Nonlinear Time Series

Theory, Methods, and Applications with R Examples

CHAPMAN & HALL/CRC
Texts in Statistical Science Series

Series Editors
Francesca Dominici, *Harvard School of Public Health, USA*
Julian J. Faraway, *University of Bath, UK*
Martin Tanner, *Northwestern University, USA*
Jim Zidek, *University of British Columbia, Canada*

Statistical Theory: A Concise Introduction
F. Abramovich and Y. Ritov

Practical Multivariate Analysis, Fifth Edition
A. Afifi, S. May, and V.A. Clark

Practical Statistics for Medical Research
D.G. Altman

Interpreting Data: A First Course in Statistics
A.J.B. Anderson

Introduction to Probability with R
K. Baclawski

Linear Algebra and Matrix Analysis for Statistics
S. Banerjee and A. Roy

Statistical Methods for SPC and TQM
D. Bissell

Bayesian Methods for Data Analysis, Third Edition
B.P. Carlin and T.A. Louis

Second Edition
R. Caulcutt

The Analysis of Time Series: An Introduction, Sixth Edition
C. Chatfield

Introduction to Multivariate Analysis
C. Chatfield and A.J. Collins

Problem Solving: A Statistician's Guide, Second Edition
C. Chatfield

Statistics for Technology: A Course in Applied Statistics, Third Edition
C. Chatfield

Bayesian Ideas and Data Analysis: An Introduction for Scientists and Statisticians
R. Christensen, W. Johnson, A. Branscum, and T.E. Hanson

Modelling Binary Data, Second Edition
D. Collett

Modelling Survival Data in Medical Research, Second Edition
D. Collett

Introduction to Statistical Methods for Clinical Trials
T.D. Cook and D.L. DeMets

Applied Statistics: Principles and Examples
D.R. Cox and E.J. Snell

Multivariate Survival Analysis and Competing Risks
M. Crowder

Statistical Analysis of Reliability Data
M.J. Crowder, A.C. Kimber, T.J. Sweeting, and R.L. Smith

An Introduction to Generalized Linear Models, Third Edition
A.J. Dobson and A.G. Barnett

Nonlinear Time Series: Theory, Methods, and Applications with R Examples
R. Douc, E. Moulines, and D. Stoffer

Introduction to Optimization Methods and Their Applications in Statistics
B.S. Everitt

Extending the Linear Model with R: Generalized Linear, Mixed Effects and Nonparametric Regression Models
J.J. Faraway

A Course in Large Sample Theory
T.S. Ferguson

Multivariate Statistics: A Practical Approach
B. Flury and H. Riedwyl

Readings in Decision Analysis
S. French

Markov Chain Monte Carlo: Stochastic Simulation for Bayesian Inference, Second Edition
D. Gamerman and H.F. Lopes

Bayesian Data Analysis, Third Edition
A. Gelman, J.B. Carlin, H.S. Stern, D.B. Dunson, A. Vehtari, and D.B. Rubin

Multivariate Analysis of Variance and Repeated Measures: A Practical Approach for Behavioural Scientists
D.J. Hand and C.C. Taylor

Practical Data Analysis for Designed Practical Longitudinal Data Analysis
D.J. Hand and M. Crowder

Logistic Regression Models
J.M. Hilbe

Richly Parameterized Linear Models: Additive, Time Series, and Spatial Models Using Random Effects
J.S. Hodges

Statistics for Epidemiology
N.P. Jewell

Stochastic Processes: An Introduction, Second Edition
P.W. Jones and P. Smith

The Theory of Linear Models
B. Jørgensen

Principles of Uncertainty
J.B. Kadane

Graphics for Statistics and Data Analysis with R
K.J. Keen

Mathematical Statistics
K. Knight

Nonparametric Methods in Statistics with SAS Applications
O. Korosteleva

Modeling and Analysis of Stochastic Systems, Second Edition
V.G. Kulkarni

Exercises and Solutions in Biostatistical Theory
L.L. Kupper, B.H. Neelon, and S.M. O'Brien

Exercises and Solutions in Statistical Theory
L.L. Kupper, B.H. Neelon, and S.M. O'Brien

Design and Analysis of Experiments with SAS
J. Lawson

A Course in Categorical Data Analysis
T. Leonard

Statistics for Accountants
S. Letchford

Introduction to the Theory of Statistical Inference
H. Liero and S. Zwanzig

Statistical Theory, Fourth Edition
B.W. Lindgren

Stationary Stochastic Processes: Theory and Applications
G. Lindgren

The BUGS Book: A Practical Introduction to Bayesian Analysis
D. Lunn, C. Jackson, N. Best, A. Thomas, and D. Spiegelhalter

Introduction to General and Generalized Linear Models
H. Madsen and P. Thyregod

Time Series Analysis
H. Madsen

Pólya Urn Models
H. Mahmoud

Randomization, Bootstrap and Monte Carlo Methods in Biology, Third Edition
B.F.J. Manly

Introduction to Randomized Controlled Clinical Trials, Second Edition
J.N.S. Matthews

Statistical Methods in Agriculture and Experimental Biology, Second Edition
R. Mead, R.N. Curnow, and A.M. Hasted

Statistics in Engineering: A Practical Approach
A.V. Metcalfe

Beyond ANOVA: Basics of Applied Statistics
R.G. Miller, Jr.

A Primer on Linear Models
J.F. Monahan

Applied Stochastic Modelling, Second Edition
B.J.T. Morgan

Elements of Simulation
B.J.T. Morgan

Probability: Methods and Measurement
A. O'Hagan

Introduction to Statistical Limit Theory
A.M. Polansky

Applied Bayesian Forecasting and Time Series Analysis
A. Pole, M. West, and J. Harrison

Statistics in Research and Development, Time Series: Modeling, Computation, and Inference
R. Prado and M. West

Introduction to Statistical Process Control
P. Qiu

Sampling Methodologies with Applications
P.S.R.S. Rao

A First Course in Linear Model Theory
N. Ravishanker and D.K. Dey

Essential Statistics, Fourth Edition
D.A.G. Rees

Stochastic Modeling and Mathematical Statistics: A Text for Statisticians and Quantitative
F.J. Samaniego

Statistical Methods for Spatial Data Analysis
O. Schabenberger and C.A. Gotway

Large Sample Methods in Statistics
P.K. Sen and J. da Motta Singer

Decision Analysis: A Bayesian Approach
J.Q. Smith

Analysis of Failure and Survival Data
P. J. Smith

Applied Statistics: Handbook of GENSTAT Analyses
E.J. Snell and H. Simpson

Applied Nonparametric Statistical Methods, Fourth Edition
P. Sprent and N.C. Smeeton

Data Driven Statistical Methods
P. Sprent

Generalized Linear Mixed Models: Modern Concepts, Methods and Applications
W. W. Stroup

Survival Analysis Using S: Analysis of Time-to-Event Data
M. Tableman and J.S. Kim

Applied Categorical and Count Data Analysis
W. Tang, H. He, and X.M. Tu

Elementary Applications of Probability Theory, Second Edition
H.C. Tuckwell

Introduction to Statistical Inference and Its Applications with R
M.W. Trosset

Understanding Advanced Statistical Methods
P.H. Westfall and K.S.S. Henning

Statistical Process Control: Theory and Practice, Third Edition
G.B. Wetherill and D.W. Brown

Generalized Additive Models: An Introduction with R
S. Wood

Epidemiology: Study Design and Data Analysis, Third Edition
M. Woodward

Experiments
B.S. Yandell

Texts in Statistical Science

Nonlinear Time Series

Theory, Methods, and Applications with R Examples

Randal Douc
Telecom SudParis
Evry, France

Eric Moulines
Telecom ParisTech
Paris, France

David S. Stoffer
University of Pittsburgh
Pennsylvania, USA

CRC Press
Taylor & Francis Group
Boca Raton London New York

CRC Press is an imprint of the
Taylor & Francis Group, an **informa** business
A CHAPMAN & HALL BOOK

CRC Press
Taylor & Francis Group
6000 Broken Sound Parkway NW, Suite 300
Boca Raton, FL 33487-2742

ISBN 13: 978-1-4665-0225-3 (hbk)

Visit the Taylor & Francis Web site at
http://www.taylorandfrancis.com

and the CRC Press Web site at
http://www.crcpress.com

Contents

Preface

This book is designed for researchers and students who want to acquire advanced skills in nonlinear time series analysis and their applications. Before reading this text, we suggest a solid knowledge of linear Gaussian time series, for which there are many texts. At the advanced level, texts that cover both the time and frequency domains are Anderson (1994), Brockwell and Davis (1991), and Fuller (1996). At the intermediate level, we mention Hamilton (1994), Kitagawa (2010), and Shumway and Stoffer (2011), which cover both the time and frequency domains, and Box and Jenkins (1970), which covers primarily the time domain. Hannan and Deistler (2012) is an advanced text on the statistical theory of linear state space systems. There are a number of texts that cover time series at a more introductory level, but the material covered in this text requires at least an intermediate level of understanding of the time domain.

While it is not sensible to view statistics simply as a branch of mathematics, we believe that statistical modeling and inference need to be firmly grounded in theory. Although we avoid delving into sophisticated mathematical derivations, most of the statements of the book are rigorously established. The reader is therefore expected to have some background in measure theory (covering the construction of the measure and Lebesgue integrals), and in probability theory (including conditional expectations, the construction of discrete time stochastic processes and martingales). Examples of courses covering this material are Doob (1953), Billingsley (1995), Shiryaev (1996), and Durrett (2010), among many others. Although we constantly use measure-theoretic concepts and notations, nothing excessively deep is used. An introduction to discrete state space Markov chains is clearly a plus, but is not needed.

The book represents a biased selection of topics in nonlinear time series, reflecting our own inclinations toward state-space representations. Our focus on principles is intended to provide readers with a solid background to craft their own stochastic models, numerical methods, and software, and to be able to assess the advantages and disadvantages of different approaches. We do not believe in pulling mathematical formulas out of thin air, or establishing a catalog of models and methods. Of course, this attitude reflects our mathematical orientation and our willingness to postpone the statistical discussion to pay attention to rigorous theoretical foundations.

There are a number of texts that cover nonlinear and non-Gaussian models from a variety of points-of-view. Because financial series tend to be nonlinear, there are many texts that focus primarily on finance such as Chan (2002), Franses and Van Dijk (2000), and Tsay (2005). The text by Teräsvirta et al. (2011), while focusing primarily on finance, is a rather comprehensive and approachable text on the subject. Other texts that present general statistical approaches to nonlinear time series models are Fan and Yao (2003) and Gao (2007), which take a nonparametric or semiparametric smoothing approach, Priestley (1988), which focuses on nonlinear models and spectral analysis for nonstationary processes, and Tong (1983), which introduces threshold models. Tong (1990) and Kantz and Schreiber (2004) take a dynamical systems approach and present a wide array of nonlinear time series models. Two other texts that focus primarily on a state-space approach to nonlinear and non-Gaussian time series are Kitagawa and Gersch (1996) and the second part of Durbin and Koopman (2012). MacDonald and Zucchini (2009), Fraser (2008) and Kitagawa (2010) present a state-space approach to modeling linear and nonlinear time series at an introductory level. Kitagawa (2010) could serve as a supplement for readers seeking a more gentle initial approach to the subject.

We are agnostic about the nature of statistical inference. The reader must definitely look elsewhere for a philosophical discussion of the relative merits of frequentist versus Bayesian inference. Our belief is that nonlinear time series generally benefit from analysis using a variety of frequentist and Bayesian methods. These different perspectives strengthen the conclusions rather than contradict one another.

Our hope is to acquaint readers with the main principles behind nonlinear time series models without overwhelming them with difficult mathematical developments. To keep the book length within acceptable limits, we have avoided the use of sophisticated probabilistic arguments, which underlie most of the recent developments of continuous state space Markov chains and sequential Monte Carlo methods. For the statistical part, we cover mostly the basics; other important concepts like the local asymptotic normality (e.g., Taniguchi and Kakizawa, 2000), empirical process techniques (e.g., Dehling et al., 2002), cointegration (e.g., Fuller, 1996), multivariate time series (e.g., Reinsel, 2003, Lütkepohl, 2005), model selection, semiparametric and nonparametric inference (e.g., Fan and Yao, 2003), and so on, may be found in other texts. We are, of course, responsible for any and all mistakes, misconceptions and omissions.

Although there is a logical progression through the chapters, the three parts can be studied independently. Some chapters within each part may also be read as independent surveys. Several chapters highlight recent developments such as explicit rate of convergence of Markov chains (we use the techniques outlined in Hairer and Mattingly, 2011 to discuss the geometric ergodicity of Markov chains), or sequential Monte Carlo techniques (covering for example the recently introduced particle Markov chain Monte Carlo methods found in Andrieu et al., 2010).

Any instructor contemplating a one-semester course based on this book will have to decide which chapters to cover and which to omit. The first part can be seen as a crash course on "classical" time series, with a special emphasis on linear state space models and a rather detailed coverage on random coefficient autoregressions, covering both ARCH and GARCH models. The second part is a self-contained introduction to Markov chain, discussing stability, the existence of a stationary distribution, ergodicity, limit theorems and statistical inference. Many examples are provided with the objective to develop empirical skills. We have already covered parts I and II in a fast-paced one semester advanced master level course. Part III is a self-contained account of nonlinear state space and sequential Monte Carlo methods. It is an elementary introduction to nonlinear state space modeling and sequential Monte Carlo, but it touches on many current topics in this field, from the theory of statistical inference to advanced computational methods. This has been used as a support to an advanced course on these methods, and can be used by readers who want to have an introduction to this field before studying more specialized texts such as Del Moral (2004) or Del Moral et al. (2010).

As with any textbook, the exercises are nearly as important as the main text. Statistics is not a spectator sport, so the book contains more than 200 exercises to challenge the readers. Most problems merely serve to strengthen intellectual muscles strained by the introduction of new theory; some problems extend the theory in significant ways.

We acknowledge the help of Julien Cornebise and Fredrik Lindsten who participated in the writing of the text and contributed to Chapter 11 and Chapter 12 in Part III. Julien also helped us considerably in the development of R code. We are also indebted to Pierre Priouret for suggesting various forms of improvement in the presentation, layout, and so on, as well as helping us track typos and errors. We are grateful to Christophe Andrieu, Pierre Del Moral, and Arnaud Doucet, who generously gave us some of their time to help to decipher many of the intricacies of the particle filters. This work would have not been possible without the continuous support of our colleagues and friends Olivier Cappé, Gersende Fort, Jimmy Olsson, François Roueff, and Philippe Soulier, who provided various helpful insights and comments. We also acknowledge Hedibert Freitas Lopes and Fredrik Lindsten for distributing code that became the basis of some of the R scripts used in Section 12.2. Finally, we thank John Kimmel for his support and enduring patience.

R. Douc thanks Telecom SudParis for a six month sabbatical in 2012 to work on the text. E. Moulines thanks Telecom ParisTech for giving him the opportunity to work on this project. D.S. Stoffer thanks the U.S. National Science Foundation and Telecom ParisTech for partial support during the preparation of this manuscript. In addition, most of the time spent on the text was during a visiting professorship at the Booth School of Business, University of Chicago, and their support is gratefully acknowledged.

The webpage for the text, www.stat.pitt.edu/stoffer/nltsa, contains the R scripts used in the examples and other useful information such as errata.

Paris	*Randal Douc*
Paris	*Eric Moulines*
Chicago	*David S. Stoffer*
September, 2013	

Frequently Used Notation

Sets and Numbers

- $\mathbb{N}$: the set of natural numbers including zero, $\mathbb{N} = \{0, 1, 2, \ldots\}$.
- $\mathbb{N}^*$: the set of natural numbers excluding zero, $\mathbb{N}^* = \{1, 2, \ldots\}$.
- $\mathbb{Z}$: the set of relative integers, $\mathbb{Z} = \{0, \pm 1, \pm 2, \ldots\}$.
- $\mathbb{R}$: the set of real numbers.
- $\mathbb{R}^d$: Euclidean space consisting of all column vectors $x = (x_1, \ldots, x_d)'$.
- $\bar{\mathbb{R}}$: the extended real line, i.e., $\mathbb{R} \cup \{-\infty, +\infty\}$.
- $\mathbb{C}$: the set of complex numbers.
- $\cap, \bigcap$: intersection.
- $\cup, \bigcup$: union.
- A^c: the complement of A.
- $B \setminus A$: $B \cap A^c$, the relative complement of B in A or set difference.
- $A \ominus B$: symmetric difference of sets; $(B \setminus A) \cup (A \setminus B)$.
- $\bar{z}$ or z^*: the complement of $z \in \mathbb{C}$.
- $\lceil x \rceil$: the smallest integer bigger than or equal to x.
- $\lfloor x \rfloor$: the largest integer smaller than or equal to x.
- $a * b$: convolution; for sequences $a = \{a(n), n \in \mathbb{Z}\}$ and $b = \{b(n), n \in \mathbb{Z}\}$, $a * b$ denotes the convolution of a and b, defined formally by $a * b = \sum_k a(k) b(n-k)$.

Metric space

- (X, d): a metric space.
- $\mathrm{B}(x, r)$: the open ball of radius $r > 0$ centerd in x, $\mathrm{B}(x, r) = \{y \in \mathsf{X} : d(x, y) < r\}$.
- $\overline{U}$: closure of the set $U \subset \mathsf{X}$.
- U^o: interior of the set $U \subset \mathsf{X}$.
- ∂U: boundary of the set $U \subset \mathsf{X}$.

Binary relations

- $a \wedge b$: the minimum of a and b.
- $a \vee b$: the maximum of a and b.

- $a(n) \asymp b(n)$: the ratio of the two sides is bounded from above and below by positive constants that do not depend on n.
- $a(n) \sim b(n)$: the ratio of the two sides converges to one.

Vectors, matrices

- $\mathbb{M}_d(\mathbb{R})$ (resp. $\mathbb{M}_d(\mathbb{C})$): the set of $d \times d$ matrices with real (resp. complex) coefficients.
- $|||M|||$: operator norm; for $M \in \mathbb{M}_d(\mathbb{C})$, and $\|\cdot\|$ any norm on $\mathbb{C}^d$,

$$|||M||| = \sup\left\{\frac{\|Mx\|}{\|x\|}, \, x \in \mathbb{C}^d, \, x \neq 0\right\} .$$

- I_d: $d \times d$ identity matrix.
- $\mathbf{1} = (1, \ldots, 1)'$: $d \times 1$ vector whose entries are all equal to 1.
- $A \otimes B$: Kronecker product; let A and B be $m \times n$ and $p \times q$ matrices, respectively. The Kronecker product of A with B is the $mp \times nq$ matrix whose (i, j)th block is the $p \times q$ $A_{i,j}B$, where $A_{i,j}$ is the (i, j)th element of A. Note that the Kronecker product is associative $(A \otimes B) \otimes C = A \otimes (B \otimes C)$ and $(A \otimes B)(C \otimes B) = (AC \otimes BD)$ (for matrices with compatible dimensions).
- $\mathrm{Vec}(A)$: vectorization of a matrix; let A be an $m \times n$ matrix, then $\mathrm{Vec}(A)$ is the $(mn \times 1)$ vector obtained from A by stacking the columns of A (from left to right). Note that $\mathrm{Vec}(ABC) = (C^T \otimes A)\mathrm{Vec}(B)$.

Functions

- $\mathbb{1}_A$: indicator function with $\mathbb{1}_A(x) = 1$ if $x \in A$ and 0 otherwise. $\mathbb{1}\{A\}$ is used if A is a composite statement
- f^+: the positive part of the function f, i.e., $f^+(x) = f(x) \vee 0$,
- f^-: the negative part of the function f, i.e., $f^-(x) = -(f(x) \wedge 0)$.
- $f^{-1}(A)$: inverse image of the set A by f.
- $\mathrm{osc}\,(f)$: the oscillation seminorm; for f a real valued function on X, $|f|_\infty = \sup\{f(x) : x \in \mathsf{X}\}$ it is the supremum norm defined as

$$\mathrm{osc}\,(f) = \sup_{(x,y) \in \mathsf{X} \times \mathsf{X}} |f(x) - f(y)| = 2 \inf_{c \in \mathbb{R}} |f - c|_\infty .$$

Measures

Let $(\mathsf{X}, \mathcal{X})$ be a measurable space.

- If X is a topological space (in particular a metric space) then $\mathcal{X}$ is always taken to be the Borel sigma-field generated by the topology of X. If $\mathsf{X} = \bar{\mathbb{R}}^d$, its Borel sigma-field is denoted by $\mathcal{F}_b\left(\bar{\mathbb{R}}^d\right)$.

- δ_x: Dirac measure with mass concentrated on x, i.e., $\delta_x(A) = 1$ if $x \in A$ and 0 otherwise.
- Leb: Lebesgue measure on $\mathbb{R}^d$.
- $\mathbb{M}(\mathcal{X})$: the set of finite signed measures on the measurable space $(\mathsf{X}, \mathcal{X})$.
- $\mathbb{M}_+(\mathcal{X})$: the set of measures on the measurable space $(\mathsf{X}, \mathcal{X})$.
- $\mathbb{M}_1(\mathcal{X})$: the set of probability measures on $(\mathsf{X}, \mathcal{X})$.
- $\mathbb{M}_0(\mathcal{X})$: the set of finite signed measures ξ on $(\mathsf{X}, \mathcal{X})$ satisfying $\xi(\mathsf{X}) = 0$.
- $\mu \ll \nu$: μ is absolutely continuous with respect to ν.

Function spaces

Let $(\mathsf{X}, \mathcal{X})$ be a measurable space.

- $\mathbb{F}(\mathsf{X}, \mathcal{X})$: the vector space of measurable functions from $(\mathsf{X}, \mathcal{X})$ to $(-\infty, \infty)$.
- $\mathbb{F}_+(\mathsf{X}, \mathcal{X})$: the cone of measurable functions from $(\mathsf{X}, \mathcal{X})$ to $[0, \infty]$.
- $\mathbb{F}_b(\mathsf{X}, \mathcal{X})$: the subset of $\mathbb{F}(\mathsf{X}, \mathcal{X})$ of bounded functions.
- $\xi(f)$: for any $\xi \in \mathbb{M}(\mathcal{X})$ and $f \in \mathbb{F}_b(\mathsf{X}, \mathcal{X})$, $\xi(f) = \int f \, d\xi$.
- $\xi \in \mathbb{M}(\mathcal{X})$: defines a linear functional on the Banach space $(\mathbb{F}_b(\mathsf{X}, \mathcal{X}), |\cdot|_\infty)$. We use the same notation for the measure and for the functional.
- $\mathrm{C}_b(\mathsf{X})$: the space of all bounded continuous real functions defined on a topological space X.
- $\mathcal{L}^p(\mu)$: the space of measurable functions f such that $\int |f|^p d\mu < \infty$.
- $\mathrm{L}^p(\mu)$: the space of classes of μ-equivalent functions in $\mathcal{L}^p(\mu)$. If $\mathsf{f} \in \mathrm{L}^p(\mu)$, $\|\mathsf{f}\|_p = (\int |f|^p d\mu)^{1/p}$ where $f \in \mathsf{f}$. When no confusion is possible, we will identify f and any $f \in \mathsf{f}$.

Probability space

Let $(\Omega, \mathcal{A}, \mathbb{P})$ be a probability space.

- $\mathbb{E}[X]$: expectation of random variable X with respect to the probability measure $\mathbb{P}$.
- $\mathrm{Var}[X]$ variance of random variable X with respect to the probability measure $\mathbb{P}$
- $\mathrm{Cov}(X, Y)$: covariance of the random variables X and Y.
- $\mathbb{P}(A \,|\, \mathcal{F})$: conditional probability of A given $\mathcal{F}$, a sub-σ-field $\mathcal{F}$, and $A \in \mathcal{A}$.
- $\mathbb{E}\left[X \mid \mathcal{F}\right]$: conditional expectation of X given $\mathcal{F}$ as defined above.
- $\mathcal{L}_{\mathbb{P}}(X)$: the law of X under $\mathbb{P}$.
- $X_n \overset{\mathbb{P}}{\Longrightarrow} X$ or $X_n \Rightarrow_{\mathbb{P}} X$: the sequence of random variables $\{X_n\}$ converges to X in distribution under $\mathbb{P}$.
- $X_n \overset{\mathbb{P}}{\longrightarrow} X$ or $X_n \to_{\mathbb{P}} X$: the sequence of random variables $\{X_n\}$ converges to X in probability under $\mathbb{P}$.

- $X_n \xrightarrow{\mathbb{P}\text{-a.s.}} X$ or $X_n \to_{\mathbb{P}\text{-a.s.}} X$: the sequence of random variables $\{X_n\}$ converges to X $\mathbb{P}$-almost surely ($\mathbb{P}$-a.s.).
- $X \stackrel{\mathrm{d}}{=} Y$ or $X =_{\mathrm{d}} Y$: X is stochastically equal to Y; i.e., $\mathcal{L}_{\mathbb{P}}(X) = \mathcal{L}_{\mathbb{P}}(Y)$

Usual distributions

- $\mathrm{N}(\mu, \sigma^2)$: Normal distribution with mean μ and variance σ^2.
- $\mathfrak{g}(x; \mu, \Sigma)$: The Gaussian density in variable x with mean μ and (co)variance Σ; see (1.36).
- $\mathrm{U}(a, b)$: uniform distribution on $[a, b]$.
- χ^2: chi-square distribution.
- χ_n^2: chi-square distribution with n degrees of freedom.

White noise types

- $Z_t \sim \mathrm{WN}(0, \sigma^2)$: The sequence $\{Z_t,\ t \in \mathcal{T}\}$ is an uncorrelated sequence with mean zero and variance σ^2. The actual set $\mathcal{T}$ will be apparent from the context. Called *second-order white noise*, *second-order noise*, *weak white noise*, *weak noise*, or simply *white noise*.
- $Z_t \sim \mathrm{iid}\,(0, \sigma^2)$: The sequence $\{Z_t,\ t \in \mathcal{T}\}$ is an i.i.d. sequence with mean zero and variance σ^2. The actual set $\mathcal{T}$ will be apparent from the context. Called *strong white noise*, *strong noise*, *i.i.d. noise*, or *independent white noise*.
- $Z_t \sim \mathrm{iid}\,\mathrm{N}(0, \sigma^2)$: The sequence $\{Z_t,\ t \in \mathcal{T}\}$ is an i.i.d. sequence of Gaussian random variables with mean zero and variance σ^2. The actual set $\mathcal{T}$ will be apparent from the context. Called *Gaussian noise* or *normal noise*.

Part I

Foundations

Chapter 1

Linear Models

In this chapter, we briefly review some aspects of stochastic processes and linear time series models in both the time and frequency domains. The chapter can serve as a review of stationarity, linearity and Gaussianity that provides a foundation on which to build a course on nonlinear time series analysis. Our discussions are brief and are meant only to establish a baseline of material that is necessary for comprehension of the material presented in this text.

1.1 Stochastic processes

The primary objective of time series analysis is to develop mathematical models that allow plausible descriptions for sample data. In order to provide a statistical setting for describing the character of data that seemingly fluctuate in a random fashion over time, we assume a time series can be defined as a collection of random elements, defined on some probability space $(\Omega, \mathcal{F}, \mathbb{P})$ and taking value in some state-space X, indexed according to the order they are obtained in time,

$$\{X_t,\ t \in \mathcal{T}\}\,. \tag{1.1}$$

Time t takes values in set $\mathcal{T}$, which can be discrete or continuous. In this book, we will take $\mathcal{T}$ to be the integers $\mathbb{Z} = \{0, \pm 1, \pm 2, \ldots,\}$ or some subset of the integers such as the non-negative integers $\mathbb{N} = \{0, 1, 2, \ldots\}$. For example, we may consider a time series as a sequence of random variables, $\{X_t,\ t \in \mathbb{N}\} = \{X_0, X_1, X_2, \ldots\}$, where the random variable X_0 denotes the initial value taken by the series, X_1 denotes the value at the first time period, X_2 denotes the value for the second time period, and so on.

The *state-space* X is the space in which the time series takes its values. Formally, the state-space should be a measurable space, $(\mathsf{X}, \mathcal{X})$ where $\mathcal{X}$ is a σ-field. In many instances, $\mathsf{X} = \mathbb{R}$ or $\mathsf{X} = \mathbb{R}^d$, if the observations are scalar or vector-valued. In this case, $\mathcal{X} = \mathcal{B}(\mathbb{R})$ or $\mathcal{X} = \mathcal{B}(\mathbb{R}^d)$, the Borel σ-fields. In some examples, the observations can be discrete; for example, the observations are integers $\mathsf{X} = \mathbb{N}$ if we are dealing with time-series of counts.

Definition 1.1 (Stochastic process). *A collection of random elements $X = \{X_t,\ t \in \mathcal{T}\}$ defined on a probability space $(\Omega, \mathcal{F}, \mathbb{P})$ and taking value in a measurable space $(\mathsf{X}, \mathcal{X})$ is referred to as a stochastic process.*

A stochastic process is therefore a function of two arguments $X: \mathcal{T} \times \Omega \to \mathsf{X}$, $(t,\omega) \mapsto X_t(\omega)$. For each $t \in \mathcal{T}$, $X_t: \omega \mapsto X_t(\omega)$ is a straightforward $\mathcal{F}/\mathcal{X}$-measurable random element from Ω to X, which induces a probability measure on $(\mathsf{X},\mathcal{X})$ (referred to as the the law of X_t). For each $\omega \in \Omega$, $X(\omega): t \mapsto X_t(\omega)$ is a function from $\mathcal{T}$ to the state-space X.

Definition 1.2 (Trajectory and path). *To every $\omega \in \Omega$ is associated a collection of numbers (or vector of numbers), indexed by $\mathcal{T}$, $t \mapsto X_t(\omega)$ representing a* realization *of the stochastic process (sometimes referred to as a* path*).*

Because it will be clear from the context of our discussions, we use the term time series whether we are referring generically to the process or to a particular realization and make no notational distinction between the two concepts.

The distribution of a stochastic process is usually described in terms of its *finite-dimensional distributions.*

Definition 1.3 (Finite-dimensional distributions). *The finite-dimensional distributions of the stochastic process X are the set of all the joint distributions of the random elements $X_{t_1}, X_{t_2}, \ldots, X_{t_n}$ for all integer n and all n-tuples $(t_1,t_2,\ldots t_n) \in \mathcal{T}^n$ of distinct indices (i.e., $t_i \neq t_j$ for $i \neq j$),*

$$\mu^X_{t_1,t_2,\ldots,t_n}(H) = \mathbb{P}\left[(X_{t_1},X_{t_2},\ldots,X_{t_n}) \in H\right] , \quad H \in \mathcal{X}^{\otimes n} . \tag{1.2}$$

Finite dimensional distributions are enough to compute the distribution of measurable functions involving a finite number of random elements $(X_{t_1},X_{t_2},\ldots,X_{t_n})$, such as, e.g., the product of $X_{t_1}X_{t_2}\ldots X_{t_k}$ or the maximum, $\max(X_{t_1},X_{t_2},\ldots,X_{t_k})$. But in the sequel we will have to consider quantities involving an infinite number of random variables, like the distribution of the $\max_{t\in\mathcal{T}} X_t$ or limits such as $\lim_{t\to\infty} X_t$. Of course, a priori, computing the distribution of such quantities require us to go beyond any *finite* dimensional distributions. The following theorem states that the finite-dimensional distributions specify the infinite-dimensional distribution uniquely.

Theorem 1.4 (Stochastic processes / finite-dimensional distributions). *Let X and Y be two X-valued stochastic processes indexed by $\mathcal{T}$. Then X and Y have the same distribution if and only if all their finite-dimensional distributions agree.*

The finite-dimensional distributions of a given stochastic process X are not an arbitrary set of distributions. Consider a collection of distinct indices t_1, $t_2, \ldots t_n \in \mathcal{T}$, and corresponding measurable sets $B_1, B_2, \ldots, B_n \in \mathcal{X}$. Then, for any further index $t_{n+1} \in \mathcal{T}$ distinct from $t_1, \ldots, t_n$,

$$\mu^X_{t_1,t_2,\ldots,t_n}(B_1 \times B_2 \times \cdots \times B_n) = \mu^X_{t_1,t_2,\ldots,t_n,t_{n+1}}(B_1 \times B_2 \times \cdots \times B_n \times \mathsf{X}) . \tag{1.3}$$

Similarly, if σ is a permutation of the set $\{1,\ldots,k\}$,

$$\mu^X_{t_1,t_2,\ldots,t_n}(B_1 \times B_2 \times \cdots \times B_n) = \mu^X_{t_{\sigma(1)},t_{\sigma(2)},\ldots,t_{\sigma(n)}}(B_{\sigma(1)} \times B_{\sigma(2)} \times \cdots \times B_{\sigma(n)}) . \tag{1.4}$$

Such relations are referred to as *consistency* relations. The finite dimensional distributions of a stochastic process necessarily satisfy (1.3) and (1.4). The *Kolmogorov Existence Theorem* basically states the converse.

Theorem 1.5 (Kolmogorov existence theorem). *Let*

$$\mathcal{M} = \left\{\mu_{t_1,t_2,\ldots,t_n} : n \in \mathbb{N}^*; (t_1,t_2,\ldots,t_n) \in \mathcal{T}^n, \text{with } t_i \neq t_j \text{ for } i \neq j\right\} \tag{1.5}$$

be a system of finite-dimensional distributions satisfying the two consistency conditions (1.3) and (1.4). Then, there exist a probability space $(\Omega,\mathcal{F},\mathbb{P})$ and an X-valued stochastic process X, having $\mathcal{M}$ as its finite dimensional distributions.

Proof. See for example Billingsley (1999, Theorem 36.2). ■

The proof is constructive. We can set $\Omega = \mathsf{X}^{\mathcal{T}} = \prod_{t\in\mathcal{T}} \mathsf{X}$ the set of all possible paths $\{x_t,\ t \in \mathcal{T}\}$ (for all $t \in \mathcal{T}$, the coordinate $x_t \in \mathsf{X}$). The σ-algebra $\mathcal{F}$ can be identified with σ-algebra generated by the cylinders, defined as

$$\mathcal{C} = \left\{\{x_t,\ t \in \mathcal{T}\} : (x_{t_1},x_{t_2},\ldots,x_{t_k}) \in H\right\}, \quad H \in \mathcal{X}^k,\ k \in \mathbb{N}. \tag{1.6}$$

In this case, the random elements X_t are the coordinate projections, for $\omega = \{x_t,\ t \in \mathcal{T}\}$, $X_t(\omega) = x_t$.

Definition 1.6 (Strict stationarity). *A process is said to be* strictly stationary *if*

$$\{X_{t_1},X_{t_2},\ldots,X_{t_n}\} \stackrel{\mathrm{d}}{=} \{X_{t_1+h},X_{t_2+h},\ldots,X_{t_n+h}\}, \tag{1.7}$$

for all $n \in \mathbb{N}^*$, all time points $(t_1,t_2,\ldots,t_n) \in \mathcal{T}^n$, and all time shifts $h \in \mathbb{Z}$.

A trivial example of a strictly stationary process is one where $\{X_t,\ t \in \mathbb{Z}\}$ is i.i.d. (where i.i.d. stands for independent and identically distributed). In addition, it is easily shown, using characteristic functions, that if $\{X_t\}$ is i.i.d., then a finite *time invariant linear filter* of X_t, say $Y_t = \sum_{j=-k}^{k} a_j X_{t-j}$, where $(a_0,a_{\pm 1},\ldots,a_{\pm k}) \in \mathbb{R}^{2k+1}$, is strictly stationary (see Exercise 1.1).

1.2 The covariance world

1.2.1 *Second-order stationary processes*

As mentioned above, a stochastic process is strictly stationary if its finite-dimensional distributions are invariant under time-shifts. Rather than imposing conditions on all possible distributions, a milder version imposes conditions only on the first two moments of the series. First, we make the following definitions. Unless stated otherwise, the state space is taken to be $\mathsf{X} = \mathbb{R}$.

Definition 1.7 (Mean function). *The* mean function *is defined as*

$$\mu_t = \mathbb{E}[X_t] \tag{1.8}$$

provided that $\mathbb{E}[|X_t|] < \infty$, where $\mathbb{E}$ denotes the expectation operator.

Definition 1.8 (Autocovariance function). *Assume that for all t, $\mathbb{E}\left[X_t^2\right] < \infty$. The* autocovariance function *is defined as the second moment product*

$$\gamma(s,t) = \mathrm{Cov}(X_s,X_t) = \mathbb{E}[(X_s-\mu_s)(X_t-\mu_t)], \tag{1.9}$$

for all s and t. Note that $\gamma(s,t) = \gamma(t,s)$.

If a time series is strictly stationary, then all of the multivariate distribution functions for subsets of variables must agree with their counterparts in the shifted set for all values of the shift parameter h. For example, when $n = 1$, (1.7) implies that $\mathbb{P}\{X_s \leq x\} = \mathbb{P}\{X_t \leq x\}$ for any time points s and t. If, in addition, the mean function, μ_t, exists, (1.7) implies that $\mu_s = \mu_t$ for all s and t, and hence μ_t must be constant. When $n = 2$, we can write (1.7) as $\mathbb{P}\{X_s \leq x_1, X_t \leq x_2\} = \mathbb{P}\{X_{s+h} \leq x_1, X_{t+h} \leq x_2\}$ for any time points s and t and shift h. Thus, if the variance function of the process exists, this implies that the autocovariance function of the series X_t satisfies $\gamma(s,t) = \gamma(s+h, t+h)$ for all s and t and h. We may interpret this result by saying the autocovariance function of the process depends only on the time difference between s and t, and not on the actual times. These considerations lead to the following definition.

Definition 1.9 (Second-order stationarity). *A second-order or weakly* stationary *time series, $\{X_t,\ t \in \mathbb{Z}\}$, is a finite variance process such that*

(i) the mean value function, μ_t, defined in (1.8) is constant and does not depend on time t, and

(ii) the autocovariance function, $\gamma(s,t)$, defined in (1.9) depends on s and t only through their difference $|s-t|$.

For brevity, a covariance or weakly stationary series is simply called a *stationary* time series. It should be clear from the discussion following Definition 1.8 that a strictly stationary, finite variance time series is also stationary. The converse is not true unless there are further conditions. One important case where weak stationarity implies strict stationarity is if the time series is Gaussian (meaning all finite distributions of the series are Gaussian; see Definition 1.28). In a linear, Gaussian world, these conditions are sufficient for inference. The idea is that one only needs to specify the mean and covariance relationships of a process to specify all finite Gaussian distributions ala Kolmogorov's existence theorem.

Because the mean function, μ_t, of a stationary time series is independent of time t, we drop the subscript and write $\mu_t = \mu$. Also, because the autocovariance function, $\gamma(s,t)$, of a stationary time series depends on s and t only through their difference $|s-t|$, the notation can be simplified. Let $s = t+h$, where h represents the time shift or lag. Then

$$\gamma(t+h,t) = \mathrm{Cov}(X_{t+h}, X_t) = \mathrm{Cov}(X_h, X_0) = \gamma(h,0)$$

because the time difference between times $t+h$ and t is the same as the time difference between times h and 0. Thus, the autocovariance function of a stationary time series does not depend on the time argument t. Hence, for convenience, the second argument of $\gamma(h,0)$ is dropped and we write

$$\gamma(h) := \mathrm{Cov}(X_{t+h}, X_t)\,. \tag{1.10}$$

Proposition 1.10 (Autocovariance function properties). *Let $\gamma(h)$, as defined in (1.10), be the autocovariance function of a stationary process. Then*

(i) $\gamma(h) = \gamma(-h)$ for $h \in \mathbb{Z}$, as indicated in Definition 1.8.

(ii) $|\gamma(h)| \leq \gamma(0)$, *by the Cauchy-Schwarz Inequality.*

(iii) $\gamma(h)$ *is a* non-negative definite (n.n.d.) function*; that is, for any set of constants* $(a_1,\ldots,a_n) \in \mathbb{R}^n$, *time points* $(t_1,\ldots,t_n) \in \mathbb{Z}^n$, *and any* $n \in \mathbb{N}^*$,

$$\sum_{i=1}^{n}\sum_{j=1}^{n} a_i \gamma(t_i - t_j) a_j \geq 0\,. \tag{1.11}$$

That (1.11) *holds follows simply from the fact that for any sample* $\{X_{t_1},\ldots,X_{t_n}\}$ *from a stationary process* $\{X_t,\, t \in \mathbb{Z}\}$, *we have* $\mathbb{E}\,|\sum_{i=1}^n a_i(X_{t_i} - \mu)|^2 \geq 0$.

It follows immediately from Proposition 1.10 that the covariance matrix, Γ_n, of a sample of size n, $\{X_{t_1},\ldots,X_{t_n}\}$, from a stationary time series is symmetric and n.n.d., where

$$\Gamma_n = \mathrm{Cov}(X_{t_1},\ldots,X_{t_n}) = \begin{bmatrix} \gamma(0) & \gamma(t_1-t_2) & \cdots & \gamma(t_1-t_n) \\ \gamma(t_2-t_1) & \gamma(0) & \cdots & \gamma(t_2-t_n) \\ \vdots & \vdots & \ddots & \vdots \\ \gamma(t_n-t_1) & \gamma(t_n-t_2) & \cdots & \gamma(0) \end{bmatrix}. \tag{1.12}$$

As in classical statistics, it is often convenient to deal with correlation, and this leads to the following definition.

Definition 1.11 (Autocorrelation function). *The* autocorrelation function (ACF) *is defined as*

$$\rho(s,t) = \frac{\gamma(s,t)}{\sqrt{\gamma(s,s)}\sqrt{\gamma(t,t)}}. \tag{1.13}$$

The ACF measures the linear predictability of the series at time t, say X_t, using only the value X_s. We can show easily that $-1 \leq \rho(s,t) \leq 1$ using the Cauchy–Schwarz inequality. If we can predict X_t perfectly from X_s through a linear relationship, $X_t = \alpha + \beta X_s$, then the correlation will be $+1$ when $\beta > 0$, and -1 when $\beta < 0$.

If the process is stationary, the ACF is given by

$$\rho(h) = \mathrm{Cor}(X_{t+h}, X_t) = \frac{\gamma(h)}{\gamma(0)}. \tag{1.14}$$

Example 1.12 (Estimation of ACF in the stationary case). If a time series is stationary, the mean function, $\mu_t = \mu$ in (1.8), is constant so that we can estimate it by the sample mean, $\overline{X} = n^{-1}\sum_{t=1}^n X_t$. The theoretical autocovariance function, (1.9), is estimated by the sample autocovariance function defined as follows:

$$\widehat{\gamma}(h) = n^{-1}\sum_{t=1}^{n-h}(X_{t+h} - \overline{X})(X_t - \overline{X}), \tag{1.15}$$

with $\widehat{\gamma}(-h) = \widehat{\gamma}(h)$ for $h = 0,1,\ldots,n-1$. The sum in (1.15) runs over a restricted range because X_{t+h} is not available for $t+h > n$. The estimator in (1.15) is preferred to the one that would be obtained by dividing by $n-h$ because (1.15) is a non-negative definite function; see (1.11).

The *sample autocorrelation function* is defined, analogously to (1.14), as

$$\widehat{\rho}(h) = \frac{\widehat{\gamma}(h)}{\widehat{\gamma}(0)}. \tag{1.16}$$

◇

Example 1.13 (White noise). A basic building block of time series models is *white noise*, which we denote by Z_t. At a basic level, $\{Z_t,\ t \in \mathbb{Z}\}$ is a process of uncorrelated random variables with mean 0 and variance σ_z^2; in this case we will write $Z_t \sim \mathrm{WN}(0, \sigma_z^2)$. Note that Z_t is stationary with ACF $\rho_z(h) = \delta_h(0)$.

We will, at times, also require the noise to be i.i.d. random variables with mean 0 and variance σ_z^2. We shall distinguish this case by saying *independent white noise* or *strong white noise*, or by writing $Z_t \sim \mathrm{iid}\,(0, \sigma_z^2)$. A particularly useful white noise series is *Gaussian white noise*, wherein the Z_t are independent normal random variables with mean 0 and variance σ_z^2, or more succinctly, $Z_t \sim \mathrm{iid}\ \mathrm{N}(0, \sigma_z^2)$. ◇

Example 1.14 (Finite time invariant linear filter). Although white noise is an uncorrelated process, correlation can be introduced by filtering the noise. For example, suppose X_t is a finite time invariant linear filter of white noise given by $X_t = \sum_{j=-k}^{k} a_j Z_{t-j}$, where $(a_0, a_{\pm 1}, \dots, a_{\pm k}) \in \mathbb{R}^{2k+1}$, then $\mu_t = \mathbb{E}(X_t) = 0$ and

$$\gamma(h) = \mathrm{Cov}(X_{t+h}, X_t) = \sum_{i,j=-k}^{k}\sum a_i a_j \mathrm{Cov}(Z_{t+h-i}, Z_{t-j}) = \sigma_z^2 \sum_{|j| \le k-h} a_j a_{j+h}\,.$$

The basic idea may be extended to a filter of a general process X_t, say $Y_t = \sum_{j=-k}^{k} a_j X_{t-j}$; we leave it as an exercise to show that if X_t is stationary, then Y_t is stationary (see Exercise 1.2). Some important filters are the *difference filter* to remove trend, $Y_t = X_t - X_{t-1}$, where $k = 1$ and $a_{-1} = 0$, $a_0 = 1$, $a_1 = -1$, and a filter used to *seasonally adjust*, or remove the annual cycle in monthly data, $Y_t = \sum_{j=-6}^{6} a_j X_{t-j}$ where $a_j = 1/12$ for $j = 0, \pm 1, \dots, \pm 5$ and $a_{\pm 6} = 1/24$. ◇

The concept of a finite linear filter may be extended to an infinite linear filter of a stationary process via the L^2 completeness theorem known as the *Riesz–Fischer Theorem*.

Theorem 1.15 (Riesz–Fischer). *Let $\{U_n,\ n \in \mathbb{N}\}$ be a sequence in L^2. Then, there exists a U in L^2 such that $U_n \xrightarrow{m.s.} U$ if and only if*

$$\lim_{m\to\infty} \sup_{n \ge m} \mathbb{E}|U_n - U_m|^2 = 0\,.$$

We now address the notion of an infinite linear filter in its generality, which we state in the following proposition.

Proposition 1.16 (Time invariant linear filter). *Consider a time-invariant linear filter defined as a convolution of the form*

$$Y_t = \sum_{j=-\infty}^{\infty} a_j X_{t-j}\,, \qquad \sum_{j=-\infty}^{\infty} |a_j| < \infty, \tag{1.17}$$

for each $t \in \mathbb{Z}$. If $\{X_t,\ t \in \mathbb{Z}\}$ is a sequence of random variables such that $\sup_{t\in\mathbb{Z}} \mathbb{E}\left[|X_t|\right] < \infty$, then the series converges absolutely $\mathbb{P}$-a.s. If, in addition, $\sup_{t\in\mathbb{Z}} \mathbb{E}\left[|X_t|^2\right] < \infty$, the series also converges in mean square to the same limit. In particular, if $\{X_t,\ t \in \mathbb{Z}\}$ is stationary, these properties hold.

Proof. Defining the finite linear filter as

$$Y_t^n = \sum_{j=-n}^{n} a_j X_{t-j}\,, \tag{1.18}$$

$n \in \mathbb{N}$, to establish mean square convergence, we need to show that, for each $t \in \mathbb{Z}$, $\{Y_t^n, n \in \mathbb{N}\}$ has a mean square limit. By Theorem 1.15, it is enough to show

$$\lim_{m\to\infty} \sup_{n\geq m} \mathbb{E}\left|Y_t^n - Y_t^m\right|^2 = 0\,. \tag{1.19}$$

For $n > m > 0$,

$$\begin{aligned}\mathbb{E}\left|Y_t^n - Y_t^m\right|^2 = \mathbb{E}\left|\sum_{m<|j|\leq n} a_j X_{t-j}\right|^2 &\leq \sum_{m<|j|\leq n}\ \sum_{m\leq|k|\leq n} |a_j||a_k||\mathbb{E}(X_{t-j}X_{t-k})| \\ &\leq \sup_{t\in\mathbb{Z}} \mathbb{E}\left[|X_t|^2\right]\Big(\sum_{m\leq|j|\leq n} |a_j|\Big)^2,\end{aligned}$$

which implies (1.19), because $\sup_{t\in\mathbb{Z}} \mathbb{E}\left[|X_t|^2\right] < \infty$ and $\{a_j\}$ is absolutely summable (the second inequality follows from Cauchy–Schwarz).

Although we know that the sequence $\{Y_t^n,\ n \in \mathbb{N}\}$ given by (1.18) converges in mean square, we have not established its mean square limit. If $\tilde{Y}_t$ denotes the mean square limit of Y_t^n, then using Fatou's Lemma, $\mathbb{E}|\tilde{Y}_t - Y_t|^2 = \mathbb{E}\liminf_{n\to\infty}|\tilde{Y}_t - Y_t^n|^2 \leq \liminf_{n\to\infty}\mathbb{E}|\tilde{Y}_t - Y_t^n|^2 = 0$, which establishes that Y_t is the mean square limit of Y_t^n.

It is also fairly easy to establish the fact that (1.17) exists $\mathbb{P}$-a.s. We have, by the Fubini-Tonelli theorem,

$$\mathbb{E}\left[\sum_{k\in\mathbb{Z}} |a_k X_{t-k}|\right] = \sum_{k\in\mathbb{Z}} |a_k|\,\mathbb{E}\left[|X_{t-k}|\right] \leq \sup_{t\in\mathbb{Z}} \mathbb{E}\left[|X_t|\right] \sum_{k\in\mathbb{Z}} |a_k|\,,$$

which is finite because $\sup_{t\in\mathbb{Z}} \mathbb{E}\left[|X_t|\right] < \infty$ and $\{a_j\}$ is absolutely summable. Hence $\sum_{j=-\infty}^{\infty} |a_j||X_{t-j}| < \infty$, $\mathbb{P}$-a.s. Therefore, the sequence $\{Y_t^n, n \in \mathbb{N}\}$ converges $(n \to \infty)$ absolutely to Y_t, $\mathbb{P}$-a.s. ■

1.2.2 *Spectral representation*

It is often advantageous to analyze the repetitive or regular behavior of a linear, stationary process based on its harmonic or oscillatory nature. This idea is the foundation of spectral analysis. In this case, it is easier to work with complex time series, say $X_t = X_{t1} + \mathrm{i}X_{t2}$ where $X_{t1} = \mathrm{Re}\,X_t$ and $X_{t2} = \mathrm{Im}\,X_t$; i.e., the state-space is $\mathsf{X} = \mathbb{C}$.

Stationarity follows as in the real case. Using obvious notation, X_t is stationary if $\mu = \mathbb{E}X_t = \mu_1 + \mathrm{i}\mu_2$ is independent of t and $\gamma(h) = \mathrm{Cov}(X_{t+h}, X_t) =$

$\mathbb{E}[(X_{t+h}-\mu)(X_t-\mu)^*]$, is independent of time t, where * indicates conjugation. In this case, $\gamma(h)$ may be a complex-valued function, but Proposition 1.10 still applies with appropriate changes. That is, (i) $\gamma(h)$ is a Hermitian function: $\gamma(h)=\gamma^*(-h)$, (ii) $0\le|\gamma(h)|\le\gamma(0)\in\mathbb{R}$, and (iii) $\gamma(h)$ is n.n.d. in the sense that, for any set of constants $(a_1,\dots,a_n)\in\mathbb{C}^n$, time points $(t_1,\dots,t_n)\in\mathbb{Z}^n$, and any $n\in\mathbb{N}^*$,

$$\sum_{i=1}^{n}\sum_{j=1}^{n} a_i^*\gamma(t_i-t_j)a_j \ge 0\,. \tag{1.20}$$

The following theorem states that the autocovariance function of a weakly stationary process $\{X_t,\, t\in\mathbb{Z}\}$ is entirely determined by a finite nonnegative measure on $(-\pi,\pi]$. This measure is called the *spectral measure* of $\{X_t,\, t\in\mathbb{Z}\}$.

Theorem 1.17 (Herglotz). *A sequence, $\{\gamma(h), h\in\mathbb{Z}\}$, is a nonnegative definite Hermitian sequence in the sense of* (1.20) *if and only if there exists a finite nonnegative measure ν on $(-\pi,\pi]$ such that*

$$\gamma(h)=\int_{-\pi}^{\pi} \mathrm{e}^{\mathrm{i}h\omega}\nu(\mathrm{d}\omega)\,, \quad \textit{for all } h\in\mathbb{Z}\,. \tag{1.21}$$

This relation defines ν uniquely.

Remark 1.18. By Proposition 1.10, Theorem 1.17 applies to all $\gamma(h)$ that are autocovariance functions of second-order stationary processes. If $\{X_t,\, t\in\mathbb{Z}\}$ is stationary, then ν (or ν_x if we need to identify the process) is called the *spectral measure* of $\{X_t,\, t\in\mathbb{Z}\}$. The *spectral distribution function* defined by $F_x(\omega)=\nu_x(-\pi,\omega]$ is right-continuous, non-decreasing and bounded on $[-\pi,\pi]$ with $F_x(-\pi)=0$ and $F_x(\pi)=\mathrm{Var}[X_t]$. If ν_x admits a density f_x with respect to the Lebesgue measure on $(-\pi,\pi]$, then f_x is referred to as the *spectral density function* of the process $\{X_t,\, t\in\mathbb{Z}\}$. In this case, $dF_x(\omega)=f_x(\omega)d\omega$.

An important situation that simplifies matters is the case where the autocovariance of a stationary process is absolutely summable. In this case, the spectral measure admits a density with respect to the Lebesgue measure.

Proposition 1.19 (Spectral Density). *If the autocovariance function $\{\gamma(h), h\in\mathbb{Z}\}$, of a stationary process satisfies*

$$\sum_{h=-\infty}^{\infty}|\gamma(h)|<\infty\,, \tag{1.22}$$

then it has the representation

$$\gamma(h)=\int_{-\pi}^{\pi} f(\omega)\mathrm{e}^{\mathrm{i}\,\omega h}\,d\omega\,, \qquad h=0,\pm1,\pm2,\dots, \tag{1.23}$$

as the inverse transform of the spectral density, which has the representation

$$f(\omega)=\frac{1}{2\pi}\sum_{h=-\infty}^{\infty}\gamma(h)\,\mathrm{e}^{-\mathrm{i}\,\omega h}\ge 0\,, \qquad -\pi\le\omega\le\pi\,. \tag{1.24}$$

The proofs of Theorem 1.17 and Proposition 1.19 may be found in advanced texts on time series such as Brillinger (2001), Brockwell and Davis (1991), Hannan (1970), Priestley (1981), or Shumway and Stoffer (2011). The spectral density is the analogue of the probability density function; the fact that $\gamma(h)$ is non-negative definite ensures $f(\omega) \geq 0$ for all $\omega \in [-\pi, \pi]$. It follows immediately from (1.24) that $f(\omega) = f(-\omega)$. In addition, putting $h = 0$ in (1.23) yields

$$\gamma(0) = \mathrm{Var}(X_t) = \int_{-\pi}^{\pi} f(\omega)\, d\omega\,,$$

which expresses the total variance as the integrated spectral density over all of the frequencies.

Example 1.20 (Spectral density of white noise). As discussed in Example 1.13, if $\{Z_t,\, t \in \mathbb{Z}\}$ is white noise, $\mathrm{WN}(0, \sigma_z^2)$, then its autocovariance function is $\gamma_z(h) = \sigma_z^2 \delta_h(0)$. The absolute summability condition of Proposition 1.19 is met, and by (1.24), its spectral density function is given by

$$f_z(\omega) = \frac{\sigma_z^2}{2\pi}\,,$$

for $\omega \in [-\pi, \pi]$, which is a uniform density. Hence the process contains equal power at all frequencies. In fact, the name *white noise* comes from the analogy to white light, which contains all frequencies in the color spectrum at the same level of intensity. ◇

Example 1.21 (Spectral measure of a harmonic process). Consider the complex-valued harmonic process $\{X_t,\, t \in \mathbb{Z}\}$ given by

$$X_t = \sum_{j=1}^{n} A_j \mathrm{e}^{\mathrm{i}t\omega_j}, \quad -\pi < \omega_1 < \cdots < \omega_n < \pi\,, \tag{1.25}$$

where the A_j are uncorrelated complex-valued random variables such that $\mathbb{E}[A_j] = 0$ and $\mathbb{E}\left[|A_j|^2\right] = \sigma_j > 0$. It follows that the autocovariance function of $\{X_t,\, t \in \mathbb{Z}\}$ is given by

$$\gamma_x(h) = \mathbb{E}[X_{t+h} X_t^*] = \sum_{j=1}^{n} \sigma_j^2\, \mathrm{e}^{\mathrm{i}h\omega_j}\,. \tag{1.26}$$

Note that the total variance of the process is the sum of the variances of the individual components, $\mathrm{Var}[X_t] = \gamma_x(0) = \sum_{j=1}^{n} \sigma_j^2$. The autocovariance function does not satisfy the absolute summability condition of Proposition 1.19, but we may write

$$\gamma_x(h) = \int_{-\pi}^{\pi} \sigma_j^2\, \mathrm{e}^{\mathrm{i}h\lambda}\, \delta_{\omega_j}(\mathrm{d}\lambda)\,, \tag{1.27}$$

where δ_{ω_j} denotes the Dirac mass at point $\omega_j \in (-\pi, \pi)$. Consequently, the spectral measure of $\{X_t,\, t \in \mathbb{Z}\}$ is given by

$$\nu_x = \sum_{j=1}^{n} \sigma_j^2\, \delta_{\omega_j}\,, \tag{1.28}$$

which is a sum of point masses with weights σ_j^2 located at the frequencies of the harmonic functions.

If the process $\{X_t,\, t \in \mathbb{Z}\}$ is real, then n must be even and there must be conjugate pairs in (1.25). For example, with $n = 2m$, for $1 \le j \le m$, let $\omega_{j+m} = -\omega_j$, write $A_j = (B_j - \mathrm{i}C_j)/2$ and $A_{j+m} = A_j^*$, with $\{B_j\}$ and $\{C_j\}$ being uncorrelated random variables with mean-zero and variance σ_j^2. In this case, $\mathbb{E}\left[|A_j|^2\right] = \sigma_j^2/2$, so that the spectral measure, (1.28), is now the sum of point masses with weights $\sigma_j^2/2$ at $\pm\omega_j$, for $j = 1, \ldots, m$. Note that

$$X_t = \sum_{j=1}^{n} A_j \mathrm{e}^{\mathrm{i}t\omega_j} = \sum_{j=1}^{m} \left(\frac{B_j - \mathrm{i}C_j}{2} \mathrm{e}^{\mathrm{i}t\omega_j} + \frac{B_j + \mathrm{i}C_j}{2} \mathrm{e}^{-\mathrm{i}t\omega_j} \right)$$

$$= \sum_{j=1}^{m} B_j \left(\frac{\mathrm{e}^{\mathrm{i}t\omega_j} + \mathrm{e}^{-\mathrm{i}t\omega_j}}{2} \right) - \mathrm{i}C_j \left(\frac{\mathrm{e}^{\mathrm{i}t\omega_j} - \mathrm{e}^{-\mathrm{i}t\omega_j}}{2} \right) = \sum_{j=1}^{m} B_j \cos(\omega_j t) + C_j \sin(\omega_j t)\,.$$

Because $B\cos(\omega t) + C\sin(\omega t)$ may be written as $D\cos(\omega t + \phi)$, where $B = D\cos(\phi)$ and $C = -D\sin(\phi)$, we may also write $X_t = \sum_{j=1}^{m} D_j \cos(\omega_j t + \phi_j)$, where ϕ_j is called the (random) *phase* of component j. Also, in the real case, we have $\gamma_x(h) = \sum_{j=1}^{m} \sigma_j^2 \cos(\omega_j h)$. ◇

In Proposition 1.16, we introduced the concept of a time invariant linear filter. In general, a linear filter uses a set of specified coefficients $\{a_j,\ j \in \mathbb{Z}\}$, to transform an input series, $\{X_t,\, t \in \mathbb{Z}\}$, producing an output series, $\{Y_t,\, t \in \mathbb{Z}\}$, of the form

$$Y_t = \sum_{j=-\infty}^{\infty} a_j X_{t-j}\,, \qquad \sum_{j=-\infty}^{\infty} |a_j| < \infty\,. \tag{1.29}$$

The coefficients $\{a_j\}$ are called the *impulse response function*, and the Fourier transform

$$A(\mathrm{e}^{-\mathrm{i}\omega}) = \sum_{j=-\infty}^{\infty} a_j \mathrm{e}^{-\mathrm{i}\omega j}\,, \tag{1.30}$$

is called the *frequency response function*.

The importance of the linear filter stems from its ability to enhance certain parts of the spectrum of the input series. This fact is expressed in the following proposition; see Exercise 1.8.

Proposition 1.22. *If, in* (1.29), $\{X_t,\, t \in \mathbb{Z}\}$ *has spectral density* f_x, *then*

$$f_y(\omega) = |A(\mathrm{e}^{-\mathrm{i}\omega})|^2 f_x(\omega)\,, \tag{1.31}$$

where f_y *is the spectral density of* $\{Y_t,\, t \in \mathbb{Z}\}$ *and the frequency response function* $A(\mathrm{e}^{-\mathrm{i}\omega})$ *is defined in* (1.30).

An important result that allows us to think of a stationary time series as being approximately a random superposition of sines and cosines is the celebrated Cramér spectral representation theorem. The result is based on an orthogonal increment process, $Z(\omega)$, by which is meant a mean-zero, finite variance, continuous-time stochastic process for which events occurring in non-overlapping intervals are uncorrelated.

Theorem 1.23 (Cramér). *If* $\{X_t,\ t \in \mathbb{Z}\}$ *is a mean-zero stationary process, with spectral measure as given in Theorem 1.17, then there exists a complex-valued orthogonal increment process* $\{Z(\omega), \omega \in (-\pi, \pi]\}$ *such that* $\{X_t,\ t \in \mathbb{Z}\}$ *can be written as the stochastic integral*

$$X_t = \int_{-\pi}^{\pi} \mathrm{e}^{\mathrm{i}t\omega}\, \mathrm{d}Z(\omega)\,,$$

where, for $-\pi < \omega_1 \leq \omega_2 \leq \pi$, $\mathrm{Var}\{Z(\omega_2) - Z(\omega_1)\} = \nu((\omega_1, \omega_2])$.

In this text, we focus primarily on time domain approaches. More details on the frequency domain approach to time series may be found in numerous texts such as Brillinger (2001), Brockwell and Davis (1991), Hannan (1970), Priestley (1981), or Shumway and Stoffer (2011).

1.2.3 Wold decomposition

The linear regression approach to modeling time series is generally implied by the assumption that the dependence between adjacent values in time is best explained in terms of a regression of the current values on the past values. This assumption is partially justified, in theory, by the Wold decomposition. First, for completeness, we state the following result for zero-mean, finite variance random variables.

Theorem 1.24 (L^2 **projection theorem).** *Let* $X \in \mathrm{L}^2$ *and suppose* $\mathcal{M}$ *is a closed subspace of* L^2. *Then* X *can be uniquely represented (a.e.) as*

$$X = \widehat{X} + Z$$

where $\widehat{X} \in \mathcal{M}$ *and* Z *is orthogonal to* $\mathcal{M}$*; i.e.,* $\mathbb{E}[ZW]=0$ *for all* W *in* $\mathcal{M}$. *Furthermore, the point* $\widehat{X}$ *is the closest to* X *in the sense that, for any* $W \in \mathcal{M}$, $\mathbb{E}[X - W]^2 \geq \mathbb{E}[X - \widehat{X}]^2$, *where equality holds if and only if* $W = \widehat{X}$ *(a.e.).*

We call the mapping $\mathsf{P}_{\mathcal{M}} X = \widehat{X}$, for $X \in \mathrm{L}^2$, the orthogonal projection mapping of X onto $\mathcal{M}$. Recall that the closed span of set $\{X_1, \ldots, X_n\}$ of elements in L^2 is defined to be the set of all linear combinations $W = a_1 X_1 + \cdots + a_n X_n$, where $a_1, \ldots, a_n$ are scalars. This subspace of L^2 is denoted by $\mathcal{M} = \overline{\mathrm{sp}}\{X_1, \ldots, X_n\}$.

For a stationary, zero-mean process $\{X_t,\ t \in \mathbb{Z}\}$, we define

$$\mathcal{M}_n^x = \overline{\mathrm{sp}}\{X_t,\ -\infty < t \leq n\}, \quad \text{with} \quad \mathcal{M}_{-\infty}^x = \bigcap_{n=-\infty}^{\infty} \mathcal{M}_n^x\,,$$

and

$$\sigma_x^2 = \mathbb{E}\left[X_{n+1} - \mathsf{P}_{\mathcal{M}_n^x} X_{n+1}\right]^2\,,$$

where $\mathsf{P}_{\mathcal{M}_n^x}(\cdot)$ denotes orthogonal projection onto the space $\mathcal{M}_n^x$. We say that $\{X_t,\ t \in \mathbb{Z}\}$ is a deterministic process if and only if $\sigma_x^2 = 0$. That is, a deterministic process is one in which its future is perfectly predictable from its past; e.g., $X_t = A\cos(\pi t)$, where $A \in \mathrm{L}^2$. We are now ready to present the decomposition.

Theorem 1.25 (The Wold decomposition). *Under the conditions and notation previously mentioned, if* $\sigma_x^2 > 0$, *then* X_t *can be expressed as*

$$X_t = \sum_{j=0}^{\infty} \psi_j Z_{t-j} + V_t$$

where

(i) $\sum_{j=0}^{\infty} \psi_j^2 < \infty$ *(*$\psi_0 = 1$*)*

(ii) $\{Z_t,\ t \in \mathbb{Z}\} \sim \mathrm{WN}(0, \sigma_z^2)$

(iii) $Z_t \in \mathcal{M}_t^x$

(iv) $\mathrm{Cov}(Z_s, V_t) = 0$ *for all* $s,t = 0, \pm 1, \pm 2, \ldots$

(v) $V_t \in \mathcal{M}_{-\infty}^x$

(vi) $\{V_t,\ t \in \mathbb{Z}\}$ *is deterministic.*

The proof of the decomposition follows from Theorem 1.24 by defining the unique sequences

$$Z_t = X_t - \mathsf{P}_{\mathcal{M}_{t-1}^x} X_t\,, \quad \psi_j = \sigma_z^{-2}\mathbb{E}\left[X_t Z_{t-j}\right], \quad V_t = X_t - \sum_{j=0}^{\infty} \psi_j Z_{t-j}\,.$$

Details may be found in Brockwell and Davis (1991). Although every stationary process can be represented by the Wold decomposition, it does not mean that the decomposition is the best way to describe the process. In addition, there may be some dependence structure among the $\{Z_t\}$; we are only guaranteed that the sequence is an uncorrelated sequence. The theorem, in its generality, falls short in that we would prefer the noise process, $\{Z_t\}$, to be i.i.d. noise.

By imposing extra structure, *conditional expectation* can be defined as a projection mapping for random variables in L^2 with the equivalence relation that, for $X, Y \in \mathrm{L}^2$, $X = Y$ if $\mathbb{P}(X = Y) = 1$. In particular, for $Y \in \mathrm{L}^2$, if $\mathcal{M}$ is a closed subspace of L^2 containing 1, the conditional expectation of Y given $\mathcal{M}$ is defined to be the projection of Y onto $\mathcal{M}$, namely, $\mathbb{E}_{\mathcal{M}}[Y] \equiv \mathbb{E}[Y|\mathcal{M}] = \mathsf{P}_{\mathcal{M}} Y$. This means that conditional expectation, $\mathbb{E}_{\mathcal{M}}$, must satisfy the orthogonality principle of the Projection Theorem, Theorem 1.24. If we let $\mathcal{M}(X)$ denote the closed subspace of all random variables in L^2 that can be written as a measurable function of X, then we may define, for $X, Y \in \mathrm{L}^2$, the *conditional expectation of* Y *given* X as $\mathbb{E}[Y|X] := \mathbb{E}[Y|\mathcal{M}(X)]$. This idea may be generalized in an obvious way to define the conditional expectation of Y given $\boldsymbol{X} = (X_1, \ldots, X_n)$; that is $\mathbb{E}[Y|\boldsymbol{X}] := \mathbb{E}[Y|\mathcal{M}(\boldsymbol{X})]$. Of particular interest to us is the following result, which states that, in the Gaussian case, conditional expectation and linear prediction are equivalent.

Theorem 1.26. *If* $(Y, X_1, \ldots, X_n)$ *is multivariate normal, then*

$$\mathbb{E}\left[Y \mid X_1, \ldots, X_n\right] = \mathsf{P}_{\overline{\mathrm{sp}}\{1, X_1, \ldots, X_n\}} Y\,.$$

Proof. It follows from Theorem 1.24 that the conditional expectation of Y given $\boldsymbol{X} = (X_1, \ldots, X_n)$ is the unique element $\mathbb{E}_{\mathcal{M}(\boldsymbol{X})} Y$ that satisfies the orthogonality principle,

$$\mathbb{E}\left[\left(Y - \mathbb{E}_{\mathcal{M}(\boldsymbol{X})} Y\right) W\right] = 0 \quad \text{for all } W \in \mathcal{M}(\boldsymbol{X})\,.$$

We must show that $\widehat{Y} = P_{\overline{\text{sp}}\{1,X_1,\dots,X_n\}}Y$ is that element. In fact, by the Projection Theorem (Theorem 1.24), $\widehat{Y}$ satisfies

$$\mathbb{E}\left[(Y-\widehat{Y})X_i\right] = 0 \quad \text{for } i = 0,1,\dots,n,$$

where we have set $X_0 = 1$. But $\mathbb{E}[(Y-\widehat{Y})X_i] = \text{Cov}(Y-\widehat{Y},X_i) = 0$, implying that $Y-\widehat{Y}$ and $\{X_1,\dots,X_n\}$ are independent because the vector $(Y-\widehat{Y},X_1,\dots,X_n)'$ is multivariate normal. Thus, if $W \in \mathcal{M}(\boldsymbol{X})$, then W and $Y-\widehat{Y}$ are independent and, hence, $\mathbb{E}\{(Y-\widehat{Y})W\} = \mathbb{E}(Y-\widehat{Y})\mathbb{E}(W) = 0$, recalling that $\mathbb{E}(Y-\widehat{Y}) = 0$. ∎

In the Gaussian case, conditional expectation has a simple explicit form. Let $\boldsymbol{Y} = (Y_1,\dots,Y_m)'$, $\boldsymbol{X} = (X_1,\dots,X_n)'$, and suppose the $(m+n)\times 1$ vector $(\boldsymbol{Y}',\boldsymbol{X}')'$ is normal:

$$\begin{pmatrix} \boldsymbol{Y} \\ \boldsymbol{X} \end{pmatrix} \sim \text{N}\left[\begin{pmatrix} \boldsymbol{\mu}_y \\ \boldsymbol{\mu}_x \end{pmatrix}, \begin{pmatrix} \Sigma_{yy} & \Sigma_{yx} \\ \Sigma_{xy} & \Sigma_{xx} \end{pmatrix}\right],$$

then $\boldsymbol{Y}|\boldsymbol{X}$ is normal with

$$\boldsymbol{\mu}_{y|x} = \boldsymbol{\mu}_y + \Sigma_{yx}\Sigma_{xx}^{-1}(\boldsymbol{X}-\boldsymbol{\mu}_x) \tag{1.32}$$

$$\Sigma_{y|x} = \Sigma_{yy} - \Sigma_{yx}\Sigma_{xx}^{-1}\Sigma_{xy}\,, \tag{1.33}$$

where Σ_{xx} is assumed to be nonsingular.

1.3 Linear processes

1.3.1 What are linear Gaussian processes?

The concept of second-order stationarity forms the basis for much of the analysis performed with linear Gaussian time series. Most models used in this situation are special cases of the linear process.

Definition 1.27 (Linear process). *A* linear process, $\{X_t,\ t\in\mathbb{Z}\}$, *is defined to be a linear combination of white noise variates* $Z_t \sim \text{WN}(0,\sigma_z^2)$, *and is given by*

$$X_t = \mu + \sum_{j=-\infty}^{\infty} \psi_j Z_{t-j}\,, \qquad \sum_{j=-\infty}^{\infty} \psi_j^2 < \infty\,. \tag{1.34}$$

The stronger condition of absolute summability, i.e., $\sum_{j=-\infty}^{\infty}|\psi_j| < \infty$, is often required and, when needed, we will make this distinction explicit. For example, the time invariant linear filter (see Proposition 1.16) is defined in such a way. Another frequent additional condition is that $Z_t \sim \text{iid}(0,\sigma_z^2)$; again, we will make this distinction when necessary. For a linear process, note that $X_t \in \text{L}^2$; also, direct calculation yields the autocovariance function,

$$\gamma_x(h) = \sigma_z^2 \sum_{j=-\infty}^{\infty} \psi_j \psi_{j+|h|}\,, \tag{1.35}$$

for $h \in \mathbb{Z}$.

Definition 1.28 (Gaussian process). *A process, $\{X_t,\ t \in \mathbb{Z}\}$, is said to be a* Gaussian process *if the n-dimensional vectors $\boldsymbol{X} = (X_{t_1}, X_{t_2}, \dots, X_{t_n})'$, for every collection of distinct time points $t_1, t_2, \dots, t_n$, and every positive integer n, have a nonsingular multivariate normal distribution.*

Defining the $n \times 1$ mean vector $\mathbb{E}(\boldsymbol{X}) \equiv \boldsymbol{\mu} = (\mu_{t_1}, \mu_{t_2}, \dots, \mu_{t_n})'$ and the $n \times n$ variance-covariance matrix as $\mathrm{Cov}(\boldsymbol{X}) \equiv \Gamma = \{\gamma(t_i, t_j);\ i, j = 1, \dots, n\}$, which is assumed to be positive definite, the multivariate normal density function is

$$g(\boldsymbol{x}; \boldsymbol{\mu}, \Gamma) = (2\pi)^{-n/2} |\Gamma|^{-1/2} \exp\left\{-\tfrac{1}{2}(\boldsymbol{x} - \boldsymbol{\mu})' \Gamma^{-1} (\boldsymbol{x} - \boldsymbol{\mu})\right\}, \tag{1.36}$$

where $|\cdot|$ denotes the determinant and $\boldsymbol{x} \in \mathbb{R}^n$. This distribution forms the basis for solving problems involving statistical inference for linear Gaussian time series. If a Gaussian time series is weakly stationary, then $\mu_t = \mu$ and $\gamma(t_i, t_j) = \gamma(|t_i - t_j|)$, so that the vector $\boldsymbol{\mu}$ and the matrix Γ are independent of time, t. These facts imply that all the finite distributions, (1.36), of the series $\{X_t\}$ depend only on time lag and not on the actual times, and hence the series must be strictly stationary.

If, in Definition 1.27, we assume that $Z_t \sim \text{iid N}(0, \sigma_z^2)$, then we are in the friendly world of stationary linear Gaussian processes. Many real processes can be approximated (i.e., modeled) under these assumptions, sometimes after coercion via simple transformations such as differencing or logging, and this forms the basis of classical, or second-order, time series analysis. However, many interesting processes are patently not linear or Gaussian, and that is the focus of this text. We note that linear processes need not be Gaussian, but by Theorem 1.25, (non-deterministic) stationary Gaussian processes are linear.

1.3.2 ARMA models

We briefly describe autoregressive-moving average (ARMA) models that were popularized by the work of Box and Jenkins (1970). There are many other texts that present an exhaustive treatment of these and associated models, e.g., Brockwell and Davis (1991), Fuller (1996), or Shumway and Stoffer (2011).

Definition 1.29 (ARMA model). *A time series $\{X_t,\ t \in \mathbb{Z}\}$ is* ARMA(p, q) *if it is stationary and*

$$X_t = \phi_1 X_{t-1} + \cdots + \phi_p X_{t-p} + Z_t + \theta_1 Z_{t-1} + \cdots + \theta_q Z_{t-q}\,, \tag{1.37}$$

with $\phi_p \neq 0$, $\theta_q \neq 0$, and $Z_t \sim \mathrm{WN}(0, \sigma_z^2 > 0)$. The parameters p and q are called the autoregressive and the moving average orders, respectively. If X_t has a nonzero mean μ, we set $\phi_0 = \mu(1 - \phi_1 - \cdots - \phi_p)$ and write the model as

$$X_t = \phi_0 + \phi_1 X_{t-1} + \cdots + \phi_p X_{t-p} + Z_t + \theta_1 Z_{t-1} + \cdots + \theta_q Z_{t-q}\,. \tag{1.38}$$

Although it is not necessary for the definition, it is typically assumed for the sake of inference and prediction that Z_t is Gaussian white noise. When $q = 0$, the model is called an autoregressive model of order p, AR(p), and when $p = 0$, the model is called a moving average model of order q, MA(q). It is sometimes advantageous to write the ARMA(p,q) model in concise form as

$$\phi(B)X_t = \theta(B)Z_t \,, \tag{1.39}$$

where B is the backshift operator

$$BX_t = X_{t-1} \,,$$

with $B^kX_t = B^{k-1}(BX_t) = X_{t-k}$, and where the *autoregressive operator* is defined to be

$$\phi(B) = 1 - \phi_1 B - \phi_2 B^2 - \cdots - \phi_p B^p \,, \tag{1.40}$$

and the *moving average operator* is

$$\theta(B) = 1 + \theta_1 B + \theta_2 B^2 + \cdots + \theta_q B^q \,. \tag{1.41}$$

Definition 1.30 (Causal/nonanticipative and invertible). *A linear process,* $\{X_t, t \in \mathbb{Z}\}$*, is said to be* causal *or* nonanticipative *if it can be written as a one-sided linear process,*

$$X_t = \sum_{j=0}^{\infty} \psi_j Z_{t-j} = \psi(B)Z_t \,, \tag{1.42}$$

where $\psi(B) = \sum_{j=0}^{\infty} \psi_j B^j$*, and* $\sum_{j=0}^{\infty} |\psi_j| < \infty$*; we set* $\psi_0 = 1$*. A linear process is said to be* invertible *if it can be written as*

$$\pi(B)X_t = \sum_{j=0}^{\infty} \pi_j X_{t-j} = Z_t \,, \tag{1.43}$$

where $\pi(B) = \sum_{j=0}^{\infty} \pi_j B^j$*, and* $\sum_{j=0}^{\infty} |\pi_j| < \infty$*; we set* $\pi_0 = 1$*.*

More generally, a process $\{X_t,\ t \in \mathbb{Z}\}$ *is said to be* causal *or* nonanticipative *if* $X_t = G(Z_s,\ s \le t)$ *where G is a measurable function such that* X_t *is a properly defined random variable. Likewise, if* $Z_t = G(X_s,\ s \le t)$*, the process* $\{X_t,\ t \in \mathbb{Z}\}$ *is said to be* invertible.

Causality (nonanticipativity) assures that the process depends only on the present and past innovations Z_t, and not on future errors. Invertibility is a similar useful property of the noise process; for example, see Exercise 1.3. We now state the conditions for which an ARMA model is both causal and invertible.

Proposition 1.31 (Causality and invertibility of ARMA). *Let* $z \in \mathbb{C}$ *and assume that there is no z for which* $\phi(z) = \theta(z) = 0$*, where* $\phi(\cdot)$ *and* $\theta(\cdot)$ *are defined in (1.40) and (1.41), respectively. An ARMA(p,q) model is causal if and only if* $\phi(z) \neq 0$ *for* $|z| \le 1$*. The coefficients of the linear process given in (1.42) can be determined by solving*

$$\psi(z) = \sum_{j=0}^{\infty} \psi_j z^j = \frac{\theta(z)}{\phi(z)}, \quad |z| \le 1 \,.$$

An ARMA(p,q) model is invertible if and only if $\theta(z) \neq 0$ for $|z| \leq 1$. The coefficients π_j of $\pi(B)$ given in (1.43) can be determined by solving

$$\pi(z) = \sum_{j=0}^{\infty} \pi_j z^j = \frac{\phi(z)}{\theta(z)}, \quad |z| \leq 1 .$$

Example 1.32 (ACF and PACF of ARMA). In view of causality, the autocovariance function of an ARMA(p,q) process is given by (1.35). Consequently, the ACF is given by

$$\rho(h) = \frac{\sum_{j=0}^{\infty} \psi_j \psi_{j+|h|}}{\sum_{j=0}^{\infty} \psi_j^2} . \tag{1.44}$$

If the process is pure MA(q), then $\psi_0 = 1$, $\psi_j = \theta_j$ for $j = 1,\ldots,q$, and $\psi_j = 0$ for $j > q$. Thus, for a pure MA process, $\rho(h) = 0$ for $|h| > q$, in which case the ACF can be used to identify the order q. For an AR(1), $\phi(z) = 1 - \phi z$, so that using Proposition 1.31, it is seen that $\psi_j = \phi^j$. In this case, $\rho(h) = \phi^{|h|}$, so that the ACF does not zero-out at any lag h. In general, for an AR(p) or ARMA(p,q), the ACF is not helpful in determining the orders because it does not zero-out as in the MA case.

A useful measure for determining the order of an AR(p) model is *partial autocorrelation function* (PACF), which extends the idea of partial correlation to time series. Recall that if $(X,Y,\mathbf{Z})$ is multivariate normal, then $\mathrm{Cor}(X,Y \mid \mathbf{Z})$ is the correlation coefficient between X and Y in the bivariate normal conditional distribution of X and Y given $\mathbf{Z} = (Z_1,\ldots,Z_k)$. This value is seen as the correlation between X and Y with the effect of $\mathbf{Z}$ removed (or partialled out). The definition can be extended to non-normal variables by defining $\mathrm{Cor}(X,Y \mid \mathbf{Z})$ to be $\mathrm{Cor}(X - \mathsf{P}_z X,\, Y - \mathsf{P}_z Y)$, where P_z denotes projection (or regression) onto $\overline{\mathrm{sp}}(1, Z_1,\ldots,Z_k)$. This definition is equivalent to the original definition in the Gaussian case. For a stationary time series, X_t, the PACF, denoted ϕ_{hh}, for $h = 1,2,\ldots$, is defined to be

$$\phi_{11} = \mathrm{Cor}(X_{t+1}, X_t) = \rho(1) \tag{1.45}$$

and

$$\phi_{hh} = \mathrm{Cor}(X_{t+h} - \mathsf{P}_{h-1}X_{t+h},\, X_t - \mathsf{P}_{h-1}X_t), \quad h \geq 2 , \tag{1.46}$$

where P_{h-1} denotes projection onto $\overline{\mathrm{sp}}\{1, X_{t+1}, X_{t+2},\ldots,X_{t+h-1}\}$. Thus, the PACF is seen as the correlation between X_{t+h} and X_t with the effect of everything in the middle being partialled out. For example, for a zero-mean AR(1) process, $X_t = \phi X_{t-1} + Z_t$, we have $\phi_{11} = \rho(1) = \phi$ and

$$\begin{aligned}\phi_{22} &= \mathrm{Cor}(X_{t+2} - \mathsf{P}_1 X_{t+2},\, X_t - \mathsf{P}_1 X_t) = \mathrm{Cor}(X_{t+2} - \phi X_{t+1},\, X_t - \mathsf{P}_1 X_t) \\ &= \mathrm{Cor}(Z_{t+2},\, X_t - \mathsf{P}_1 X_t) = 0\end{aligned}$$

because $X_t - \mathsf{P}_1 X_t$ is a function of $Z_t, Z_{t-1},\ldots$, which are uncorrelated with Z_{t+2}. Similar calculations will show $\phi_{hh} = 0$ for $h > 1$. Similarly, for a general AR(p), we have $\phi_{hh} = 0$ for $h > p$, which can be used to identify the order of the AR. ◇

Example 1.33 (Spectrum of ARMA). Using the results and notation of Proposition 1.22 and Proposition 1.31, the spectral density of an ARMA(p,q) process is given by

$$f_x(\omega) = f_z(\omega)\,|\psi(\mathrm{e}^{-\mathrm{i}\omega})|^2 = \frac{\sigma_z^2}{2\pi}\frac{|\theta(\mathrm{e}^{-\mathrm{i}\omega})|^2}{|\phi(\mathrm{e}^{-\mathrm{i}\omega})|^2}, \tag{1.47}$$

recalling from Example 1.20 that $f_z(\omega) = \sigma_z^2/2\pi$. ◇

1.3.3 Prediction

In prediction or forecasting, the goal is to predict future values of a time series, X_{n+m}, $m = 1, 2, \ldots$, based on the data collected to the present, $X_{1:n} = \{X_1, \ldots, X_n\}$. Let $\mathcal{F}_n = \sigma(X_1, \ldots, X_n)$, then the minimum mean square error predictor of X_{n+m} is

$$X_{n+m|n} := \mathbb{E}\left[X_{n+m} \mid \mathcal{F}_n\right] \tag{1.48}$$

because the conditional expectation minimizes the mean square error

$$\mathbb{E}\left[X_{n+m} - g(X_{1:n})\right]^2, \tag{1.49}$$

where $g : \mathbb{R}^n \to \mathbb{R}$ is a function of the observations.

Except in the Gaussian case, it is typically difficult to perform optimal prediction because all the finite dimensional distributions must be known to compute (1.48). Typically, one makes a compromise and works with best linear prediction; that is, predictors of the form

$$X_{n+m|n} = \phi_{n0}^{(m)} + \sum_{j=1}^{n} \phi_{nj}^{(m)} X_{n+1-j}, \tag{1.50}$$

where $\phi_{n0}^{(m)}, \phi_{n1}^{(m)}, \ldots, \phi_{nn}^{(m)}$ are real numbers. Linear predictors of the form (1.50) that minimize the mean square prediction error (1.49) are called *best linear predictors* (BLPs). Linear prediction depends only on the second-order moments of the process, which are easy to estimate from the data in the stationary case. If, in addition, the process is Gaussian, minimum mean square error predictors and best linear predictors are the same. We note that in the stationary case we can eliminate $\phi_{n0}^{(m)}$ by centering the process; i.e., in (1.50), we replace $\phi_{n0}^{(m)}$ by μ and X_t by $X_t - \mu$; consequently, we drop $\phi_{n0}^{(m)}$ from the discussion. The following property, which is based on Theorem 1.24, is a key result; e.g., see Shumway and Stoffer (2011).

Proposition 1.34 (Best linear prediction for stationary process). *Given observations $X_1, \ldots, X_n$, the (a.e. unique) best linear predictor, $X_{n+m|n} = \sum_{j=1}^{n} \phi_{nj}^{(m)} X_{n+1-j}$, of X_{n+m}, for $m \geq 1$, is found by solving*

$$\mathbb{E}\left[(X_{n+m} - X_{n+m|n})X_t\right] = 0, \quad t = 1, \ldots, n, \tag{1.51}$$

for $\phi_{n1}^{(m)}, \ldots, \phi_{nn}^{(m)}$. The prediction equations (1.51) can be written in matrix notation as

$$\Gamma_n \phi_n^{(m)} = \gamma_n^{(m)}, \tag{1.52}$$

where $\phi_n^{(m)} = (\phi_{n1}^{(m)}, \dots, \phi_{nn}^{(m)})'$ *and* $\gamma_n^{(m)} = (\gamma(m), \dots, \gamma(m+n-1))'$ *are* $n \times 1$ *vectors, and* $\Gamma_n = \{\gamma(i-j)\}_{i,j=1}^n$ *is an* $n \times n$ *matrix, If* Γ_n *is non-singular, then the* ϕ*s are unique and may be obtained as*

$$\phi_n^{(m)} = \Gamma_n^{-1} \gamma_n^{(m)}. \tag{1.53}$$

In addition, the m-step-ahead mean square prediction error can be computed as

$$P_{n+m|n} := \mathbb{E}[X_{n+m} - X_{n+m|n}]^2 = \gamma(0) - \gamma_n^{(m)\prime} \Gamma_n^{-1} \gamma_n^{(m)}. \tag{1.54}$$

Moreover, $\phi_{hh}^{(1)}$ *as defined here, is equivalent to the PACF,* ϕ_{hh}*, defined in (1.45)–(1.46).*

The Durbin-Levinson algorithm, Levinson (1947) and Durbin (1960), can be used to obtain the components of $\phi_n^{(1)}$ recursively, $n = 1, 2, \dots,$ without having to invert Γ_n. Most texts on linear time series discuss the algorithm. We also note that Proposition 1.34 implies that the PACF, ϕ_{hh}, is a function of $\gamma(h)$ alone, and consequently it may be estimated easily from the data via (1.15).

Example 1.35 (BLP versus minimum mean square prediction). Suppose X and Z are independent standard normal random variables and let

$$Y = X^2 + Z\,.$$

We wish to predict Y based on the data X. The minimum mean square predictor is given by $\mathbb{E}[Y|X] = X^2$ with mean square prediction error (MSPE)

$$\mathbb{E}[Y - \mathbb{E}(Y|X)]^2 = \mathbb{E}[Y - X^2]^2 = \mathbb{E}Z^2 = 1\,.$$

Now, let $g(X) = a + bX$ be the BLP. Then using Proposition 1.34, $g(X)$ satisfies

$$\mathbb{E}\left[Y - g(X)\right] = 0 \quad \text{and} \quad \mathbb{E}\left[(Y - g(X))X\right] = 0\,,$$

or

$$\mathbb{E}[Y] = \mathbb{E}\left[a + bX\right] \quad \text{and} \quad \mathbb{E}\left[XY\right] = \mathbb{E}\left[(a + bX)X\right]\,.$$

From the first equation we have $a + b\mathbb{E}[X] = \mathbb{E}[Y]$, but $\mathbb{E}\left[X\right] = 0$ and $\mathbb{E}\left[Y\right] = 1$, so $a = 1$. From the second equation we have $a\mathbb{E}[X] + b\mathbb{E}[X]^2 = \mathbb{E}[XY]$, or $b = \mathbb{E}\left[X\left(X^2 + Z\right)\right] = \mathbb{E}[X^3] + \mathbb{E}[XZ] = 0 + 0$. Consequently, the BLP is $g(X) = 1$, independent of the data X, and thus the MSPE is

$$\mathbb{E}\left[Y - 1\right]^2 = \mathbb{E}\left[Y^2\right] - 1 = \mathbb{E}[X]^4 + \mathbb{E}[Z]^2 - 1 = 3 + 1 - 1 = 3\,,$$

noting that the fourth moment of a standard normal is 3.

We see that using linear prediction in a nonlinear setting can give strange results; here, the predictor does not even rely on the data and the BLP has three times the error of the optimal predictor (conditional expectation). Three times the error may not seem that large at first, but we forgot to mention at the start of this example that the units of the problem are in trillions of dollars. ◇

As previously mentioned, BLP and minimum mean square prediction are the equivalent in the Gaussian case. For example, given data vector $\boldsymbol{X}_n = (X_n, \ldots, X_1)'$, suppose we are interested in the one-step-ahead predictor, $X_{n+1|n}$, and the corresponding MSPE, $P_{n+1|n}$. Then, using standard multivariate normal distribution theory, we may write

$$X_{n+1|n} = \mu + \boldsymbol{\gamma}_n' \Gamma_n^{-1} (\boldsymbol{X}_n - \boldsymbol{\mu}_n) \tag{1.55}$$

$$P_{n+1|n} = \gamma(0) - \boldsymbol{\gamma}_n' \Gamma_n^{-1} \boldsymbol{\gamma}_n \,, \tag{1.56}$$

where $\mu = \mathbb{E}X_t$, $\boldsymbol{\mu}_n = \mathbb{E}\boldsymbol{X}_n = (\mu, \ldots, \mu)'$ and $\boldsymbol{\gamma}_n = \text{Cov}\,(X_{n+1}, \boldsymbol{X}_n) = (\gamma(1), \ldots, \gamma(n))'$ are $n \times 1$ vectors, and $\Gamma_n = \text{Cov}\,(\boldsymbol{X}_n) = \{\gamma(i-j)\}_{i,j=1}^n$ is an $n \times n$ matrix.

For ARMA models, many simplifications are available. For example, if X_t is AR(1), then, for $n \geq 1$, $X_{n+1|n} = \phi_0 + \phi_1 X_n$ and $P_{n+1|n} = \mathbb{E}[X_{n+1} - (\phi_0 + \phi_1 X_n)]^2 = \mathbb{E}\left[Z_t^2\right] = \sigma_z^2$. In addition, $X_{1|0} = \mathbb{E}\left[X_1\right] = \mu = \phi_0/(1-\phi_1)$ and $P_{1|0} = \text{Var}[X_1] = \sigma_z^2/(1-\phi_1)^2$; see Exercise 1.5. It should be evident that prediction for pure AR(p) models, when $n \geq p$, parallels the AR(1) case; i.e.,

$$X_{n+1|n} = \phi_0 + \phi_1 X_{n+1-1} + \phi_2 X_{n+1-2} + \cdots + \phi_p X_{n+1-p} \,, \tag{1.57}$$

and $P_{n+1|n} = \sigma_z^2$. For $n < p$, the prediction equations (1.55) – (1.56) can be used.

For general ARMA(p,q), one can simplify matters by assuming that a complete history is available; i.e., the data are $\{X_n, \ldots, X_1, X_0, X_{-1}, \ldots\}$. Using invertibility, we can write $X_{n+1} = \sum_{j=1}^{\infty} \pi_j X_{n+1-j} + Z_t$. Consequently,

$$X_{n+1|n} = \sum_{j=1}^{\infty} \pi_j X_{n+1-j} \,. \tag{1.58}$$

Of course, only the data $X_1, \ldots, X_n$ are available, so (1.58) is truncated to $X_{n+1|n} \approx \sum_{j=1}^{n} \pi_j X_{n+1-j}$, the idea being that if n is sufficiently large, there is negligible difference between (1.58) and the truncated version. From (1.58), it is also evident that $P_{n+1|n} = \mathbb{E}\left[(X_{n+1} - X_{n+1|n})^2\right] = \mathbb{E}\left[Z_{n+1}^2\right] = \sigma_z^2$, and this approximation is used in the truncated case. Forecasting ARMA processes m steps ahead uses similar approximations, see, e.g., Shumway and Stoffer (2011, Chapter 3).

1.3.4 Estimation

Time domain

Estimation for pure AR(p) models is fairly easy because they are essentially linear regression models. For example, for the AR(1) model, $X_t = \phi_0 + \phi_1 X_{t-1} + Z_t$, given a realization $x_1, \ldots, x_n$, one can perform ordinary least squares ("$Y = \beta_0 + \beta_1 x + \varepsilon$") on the $n-1$ data pairs $\{(y,x): (x_2,x_1), (x_3,x_2), \ldots, (x_n,x_{n-1})\}$. The approach for an AR(p) is similar, and the technique is efficient. An alternate method that is also efficient is Yule–Walker estimation, which is method of moments estimation (MME). Recall that in MME, we equate population moments to sample moments and then solve for the parameters in terms of the sample moments. Because $\mathbb{E}\left[X_t\right] = \mu$, the

MME of μ is the sample average, $\overline{X}$; thus to ease the notation, we will assume $\mu = 0$. For an AR(p) process,

$$X_t = \phi_1 X_{t-1} + \cdots + \phi_p X_{t-p} + Z_t \,,$$

multiply through by X_{t-h}, for $h \geq 0$ and take expectation to obtain the following $p+1$ Yule–Walker equations (see Exercise 1.4),

$$\gamma(h) = \phi_1 \gamma(h-1) + \cdots + \phi_p \gamma(h-p)\,, \quad h = 1, 2, \ldots, p, \tag{1.59}$$

$$\sigma_z^2 = \gamma(0) - \phi_1 \gamma(1) - \cdots - \phi_p \gamma(p)\,. \tag{1.60}$$

Next, replace $\gamma(h)$ with $\widehat{\gamma}(h)$ (see (1.15)) and solve for ϕs and σ_z^2.

Maximum likelihood estimation for normal ARMA(p,q) models proceeds as follows. Let $\beta = (\mu, \phi_1, \ldots, \phi_p, \theta_1, \ldots, \theta_q)'$ be the $(p+q+1)$-dimensional vector of the model parameters. The likelihood can be written as

$$\mathrm{L}(\beta, \sigma_z^2; X_{1:n}) = \prod_{t=1}^{n} \mathrm{p}^{\beta, \sigma_z^2}(X_t \mid X_{t-1}, \ldots, X_1)\,,$$

where $\mathrm{p}^{\beta,\sigma_z^2}(\cdot \mid \cdot)$ is the $\mathrm{N}(X_{t|t-1}, P_{t|t-1})$ distribution. For the ARMA model, the noise variance, σ_z^2, may be factored out and we can write $P_{t|t-1} = \sigma_z^2 r_t(\beta)$, where $r_t(\cdot)$ depends only on β. This fact can be seen from (1.56) by factoring out σ_z^2 from the $\gamma(h)$ terms; recall (1.35). The likelihood of the data can now be written as

$$\mathrm{L}(\beta, \sigma_z^2; X_{1:n}) = (2\pi\sigma_z^2)^{-n/2} \left[r_1(\beta) r_2(\beta) \cdots r_n(\beta)\right]^{-1/2} \exp\left[-\frac{S(\beta)}{2\sigma_z^2}\right], \tag{1.61}$$

where

$$S(\beta) = \sum_{t=1}^{n} \left\{ \frac{\left[X_t - X_{t|t-1}(\beta)\right]^2}{r_t(\beta)} \right\}. \tag{1.62}$$

Note that $X_{t|t-1}$ is also a function of β alone, and we make that fact explicit in (1.61) – (1.62). Given a realization $x_1, \ldots, x_n$, and values for β and σ_z^2, the likelihood may be evaluated using the prediction techniques of the previous subsection.

Maximum likelihood estimation would now proceed by maximizing (1.61) with respect to β and σ_z^2. Note that

$$\widehat{\sigma}_z^2 = n^{-1} S(\widehat{\beta})\,, \tag{1.63}$$

where $\widehat{\beta}$ is the value of β that minimizes the concentrated or profile likelihood,

$$l(\beta) = \ln\left[n^{-1} S(\beta)\right] + n^{-1} \sum_{t=1}^{n} \ln r_t(\beta)\,, \tag{1.64}$$

where $l(\beta) \propto -2 \ln \mathrm{L}(\beta, \widehat{\sigma}_z^2)$.

Because $l(\beta)$ is a complicated function of $p+q+1$ parameters, optimization routines are employed. Numerical methods for accomplishing maximum likelihood estimation are discussed in Chapter 8. When fitting ARMA models to data, one often considers multiple models. In such a cases, we often rely on the theory of parsimony to choose the best model. The choice is typically aided by the use of various information theoretic criteria that take the general form

$$\Delta_{k,n} = -2\ln \mathrm{L}_k + C_{k,n} , \tag{1.65}$$

where L_k is the value of the maximized likelihood, k is the number of regression parameters in the model, n is the sample size and $C_{k,n}$ is a penalty for adding parameters. The most used are (i) Akaike's Information Criterion (AIC), wherein $C_{k,n} = 2k$, Akaike (1973, 1974); (ii) Corrected AIC (AICc), wherein $C_{k,n} = \frac{n+k}{n-k-2}$, Hurvich and Tsai (1993); (iii) Bayesian (or Schwarz's) Information Criterion (BIC), wherein $C_{k,n} = k\ln(n)$, Schwarz (1978); and (iv) Hannan-Quinn (HQ), wherein $C_{k,n} = 2k\ln\ln(n)$, Hannan and Quinn (1979). Most texts on time series analysis provide further discussion of selection criteria; see McQuarrie and Tsai (1998) for a general reference.

Frequency domain

There are two basic methods for estimating a spectral density; parametric estimation and nonparametric estimation. In Example 1.33, we exhibited the spectrum of an ARMA process and we might consider basing a spectral estimator on this function, substituting the parameter estimates from an ARMA(p,q) fit on the data into the formula for the spectral density $f_x(\omega)$ given in (1.47). Such an estimator is called a parametric spectral estimator. For convenience, a parametric spectral estimator is obtained by fitting an AR(p) to the data, where the order p is determined by one of the model selection criteria. Parametric autoregressive spectral estimators will often have superior resolution in problems when several closely spaced narrow spectral peaks are present and are preferred by engineers for a broad variety of problems; e.g., see Kay (1988). The development of autoregressive spectral estimators has been summarized by Parzen (1983).

If $\widehat{\phi}_1, \widehat{\phi}_2, \ldots, \widehat{\phi}_p$ and $\widehat{\sigma}_z^2$ are the estimates from an AR(p) fit to observations $X_1, \ldots, X_n$, then a parametric spectral estimate of $f_x(\omega)$ is attained by substituting these estimates into (1.47), that is,

$$\widehat{f}_x(\omega) = \frac{\widehat{\sigma}_z^2}{2\pi} \left|\widehat{\phi}\left(\mathrm{e}^{-\mathrm{i}\omega}\right)\right|^{-2}, \tag{1.66}$$

where

$$\widehat{\phi}(z) = 1 - \widehat{\phi}_1 z - \widehat{\phi}_2 z^2 - \cdots - \widehat{\phi}_p z^p . \tag{1.67}$$

An interesting fact about rational spectra of the form (1.47) is that any spectral density can be approximated, arbitrarily close, by the spectrum of an AR process; e.g., see Brockwell and Davis (1991) or Fuller (1996). In a sense, we can say that the spectral densities of AR processes are dense in the space of continuous spectral densities.

Nonparametric spectral estimation is a little more involved and for full details, we refer the reader to other texts such as Brockwell and Davis (1991) or Shumway and Stoffer (2011). The basic building block is the *discrete Fourier transform* (DFT). Given data $X_1,\ldots,X_n$, the DFT is defined to be

$$d(\omega_j) = (2\pi n)^{-1/2} \sum_{t=1}^{n} X_t e^{-i\omega_j t} \tag{1.68}$$

for $j = 0, 1, \ldots, n-1$, where the frequencies $\omega_j = 2\pi j/n$ are called the *Fourier* or *fundamental frequencies*. The *periodogram* is then defined as the squared modulus of the DFT,

$$I(\omega_j) = \left|d(\omega_j)\right|^2 \tag{1.69}$$

for $j = 0, 1, 2, \ldots, n-1$. Replacing X_t by $X_t - \overline{X}$ in (1.68), it is easily shown (e.g., Shumway and Stoffer, 2011) that

$$I(\omega_j) = \frac{1}{2\pi} \sum_{h=-(n-1)}^{n-1} \widehat{\gamma}(h)\, e^{-i\omega_j h},$$

for $\omega_j \neq 0, \frac{1}{2}$, which, in view of (1.24), shows that the periodogram may be considered the "sample" spectral density. One can extend the peridogram to all $\omega \in [-\pi, \pi]$ by defining $I(\omega) = I(\omega_j)$ for $|\omega_j - \omega| \leq 1/2n$ and $I(-\omega) = I(\omega)$. Although

$$\mathbb{E}\left[I(\omega)\right] \to f(\omega)$$

as $n \to \infty$, it can be shown that under mild conditions, $\text{Var}[I(\omega)] \to f^2(\omega)$, so that the periodogram is not a consistent estimator of the spectral density $f(\omega)$. This problem is overcome by local smoothing of the periodogram in a neighborhood of a frequency of interest, $\{\omega_j - \frac{m_n}{n} \leq \omega \leq \omega_j + \frac{m_n}{n}\}$, say

$$\widehat{f}(\omega) = \sum_{k=-m_n}^{m_n} h_k I(\omega_j + k/n), \tag{1.70}$$

where $m_n \to \infty$ but $m_n/n \to 0$ as $n \to \infty$, and the weights $h_k > 0$ satisfy $(n \to \infty)$

$$\sum_{k=-m_n}^{m_n} h_k = 1 \quad \text{and} \quad \sum_{k=-m_n}^{m_n} h_k^2 \to 0.$$

In essence, one may think of the periodogram as a histogram, and (1.70) as smoothing the histogram to obtain an estimate of the smooth density $f(\omega)$. A simple average corresponds to the case where $h_q = 1/(2m+1)$ for $q = -m,...,0,...,m$. The number m is chosen to obtain a desired degree of smoothness. Larger values of m lead to smoother estimates, but one has to be careful not to smooth away significant peaks (this is the bias-variance tradeoff problem). Experience and trial-and-error can be used to select good values of m and the set of weights $\{h_q\}$. Another consideration is that of tapering the data prior to a spectral analysis; i.e. rather than work with the data

X_t directly, one can improve the estimation of spectra by working with tapered data, say $Y_t = a_t X_t$, where tapers $\{a_t\}$ generally have a shape that enhances the center of the data relative to the extremities, such as a cosine bell, $a_t = .5[1+\cos(2\pi t'/n)]$ where $t' = t-(n+1)/2$, favored by Blackman and Tukey (1959). Another related approach is window spectral estimation. Specifically, consider a window function $H(\alpha)$, $-\infty < \alpha < \infty$, that is real-valued, even, of bounded variation, with $\int_{-\infty}^{\infty} H(\alpha)d\alpha = 1$, and $\int_{-\infty}^{\infty} |H(\alpha)|d\alpha < \infty$. The window spectral estimator is

$$\widehat{f}(\omega) = n^{-1} \sum_{q=1}^{n-1} H_n(\omega - q/n) I(q/n) , \tag{1.71}$$

where $H_n(\alpha) = B_n^{-1} \sum_{j=-\infty}^{\infty} H(B_n^{-1}[\alpha + j])$ and B_n is a bounded sequence of non-negative scale parameters such that $B_n \to 0$ and $nB_n \to \infty$ as $n \to \infty$. Estimation of the spectral density requires special attention to the issues of leakage and of the variance-bias tradeoff typically associated with the estimation of density functions. Readers who are unfamiliar with this material may consult one of the many texts on the spectral domain analysis of time series; e.g., Brillinger (2001), Bloomfield (2004), or Shumway and Stoffer (2011, Chapter 4).

1.4 The multivariate cases

1.4.1 Time domain

In this situation, we are interested in modeling and forecasting $k \times 1$ vector-valued time series $X_t = (X_{t1}, \ldots, X_{tk})'$, for $t = 0, \pm 1, \pm 2, \ldots$. Unfortunately, extending univariate ARMA models to the multivariate case is not so simple. The multivariate autoregressive model, however, is a straight-forward extension of the univariate AR model.

A $k \times 1$ vector-valued time series X_t, for $t = 0, \pm 1, \pm 2, \ldots$, is said to be VARMA$(p,q)$ if X_t is stationary and

$$X_t = \Phi_0 + \Phi_1 X_{t-1} + \cdots + \Phi_p X_{t-p} + Z_t + \Theta_1 Z_{t-1} + \cdots + \Theta_q Z_{t-q} , \tag{1.72}$$

with $\Phi_p \neq 0$, $\Theta_q \neq 0$, and the *vector white noise* process Z_t, typically taken to be multivariate normal with mean-zero and variance-covariance matrix $\Sigma_z > 0$ (that is, Σ_z is positive definite). The coefficient matrices Φ_j for $j = 1, \ldots, p$ and Θ_j for $j = 1, \ldots, q$ are $k \times k$ matrices. If $q = 0$, the model is a pure VAR(p) model, and if $p = 0$, the model is a pure VMA(q) model. If X_t has mean μ then $\Phi_0 = (I - \Phi_1 - \cdots - \Phi_p)\mu$. As in the univariate case, a number of conditions must be placed on the multivariate ARMA model to ensure the model is unique and has desirable properties such as causality. The special form assumed for the constant component Φ_0 can be generalized to include a fixed $r \times 1$ vector of inputs, U_t. That is, we could have proposed the *vector ARMAX model*,

$$X_t = \Upsilon U_t + \sum_{j=1}^{p} \Phi_j X_{t-j} + \sum_{k=1}^{q} \Theta_k Z_{t-k} + Z_t , \tag{1.73}$$

where Υ is a $k \times r$ parameter matrix.

In the multivariate case, the *autoregressive operator* is

$$\Phi(B) = I - \Phi_1 B - \cdots - \Phi_p B^p , \tag{1.74}$$

and the *moving average operator* is

$$\Theta(B) = I + \Theta_1 B + \cdots + \Theta_q B^q , \tag{1.75}$$

The zero-mean VARMA(p,q) model is then written in the concise form as

$$\Phi(B) X_t = \Theta(B) Z_t . \tag{1.76}$$

The model is said to be *causal* if the roots of $\Phi(z)$ lie outside the unit circle, i.e., $\det\{\Phi(z)\} \neq 0$ for any value z such that $|z| \leq 1$. In this case, we can write

$$X_t = \Psi(B) Z_t ,$$

where $\Psi(B) = \sum_{j=0}^{\infty} \Psi_j B^j$, $\Psi_0 = I$, and $\sum_{j=0}^{\infty} ||\Psi_j|| < \infty$. The model is said to be *invertible* if the roots of $\Theta(z)$ lie outside the unit circle. Then, we can write

$$Z_t = \Pi(B) X_t ,$$

where $\Pi(B) = \sum_{j=0}^{\infty} \Pi_j B^j$, $\Pi_0 = I$, and $\sum_{j=0}^{\infty} ||\Pi_j|| < \infty$. Analogous to the univariate case, we can determine the matrices Ψ_j by solving $\Psi(z) = \Phi(z)^{-1}\Theta(z)$, for $|z| \leq 1$, and the matrices Π_j by solving $\Pi(z) = \Theta(z)^{-1}\Phi(z)$, for $|z| \leq 1$.

For a causal model, we can write $X_t = \Psi(B) Z_t$ so the general autocovariance structure of an ARMA(p,q) model is

$$\Gamma(h) = \mathrm{Cov}(X_{t+h}, X_t) = \sum_{j=0}^{\infty} \Psi_{j+h} \Sigma_z \Psi_j' . \tag{1.77}$$

and $\Gamma(-h) = \Gamma(h)'$. For pure MA(q) processes, (1.77) becomes

$$\Gamma(h) = \sum_{j=0}^{q-h} \Theta_{j+h} \Sigma_z \Theta_j' , \tag{1.78}$$

where $\Theta_0 = I$. Of course, (1.78) implies $\Gamma(h) = 0$ for $h > q$.

For pure VAR(p) models, the autocovariance structure leads to the multivariate version of the *Yule–Walker equations*

$$\Gamma(h) = \sum_{j=1}^{p} \Phi_j \Gamma(h-j), \quad h = 1, 2, ..., \tag{1.79}$$

$$\Gamma(0) = \sum_{j=1}^{p} \Phi_j \Gamma(-j) + \Sigma_z . \tag{1.80}$$

where $\Gamma(h) = \mathrm{Cov}(X_{t+h}, X_t)$ is a $k \times k$ matrix. Analogous to the univariate case, these equations can be used to obtain method of moment estimators of the model parameters by replacing $\Gamma(h)$ with the corresponding sample moments, and solving for the Φs and Σ_z. For moment estimation of the autocovariance matrix, we set $\overline{X} = n^{-1}\sum_{t=1}^{n} X_t$, as an estimate of $\mu = \mathbb{E}X_t$,

$$\widehat{\Gamma}(h) = n^{-1}\sum_{t=1}^{n-h}(X_{t+h}-\overline{X})(X_t-\overline{X})', \quad h = 0,1,2,..,n-1, \tag{1.81}$$

and $\widehat{\Gamma}(-h) = \widehat{\Gamma}(h)'$.

As previously mentioned, there are many problems with the general VARMA model that have to do with parameter uniqueness and estimation. For further details, the reader is referred to Lütkepohl (2005), Reinsel (2003), or Shumway and Stoffer (2011). These problems may be avoided by using only pure models, and typically that means relying on VAR(p) models. In this case, estimation is fairly straight-forward via the Yule–Walker equations, or by realizing that conditional on a few stating values, $X_1,\ldots,X_p$, the model is multivariate linear regression. Maximum likelihood estimation in the multivariate setting can also be performed numerically as in the scalar case, but in this case, Σ_z cannot be factored out of the likelihood.

1.4.2 Frequency domain

As in Section 1.2.2, it is convenient to work with complex-valued time series. A $p\times 1$ complex-valued time series can be represented as $X_t = X_{1t} - \mathrm{i}X_{2t}$, where X_{1t} is the real part and X_{2t} is the imaginary part of X_t. The process is said to be stationary if $\mathbb{E}(X_t)$ and $\mathbb{E}(X_{t+h}X_t^*)$ exist and are independent of time t, where * denotes conjugation and transposition. The $p\times p$ autocovariance function,

$$\Gamma(h) = \mathbb{E}(X_{t+h}X_t^*) - \mathbb{E}(X_{t+h})\mathbb{E}(X_t^*)\,,$$

of X_t satisfies conditions similar to those of the real-valued case. Writing $\Gamma(h) = \{\gamma_{ij}(h)\}$, for $i,j = 1,\ldots,p$, we have (i) $\gamma_{ii}(0)\ge 0$ is real, (ii) $|\gamma_{ij}(h)|^2 \le \gamma_{ii}(0)\gamma_{jj}(0)$ for all integers h, and (iii) $\Gamma(h)$ is a non-negative definite function. The spectral theory of complex-valued vector time series is analogous to the real-valued case. For example, if $\sum_h ||\Gamma(h)|| < \infty$, the spectral density matrix of the complex series X_t is given by

$$f(\omega) = \frac{1}{2\pi}\sum_{h=-\infty}^{\infty}\Gamma(h)\mathrm{e}^{-\mathrm{i}h\omega}\,.$$

The off-diagonal elements of $f(\omega)$, say $f_{ij}(\omega)$, for $i\ne j = 1,\ldots,p$ are called *cross-spectra*. Typically one is interested in *squared-coherence*, which is a frequency based measure of squared correlation, defined as

$$\rho_{ij}^2(\omega) = \frac{|f_{ij}(\omega)|^2}{f_{ii}(\omega)f_{jj}(\omega)}\,. \tag{1.82}$$

As with squared correlation, $0\le \rho_{ij}^2(\omega)\le 1$, with values close to one indicating a strong linear relationship between X_{it} and X_{jt} at frequency ω.

Estimation can be achieved, as in the univariate case, by parametric or nonparametric methods. If a VAR(p) model is assumed for the process, then the spectral density is

$$f(\omega) = \frac{1}{2\pi}\Phi^{-1}(e^{i\omega})\Sigma_z\Phi'^{-1}(e^{-i\omega}),$$

where $\Phi(z)$ is given in (1.74). Once the VAR parameters are estimated, one simply substitutes the parameters for their estimates in the above. In the nonparametric case, the vector DFT is calculated,

$$d(\omega_j) = (2\pi n)^{-1/2}\sum_{t=1}^{n} X_t e^{-i\omega_j t},$$

in which case the periodogram is a $p \times p$ matrix, $I(\omega_j) = d(\omega_j)d(\omega_j)^*$. Consistency and non-negative definiteness are still concerns, so a smoothed estimator is used,

$$\widehat{f}(\omega) = \sum_{k=-m_n}^{m_n} h_k I(\omega_j + k/n).$$

Here, the h_k are scalars and satisfy the conditions given in the univariate case.

1.5 Numerical examples

In this section we provide brief numerical examples on fitting ARMA models and performing spectral analysis on an R data set. In both examples we use the R data set `sunspot.year`, which are the annual sunspot numbers for the years 1700–1988. Because there are extreme values, we first transform the data by taking the square root. The transformed data are displayed in Figure 1.1 using the R code displayed in the first two lines of Example 1.36. It has been noted by many authors that the sunspot series exhibits nonlinear and non-normal behavior. We will discuss this problem in more detail throughout the text, but as a quick check, note that in Figure 1.1, the sunspot numbers tend to rise quickly to a peak and then decline slowly to a trough (↑↘); thus the series resembles a reverse, or inverse, saw tooth wave. Thus, if put in reverse-time order, the data will rise slowly and decline quickly. A linear Gaussian process cannot have this behavior because the distribution of $\{X_1,\ldots,X_n\}$ must be the same as $\{X_n,\ldots,X_1\}$; i.e., both have the same mean vector, μ, and variance-covariance matrix, Γ, as defined in (1.36). Nevertheless, for the sake of demonstration, we will apply linear models to the series.

Example 1.36 (Sunspots – ARMA). Continuing with the sunspot data, plotting the sample ACF and PACF (not shown) suggest fitting an autoregression of at least order $p = 2$. We then used the R command `ar()` to search for the best autoregressive model based on AIC when the parameters are estimated by Yule-Walker. This method suggests an AR(9) model, which is then fit using maximum likelihood. The final results are:

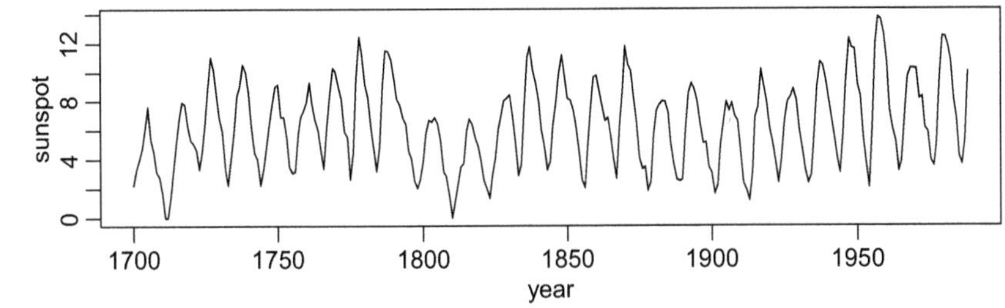

Figure 1.1 *The square root of the* R *data set* sunspots.year, *which are the annual sunspot numbers for the years 1700–1988.*

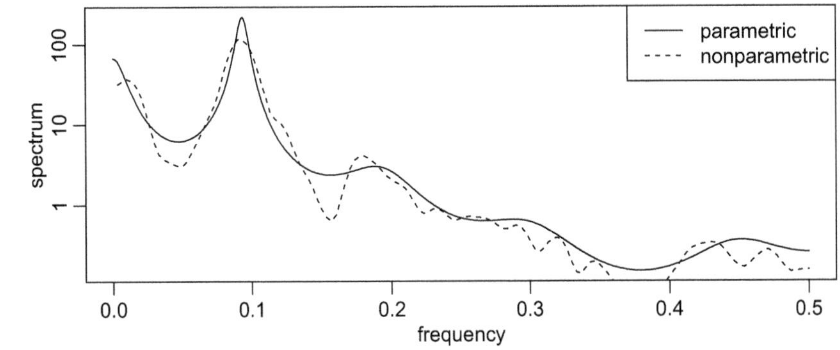

Figure 1.2: *Results of parametric and nonparametric spectral analysis on the sunspot data.*

```
> sunspot = sqrt(sunspot.year)
> arima(sunspot, order=c(9,0,0))  # final model fit via MLE
Coefficients:
       ar1     ar2     ar3    ar4     ar5    ar6    ar7     ar8    ar9  intrcpt
     1.219  -0.479  -0.142  0.270  -0.243  0.017  0.168  -0.206  0.297     6.40
se: 0.057   0.091   0.094  0.095   0.095  0.096  0.096   0.092  0.057     0.58
sigma^2 estimated as 1.08:  log likelihood = -424,  aic = 870
```

Note that R reports the intercept as `6.40` when, in fact, it is the estimate of the mean $\mu_x = \mathbb{E}[X_t]$. Unfortunately, this is a misleading label that the R core group refuses to fix. ◇

Example 1.37 (Sunspots – spectral analysis). Here, we perform parametric and nonparametric spectral analysis on the sunspot data. The R program `spec.ar()` can be used to fit the best (via AIC) AR spectrum to a data set, whereas `spectrum()` may be used for the nonparametric analysis. In this case, a smoothing window must be given and we refer the reader to Shumway and Stoffer (2011) for more details. The results are displayed in Figure 1.2. The major peak appears at approximately the 11 year cycle ($\omega \approx .09$). ◇

Exercises

1.1. Suppose $\{X_t,\ t \in \mathbb{Z}\}$ is i.i.d. with characteristic function $\varphi_x(\xi)$ and consider the finite linear filter given by $Y_t = X_t + \theta X_{t-1}$, where $\theta \in \mathbb{R}$. Express the joint characteristic function of $\{Y_{t_1}, Y_{t_2}, \ldots, Y_{t_n}\}$, say, $\varphi_y(\xi_1, \xi_2, \ldots, \xi_n)$, in terms of $\varphi_x(\cdot)$, and use it to deduce that $\{Y_t,\ t \in \mathbb{Z}\}$ is strictly stationary. Is Y_t weakly stationary?

1.2. Suppose $\{X_t,\ t \in \mathbb{Z}\}$ is a stationary process with mean μ_x and autocovariance function $\gamma_x(h)$. Show that $Y_t = \sum_{j=-k}^{k} a_j X_{t-j}$ is stationary.

1.3. (a) Consider two pure MA(1) processes,

$$X_t = Z_t + \tfrac{1}{5} Z_{t-1}, \quad \{Z_t,\ t \in \mathbb{Z}\} \sim \text{iid N}(0, \sigma_z = 5)$$
$$Y_t = V_t + 5V_{t-1}, \quad \{V_t,\ t \in \mathbb{Z}\} \sim \text{iid N}(0, \sigma_v = 1).$$

Show that $\{X_{t_1}, X_{t_2}, \ldots, X_{t_n}\} \stackrel{\mathrm{d}}{=} \{Y_{t_1}, Y_{t_2}, \ldots, Y_{t_n}\}$ for all $n = 1, 2, \ldots$, all time points $t_1, t_2, \ldots, t_n$, and consequently, based solely on data, it is impossible to distinguish between the two models. Which process is the invertible process?

(b) Consider two pure AR(1) processes,

$$X_t = \tfrac{1}{2} X_{t-1} + Z_t, \quad \{Z_t,\ t \in \mathbb{Z}\} \sim \text{iid N}(0, \sigma_z = \tfrac{1}{2});$$
$$Y_t = 2Y_{t-1} + V_t, \quad \{V_t,\ t \in \mathbb{Z}\} \sim \text{iid N}(0, \sigma_v = 1)\ .$$

Show that $\{X_{t_1}, X_{t_2}, \ldots, X_{t_n}\} \stackrel{\mathrm{d}}{=} \{Y_{t_1}, Y_{t_2}, \ldots, Y_{t_n}\}$ for all $n = 1, 2, \ldots$, all time points $t_1, t_2, \ldots, t_n$, and consequently, based solely on data, it is impossible to distinguish between the two models. Which process is the causal process?

1.4. Prove (1.59) and (1.60) to verify the Yule-Walker equations.

1.5. Show that if $\{X_t,\ t \in \mathbb{Z}\}$ is a causal AR(1), $X_t = \phi_0 + \phi_1 X_{t-1} + Z_t$, with $|\phi_1| < 1$, then $\mathbb{E}[X_t] = \phi_0/(1-\phi_1)$, $\operatorname{Var}(X_t) = \sigma_z^2/(1-\phi_1^2)$, and $\rho(h) = \phi^{|h|}$.

1.6. Show that the autocovariance function of an ARMA(1,1) model, $X_t = \phi X_{t-1} + Z_t + \theta Z_{t-1}$, is given by

$$\gamma_x(0) = \sigma_z^2 \frac{1 + 2\theta\phi + \theta^2}{1 - \phi^2},$$

and, for $h \neq 0$,

$$\gamma_x(h) = \sigma_z^2 \frac{(1 + \theta\phi)(\phi + \theta)}{1 - \phi^2} \phi^{|h|-1}.$$

Finally, show, for $h \neq 0$, the ACF is

$$\rho_x(h) = \frac{(1 + \theta\phi)(\phi + \theta)}{1 + 2\theta\phi + \theta^2} \phi^{|h|-1}.$$

1.7. Show that the function γ defined by $\gamma(0) = 1$, $\gamma(1) = \gamma(-1) = \rho$ and $\gamma(k) = 0$ otherwise, is a covariance function if and only if $|\rho| \leq \frac{1}{2}$.

1.8. Use Proposition 1.19 to prove Proposition 1.22.

1.9. Let $\{Y_t,\ t \in \mathbb{Z}\}$ be a weakly stationary process with spectral density f such that $0 \le m \le f(\lambda) \le M < \infty$ for all $\lambda \in [-\pi, \pi]$. For $n \ge 1$, denote by Γ_n the covariance matrix of $[Y_1, \ldots, Y_n]'$. Show that the eigenvalues of Γ_n belong to the interval $[2\pi m, 2\pi M]$.

1.10. Let $\{X_t^{(i)},\ t \in \mathbb{Z}\}$, for $i = 1, 2$, denote two uncorrelated MA processes such that

$$X_t^{(1)} = Z_t^{(1)} + \theta_1 Z_{t-1}^{(1)} + \cdots + \theta_q Z_{t-q}^{(1)}$$
$$X_t^{(2)} = Z_t^{(2)} + \rho_1 Z_{t-1}^{(2)} + \cdots + \rho_p Z_{t-p}^{(2)}$$

where $\{Z_t^{(i)},\ t \in \mathbb{Z}\}$ for $i = 1, 2$, are white noise processes with variances σ_i^2. Define $Y_t = X_t^{(1)} + X_t^{(2)}$.

(a) Show that $\{Y_t,\ t \in \mathbb{Z}\}$ is an ARMA process.

(b) Assuming that $q = p = 1$ and $0 < \theta_1, \rho_1 < 1$, compute the variance of the innovation process, $Y_t - Y_{t|t-1}$, of $\{Y_t,\ t \in \mathbb{Z}\}$.

1.11. Let $\{X_t^{(1)},\ t \in \mathbb{Z}\}$ and $\{X_t^{(2)},\ t \in \mathbb{Z}\}$ denote two uncorrelated AR(1) processes,

$$X_t^{(1)} = aX_{t-1}^{(1)} + Z_t^{(1)}$$
$$X_t^{(2)} = bX_{t-1}^{(2)} + Z_t^{(2)}$$

where $\{Z_t^{(1)},\ t \in \mathbb{Z}\}$ and $\{Z_t^{(2)},\ t \in \mathbb{Z}\}$ have variances $\sigma^2_{Z^{(1)}}$ and $\sigma^2_{Z^{(2)}}$, respectively, and $0 < a, b < 1$. Define $Y_t = X_t^{(1)} + X_t^{(2)}$.

(a) Show that there exists a white noise $\{Z_t,\ t \in \mathbb{Z}\}$ with variance σ^2 and θ with $|\theta| < 1$ such that

$$Y_t - (a+b)Y_{t-1} + abY_{t-2} = Z_t - \theta Z_{t-1}\,.$$

(b) Check that

$$Z_t = Z_t^{(1)} + (\theta - b)\sum_{k=0}^{\infty} \theta^k Z_{t-1-k}^{(1)} + Z_t^{(2)} + (\theta - a)\sum_{k=0}^{\infty} \theta^k Z_{t-1-k}^{(2)}$$

(c) Determine the best predictor of Y_{t+1} when $(X_s^{(1)})$ and $(X_s^{(2)})$ are known up to time $s = t$.

(d) Determine the best predictor of Y_{t+1} when (Y_s) is known up to time $s = t$.

(e) Compare the variances of the prediction errors corresponding to the two predictors defined above.

1.12. Let $\{X_t,\ t \in \mathbb{Z}\}$ be a zero-mean second-order stationary process with covariance function γ. For $t \in \mathbb{N}$, denote by Γ_t the $(t \times t)$-matrix given by

$$\Gamma_t = [\gamma(i-j)]_{1 \le i,j \le t} \qquad \text{for all } t \ge 1.$$

In addition, for n successive observations $X_1,\ldots,X_n$ of $\{X_t,\ t\in\mathbb{Z}\}$, define the *empirical covariance function* by $\hat{\gamma}$

$$\hat{\gamma}(k)=\begin{cases} n^{-1}\sum_{t=1}^{n-|k|}(X_t-\bar{X}_n)(X_{t+|k|}-\bar{X}_n) & \text{if } |k|\le n-1 \\ 0 & \text{otherwise}\end{cases}$$

where $\bar{X}_n = 1/n\sum_{t=1}^{n} X_t$.

(a) Assume that there exists $k\ge 1$ such that Γ_k is invertible but Γ_{k+1} is not. Show that we can then write X_n as $\sum_{t=1}^{k}\alpha_t^{(n)}X_t$, where $\alpha^{(n)}\in\mathbb{R}^k$, for all $n\ge k+1$.

(b) Show that the vectors $\alpha^{(n)}$ are bounded independently of n.

(c) If we now assume $\gamma(0)>0$ and $\gamma(t)\to 0$ as $t\to\infty$, can we find $t\ge 1$ such that Γ_t is singular?

(d) Show that $\hat{\gamma}$ is a covariance function. Deduce that the empirical covariance matrices $\hat{\Gamma}_k=(\hat{\gamma}(i-j))_{1\le i,j\le k}$ are invertible for all $k\ge 1$ under a simple condition on $X_1,\ldots,X_n$.

Chapter 2

Linear Gaussian State Space Models

A very general model that subsumes a whole class of special cases of interest in much the same way that linear regression does is the state-space model or the *dynamic linear model* (DLM), which was presented in Kalman (1960) and Kalman and Bucy (1961). The model arose in the space tracking setting, where the state equation defines the motion equations for the position or *state* of a spacecraft with location $\{X_t,\ t \in \mathbb{N}\}$ and the data $\{Y_t,\ t \in \mathbb{N}\}$ reflect information that can be observed from a tracking device such as velocity and azimuth. According to various accounts (e.g., Cipra, 1993), although not interested at first, the researchers at NASA eventually latched onto the Kalman filter as a way of dealing with problems in satellite orbit determination and Kalman filtering became a mainstay of aerospace engineering. For example, it was used in the Ranger, Mariner, and Apollo missions of the 1960s. In particular, the on-board computer that guided the descent of the Apollo 11 lunar module to the moon had a Kalman filter.

It should be noted that the Kalman filter was derived in 1880 by T. N. Thiele in a paper on a problem in astronomical geodesy; see Lauritzen (1981) for details. However, it seems that paper had been overlooked. When introduced by Kalman and Bucy in the 1960s, the model was proposed as a method primarily for use in aerospace-related research. Since then, it has been applied to modeling data from many disciplines including economics, biology, psychology, and medicine. Because of its broad applicability, the idea has been repackaged and claimed by researchers who were perhaps not even ten years old when the methodology was successfully being used for manned missions to the Moon.

In this chapter, we focus solely on the linear Gaussian state space model. Here, we give a brief account of the general methodology in order to establish a foundation on which to proceed to more complex ideas. Nonlinear and non-Gaussian cases are treated in Chapter 9.

2.1 Model basics

The linear state-space model or DLM, in its basic form, employs an order one, vector autoregression as the *state equation*,

$$X_t = \Phi X_{t-1} + \Upsilon U_t + W_t\,, \tag{2.1}$$

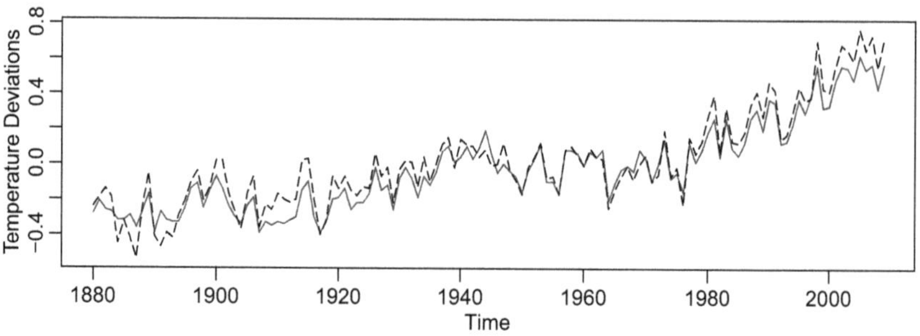

Figure 2.1 *Annual global temperature deviation series, measured in degrees centigrade, 1880–2009. The solid line is the land-marine series whereas the dashed line shows the land-based series.*

where the state equation determines the rule for the generation of the $p \times 1$ state vector X_t from the past $p \times 1$ state X_{t-1}, for time points $t = 1, \ldots, n$. In this chapter, we assume that $\{W_t,\, t \in \mathbb{N}\}$ are $p \times 1$ independent and identically distributed, zero-mean normal vectors with covariance matrix Q. In the DLM, we assume the initial state X_0 is Gaussian with mean μ_0 and covariance matrix Σ_0, and is independent of $\{W_t,\, t \in \mathbb{N}\}$. Here, $\{U_t,\, t \in \mathbb{N}\}$ denotes an exogenous input series, and Υ is a $p \times r$ matrix of parameters, which is possibly the zero matrix.

The DLM adds an additional component to the model in assuming we do not observe the state sequence $\{X_t,\, t \in \mathbb{N}\}$ directly, but only a linear transformed version of it with noise added,

$$Y_t = A_t X_t + \Gamma U_t + V_t \;, \tag{2.2}$$

where A_t is a $q \times p$ *measurement* or *observation matrix*; equation (2.2) is called the *observation equation.*

The observation vector, Y_t, is q-dimensional, which can be larger than or smaller than p, the state dimension. The additive observation noise $\{V_t,\, t \in \mathbb{N}\}$ is assumed to be white and Gaussian with $q \times q$ covariance matrix R and is independent of the state noise $\{W_t,\, t \in \mathbb{N}\}$ and the initial state X_0. The matrix Γ is a $q \times r$ regression matrix which may be the zero matrix. In addition, we initially assume that the initial state X_0, the state noise $\{W_t,\, t \in \mathbb{N}\}$ and the observation noise $\{V_t,\, t \in \mathbb{N}\}$ are uncorrelated; this assumption is not necessary, but it helps in the explanation of first concepts. The case of correlated noises is discussed in Section 2.8.

As previously mentioned, the model is quite general and can be used in a number of problems from a broad class of disciplines. We will see a few examples in this chapter.

Example 2.1 (Climate change). Figure 2.1 shows two different estimates of the global temperature deviations from 1880 to 2009. The solid line shows the global mean land-ocean temperature index data. The second series (dashed line) shows the surface air temperature index data using only land based meteorological station data. Conceptually, both series should be measuring the same underlying climatic signal, and we may consider the problem of extracting this underlying signal.

We suppose both series are observing the same signal with different noises; that is,

$$Y_{1,t} = X_t + V_{1,t} \quad \text{and} \quad Y_{2,t} = X_t + V_{2,t} ,$$

or more compactly, the observation equation can be written as

$$\begin{pmatrix} Y_{1,t} \\ Y_{2,t} \end{pmatrix} = \begin{pmatrix} 1 \\ 1 \end{pmatrix} X_t + \begin{pmatrix} V_{1,t} \\ V_{2,t} \end{pmatrix} , \tag{2.3}$$

where

$$R = \text{Cov} \begin{pmatrix} V_{1,t} \\ V_{2,t} \end{pmatrix} = \begin{pmatrix} r_{11} & r_{12} \\ r_{21} & r_{22} \end{pmatrix} .$$

It is reasonable to suppose that the unknown common signal, X_t, can be modeled as a random walk with drift, so the state equation is of the form

$$X_t = \delta + X_{t-1} + W_t , \tag{2.4}$$

where $\{W_t,\, t \in \mathbb{N}\} \sim \text{iid}\, \text{N}(0, Q)$. In this example, $p = 1$, $q = 2$, $\Phi = 1$, $\Upsilon = \delta$ with $U_t \equiv 1$, and $\Gamma = 0$. ◇

The questions of general interest for the dynamic linear model (2.1) and (2.2) relate to estimating the unknown parameters contained in $\Phi, \Upsilon, Q, \Gamma, A_t$, and R, that define the particular model, and estimating or forecasting values of the underlying unobserved process $\{X_t,\, t \in \mathbb{N}\}$. The advantages of the state-space formulation are in the ease with which we can treat various missing data configurations and in the incredible array of models that can be generated from the model. The analogy between the observation matrix A_t and the design matrix in the usual regression and analysis of variance setting is a useful one. We can generate fixed and random effect structures that are either constant or vary over time simply by making appropriate choices for the matrix A_t and the transition structure Φ. We give a few examples in this chapter; for further examples, see Durbin and Koopman (2012), Harvey (1993), Jones (1993), Kitagawa (2010), or Shumway and Stoffer (2011) to mention a few.

Before continuing our investigation of the more complex model, it is instructive to consider a simple univariate state-space model wherein an AR(1) process is observed using a noisy instrument.

Example 2.2 (Noisy AR(1) – AR with observational noise). Consider a univariate state-space model where the observations are noisy,

$$Y_t = X_t + V_t , \tag{2.5}$$

and the signal (state) is a stationary AR(1) process,

$$X_t = \phi X_{t-1} + W_t , \tag{2.6}$$

for $t \in \mathbb{N}^*$, where $|\phi| < 1$ and $\{V_t,\, t \in \mathbb{N}\} \sim \text{iid}\, \text{N}(0, \sigma_v^2)$, $\{W_t,\, t \in \mathbb{N}\} \sim \text{iid}\, \text{N}(0, \sigma_w^2)$, $\{V_t,\, t \in \mathbb{N}\}$, $\{W_t,\, t \in \mathbb{N}\}$, and $X_0 \sim \text{N}(0, \sigma_w^2/(1-\phi^2))$ are independent.

In Chapter 1, Exercise 1.5, we investigated the properties of the state, $\{X_t,\, t \in \mathbb{N}\}$, because it is a stationary AR(1) process. For example, we know the autocovariance function of $\{X_t,\ t \in \mathbb{N}\}$ is

$$\gamma_x(h) = \frac{\sigma_w^2}{1-\phi^2}\,\phi^{|h|}, \quad h = 0, \pm 1, \pm 2, \ldots \tag{2.7}$$

But here, we must investigate how the addition of observation noise affects the dynamics. Although it is not a necessary assumption, we have assumed in this example that $\{X_t,\ t \in \mathbb{N}\}$ is stationary. In this case, the observations are also stationary because $\{Y_t,\ t \in \mathbb{N}\}$ is the sum of two independent stationary components $\{X_t,\ t \in \mathbb{N}\}$ and $\{V_t,\ t \in \mathbb{N}\}$. We have

$$\gamma_y(0) = \operatorname{Var}(Y_t) = \operatorname{Var}(X_t + V_t) = \frac{\sigma_w^2}{1-\phi^2} + \sigma_v^2\,, \tag{2.8}$$

and, when $h \neq 0$,

$$\gamma_y(h) = \operatorname{Cov}(Y_t, Y_{t-h}) = \operatorname{Cov}(X_t + V_t, X_{t-h} + V_{t-h}) = \gamma_x(h)\,. \tag{2.9}$$

Consequently, for $h \neq 0$, the ACF of the observations is

$$\rho_y(h) = \frac{\gamma_y(h)}{\gamma_y(0)} = \left(1 + \frac{\sigma_v^2}{\sigma_w^2}(1-\phi^2)\right)^{-1} \phi^{|h|}\,. \tag{2.10}$$

It should be clear from (2.10) that the observations, $\{Y_t,\ t \in \mathbb{N}\}$, are not AR(1) unless $\sigma_v^2 = 0$. In addition, it is easy to show that the autocorrelation structure of Y_t is identical to the autocorrelation structure of an ARMA(1,1) process; see Exercise 1.6. ◇

We will see in Section 2.8 that an equivalence exists between stationary ARMA models and stationary state-space models. Nevertheless, it is sometimes easier to work with one form than another. As previously mentioned, in the case of missing data, complex multivariate systems, mixed effects, and certain types of nonstationarity, it is easier to work in the framework of state-space models.

2.2 Filtering, smoothing, and forecasting

From a practical view, the primary aims of any analysis involving the state-space model as defined by (2.1)-(2.2), would be to produce estimators for the underlying unobserved signal X_t, given the data $Y_{1:s} = \{Y_1, \ldots, Y_s\}$, to time s. When $s < t$, the problem is called *forecasting* or *prediction*. When $s = t$, the problem is called *filtering*, and when $s > t$, the problem is called *smoothing*. In addition to these estimates, we would also want to measure their precision. The solution to these problems is accomplished via the *Kalman filter and smoother* and is the focus of this section.

Throughout this chapter, we will use the following definitions,

$$X_{t|s} := \mathbb{E}\left[X_t \mid Y_{1:s}\right] \tag{2.11}$$

and

$$P_{t_1,t_2|s} := \mathbb{E}\left\{(X_{t_1} - X_{t_1|s})(X_{t_2} - X_{t_2|s})'\right\} . \tag{2.12}$$

When $t_1 = t_2$ ($=t$ say) in (2.12), we will write $P_{t|s}$ for convenience.

In obtaining the filtering and smoothing equations in this chapter, we will rely heavily on the Gaussian assumption. When we assume, as in this chapter, the processes are Gaussian, (2.12) is also the conditional error covariance; that is,

$$P_{t_1,t_2|s} = \mathbb{E}\left[(X_{t_1} - X_{t_1|s})(X_{t_2} - X_{t_2|s})' \mid Y_{1:s}\right] .$$

This fact can be seen, for example, by noting the covariance matrix between $(X_t - X_{t|s})$ and $Y_{1:s}$, for any t and s, is zero; we could say they are orthogonal in the sense of Section 1.2.3. This result implies that $(X_t - X_{t|s})$ and $Y_{1:s}$ are independent (because of the normality), and hence, the conditional distribution of $(X_t - X_{t|s})$ given $Y_{1:s}$ is the unconditional distribution of $(X_t - X_{t|s})$. Derivations of the filtering and smoothing equations from a Bayesian perspective are given in Meinhold and Singpurwalla (1983) and West and Harrison (1997); more traditional approaches based on the concept of projection and on multivariate normal distribution theory are given in Jazwinski (1970) and Anderson and Moore (1979).

First, we present the Kalman filter, which gives the filtering and forecasting equations. The name filter comes from the fact that $X_{t|t}$ is a linear filter of the observations $Y_1, \ldots, Y_t$; that is, $X_{t|t} = \sum_{s=1}^{t} B_{t,s} Y_s$ for suitably chosen $p \times q$ matrices $\{B_{t,s}\}_{s=1}^{t}$. The advantage of the Kalman filter is that it specifies how to update the filter from $X_{t-1|t-1}$ to $X_{t|t}$ once a new observation Y_t is obtained, without having to reprocess the entire data set $Y_1, \ldots, Y_t$.

Proposition 2.3 (The Kalman filter). *For the state-space model specified in (2.1) and (2.2), with initial conditions $X_{0|0} = \mu_0$ and $P_{0|0} = \Sigma_0$, for $t = 1, \ldots, n$,*

$$X_{t|t-1} = \Phi X_{t-1|t-1} + \Upsilon U_t , \tag{2.13}$$

$$P_{t|t-1} = \Phi P_{t-1|t-1} \Phi' + Q , \tag{2.14}$$

with

$$X_{t|t} = X_{t|t-1} + K_t (Y_t - A_t X_{t|t-1} - \Gamma U_t) , \tag{2.15}$$

$$P_{t|t} = [I - K_t A_t] P_{t|t-1} , \tag{2.16}$$

$$K_t = P_{t|t-1} A_t' [A_t P_{t|t-1} A_t' + R]^{-1} . \tag{2.17}$$

K_t is often called the gain matrix. Prediction for $t > n$ is accomplished via (2.13) and (2.14) with initial conditions $X_{n|n}$ and $P_{n|n}$. Important byproducts of the filter are the innovations (prediction errors)

$$\varepsilon_t = Y_t - Y_{t|t-1} = Y_t - A_t X_{t|t-1} - \Gamma U_t , \tag{2.18}$$

and the corresponding variance-covariance matrices

$$\Sigma_t := \operatorname{Cov}(\varepsilon_t) = \operatorname{Cov}(A_t (X_t - X_{t|t-1}) + V_t) = A_t P_{t|t-1} A_t' + R \tag{2.19}$$

for $t = 1, \ldots, n$.

Proof. The derivations of (2.13) and (2.14) follow from straight-forward calculations, because from (2.1) we have

$$X_{t|t-1} = \mathbb{E}\left[X_t \mid Y_{1:t-1}\right] = \mathbb{E}\left[\Phi X_{t-1} + \Upsilon U_t + W_t \mid Y_{1:t-1}\right] = \Phi X_{t-1|t-1} + \Upsilon U_t \,,$$

and thus

$$\begin{aligned} P_{t|t-1} &= E\left\{(X_t - X_{t|t-1})(X_t - X_{t|t-1})'\right\} \\ &= E\left\{\left[\Phi(X_{t-1} - X_{t-1|t-1}) + W_t\right]\left[\Phi(X_{t-1} - X_{t-1|t-1}) + W_t\right]'\right\} \\ &= \Phi P_{t-1|t-1}\Phi' + Q \,. \end{aligned}$$

To derive (2.15), we note that the matrix $\mathrm{Cov}\,(\varepsilon_t, Y_s) = 0$ for $s < t$, which in view of the fact that the innovation sequence is a Gaussian process, implies that the innovations are independent of the past observations. Furthermore, the conditional covariance between X_t and ε_t given $Y_{1:t-1}$ is

$$\begin{aligned} \mathrm{Cov}\,(X_t,\ \varepsilon_t \mid Y_{1:t-1}) &= \mathrm{Cov}\,\left(X_t,\ Y_t - A_t X_{t|t-1} - \Gamma U_t \mid Y_{1:t-1}\right) \\ &= \mathrm{Cov}\,\left(X_t - X_{t|t-1},\ Y_t - A_t X_{t|t-1} - \Gamma U_t \mid Y_{1:t-1}\right) \\ &= \mathrm{Cov}\,\left(X_t - X_{t|t-1},\ A_t(X_t - X_{t|t-1}) + V_t\right) \\ &= P_{t|t-1}A_t' \,. \end{aligned} \tag{2.20}$$

Using these results we have that the joint conditional distribution of X_t and ε_t given $Y_{1:t-1}$ is normal,

$$\begin{pmatrix} X_t \\ \varepsilon_t \end{pmatrix} \Bigg| \; Y_{1:t-1} \sim \mathrm{N}\left(\begin{bmatrix} X_{t|t-1} \\ 0 \end{bmatrix}, \begin{bmatrix} P_{t|t-1} & P_{t|t-1}A_t' \\ A_t P_{t|t-1} & \Sigma_t \end{bmatrix}\right). \tag{2.21}$$

Thus, using (1.32), we can write

$$X_{t|t} = \mathbb{E}\left[X_t \mid Y_1, \ldots, Y_{t-1}, Y_t\right] = \mathbb{E}\left[X_t \mid Y_{1:t-1}, \varepsilon_t\right] = X_{t|t-1} + K_t\varepsilon_t, \tag{2.22}$$

where

$$K_t = P_{t|t-1}A_t'\Sigma_t^{-1} = P_{t|t-1}A_t'(A_t P_{t|t-1}A_t' + R)^{-1} \,.$$

The evaluation of $P_{t-1|t-1}$ is easily computed from (2.21) [see (1.33)] as

$$P_{t|t} = \mathrm{Cov}\,(X_t \mid Y_{t-1}, \varepsilon_t) = P_{t|t-1} - P_{t|t-1}A_t'\Sigma_t^{-1}A_t P_{t|t-1} \,,$$

which simplifies to (2.16). ∎

Remark 2.4 (The Kalman filter as BLP). In view of Theorem 1.26, we may also think of $X_{t|s}$ as linear projection of X_t onto $\mathcal{M} = \overline{\mathrm{sp}}\{1, Y_1, \ldots, Y_s\}$. In particular, in the non-Gaussian case, the Kalman filter provides a method to recursively generate the best linear predictors and the corresponding MSPEs; also, see Exercise 2.3.

Remark 2.5 (Robustness). Note that the filter, $X_{t|t}$, in (2.15) is an unbounded function of the new observation, Y_t. This means the filter is not robust to outliers. For example, if the t-th observation is actually 4.20, but 420 is recorded instead, the filter value would jump off its trajectory by a large amount. In addition, the error covariance matrices $P_{t|t-1}$ and $P_{t|t}$ in Proposition 2.3 may be computed without obtaining any observations. Thus, if the data become more variable over time (i.e., the model is not correct), the error variances will not change with that information.

Nothing in the proof of Proposition 2.3 precludes the cases where some or all of the parameters vary with time, or where the observation dimension changes with time, which leads to the following corollary.

Corollary 2.6 (Kalman filter: The time-varying case). *If, in the DLM,* (2.1) *and* (2.2), *any or all of the parameters are time dependent,* $\Phi = \Phi_t$, $\Upsilon = \Upsilon_t$, $Q = Q_t$ *in the state equation or* $\Gamma = \Gamma_t$, $R = R_t$ *in the observation equation, or the dimension of the observational equation is time dependent,* $q = q_t$, *Proposition 2.3 holds with the appropriate substitutions.*

Next, we consider the problem of obtaining estimators for X_t based on an entire set of observations, $Y_1, \dots, Y_n$, where $t \le n$, namely, $X_{t|n}$. These estimators are called smoothers because a time plot of the sequence $\{X_{t|n};\ t = 1, \dots, n\}$ is typically smoother than the forecasts $\{X_{t|t-1};\ t = 1, \dots, n\}$ or the filters $\{X_{t|t};\ t = 1, \dots, n\}$. We note that there is a relationship between the Kalman smoother and smoothing splines; this relationship is explored in Section 2.4.

Proposition 2.7 (The Kalman smoother). *For the state-space model specified in* (2.1) *and* (2.2), *with initial conditions* $X_{n|n}$ *and* $P_{n|n}$ *obtained via Proposition 2.3, for* $t = n, n-1, \dots, 1$,

$$X_{t-1|n} = X_{t-1|t-1} + J_{t-1}\left(X_{t|n} - X_{t|t-1}\right), \tag{2.23}$$

$$P_{t-1|n} = P_{t-1|t-1} + J_{t-1}\left(P_{t|n} - P_{t|t-1}\right)J_{t-1}', \tag{2.24}$$

where

$$J_{t-1} = P_{t-1|t-1}\Phi' P_{t|t-1}^{-1}. \tag{2.25}$$

Proof. The smoother can be derived in many ways. Here we provide a proof that was given in Ansley and Kohn (1982). First, for $1 \le t \le n$, define

$$Y_{1:t-1} = \{Y_1, \dots, Y_{t-1}\} \quad \text{and} \quad \mathcal{E}_{t:n} = \{V_t, \dots, V_n; W_{t+1}, \dots, W_n\},$$

with $Y_{1:0}$ being empty, and let

$$\widetilde{X}_{t-1} = \mathbb{E}\left[X_{t-1} \,\middle|\, Y_{1:t-1}, X_t - X_{t|t-1}, \mathcal{E}_{t:n}\right].$$

Then, because $Y_{1:t-1}$, $\{X_t - X_{t|t-1}\}$, and $\mathcal{E}_{t:n}$ are mutually independent, and X_{t-1} and $\mathcal{E}_{t:n}$ are independent, using (1.32) we have

$$\widetilde{X}_{t-1} = X_{t-1|t-1} + J_{t-1}(X_t - X_{t|t-1}), \tag{2.26}$$

where

$$J_{t-1} = \mathrm{Cov}(X_{t-1}, X_t - X_{t|t-1})P_{t|t-1}^{-1} = P_{t-1|t-1}\Phi' P_{t|t-1}^{-1}.$$

Finally, because $Y_{1:t-1}$, $X_t - X_{t|t-1}$, and $\mathcal{E}_{t:n}$ generate $Y_{1:n} = \{Y_1, \ldots, Y_n\}$,

$$X_{t-1|n} \equiv \mathbb{E}\left[X_{t-1} \mid Y_{1:n}\right] = \mathbb{E}\left[\widetilde{X}_{t-1} \mid Y_{1:n}\right] = X_{t-1|t-1} + J_{t-1}(X_{t|n} - X_{t|t-1}),$$

which establishes (2.23).

The recursion for the error covariance, $P_{t-1|n}$, is obtained by straight-forward calculation. Using (2.23) we obtain

$$X_{t-1} - X_{t-1|n} = X_{t-1} - X_{t-1|t-1} - J_{t-1}\left(X_{t|n} - \Phi X_{t-1|t-1}\right),$$

or

$$\left(X_{t-1} - X_{t-1|n}\right) + J_{t-1}X_{t|n} = \left(X_{t-1} - X_{t-1|t-1}\right) + J_{t-1}\Phi X_{t-1|t-1}. \tag{2.27}$$

Multiplying each side of (2.27) by the transpose of itself and taking expectation, we have

$$P_{t-1|n} + J_{t-1}\mathbb{E}[X_{t|n}X'_{t|n}]J'_{t-1} = P_{t-1|t-1} + J_{t-1}\Phi\mathbb{E}[X_{t-1|t-1}X'_{t-1|t-1}]\Phi' J'_{t-1}, \tag{2.28}$$

using the fact the cross-product terms are zero. But,

$$\mathbb{E}[X_{t|n}X'_{t|n}] = \mathbb{E}\left[X_t X'_t\right] - P_{t|n} = \Phi\mathbb{E}\left[X_{t-1}X'_{t-1}\right]\Phi' + Q - P_{t|n},$$

and

$$\mathbb{E}[X_{t-1|t-1}X'_{t-1|t-1}] = \mathbb{E}\left[X_{t-1}X'_{t-1}\right] - P_{t-1|t-1},$$

so (2.28) simplifies to (2.24). ∎

Example 2.8 (A local level model). In this example, we suppose that we observe a univariate series Y_t that consists of a trend component, μ_t, and a noise component, V_t, where

$$Y_t = \mu_t + V_t \tag{2.29}$$

and $V_t \sim$ iid $\mathrm{N}(0, \sigma_v^2)$. In particular, we assume the trend is a random walk given by

$$\mu_t = \mu_{t-1} + W_t \tag{2.30}$$

where $W_t \sim$ iid $\mathrm{N}(0, \sigma_w^2)$ is independent of $\{V_t\}$. Recall Example 2.1, where we suggested this type of trend model for the global temperature series.

The model, which is sometimes referred to as a *structural model* (see Section 2.7), is a state-space model with (2.29) being the observation equation, and (2.30) being the state equation. Consequently, Proposition 2.3 and Proposition 2.7 can be used to obtain forecasts, filters, and smoothers. For this example, we simulated $n = 10$ observations from the local level trend model. The sequences $\{W_t\}$, $\{V_t\}$ and μ_0 were generated independently as standard normals. We then ran the Kalman filter and smoother using the actual parameters.

Table 2.1 shows 10 observations as well as the corresponding state values, the predictions, filters and smoothers. Note that in Table 2.1 one-step-ahead prediction is more uncertain than the corresponding filtered value, which, in turn, is more uncertain than the corresponding smoother value (that is $P_{t|t-1} \geq P_{t|t} \geq P_{t|n}$). Also, in each case, the error variances stabilize quickly, virtually by $t = 4$. Figure 2.2 shows the trend μ_t for $t = 1, \ldots, 10$ as points along with the smoothed values $\mu_{t|n}$ with error bounds $\pm 2\sqrt{P_{t|n}}$. ◇

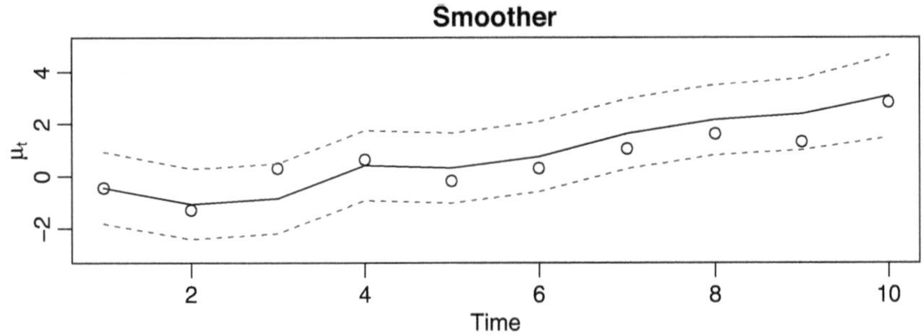

Figure 2.2 *The trend μ_t shown as points along with the smoother $\mu_{t|n}$ obtained via the Kalman smoother shown as a line. Error bounds, $\mu_{t|n} \pm 2\sqrt{P_{t|n}}$, are shown as dashed lines.*

Table 2.1: *Forecasts, Filters, and Smoothers for Example 2.8*

t	y_t	μ_t	$\mu_{t\|t-1}$	$P_{t\|t-1}$	$\mu_{t\|t}$	$P_{t\|t}$	$\mu_{t\|n}$	$P_{t\|n}$
0	—	−0.63	—	—	0.00	1.00	−0.22	0.83
1	−0.05	−0.44	0.00	2.00	−0.04	0.67	−0.45	0.47
2	−1.90	−1.28	−0.04	1.67	−1.20	0.63	−1.07	0.45
3	−1.90	0.32	−1.20	1.63	−1.63	0.62	−0.85	0.45
4	1.77	0.65	−1.63	1.62	0.47	0.62	0.41	0.45
5	−0.22	−0.17	0.47	1.62	0.04	0.62	0.31	0.45
6	0.30	0.31	0.04	1.62	0.20	0.62	0.75	0.45
7	2.00	1.05	0.20	1.62	1.31	0.62	1.63	0.45
8	2.45	1.63	1.31	1.62	2.01	0.62	2.15	0.45
9	1.92	1.32	2.01	1.62	1.95	0.62	2.38	0.47
10	3.75	2.83	1.95	1.62	3.07	0.62	3.07	0.62

When we discuss maximum likelihood estimation via the EM algorithm in the next section, we will need a set of recursions for obtaining $P_{t,t-1|n}$, as defined in (2.12). We give the necessary recursions in the following property, the proof of which may be found in Shumway and Stoffer (2011, Chapter 6).

Proposition 2.9 (The lag-one covariance smoother). *Given the state-space model specified in* (2.1) *and* (2.2)*, with K_t, J_t ($t = 1,\ldots,n$), and $P_{n|n}$ obtained from Proposition 2.3 and Proposition 2.7, and with initial condition*

$$P_{n,n-1|n} = (I - K_n A_n)\Phi P_{n-1|n-1}, \tag{2.31}$$

for $t = n, n-1, \ldots, 2$,

$$P_{t-1,t-2|n} = P_{t-1|t-1}J_{t-2}' + J_{t-1}\left(P_{t,t-1|n} - \Phi P_{t-1|t-1}\right)J_{t-2}'. \tag{2.32}$$

Remark 2.10. Although it would be computationally inconvenient, the lag-one covariance smoothers, $P_{t-1,t-2|n}$, could also be obtained by stacking the state vectors and observation vectors as $X(t) = (X_t', X_{t-1}')'$ and $Y(t) = (Y_t', Y_{t-1}')'$, respectively, and then running the Kalman filter and smoother on these vectors with the parameter matrices appropriately reconfigured. In this case, the smoother variance of the stacked data would have the form

$$P_{(t)|(n)} = \begin{pmatrix} P_{t|n} & P_{t,t-1|n} \\ P'_{t,t-1|n} & P_{t-1|n} \end{pmatrix},$$

where subscripts (t) and (n) refer to operations on the stacked values.

2.3 Maximum likelihood estimation

2.3.1 Newton–Raphson

The estimation of the parameters that specify the state-space model, (2.1) and (2.2), is quite involved. We use θ to represent the collection of parameters contained as the elements of the initial mean and covariance μ_0 and Σ_0, the transition matrix Φ, and the state and observation covariance matrices Q and R and the input coefficient matrices, Υ and Γ. We use maximum likelihood under the assumption that the initial state vector is normal, $X_0 \sim \mathrm{N}(\mu_0, \Sigma_0)$, and the error vectors, $\{W_1, \ldots, W_n\}$ and $\{V_1, \ldots, V_n\}$, are each Gaussian white noise. We continue to assume, for simplicity, that X_0, $\{W_t\}$ and $\{V_t\}$ are uncorrelated.

The likelihood is computed using the *innovations* $\varepsilon_1, \varepsilon_2, \ldots, \varepsilon_n$, defined by (2.18),

$$\varepsilon_t = Y_t - A_t X_{t|t-1} - \Gamma U_t.$$

The innovations form of the likelihood function, which was first given by Schweppe (1965), is obtained by noting that the innovations are independent Gaussian random vectors with zero means and, as shown in (2.19), covariance matrices

$$\Sigma_t = A_t P_{t|t-1} A_t' + R. \tag{2.33}$$

Hence, ignoring a constant, we may write the likelihood, $\mathrm{L}(\theta; Y_{1:n})$ as

$$-\ln \mathrm{L}(\theta; Y_{1:n}) = \frac{1}{2}\sum_{t=1}^{n} \ln |\Sigma_t(\theta)| + \frac{1}{2}\sum_{t=1}^{n} \varepsilon_t(\theta)' \Sigma_t(\theta)^{-1} \varepsilon_t(\theta), \tag{2.34}$$

where we have emphasized the dependence of the innovations on the parameters θ. Of course, (2.34) is a highly nonlinear and complicated function of the unknown parameters. The usual procedure is to fix μ_0 and Σ_0, and then develop a set of recursions for the log likelihood function and its first two derivatives; see Exercise 2.11 and Exercise 2.12, also Gupta and Mehra (1974a,b). Then, a Newton–Raphson algorithm (see Section 1.3.4) can be used successively to update the parameter values until the negative of the log likelihood is minimized.

The steps involved in performing a Newton–Raphson estimation procedure are as follows.

(i) Select initial values for the parameters, say, $\theta^{(0)}$.

(ii) Run the Kalman filter, Proposition 2.3, using the initial parameter values, $\theta^{(0)}$, to obtain a set of innovations and error covariances, say, $\{\varepsilon_t^{(0)}; t = 1, \ldots, n\}$ and $\{\Sigma_t^{(0)}; t = 1, \ldots, n\}$.

(iii) Run one iteration of a Newton–Raphson procedure with $-\ln \mathrm{L}(\theta; Y_{1:n})$ as the criterion function to obtain a new set of estimates, say $\theta^{(1)}$.

(iv) At iteration j, $(j = 1, 2, \ldots)$, repeat step 2 using $\theta^{(j)}$ in place of $\theta^{(j-1)}$ to obtain a new set of innovation values $\{\varepsilon_t^{(j)};\ t = 1, \ldots, n\}$ and $\{\Sigma_t^{(j)};\ t = 1, \ldots, n\}$. Then repeat step 3 to obtain a new estimate $\theta^{(j+1)}$. Stop when the estimates or the likelihood stabilize.

Example 2.11 (Newton–Raphson for Example 2.2). In this example, we generated $n = 100$ observations, $Y_1, \ldots, Y_{100}$, using the model in Example 2.2, to perform a Newton–Raphson estimation of the parameters ϕ, σ_w^2, and σ_v^2. In the notation of Section 2.2, we would have $\Phi = \phi$, $Q = \sigma_w^2$ and $R = \sigma_v^2$. The actual values of the parameters are $\phi = .8$, $\sigma_w^2 = \sigma_v^2 = 1$.

Newton–Raphson estimation was accomplished using the R program `optim`. In that program, we must provide an evaluation of the function to be minimized, namely, $-\ln \mathrm{L}(\theta; Y_{1:n})$. In this case, the function call combines steps (ii) and (iii), using the current values of the parameters, $\theta^{(j-1)}$, to obtain first the filtered values, then the innovation values, and then calculates the criterion function, $-\ln \mathrm{L}(\theta^{(j-1)}; Y_{1:n})$, to be minimized. We can also provide analytic forms of the gradient or *score vector*, $-\partial \ln \mathrm{L}(\theta; Y_{1:n})/\partial \theta$, and the *Hessian matrix*, $-\partial^2 \ln \mathrm{L}(\theta; Y_{1:n})/\partial \theta\ \partial \theta'$, in the optimization routine, or allow the program to calculate these values numerically. In this example, we let the program proceed numerically and we note the need to be cautious when calculating gradients numerically. For better stability, we can also provide an iterative solution for obtaining analytic gradients and Hessians of the log likelihood function.

The final estimates, along with their standard errors (in parentheses), are $\widehat{\phi} = .81_{(.08)}$, $\widehat{\sigma}_w = .85_{(.18)}$, $\widehat{\sigma}_v = .87_{(.14)}$. The standard errors are a byproduct of the estimation procedure, and we will discuss their evaluation later in this section. ◇

2.3.2 *EM algorithm*

In addition to Newton–Raphson, Shumway and Stoffer (1982) presented a conceptually simpler estimation procedure based on the EM (*expectation-maximization*) algorithm (Dempster et al., 1977).

The EM algorithm is an iterative method for finding the modes of the likelihood function, which has been found to be extremely useful for models where the direct maximization of likelihood function is difficult, but for which it is easy to work with the "complete" data model. The EM algorithm formalizes a relatively old idea for handling maximum likelihood estimation in missing data problems. Details may be found in Appendix D.

For the sake of brevity, we consider the model, (2.1)–(2.2), without inputs. The basic idea is that if we could observe the states, $X_{0:n} = \{X_0, X_1, \ldots, X_n\}$, in addition to the observations $Y_{1:n} = \{Y_1, \ldots, Y_n\}$, then we would consider $\{X_{0:n}, Y_{1:n}\}$ as the *complete data*, with the joint density

$$\mathrm{L}(\theta; X_{0:n}, Y_{1:n}) = \mathrm{p}^{\mu_0, \Sigma_0}(X_0) \prod_{t=1}^{n} \mathrm{p}^{\Phi, Q}(X_t | X_{t-1}) \prod_{t=1}^{n} \mathrm{p}^{R}(Y_t | X_t); \tag{2.35}$$

the values in the likelihood are kept random for a purpose to be revealed in (2.37). Under the Gaussian assumption and ignoring constants, the complete data likelihood, (2.35), can be written as

$$\begin{aligned} -2\ln \mathrm{L}(\theta; X_{0:n}, Y_{1:n}) &= \ln|\Sigma_0| + (X_0 - \mu_0)'\Sigma_0^{-1}(X_0 - \mu_0) \\ &\quad + n\ln|Q| + \sum_{t=1}^{n}(X_t - \Phi X_{t-1})'Q^{-1}(X_t - \Phi X_{t-1}) \\ &\quad + n\ln|R| + \sum_{t=1}^{n}(Y_t - A_tX_t)'R^{-1}(Y_t - A_tX_t). \end{aligned} \tag{2.36}$$

Thus, in view of (2.36), if we did have the complete data, we could then use the results from multivariate normal theory to easily obtain the MLEs of θ. We do not have the complete data; however, the EM algorithm gives us an iterative method for finding the MLEs of θ based on the *incomplete data*, $Y_{1:n}$, by successively maximizing the conditional expectation of the complete data likelihood. To implement the EM algorithm, we write, at iteration j (for $j = 1, 2, \ldots$),

$$\mathcal{Q}(\theta; \theta^{(j-1)}) := \mathbb{E}\left[-2\ln \mathrm{L}(\theta; X_{0:n}, Y_{1:n}) \mid Y_{1:n}, \theta^{(j-1)}\right]. \tag{2.37}$$

Calculation of (2.37) is the *expectation step*. Of course, given the data $Y_{1:n}$ and the current value of the parameters, $\theta^{(j-1)}$, we can use Proposition 2.7 to obtain the desired conditional expectations as smoothers. This property yields

$$\begin{aligned} \mathcal{Q}(\theta; \theta^{(j-1)}) &= \ln|\Sigma_0| + \mathrm{tr}\left\{\Sigma_0^{-1}[P_{0|n} + (X_{0|n} - \mu_0)(X_{0|n} - \mu_0)']\right\} \\ &\quad + n\ln|Q| + \mathrm{tr}\left\{Q^{-1}[S_{00} - S_{01}\Phi' - \Phi S_{10} + \Phi S_{11}\Phi']\right\} \\ &\quad + n\ln|R| + \mathrm{tr}\left\{R^{-1}\sum_{t=1}^{n}[(Y_t - A_tX_{t|n})(Y_t - A_tX_{t|n})' + A_tP_{t|n}A_t']\right\}, \end{aligned} \tag{2.38}$$

where, for $k, \ell = 0, 1$,

$$S_{k,\ell} = \sum_{t=1}^{n}(X_{t-k|n}X'_{t-\ell|n} + P_{t-k,t-\ell|n}). \tag{2.39}$$

In (2.38) and (2.39), the smoothers are calculated under the current value of the parameters $\theta^{(j-1)}$; for simplicity, we have not explicitly displayed this fact.

Minimizing (2.38) with respect to the parameters, at iteration j, constitutes the *maximization step*, and is analogous to the usual multivariate regression approach, which yields the updated estimates

$$\Phi^{(j)} = S_{01}S_{11}^{-1}, \tag{2.40}$$

$$Q^{(j)} = n^{-1}\left(S_{00} - S_{01}S_{11}^{-1}S_{10}\right), \tag{2.41}$$

$$R^{(j)} = n^{-1}\sum_{t=1}^{n}[(Y_t - A_tX_{t|n})(Y_t - A_tX_{t|n})' + A_tP_{t|n}A_t']. \tag{2.42}$$

The updates for the initial mean and variance–covariance matrix are

$$\mu_0^{(j)} = X_{0|n} \quad \text{and} \quad \Sigma_0^{(j)} = P_{0|n} \tag{2.43}$$

obtained from minimizing (2.38).

The overall procedure can be regarded as simply alternating between the Kalman filtering and smoothing recursions and the multivariate normal maximum likelihood estimators, as given by (2.40)–(2.43). Convergence results for the EM algorithm under general conditions can be found in Wu (1983); also see Appendix D. We summarize the iterative procedure as follows.

(i) For $\theta = \{\mu_0, \Sigma_0, \Phi, Q, R\}$, initialize the procedure by selecting starting values for the parameters, say $\theta^{(0)}$.

On iteration j, ($j = 1, 2, \ldots$):

(ii) Perform the E-Step: Use Proposition 2.3, Proposition 2.7, and Proposition 2.9 to obtain the smoothed values $X_{t|n}, P_{t|n}$ and $P_{t,t-1|n}$, for $t = 1, \ldots, n$, using the parameters $\theta^{(j-1)}$. Use the smoothed values to calculate $S_{k,\ell}$ for $k, \ell = 0, 1$, given in (2.39).

(iii) Perform the M-Step: Update the estimates using (2.40)–(2.43), to obtain $\theta^{(j)}$.

(iv) Repeat Steps (ii) – (iv) to convergence.

Example 2.12 (EM algorithm for Example 2.2). Using the same data generated in Example 2.11, we performed an EM algorithm estimation of the parameters ϕ, σ_w^2 and σ_v^2 as well as the initial parameters μ_0 and Σ_0 using the script `EM0` from the R package `astsa`. The convergence rate of the EM algorithm compared with the Newton–Raphson procedure is slow. In this example, with convergence being claimed when the relative change in the log likelihood is less than .00001, convergence was attained after 33 iterations.

The results were $\widehat{\phi} = 0.79_{(0.08)}$, $\widehat{\sigma}_w = 0.90_{(0.17)}$, $\widehat{\sigma}_v = 0.82_{(0.14)}$, with $\widehat{\mu}_0 = -2.49$ and $\widehat{\Sigma}_0 = 0.06$. We note that the results from the EM algorithm are close to the results from the Newton procedure discussed in Example 2.11. ◇

2.4 Smoothing splines and the Kalman smoother

There is a direct connection between smoothing splines (e.g., Eubank, 1999, Green and Silverman, 1993, Wahba, 1990) and state space models. The basic idea of smoothing splines in discrete time is we suppose that data Y_t are generated by $Y_t = \mu_t + \varepsilon_t$ for $t = 1, \ldots, n$, where μ_t is a smooth function of t, and ε_t is white noise. In cubic smoothing with knots at the time points t, μ_t is estimated by minimizing

$$\sum_{t=1}^{n} [Y_t - \mu_t]^2 + \lambda \sum_{t=1}^{n} \left(\nabla^2 \mu_t\right)^2 \tag{2.44}$$

with respect to μ_t, where $\lambda > 0$ is a smoothing parameter. The parameter λ controls the degree of smoothness, with larger values yielding smoother estimates. For example, if $\lambda = 0$, then the minimizer is the data itself $\widehat{\mu}_t = Y_t$; consequently, the estimate will not be smooth. If $\lambda = \infty$, then the only way to minimize (2.44) is to choose the second term to be zero, i.e., $\nabla^2 \mu_t = 0$, in which case it is of the form $\mu_t = \alpha + \beta t$, and we are in the setting of linear regression.[1] Hence, the choice of $\lambda > 0$ is seen as a trade-off between fitting a line that goes through all the data points and linear regression.

Now, consider the model given by

$$\nabla^2 \mu_t = W_t \quad \text{and} \quad Y_t = \mu_t + V_t, \tag{2.45}$$

where W_t and V_t are independent white noise processes with $\text{Var}[W_t] = \sigma_w^2$ and $\text{Var}[V_t] = \sigma_v^2$. Rewrite (2.45) as

$$\begin{pmatrix} \mu_t \\ \mu_{t-1} \end{pmatrix} = \begin{bmatrix} 2 & -1 \\ 1 & 0 \end{bmatrix} \begin{pmatrix} \mu_{t-1} \\ \mu_{t-2} \end{pmatrix} + \begin{bmatrix} 1 \\ 0 \end{bmatrix} W_t \quad \text{and} \quad Y_t = \begin{bmatrix} 1 & 0 \end{bmatrix} \begin{pmatrix} \mu_t \\ \mu_{t-1} \end{pmatrix} + V_t, \tag{2.46}$$

or equivalently, with $X_t = (\mu_t, \mu_{t-1})'$,

$$X_t = \Phi X_{t-1} + W_t^* \quad \text{and} \quad Y_t = A X_t + V_t, \tag{2.47}$$

in obvious notation. It is clear from (2.46) or (2.47) that (2.45) specifies a state space model, (2.1)-(2.2). Note that, in this random signal case, the model is similar to the local level model discussed in Example 2.8. In particular, the state process could be written as $\mu_t = \mu_{t-1} + \eta_t$, where $\eta_t = \eta_{t-1} + W_t$. An example of such a trajectory can be seen in Figure 2.3; note that data in Figure 2.3 look like the global temperature data in Figure 2.1.

Next, examine the problem of estimating the states, X_t, when the model parameters, σ_w^2 and σ_v^2, are specified. For ease, we assume X_0 is fixed. Then using the notation surrounding equations (2.35)-(2.36), the goal is to find the MLE of $X_{1:n}$ given $Y_{1:n}$; i.e. to maximize $\ln p(X_{1:n} \mid Y_{1:n})$ with respect to the states. Because of the Gaussianity, the maximum (or mode) of the distribution is when the states are estimated by $X_{t|n}$, the conditional means. These values are, of course, the smoothers obtained via Proposition 2.7.

But $\ln p(X_{1:n} \mid Y_{1:n}) = \ln p(X_{1:n}, Y_{1:n}) - \ln p(Y_{1:n})$, so maximizing $\ln p(X_{1:n}, Y_{1:n})$ with respect to $X_{1:n}$, is an equivalent problem. Writing (2.36) in the notation of (2.45), we have,

$$-2 \ln p(X_{1:n}, Y_{1:n}) \propto \sigma_w^{-2} \sum_{t=1}^{n} \left(\nabla^2 \mu_t\right)^2 + \sigma_v^{-2} \sum_{t=1}^{n} (Y_t - \mu_t)^2, \tag{2.48}$$

where we have kept only the terms involving the states, μ_t. If we set $\lambda = \sigma_v^2 / \sigma_w^2$, we can write

$$-2 \ln p(X_{1:n}, Y_{1:n}) \propto \lambda \sum_{t=1}^{n} \left(\nabla^2 \mu_t\right)^2 + \sum_{t=1}^{n} (Y_t - \mu_t)^2, \tag{2.49}$$

so that maximizing $p(X_{1:n}, Y_{1:n})$ with respect to the states is equivalent to minimizing (2.49), which is the original problem stated in (2.44).

[1]That the unique general solution to $\nabla^2 \mu_t = 0$ is of the form $\mu_t = \alpha + \beta t$ follows from difference equation theory; e.g., see Mickens (1990).

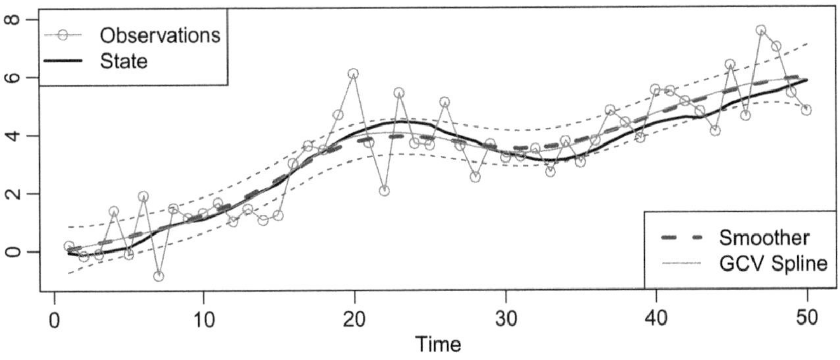

Figure 2.3 *Display for Example 2.13: Simulated state process,* μ_t *and observations* y_t *from the model* (2.45) *with* $n = 50$, $\sigma_w = .1$ *and* $\sigma_v = 1$. *Estimated smoother (dashed lines):* $\widehat{\mu}_{t|n}$ *and corresponding 95% confidence band.* GCV *smoothing spline (thin solid line).*

In the general state space setting, we would estimate σ_w^2 and σ_v^2 via maximum likelihood as described in Section 2.3, and then obtain the smoothed state values by running Proposition 2.7 with the estimated variances, say $\widehat{\sigma}_w^2$ and $\widehat{\sigma}_v^2$. In this case, the estimated value of the smoothing parameter would be given by $\widehat{\lambda} = \widehat{\sigma}_v^2 / \widehat{\sigma}_w^2$.

Example 2.13. In this example, we generated the signal, or state process, μ_t and observations y_t from the model (2.45) with $n = 50$, $\sigma_w = .1$ and $\sigma_v = 1$. The state is displayed in Figure 2.3 as a thick solid line, and the observations are displayed as points. We then estimated σ_w and σ_v using Section 2.3.1 techniques and obtained $\widehat{\sigma}_w = .08$ and $\widehat{\sigma}_v = .94$. We then used Proposition 2.7 to generate the estimated smoothers, say, $\widehat{\mu}_{t|n}$, and those values are displayed in Figure 2.3 as a thick dashed line along with a corresponding 95% (pointwise) confidence band as thin dashed lines. Finally, we used the R function `smooth.spline` to fit a smoothing spline to the data based on the method of generalized cross-validation (GCV). The fitted spline is displayed in Figure 2.3 as a thin solid line, which is nearly identical to $\widehat{\mu}_{t|n}$.

2.5 Steady state and the asymptotic distribution of the MLE

The asymptotic distribution of estimators of the model parameters, say, $\widehat{\theta}_n$, was studied extensively in Caines (1988, Chapters 7, 8), and in Hannan and Deistler (1988, Chapter 4). In those references, and as well as in Chapter 13, the consistency and asymptotic normality of the estimators is established under general conditions. Although we will only state the basic result here, some crucial elements are needed to establish large sample properties of the estimators. An essential condition is the stability of the filter. Stability of the filter assures that, for large t, the innovations ε_t are basically copies of each other (that is, independent and identically distributed) with a stable covariance matrix Σ that does not depend on t and that, asymptotically, the innovations contain all of the information about the unknown parameters. Although it is not necessary, for simplicity, we shall assume here that $A_t \equiv A$ for all t. Details on departures from this assumption can be found in Jazwinski (1970, Chapter 7). We also drop the inputs from the model (2.1) and (2.2).

For stability of the filter, we assume the eigenvalues of Φ are less than one in absolute value; this assumption can be weakened (e.g., see Harvey 1991, Section 4.3), but we retain it for simplicity. This assumption is enough to ensure the stability of the filter in that, as $t \to \infty$, the filter error covariance matrix $P_{t|t}$ converges to P, the steady-state error covariance matrix, the gain matrix K_t converges to K, the steady-state gain matrix, from which it follows that the innovation variance–covariance matrix Σ_t converges to Σ, the steady-state variance–covariance matrix of the stable innovations; details can be found in Jazwinski (1970, Chapter 7) and Anderson and Moore (1979, Chapter 4). In particular, the steady-state filter error covariance matrix, P, satisfies the Riccati equation:

$$P = \Phi[P - PA'(APA' + R)^{-1}AP]\Phi' + Q;$$

the steady-state gain matrix satisfies $K = PA'[APA' + R]^{-1}$. In Example 2.8, for all practical purposes, stability was reached by the fourth observation.

When the process is in steady-state, we may consider $X_{t+1|t}$ as the steady-state predictor and interpret it as being based on the infinite past, i.e., $X_{t+1|t} = \mathbb{E}\left[X_{t+1} \mid Y_t, Y_{t-1}, \ldots\right]$. As can be seen from (2.13) and (2.15), the steady-state predictor can be written as

$$\begin{aligned} X_{t+1|t} &= \Phi[I - KA]X_{t|t-1} + \Phi K Y_t \\ &= \Phi X_{t|t-1} + \Phi K \varepsilon_t, \end{aligned} \tag{2.50}$$

where ε_t is the steady-state innovation process given by

$$\varepsilon_t = Y_t - \mathbb{E}\left[Y_t \mid Y_{t-1}, Y_{t-2}, \ldots\right] .$$

In the Gaussian case, $\varepsilon_t \sim \text{iid N}(0, \Sigma)$, where $\Sigma = APA' + R$. In steady-state, the observations can be written as

$$Y_t = AX_{t|t-1} + \varepsilon_t. \tag{2.51}$$

Together, (2.50) and (2.51) make up the *steady-state innovations form* of the dynamic linear model.

In the following property, we assume the Gaussian state-space model, (2.1) and (2.2), is time invariant, i.e., $A_t \equiv A$, the eigenvalues of Φ are within the unit circle and the model has the smallest possible dimension (see Hannan and Deistler, 1988, 2012, Section 2.3 for details). We denote the true parameters by $\theta_0 \in \Omega$, and we assume they are identifiable and not on the boundary of Ω. Although it is not necessary to assume W_t and V_t are Gaussian, certain additional conditions would have to apply and adjustments to the asymptotic covariance matrix would have to be made (e.g., see Caines, 1988, Chapter 8).

Proposition 2.14. *Under the conditions previously stated, let $\widehat{\theta}_n$ be the estimator of θ_0 obtained by maximizing the innovations likelihood,* $\mathrm{L}(\theta; Y_{1:n})$*, as given in (2.34). Then, as $n \to \infty$,*

$$\sqrt{n}\left(\widehat{\theta}_n - \theta_0\right) \stackrel{\mathbb{P}}{\Longrightarrow} \mathrm{N}\left[0,\, \mathcal{I}(\theta_0)^{-1}\right],$$

where $\mathcal{I}(\theta)$ is the asymptotic information matrix given by

$$\mathcal{I}(\theta) = \lim_{n\to\infty} n^{-1}\mathbb{E}\left[-\partial^2 \ln \mathrm{L}(\theta; Y_{1:n})/\partial\theta\, \partial\theta'\right].$$

Precise details and the proof of Proposition 2.14 are given in Chapter 13; also, see Caines (1988, Chapter 7) or Hannan and Deistler (1988, Chapter 4). For a Newton procedure, the Hessian matrix (as described in Example 2.11) at the time of convergence can be used as an estimate of $n\mathcal{I}(\theta_0)$ to obtain estimates of the standard errors. In the examples of this section, the estimated standard errors were obtained from the numerical Hessian matrix of $-\ln \mathrm{L}(\widehat{\theta}; Y_{1:n})$, where $\widehat{\theta}$ is the vector of parameters estimates at the time of convergence.

We briefly mention computation of standard errors for the EM algorithm. In Example 2.12, we had the luxury of being able to compute the likelihood of the data, and we used that ability to obtain estimates of the standard errors from the inverse of observed information. In the general case, the observed information is not known directly, but standard errors can be estimated using Louis's identity, which is specified in Proposition D.4.

2.6 Missing data modifications

An attractive feature available within the state-space framework is its ability to treat time series that have been observed irregularly over time. For example, Jones (1980) used the state-space representation to fit ARMA models to series with missing observations. Shumway and Stoffer (1982) described the modifications necessary to fit multivariate state-space models via the EM algorithm when data are missing. We will discuss the procedure in detail in this section. Throughout this section, for notational simplicity, we drop the inputs from the model, (2.1) and (2.2).

Suppose, at a given time t, we define the partition of the $q \times 1$ observation vector $Y_t = (Y_t^{(1)\prime}, Y_t^{(2)\prime})'$, where the first $q_{1t} \times 1$ component is observed and the second $q_{2t} \times 1$ component is unobserved, $q_{1t} + q_{2t} = q$. Then, write the partitioned observation equation

$$\begin{pmatrix} Y_t^{(1)} \\ Y_t^{(2)} \end{pmatrix} = \begin{bmatrix} A_t^{(1)} \\ A_t^{(2)} \end{bmatrix} X_t + \begin{pmatrix} V_t^{(1)} \\ V_t^{(2)} \end{pmatrix}, \tag{2.52}$$

where $A_t^{(1)}$ and $A_t^{(2)}$ are, respectively, the $q_{1t} \times p$ and $q_{2t} \times p$ partitioned observation matrices, and

$$\mathrm{Cov}\begin{pmatrix} V_t^{(1)} \\ V_t^{(2)} \end{pmatrix} = \begin{bmatrix} R_{11t} & R_{12t} \\ R_{21t} & R_{22t} \end{bmatrix} \tag{2.53}$$

denotes the covariance matrix of the measurement errors between the observed and unobserved parts.

In the missing data case where $Y_t^{(2)}$ is not observed, we may modify the observation equation so that the model is

$$X_t = \Phi X_{t-1} + W_t \quad \text{and} \quad Y_t^{(1)} = A_t^{(1)} X_t + V_t^{(1)}, \tag{2.54}$$

where now, the observation equation is q_{1t}-dimensional at time t. In this case, it follows directly from Corollary 2.6 that the filter equations hold with the appropriate notational substitutions. If there are no observations at time t, then set the gain matrix, K_t, to the $p \times q$ zero matrix in Proposition 2.3, in which case $X_{t|t} = X_{t|t-1}$ and $P_{t|t} = P_{t|t-1}$.

Rather than deal with varying observational dimensions, it is computationally easier to modify the model by zeroing out certain components and retaining a q-dimensional observation equation throughout. In particular, Corollary 2.6 holds for the missing data case if, at update t, we substitute

$$Y_{(t)} = \begin{pmatrix} Y_t^{(1)} \\ 0 \end{pmatrix}, \quad A_{(t)} = \begin{bmatrix} A_t^{(1)} \\ 0 \end{bmatrix}, \quad R_{(t)} = \begin{bmatrix} R_{11t} & 0 \\ 0 & I_{22t} \end{bmatrix}, \tag{2.55}$$

for Y_t, A_t, and R, respectively, in (2.15)–(2.17), where I_{22t} is the $q_{2t} \times q_{2t}$ identity matrix. With the substitutions (2.55), the innovation values (2.18) and (2.19) will now be of the form

$$\varepsilon_{(t)} = \begin{pmatrix} \varepsilon_t^{(1)} \\ 0 \end{pmatrix}, \quad \Sigma_{(t)} = \begin{bmatrix} A_t^{(1)} P_t^{t-1} A_t^{(1)\prime} + R_{11t} & 0 \\ 0 & I_{22t} \end{bmatrix}, \tag{2.56}$$

so that the innovations form of the likelihood given in (2.34) is correct for this case. Hence, with the substitutions in (2.55), maximum likelihood estimation via the innovations likelihood can proceed as in the complete data case.

Once the missing data filtered values have been obtained, it can be shown that the smoother values can be processed using Proposition 2.7 with the values obtained from the missing data-filtered values. Missing data examples may be found in Shumway and Stoffer (2011, Chapter 6); see also Shumway and Stoffer (1982). The following example examines a univariate missing data example.

Example 2.15 (Longitudinal biomedical data). Suppose we consider the problem of monitoring the level of a biomedical marker after a cancer patient undergoes a bone marrow transplant. The data in Figure 2.4, used by Jones (1980), are measurements made for 91 days on log(white blood count) [WBC]. Approximately 40% of the values are missing, with missing values occurring primarily after the 35th day. The main objectives are to model WBC using the state-space approach, and to estimate the missing values.

We fit the model (2.54) using (2.55) to the WBC data using the EM algorithm, which converged after 54 iterations. The maximum likelihood procedure yielded the estimators $\widehat{\Phi} = 1.003$, $\widehat{Q} = 0.021$, $\widehat{R} = 0.003$, $\widehat{\mu}_0 = 2.264$, and $\widehat{\Sigma}_0 = 0.0005$.

Byproducts of the procedure are the estimated trajectory and the respective prediction intervals. In particular, Figure 2.4 shows the data as points, the estimated smoothed values $\widehat{X}_{t|n}$ as solid lines, and error bounds, $\widehat{X}_{t|n} \pm 3\sqrt{\widehat{P}_{t|n}}$, as a gray area, for critical post-transplant white blood cell count. ◇

2.7 Structural component models

Structural component models have been advocated by Harvey (1991) and others. The recent text Kitagawa (2010) has a number of interesting examples including

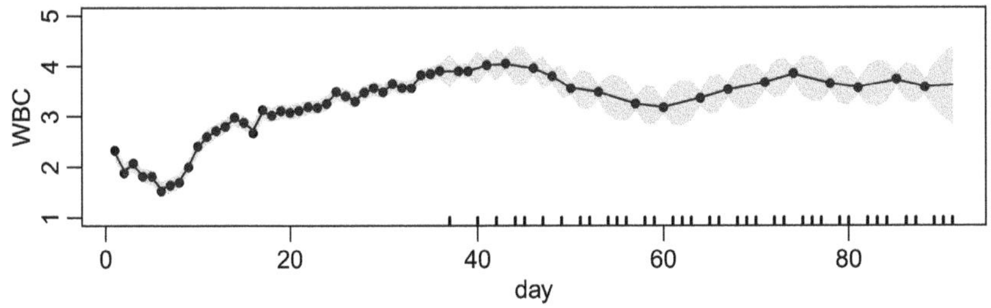

Figure 2.4 *Smoothed values for the white blood count [WBC] tracking problem. The actual data are shown as points, the smoothed values are shown as a solid line, and ± 3 standard error bounds are shown as a gray area. Days where no data are recorded are marked by hash marks.*

nonlinear problems. In the basic case, the components of the model are taken as linear processes that can be adapted to represent fixed and disturbed trends and periodicities as well as classical autoregressions. The observed series is regarded as being a sum of component signal series, such as trend and seasonal components. For example, the trend models in Example 2.8 and Example 2.13 are simple cases of structural models. We illustrate the technique via a more complex example that shows how to fit a sum of trend, seasonal, and irregular components to the quarterly earnings data.

Example 2.16 (Johnson & Johnson quarterly earnings). Consider the quarterly earnings series from the U.S. company Johnson & Johnson as given in Figure 2.5. The series is highly nonstationary, and there is both a trend signal that is gradually increasing over time and a seasonal component that cycles every four quarters or once per year. The seasonal component is getting larger over time as well. Transforming into logarithms or even taking the n-th root does not seem to make the series stationary, as there is a slight bend to the transformed curve. Suppose, however, we consider the series to be the sum of a trend component, a seasonal component, and a white noise. That is, let the observed series be expressed as

$$Y_t = T_t + S_t + V_t, \tag{2.57}$$

where T_t is trend and S_t is the seasonal component. Suppose we allow the trend to increase exponentially; that is,

$$T_t = \phi T_{t-1} + W_{t1}, \tag{2.58}$$

where the coefficient $\phi > 1$ characterizes the increase. Let the seasonal component be modeled as

$$S_t + S_{t-1} + S_{t-2} + S_{t-3} = W_{t2}, \tag{2.59}$$

which corresponds to assuming the seasonal component is expected to sum to zero over a complete period or four quarters. To express this model in state-space form, let $X_t = (T_t, S_t, S_{t-1}, S_{t-2})'$ be the state vector so the observation equation can be written

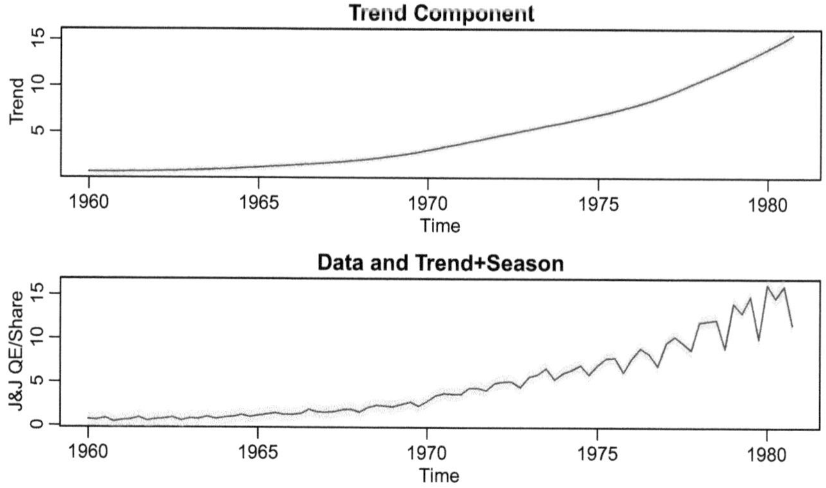

Figure 2.5 *Estimated trend component, $T_{t|n}$ with error bounds (top). The data (solid line) with error bounds for $T_{t|n}+S_{t|n}$ superimposed as a gray area (bottom).*

as

$$Y_t = \begin{pmatrix}1 & 1 & 0 & 0\end{pmatrix}\begin{pmatrix}T_t\\ S_t\\ S_{t-1}\\ S_{t-2}\end{pmatrix} + V_t,$$

with the state equation written as

$$\begin{pmatrix}T_t\\ S_t\\ S_{t-1}\\ S_{t-2}\end{pmatrix} = \begin{pmatrix}\phi & 0 & 0 & 0\\ 0 & -1 & -1 & -1\\ 0 & 1 & 0 & 0\\ 0 & 0 & 1 & 0\end{pmatrix}\begin{pmatrix}T_{t-1}\\ S_{t-1}\\ S_{t-2}\\ S_{t-3}\end{pmatrix} + \begin{pmatrix}W_{t1}\\ W_{t2}\\ 0\\ 0\end{pmatrix},$$

where $R = r_{11}$ and

$$Q = \begin{pmatrix}q_{11} & 0 & 0 & 0\\ 0 & q_{22} & 0 & 0\\ 0 & 0 & 0 & 0\\ 0 & 0 & 0 & 0\end{pmatrix}.$$

The model reduces to state-space form, (2.1) and (2.2), with $p = 4$ and $q = 1$. The parameters to be estimated are r_{11}, the noise variance in the measurement equations, q_{11} and q_{22}, the model variances corresponding to the trend and seasonal components and ϕ, the transition parameter that models the growth rate. Growth is about 3% per year, and we began with $\phi = 1.03$. The initial mean was fixed at $\mu_0 = (.7, 0, 0, 0)'$, with uncertainty modeled by the diagonal covariance matrix with $\Sigma_{0ii} = .04$, for $i = 1, \dots, 4$. Initial state covariance values were taken as $q_{11} = .01, q_{22} = .01$. The measurement error covariance was started at $r_{11} = .25$.

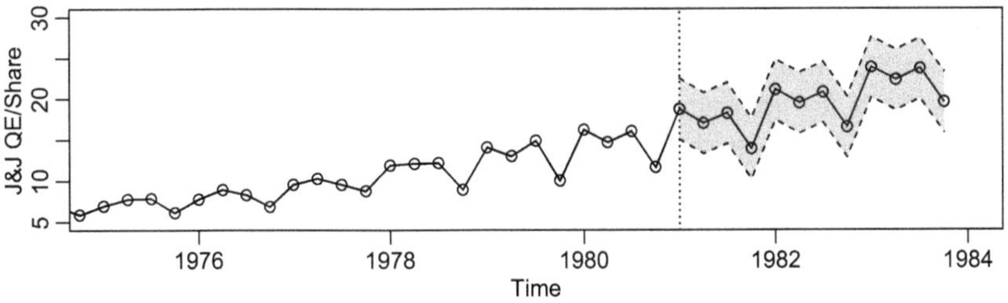

Figure 2.6 *A 12-quarter forecast for the Johnson & Johnson quarterly earnings series. The forecasts are shown as a continuation of the data (points connected by a solid line). The dashed lines indicate the upper and lower 95% prediction intervals.*

After about 20 iterations of a Newton–Raphson, the transition parameter estimate was $\widehat{\phi} = 1.035_{(.003)}$, corresponding to exponential growth with inflation at about 3.5% per year. The measurement uncertainty was small at $\sqrt{\widehat{r}_{11}} = .0005_{(.242)}$, compared with the model uncertainties $\sqrt{\widehat{q}_{11}} = .140_{(.022)}$ and $\sqrt{\widehat{q}_{22}} = .221_{(.024)}$. Note that standard errors are shown in parentheses. Figure 2.5 shows the smoothed trend estimate and the exponentially increasing seasonal components. We may also consider forecasting the Johnson & Johnson series, and the result of a 12-quarter forecast is shown in Figure 2.6 as basically an extension of the latter part of the observed data. The R code for this example may be found in its entirety in Shumway and Stoffer (2011, Example 6.10) and on the website for the text, www.stat.pitt.edu/stoffer/tsa3. ◇

2.8 State-space models with correlated errors

Sometimes it is advantageous to write the state-space model in a slightly different way, as is done by numerous authors; for example, Anderson and Moore (1979) and Hannan and Deistler (1988, 2012). Here, we write the state-space model as

$$X_{t+1} = \Phi X_t + \Upsilon U_{t+1} + \Theta W_t \qquad t = 0, 1, \dots, n \tag{2.60}$$

$$Y_t = A_t X_t + \Gamma U_t + V_t \qquad t = 1, \dots, n \tag{2.61}$$

where, in the state equation, $X_0 \sim \mathrm{N}_p(\mu_0, \Sigma_0)$, Φ is $p \times p$, and Υ is $p \times r$, Θ is $p \times m$ and $W_t \sim$ iid $\mathrm{N}_m(0, Q)$. In the observation equation, A_t is $q \times p$ and Γ is $q \times r$, and $V_t \sim$ iid $\mathrm{N}_q(0, R)$. In this model, while W_t and V_t are still white noise series (both independent of X_0), we also allow the state noise and observation noise to be correlated at time t; that is,

$$\mathrm{Cov}\,(W_s, V_t) = S\,\delta_s^t, \tag{2.62}$$

where δ_s^t is Kronecker's delta. Note that S is an $m \times q$ matrix. The major difference between this form of the model and the one specified by (2.1)-(2.2) is that this model starts the state noise process at $t = 0$ in order to ease the notation related to the concurrent covariance between W_t and V_t. Also, the inclusion of the matrix Θ allows us to avoid using a singular state noise process as was done in Example 2.16.

To obtain the innovations, $\varepsilon_t = Y_t - A_t X_{t|t-1} - \Gamma U_t$, and the innovation variance $\Sigma_t = A_t P_{t|t-1} A_t' + R$, in this case, we need the one-step-ahead state predictions. Of course, the filtered estimates will also be of interest, and they will be needed for smoothing; Proposition 2.7 still holds. The following property generates the predictor $X_{t+1|t}$ from the past predictor $X_{t|t-1}$ when the noise terms are correlated and exhibits the filter update.

Proposition 2.17 (The Kalman filter with correlated noise). *For the state-space model specified in (2.60) and (2.61), with initial conditions $X_{1|0}$ and $P_{1|0}$, for $t = 1,\dots,n$,*

$$X_{t+1|t} = \Phi X_{t|t-1} + \Upsilon U_{t+1} + K_t \varepsilon_t \tag{2.63}$$

$$P_{t+1|t} = \Phi P_{t|t-1} \Phi' + \Theta Q \Theta' - K_t \Sigma_t K_t' \tag{2.64}$$

where $\varepsilon_t = Y_t - A_t X_{t|t-1} - \Gamma U_t$ and the gain matrix is given by

$$K_t = [\Phi P_{t|t-1} A_t' + \Theta S][A_t P_{t|t-1} A_t' + R]^{-1}. \tag{2.65}$$

The filter values are given by

$$X_{t|t} = X_{t|t-1} + P_{t|t-1} A_t' \left[A_t P_{t|t-1} A_t' + R\right]^{-1} \varepsilon_{t+1}, \tag{2.66}$$

$$P_{t|t} = P_{t|t-1} - P_{t|t-1} A_{t+1}' \left[A_t P_{t|t-1} A_t' + R\right]^{-1} A_t P_{t|t-1}. \tag{2.67}$$

The derivation of Proposition 2.17 is similar to the derivation of the Kalman filter in Proposition 2.3; we note that the gain matrix K_t differs in the two properties. The filter values, (2.66)–(2.67), are symbolically identical to (2.13) and (2.14). To initialize the filter, we note that

$$X_{1|0} = \mathbb{E}[X_1] = \Phi \mu_0 + \Upsilon U_1, \quad \text{and} \quad P_{1|0} = \operatorname{Var}[X_1] = \Phi \Sigma_0 \Phi' + \Theta Q \Theta'.$$

In the next two subsections, we show how to use the model (2.60)-(2.61) for fitting ARMAX models and for fitting (multivariate) regression models with autocorrelated errors. To put it succinctly, for ARMAX models, the inputs enter in the state equation and for regression with autocorrelated errors, the inputs enter in the observation equation. It is, of course, possible to combine the two models.

2.8.1 ARMAX models

Consider a k-dimensional ARMAX model given by

$$Y_t = \Upsilon U_t + \sum_{j=1}^{p} \Phi_j Y_{t-j} + \sum_{k=1}^{q} \Theta_k V_{t-k} + V_t. \tag{2.68}$$

The observations Y_t are a k-dimensional vector process, the Φs and Θs are $k\times k$ matrices, Υ is $k\times r$, U_t is the $r\times 1$ input, and V_t is a $k\times 1$ white noise process; in fact, (2.68) and (1.73) are identical models, but here, we have written the observations as Y_t and the Gaussian noise as V_t. We now have the following property.

Proposition 2.18 (A state-space form of ARMAX). *For $p\ge q$, let*

$$F=\begin{bmatrix}\Phi_1 & I_k & 0 & \cdots & 0\\ \Phi_2 & 0 & I_k & \cdots & 0\\ \vdots & \vdots & \vdots & \ddots & \vdots\\ \Phi_{p-1} & 0 & 0 & \cdots & I_k\\ \Phi_p & 0 & 0 & \cdots & 0\end{bmatrix}\quad G=\begin{bmatrix}\Theta_1+\Phi_1\\ \vdots\\ \Theta_q+\Phi_q\\ \Phi_{q+1}\\ \vdots\\ \Phi_p\end{bmatrix}\quad H=\begin{bmatrix}\Upsilon\\ 0\\ \vdots\\ 0\end{bmatrix} \tag{2.69}$$

where I_k is the $k\times k$ identity matrix, F is $kp\times kp$, G is $kp\times k$, and H is $kp\times r$. Then, the state-space model given by

$$X_{t+1}=FX_t+HU_{t+1}+GV_t, \tag{2.70}$$

$$Y_t=AX_t+V_t, \tag{2.71}$$

where $A=\left[I_k,0,\cdots,0\right]$ is $k\times pk$, implies the ARMAX model (2.68). If $p<q$, set $\Phi_{p+1}=\cdots=\Phi_q=0$, in which case $p=q$ and (2.70)–(2.71) still apply. Note that the state process is kp-dimensional, whereas the observations are k-dimensional.

Example 2.19 (Univariate ARMAX$(1,1)$ in state-space form). Consider the univariate ARMAX$(1,1)$ model

$$Y_t=\alpha_t+\phi Y_{t-1}+\theta V_{t-1}+V_t,$$

where $\alpha_t=\Upsilon U_t$ to ease the notation. For a simple example, if $\Upsilon=(\beta_0,\beta_1)$ and $U_t=(1,t)'$, the model for Y_t would be ARMA(1,1) with linear trend, $Y_t=\beta_0+\beta_1 t+\phi Y_{t-1}+\theta V_{t-1}+V_t$. Using Proposition 2.18, we can write the model as

$$X_{t+1}=\phi X_t+\alpha_{t+1}+(\theta+\phi)V_t, \tag{2.72}$$

and

$$Y_t=X_t+V_t. \tag{2.73}$$

In this case, (2.72) is the state equation with $W_t\equiv V_t$ and (2.73) is the observation equation. Consequently, $\mathrm{Cov}(W_t,V_t)=\mathrm{Var}(V_t)=R$, and $\mathrm{Cov}(W_t,V_s)=0$ when $s\neq t$, so Proposition 2.17 would apply. To verify that (2.72) and (2.73) specify an ARMAX$(1,1)$ model, we have

$$\begin{aligned} Y_t &= X_t+V_t && \text{from (2.73)}\\ &= [\phi X_{t-1}+\alpha_t+(\theta+\phi)V_{t-1}]+V_t && \text{from (2.72)}\\ &= \alpha_t+\phi(X_{t-1}+V_{t-1})+\theta V_{t-1}+V_t && \text{rearrange terms}\\ &= \alpha_t+\phi Y_{t-1}+\theta V_{t-1}+V_t, && \text{from (2.73).} \end{aligned}$$

◇

Together, Proposition 2.17 and Proposition 2.18 can be used to accomplish maximum likelihood estimation as described in Section 2.3 for ARMAX models. The ARMAX model is only a special case of the model (2.60)–(2.61), which is quite rich, as will be reinforced in the next subsection.

2.8.2 *Regression with autocorrelated errors*

In (multivariate) regression with autocorrelated errors, we are interested in fitting the regression model

$$Y_t = \Gamma U_t + \varepsilon_t \tag{2.74}$$

to a $k \times 1$ vector process, Y_t, with r regressors $U_t = (U_{t1}, \ldots, U_{tr})'$ where ε_t is vector ARMA(p,q) and Γ is a $k \times r$ matrix of regression parameters. We note that the regressors do not have to vary with time (e.g., $U_{t1} \equiv 1$ includes a constant in the regression).

To put the model in state-space form, we simply notice that $\varepsilon_t = Y_t - \Gamma U_t$ is a k-dimensional ARMA(p,q) process. Thus, if we set $H = 0$ in (2.70), and include ΓU_t in (2.71), we obtain

$$X_{t+1} = FX_t + GV_t, \tag{2.75}$$

$$Y_t = \Gamma U_t + AX_t + V_t, \tag{2.76}$$

where the model matrices A, F, and G are defined in Proposition 2.18. The fact that (2.75)–(2.76) is multivariate regression with autocorrelated errors follows directly from Proposition 2.18 by noticing that together, $X_{t+1} = FX_t + GV_t$ and $\varepsilon_t = AX_t + V_t$ imply $\varepsilon_t = Y_t - \Gamma U_t$ is vector ARMA(p,q).

As in the case of ARMAX models, regression with autocorrelated errors is a special case of the state-space model, and the results of Proposition 2.17 can be used to obtain the innovations form of the likelihood for parameter estimation. Numerical examples of both types may be found in Shumway and Stoffer (2011, Example 6.12).

Exercises

2.1. Consider a system process given by

$$X_t = -.9X_{t-2} + W_t \quad t = 1, 2, \ldots,$$

where $X_0 \sim \mathrm{N}(0, \sigma_0^2)$, $X_{-1} \sim N(0, \sigma_1^2)$, and $\{W_t,\, t \in \mathbb{N}^*\}$ is Gaussian white noise with variance σ_w^2. The system process is observed with noise, say,

$$Y_t = X_t + V_t,$$

where $\{V_t,\, t \in \mathbb{N}^*\}$ is Gaussian white noise with variance σ_v^2. Further, suppose X_0, X_{-1}, $\{W_t,\, t \in \mathbb{Z}\}$ and $\{V_t,\, t \in \mathbb{Z}\}$ are independent.

(a) Write the system and observation equations in the form of a state space model.

(b) Find the values of σ_0^2 and σ_1^2 that make the observations, Y_t, stationary.

2.2. Consider the AR(1) process with observational noise model presented in Example 2.2. Let $X_{t|t-1} = \mathbb{E}\left[X_t | Y_{t-1}, \ldots, Y_1\right]$ and let $P_{t|t-1} = \mathbb{E}[X_t - X_{t|t-1}]^2$. The innovations are $\varepsilon_t = Y_t - Y_{t|t-1}$, where $Y_{t|t-1} = \mathbb{E}\left[Y_t | Y_{t-1}, \ldots, Y_1\right]$. Find $\mathrm{Cov}(\varepsilon_s, \varepsilon_t)$ in terms of $X_{t|t-1}$ and $P_{t|t-1}$ for (a) $s \neq t$ and (b) $s = t$.

2.3 (Projection theorem derivation of Proposition 2.3). Suppose the vector $Z = (X', Y')'$, where X $(p \times 1)$ and Y $(q \times 1)$ are jointly distributed with mean vectors μ_x and μ_y and with covariance matrix

$$\mathrm{Cov}_Z = \begin{pmatrix} \Sigma_{xx} & \Sigma_{xy} \\ \Sigma_{yx} & \Sigma_{yy} \end{pmatrix}$$

with $\Sigma_{yy} > 0$. Consider projecting X on $\mathcal{M} = \overline{\mathrm{sp}}\{1, Y\}$, say, $\mathsf{P}_{\mathcal{M}} X = b + BY$.

(a) Show the orthogonality conditions of Proposition 1.34, (1.51), can be written as

$$\mathbb{E}(X - b - BY) = 0, \quad \text{and} \quad \mathbb{E}\left[(X - b - BY)Y'\right] = 0,$$

leading to the solutions

$$b = \mu_x - B\mu_y \quad \text{and} \quad B = \Sigma_{xy}\Sigma_{yy}^{-1}.$$

(b) Prove the mean square error matrix is

$$MSE = \mathbb{E}\left[(X - b - BY)X'\right] = \Sigma_{xx} - \Sigma_{xy}\Sigma_{yy}^{-1}\Sigma_{yx}.$$

(c) How can these results be used to justify the claim that, in the absence of normality, Proposition 2.3 yields the best linear estimate of the state X_t given the data $\{Y_1, \ldots, Y_t\}$, say, $X_{t|t}$, and its corresponding MSE, $P_{t|t}$?

2.4 (Projection theorem derivation of Proposition 2.7). Let $\mathcal{Y}_t = \overline{\mathrm{sp}}\{Y_1, \ldots, Y_t\}$, and $\mathcal{E}_t = \overline{\mathrm{sp}}\{Y_t - Y_{t|t-1}\}$, for $t = 1, \ldots, n$, where $Y_{t|t-1}$ is the projection of Y_t on $\mathcal{Y}_{t-1}$ ($\mathcal{Y}_0$ is empty). Note that, $\mathcal{Y}_t = \mathcal{Y}_{t-1} \oplus \mathcal{E}_t$. We assume $P_{0|0} > 0$ and $R > 0$.

(a) Show the projection of X_{t-1} on $\mathcal{Y}_t$, that is, $X_{t-1|t}$, is given by

$$X_{t-1|t} = X_{t-1|t-1} + H_t(Y_t - Y_{t|t-1}),$$

where H_t can be determined by the orthogonality property

$$\mathbb{E}\left\{\left(X_{t-1} - H_t(Y_t - Y_{t|t-1})\right)\left(Y_t - Y_{t|t-1}\right)'\right\} = 0.$$

Show

$$H_t = P_{t-1|t-1}\Phi' A_t' \left[A_t P_{t|t-1} A_t' + R\right]^{-1}.$$

(b) Define $J_{t-1} = P_{t-1|t-1}\Phi' P_{t|t-1}^{-1}$, and show

$$X_{t-1|t} = X_{t|t} + J_{t-1}(X_{t|t} - X_{t|t-1}).$$

(c) Repeating the process, show

$$X_{t-1|t+1} = X_{t-1|t-1} + J_{t-1}(X_{t|t+1} - X_{t|t-1}) + H_{t+1}(Y_{t+1} - Y_{t+1|t}),$$

solving for H_{t+1}. Simplify and show

$$X_{t-1|t+1} = X_{t-1|t-1} + J_{t-1}(X_{t|t+1} - X_{t|t-1}).$$

(d) Using induction, conclude

$$X_{t-1|n} = X_{t-1|t-1} + J_{t-1}(X_{t|n} - X_{t|t-1}),$$

which yields the smoother for $t = 1, \ldots, n$.

2.5. Consider the univariate state-space model given by state conditions $X_0 = W_0$, $X_t = X_{t-1} + W_t$ and observations $Y_t = X_t + V_t$, for $t = 1, 2, \ldots$, where $\{W_t,\, t \in \mathbb{N}^*\}$ and $\{V_t,\, t \in \mathbb{N}^*\}$ are independent, Gaussian, white noise processes with $\text{Var}(W_t) = \sigma_w^2$ and $\text{Var}(V_t) = \sigma_v^2$. Show that Y_t follows an IMA(1,1) model, that is, ∇Y_t follows an MA(1) model.

2.6 (Estimating trend of global temperature). Let Y_t represent the global temperature series (`gtemp` from the R package `astsa`) shown in Figure 2.1 and discussed in Example 2.1.

(a) Use R to fit a smoothing spline using GCV (the default) to Y_t. Repeat the fit using `spar=.7`; the GCV method yields `spar=.5` approximately (in R, see the help file for `smooth.spline`). Save the results for part (c).

(b) Consider the model $Y_t = X_t + V_t$ with $\nabla^2 X_t = W_t$ with $\text{Var}[W_t] = \sigma_w^2$ and $\text{Var}[V_t] = \sigma_v^2$. Using Example 2.13 as a guide, fit this model to Y_t, and compute the estimated smoother, say $\widehat{X}_{t|n}$, and the corresponding estimated MSE, $\widehat{P}_{t|n}$. Save the results for part (c).

(c) Superimpose all the fits from parts (a) and (b) [include error bounds with (b)] on the data and briefly compare and contrast the results.

2.7. Consider data $\{Y_1, \ldots, Y_n\}$ from the model

$$Y_t = X_t + V_t,$$

where $\{V_t,\, t \in \mathbb{N}^*\}$ is Gaussian white noise with variance σ_v^2, $\{X_t,\, t \in \mathbb{N}^*\}$ are independent Gaussian random variables with mean zero and $\text{Var}(X_t) = r_t \sigma_x^2$ with $\{X_t,\, t \in \mathbb{N}^*\}$ independent of $\{V_t,\, t \in \mathbb{N}^*\}$, and $r_1, \ldots, r_n$ are known constants. Show that applying the EM algorithm to the problem of estimating σ_x^2 and σ_v^2 leads to updates (represented by hats)

$$\widehat{\sigma}_x^2 = \frac{1}{n} \sum_{t=1}^{n} \frac{\sigma_t^2 + \mu_t^2}{r_t} \quad \text{and} \quad \widehat{\sigma}_v^2 = \frac{1}{n} \sum_{t=1}^{n} [(Y_t - \mu_t)^2 + \sigma_t^2],$$

where, based on the current estimates (represented by tildes),

$$\mu_t = \frac{r_t \widetilde{\sigma}_x^2}{r_t \widetilde{\sigma}_x^2 + \widetilde{\sigma}_v^2} Y_t \quad \text{and} \quad \sigma_t^2 = \frac{r_t \widetilde{\sigma}_x^2 \widetilde{\sigma}_v^2}{r_t \widetilde{\sigma}_x^2 + \widetilde{\sigma}_v^2}.$$

2.8. To explore the stability of the filter, consider a univariate ($p = q = 1$) state-space model, (2.1) and (2.2), without inputs. That is, for $t = 1, 2, \ldots$, the observations are $Y_t = X_t + V_t$ and the state equation is $X_t = \phi X_{t-1} + W_t$, where $Q = R = 1$ and $|\phi| < 1$. The initial state, X_0, has zero mean and variance one.

(a) Exhibit the recursion for $P_{t|t-1}$ in Proposition 2.3 in terms of $P_{t-1|t-2}$.

(b) Use the result of (a) to verify that $P_{t|t-1}$ approaches a limit ($t \to \infty$) P that is the positive solution of $P^2 - \phi^2 P - 1 = 0$.

(c) With $K = \lim_{t\to\infty} K_t$ as given in Proposition 2.3, (2.17), show $|1-K| < 1$.

(d) Show that in steady-state, $Y_{t+1|t} = \mathbb{E}\left[Y_{t+1} \mid Y_t, Y_{t-1}, \ldots\right]$, the one-step-ahead predictor of a future observation, satisfies

$$Y_{t+1|t} = \sum_{j=0}^{\infty} \phi^j K(1-K)^{j-1} Y_{t+1-j}.$$

2.9. As an example of the way the state-space model handles the missing data problem, suppose the first-order autoregressive process

$$X_t = \phi X_{t-1} + W_t,$$

$|\phi| < 1$, has an observation missing at $t = m$, leading to the observations $Y_t = A_t X_t$, where $A_t = 1$ for all t, except $t = m$ wherein $A_t = 0$. Assume $X_0 \sim \mathrm{N}(0, \sigma_w^2/(1-\phi^2))$, where the variance of $\{W_t,\ t \in \mathbb{N}^*\}$ is σ_w^2. Show the Kalman smoother estimators in this case are

$$X_{t|n} = \begin{cases} \phi Y_1 & t = 0, \\ \frac{\phi}{1+\phi^2}(Y_{m-1} + Y_{m+1}) & t = m, \\ Y_t, & t \neq 0, m, \end{cases}$$

with mean square covariances determined by

$$P_{t|n} = \begin{cases} \sigma_w^2 & t = 0, \\ \sigma_w^2/(1+\phi^2) & t = m, \\ 0 & t \neq 0, m. \end{cases}$$

2.10. Use Proposition 2.18 to complete the following exercises.

(a) Write a univariate AR(1) model, $Y_t = \phi Y_{t-1} + V_t$, in state-space form. Verify your answer is indeed an AR(1).

(b) Repeat (a) for an MA(1) model, $Y_t = V_t + \theta V_{t-1}$.

(c) Write an IMA(1,1) model, $Y_t = Y_{t-1} + V_t + \theta V_{t-1}$, in state-space form.

2.11. In Section 2.3, we discussed that it is possible to obtain a recursion for the gradient or score vector, $-\partial \ln \mathrm{L}(\theta; Y_{1:n})/\partial\theta$. Assume the model is given by (2.1) and (2.2) and A_t is a known design matrix that does not depend on θ, in which case Proposition 2.3 applies. For the gradient vector, show

$$\begin{aligned} \partial \ln \mathrm{L}(\theta; Y_{1:n})/\partial\theta_i &= \sum_{t=1}^{n} \Bigg\{ \varepsilon_t' \Sigma_t^{-1} \frac{\partial \varepsilon_t}{\partial \theta_i} - \frac{1}{2} \varepsilon_t' \Sigma_t^{-1} \frac{\partial \Sigma_t}{\partial \theta_i} \Sigma_t^{-1} \varepsilon_t \\ &\qquad + \frac{1}{2} \mathrm{tr}\left(\Sigma_t^{-1} \frac{\partial \Sigma_t}{\partial \theta_i} \right) \Bigg\}, \end{aligned}$$

where the dependence of the innovation values on θ is understood. In addition, with the general definition $\partial_i g := \partial g(\theta)/\partial\theta_i$, show that the following recursions, for $t = 2,\dots,n$, apply:

(i) $\partial_i\varepsilon_t = -A_t\,\partial_i X_{t|t-1}$,

(ii) $\partial_i X_{t|t-1} = \partial_i\Phi\,X_{t-1|t-2} + \Phi\,\partial_i X_{t-1|t-2} + \partial_i K_{t-1}\,\varepsilon_{t-1} + K_{t-1}\,\partial_i\varepsilon_{t-1}$,

(iii) $\partial_i\Sigma_t = A_t\,\partial_i P_{t|t-1}A_t' + \partial_i R$,

(iv) $\partial_i K_t = \left[\partial_i\Phi\,P_{t|t-1}A_t' + \Phi\,\partial_i P_{t|t-1}\,A_t' - K_t\,\partial_i\Sigma_t\right]\Sigma_t^{-1}$,

(v)
$$\partial_i P_{t|t-1} = \partial_i\Phi\,P_{t-1|t-2}\Phi' + \Phi\,\partial_i P_{t-1|t-2}\,\Phi' + \Phi\,P_{t-1|t-2}\,\partial_i\Phi' + \partial_i Q - \partial_i K_{t-1}\,\Sigma_t K_{t-1}' - K_{t-1}\,\partial_i\Sigma_t\,K_{t-1}' - K_{t-1}\Sigma_t\,\partial_i K_{t-1}',$$

using the fact that $P_{t|t-1} = \Phi P_{t-1|t-2}\Phi' + Q - K_{t-1}\Sigma_t K_{t-1}'$.

2.12. Continuing with the previous problem, consider the evaluation of the Hessian matrix and the numerical evaluation of the asymptotic variance–covariance matrix of the parameter estimates. The information matrix satisfies

$$\mathbb{E}\left\{-\frac{\partial^2 \ln \mathrm{L}(\theta;Y_{1:n})}{\partial\theta\,\partial\theta'}\right\} = \mathbb{E}\left\{\left(\frac{\partial \ln \mathrm{L}(\theta;Y_{1:n})}{\partial\theta}\right)\left(\frac{\partial \ln \mathrm{L}(\theta;Y_{1:n})}{\partial\theta}\right)'\right\};$$

see Anderson (2003, Section 3.4), for example. Show the (i,j)-th element of the information matrix, say, $\mathcal{I}_{ij}(\theta) = \mathbb{E}\left\{-\partial^2 \ln \mathrm{L}(\theta;Y_{1:n})/\partial\theta_i\,\partial\theta_j\right\}$, is

$$\begin{aligned}\mathcal{I}_{ij}(\theta) &= \sum_{t=1}^{n}\mathbb{E}\Big\{\,\partial_i\varepsilon_t'\,\Sigma_t^{-1}\,\partial_j\varepsilon_t + \tfrac{1}{2}\mathrm{tr}\big(\Sigma_t^{-1}\,\partial_i\Sigma_t\,\Sigma_t^{-1}\,\partial_j\Sigma_t\big)\\ &\qquad + \tfrac{1}{4}\mathrm{tr}\big(\Sigma_t^{-1}\,\partial_i\Sigma_t\big)\,\mathrm{tr}\big(\Sigma_t^{-1}\,\partial_j\Sigma_t\big)\Big\}.\end{aligned}$$

Consequently, an approximate Hessian matrix can be obtained from the sample by dropping the expectation, $\mathbb{E}$, in the above result and using only the recursions needed to calculate the gradient vector.

Chapter 3

Beyond Linear Models

The goal of this chapter is to to provide a cursory introduction to nonlinear processes and models that may be used for data analysis. We motivate the need for nonlinear and non-Gaussian models through real data examples, discussing why there is a need for such models. We give some examples that may help explain some of the similarities seen in nonlinear or non-Gaussian process from many different disciplines. Then we exhibit some of the models used to analyze such processes and briefly discuss their properties. Our intention is not to be exhaustive in covering these topics, but rather to give a sampling of various situations and approaches to modeling nonlinear and non-Gaussian processes.

As discussed in Chapter 1, the main goal of classical time series is the analysis of the second order structure of stationary processes. This structure is fully determined by the autocovariance or autocorrelation functions, or alternatively by the associated spectral measure. The second order structure of the process fully determines the structure of stationary Gaussian processes. It is known from the Wold decomposition (see Theorem 1.25) that a regular second order stationary process $\{X_t,\, t \in \mathbb{Z}\}$ may be represented as

$$X_t = \sum_{j=0}^{\infty} \psi_j Z_{t-j}\,, \tag{3.1}$$

where $\sum_{j=0}^{\infty} \psi_j^2 < \infty$ and $\{Z_t,\, t \in \mathbb{Z}\} \sim \mathrm{WN}(0, \sigma_z^2)$ is white noise. In this case, the spectral measure of the process $\{X_t,\, t \in \mathbb{Z}\}$ has a density $f_x(\omega) = \frac{\sigma_z^2}{2\pi} \left|\psi(\mathrm{e}^{-\mathrm{i}\omega})\right|^2$, where $\psi(\mathrm{e}^{-\mathrm{i}\omega})$ is the transfer function associated to the impulse response $\{\psi_j,\, j \in \mathbb{N}\}$; see Example 1.33. If we are only interested in the second order structure, $\{X_t,\, t \in \mathbb{Z}\}$ is equivalent to the (strong sense) causal linear process $\{\tilde{X}_t,\, t \in \mathbb{Z}\}$ given by

$$\tilde{X}_t = \sum_{j=0}^{\infty} \psi_j \tilde{Z}_{t-j}\,, \tag{3.2}$$

where $\{\tilde{Z}_t,\, t \in \mathbb{Z}\}$ is strong (i.i.d.) white noise with variance σ_z^2. The structure of a linear process is therefore intimately related to the properties of causal linear systems.

For simplicity, assume that $\sum_{j=0}^{\infty} |\psi_j| < \infty$. First, if the input of a linear system is a sine wave of pulsation ω_0, i.e., $Z_t = A\cos(\omega_0 t + \varphi)$, then the output, $X_t = \sum_{j=0}^{\infty} \psi_j Z_{t-j}$, is a sine wave of same frequency ω_0 but with the amplitude scaled by $|\psi(e^{-i\omega_0})|$ and phase shifted by $\arg[\psi(e^{-i\omega_0})]$; see Exercise 3.1. This property is typically lost in nonlinear systems. If we input a sine wave into a nonlinear system, then the output (provided it is well defined) contains not only a component at the fundamental frequency ω_0 but also components at the harmonics, i.e., multiples of the fundamental frequencies $2\omega_0, 3\omega_0$, and so on. Second, a linear process satisfies a superposition principle, i.e., if we input a sum $\{Z_t^{(1)},\ t \in \mathbb{Z}\}$ and $\{Z_t^{(2)},\ t \in \mathbb{Z}\}$ into a linear system, then the output will be the sum $X_t = X_t^{(1)} + X_t^{(2)}$, where $X_t^{(i)} = \sum_{j=0}^{\infty} \psi_j Z_{t-j}^{(i)}$, $i = 1, 2$. This property clearly extends to an arbitrary number of components and explains why the process $\{X_t,\ t \in \mathbb{Z}\}$ may be represented as

$$X_t = \int_{-\pi}^{\pi} e^{i\omega t} \psi(e^{-i\omega}) dZ(\omega) ,$$

where $Z(\omega)$ is the spectral field associated with $\{Z_t,\ t \in \mathbb{Z}\}$, i.e., $Z_t = \int_{-\pi}^{\pi} e^{i\omega t} dZ(\omega)$. In the linear world, there is a kind of natural duality between the time-domain and the frequency-domain, since linear transformation preserves the frequencies and obeys a superposition principle. In the nonlinear world, there is no such thing as *impulse response* or *transfer function* and there is no longer a nice correspondence between time and frequency domains.

When a linear process is invertible, the innovations $\{Z_t,\ t \in \mathbb{Z}\}$ can be expressed in terms of the process $\{X_t,\ t \in \mathbb{Z}\}$. If the process is causally invertible, then there exists a sequence $\{\pi_j,\ j \in \mathbb{N}\}$ such that

$$Z_t = \sum_{j=0}^{\infty} \pi_j X_{t-j} ;$$

see Definition 1.30. If $Z_t \sim \text{iid}(0, \sigma_z^2)$, then, according to (3.1), for any $t \in \mathbb{Z}$, X_t is a (linear function) of Z_s, for $s \leq t$. Hence, X_t is $\mathcal{F}_t^Z$ measurable, where $\mathcal{F}_t^Z = \sigma(Z_s, s \leq t)$ is the past history at time t of the process $\{Z_t,\ t \in \mathbb{Z}\}$; this implies that $\mathcal{F}_t^X \subset \mathcal{F}_t^Z$, where $\mathcal{F}_t^X = \sigma(X_s,\ s \leq t)$. Thus, for any $t \in \mathbb{Z}$, Z_t is independent of $\mathcal{F}_{t-1}^X \subset \mathcal{F}_{t-1}^Z$. Therefore $\mathbb{E}\left[Z_t \mid \mathcal{F}_{t-1}^X\right] = 0$ showing that the conditional expectation of X_t of the process given the past $\mathcal{F}_{t-1}^X$ can be linearly expressed as a function of its past values, the optimal predictor is linear.

3.1 Nonlinear non-Gaussian data

In Chapter 1, we indicated that linear Gaussian models can handle a broad range of problems, but that it is often necessary to go beyond these models. One might say that in the linear Gaussian world, "$\infty = 2$". In Section 1.5, however, we argued that the annual sunspot numbers were not a linear Gaussian process because the data are not time reversible, i.e., the data plotted in time order as $X_{1:n} = \{X_1, X_2, \ldots, X_n\}$ does not look the same as the data plotted in reverse time order $X_{n:1} = \{X_n, X_{n-1}, \ldots, X_1\}$. In that section we pointed out that such an occurrence will not happen for a linear Gaussian process because, in that case, $X_{1:n}$ and $X_{n:1}$ have the same distribution.

Trying to model something that is *not* linear or *not* Gaussian might seem like a

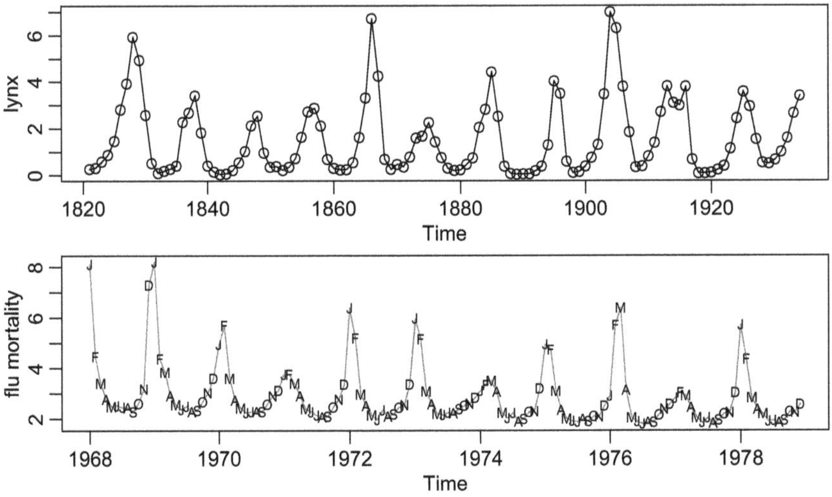

Figure 3.1 *Top: Annual numbers ($\div$1000) of lynx trappings for 1821–1934 in Canada. Bottom: Monthly rates of pneumonia and influenza deaths in the United States for 1968–1978.*

daunting task at first, because how does one model something in the negative? That is, how do you model a process that is *not* linear or *not* Gaussian; there are so many different ways to be not something. Fortunately, there are patterns of nonlinearity and non-Gaussianity that are common to many processes. In recognizing these similarities, we are able to develop general strategies and models to cover a wide range of processes observed in diverse disciplines. The following examples will help in explaining some of the commonalities of processes that are in the complement of linear Gaussian processes.

Example 3.1 (Sunspots, felines, and flu). The irreversiblity of the sunspot data (Figure 1.1) is a trait that is observed in a variety of processes. For example, the data shown in Figure 3.1 are typical of predator-prey relationships; the data are the annual numbers of lynx trappings for 1821–1934 in Canada; see Campbell and Waker (1977). Such relationships are often modeled by the Lotka-Volterra equations, which are a pair of simple nonlinear differential equations used to describe the interaction between the size of predator and prey populations; e.g., see Edelstein-Keshet (2005, Ch. 6). Note that, as opposed to the sunspot data set, the lynx data tend to increase slowly to a peak and then decline quickly to a trough ($\nearrow\Downarrow$). Another process that has a similar pattern is the influenza data also shown in Figure 3.1. These data are taken from Shumway and Stoffer (2011) and are monthly pneumonia and influenza deaths per 1,000 people in the United States for 11 years. ◇

Example 3.2 (EEG, S&P500, and explosions). The data shown in the top of Figure 3.2 are a single channel EEG signal taken from the epileptogenic zone of a subject with epilepsy, but during a seizure free interval of 23.6 seconds, and is series (d) shown in Andrzejak et al. (2001, Figure 3). The bottom of Figure 3.2 shows the

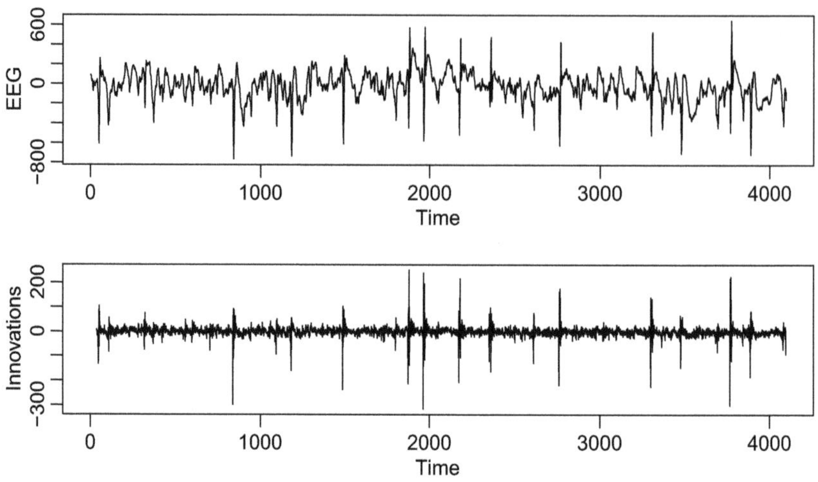

Figure 3.2 *Top: A single channel EEG signal taken from the epileptogenic zone of a subject with epilepsy during a seizure free interval of 23.6 seconds; see Andrzejak et al. (2001). Bottom: The innovations after removal of the signal using an autoregression based on AIC.*

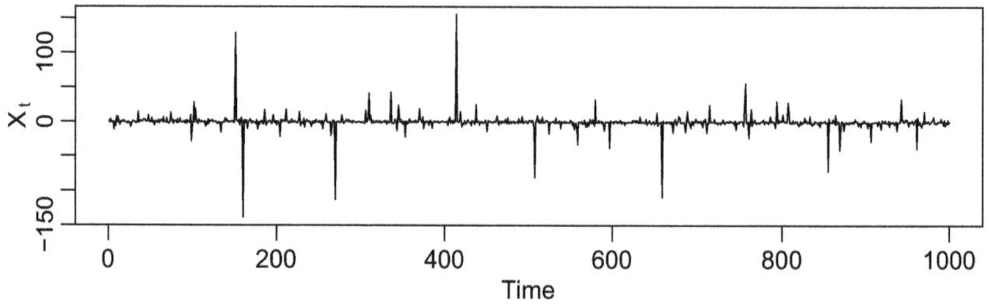

Figure 3.3: *Simulated infinite variance series generated as i.i.d. standard Cauchy errors.*

innovations (residuals) after the signal has been removed based on fitting an AR(p) using AIC to determine the order.

Due to the large spikes in the EEG trace, it is apparent that the data are not normal. In fact, the innovations in Figure 3.2 look more like the simulated infinite variance noise series shown in Figure 3.3, which were generated from i.i.d. standard Cauchy errors.

Moreover, the left side of Figure 3.4 shows the sample ACF of the EEG innovations. The fact that the values are small indicates that the innovations are white noise. However, the right side of Figure 3.4 shows the sample ACF of the squared EEG innovations, where we clearly see significant autocorrelation. Thus, while the innovations appear to be white, the are clearly not independent, and hence not Gaussian.

The behavior seen in the EEG trace is not particular to EEGs, and in fact is quite common in financial data. For example, the top of Figure 3.5 shows the daily

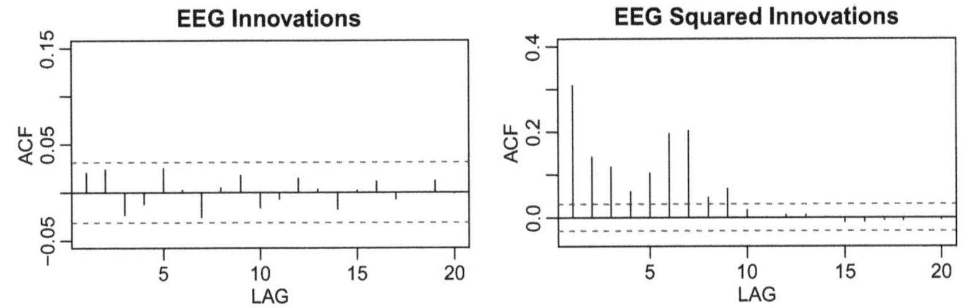

Figure 3.4 *The sample ACF of the EEG innovations (left) and the squared innovations (right); the EEG innovations series is shown in Figure 3.2.*

returns[1] of the S&P 500 from 2001 to the end of 2011. There, the data exhibit what is called volatility clusters, that is, regions of highly volatile periods tend to be clustered together. As with the EEG series, the data have very little autocorrelation, but the squares of the returns have significant autocorrelation; this is demonstrated in the bottom part of Figure 3.5.

Figure 3.6 shows the two phases or arrivals (the P-wave and then the S-wave) along the surface of an explosion at a seismic recording station. The recording instruments are observing earthquakes and mining explosions with the general problem of interest being to distinguish between waveforms generated by earthquakes and those generated by explosions. This distinction is key in enforcing a comprehensive nuclear test ban treaty. The general problem of interest, which is discussed in more detail in Shumway and Stoffer (2011, Chapter 7), is in discriminating between waveforms generated by earthquakes and those generated by explosions. The data behave in a similar fashion to the EEG trace and the S&P500 returns; see Exercise 3.2. ◇

Example 3.3 (Great discoveries and polio). Another situation in which normality is an unreasonable assumption is when the data are discrete-valued and small. Two such process are the numbers of "great" inventions and scientific discoveries in each year from 1860 to 1959, shown in Figure 3.7, and the number of poliomyelitis cases reported to the U.S. Centers for Disease Control for the years 1970 to 1983, displayed in Figure 3.8.

These two unrelated processes have striking similarities in that the marginal distributions appear to be Possion, or more specifically generalized Poisson or negative binomial (which are a mixture of Poissons; this is often used to account for over- or under-dispersion, where the mean and the standard deviation are not equal; e.g., see Joe and Zhu (2005). Moreover, we see that the ACFs of each process seems to imply a simple autocorrelation structure, which might be modeled as a simple non-Gaussian AR(1) type of model.

[1] If X_t is the price of an asset at time t, the *return* or *growth rate* of that asset, at time t, is $R_t = (X_t - X_{t-1})/X_{t-1}$. Alternately, we may write $X_t = (1+R_t)X_{t-1}$, or $\nabla \ln X_t = \ln(1+R_t)$. But $\ln(1+R_t) = R_t - R_t^2/2 + R_t^3/3 - \cdots$ for $-1 < R_t \leq 1$. If R_t is a small percentage, then the higher order terms are negligible, and $\ln(1+R_t) \approx R_t$. It is easier to program $\nabla \ln X_t$, so this is often used instead of calculating R_t directly. Although it is a misnomer, $\nabla \ln X_t$ is often called the *log-return*.

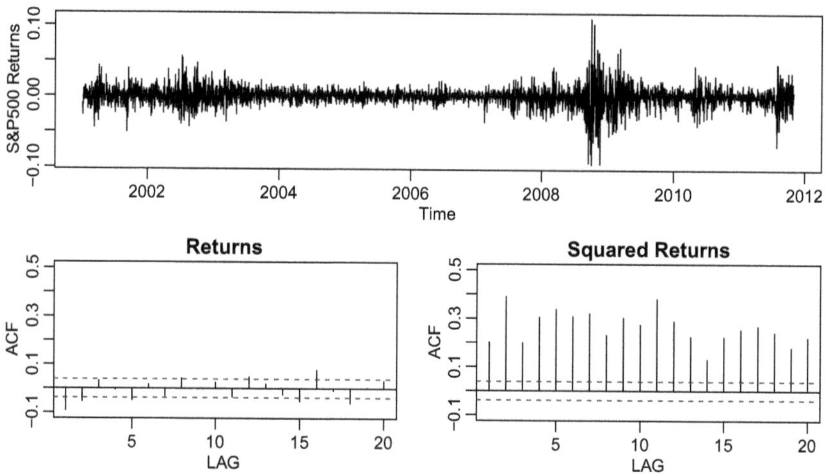

Figure 3.5 *Top: Daily returns of the S&P 500 from 2001 to the end of 2011. Bottom: The sample ACF of the returns and of the squared returns.*

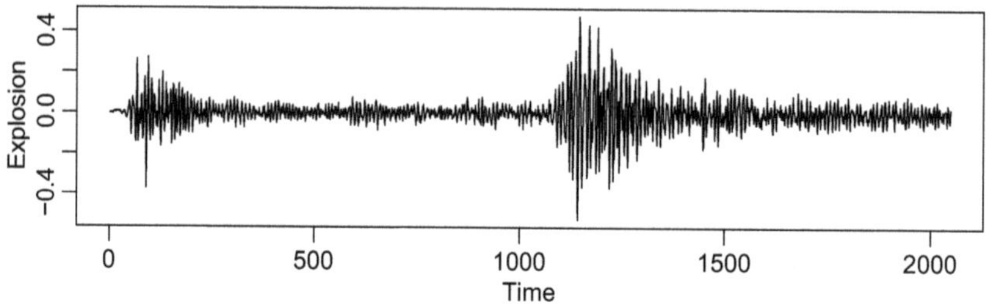

Figure 3.6 *Two phases or arrivals along the surface of an explosion at a seismic recording station. Compressional waves, also known as primary or P-waves, travel fastest, at speeds between 1.5 and 8 kilometers per second in the Earth's crust. Shear waves, also known as secondary or S-waves, travel more slowly, usually at 60% to 70% of the speed of P-waves.*

These data sets make it clear that, in addition to the problems discussed in the previous examples, there is a need to have non-Gaussian time series models that can take into account processes that produce discrete-valued observations that may have an autocorrelation structure similar to what is seen in ARMA models. The data set `discoveries` is an R data set that was taken from McNeil (1977). The polio data set is taken from Zeger (1988) and can be found in the R package `gamlss.data`. We test if the marginal number of reported polio cases is Poisson or negative binomial using `goodfit` from the R package `vcd`.

```
> summary(goodfit(as.integer(polio)))                 # Poisson
 Goodness-of-fit test for poisson distribution
                        X^2 df      P(> X^2)
  Likelihood Ratio 78.04415  9  3.949539e-13
```

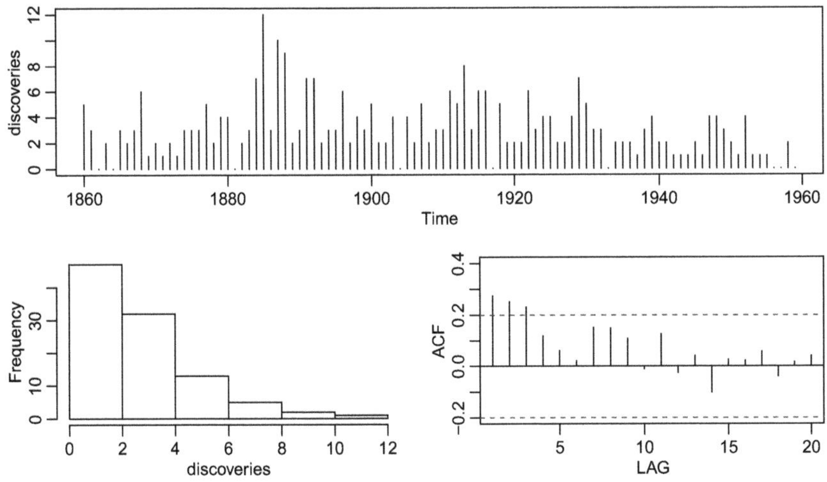

Figure 3.7 *The numbers of "great" inventions and scientific discoveries in each year from 1860 to 1959. Source: The World Almanac and Book of Facts, 1975 Edition.*

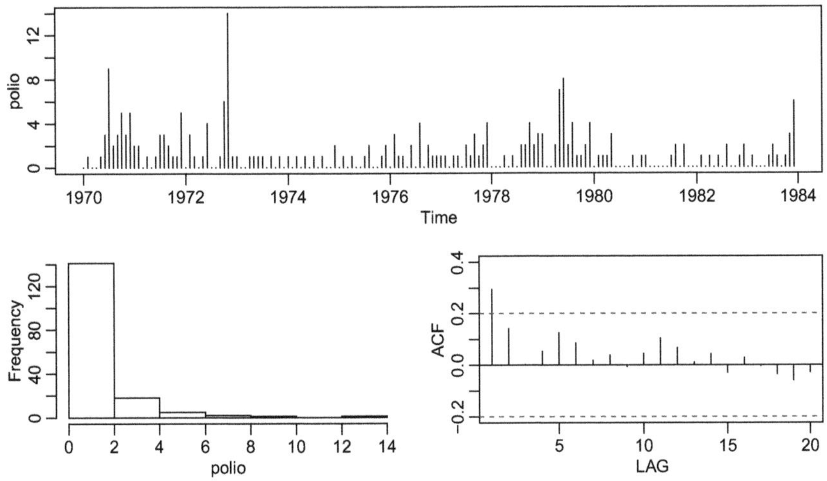

Figure 3.8 *Poliomyelitis cases reported to the U.S. Centers for Disease Control for the years 1970 to 1983.*

```
> summary(goodfit(as.integer(polio), "nbinomial")) # Neg Binomial
  Goodness-of-fit test for nbinomial distribution
                        X^2  df  P(> X^2)
  Likelihood Ratio 12.8793   8  0.116070
```

Clearly the Poisson distribution does not fit, while the negative binomial appears to be satisfactory. ◇

The essential points of Examples 3.1, 3.2 and 3.3 are that (i) linear or Gaussian time series models are limited, and not all situations can be handled by such models even after transformation; (ii) similar types of departures from the linear or Gaussian process are observed in data from many diverse fields, and in varied and unrelated situations. Hence, the development of general nonlinear models has played a prominent role in the field of time series for decades.

3.2 Volterra series expansion

A natural idea for going beyond the linear structure of (3.1) is to consider the following Volterra series. An M-th order Volterra process is given by

$$X_t = \sum_{i=1}^{M} \sum_{m_1=0}^{\infty} \cdots \sum_{m_i=0}^{\infty} \psi^{(i)}_{m_1,\ldots,m_i} \prod_{j=1}^{i} Z_{t-m_j}, \tag{3.3}$$

where $\{Z_t,\, t \in \mathbb{Z}\}$ is an i.i.d. sequence. The coefficients $\{\psi^{(i)}_{m_1,\ldots,m_i},\, (m_1,\ldots,m_i) \in \mathbb{N}^i\}$ are the coefficients determining the i-th order Volterra kernel. For simplicity, it is assumed that $\sum |\psi^{(i)}_{m_1,\ldots,m_i}| < \infty$, but of course, this assumption may be weakened. These types of expansions were first considered by Wiener (1958), where the concern was with the case when both the input $\{Z_t,\, t \in \mathbb{Z}\}$ and the output $\{X_t,\, t \in \mathbb{Z}\}$ were observable. In the context of time series, only $\{X_t,\, t \in \mathbb{Z}\}$ is observable. The first term $H_1[Z_s,\, s \le t] = \sum_{m=0}^{\infty} \psi^{(1)}_m Z_{t-m}$ is a linear model. The second term $H_2[Z_s,\, s \le t] = \sum_{m_1,m_2=0}^{\infty} \psi^{(2)}_{m_1,m_2} Z_{t-m_1} Z_{t-m_2}$ is a linear combination of quadratic terms. The higher order terms may be called the cubic component, the quartic component, and so on. This expansion might be seen as the M-th order principal part of the multidimensional Taylor expansion of the generic nonlinear model $X_t = g(Z_s,\, s \le t)$ (assuming that the operator g is analytic). By the Stone-Weierstrass Theorem, any continuous function $g\colon (z_1,\ldots,z_m) \mapsto g(z_1,\ldots,z_m)$ on a compact set of $\mathbb{R}^m$ can be approximated with an arbitrary precision in the topology of uniform convergence by a polynomial $p(z_1,\ldots,z_m)$. Hence, it is not difficult to guess that, under rather weak assumptions, an arbitrary finite memory nonlinear system $X_t = g(Z_t, Z_{t-1}, \ldots, Z_{t-m+1})$ can be approximated arbitrarily well by a Volterra series expansion. Infinite memory processes $X_t = g(Z_s, s \le t)$ can also be approximated arbitrarily well by a finite order Volterra series provided that the infinite memory possesses some forgetting property (roughly speaking, the influence of the infinite past should fade away in some appropriate sense).

Even simple Volterra series expansions display properties that are markedly different from linear processes. For example, consider the process $\{X_t,\, t \in \mathbb{Z}\}$ defined by,

$$X_t = Z_t + \beta Z_{t-1} Z_{t-2} \tag{3.4}$$

where $\{Z_t,\, t \in \mathbb{Z}\}$ is a strong (i.i.d.) white noise sequence with zero mean and constant variance. It follows immediately that $\{X_t,\, t \in \mathbb{Z}\}$ has zero mean, constant variance, and autocovariance function given by,

$$\mathbb{E}[X_t X_{t+h}] =$$
$$\mathbb{E}\left[Z_t Z_{t+h} + \beta Z_{t-1} Z_{t-2} Z_{t+h} + \beta Z_t Z_{t+h-1} Z_{t+h-2} + \beta^2 Z_{t-1} Z_{t-2} Z_{t+h-1} Z_{t+h-2}\right] .$$

For all $h \neq 0$, each of the terms on the right-hand side is zero because Z_t is a strong white noise. Thus, as far as its second order properties are concerned, $\{X_t,\ t \in \mathbb{Z}\}$ behaves just like a white noise process. However, given observations up to time t, one can clearly construct a non-trivial prediction of X_{t+h}. Specifically, if we adopt the mean square error criterion, the optimal forecast of a future observation, X_{t+h}, is its conditional expectation, $X_{t+h|t} = \mathbb{E}\left[X_{t+h} \mid \mathcal{F}_t^X\right]$, where $\mathcal{F}_t^X = \sigma(X_s, s \leq t)$. Computing this conditional expectation is not entirely trivial as we shall see. Assume that the system is invertible, i.e., that there exists a measurable non-anticipative function such that $Z_t = g(X_s, s \leq t)$. In this case, for any $t \in \mathbb{Z}$, Z_t belongs to $\mathcal{F}_t^X$ and therefore

$$\mathbb{E}\left[X_{t+1} \mid \mathcal{F}_t^X\right] = \beta g(X_s, s \leq t) \times g(X_s, s \leq t-1) \quad \mathbb{P}\text{-a.s.}$$

Note that such inverse does not always exist; in this case, more delicate arguments are used to compute forecasts.

There is a substantial literature on the theoretical properties of these models, which plays an important role in nonlinear system theory. Estimating the coefficients of the Volterra kernels individually is difficult for two reasons. First, the kernels of the Volterra series are strongly dependent. A direct approach leads to the problem of simultaneously solving a strongly coupled set of nonlinear equations for the kernel coefficients. Second, the canonical representation (3.3) contains, in general, far too many parameters to estimate efficiently from a finite set of observations. To alleviate this problem, following the original suggestion by Wiener, the estimation of the Volterra kernels is generally performed by developing the coefficients on appropriately chosen orthogonal basis function, such as the Laguerre and Kautz functions or generalized orthonormal basis functions (GOBFs); see Campello et al. (2004) and the references therein. This technique requires us to assume that the coefficients in the expansion may be expressed as known functions of some relatively small number of other parameters.

3.3 Cumulants and higher-order spectra

We have seen that in the linear Gaussian world, it is sufficient to work with second-order statistics. Now, reconsider Example 1.35 where X, Y, Z are i.i.d. N(0,1) random variables with $Y = X^2 + Z$. This could be a model (appropriately parameterized) for automobile fuel consumption Y versus speed X; i.e., fuel consumption is lowest at moderate speeds, but is highest at very low and very high speeds. In that example, we saw that the BLP ($\widehat{Y} = 1$) was considerably worse than the minimum mean square predictor ($\widehat{Y} = X^2$). If, however, we consider linear prediction on $\mathcal{M} = \overline{\text{sp}}\{1, X, X^2\}$, then from Proposition 1.34, the prediction equations are

$$(i)\ \mathbb{E}[Y - P_{\mathcal{M}}Y] = 0; \quad (ii)\ \mathbb{E}[Y - P_{\mathcal{M}}Y]X = 0; \quad (iii)\ \mathbb{E}[Y - P_{\mathcal{M}}Y]X^2 = 0 \qquad (3.5)$$

where $P_{\mathcal{M}}Y = a + bX + cX^2$. Solving these equations will yield $a = b = 0$, and $c = 1$ (see Exercise 3.3) so that $P_{\mathcal{M}}Y = X^2$, which was also the optimal predictor $\mathbb{E}\left[Y \mid X\right]$. The problem with the BLP in Example 1.35 was that it only considered moments up to order 2, e.g., $\mathbb{E}[YX]$ and $\mathbb{E}\left[X^2\right]$. But here, we have improved the predictor by considering slightly higher-order moments such as $\mathbb{E}\left[YX^2\right]$ and $\mathbb{E}\left[X^4\right]$.

For a collection of random variables $\{X_1, \ldots, X_k\}$, let $\varphi(\xi_1, \ldots, \xi_k) = \varphi(\xi)$ be the corresponding joint characteristic function,

$$\varphi(\xi) = \mathbb{E}\left[\exp\left\{\mathrm{i}\sum_{j=1}^{k} \xi_j X_j\right\}\right]. \tag{3.6}$$

For $r = (r_1, \ldots r_k)$, if the moments $\mu_r = \mathbb{E}\left[X_1^{r_1} \cdots X_k^{r_k}\right]$ exist up to a certain order $|r| := \sum_{j=1}^{k} r_j \leq n$, then they are the coefficients in the expansion of $\varphi(\xi)$ around zero,

$$\varphi(\xi) = \sum_{|r| \leq n} (\mathrm{i}\xi)^r \mu_r / r! + o(|\xi|^n), \tag{3.7}$$

where $r! = \prod_{j=1}^{k} r_j!$ and $\xi^r = \xi_1^{r_1} \ldots \xi_k^{r_k}$. Similarly, the joint cumulants $\kappa_r \equiv \mathrm{cum}[X_1^{r_1} \cdots X_k^{r_k}]$ are the coefficients in the expansion of the *cumulant generating function*, defined as the logarithm of the characteristic function:

$$\ln \varphi(\xi) = \sum_{|r| \leq n} (\mathrm{i}\xi)^r \kappa_r / r! + o(|\xi|^n). \tag{3.8}$$

A special case of (3.8) is $X_j = X$ for $j = 1, \ldots, k$, in which one obtains the r-th cumulant of X. If $X \sim \mathrm{N}(\mu, \sigma^2)$, then $\ln \varphi(\xi) = \mathrm{i}\mu\xi - \frac{1}{2}\sigma^2\xi^2$, so that $\kappa_1 = \mu$, $\kappa_2 = \sigma^2$, and $\kappa_r = 0$, for $r > 2$. In fact, the normal distribution is the only distribution for which this is true (i.e., there are a finite number of non-zero cumulants, Marcinkiewicz, 1939). Another interesting case is the Poisson(λ) distribution wherein $\ln \varphi(\xi) = \lambda(\mathrm{e}^{\xi} - 1)$ and consequently $\kappa_r = \lambda$ for all r.

Some special properties of cumulants are:

- The cumulant is invariant with respect to the permutations: $\mathrm{cum}(X_1, \ldots, X_k) = \mathrm{cum}(X_{\sigma(1)}, \ldots, X_{\sigma(k)})$ where σ is any permutation on $\{1, \ldots, k\}$.
- For every $(a_1, \ldots, a_k) \in \mathbb{R}^k$, $\mathrm{cum}(a_1 X_1, \ldots, a_k X_k) = a_1 \cdots a_k \, \mathrm{cum}(X_1, \ldots, X_k)$.
- The cumulant is multilinear:

$$\mathrm{cum}(X_1 + Y_1, X_2, \ldots, X_k) = \mathrm{cum}(X_1, X_2, \ldots, X_k) + \mathrm{cum}(Y_1, X_2, \ldots, X_k).$$

- If $\{X_1, \ldots, X_k\}$ can be partitioned into two disjoint sets that are independent of each other, then $\mathrm{cum}(X_1, \ldots, X_k) = 0$.
- If $\{X_1, \ldots, X_k\}$ and $\{Y_1, \ldots, Y_k\}$ are independent, then $\mathrm{cum}(X_1 + Y_1, \ldots X_k + Y_k) = \mathrm{cum}(X_1, \ldots, X_k) + \mathrm{cum}(Y_1, \ldots, Y_k)$.
- $\mathrm{cum}(X) = \mathbb{E}[X]$ and $\mathrm{cum}(X, Y) = \mathrm{Cov}(X, Y)$.
- $\mathrm{cum}(X, Y, Z) = \mathbb{E}[XYZ]$ if the means of the random variables are zero.

Further information on the properties of cumulants may be found in Brillinger (2001), Leonov and Shiryaev (1959), and Rosenblatt (1983).

Using Theorem 1.23 for a zero-mean stationary time series $\{X_t,\, t \in \mathbb{Z}\}$, and defining $\kappa_x(r) = \mathrm{cum}(X_{t+r}, X_t) = \gamma_x(r)$ we may write

$$\begin{aligned}\kappa_x(r) = \mathbb{E}\left[X_{t+r}X_t\right] &= \iint_{-\pi}^{\pi} \mathrm{e}^{\mathrm{i}(t+r)\omega}\, \mathrm{e}^{\mathrm{i}t\lambda}\, \mathbb{E}\left[\mathrm{d}Z(\omega)\, \mathrm{d}Z(\lambda)\right] \\ &= \iint_{-\pi}^{\pi} \mathrm{e}^{\mathrm{i}t(\omega+\lambda)}\, \mathrm{e}^{\mathrm{i}r\omega}\, \mathbb{E}\left[\mathrm{d}Z(\omega)\, \mathrm{d}Z(\lambda)\right].\end{aligned} \tag{3.9}$$

Because the left-hand side of (3.9) does not depend on t, the right-hand side cannot depend on t. Thus, it must be the case that $\mathbb{E}\left[\mathrm{d}Z(\omega)\, \mathrm{d}Z(\lambda)\right] = 0$ unless $\lambda = -\omega$, and consequently, as pointed out in Theorem 1.23, $\mathbb{E}\left[\mathrm{d}Z(-\omega)\, \mathrm{d}Z(\omega)\right] = \mathbb{E}\left[|\mathrm{d}Z(\omega)|^2\right] = \mathrm{d}F(\omega)$. If $\kappa_x(r) = \gamma_x(r)$ is absolutely summable, then by Proposition 1.19, $\mathrm{d}F(\omega) = f(\omega)\mathrm{d}\omega$, where $f(\omega)$ is the spectral density of the process.

This concept may be applied to higher order moments. For example, suppose the cumulant $\kappa_x(r_1, r_2) = \mathrm{cum}(X_{t+r_1}, X_{t+r_2}, X_t) = \mathbb{E}\left[X_{t+r_1} X_{t+r_2} X_t\right]$ exists and does not depend on t. Then,

$$\begin{aligned}\kappa_x(r_1, r_2) &= \iiint_{-\pi}^{\pi} \mathrm{e}^{\mathrm{i}(t+r_1)\omega_1}\, \mathrm{e}^{\mathrm{i}(t+r_2)\omega_2}\, \mathrm{e}^{\mathrm{i}t\lambda}\, \mathbb{E}\left[\mathrm{d}Z(\omega_1)\, \mathrm{d}Z(\omega_2)\, \mathrm{d}Z(\lambda)\right] \\ &= \iiint_{-\pi}^{\pi} \mathrm{e}^{\mathrm{i}t(\omega_1+\omega_2+\lambda)}\, \mathrm{e}^{\mathrm{i}r_1\omega_1}\, \mathrm{e}^{\mathrm{i}r_2\omega_2}\, \mathbb{E}\left[\mathrm{d}Z(\omega_1)\, \mathrm{d}Z(\omega_2)\, \mathrm{d}Z(\lambda)\right].\end{aligned} \tag{3.10}$$

Because $\kappa_x(r_1, r_2)$ does not depend on t, it must be that $\mathbb{E}\left[\mathrm{d}Z(\omega_1)\, \mathrm{d}Z(\omega_2)\, \mathrm{d}Z(\lambda)\right] = 0$ unless $\omega_1 + \omega_2 + \lambda = 0$. Consequently, we may write

$$\kappa_x(r_1, r_2) = \iint_{-\pi}^{\pi} \mathrm{e}^{\mathrm{i}r_1\omega_1}\, \mathrm{e}^{\mathrm{i}r_2\omega_2} \mathbb{E}\left[\mathrm{d}Z(\omega_1)\, \mathrm{d}Z(\omega_2)\, \mathrm{d}Z(-[\omega_1 + \omega_2])\right]. \tag{3.11}$$

Hence, the *bispectral distribution* may be defined as

$$\mathrm{d}F(\omega_1, \omega_2) = \mathbb{E}\left[\mathrm{d}Z(\omega_1)\, \mathrm{d}Z(\omega_2)\, \mathrm{d}Z(-[\omega_1 + \omega_2])\right]. \tag{3.12}$$

Following Proposition 1.19, under absolute summability conditions, we may define the bispectral density or *bispectrum* as $f(\omega_1, \omega_2)$ where

$$\kappa_x(r_1, r_2) = \iint_{-\pi}^{\pi} \mathrm{e}^{\mathrm{i}r_1\omega_1}\, \mathrm{e}^{\mathrm{i}r_2\omega_2}\, f(\omega_1, \omega_2)\, \mathrm{d}\omega_1 \mathrm{d}\omega_2 \tag{3.13}$$

and

$$f(\omega_1, \omega_2) = (2\pi)^{-2} \sum\sum_{-\infty < r_1, r_2 < \infty} \kappa_x(r_1, r_2)\, \mathrm{e}^{-\mathrm{i}r_1\omega_1}\, \mathrm{e}^{-\mathrm{i}r_2\omega_2}. \tag{3.14}$$

If $\{X_t\}$ is Gaussian, then $\kappa_x(r_1, r_2) = 0$ for all $(r_1, r_2) \in \mathbb{Z}^2$, and thus the bispectrum $f(\omega_1, \omega_2) \equiv 0$ for $(\omega_1, \omega_2) \in [-\pi, \pi]^2$. Consequently, tests of linearity and Gaussianity may rely on the bispectrum; see Exercise 3.4.

Finally, higher order cumulant spectra may be defined analogously to the bispectrum. That is, let $\kappa_x(r) = \kappa_x(r_1, \dots, r_k) = \mathrm{cum}(X_{t+r_1}, \dots, X_{t+r_k}, X_t)$ and assume that

$$\sum_{-\infty<r<\infty} \cdots \sum |\kappa_x(r)| < \infty.$$

Then, the $k+1$-*st order cumulant spectrum* is defined by

$$f_x(\omega_1, \dots, \omega_k) = \sum_{-\infty<r<\infty} \cdots \sum \kappa_x(r) \exp\Big\{-\mathrm{i} \sum_{j=1}^{k} r_j \omega_j\Big\}. \tag{3.15}$$

We note that higher-order spectra are generally complex-valued. The inverse relationship is,

$$\kappa_x(r_1, \dots, r_k) = \int_{-\pi}^{\pi} \cdots \int f_x(\omega_1, \dots, \omega_k) \exp\Big\{\mathrm{i} \sum_{j=1}^{k} r_j \omega_j\Big\} \mathrm{d}\omega_1 \dots \mathrm{d}\omega_k. \tag{3.16}$$

For further details, the reader is referred to Brillinger (1965, 2001) and Rosenblatt (1983).

3.4 Bilinear models

In Example 3.2, we saw that time series data may exhibit simple or no autocorrelation structure, but still be highly dependent; this dependence was seen in the squares of the observations. For example, the EEG innovations shown in Figure 3.2 and the autocorrelation structure shown in Figure 3.4 suggest that the innovations are white (i.e., uncorrelated noise), but the squared innovations indicate that there is still a dependence structure. This is a common occurrence, especially in financial time series. For example, in Figure 3.5, the daily returns of the S&P 500 exhibit obvious dependence, whereas the sample ACF indicates only a small correlation structure. The squares of the process, however, indicate a strong correlation structure is present. Exercise 3.2 explores the fact that the innovations of the explosion series shown in Figure 3.6 also have this property.

One early exploration of models for this type of behavior was the bilinear model developed for the statistical analysis of time series by Granger and Andersen (1978) and explored further by Subba Rao (1981) and others. The basic idea is that of using higher order terms of a Volterra expansion of the noise. For example, think of an ARMA model, $X_t = \sum_{j=0}^{\infty} \psi_j Z_{t-i}$, as a first order (linear) approximation of the Volterra expansion (3.3). In the nonlinear case, it seems reasonable to, at least, use a second order approximation, and that is the idea behind the bilinear model,

$$X_t = \sum_{j=1}^{p} \phi_j X_{t-j} + \sum_{j=1}^{q} \theta_j Z_{t-j} + \sum_{i=1}^{P} \sum_{j=1}^{Q} b_{ij} X_{t-i} Z_{t-j} + Z_t, \tag{3.17}$$

which is denoted as BL(p,q,P,Q). In this case, the process $\{X_t,\ t \in \mathbb{Z}\}$ is said to be causal (or nonanticipative) if for all $t \in \mathbb{Z}$, X_t is measurable with respect to $\mathcal{F}_t^Z = \sigma(Z_s, s \le t)$, i.e., X_t can be expressed as a measurable (but nonlinear) function of Z_s, for $s \le t$.

While the model appears to be a simple extension of the ARMA model to include nonlinearity, the existence, stationarity and invertibility of bilinear processes is a delicate topic. In fact, the model is too complicated to be examined in full generality. Consequently, investigations such as Granger and Andersen (1978), Pham and Tran (1981), Subba Rao (1981) and Subba Rao and Gabr (1984) focus on restricted models. Although the model has the desired property of exhibiting ARMA-type correlation structure for X_t with dependent innovations, the problem of examining the model in its generality seems to be the reason the model lost favor as a procedure to analyze nonlinear time series. Priestley (1988) has a nice discussion of the model and its history.

As a simple illustration of the properties of the model, consider the following bilinear model, BL(0,0,2,1),

$$X_t = bZ_{t-1}X_{t-2} + Z_t, \tag{3.18}$$

where $Z_t \sim \text{iid}\,(0, \sigma_z^2)$ with Z_t independent of X_s for $s < t$. If we assume that $\mathbb{E}\left[Z_t^4\right] < \infty$ and $b^2\sigma_z^2 < 1$, we can show that X_t is stationary using the techniques of Chapter 4; see Exercise 4.11. Let $\mathcal{F}_t = \sigma(X_t, X_{t-1}, \dots)$, then direct calculation (see Exercise 3.6) establishes that $\mathbb{E}\left[X_t\right] = 0$ and

$$\mathbb{E}\left[X_t X_{t-h} \mid \mathcal{F}_{t-2}\right] = \begin{cases} \sigma_z^2 + b^2\sigma_z^2 X_{t-2}^2, & h = 0, \\ b\sigma_z^2 X_{t-2}, & h = 1, \\ 0, & h \ge 2, \end{cases} \tag{3.19}$$

with probability one. From these facts we can establish that X_t is white noise, but X_t^2 is predictable from its history. Such a model could be used to describe the innovations of the EEG data set shown in Figure 3.2. Recall Figure 3.4 where the innovations are white, but the squares of the innovations are correlated.

3.5 Conditionally heteroscedastic models

An autoregressive conditional heteroscedastic model of order p, ARCH(p), is defined as

$$X_t = \sigma_t\,\varepsilon_t\,, \tag{3.20}$$

$$\sigma_t^2 = \alpha_0 + \alpha_1 X_{t-1}^2 + \cdots + \alpha_p X_{t-p}^2\,, \tag{3.21}$$

where the coefficients $\alpha_j \ge 0$, $j \in \{0,\dots,p\}$ are non-negative and $\varepsilon_t \sim \text{iid}\,(0,1)$ is the driving noise. If the driving noise is assumed to be Gaussian, the model implies that the conditional distribution of X_t given $X_{t-1},\dots,X_{t-p}$ is Gaussian,

$$X_t \mid X_{t-1},\dots,X_{t-p} \sim \mathrm{N}(0, \alpha_0 + \alpha_1 X_{t-1}^2 + \cdots + \alpha_p X_{t-p}^2). \tag{3.22}$$

Often, the driving noise is not normal and other distributions, such as the t-distribution, are used to model the noise.

To explore the properties of the model, first define $\mathcal{F}_t^X = \sigma(X_s, s \le t)$. In general, we are mainly interested in causal (or nonanticipative) solutions in which X_t is measurable with respect to $\mathcal{F}_t^\varepsilon = \sigma(\varepsilon_s, s \le t)$. This implies that $\mathcal{F}_t^X \subseteq \mathcal{F}_t^\varepsilon$ for all $t \in \mathbb{Z}$. Assume that the parameters $\{\alpha_i; i = 0, \dots, p\}$ are chosen in such a way that there exists a nonanticipative second-order stationary solution to (3.20)-(3.21). It then follows from (3.20) that

$$\begin{aligned}\mathbb{E}\left[X_t \mid \mathcal{F}_{t-1}^X\right] &= \mathbb{E}\left[\sigma_t \varepsilon_t \mid \mathcal{F}_{t-1}^X\right] \overset{(1)}{=} \sigma_t \mathbb{E}\left[\varepsilon_t \mid \mathcal{F}_{t-1}^X\right] \\ &\overset{(2)}{=} \sigma_t \mathbb{E}\left[\mathbb{E}\left[\varepsilon_t \mid \mathcal{F}_{t-1}^\varepsilon\right] \mid \mathcal{F}_{t-1}^X\right] \overset{(3)}{=} 0 \quad \mathbb{P}\text{-a.s.},\end{aligned} \tag{3.23}$$

where we have used that for all $t \in \mathbb{Z}$: (1) σ_t is $\mathcal{F}_{t-1}^X$-measurable (the process $\{\sigma_t^2, t \in \mathbb{Z}\}$ is previsible); (2) that $\mathcal{F}_{t-1}^X \subset \mathcal{F}_{t-1}^\varepsilon$ (the process $\{X_t, t \in \mathbb{Z}\}$ is nonanticipative); (3) $\mathbb{E}\left[\varepsilon_t \mid \mathcal{F}_{t-1}^\varepsilon\right] = \mathbb{E}(\varepsilon_t) = 0$. Note that, as a consequence of (3.23), we have $\mathbb{E}[X_t] = \mathbb{E}\left[\mathbb{E}\left[X_t \mid \mathcal{F}_{t-1}^X\right]\right] = 0$.

Because $\mathbb{E}\left[X_t \mid \mathcal{F}_{t-1}^X\right] = 0$ $\mathbb{P}$-a.s., for all $t \in \mathbb{Z}$, the process $\{X_t, t \in \mathbb{Z}\}$ is said to be a *martingale difference* or *increment process* (see Appendix B). Assume that $\mathbb{E}\left[X_t^2\right] < \infty$. The fact that $\{X_t, t \in \mathbb{Z}\}$ is a martingale difference implies that it is also an uncorrelated sequence. To see this, let $h > 0$, then

$$\begin{aligned}\mathrm{Cov}(X_{t+h}, X_t) &= \mathbb{E}\left[X_t X_{t+h}\right] = \mathbb{E}\left[\mathbb{E}\left[X_t X_{t+h} \mid \mathcal{F}_{t+h-1}^X\right]\right] \\ &= \mathbb{E}\left[X_t \mathbb{E}\left[X_{t+h} \mid \mathcal{F}_{t+h-1}^X\right]\right] = 0.\end{aligned} \tag{3.24}$$

The last line of (3.24) follows because X_t is $\mathcal{F}_{t+h-1}^X$-measurable for $h > 0$, and $\mathbb{E}\left[X_{t+h} \mid \mathcal{F}_{t+h-1}\right] = 0$ $\mathbb{P}$-a.s., as determined in (3.23).

While (3.24) implies that the ARCH process is white noise, it is still a dependent sequence. In fact, it is possible to write the ARCH(p) model as a non-Gaussian AR(p) model in the squares, X_t^2. First, square (3.20), $X_t^2 = \sigma_t^2 \varepsilon_t^2$, and then subtract (3.21), to obtain

$$X_t^2 - (\alpha_0 + \alpha_1 X_{t-1}^2 + \cdots + \alpha_p X_{t-p}^2) = Z_t\,, \tag{3.25}$$

where $Z_t = \sigma_t^2(\varepsilon_t^2 - 1)$. If the driving noise is Gaussian, then ε_t^2 is the square of a standard normal random variable, and $\varepsilon_t^2 - 1$ is a shifted (to have mean-zero) χ_1^2 random variable. The fact that $\{Z_t, t \in \mathbb{Z}\}$ is white noise follows from the fact that it is a martingale difference, $\mathbb{E}\left[Z_t \mid \mathcal{F}_{t-1}^X\right] = \sigma_t^2 \mathbb{E}\left[\varepsilon_t^2 - 1\right] = 0$, $\mathbb{P}$-a.s., noting that $\sigma_t^2 \in \mathcal{F}_{t-1}^X$.

ARCH models were introduced by Engle (1982) to model the varying (conditional) variance or volatility of time series. It is often found in economics and finance that the larger values of time series (shocks) also cause instability at later times (i.e., larger variances); this phenomenom is referred to as *conditional heteroscedasticity*. For example, as illustrated in Figure 3.5 the returns of the S&P 500 exhibit largest variance after shocks. Allowing the conditional variance of X_t to depend on $X_{t-1}^2, \dots, X_{t-p}^2$ is a first step in this direction.

The limitation of the ARCH model is that the squared process admits an AR correlation structure, which is not always the case. Bollerslev (1986) generalized the ARCH model by allowing the conditional variance $\mathbb{E}\left[X_t^2 \mid \mathcal{F}_{t-1}^X\right]$ to depend not only on the lagged squared returns $(X_{t-1}^2, \ldots, X_{t-p}^2)$ but also on the lagged conditional variances, leading to the *generalized autoregressive conditional heteroscedastic* model, GARCH(p,q), where (3.20) still holds, but now

$$\sigma_t^2 = \alpha_0 + \alpha_1 X_{t-1}^2 + \cdots + \alpha_p X_{t-p}^2 + \beta_1 \sigma_{t-1}^2 + \cdots + \beta_q \sigma_{t-q}^2 , \tag{3.26}$$

where the coefficients α_j, $j \in \{0, \ldots, p\}$ and β_j, $j \in \{1, \ldots, q\}$ are nonnegative (although this assumption can be relaxed). The extension of the ARCH process to the GARCH process bears much similarity to the extension of standard AR models to ARMA models described in Section 1.3.2. To see this, consider a GARCH$(1,1)$ model (for ease of notation). As with the ARCH model, square (3.20), $X_t^2 = \sigma_t^2 \varepsilon_t^2$, and then subtract σ_t^2 to obtain

$$X_t^2 - \sigma_t^2 = \sigma_t^2(\varepsilon_t^2 - 1) := Z_t . \tag{3.27}$$

Consequently,

$$\beta_1 (X_{t-1}^2 - \sigma_{t-1}^2) = \beta_1 Z_{t-1} , \tag{3.28}$$

and thus subtracting (3.28) from (3.27), we obtain

$$(X_t^2 - \sigma_t^2) - \beta_1 (X_{t-1}^2 - \sigma_{t-1}^2) = Z_t - \beta_1 Z_{t-1} ,$$

or

$$X_t^2 - \beta_1 X_{t-1}^2 - (\sigma_t^2 - \beta_1 \sigma_{t-1}^2) = Z_t - \beta_1 Z_{t-1} .$$

But $\sigma_t^2 - \beta_1 \sigma_{t-1}^2 = \alpha_0 + \alpha_1 X_{t-1}^2$, so finally

$$X_t^2 - (\alpha_1 + \beta_1) X_{t-1}^2 = Z_t - \beta_1 Z_{t-1} , \tag{3.29}$$

implying $\{X_t^2,\ t \in \mathbb{Z}\}$ is a non-Gaussian ARMA$(1,1)$. We note that this technique generalizes to any GARCH(p,q) by writing it as a GARCH(m,m) model where $m = \max(p,q)$ and setting any additional coefficients to zero, i.e., $\alpha_{p+1} = \cdots = \alpha_q = 0$ if $p < q$ or $\beta_{q+1} = \cdots = \beta_p = 0$ if $p > q$; see Exercise 3.7.

Summarizing, in general, if $\{X_t,\ t \in \mathbb{Z}\}$ is GARCH(p,q), then it is a martingale difference, $\mathbb{E}\left[X_t \mid \mathcal{F}_{t-1}^X\right] = 0$ $\mathbb{P}$-a.s., and consequently is white noise. In addition, $\{X_t^2,\ t \in \mathbb{Z}\}$ is a non-Gaussian ARMA(p,q) process. This type of result was the goal of the bilinear model presented in Section 3.4, but as opposed to the bilinear model, the correlation structure of the GARCH model easily generalizes.

Another reason for the popularity of GARCH(p,q) models is that parameter estimation is straight-forward by conditioning on initial values. That is, the conditional likelihood of the data $X_{p+1}, \ldots, X_n$ given $X_1, \ldots, X_p$, and $\sigma_p^2 = \cdots = \sigma_{p+1-q}^2 = 0$ (if $q > 0$) is

$$\mathrm{L}(\theta; X_1, \ldots, X_p, \sigma_p^2 = \cdots = \sigma_{p+1-q}^2 = 0) = \prod_{t=p+1}^{n} \mathrm{p}^\theta(X_t \mid X_{t-1}, \ldots, X_1), \tag{3.30}$$

where

$$\theta = \{\alpha_0, \alpha_1, \dots, \alpha_p, \beta_1, \dots, \beta_q\}.$$

If $\varepsilon_t \sim \text{iid N}(0,1)$, then the conditional densities $\text{p}^\theta(\cdot|\cdot)$ in (3.30) are Gaussian, i.e.,

$$X_t \mid X_{t-1}, \dots, X_1 \sim \text{N}(0, \alpha_0 + \alpha_1 X_{t-1}^2 + \cdots + \alpha_p X_{t-p}^2 + \beta_1 \sigma_{t-1}^2 + \cdots + \beta_q \sigma_{t-q}^2),$$

for $t = p+1, \dots, n$, with $\sigma_p^2 = \cdots = \sigma_{p+1-q}^2 = 0$. The sample size is typically large in financial applications so that conditioning on a few initial values is not problematic. Because the conditional likelihood is easily evaluated for a specified θ, a numerical routine such as Newton–Raphson (Section 1.3.4) is typically employed. In addition, the gradient of the likelihood is easily evaluated; see Exercise 3.8.

Some drawbacks of the GARCH model are that the likelihood tends to be flat unless n is very large, and the model tends to overpredict volatility because it responds slowly to large isolated returns. Returns are rarely conditionally normal or symmetric, so various extensions to the basic model have been developed to handle the various situations noticed empirically. Interested readers might find the general discussions in Bollerslev et al. (1994) and Shephard (1996) worthwhile reading. Also, Gouriéroux (1997) gives a detailed presentation of ARCH and related models with financial applications and contains an extensive bibliography. Excellent texts on financial time series analysis are Chan (2002), Teräsvirta et al. (2011), and Tsay (2005).

Finally, we briefly mention *stochastic volatility models*; a detailed treatment of these models is given in Chapter 9. The volatility component, σ_t^2, in the GARCH model is conditionally nonstochastic. For example, in the ARCH(1) model, any time the previous return is zero, i.e., $X_{t-1} = 0$, it must be the case that $\sigma_t^2 = \alpha_0$, and so on. This assumption seems a bit unrealistic in that one would expect some variability in this outcome. The stochastic volatility model adds a stochastic component to the volatility in the following way. The GARCH model assumes $X_t = \sigma_t \varepsilon_t$, or equivalently,

$$\ln X_t^2 = \ln \sigma_t^2 + \ln \varepsilon_t^2. \tag{3.31}$$

Thus, the observations, $\ln X_t^2$, are generated by two components, the unobserved volatility $\ln \sigma_t^2$ and the unobserved non-Gaussian noise $\ln \varepsilon_t^2$. While, for example, the GARCH$(1,1)$ models volatility without error, $\sigma_{t+1}^2 = \alpha_0 + \alpha_1 X_t^2 + \beta_1 \sigma_t^2$, the basic stochastic volatility model assumes the latent variable is an autoregressive process,

$$\ln \sigma_{t+1}^2 = \phi_0 + \phi_1 \ln \sigma_t^2 + Z_t \tag{3.32}$$

where $Z_t \sim \text{iid N}(0, \sigma_z^2)$. The introduction of the noise term Z_t makes the latent volatility process stochastic.

Together (3.31) and (3.32) comprise the stochastic volatility model. In fact, the model is a non-Gaussian state space model. Let $h_t = \ln \sigma_t^2$, $Y_t = \ln X_t^2$, and $V_t = \ln \varepsilon_t^2$, then the basic stochastic volatility model may be written as

$$\begin{aligned} h_{t+1} &= \phi_0 + \phi_1 h_t + Z_t \quad &\text{(state)} \\ Y_t &= h_t + V_t \quad &\text{(observation)} \end{aligned}$$

where Z_t is a Gaussian process, but V_t is not a Gaussian process. Given n observations, the goals are to estimate the parameters ϕ_0, ϕ_1 and σ_z^2, and then predict future volatility. Further details and extensions are discussed in Chapter 9.

3.6 Threshold ARMA models

Self-exciting threshold ARMA (SETARMA or TARMA) models, introduced by Tong (1978, 1983, 1990), have been widely employed as a model for nonlinear time series. Threshold models are piecewise linear ARMA models for which the linear relationship varies according to delayed values of the process (hence the term *self-exciting*). In this class of models, it is hypothesized that different autoregressive processes may operate and that the change between the various ARMA is governed by threshold values and a time lag. A k-regimes TARMA model has the form

$$X_t = \begin{cases} \phi_0^{(1)} + \sum_{i=1}^{p_1} \phi_i^{(1)} X_{t-i} + Z_t^{(1)} + \sum_{j=1}^{q_1} \theta_j^{(1)} Z_{t-j}^{(1)} & \text{if } X_{t-d} \le r_1, \\ \phi_0^{(2)} + \sum_{i=1}^{p_2} \phi_i^{(2)} X_{t-i} + Z_t^{(2)} + \sum_{j=1}^{q_2} \theta_j^{(2)} Z_{t-j}^{(2)} & \text{if } r_1 < X_{t-d} \le r_2, \\ \vdots & \vdots \\ \phi_0^{(k)} + \sum_{i=1}^{p_k} \phi_i^{(k)} X_{t-i} + Z_t^{(k)} + \sum_{j=1}^{q_k} \theta_j^{(k)} Z_{t-j}^{(k)} & \text{if } r_{k-1} < X_{t-d}, \end{cases} \tag{3.33}$$

where $Z_t^{(j)} \sim \text{iid N}(0, \sigma_j^2)$, for $j = 1, \ldots, k$, the positive integer d is a specified delay, and $-\infty < r_1 < \cdots < r_{k-1} < \infty$ is a partition of $\mathsf{X} = \mathbb{R}$. These models allow for changes in the ARMA coefficients over time, and those changes are determined by comparing previous values (back-shifted by a time lag equal to d) to fixed threshold values. Each different ARMA model is referred to as a *regime*. In the definition above, the values (p_j, q_j) of the order of ARMA models can differ in each regime, although in many applications, they are equal. Stationarity and invertibility are obvious concerns when fitting time series models. For the threshold time series models, such as TAR, TMA and TARMA models, however, the stationary and invertible conditions in the literature are less well-known in general. If known, often they are restricted to TAR or TMA processes with order one, and/or only sufficient conditions for higher orders; see e.g., Petruccelli and Woolford (1984), Brockwell et al. (1992), Ling (1999), and Ling et al. (2007).

The model can be generalized to include the possibility that the regimes depend on a collection of the past values of the process, or that the regimes depend on an exogenous variable (in which case the model is not self-exciting). For example, in the case such as that of the lynx, its prey varies from small rodents to deer, with the Snowshoe Hare being its overwhelmingly favored prey. In fact, in certain areas the lynx is so closely tied to the Snowshoe that its population rises and falls with that of the hare, even though other food sources may be abundant. In this case, it seems reasonable to replace X_{t-d} in (3.33) with say Y_{t-d}, where Y_t is the size of the Snowshoe Hare population.

The popularity of TAR models is due to their being relatively simple to specify, estimate, and interpret as compared to many other nonlinear time series models. In addition, despite its apparent simplicity, the class of TAR models could reproduce many nonlinear phenomena such as stable and unstable limit cycles, jump resonance, harmonic distortion, modulation effects, chaos and so on.

As a simple example, Tong (1990, p. 377) fit the following TAR model with two regimes with delay variable $d = 2$ to the logarithm (base 10) of the lynx data,

$$X_t = \begin{cases} 0.62 + 1.25X_{t-1} - 0.43X_{t-2} + Z_t^{(1)}, & X_{t-2} \leq 3.25\,, \\ 2.25 + 1.52X_{t-1} - 1.24X_{t-2} + Z_t^{(2)}, & X_{t-2} > 3.25\,, \end{cases} \tag{3.34}$$

although more complicated models were also fit to these data. Tong and Lim (1980) fit a two-regime TAR(11) to the sunspot data, the square root of which is shown in Figure 1.1. Also, Shumway and Stoffer (2011, Section 5.5) fit a threshold model to the differenced influenza and pneumonia mortality data set shown at the bottom of Figure 3.1.

3.7 Functional autoregressive models

In its basic form, a functional AR(p) model is written as

$$X_t = f(X_{t-1}, \ldots, X_{t-p}) + Z_t \tag{3.35}$$

where $\{Z_t,\ t \in \mathbb{N}\}$ is a strong white noise and is independent of X_s for $s < t$. The function $f(\cdot)$ is understood to be the conditional expectation, $f(X_{t-1}, \ldots, X_{t-p}) = \mathbb{E}\left[X_t \mid X_{t-1}, \ldots, X_{t-p}\right]$, and can be left unspecified but with various smoothness conditions on $f(\cdot)$ and often under weak-dependence conditions on the process $\{X_t\}$. Sometimes the noise process is written as

$$Z_t = h(X_{t-1}, \ldots, X_{t-p})\, \varepsilon_t, \tag{3.36}$$

where $\varepsilon_t \sim \mathrm{iid}\,(0,1)$. The function $h(\cdot)$ represents the possibility of conditionally heteroscedastic variance, with $h(\cdot) \equiv \sigma_z$ representing the homoscedastic case.

The basic goal is to estimate $f(\cdot)$, often via nonparametric methods, and then use the estimated relationship for prediction. We note, however, that many parametric models fit into this general model. For example, the TAR model in Section 3.6 would be considered a parametric form of the model with $f(\cdot)$ being an AR(q_j) with parameters depending on X_{t-d} as specified in (3.33), and with $\sigma(\cdot) = \sigma_j$ also depending on the observed value of X_{t-d}, with d specified as in (3.33). The model with (3.36) added clearly includes various forms of the ARCH model. As another example, we mention the amplitude-dependent exponential autoregressive (EXPAR) model introduced in Haggan and Ozaki (1981), which assumes that

$$f(x_1, \ldots, x_p) = (\phi_1 + \pi_1 \mathrm{e}^{-\gamma x_1^2})x_1 + \cdots + (\phi_p + \pi_p \mathrm{e}^{-\gamma x_1^2})x_p\,. \tag{3.37}$$

In this case, the autoregressive part retains an additive form, but the coefficients entering the regression are made to change instantaneously with x_1^2.

In more recent works, estimation of f or h is performed using some of the same tools used in non- or semi-parametric estimation of regression functions. Note, however, that some care should be exercised in controlling the functions f and h in such a way that there exists a (strict-sense or second-order) stationary solution for (3.35)–(3.36). This of course involves some rather non-trivial conditions on the behavior of f and h; we will have to wait until Chapter 4 to develop the tools required to show the existence of such solutions. Various versions of the non- or semi-parametric approach have been explored. For example, Hastie and Tibshirani (1990) examined the additive model,

$$X_t = f_1(X_{t-1}) + \cdots + f_p(X_{t-p}) + Z_t \tag{3.38}$$

and Chen and Tsay (1993) explored the functional coefficient AR model, which, in its simplest form, is written as

$$X_t = f_1(X_{t-d})X_{t-1} + \cdots + f_p(X_{t-d})X_{t-p} + Z_t \tag{3.39}$$

where $d > 0$ is some specified delay. Another interesting model is the *partially linear model*, where for example, we might have

$$f(X_{t-1}, \ldots, X_{t-p}) = \mu(t) + \sum_{j=1}^{p} \phi_j X_{t-j} \tag{3.40}$$

where $\mu(t)$ is a local trend function of time t that we do not wish to model parametrically. For example, the rates of pneumonia and influenza mortality series shown at the bottom of Figure 3.1 exhibits some negative, but not necessarily linear trend over the decade (e.g., it appears that the decline in the average annual mortality is more pronounced over the first part of the series than at the end of the series). In this case, we may wish to fit $f(\cdot)$ via semiparametric methods.

Semiparametric and nonparametric estimation for time series models in various forms runs the gamut of the methods used for independent data. These mainly involve some type of local smoothing such as running means or medians, kernel smoothing, local polynomial regression, smoothing splines, and backfitting algorithms such as the ACE algorithm. There are a number of excellent modern expositions on this topic and we refer the reader to texts by Fan and Yao (2003) and by Gao (2007). In addition, the comprehensive review by Härdle et al. (1997) provides an accessible overview of the field.

3.8 Linear processes with infinite variance

In Example 3.2, we argued that the EEG data shown at the top of Figure 3.2 may be best described as having infinite variance, and we compared the innovations after an AR(p) fit to the data to Cauchy noise, a realization of which is shown in Figure 3.3. Such models have been used in a variety of situations, for example Fama (1965) used them to examine stock market prices.

An important property of Gaussian random variables is that the sum of two of them is itself a normal random variable. One consequence of this is that if Z is normal, then for Z_1 and Z_2 independent copies of Z and any positive constants a and b, $aZ_1 + bZ_2 =_d cZ + d$, for some positive c and some $d \in \mathbb{R}$. (The symbol $=_d$ means equality in distribution). In other words, the shape of Z is preserved (up to scale and shift) under addition. One typically defines an infinite variance linear process via symmetric (about zero) stable innovations.

Definition 3.4 (Stable law). *A random variable Z is said to be* stable, *or have a stable distribution, if Z_1 and Z_2 are two independent copies of Z, and for any positive constants a and b, the linear combination $aZ_1 + bZ_2$ has the same distribution as $cZ + d$, for some positive c and $d \in \mathbb{R}$. A random variable is said to be strictly stable if $d = 0$ for all positive a and b. A random variable is symmetric stable if X and $-X$ have the same stable distribution.*

Equivalently, the random variable Z is stable if for every $n \in \mathbb{N}^*$, there exists constants $a_n > 0$ and b_n such that the sum of i.i.d. copies, $Z_1 + \cdots + Z_n$, has the same distribution as $a_n Z + b_n$. We say that Z is strictly stable if $b_n = 0$.

Remark 3.5. It is possible to show that the only possible choice for the scaling constant a_n is $a_n = n^{1/\alpha}$ for some $\alpha \in (0,2]$.

Remark 3.6. The addition rule for independent random variables says that the mean of the sum is the sum of the means and the variance of the sum is the sum of the variances. Suppose $Z \sim \mathrm{N}(\mu,\ \sigma^2)$. Let Z_1 and Z_2 be two independent copies of Z. Then, $aZ_1 \sim \mathrm{N}(a\mu,\ (a\sigma)^2)$, $bZ_2 \sim \mathrm{N}(b\mu,\ (b\sigma)^2)$, and $cZ + d \sim \mathrm{N}(c\mu + d,\ (c\sigma)^2)$. The addition rule implies that $c^2 = a^2 + b^2$ and $d = (a + b - c)\mu$.

The most effective way to define the set of stable distributions is through their characteristic functions; see Billingsley (1995, Chapters 5, 26).

Theorem 3.7. *A random variable X is* stable *if and only if $X =_d aZ + b$, where $a > 0$, $b \in \mathbb{R}$ and Z is a random variable with characteristic function*

$$\varphi(\xi) = \mathbb{E}\exp(\mathrm{i}\xi Z) = \begin{cases} \exp\left(-|\xi|^\alpha[1 - \mathrm{i}\beta \tan\frac{\pi\alpha}{2}\,(\operatorname{sign}\xi)]\right) & \alpha \neq 1 \\ \exp\left(-|\xi|[1 + \mathrm{i}\beta\frac{2}{\pi}\,(\operatorname{sign}\xi)\ln|\xi|]\right) & \alpha = 1 \end{cases},$$

where $0 < \alpha \leq 2$, $-1 \leq \beta \leq 1$, and sign *is the sign function given by* $\operatorname{sign}(\xi) = -1$ *if $\xi < 0$,* $\operatorname{sign}(\xi) = 0$ *if $\xi = 0$ and* $\operatorname{sign}(\xi) = 1$ *if $\xi > 0$.*

When $\beta = 0$ and $b = 0$, these distributions are symmetric around zero, in which case the characteristic function of aZ has the simpler form

$$\varphi(\xi) = \mathrm{e}^{-a^\alpha|\xi|^\alpha}$$

Remark 3.8. The Gaussian distribution is stable with parameters $\alpha = 2$, $\beta = 0$. The Cauchy distribution is stable with parameters $\alpha = 1$, $\beta = 0$. A random variable Z is said to be Lévy$(\gamma,\ \delta)$ if it has density

$$f(x) = \sqrt{\frac{\gamma}{2\pi}}\frac{1}{(x-\delta)^{3/2}}\exp\left(-\frac{\gamma}{2(x-\delta)}\right), \quad \delta < x < \infty.$$

The Lévy distribution is stable with parameters $\alpha = 1/2$, $\beta = 1$.

Remark 3.9. Both the Gaussian and Cauchy distributions are symmetric and bell-shaped, but the Cauchy distribution has much heavier tails. If Z is standard normal, $\mathbb{P}(Z \geq 3)$ is 1.310^{-3}, whereas if Z is standard Cauchy (equivalently, a t-distribution with 1 degree of freedom), $\mathbb{P}(Z \geq 3) = 10^{-1}$. In a sample from these two distributions, there will be (on average) more than 100 times as many values above 3 in the Cauchy case than in the normal case. This is the reason stable distributions are called heavy tailed. In contrast to the normal and Cauchy distributions, the Lévy distribution is highly skewed. The distribution is concentrated on $x > 0$, and it has even heavier tails than the Cauchy.

Remark 3.10. For non-normal stable random variables Z (i.e., $\alpha < 2$), it can be shown that $\mathbb{E}\left[|Z|^\delta\right] < \infty$ only for $0 < \delta < \alpha$. Consequently, $\operatorname{Var} Z = \infty$ for $0 < \alpha < 2$ and $\mathbb{E}\left[|Z|\right] = \infty$ for $0 < \alpha \leq 1$.

It is possible to define an ARMA-type model with stable innovations. That is, we may define a process $\{X_t,\ t \in \mathbb{Z}\}$ such that $\mathbb{P}$-a.s.,

$$X_t = \sum_{j=-\infty}^{\infty} \psi_j Z_{t-j},$$

where $\{Z_t,\ t \in \mathbb{Z}\}$ is a sequence of i.i.d. stable random variables, and $\sum_j |\psi_j|^\delta < \infty$ for some $\delta \in (0, \alpha) \cap [0, 1]$. Moreover, it is possible to write the process as

$$\phi(B) X_t = \theta(B) Z_t,$$

where $\{X_t,\ t \in \mathbb{Z}\}$ is strictly stationary (but, of course, not covariance stationary unless $\alpha = 2$) and $\phi(B)$ and $\theta(B)$ are as in Section 1.3.2. These models are described in a fair amount of detail in Brockwell and Davis (1991, §12.5), who also discuss fitting these models to data.

3.9 Models for counts

In Example 3.3, we presented two time series that are discrete-valued and take on small values. These series should be contrasted with the series discussed in Section 1.5 and Example 3.1, which are also counts (the number of sunspots in Figure 1.1; the number of lynx trappings and the number of flu deaths in Figure 3.1), but are quite different in that one could use, for example, a TAR model with Gaussian noise as a reasonable approximation in the latter cases, but any use of normality is out of the question for the great discoveries series displayed in Figure 3.7 and for the cases of polio time series displayed in Figure 3.8.

There are two basic approaches to the problem. One approach is to develop models that produce integer-valued outcomes, and the other is to develop generalized linear models for dependent data. We briefly describe each approach in the following sections. Our presentation is very brief, the texts by MacDonald and Zucchini (1997) and by Kedem and Fokianos (2002) present rather extensive discussions of these models. In addition, the second part of Durbin and Koopman (2012) details the generalized linear model approach to the problem.

3.9.1 *Integer valued models*

In the late 1970s and through the 1980s, there were a number of researchers who worked on models with specific non-Gaussian marginals. The driving force behind these models is that discrete-valued time series can have ARMA-type autocorrelation structures along with marginal distributions that follow standard distributions such as Poisson, negative binomial, and so on. For a linear model to have marginals that match the innovations, the distributions must be stable; see Section 3.8. However, a number of researchers focused on random mixing or random summation as a method to obtain models that admit desired marginals and correlation structures.

For example, Jacobs and Lewis (1978a,b, 1983) developed DARMA, or discrete ARMA, models via mixing. For example, a DAR(1) model is of the form

$$X_t = V_t X_{t-1} + (1 - V_t) Z_t \tag{3.41}$$

where V_t is i.i.d. Bernoulli with $\Pr\{V_t = 1\} = 1 - \Pr\{V_t = 0\} = \rho$, and $\{Z_t\}$ is i.i.d. according to some specified discrete-valued distribution. Clearly, the support of X_t is the support of the noise Z_t. It is easy to show that (3.41) has the ACF structure of an AR(1), i.e., $\rho(h) = \rho^h$ for $h \in \mathbb{N}$; see Exercise 3.9. The authors also developed a 'new' DARMA, or NDARMA model with similar properties. These types of models are discussed further in Example 4.22 of Chapter 4.

Langberg and Stoffer (1987), and Block et al. (1988, 1990) developed (bivariate) exponential and geometric time series with ARMA correlation structure. We briefly discuss the univariate aspects of the geometric model; the bivariate model was called the BGARMA model. The authors developed the model using both random mixing and random summation, and then showed that the two methods are equivalent. The basic idea is as follows. If $X \sim \mathrm{G}(p)$ and $Z \sim \mathrm{G}(p/\pi)$ are independent geometric random variables [e.g., $\Pr(X = k) = p(1-p)^{k-1}$ for $k \in \mathbb{N}^*$], independent of I, which is Bernoulli$(1 - \pi)$, then $X' = IX + Z$, has the same distribution as X. For random summation, suppose $N \sim G(\pi)$ independent of $Z_j \sim \text{iid}\, \mathrm{G}(p/\pi)$, then $X' = \sum_{j=1}^{N} Z_j$ has the representation $X' = IX + Z$. This basic idea can be extended and used to formulate various non-Gaussian multivariate processes, and we refer the reader to Block et al. (1988) for a thorough presentation. As a simple example, let $X_0 \sim \mathrm{G}(p)$ and define, for $t \in \mathbb{N}$,

$$X_t = I_t X_{t-1} + Z_t \tag{3.42}$$

where $I_t \sim \text{iid Bernoulli}(1 - \pi)$, and $Z_t \sim \text{iid}\, \mathrm{G}(p/\pi)$ is the noise process. Then, $\{X_t,\, t \in \mathbb{N}\}$ is a process with the autocorrelation structure of an AR(1), and where $X_t \sim \mathrm{G}(p)$; see Exercise 3.9.

Finally we mention some models that are based on the notion of *thinning* that was discussed in Steutel and Van Harn (1979) as an integer-valued analog to stability for continuous-valued random variables. The idea is closely related to the random summation concept in Block et al. (1988), but in this case, the decomposition is given by $X' = \alpha \circ X + X_\alpha$ where X and X_α are independent, and $\alpha \circ X = \sum_{j=1}^{X} N_j$ where $N_j \sim \text{iid Bernoulli}(\alpha)$. Under this decomposition, X' and X have the same distribution, and such processes are called discrete stable; Steutel and Van Harn (1979) show, for example, that the Poisson distribution is discrete stable. McKenzie (1986), Al-Osh and Alzaid (1987) and others used the idea of thinning to obtain the INAR, or integer-valued AR, model. For example, an INAR(1) has the form

$$X_t = \alpha \circ X_{t-1} + Z_t \tag{3.43}$$

for $\alpha \in [0,1)$, where Z_t is an i.i.d. sequence of integer-valued random variables such as Poisson(λ) wherein the marginal of X_t is also Poisson with rate $\lambda/1-\alpha$. Moreover, the ACF of X_t is like an AR(1) and is given by $\rho(h) = \alpha^h$ for $h \in \mathbb{N}$; see Exercise 3.9.

3.9.2 Generalized linear models

The basic idea here is to extend the theory of generalized linear models to dependent data. This approach is apparently more successful for data analysis than the integer-valued models discussed in the previous subsection because there seem to be fewer pathologies in this setup. This topic is best discussed in more generality than is done in this section, and should be presented after the material in Part III on nonlinear state space models. Our brief discussion here can be supplemented with the texts mentioned in the introduction to this section.

We restrict attention to the univariate case. Let U_t be a vector of deterministic exogenous inputs or covariates, let $\mathcal{F}_t = \sigma(X_t, X_{t-1}, \ldots; U_t)$, and denote the conditional mean and variance by $\mu_t = \mathbb{E}\left[X_t \mid \mathcal{F}_{t-1}\right]$ and $\sigma_t^2 = \mathrm{Var}(X_t \mid \mathcal{F}_{t-1})$. Assume that conditionally, the observations are from an exponential family,

$$f(x_t \mid \theta_t, \mathcal{F}_{t-1}) = \exp\left[\frac{x_t\theta_t - b(\theta_t)}{\phi} + c(x_t;\phi)\right]. \tag{3.44}$$

The parameter ϕ is called the dispersion or scale parameter. It is assumed that $b(\theta_t)$ is twice differentiable, $c(x_t;\phi)$ does not involve θ_t, and θ_t is the (monotone) canonical link function. For this family, it can be shown that $\mu_t = b'(\theta_t)$ and $\sigma_t^2 = \phi\, b''(\theta_t)$. As an example, consider the Poisson distribution with mean function μ_t, in which case $\phi = 1$, $\theta_t = \ln \mu_t$ is the canonical link, $b(\theta_t) = \exp(\theta_t)$, and $c(x_t;\phi) = -\ln(x_t!)$; see Exercise 3.9. In the basic overdispersed or quasi-Poisson model, the scale parameter ϕ is left unspecified and estimated from the data rather than fixing it at 1. Typically, an estimating function is used and a quasi-Poisson model does not correspond to models with a fully specified likelihood.

Various approaches to modeling θ_t exist and include non- and semi-parametric methods, observation-driven models (i.e., μ_t is driven by the past data) and parameter driven models (i.e., μ_t is driven by the past parameter values) and various combinations of these models. In these settings, we have a link function,

$$\theta_t := \theta_t(\mu_t) = h(U_t, X^{t-1}, \mu^{t-1}, \varepsilon_t) \tag{3.45}$$

where $X^{t-1} = \{X_{t-1}, X_{t-2}, \ldots\}$ represents the data history, $\mu^{t-1} = \{\mu_{t-1}, \mu_{t-2}, \ldots\}$, and ε_t represents a vector of latent processes.

For time series of counts, the Poisson distribution is used most often. In the case of time series, it is typically necessary to account for over-dispersion and autocorrelation found in the data. For example, in Example 3.3 we saw overdispersion in that the data seem to have negative binomial marginals, and in Figure 3.7 and Figure 3.8, where autocorrelation is evident. Static models have $h(\cdot) = \beta' U_t$ where β is a vector of regression parameters and U_t is a vector of fixed inputs as previously explained. It

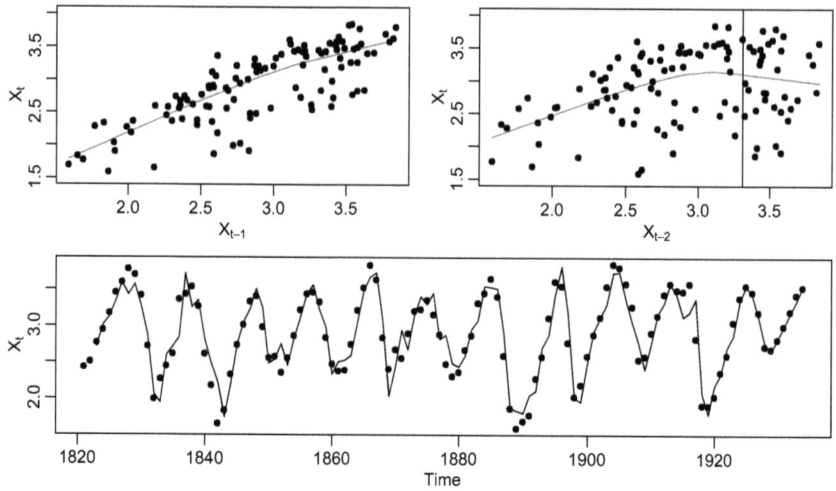

Figure 3.9 *Display for Example 3.11. Top: Lag plots of the Lynx series with a lowess fit emphasizing nonlinearity. The vertical line shows the threshold value of 3.31. Bottom: The logged Lynx series as points, and the one-step-ahead predictions as a solid line.*

is also possible to have $h(\cdot)$ be non- or semi-parametric, particularly to evaluate nonlinear trend; see Example 3.13. An extension that is treated in Kedem and Fokianos (2002) and reviewed in Fokianos (2009) is the case where $h(\cdot) = \beta' U_t + \sum_{j=1}^{p} \phi_j X_{t-j}$. Zeger (1988) introduced a stochastic element via a stationary latent process; i.e., $h(\cdot) = \beta' U_t + \varepsilon_t$. Here, overdispersion is introduced via the latent variable. Davis et al. (2003) and Shephard (1995) extended this idea to the *generalized linear ARMA*, or GLARMA, model by writing $\varepsilon_t = \sum_{j=1}^{p} \alpha_j(\varepsilon_{t-j} + e_{t-j}) + \sum_{j=1}^{q} \beta_j e_{t-j}$, where $e_t = (X_t - \mu_t)/\sqrt{\mu_t}$, and $\varepsilon_t = e_t = 0$ for $t \leq 0$. Finally, we mention GARCH-type Poisson models wherein $\mu_t = \alpha_0 + \sum_{j=1}^{p} \alpha_j X_{t-j} + \sum_{k=1}^{q} \beta_j \mu_{t-j}$; see Engle and Russell (1998). There is a considerable amount of literature on this topic, and we have only presented a few approaches. For an extensive and up-to-date review, see Jung and Tremayne (2011), which also presents an empirical comparison of various methods for analyzing discrete-valued time series.

3.10 Numerical examples

In this section, we use some of the models presented in this chapter to analyze a few of the data sets presented in Section 3.1. In particular, we will fit a SETAR model to the lynx data set using the R package `tsDyn`, an asymmetric GARCH-type model to the S&P 500 data set using the package `fGarch`, and overdispersed Poisson models to the polio data set using the packages `dyn`, `mgcv`, and `glm`. We also present some tests for detecting nonlinearity.

Example 3.11 (SETAR model). We used the `tsDyn` package to fit the SETAR model specified in (3.34) to the logarithm (base 10) of the lynx data. However, we allow the program to choose the optimal value of the threshold, rather than the one

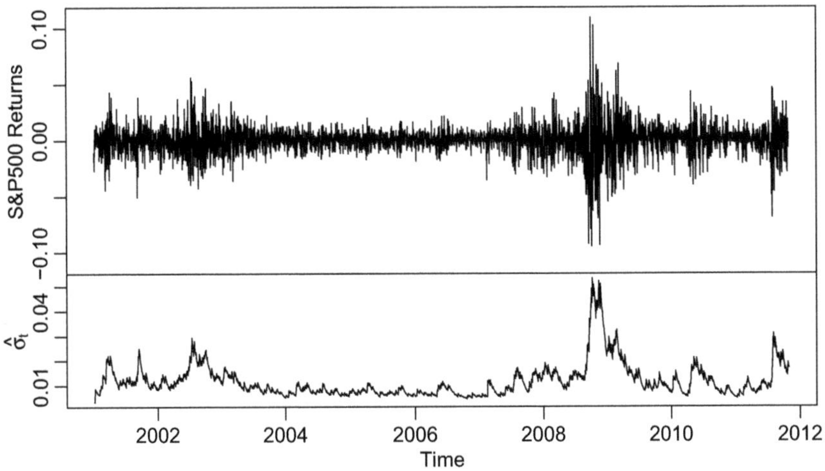

Figure 3.10 *Display for Example 3.12. Top: The S&P 500 returns. Bottom: The predicted one-step-ahead volatility from an Asymmetric Power ARCH fit.*

specified in (3.34), namely, 3.25. The top of Figure 3.9 shows two lag plots with lowess fits superimposed; these plots clearly indicate nonlinear behavior. The vertical line in the lag 2 plot indicates the optimal threshold value for the model, which is 3.31. The fitted model is similar to the fitted model displayed in (3.34). The results of the fit are as follows.

```
SETAR model ( 2 regimes)
Coefficients:
 Low regime:                          High regime:
    phiL.1    phiL.2    const L          phiH.1    phiH.2    const H
  1.264279 -0.428429  0.588437        1.599254 -1.011575  1.165692
Threshold:
-Variable: Z(t) = + (0) X(t)+ (1)X(t-1)
-Value: 3.31
Proportion of points in low regime: 69.64%   High regime: 30.36%
```

Finally, we note that it is not necessary to use a special package to fit a SETAR model. The model can be fit using piecewise linear regressions, `lm` in R, once a threshold value has been determined. ◇

Example 3.12 (Asymmetric power ARCH). The R package `fGarch` was used to fit a model to the S&P 500 returns discussed in Example 3.2. The data are displayed in Figure 3.5, where a small amount of autocorrelation is noticed. Hence, we include an AR(1) in the model to account for the conditional mean. For the conditional variance, we fit an Asymmetric Power ARCH (APARCH) model to the data; see Exercise 8.17 for details. In this case, the model is $X_t = \mu_t + \varepsilon_t$ where μ_t is an AR(1), and ε_t is GARCH-type noise where the conditional variance is modeled as

$$\sigma_t^\delta = \alpha_0 + \sum_{j=1}^{p} \alpha_j f_j(\varepsilon_{t-j}) + \sum_{j=1}^{q} \beta_j \sigma_{t-j}^\delta , \tag{3.46}$$

where

$$f_j(\varepsilon) = (|\varepsilon| - \gamma_j \varepsilon)^\delta . \tag{3.47}$$

Note that the model is GARCH when $\delta = 2$ and $\gamma_j = 0$, $j \in \{1, \dots, p\}$. The parameters γ_j ($|\gamma_j| \leq 1$) are the *leverage* parameters, which are a measure of asymmetry, and $\delta > 0$ is the parameter for the power term. A positive (resp. negative) value of γ_j's means that past negative (resp. positive) shocks have a deeper impact on current conditional volatility than past positive shocks (Black, 1976). This model couples the flexibility of a varying exponent with the asymmetry coefficient to take the *leverage effect* into account. Further, to guarantee that $\sigma_t > 0$, we assume that $\alpha_0 > 0$, $\alpha_j \geq 0$ with at least one $\alpha_j > 0$, and $\beta_j \geq 0$.

We fit an AR(1)-APARCH(1,1) model to the data. The (partial) results of the fit are outlined below, and Figure 3.10 displays the returns as well as the estimated one-step-ahead predicted volatilty, $\widehat{\sigma}_t$.

```
        Estimate  Std. Error  t value    Pr(>t)
mu      5.456e-05   1.685e-04    0.324  0.74605
ar1    -6.409e-02   1.989e-02   -3.221  0.00128
alpha0  1.596e-05   3.419e-06    4.668 3.04e-06
alpha1  4.676e-02   8.193e-03    5.708 1.15e-08
gamma1  1.000e+00   4.319e-02   23.156  < 2e-16
beta1   9.291e-01   7.082e-03  131.207  < 2e-16
delta   1.504e+00   2.054e-01    7.323 2.42e-13

Standardised Residuals Tests:
                                Statistic p-Value
 Shapiro-Wilk Test  R    W      0.9810958 0
 Ljung-Box Test     R    Q(20)  19.58712  0.4840092
 Ljung-Box Test     R^2  Q(20)  25.55894  0.1808778
```

Finally, we mention that ACF of the squared returns shown in Figure 3.5 indicates persistent volatility, and it seems reasonable that some type of integrated, or IGARCH(1,1) model could be fit to the data. In this case, (3.26) would be fit but with $\alpha_1 + \beta_1 \equiv 1$; recall the discussion following (3.29). ◇

Example 3.13 (Overdispersed Poisson model). At this point, we do not have all the tools necessary to fit complex models to dependent count data, so we use some existing R packages for independent data to fit simple time series models. In particular, we fit two overdispersed Poisson models to the polio data set displayed in Figure 3.8. In both cases, we followed Zeger (1988) by adding sinusoidal terms to account for seasonal behavior, namely, $C_{kt} = \cos(2\pi tk/12)$ and $S_{kt} = \sin(2\pi tk/12)$, for $k = 1, 2$. The first model is fully parametric, while the second model is semi-parametric. The link functions, (3.45), for the two models are:

$$Model\ 1: \ \ln(\mu_t) = \alpha_0 + \alpha_1 t + \beta_1 C_{1t} + \beta_2 S_{1t} + \beta_3 C_{2t} + \beta_4 S_{2t} + \varphi X_{t-1} \tag{3.48}$$

$$Model\ 2: \ \ln(\mu_t) = \alpha_0 + \mathrm{sm}(t) + \beta_1 C_{1t} + \beta_2 S_{1t} + \beta_3 C_{2t} + \beta_4 S_{2t} \tag{3.49}$$

A lagged value of the series was included in Model 2 in a first run, but it was not needed when the semi-parametrically fit smooth trend term, $\mathrm{sm}(t)$, was included in the model. In each case, a scale parameter ϕ is estimated. The results are displayed in Figure 3.11. ◇

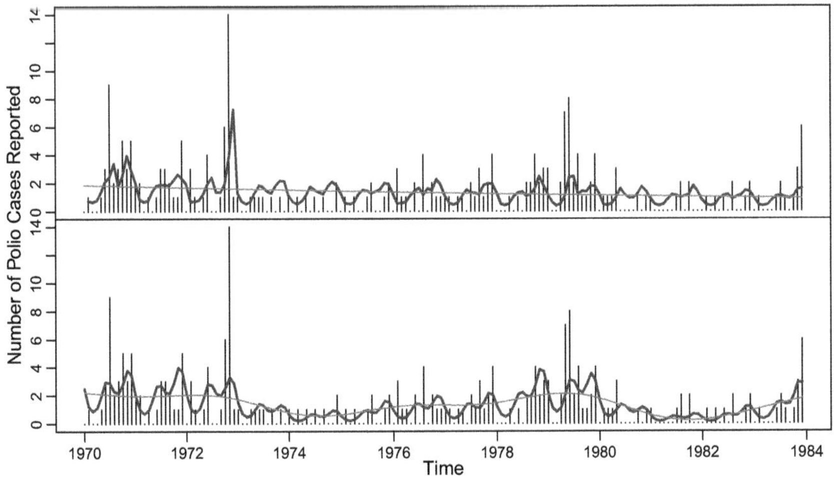

Figure 3.11 *Display for Example 3.13. In both cases, an overdispersed Poisson model is used. Top: The result of the Model 1, (3.48), fit superimposed on the polio data set. Displayed are the estimated mean function and linear trend lines. Bottom: The result of the Model 2, (3.49), fit superimposed on the polio data set. Displayed are the estimated mean function and trend lines; the assumed smooth trend is estimated via semi-parametric techniques.*

Example 3.14 (BiSpectrum). A number of researchers have suggested tests of nonlinearity based on the bispectrum given in 3.14. Given the results displayed in Exercise 3.4, it is clear that an estimate of

$$B(\omega_1, \omega_2) = \frac{|f(\omega_1, \omega_2)|^2}{f(\omega_1)f(\omega_2)f(\omega_1 + \omega_2)}$$

could be used to determine whether or not a process is linear. Because the value of $B(\omega_1, \omega_2)$ is unbounded, some researchers have proposed normalizing it so that, like squared coherence given in (1.82), it lives on the unit interval, with larger values indicating nonlinear (specifically, quadratic) dynamics.

The method proposed in Hinich and Wolinsky (2005) uses "frame averaging" wherein one first partitions time into blocks. Then, DFTs are calculated in each block and their averages are used to estimate the spectrum and bispectrum and to form an estimate of $B(\omega_1, \omega_2)$. This estimate is then transformed using a normalization based on a noncentral chi-squared distribution under the null hypothesis that the process is linear. For details, we refer the reader to Hinich and Wolinsky (2005). We provide an R script, `bi.coh`, that can be used to estimate and plot the normalized bispectrum.

Figure 3.12 shows the graphic produced by the script for the S&P 500 returns discussed in Example 3.12. Note that values over .95 are dark (or pink if color is used); numerous dark values indicate nonlinearity. ◇

Example 3.15 (Time domain tests for nonlinearity). There are a number of time domain tests for nonlinearity in the conditional mean, and many of them are discussed and compared in Lee et al. (1993). An obvious approach is to assess whether

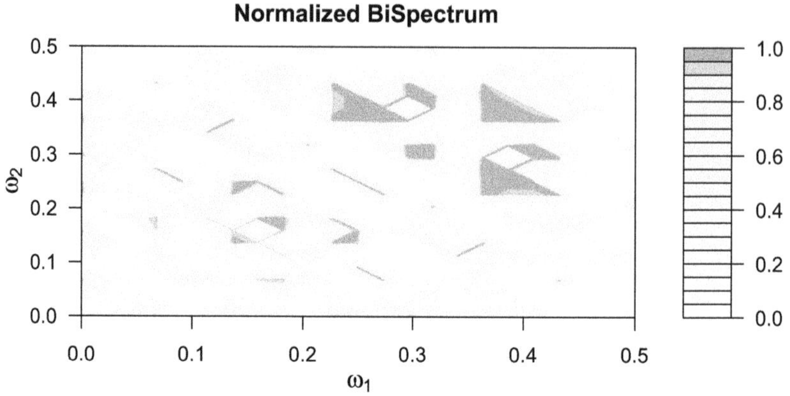

Figure 3.12 *Display for Example 3.14. Estimated normalized bispectrum of the S&P 500 returns displayed in Figure 3.5, with highlighted regions indicating departure from the linearity assumption.*

the coefficients of the higher-order ($M \geq 2$) terms in the Volterra series (3.3) are zero. For practical purposes, given a finite set of data, tests typically focus on whether or not there is the existence of second-order terms.

Keenan (1985) developed a one-degree-of-freedom test by first fitting a linear AR(p) model, where p is chosen arbitrarily or by some model choice criterion such as those described in (1.65). Given data, $\{X_1, \ldots, X_n\}$, a model is fit, and the one-step-ahead predictions,

$$\widehat{X}_{t|t-1} = \widehat{\phi}_0 + \widehat{\phi}_1 X_{t-1} + \cdots + \widehat{\phi}_p X_{t-p},$$

for $t = p+1, \ldots, n$ are calculated. Then, the AR(p) model is fit again, but now with the squared predictions included in the model. That is, the model

$$X_t = \phi_0 + \phi_1 X_{t-1} + \cdots + \phi_p X_{t-p} + \theta \widehat{X}^2_{t|t-1} + Z_t$$

is fit for $t = p+1, \ldots, n$, and the null hypothesis that $\theta = 0$ is tested against the alternative hypothesis that $\theta \neq 0$, in the usual fashion. In a sense, one is testing if the squared forecasts have additional predictive ability.

Tsay (1986) extended this idea by testing whether any of the second-order terms are additionally predictive. That is, the model,

$$X_t = \phi_0 + \phi_1 X_{t-1} + \cdots + \phi_p X_{t-p} + \sum_{1 \leq i \leq j \leq p}\sum \theta_{i,j} X_{t-i} X_{t-j} + Z_t,$$

is fit to the data and the null hypothesis $\theta_{i,j} = 0$ for all $1 \leq i \leq j \leq p$ is tested against the alternative hypothesis that at least one $\theta_{i,j} \neq 0$.

Both of these tests are available in the R package TSA, and we perform both tests on the logged Lynx data set analyzed in Example 3.11. Note that the series is fairly short.

```
> Keenan.test(log10(lynx))
  $test.stat  $p.value   $order
  [1] 11.669  [1] 0.001  [1] 11
> Tsay.test(log10(lynx))
  $test.stat  $p.value   $order
  [1]  1.316  [1] 0.226  [1] 11
```

We see that AIC chooses $p = 11$ and that the Keenan test identifies nonlinearity in the conditional mean whereas the Tsay test does not. ◇

Exercises

3.1. Suppose that $Z_t = A\cos(\omega_0 t + \varphi)$, where A, ω_0 and φ are fixed.

(a) If Z_t is the input process in (3.1), show that the output, X_t is a sinusoid of frequency ω_0 with squared amplitude $|\psi(e^{-i\omega_0})|^2$ and phase shifted by $\arg(\psi(e^{-i\omega_0}))$.

(b) Let $X_t = Z_t^2$. What are the frequency, amplitude and phase of X_t? Comment.

3.2. Using Example 3.2 as a guide, remove the signal from the explosion series shown in Figure 3.6 using the R command `ar`. Then calculate the sample ACF of the innovations of explosion series and compare it to the sample ACF of the squared innovations. How do these results compare to the results of Example 3.2 for the EEG and the S&P500 series?

3.3. Show that the solution to (3.5) is $a = b = 0$ and $c = 1$, as claimed.

3.4. Suppose $X_t = \sum_{j=0}^{\infty} \psi_j Z_{t-j}$, where $\sum_j |\psi_j| < \infty$ and $\{Z_t,\ t \in \mathbb{Z}\}$ are i.i.d. with mean-zero, variance σ_z^2 and finite third moment $\mathbb{E}\left[Z_t^3\right] = \mu_3$.

(a) Let $\kappa_x(r_1, r_2) = \text{cum}(X_{t+r_1}, X_{t+r_2}, X_t)$. Show that $\kappa_x(r_1, r_2) = \mu_3 \sum_j \psi_j \psi_{j+r_1} \psi_{j+r_2}$.

(b) Use part (a) and Proposition 1.22 (see also Example 1.33) to show that the bispectrum of $\{X_t,\ t \in \mathbb{Z}\}$ is $f(\omega_1, \omega_2) = \frac{\mu_3}{(2\pi)^2} \psi(e^{-i\omega_1}) \psi(e^{-i\omega_2}) \psi(e^{i(\omega_1+\omega_2)})$.

(c) Finally, show that
$$\frac{|f(\omega_1, \omega_2)|^2}{f(\omega_1) f(\omega_2) f(\omega_1 + \omega_2)}$$
is equal to $\mu_3^2 / 2\pi\sigma_z^6$, independent of frequency.

(d) How can the facts of this exercise be used to determine if a process $\{X_t,\ t \in \mathbb{Z}\}$ is (i) linear, and (ii) Gaussian?

3.5. Consider the process given by $X_t = Z_t + \theta Z_{t-1} Z_{t-2}$, where $\{Z_t,\ t \in \mathbb{Z}\}$ is a sequence of i.i.d. Gaussian variables with zero mean and variance σ^2.

(a) Show that $\{X_t,\ t \in \mathbb{Z}\}$ is strict-sense and weak-sense stationary.

(b) Show that $\{X_t,\ t \in \mathbb{Z}\}$ is a (weak) white-noise.

(c) Compute the bispectrum of $\{X_t,\ t \in \mathbb{Z}\}$.

3.6. For the bilinear model $\text{BL}(0,0,2,1)$ shown in (3.18),

(a) Show that $\mathbb{E}[X_t] = 0$ and then verify (3.19).

(b) Verify the statement that $\{X_t,\ t \in \mathbb{Z}\}$ is white noise but that $\{X_t^2,\ t \in \mathbb{Z}\}$ is predictable from its history.

3.7. If $\{X_t,\ t \in \mathbb{Z}\}$ is GARCH(p,q), show that $\{X_t^2,\ t \in \mathbb{Z}\}$ is non-Gaussian ARMA.

3.8. If $\{X_t,\ t \in \mathbb{Z}\}$ is ARCH(1), show that the gradient of the conditional log-likelihood $l(\alpha_0, \alpha_1 \,|\, X_1)$ is given by the 2×1 gradient vector,

$$\begin{pmatrix} \partial l/\partial \alpha_0 \\ \partial l/\partial \alpha_1 \end{pmatrix} = \sum_{t=2}^{n} \begin{pmatrix} 1 \\ X_{t-1}^2 \end{pmatrix} \times \frac{\alpha_0 + \alpha_1 X_{t-1}^2 - X_t^2}{2\left(\alpha_0 + \alpha_1 X_{t-1}^2\right)^2}.$$

3.9. The following problems are based on the material in Section 3.9.

(a) For the DAR(1) model in (3.41), show that the support of X_t and Z_t are the same, and then derive the ACF.

(b) Show that the marginal distribution of X_t defined by (3.42) is Geometric with parameter p, and then show the ACF is that of an AR(1) model.

(c) Show that the marginal of X_t in (3.43) is Poisson if Z_t is Poisson, and then derive the ACF of X_t.

(d) Suppose X_t is Poisson with conditional mean given by μ_t. Verify that the marginal of X_t is in the exponential family given by (3.44), identify the components, θ_t, ϕ, $b(\cdot)$, and $c(\cdot)$, and verify that $\mu_t = b'(\theta_t)$ and $\sigma_t^2 = \phi\, b''(\theta_t)$.

3.10. Using Section 3.10 as a guide, perform the following analyses.

(a) Fit a threshold model to the sunspots series displayed in Figure 1.1.

(b) For the mortality series, say M_t, shown at the bottom of Figure 3.1, calculate the normalized bispectrum of the series itself and then of $X_t = \nabla \ln M_t$. What is the interpretation of X_t? Comment on the difference between the results. Then, fit a threshold model to X_t. Explain why it is better to fit such a model to X_t rather than M_t.

(c) Fit a GARCH (or GARCH-type) model to the explosion series in Figure 3.6 and comment.

(d) Analyze the great discoveries and innovations series displayed in Figure 3.7.

3.11. Using Example 3.12 as guide, use the `fGarch` package to fit an AR-GARCH-type model to the returns of either (a) the CAC40, or (b) the NASDAQ. Include a complete set of residual diagnostics.

3.12. Using Example 3.14 and Example 3.15 as guides:

(a) Generate an AR(2) with $\phi_1 = 1$, $\phi_2 = -.9$, and $n = 2048$ and calculate the normalized bispectrum. Comment on the results.

(b) Calculate the normalized bispectrum of the R data set `sunspots` and comment.

(c) For the data generated in (a) and used in (b), perform the Keenan and Tsay tests for nonlinearity and comment.

Chapter 4

Stochastic Recurrence Equations

Establishing comprehensive properties of linear, Gaussian time series models is fairly straight-forward. As noted in Chapter 3, however, even slight departures from the linear model, such as a second-order Volterra series (Section 3.2), can be extremely difficult to work with. Quite often, properties of a nonlinear process, such as existence of a stationary solution, must be done on a case-by-case basis. In this chapter, we discuss *stochastic difference equations*. also referred to as *random coefficient autoregression*. This approach gives us a general framework within which to establish properties for a variety of nonlinear models.

For example, the first order autoregression, $X_t = \phi X_{t-1} + Z_t$, where $\{Z_t,\, t \in \mathbb{Z}\}$ is a strong (i.i.d.) white noise sequence, is of fundamental importance in time series analysis. The conditions for the existence of a stationary solution to the first-order autoregressive equation are simple to establish. A closely related process, but where the autoregressive coefficient is itself a stochastic process is given by

$$Y_t = A_t Y_{t-1} + B_t \,, \tag{4.1}$$

where $\{(A_t, B_t),\, t \in \mathbb{Z}\}$ is a sequence of independent random vectors in $\mathbb{R}^2$. These types of models were introduced in Andel (1976) and later developed in the seminal book by Nicholls and Quinn (1982). Despite its apparent simplicity, it is quite general and many important nonlinear models fall into the framework of (4.1). For example, consider the GARCH$(1,1)$ process discussed in Section 3.5 where $X_t = \sigma_t \varepsilon_t$, and $\sigma_t^2 = \alpha_0 + \alpha_1 X_{t-1}^2 + \beta_1 \sigma_{t-1}^2$. It then follows that $\sigma_t^2 = (\beta_1 + \alpha_1 \varepsilon_{t-1}^2)\sigma_{t-1}^2 + \alpha_0$, which fits into the framework of (4.1) where $B_t = \alpha_0$ is deterministic. As another example, consider the bilinear process BL$(1,0,1,1)$ [see Section 3.4],

$$X_t = aX_{t-1} + bX_{t-1}Z_{t-1} + Z_t \,, \tag{4.2}$$

where $\{Z_t,\, t \in \mathbb{Z}\}$ is strong white noise. We may write $\{X_t,\, t \in \mathbb{Z}\}$ in the form (4.1) by setting $A_t = a + bZ_{t-1}$ and $B_t = Z_t$.

Whereas these models are closely related to strong AR(1) processes, their properties are markedly different. Under suitable regularity conditions, we will show that there exists a unique strict-sense stationary solution to (4.1), given by

$$B_t + A_t B_{t-1} + A_t A_{t-1} B_{t-2} + A_t A_{t-1} A_{t-2} B_{t-3} + \cdots \tag{4.3}$$

We will also show that the tails of the stationary distributions are typically much heavier than the tails of $\{Z_t, t \in \mathbb{Z}\}$. The stationary distribution might fail to have a second-order moment even if $\{Z_t, t \in \mathbb{Z}\}$ is short-tailed (e.g., Gaussian). We establish such properties in Section 4.1.

It is possible to generalize (4.1). Instead of considering a first order autoregression, it is natural to consider a p-th order random coefficient autoregression,

$$X_t = \phi_1 X_{t-1} + \ldots \phi_p X_{t-p} + Z_t \,, \tag{4.4}$$

where the innovation process $\{Z_t, t \in \mathbb{Z}\}$ is a white noise. Define the lag-vector

$$\boldsymbol{Y}_t = \begin{bmatrix} X_t \\ X_{t-1} \\ \vdots \\ X_{t-p+1} \end{bmatrix} , \tag{4.5}$$

and rewrite (4.4) as a linear order 1 vector autoregressive equation,

$$\boldsymbol{Y}_t = \boldsymbol{\Phi} \boldsymbol{Y}_{t-1} + \boldsymbol{B}_t \,, \tag{4.6}$$

where $\boldsymbol{\Phi}$ is the *companion matrix* of the polynomial $\phi(z) = 1 - \phi_1 z - \cdots - \phi_p z^p$,

$$\boldsymbol{\Phi} := \begin{bmatrix} \phi_1 & \phi_2 & \phi_3 & \ldots & \phi_p \\ 1 & 0 & 0 & \ldots & 0 \\ 0 & 1 & 0 & \ldots & 0 \\ \vdots & \ddots & \ddots & \ddots & \vdots \\ 0 & 0 & \ldots & 1 & 0 \end{bmatrix} \quad \text{and} \quad \boldsymbol{B}_t := \begin{bmatrix} Z_t \\ 0 \\ \vdots \\ 0 \end{bmatrix} . \tag{4.7}$$

A natural extension consists of replacing the autoregressive coefficients by stochastic processes, leading to random coefficient autoregressive processes of order p,

$$X_t = \sum_{i=1}^{p} A_{i,t} X_{t-i} + B_t \,, \tag{4.8}$$

As for (4.4), we may rewrite (4.8) as a first order vector random coefficient autoregressive equation

$$\boldsymbol{Y}_t = A_t \boldsymbol{Y}_{t-1} + \boldsymbol{B}_t \,, \tag{4.9}$$

where

$$A_t := \begin{bmatrix} A_{1,t} & A_{2,t} & A_{3,t} & \cdots & A_{p,t} \\ 1 & 0 & 0 & \cdots & 0 \\ 0 & 1 & \ddots & \ddots & 0 \\ \vdots & \ddots & \ddots & \ddots & \vdots \\ 0 & \cdots & \cdots & 1 & 0 \end{bmatrix} , \quad \boldsymbol{B}_t := \begin{bmatrix} B_t \\ 0 \\ \vdots \\ 0 \end{bmatrix} ,$$

and the lag-vector $\boldsymbol{Y}_t$ is defined in (4.5). The vector random coefficient autoregressive model (4.8) has been host in a lot of modeling efforts. Here again, under suitable regularity conditions, (4.9) has a unique non-anticipativc solution, givcn by thc scrics

$$\boldsymbol{B}_t + A_t \boldsymbol{B}_{t-1} + A_t A_{t-1} \boldsymbol{B}_{t-2} + A_t A_{t-1} A_{t-2} \boldsymbol{B}_{t-3} + \cdots \tag{4.10}$$

The convergence of such series is more difficult to establish, because this time $A_t A_{t-1} \cdots A_{t-p}$ is a product of random matrix. We establish the conditions upon which such series converge and establish some properties of this solution in Section 4.2.

4.1 Random coefficient autoregression: The scalar case

4.1.1 Strict stationarity

In this section we consider a random coefficient autoregressive equation of the form

$$Y_t = A_t Y_{t-1} + B_t \,, \quad t \in \mathbb{Z} \,, \tag{4.11}$$

where the sequence $\{(A_t, B_t),\ t \in \mathbb{Z}\}$ is assumed to be a strictly stationary and ergodic (see Definition 7.11) process on $(\Omega, \mathcal{F}, \mathbb{P})$ taking values in $\mathbb{R}^2$. In this section, we will only need to know that, for any measurable function $f : \mathbb{R} \to \mathbb{R}$ such that $\mathbb{E}[f^+(A_0, B_0)] < \infty$, where $x^+ = \max(x, 0)$, we have, as $n \to \infty$,

$$\frac{1}{n}\sum_{t=1}^{n} f(A_t, B_t) \stackrel{\mathbb{P}\text{-a.s.}}{\longrightarrow} \mathbb{E}[f(A_0, B_0)] \quad \text{and} \quad \frac{1}{n}\sum_{t=1}^{n} f(A_{-t}, B_{-t}) \stackrel{\mathbb{P}\text{-a.s.}}{\longrightarrow} \mathbb{E}[f(A_0, B_0)] \,.$$

By the strong law of large numbers, a sequence of i.i.d. random variables is ergodic. Also, an m-dependent strict-sense stationary process is ergodic.

A natural question is the existence of strict-sense and weak-sense stationary solutions to (4.11). For ARMA processes, the conditions upon which such solutions exist are given in Proposition 1.31. It is worthwhile to extend such results to random coefficient autoregressive processes, and in Theorem 4.1, we show the existence of a strict-sense stationary solution under very weak conditions. The existence of a weak-sense stationary solution typically requires more stringent conditions.

When generating a time series by a recurrence equation of the form (4.11), it is sometimes beneficial to initialize the variable at time zero, $Y_0 = U$, where U is an arbitrary random variable, and to generate the random variables Y_1, Y_2, and so on. A straightforward recurrence shows that, for any $t \in \mathbb{N}$,

$$Y_t(U) = \sum_{j=0}^{t-1} \left(\prod_{i=t+1-j}^{t} A_i \right) B_{t-j} + \left(\prod_{i=1}^{t} A_i \right) U \,, \tag{4.12}$$

where by convention $\prod_{i=t+1}^{t} A_i = 1$. Important questions are whether or not the initial value $Y_0 = U$ affects the long-term behavior of the process $\{Y_t(U), t \in \mathbb{N}\}$ and whether or not this process attains some stationary regime as t goes to infinity. Theorem 4.1 shows that, with probability one, and for any initial condition U (which can be random or not random), the solution of (4.11) given by (4.12) converges as $t \to \infty$ (in some sense to be defined) to the strict-sense stationary solution.

Theorem 4.1. *Let $\{(A_t, B_t),\ t \in \mathbb{Z}\}$ be a strict-sense stationary and ergodic real-valued stochastic process. Assume that*

$$\mathbb{E}[\ln^+ |A_0|] < \infty, \quad -\infty \le \mathbb{E}[\ln |A_0|] < 0, \quad \text{and} \quad \mathbb{E}[\ln^+ |B_0|] < \infty\,, \tag{4.13}$$

where $x^+ = \max(0,\ x)$ for $x \in \mathbb{R}$. Then

$$\widetilde{Y}_t = \sum_{j=0}^{\infty} \left(\prod_{i=t-j+1}^{t} A_i \right) B_{t-j}\,, \quad t \in \mathbb{Z} \tag{4.14}$$

is the only strict-sense stationary solution of (4.11). *The series on the right-hand side of* (4.14) *converges absolutely almost surely. Furthermore, for all arbitrary random variables U defined on the same basic probability space as $\{(A_t, B_t),\ t \in \mathbb{Z}\}$,*

$$|Y_t(U) - \widetilde{Y}_t| \overset{\mathbb{P}\text{-a.s.}}{\longrightarrow} 0\,, \tag{4.15}$$

as $t \to \infty$, where, for any $t \in \mathbb{N}$, $Y_t(U)$ is given in (4.12). *In particular, the random variable $Y_t(U)$ converges in distribution to $\widetilde{Y}_0$, i.e., as $t \to \infty$,*

$$Y_t(U) \overset{\mathbb{P}}{\Longrightarrow} \widetilde{Y}_0\,. \tag{4.16}$$

Before proving Theorem 4.1, we introduce some notations and state and prove a useful technical lemma. For all random variable U, all $t \in \mathbb{Z}$ and all $k \in \mathbb{N}$, we denote

$$Y_t^k(U) = \sum_{j=0}^{k-1} \left(\prod_{i=t+1-j}^{t} A_i \right) B_{t-j} + \left(\prod_{i=t+1-k}^{t} A_i \right) U\,. \tag{4.17}$$

The random variable $Y_t^k(U)$ can be interpreted as the solution of the stochastic difference equation (4.11) at time t starting at time $t-k$ with the random variable U. Obviously, $Y_t(U) = Y_t^t(U), t \ge 0$, where $Y_t(U)$ is defined in (4.12). Note also that

$$\widetilde{Y}_t = \lim_{k \to \infty} Y_t^k(0) = \sum_{j=0}^{\infty} \left(\prod_{i=t-j}^{t-1} A_i \right) B_{t-j-1}\,,$$

if the limit of the right-hand side exists (pointwise). The purpose of the following lemma is to show that the series in the right-hand side of the previous equation converges absolutely, $\mathbb{P}$-a.s., in which case the process $\{\widetilde{Y}_t,\ t \in \mathbb{Z}\}$ is well-defined.

Lemma 4.2.

$$\sum_{j=0}^{\infty} \left| \left(\prod_{i=t-j}^{t-1} A_i \right) B_{t-1-j} \right| < \infty\,, \quad \mathbb{P}\text{-a.s.} \tag{4.18}$$

Proof. We will show that

$$\limsup_{j \to \infty} (|A_{t-1} \cdots A_{t-j}|\, |B_{t-1-j}|)^{1/j} < 1\,, \quad \mathbb{P}\text{-a.s.} \tag{4.19}$$

The proof then follows by application of the Cauchy root criterion for the absolute convergence of series.

We now prove (4.19) for a fixed $t \in \mathbb{Z}$. Note that the ergodicity of the process $\{(A_t, B_t),\ t \in \mathbb{Z}\}$ implies that

$$\limsup_{j\to\infty} \frac{1}{j} \sum_{i=1}^{j} \ln|A_{t-i}| < 0\,, \quad \mathbb{P}\text{-a.s.} \tag{4.20}$$

Furthermore, note that by stationarity, for any $\delta > 0$,

$$\sum_{j=1}^{\infty} \mathbb{P}(j^{-1}\ln^+|B_{t-1-j}| \geq \delta) = \sum_{j=1}^{\infty} \mathbb{P}(\delta^{-1}\ln^+|B_0| \geq j)$$
$$= \mathbb{E}\left(\sum_{j=1}^{\infty} \mathbb{1}\{j \leq \delta^{-1}\ln^+|B_0|\} \right) \leq \delta^{-1}\mathbb{E}(\ln^+|B_0|) < \infty\,,$$

and the Borel-Cantelli Lemma therefore shows that $\limsup_{j\to\infty} j^{-1}\ln^+|B_{t-1-j}| = 0$, $\mathbb{P}$-a.s. Combining this with (4.20), we get

$$\limsup_{j\to\infty} \ln\left(|A_{t-1}|\cdots|A_{t-j}||B_{t-1-j}|\right)^{1/j} < 0\,, \quad \mathbb{P}\text{-a.s.}$$

which implies (4.19) by exponentiating the two terms. ■

Proof (of Theorem 4.1). By Lemma 4.2, $\widetilde{Y}_t$ is well defined. We now prove assertion (4.15). It follows from (4.12) and (4.14) that

$$Y_t(U) - \widetilde{Y}_t = -\sum_{j=t}^{\infty} \left(\prod_{i=t+1-j}^{t} A_i \right) B_{t-j} + \left(\prod_{i=1}^{t} A_i \right) U\,.$$

Since

$$\sum_{j=t}^{\infty} \left(\prod_{i=t+1-j}^{t} A_i \right) B_{t-j} = \left(\prod_{i=1}^{t} A_i \right) \sum_{j=t}^{\infty} \left(\prod_{i=t+1-j}^{0} A_i \right) B_{t-j} = \left(\prod_{i=1}^{t} A_i \right) \widetilde{Y}_0\,,$$

we get

$$|Y_t(U) - \widetilde{Y}_t| \leqq \left(\prod_{i=1}^{t} |A_i| \right) (|\widetilde{Y}_0| + |U|)\,. \tag{4.21}$$

Using $\mathbb{E}[\ln|A_0|] < 0$ and the ergodicity of the sequence $\{(A_t, B_t),\ t \in \mathbb{Z}\}$, we obtain

$$\prod_{i=1}^{t} |A_i| = \exp\left(t \times \frac{1}{t} \sum_{i=1}^{t} \ln|A_i| \right) \overset{\mathbb{P}\text{-a.s.}}{\longrightarrow} 0\,.$$

From (4.21) this implies that $|Y_t(U) - \widetilde{Y}_t| \overset{\mathbb{P}\text{-a.s.}}{\longrightarrow} 0$ as $t \to \infty$ showing (4.15).

By construction, the process $\{\widetilde{Y}_t,\ t \in \mathbb{Z}\}$ is strictly stationary. Moreover, let h be a real bounded uniformly continuous function on $\mathbb{R}$. Then, by the dominated convergence theorem,

$$\lim_{t\to\infty} \mathbb{E}[h(Y_t(U))] = \mathbb{E}[h(\widetilde{Y}_t)] = \mathbb{E}[h(\widetilde{Y}_0)]\,,$$

where the last equality follows from the stationarity of $\{\widetilde{Y}_t,\ t \in \mathbb{Z}\}$. This implies (4.16). In addition,

$$A_t\widetilde{Y}_{t-1} + B_t = A_t \left(\sum_{j=0}^{\infty} \left(\prod_{i=t+1-(j+1)}^{t-1} A_i \right) B_{t-(j+1)} \right) + B_t$$
$$= \sum_{j=1}^{\infty} \left(\prod_{i=t+1-j}^{t-1} A_i \right) B_{t-j} + B_t = \widetilde{Y}_t\,, \quad \mathbb{P}\text{-a.s.},$$

i.e., the difference equation (4.11) is almost surely satisfied. Thus $\{\widetilde{Y}_t,\ t \in \mathbb{Z}\}$ is a strict-sense stationary solution of the difference equation (4.11).

We complete the proof by showing that $\{\widetilde{Y}_t,\ t \in \mathbb{Z}\}$ is the unique strict-sense stationary solution. Assume that $\{Y_t,\ t \in \mathbb{Z}\}$ is another stationary solution of the stochastic difference equation (4.11), i.e., $Y_{t+1} = A_{t+1}Y_t + B_{t+1}$, $\mathbb{P}$-a.s. and $\{Y_t,\ t \in \mathbb{Z}\}$ is stationary. Then,

$$\begin{aligned} |Y_t - \widetilde{Y}_t| &= |A_t|\,|Y_{t-1} - \widetilde{Y}_{t-1}| \\ &= \left|\prod_{i=0}^{k-1} A_{t-i}\right| |Y_{t-k} - \widetilde{Y}_{t-k}| \\ &\leq \left|\prod_{i=0}^{k-1} A_{t-i}\right| |Y_{t-k}| + \left|\prod_{i=0}^{k-1} A_{t-i}\right| |\widetilde{Y}_{t-k}|\,. \end{aligned} \tag{4.22}$$

Because the sequence $\{(A_t, B_t),\ t \in \mathbb{Z}\}$ is ergodic, if (4.13) is satisfied, then as $k \to \infty$,

$$\left|\prod_{i=0}^{k-1} A_{t-i}\right| = \left(\exp\left(\frac{1}{k}\sum_{i=0}^{k-1} \ln |A_{t-i}|\right)\right)^k \xrightarrow{\mathbb{P}\text{-a.s.}} 0\,.$$

Since $\{Y_t,\ t \in \mathbb{Z}\}$ and $\{\widetilde{Y}_t,\ t \in \mathbb{Z}\}$ are stationary, we have as $k \to \infty$,

$$\left|\prod_{i=1}^{k} A_{t-i}\right| |Y_{t-k}| \xrightarrow{\mathbb{P}} 0 \quad \text{and} \quad \left|\prod_{i=1}^{k} A_{t-i}\right| |\widetilde{Y}_{t-k}| \xrightarrow{\mathbb{P}} 0\,. \tag{4.23}$$

Plugging these limits into (4.22), we find $Y_t - \widetilde{Y}_t = 0$, $\mathbb{P}$-a.s., for all $t \in \mathbb{Z}$, establishing that $\{\widetilde{Y}_t,\ t \in \mathbb{Z}\}$ is the only stationary solution of (4.11). ■

Remark 4.3 (Nonanticipativity). Under the conditions stated in Theorem 4.1, (4.11) admits a unique strict-sense stationary solution. When $\{(A_t, B_t),\ t \in \mathbb{Z}\}$ is i.i.d., it is worthwhile to note that this solution is *nonanticipative*, in the sense that, for any $t \in \mathbb{Z}$, X_t is independent of the random variables $\{(A_s, B_s),\ s > t\}$.

Remark 4.4. Note also that, if $U = \widetilde{Y}_0$ (i.e., the recursion is started from the stationary solution at time $t = 0$), then for all $t \geq 0$, we get $Y_t(\widetilde{Y}_0) = \widetilde{Y}_t$, for any $t \geq 0$. That is, the nonanticipative solution of (4.11) coincides with the strict-sense stationary solution for all latter time instances. This type of behavior is classical for Markov chains, which we discuss in Part II.

Example 4.5. The discrete autoregressive process of order 1 model, DAR(1), over a state-space $(\mathsf{X},\mathcal{X})$ is given by the following stochastic recurrence equation,

$$X_t = V_t X_{t-1} + (1-V_t)Z_t \,, \tag{4.24}$$

where $\{V_t,\, t\in\mathbb{Z}\}$ is an i.i.d. sequence of binary random variables with $\mathbb{P}[V_t=1] = \alpha\in[0,1)$, $\{Z_t,\, t\in\mathbb{Z}\}$ is an i.i.d. sequence of random variables with distribution given by π on $(\mathsf{X},\mathcal{X})$, and $\{V_t,\, t\in\mathbb{Z}\}$ and $\{Z_t,\, t\in\mathbb{Z}\}$ are independent. These models were discussed in Section 3.9.

The model defines the current observation to be a mixture of two independent random variables: it is either the last observation, with probability α, or another, independent, sample from the distribution π. The DAR(1) is a very simple and very general model since π can be any distribution over a state-space, which does need to be a vector field. On the other hand, it is clear from the definition (4.24) that the dependence is obtained by runs of constant values in the sample path, and the larger the value of α the longer the runs. Such behavior makes this model not appropriate for continuous random variables; it might be of interest for discrete or categorical random variables.

By construction $\mathbb{E}\left[\ln(V_t)\right] = -\infty$; therefore, Theorem 4.1 shows that the DAR(1) model admits a unique nonanticipative solution given by

$$\widetilde{Y}_t = \sum_{j=0}^{\infty}\left(\prod_{i=t-j+1}^{t} V_i\right)(1-V_{t-j})Z_{t-j}\,, \quad t\in\mathbb{Z}\,. \tag{4.25}$$

Note that $\prod_{i=t-j+1}^{t} V_i = 1$ for $0\le j<\sigma_t$, and $\prod_{i=t-j+1}^{t} V_i = 0$ otherwise, where $\sigma_t = \inf\{j\ge 0 : V_{t-j}=0\}$; since $\mathbb{P}[\sigma_t=k] = \alpha^k(1-\alpha)$ for $k\ge 0$, $\mathbb{P}[\sigma_t<\infty]=1$, this implies that $\widetilde{Y}_t = Z_{t-\sigma_t}$, where σ_t and $\{Z_t,\, t\in\mathbb{Z}\}$ are independent. Therefore, for any $A\in\mathcal{X}$, we get

$$\mathbb{P}[\widetilde{Y}_t\in A] = \mathbb{P}[Z_{t-\sigma_t}\in A] = \sum_{k=0}^{\infty}\mathbb{P}[Z_{t-k}\in A]\mathbb{P}[\sigma_t=k] = \mathbb{P}[Z_0\in A]\,,$$

showing that the marginal distribution of the stationary solution coincides with the distribution of Z. ◇

When $\{(A_t,B_t),\, t\in\mathbb{Z}\}$ is an i.i.d. sequence, the conditions (4.13) are necessary for the almost-sure absolute convergence of the series in the right-hand side of (4.14). We make this statement precise in the following theorem.

Theorem 4.6. *Assume that $\{(A_t,B_t),\, t\in\mathbb{Z}\}$ is an i.i.d. sequence such that $\mathbb{P}(B_0 = 0)<1$ and $\mathbb{E}[\ln^+|A_0|]<\infty$. In addition, assume that for some $t\in\mathbb{Z}$, the series*

$$\sum_{j=0}^{\infty}\left(\prod_{i=t-j+1}^{t} A_i\right)B_{t-j}\,,$$

converges with probability one. Then, $-\infty\le\mathbb{E}[\ln|A_0|]<0$.

Proof. See Exercise 4.1. ■

4.1.2 Weak stationarity

We now consider conditions upon which (4.11) admits moments of order p. To keep the discussion simple, we restrict our discussion to the case where $\{(A_t, B_t),\ t \in \mathbb{Z}\}$ is an i.i.d. sequence.

Theorem 4.7. *Let $p \in (0, \infty)$. Assume that $\{(A_t, B_t),\ t \in \mathbb{Z}\}$ is an i.i.d. sequence such that $\mathbb{E}[|A_0|^p] < 1$ and $\mathbb{E}[|B_0|^p] < \infty$. Then,*

(i) The stochastic difference equation (4.11) has a unique strict-sense stationary solution $\{\widetilde{Y}_t,\ t \in \mathbb{Z}\}$ given by (4.14), and this solution satisfies $\mathbb{E}[|\widetilde{Y}_0|^p] < \infty$. In addition, if $p \geq 1$, the series on the right-hand-side of (4.14) converges in p-th norm.

(ii) The moments $\mathbb{E}[\widetilde{Y}_0^j]$, for $j = 1, 2, \ldots, \lfloor p \rfloor$, are uniquely determined by the recursive equations,

$$\mathbb{E}[\widetilde{Y}_0^j] = \sum_{k=0}^{j} \binom{j}{k} \mathbb{E}[A_0^k B_0^{j-k}] \mathbb{E}[\widetilde{Y}_0^k] . \tag{4.26}$$

(iii) Let U be a random variable such that $\mathbb{E}[|U|^p] < \infty$. For any $t \in \mathbb{N}$, $\mathbb{E}[|Y_t(U)|^p] < \infty$ where $\{Y_t(U), t \in \mathbb{N}\}$ is the solution of the stochastic difference equation (4.11) started at $t = 0$ with U as defined in (4.12) and

$$\lim_{t \to \infty} \mathbb{E}[|Y_t(U) - \widetilde{Y}_t|^p] = 0 . \tag{4.27}$$

In particular, $\lim_{t\to\infty} \|Y_t(U)\|_p = \|\widetilde{Y}_0\|_p$ where $\|V\|_p = (\mathbb{E}[V^p])^{1/p}$ for all random variable V such that $\mathbb{E}[V^p] < \infty$.

Proof. Assume first that $p \geq 1$. Since $\|A_0\|_p < 1$ and $\|B_0\|_p < \infty$, we have for all $t \in \mathbb{Z}$,

$$\sum_{j=0}^{\infty} \| \prod_{i=t-j+1}^{t} A_i B_{t-j} \|_p \leq \sum_{j=0}^{\infty} \|A_0\|_p^j \, \|B_0\|_p , \tag{4.28}$$

so that the series in the right-hand side of (4.14) converges in p-th moment. Jensen's inequality implies that $\mathbb{E}[\ln |A_0|] \leq \ln \mathbb{E}[|A_0|] < 0$ and $\mathbb{E}\left[\ln^+ |B_0|\right] < \infty$, so that (4.11) has a unique strict-sense stationary solution, $\{\widetilde{Y}_t,\ t \in \mathbb{Z}\}$, by Theorem 4.1. Since the space $\mathrm{L}^p(\Omega, \mathcal{F}, \mathbb{P})$ is complete, the relation (4.28) implies that, for any $t \in \mathbb{Z}$, the sequence $\{Y_t^k(0), k \in \mathbb{N}\}$, defined in (4.17) converges $\mathbb{P}$-a.s. and in p-th moment to $\widetilde{Y}_t$.

From (4.11), it is clear that (4.26) holds by expanding $(A_t \widetilde{Y}_{t-1} + B_t)^j$ and computing the expectations. The equations determine $\mathbb{E}[\widetilde{Y}_0^j]$ recursively for $j = 1, 2, \ldots, \lfloor p \rfloor$, since the coefficient of $\mathbb{E}[\widetilde{Y}_0^j]$ on the right-hand side of (4.26) is strictly smaller than 1. For $j \in \{1, \ldots, \lfloor p \rfloor\}$, we have $|\mathbb{E}[A_0^j]| \leq \mathbb{E}[|A_0|^j] < 1$, since $\|A_0\|_j \leq \|A_0\|_p < 1$. Eq. (4.21) shows that

$$\|Y_t(U) - \widetilde{Y}_t\|_p \leq \|\prod_{i=1}^{t} |A_i|\|_p \left(\|\widetilde{Y}_0\|_p + \|U\|_p\right) \leq \|A_0\|_p^t \left(\|\widetilde{Y}_0\|_p + \|U\|_p\right) .$$

Consider now the case $p < 1$. We have

$$\mathbb{E}[\ln|A_0|] = p^{-1}\mathbb{E}[\ln|A_0|^p] \le p^{-1}\ln\mathbb{E}[|A_0|^p] < 0$$

and since $\ln(x) \le x^p$ for $x \in \mathbb{R}^+$, $\mathbb{E}[\ln^+|B_0|] \le \mathbb{E}[|B_0|^p] < \infty$. Therefore, (4.11) has a unique strict-sense stationary solution, $\{\widetilde{Y}_t,\ t \in \mathbb{Z}\}$. The inequality $(\sum_{i=1}^{\infty} x_i)^p \le \sum_{i=1}^{\infty} x_i^p$ for any nonnegative sequence $\{x_n,\ n \in \mathbb{N}\}$ implies that

$$\begin{aligned}\mathbb{E}[|\widetilde{Y}_t|^p] &\le \sum_{j=0}^{\infty} \mathbb{E}[|\prod_{i=t-j+1}^{t} A_i|^p]\,\mathbb{E}[|B_{t-j}|^p] \\ &\le \sum_{j=0}^{\infty} (\mathbb{E}[|A_0|^p])^j\,\mathbb{E}[|B_0|^p],\end{aligned}$$

showing that the unique stationary solution has a moment of order p. Similarly, if $\mathbb{E}[|U|^p] < \infty$, then $\lim_{t\to\infty}\mathbb{E}[|Y_t(U) - \widetilde{Y}_t|^p] = 0$. ■

Remark 4.8. According to Theorem 4.7, the stationary solution $\{\widetilde{Y}_t,\ t \in \mathbb{Z}\}$ exists and admits a moment of order p provided $k(p) = \mathbb{E}[|A_0|^p] < 1$. We will show that this condition can be satisfied under the assumptions of Theorem 4.1 if we assume in addition that $k(p_\star) < \infty$ for some $p_\star > 0$. To see this, first note that $k(p) < \infty$ for all $p \in [0, p_\star]$. Indeed, by Jensen's inequality, for all such p,

$$\mathbb{E}[|A_0|^p] \le \mathbb{E}[|A_0|^{p_\star}]^{\frac{p}{p_\star}} < \infty\,.$$

Moreover, for any $(p,q) \in \mathbb{R}^+$ and $\lambda \in [0,1]$, the Hölder inequality implies that

$$\begin{aligned}\mathbb{E}[|A_0|^{\lambda p + (1-\lambda)q}] &= \left[\mathbb{E}[|A_0|^{\lambda p/\lambda}]\right]^{\lambda}\left[\mathbb{E}[|A_0|^{(1-\lambda)q/(1-\lambda)}]\right]^{1-\lambda} \\ &\le (\mathbb{E}[|A_0|^p])^{\lambda}(\mathbb{E}[|A_0|^q])^{1-\lambda}\,,\end{aligned}$$

showing that the function $p \mapsto \ln k(p)$ is convex. Moreover, $\ln k(0) = 0$. Finally, $p \mapsto k(p)$ is well-defined in $[0, p_\star]$, differentiable from the right at $p = 0$ with derivative $\mathbb{E}[\ln|A_0|] < 0$. This shows that there exists some $p > 0$ such that $k(p) = \mathbb{E}[|A_0|^p] < 1$.

Remark 4.9. Whenever $\mathbb{E}[|A_0|^j] < \infty$ and $\mathbb{E}[|B_0|^j] < \infty$, we set $\mathbb{E}[A_0^j] = a_j$ and $\mathbb{E}[B_0^j] = b_j$. Recall that in this section, we assume that $\{(A_t, B_t),\ t \in \mathbb{Z}\}$ is an i.i.d. sequence. If $\mathbb{E}[|A_0|] < 1$ and $\mathbb{E}[|B_0|] < \infty$, then $\mathbb{E}[\widetilde{Y}_0] = b_1/(1-a_1)$. Moreover, if $\mathbb{E}[|U|] < \infty$, then $\mathbb{E}[Y_t(U)] \to \mathbb{E}[\widetilde{Y}_0]$ as $t \to \infty$. Similarly, if $\mathbb{E}[A_0^2] < 1$ and $\mathbb{E}[B_0^2] < \infty$, then, by straightforward algebra,

$$\mathrm{Var}(\widetilde{Y}_0) = \frac{2b_1c(1-a_1) + b_2(1-a_1)^2 - (1-a_2)b_1^2}{(1-a_2)(1-a_1)^2}\,,$$

where $c = \mathbb{E}[A_0B_0]$. If $\mathbb{E}[U^2] < \infty$, then $\mathrm{Var}(Y_t(U)) \to \mathrm{Var}(\widetilde{Y}_0)$ as $t \to \infty$.

By combining Theorem 4.1 and Theorem 4.7, we may state conditions upon which (4.11) admits a weak-sense stationary solution.

Theorem 4.10. *Let $\{(A_t, B_t),\ t \in \mathbb{Z}\}$ be an i.i.d. sequence such that $\mathbb{E}[|A_0|^2] < 1$ and $\mathbb{E}[|B_0|^2] < \infty$, then the following hold.*

(i) The stochastic difference equation (4.11) *has a unique nonanticipative weak-sense stationary solution $\{\widetilde{Y}_t,\ t \in \mathbb{Z}\}$, given by* (4.14). *This solution is also stationary in the strict sense.*

(ii) For an arbitrary random variable U satisfying $\mathbb{E}[|U|^2] < \infty$,

$$\lim_{t\to\infty} \|Y_t(U) - \widetilde{Y}_t\|_2 = 0 ,$$

where $Y_t(U)$ is defined by (4.12). *In particular,* $\lim_{t\to\infty} \mathbb{E}[Y_t^2(U)] = \mathbb{E}[\widetilde{Y}_0^2]$.

Proof (of Theorem 4.10). Using Theorem 4.1 and Theorem 4.7, $\{\widetilde{Y}_t,\ t \in \mathbb{Z}\}$ as defined in (4.14), is a strict-sense stationary solution and $\mathbb{E}\widetilde{Y}_0^2 < \infty$. Therefore, it is also a weak-sense stationary solution. Assume now that there exists another nonanticipative weak-sense stationary solution $\{\check{Y}_t,\ t \in \mathbb{Z}\}$. Since, by construction, $\check{Y}_t = A_t\check{Y}_{t-1} + B_t$ and $\widetilde{Y}_t = A_t\widetilde{Y}_{t-1} + B_t$, we get by substracting these two identities, $\check{Y}_t - \widetilde{Y}_t = A_t(\check{Y}_{t-1} - \widetilde{Y}_{t-1})$. Iterating n-times this relation shows that, for any integer n,

$$|\check{Y}_t - \widetilde{Y}_t| = |A_t \dots A_{t-n+1}||\check{Y}_{t-n} - \widetilde{Y}_{t-n}| .$$

Noting that $\{\widetilde{Y}_t,\ t \in \mathbb{Z}\}$ and $\{\check{Y}_t,\ t \in \mathbb{Z}\}$ are both nonanticipative, and using the fact that $\mathbb{E}(X+Y)^2 \le 2[\mathbb{E}(X^2) + \mathbb{E}(Y^2)]$, we take the expectation of the square of the both sides to obtain

$$\begin{aligned}\mathbb{E}[(\check{Y}_t - \widetilde{Y}_t)^2] &= \mathbb{E}[|A_t \dots A_{t-n+1}|^2]\,\mathbb{E}[(\check{Y}_{t-n} - \widetilde{Y}_{t-n})^2] \\ &\le 2\left(\mathbb{E}[A_0^2]\right)^n \left(\mathbb{E}[\check{Y}_0^2] + \mathbb{E}[\widetilde{Y}_0^2]\right) .\end{aligned}$$

Now, by taking $n \to \infty$ in the above, we have that $\mathbb{E}[(\check{Y}_t - \widetilde{Y}_t)^2] = 0$, which implies that $\check{Y}_t = \widetilde{Y}_t$, $\mathbb{P}$-a.s. ∎

We conclude this section by studying the covariance structure of the solutions of (4.11). We suppose that Y_0, A_0 and B_0 have finite second moments and we write $a_1 = \mathbb{E}[A_0]$ and $b_1 = \mathbb{E}[B_0]$. For simplicity, we further assume that the processes $\{A_t,\ t \in \mathbb{Z}\}$ and $\{B_t,\ t \in \mathbb{Z}\}$ are independent. The recursion, for $\tau \ge 0$ and $t \in \mathbb{Z}$,

$$Y_{t+\tau} = B_{t+\tau} + A_{t+\tau}B_{t+\tau-1} + \dots + A_{t+\tau}A_{t+\tau-1}\dots A_{t+2}B_{t+1} + A_{t+\tau}A_{t+\tau-1}\cdots A_{t+1}Y_t$$

implies that

$$\mathbb{E}[Y_{t+\tau}Y_t] = b_1(1-a_1^\tau)(1-a_1)^{-1}\mathbb{E}[Y_t] + a_1^\tau\mathbb{E}[Y_t^2] .$$

If $\mathbb{E}[|A_0|] < 1$ and $Y_t = \widetilde{Y}_t$, the strict-sense stationary solution, we obtain $\mathbb{E}[\widetilde{Y}_0] = b_1/(1-a_1)$, and

$$\mathrm{Cov}(\widetilde{Y}_{t+\tau}, \widetilde{Y}_t) = a_1^\tau\,\mathrm{Var}(\widetilde{Y}_0) . \tag{4.29}$$

Note that since $|a_1| \le \mathbb{E}[|A_0|] < 1$, the covariance decreases exponentially fast with τ.

Example 4.11 (Stationarity of a bilinear model). Consider the model given by

$$X_t = aX_{t-1} + Z_t + bZ_{t-1}X_{t-1} = (a + bZ_{t-1})X_{t-1} + Z_t \,, \tag{4.30}$$

where $\{Z_t,\, t \in \mathbb{Z}\}$ is an i.i.d. sequence with zero mean, variance σ_z^2 and finite fourth moment. It is an AR(1) process with ARCH(l)-type errors. This model is useful for modeling financial time series in which the current volatility depends on the past value, including on its sign. This asymmetry has been pointed out to be a characteristic feature of financial time series. This is a $\mathrm{BL}(1,0,1,1)$ model, but it is also random coefficient autoregression with $A_t = a + bZ_{t-1}$ and $B_t = Z_t$.

We will show that (4.30) admits a weak-sense stationary solution if

$$a^2 + \sigma_z^2 b^2 < 1 \,. \tag{4.31}$$

We may not directly use Theorem 4.10 because when $A_t = a + bZ_{t-1}$ and $B_t = Z_t$, the sequence $\{(A_t, B_t),\, t \in \mathbb{Z}\}$ is not i.i.d. Instead, we use another equivalent representation. Note that

$$X_t = aX_{t-1} + bZ_{t-1}X_{t-1} + Z_t = U_{t-1} + Z_t \,,$$

with $U_t = (a + bZ_t)X_t$. Plugging $X_t = U_{t-1} + Z_t$ into the previous equation yields

$$U_t = (a + bZ_t)U_{t-1} + (a + bZ_t)Z_t \,. \tag{4.32}$$

In this representation, $\{U_t,\, t \in \mathbb{Z}\}$ is a random coefficient autoregressive sequence, with $A_t = a + bZ_t$ and $B_t = (a + bZ_t)Z_t$. In that case, $\{(A_t, B_t),\, t \in \mathbb{Z}\}$ is an i.i.d. sequence and we may apply Theorem 4.10. Consequently, under (4.31), $\mathbb{E}[A_0^2] = a^2 + b^2\sigma_z^2 < 1$ and

$$\mathbb{E}[B_0^2] = a^2\sigma_z^2 + 2ab\mathbb{E}[Z_0^3] + b^2\mathbb{E}[Z_0^4] < \infty \,.$$

so that (4.32) admits a unique nonanticipative weak-sense stationary solution. Conversely, assume that (4.30) admits a nonanticipative solution such that $\{(X_t, Z_t),\, t \in \mathbb{Z}\}$ is second-order stationary. Then, $\{U_t,\, t \in \mathbb{Z}\}$ is second-order stationary. From (4.32), we get that $\mathbb{E}[U_t] = a\mathbb{E}[U_{t-1}] + b\sigma_z^2$ and

$$U_t - \mathbb{E}[U_t] = (a + bZ_t)(U_{t-1} - \mathbb{E}[U_{t-1}]) + (a + b\mathbb{E}[U_{t-1}])Z_t + b(Z^2(t) - \sigma_z^2) \,.$$

Therefore, the stationarity implies that

$$\mathbb{E}[U_t] = \frac{b\sigma_z^2}{1-a}$$

and that

$$\mathrm{Var}(U_t) = (a^2 + b^2\sigma_z^2)\,\mathrm{Var}(U_t) + \mathrm{Var}\left(bZ_t^2 + \left[a + \frac{b^2\sigma_z^2}{1-a}\right]Z_t\right) \,.$$

Now, assume that with probability 1, Z_t^2 is not a linear combination of Z_t. Then $(1 - a^2 - b^2\sigma_z^2)\,\mathrm{Var}(U_t) > 0$, which implies that $1 - a^2 - b^2\sigma_z^2 > 0$ since $\mathrm{Var}(U_t) \geq 0$. Note that the condition (4.31) may be weakened: see Exercise 4.10. ◇

4.1.3 GARCH(1, 1)

We now specialize the results obtained in the previous section to the GARCH$(1,1)$ process.

Definition 4.12. *A process $\{X_t,\, t \in \mathbb{Z}\}$ is GARCH$(1,1)$ if it is a solution of the following recursive equations*

$$X_t = \sigma_t \varepsilon_t \tag{4.33}$$

$$\sigma_t^2 = \alpha_0 + \alpha_1 X_{t-1}^2 + \beta_1 \sigma_{t-1}^2 \,, \tag{4.34}$$

where $\{\varepsilon_t,\, t \in \mathbb{Z}\}$ is an i.i.d. sequence of random variables with zero-mean and unit-variance, and the coefficients satisfy $\alpha_0 > 0$, and $\alpha_1, \beta_1 \geq 0$.

As mentioned in Section 3.5, the conditional variance of a GARCH process is not constant. The two equations in Definition 4.12 may be combined as

$$\begin{aligned} \sigma_t^2 &= \alpha_0 + \alpha_1 \sigma_{t-1}^2 \varepsilon_{t-1}^2 + \beta_1 \sigma_{t-1}^2 \\ &= \alpha_0 + (\beta_1 + \alpha_1 \varepsilon_{t-1}^2) \sigma_{t-1}^2 \,. \end{aligned} \tag{4.35}$$

This implies that the conditional variance in a GARCH$(1,1)$ model follows a first order stochastic difference equation; i.e., $\{\sigma_t^2,\, t \in \mathbb{Z}\}$ behaves like a first-order autoregressive process, but with a random autoregressive coefficient.

Before discussing the properties of the model in more detail, in Figure 4.1, we display two simulated GARCH$(1,1)$ processes. The first process has $\alpha_0 = 0.3, \alpha_1 = 0.1$ and $\beta_1 = 0.85$; the second process is nearly an IGARCH, with $\alpha_0 = 0.3, \alpha_1 = 0.1$ and $\beta_1 = 0.899$; i.e., $\alpha_1 + \beta_1 \approx 1$. Although the parameters in both models are nearly the same, there is a remarkable difference between the two realizations. Also, it is worthwhile to compare the simulated realizations with the S&P 500 returns shown in Figure 3.5 and seismic trace of an explosion displayed in Figure 3.6.

GARCH$(1,1)$: Strict stationarity

To ease the notation in this section, we define the function $a : \mathbb{R} \to \mathbb{R}^+$ to be

$$a(z) = \beta_1 + \alpha_1 z^2 \,. \tag{4.36}$$

Theorem 4.13 (Strict-sense stationarity of a GARCH$(1,1)$ model). *Assume that*

$$-\infty \leq \gamma := \mathbb{E}[\ln(a(\varepsilon_0))] < 0 \,.$$

Then, the series $\sum_{n=1}^{\infty} \prod_{j=1}^{n} a(\varepsilon_{t-j})$ converges with probability one and the process

$$\tilde{X}_t = \tilde{\sigma}_t \varepsilon_t \tag{4.37}$$

$$\tilde{\sigma}_t^2 = \alpha_0 \left[1 + \sum_{n=1}^{\infty} \prod_{j=1}^{n} a(\varepsilon_{t-j}) \right] , \tag{4.38}$$

is the unique (strict-sense) stationary solution of (4.33) *and* (4.34)*. If $\gamma \geq 0$, the equations* (4.33) *and* (4.34) *do not admit any stationary solution.*

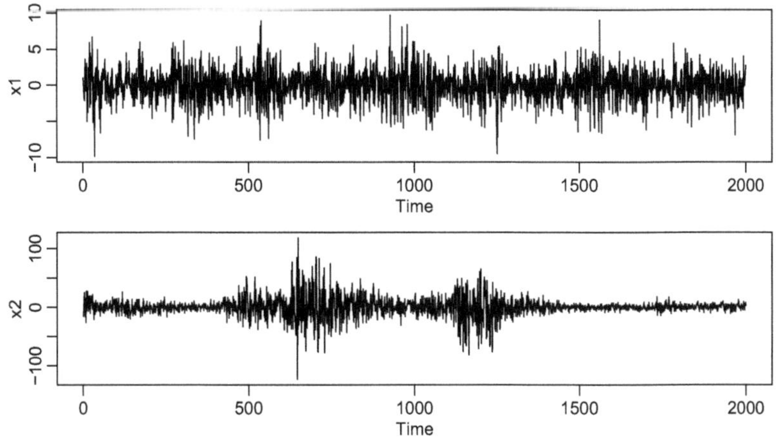

Figure 4.1 *Realizations of GARCH*$(1,1)$ *processes. In each case,* $n = 2000$ *and the innovations,* $\{\varepsilon_t,\ t \in \mathbb{Z}\}$, *are standard normal. The parameters are:* $\alpha_0 = .3, \alpha_1 = .1$ *for both plots, and* $\beta_1 = .85$ *(top),* $\beta_1 = .899$ *(bottom).*

Proof. Assume that $\gamma < 0$. The existence of a unique stationary solution is a simple application of Theorem 4.1. From (4.35) we have

$$\sigma_t^2 = \alpha_0 + (\beta_1 + \alpha_1 \varepsilon_{t-1}^2)\sigma_{t-1}^2 \,.$$

Therefore, the volatility process follows a stochastic first-order difference equation of the form (4.11), with

$$A_t = \beta_1 + \alpha_1 \varepsilon_{t-1}^2 = a(\varepsilon_{t-1}) \quad \text{and} \quad B_t = \alpha_0 \,,$$

where $a(\varepsilon) = \beta_1 + \alpha_1 \varepsilon^2$. Under the stated condition, (4.13) is satisfied.

For $\gamma \geq 0$, Theorem 4.6 shows that there is no strict-sense stationary solutions. ■

Remark 4.14. When $\alpha_0 = 0$ and $\gamma := \mathbb{E}[\ln(\beta_1 + \alpha_1 Z_0^2)] < 0$, (4.38) shows that the unique strictly stationary solution is the degenerate process $X_t = 0$, for all $t \in \mathbb{Z}$.

Remark 4.15. To establish the result of Theorem 4.13, we did not have to assume that $\mathbb{E}[\varepsilon_t] = 0$ or $\mathbb{E}[\varepsilon_t^2] = 1$, or even that $\mathbb{E}[|\varepsilon_t|]$ is finite. Assuming that $\mathbb{E}[\varepsilon_t] = 0$ and $\mathbb{E}[\varepsilon_t^2] = 1$ makes the interpretation of the model easier, but this assumption is not needed to derive the result. We may find strict-sense stationary solutions even if $\mathbb{E}[|\varepsilon_t|] = \infty$.

Remark 4.16. For a GARCH$(1,1)$, assume that $\mathbb{E}[\varepsilon_0^2] = 1$. Then, using $\ln(u) \leq u - 1$, we obtain

$$\mathbb{E}[\ln(\beta_1 + \alpha_1 \varepsilon_0^2)] \leq \mathbb{E}[\beta_1 + \alpha_1 \varepsilon_0^2 - 1] = \beta_1 + \alpha_1 - 1 \,,$$

showing that a sufficient condition for the existence of an invariant distribution is $\alpha_1 + \beta_1 < 1$.

Remark 4.17. Suppose that ε_0 is standard normally distributed. Assume first that $\beta_1 = 0$. Then

$$\mathbb{E}[\ln(\alpha_1 \varepsilon_0^2)] = \ln \alpha_1 + \frac{4}{\sqrt{2\pi}} \int_0^\infty \ln(x) \mathrm{e}^{-x^2/2} \mathrm{d}x = \ln(\alpha_1) - (\gamma + \ln(2)),$$

where here, $\gamma = \lim_{N\to\infty} \sum_{n=1}^N \frac{1}{n} - \ln(N) \approx 0.577$ is the Euler-Mascheroni constant. Hence, the ARCH(1) process with standard normal noise has a strictly stationary solution if and only if

$$\alpha_1 < 2\exp(\gamma) \approx 3.562.$$

Since $\lim_{\beta_1 \downarrow 0} \mathbb{E}[\ln(\beta_1 + \alpha_1 \varepsilon_0^2)] = \mathbb{E}[\ln(\alpha_1\ \varepsilon_0^2)]$, it follows that for all $\alpha_1 < 2\exp(\gamma)$, one may find $\overline{\beta_1} > 0$ (depending on α_1) such that the GARCH$(1,1)$ process with parameters α_0, α_1 and $\beta_1 \in \left[0,\ \overline{\beta}_1\right]$ and standard normal innovations has a unique invariant distribution. It is worthwhile to note that one may find GARCH$(1,1)$ process admitting a unique invariant distribution with $\alpha_1 + \beta_1 > 1$. Note that, while α_1 may be bigger than 1, $\beta_1 < 1$ is a necessary condition for the existence of a stationary distribution.

The quantity $\mathbb{E}[\ln(\beta_1 + \alpha_1 \varepsilon_0^2)]$ can be expressed in terms of confluent hypergeometric functions, which can be evaluated numerically.

GARCH$(1,1)$: Weak stationarity

For an ARMA process, when the innovation is a sequence of i.i.d. random variables (strong white noise) with zero mean and finite variance, the conditions guaranteeing the existence of a nonanticipative strict-sense stationary solution and second-order stationary solution are the same. For the GARCH$(1,1)$, the interplay between a strict-sense stationary solution and a second-order stationary solution of the recurrence equation is a bit trickier. The existence of a second-order stationary solution requires more stringent constraints on the parameters $\alpha_0, \alpha_1, \beta_1$. This is discussed in the following theorem.

Theorem 4.18 (Weak stationarity of GARCH$(1,1)$). *Assume that $\alpha_0 > 0$ and that $\{\varepsilon_t,\ t \in \mathbb{Z}\}$ is an i.i.d. sequence with $\mathbb{E}[\varepsilon_0] = 0$ and $\mathbb{E}[\varepsilon_0^2] = 1$. Then,*

(i) if $\alpha_1 + \beta_1 < 1$, the process $\{(\tilde{X}_t, \tilde{\sigma}_t),\ t \in \mathbb{Z}\}$ defined by (4.37)-(4.38) *is the unique second-order stationary solution of* (4.33)-(4.34). *The process $\{\tilde{X}_t,\ t \in \mathbb{Z}\}$ is weak (second-order) white noise;*

(ii) if $\alpha_1 + \beta_1 \geq 1$, there does not exist a nonanticipative and second-order stationary solution to the GARCH$(1,1)$ recursions given by (4.33) *and* (4.34).

Proof. Assume first that $\{(X_t, \sigma_t),\ t \in \mathbb{Z}\}$ is a nonanticipative stationary solution of the recursive equations (4.33)-(4.34), i.e., for all $t \in \mathbb{Z}$, ε_t is independent of $\{\sigma_s^2,\ s \leq t\}$ (the current volatility is independent of future shocks). Then,

$$\mathbb{E}[X_t^2] \stackrel{(1)}{=} \mathbb{E}[\sigma_t^2]\mathbb{E}[\varepsilon_t^2] = \alpha_0 + \alpha_1 \mathbb{E}[X_{t-1}^2] + \beta_1 \mathbb{E}[\sigma_{t-1}^2] \stackrel{(2)}{=} \alpha_0 + (\alpha_1 + \beta_1)\mathbb{E}[X_{t-1}^2],$$

where (1) follows from the fact that the process is nonanticipative, and (2) follows from

$$\mathbb{E}[X_{t-1}^2] = \mathbb{E}\left[\mathbb{E}\left[X_{t-1}^2 \mid \varepsilon_s,\, s < t-1\right]\right] = \mathbb{E}[\sigma_{t-1}^2]\,.$$

Therefore,

$$(1 - \alpha_1 - \beta_1)\mathbb{E}[X_t^2] = \alpha_0\,, \tag{4.39}$$

showing that a necessary solution for such a second-order stationary and nonanticipative solution to exist is $\alpha_1 + \beta_1 < 1$. Note that

$$\mathbb{E}[X_t^2] = \alpha_0/(1 - \alpha_1 - \beta_1) > 0\,. \tag{4.40}$$

Conversely, assume that $\alpha_1 + \beta_1 < 1$. Then, by Jensen's inequality, we get that $\mathbb{E}[\ln(\alpha_1 + \beta_1 \varepsilon_0^2)] \le \ln(\alpha_1 + \beta_1) < 0$. We know from Theorem 4.13 that $\{(\tilde{X}_t, \tilde{\sigma}_t),\, t \in \mathbb{Z}\}$ defined by (4.37)-(4.38) is the unique strict-sense stationary solution to (4.33)-(4.34). To obtain that $\{\tilde{X}_t,\, t \in \mathbb{Z}\}$ is a second-order stationary solution, it then suffices to prove that $\tilde{X}_0$, has a finite variance. Since $\tilde{\sigma}_t^2$ can be expressed as a series of non-negative random variables, the series version of the monotone convergence theorem Billingsley (1995, Theorem 16.6) implies that (putting $a(z) = \alpha_1 + \beta_1 z^2$)

$$\begin{aligned}
\mathbb{E}[\tilde{X}_t^2] = \mathbb{E}[\tilde{\sigma}_t^2] &= \left(1 + \sum_{j=1}^{+\infty} \mathbb{E}\left[\prod_{i=t-j}^{t-1} a(\varepsilon_i)\right]\right)\alpha_0 \\
&= \left(1 + \sum_{j=1}^{+\infty} \{\mathbb{E}[a(\varepsilon_0)]\}^j\right)\alpha_0 \\
&= \left(1 + \sum_{j=1}^{+\infty} (\alpha_1 + \beta_1)^j\right)\alpha_0 = \frac{\alpha_0}{1 - (\alpha_1 + \beta_1)}\,.
\end{aligned}$$

This proves that $\mathbb{E}[\tilde{X}_t^2] < \infty$ and since $\{\tilde{X}_t,\, t \in \mathbb{Z}\}$ is strict-sense stationary, finally $\{\tilde{X}_t,\, t \in \mathbb{Z}\}$ is second-order stationary. Since the process is nonanticipative, ε_t is independent of $\mathcal{F}_{t-1}^X$, which implies that the process $\{\tilde{X}_t,\, t \in \mathbb{Z}\}$ is a martingale increment sequence and since $\mathbb{E}[\tilde{X}_t^2] < \infty$, $\{\tilde{X}_t,\, t \in \mathbb{Z}\}$ is finally a weak (second order) white noise.

Finally, it remains to show the unicity of the second-order stationary solution. Denote by $\check{X}_t = \check{\sigma}_t \varepsilon_t$ another second-order and nonanticipative stationary solution. We have

$$|\tilde{\sigma}_t^2 - \check{\sigma}_t^2| = a(\varepsilon_{t-1}) \dots\, a(\varepsilon_{t-n})|\tilde{\sigma}_{t-n-1}^2 - \check{\sigma}_{t-n-1}^2|\,,$$

which implies, since both $\{\tilde{\sigma}_t,\, t \in \mathbb{Z}\}$ and $\{\check{\sigma}_t,\, t \in \mathbb{Z}\}$ are nonanticipative,

$$\begin{aligned}
\mathbb{E}[|\tilde{\sigma}_t^2 - \check{\sigma}_t^2|] &= \mathbb{E}\left[a(\varepsilon_{t-1}) \dots a(\varepsilon_{t-n})\right]\mathbb{E}[|\tilde{\sigma}_{t-n-1}^2 - \check{\sigma}_{t-n-1}^2|] \\
&= (\alpha_1 + \beta_1)^n \mathbb{E}[|\tilde{\sigma}_{t-n-1}^2 - \check{\sigma}_{t-n-1}^2|]\,.
\end{aligned}$$

Since $\mathbb{E}[|\tilde{\sigma}_{t-n-1}^2 - \check{\sigma}_{t-n-1}^2|] \le \mathbb{E}[\tilde{\sigma}_0{}^2] + \mathbb{E}[\check{\sigma}_0{}^2] < \infty$ and since $\lim_{n\to\infty}(\alpha_1 + \beta_1)^n = 0$, we get that $\mathbb{E}[|\tilde{\sigma}_t^2 - \check{\sigma}_t^2|] = 0$. ■

Remark 4.19. It is interesting to note that if $\mathbb{E}[\ln(\alpha_1+\beta_1\varepsilon_0^2)]<0$ and $\alpha_1+\beta_1\geq 1$, there exists a strict-sense stationary solution of the GARCH$(1,1)$ process but this solution has an infinite variance. The interest in financial econometric for infinite variance models was pointed out in an early paper of Mandelbrot (1963), who advocates the use of stable laws to model the absolute return of financial assets; see Section 3.8.

Assume now that $\alpha_1+\beta_1<1$. Set $Z_t=X_t^2-\sigma_t^2=\sigma_t^2(\varepsilon_t^2-1)$. Note that $\mathbb{E}[|Z_t|]=\mathbb{E}[\sigma_t^2]\mathbb{E}[|\varepsilon_t^2-1|]<\infty$ and, because the second order stationary solution is nonanticipative, $\{(Z_t,\mathcal{F}_t^\varepsilon),\ t\in\mathbb{Z}\}$ is a martingale increment sequence. Then,

$$X_t^2=\sigma_t^2\varepsilon_t^2=\sigma_t^2+Z_t=\alpha_0+\alpha_1X_{t-1}^2+\beta_1(X_{t-1}^2-Z_{t-1})+Z_t\,,$$

which shows that, as discussed in Section 3.5, the square of a GARCH$(1,1)$ process $\{X_t^2,\ t\in\mathbb{Z}\}$ can be interpreted as a non-Gaussian ARMA(1,1) process,

$$X_t^2-(\alpha_1+\beta_1)X_{t-1}^2=\alpha_0+Z_t-\beta_1Z_{t-1}\,. \tag{4.41}$$

This representation allows us to construct forecasts.

It is also interesting to note that, contrary to strong sense linear processes, the coefficients α_1 and β_1 control not only the dependence of the squared process, but also of the marginal distributions.

Proposition 4.20. *Let m be a positive integer. Assume that*

$$M_m(\alpha_1,\beta_1,\varepsilon):=\mathbb{E}[a^m(\varepsilon_0)]=\sum_{i=0}^{m}\binom{m}{i}\alpha_1^i\beta_1^{m-i}\mathbb{E}[\varepsilon_0^{2i}]<1\,. \tag{4.42}$$

Then, the process $\{\tilde{X}_t,\ t\in\mathbb{Z}\}$ defined by (4.37)–(4.38) *is a strict-sense stationary solution of* (4.33)–(4.34) *and $\mathbb{E}[\tilde{X}_0^{2m}]<\infty$. Conversely, if $M_m(\alpha_1,\beta_1,\varepsilon)\geq 1$, there is no strictly stationary solution of* (4.33)-(4.34).

Proof. Recall the notation $\|\tilde{X}\|_m=(\mathbb{E}[\tilde{X}^m])^{1/m}$. Note first that, since $\{\varepsilon_t,\ t\in\mathbb{Z}\}$ is a strong white noise,

$$\sum_{n=1}^{\infty}\|\prod_{j=1}^{n}a(\varepsilon_{t-j})\|_m=\sum_{n=1}^{\infty}\prod_{j=1}^{n}\|a(\varepsilon_{t-j})\|_m=\sum_{n=1}^{\infty}\|a(\varepsilon_0)\|_m^n\,.$$

By the Cauchy criterion, this series converges if $\|a(\varepsilon_0)\|_m<1$, which follows from (4.42).

Conversely, assume that $\|\tilde{X}_t^2\|_m<\infty$, which implies $\mathbb{E}[\tilde{X}_t^{2m}]=\mathbb{E}[\varepsilon_t^{2m}]\mathbb{E}[\tilde{\sigma}_t^{2m}]<\infty$.

For any $n \in \mathbb{N}$, we get

$$\begin{aligned}
\mathbb{E}[\tilde{\sigma}_t^{2m}] &= \mathbb{E}\left(\alpha_0 \sum_{j=0}^{n} \prod_{i=t-j+1}^{t} a(\varepsilon_i) + \alpha_0 \prod_{i=t-n}^{t} a(\varepsilon_i)\right)^m \\
&\geq \mathbb{E}\left(\alpha_0 \sum_{j=0}^{n} \prod_{i=t-j+1}^{t} a(\varepsilon_i)\right)^m \\
&\geq \alpha_0^m \sum_{j=0}^{n} \mathbb{E}\left[\prod_{i=t-j+1}^{t} a^m(\varepsilon_i)\right] \\
&= \alpha_0^m \sum_{j=0}^{n} \left(\mathbb{E}\left[a^m(\varepsilon_0)\right]\right)^j .
\end{aligned}$$

The sum on the right-hand side of the previous relation is finite only if $\mathbb{E}[a^m(\varepsilon_0)] < 1$. ■

When (4.42) is satisfied, it is possible to compute the moments recursively, by noting that, for nonanticipative solutions to (4.33)-(4.34), $\mathbb{E}[\tilde{X}_t^{2m}] = \mathbb{E}[\tilde{\sigma}_t^{2m}]\mathbb{E}[\varepsilon_t^{2m}]$ and then by expanding the relation

$$\mathbb{E}[\tilde{\sigma}_t^{2m}] = \mathbb{E}[(\alpha_0 + \alpha_1 \tilde{X}_{t-1}^2 + \beta_1 \tilde{\sigma}_{t-1}^2)^m] .$$

For the fourth-order moment, assuming $\mathbb{E}[\varepsilon_0^2] = 1$, a direct computation gives $\mathbb{E}[\tilde{X}_t^2] = \mathbb{E}[\tilde{\sigma}_t^2]$, $\mathbb{E}[\tilde{X}_t^4] = \mathbb{E}[\tilde{\sigma}_t^4]\mathbb{E}[\varepsilon_t^4]$, and

$$\mathbb{E}[\tilde{\sigma}_t^4] = \alpha_0^2 + 2\alpha_0(\alpha_1 + \beta_1)\mathbb{E}[\tilde{\sigma}_{t-1}^2] + (\beta_1^2 + 2\alpha_1\beta_1)\mathbb{E}[\tilde{\sigma}_{t-1}^4] + \alpha_1^2\mathbb{E}[\tilde{X}_{t-1}^4] .$$

Hence, setting $\mu_4 = \mathbb{E}[\varepsilon_0^4]$ and recalling that $\mathbb{E}[\tilde{\sigma}_{t-1}^2] = \alpha_0/[1-(\alpha_1+\beta_1)]$, we then get

$$\mathbb{E}[\tilde{X}_t^4] = \frac{\alpha_0^2(1+\alpha_1+\beta_1)}{(1-\mu_4\alpha_1^2-\beta_1^2-2\alpha_1\beta_1)(1-\alpha_1-\beta_1)}\mu_4 ,$$

provided that the denominator is positive.

4.2 Random coefficient autoregression: The vector case

A vector-valued stochastic difference equation is defined as

$$\boldsymbol{Y}_{t+1} = A_t \boldsymbol{Y}_t + \boldsymbol{B}_t , \quad t \in \mathbb{Z} , \tag{4.43}$$

where $\{(A_t, \boldsymbol{B}_t),\ t \in \mathbb{Z}\}$ is a sequence of i.i.d. random elements defined on $(\Omega, \mathcal{F}, \mathbb{P})$ taking values in $\mathbb{M}_d(\mathbb{R}) \times \mathbb{R}^d$, where $\mathbb{M}_d(\mathbb{R})$ is the set of $d \times d$ matrices with real coefficients.

To specify the conditions under which there exists a strict-sense stationary solution, we need to introduce some additional definitions and recall some important properties of the spectral radius.

Example 4.21 (Bilinear model). Consider the bilinear model BL$(p,0,p,1)$,

$$X_t = \sum_{j=1}^{p} \phi_j X_{t-j} + Z_t + \sum_{j=1}^{p} b_{j,1} X_{t-j} Z_{t-1} , \tag{4.44}$$

where $\{Z_t,\, t \in \mathbb{Z}\}$ is a strong white-noise. Then, let $A_{j,t} = \phi_j + b_{j,1} Z_{t-1}$, and define

$$A_t := \begin{bmatrix} A_{1,t} & A_{2,t} & A_{3,t} & \cdots & A_{p,t} \\ 1 & 0 & 0 & \cdots & 0 \\ 0 & 1 & \ddots & \ddots & 0 \\ \vdots & \ddots & \ddots & \ddots & \vdots \\ 0 & \cdots & \cdots & 1 & 0 \end{bmatrix} , \tag{4.45}$$

and $\boldsymbol{B}_t = [Z_t, 0, \ldots, 0]'$. With this notation, we may write the model (4.44) in the form

$$\boldsymbol{Y}_t = A_t \boldsymbol{Y}_{t-1} + \boldsymbol{B}_t \tag{4.46}$$

$$X_t = [1, 0, \ldots, 0]\, \boldsymbol{Y}_t , \tag{4.47}$$

where $\boldsymbol{Y}_t$ is the lag-vector $\boldsymbol{Y}_t = [X_t, X_{t-1}, \ldots, X_{t-p+1}]'$. Such a representation is sometimes referred to as the vector form of the bilinear model BL$(p,0,p,1)$; see Subba Rao and Gabr (1984, p. 149). ◇

Example 4.22 (DAR(p) model; Example 4.5 cont.). The DAR(p) model is given by the following stochastic difference equation,

$$X_t = V_t X_{t-D_t} + (1 - V_t) Z_t , \quad t \geq p , \tag{4.48}$$

where $\{V_t,\, t \in \mathbb{N}\}$ and $\{Z_t,\, t \in \mathbb{N}\}$ are defined as before and the delay $\{D_t,\, t \in \mathbb{N}\}$ are i.i.d. random variables independent of $\{V_t,\, t \in \mathbb{N}\}$ and $\{Z_t,\, t \in \mathbb{N}\}$ defined on the set $\{1, 2, \ldots,\, p\}$ with probability $\mathbb{P}(D_t = k) = \phi_k$, $k \in \{1, \ldots, p\}$ where $\phi_k \geq 0$ for any $k \in \{1, \ldots, p\}$ and $\sum_{k=1}^{p} \phi_k = 1$. The initial random variables $(X_0, X_1, \ldots, X_{p-1})$ are distributed according to ξ on $(\mathsf{X}^p, \mathcal{X}^{\otimes p})$. Thus, the current value is either one of the last p observed values, chosen stochastically, or an independent choice Z_t.

To express $\{X_t,\, t \in \mathbb{N}\}$ in the form of (4.43), we simply introduce the lag-vector

$$\boldsymbol{Y}_t = \begin{bmatrix} X_t \\ X_{t-1} \\ \vdots \\ X_{t-p+1} \end{bmatrix} , \tag{4.49}$$

and rewrite (4.48) as a linear order 1 vector recurrence equation $\boldsymbol{Y}_t = A_t \boldsymbol{Y}_{t-1} + \boldsymbol{B}_t$, where, with $A_{j,t} = V_t \mathbb{1}_{\{j\}}(D_t)$ for $j = 1, \ldots, p$, we define

$$A_t := \begin{bmatrix} A_{1,t} & A_{2,t} & A_{3,t} & \cdots & A_{p,t} \\ 1 & 0 & 0 & \cdots & 0 \\ 0 & 1 & \ddots & \ddots & 0 \\ \vdots & \ddots & \ddots & \ddots & \vdots \\ 0 & \cdots & \cdots & 1 & 0 \end{bmatrix} , \quad \boldsymbol{B}_t := \begin{bmatrix} B_t \\ 0 \\ \vdots \\ 0 \end{bmatrix} . \tag{4.50}$$

◇

4.2.1 *Strict stationarity*

Definition 4.23. *The* spectral radius $\rho(A)$ *of a matrix* $A \in \mathbb{M}_d(\mathbb{R})$ *is*

$$\rho(A) := \max\{|\lambda| : \lambda \text{ is an eigenvalue of } A\} . \tag{4.51}$$

Observe that if λ is any eigenvalue of A, then $|\lambda| \le \rho(A)$; moreover, there is at least one eigenvalue λ for which $|\lambda| = \rho(A)$. If $A\boldsymbol{x} = \lambda \boldsymbol{x}$, for $\boldsymbol{x} \neq 0$, and if $|\lambda| = \rho(A)$, consider the matrix $X \in \mathbb{M}_d(\mathbb{R})$ for which all of the columns are equal to the eigenvector $\boldsymbol{x}$, and observe that $AX = \lambda X$. Let $|||\cdot|||$ be a *matrix norm*. Recall that a matrix norm is a norm on the vector-space of matrices $\mathbb{M}_d(\mathbb{R})$, which is submultiplicative, i.e., for any $A, B \in \mathbb{M}_d(\mathbb{R})$, $|||AB||| \le |||A|||\ |||B|||$. Therefore, we get

$$|\lambda|\,|||X||| = |||\lambda X||| = |||AX||| \le |||A|||\,|||X||| ,$$

which implies that $|\lambda| = \rho(A) \le |||A|||$. Although the spectral radius ρ is not itself a matrix or vector norm on $\mathbb{M}_d(\mathbb{R})$, for each fixed $A \in \mathbb{M}_d(\mathbb{R})$, it is the greatest lower bound for the values of all matrix norms of A. We summarize the discussion above in the following proposition; see Horn and Johnson (1985, Theorem 5.6.9 and Lemma 5.6.10).

Proposition 4.24. *If* $|||\cdot|||$ *is any matrix norm and if* $A \in \mathbb{M}_d(\mathbb{R})$*, then,*

(i) $\rho(A) \le |||A|||$;

(ii) *for any* $\varepsilon > 0$, *there is a matrix norm such that* $\rho(A) \le |||A||| \le \rho(A) + \varepsilon$.

As a direct consequence of the preceding result, we may write $\rho(A)$ as the infimum of $|||A|||$ over the set of all matrix norms on $\mathbb{M}_d(\mathbb{R})$.

Lemma 4.25. *Let* $|||\cdot|||$ *be a matrix norm on* $\mathbb{M}_d(\mathbb{R})$*. Then,*

$$\rho(A) = \lim_{k\to\infty} |||A^k|||^{1/k} , \tag{4.52}$$

Proof. Since $\rho(A)^k = \rho(A^k) \le |||A^k|||$, we have that $\rho(A) \le |||A^k|||^{1/k}$ for all $k \in \mathbb{N}$. If $\varepsilon > 0$ is given, the matrix $\tilde{A} \equiv [\rho(A)+\varepsilon]^{-1}A$ has spectral radius strictly less than 1 and hence it is convergent (see Exercise 4.4–Exercise 4.5). Thus, $|||\tilde{A}^k||| \to 0$ as $k \to \infty$ and there is some $N = N(\varepsilon,\, A)$ such that $|||\tilde{A}^k||| < 1$ for all $k \ge N$. This is just the statement that $|||A^k|||^{1/k} \le [\rho(A)+\varepsilon]^k$ for all $k \ge N$, or that $|||A^k|||^{1/k} \le \rho(A)+\varepsilon$ for all $k \ge N$. Since $\rho(A) \le |||A^k|||^{1/k}$ for all k and since $\varepsilon > 0$ is arbitrary, we conclude that $\lim_{k\to\infty} |||A^k|||^{1/k}$ exists and equals $\rho(A)$. ∎

We now generalize the concept of spectral radius to sequences of matrices.

Theorem 4.26 (Top Lyapunov exponent). *Let* $\{A_t,\, t \in \mathbb{Z}\}$ *be a strictly stationary and ergodic sequence of random matrices such that* $\mathbb{E}[\ln^+(|||A_0|||)] < \infty$, *where* $|||\cdot|||$ *is any matrix norm. Then,*

$$\gamma = \lim_{k\to\infty} \frac{1}{k}\mathbb{E}[\ln |||A_k A_{k-1} \ldots A_1|||] = \inf_{k\in\mathbb{N}^*} \frac{1}{k}\mathbb{E}[\ln |||A_k A_{k-1} \ldots A_1|||] . \tag{4.53}$$

Moreover, for all $t \in \mathbb{Z}$,

$$\frac{1}{k} \ln |||A_t A_{t-1} \dots A_{t-k}||| \xrightarrow{\mathbb{P}\text{-a.s.}} \gamma , \tag{4.54}$$

as $k \to \infty$.

Proof. The proof is a simple consequence of the Kingman subadditive theorem Theorem 7.45; see Exercise 7.21. We will have to wait a little to fully appreciate this result, and we will take it for granted. ∎

The coefficient γ is referred to as the *top-Lyapunov exponent* of the sequence of matrices $\{A_t,\ t \in \mathbb{Z}\}$. An inspection of the proof of Theorem 4.1 reveals that the scalar assumption can easily be lifted.

Theorem 4.27. *Let $|||\cdot|||$ be any matrix norm. Let $\{(A_t, \boldsymbol{B}_t),\ t \in \mathbb{Z}\}$ be a stationary and ergodic sequence, where $A_t \in \mathbb{M}_d(\mathbb{R})$ and $\boldsymbol{B}_t$ is a $d \times 1$ vector. Assume that*

$$\mathbb{E}[\ln^+ |||A_0|||] < \infty \quad \text{and} \quad \mathbb{E}[\ln^+ \|\boldsymbol{B}_0\|] < \infty , \tag{4.55}$$

where $x^+ = \max(0,\ x)$ for $x \in \mathbb{R}$. Assume in addition that the top-Lyapunov exponent γ, given in (4.53), is strictly negative: $\gamma < 0$. Then, for any $t \in \mathbb{Z}$,

$$\widetilde{\boldsymbol{Y}}_t = \sum_{j=0}^{\infty} \left(\prod_{i=t-j+1}^{t} A_i \right) \boldsymbol{B}_{t-j} , \tag{4.56}$$

where $\prod_{i=t-j+1}^{t} A_i = A_t A_{t-1} \dots A_{t-j+1}$ if $j \geq 1$ and $\prod_{i=t+1}^{t} A_i = \mathrm{I}_d$ is the only strict-sense stationary solution of (4.43).

The series on the right-hand side of (4.56) is normally convergent with probability 1. Furthermore, for an arbitrary random variable $\boldsymbol{U}$ (defined on the same basic probability space as $\{(A_t, B_t),\ t \in \mathbb{Z}\}$),

$$|||\boldsymbol{Y}_t(\boldsymbol{U}) - \widetilde{\boldsymbol{Y}}_t||| \xrightarrow{\mathbb{P}\text{-a.s.}} 0 , \tag{4.57}$$

as $t \to \infty$, where

$$\boldsymbol{Y}_t(\boldsymbol{U}) = \sum_{j=0}^{t-1} \left(\prod_{i=t-j+1}^{t} A_i \right) \boldsymbol{B}_{t-1} + \left(\prod_{i=0}^{t} A_i \right) \boldsymbol{U} . \tag{4.58}$$

In particular, as $t \to \infty$, the distribution of the random variable $\boldsymbol{Y}_t(\boldsymbol{U})$ converges in distribution to $\tilde{\boldsymbol{Y}}_0$, i.e.,

$$\boldsymbol{Y}_t(\boldsymbol{U}) \xRightarrow{\mathbb{P}} \widetilde{\boldsymbol{Y}}_0 . \tag{4.59}$$

Proof. The proof is similar to that of Theorem 4.1; the elementary adaptations are left to the reader. ∎

The stability of AR(p) processes follows immediately from Theorem 4.27 and the representation (4.5)–(4.6). The characteristic polynomial of the state transition matrix $\boldsymbol{\Phi}$ given in (4.50) is equal to $z^p \phi(z^{-1})$. Hence, the eigenvalues of the companion matrix $\boldsymbol{\Phi}$ are equal to the reciprocals of the zeros of the polynomial ϕ. The autoregressive process is said to be *causal* if $\phi(z) \neq 0$ for $|z| \leq 1$. This is equivalent to the condition that the spectral radius of $\boldsymbol{\Phi}$ is strictly less than one. Therefore, we get the following result.

Theorem 4.28. *Let $\{X_t,\ t \in \mathbb{Z}\}$ be an autoregressive process of order p, $X_t = \phi_1 X_{t-1} + \cdots + \phi_p X_{t-p} + Z_t$ where $\{Z_t,\ t \in \mathbb{Z}\}$ is a stationary ergodic sequence. Assume that $\phi(z) \neq 0$ for $|z| \leq 1$, where $\phi(z) = 1 - \phi_1 z - \cdots - \phi_p z^p$ and $\mathbb{E}[\ln^+ |Z_0|] < \infty$. Then, with $\boldsymbol{Y}_t$ and $\boldsymbol{B}_t$ defined in* (4.5) *and* (4.7)*, respectively,*

$$\boldsymbol{Y}_t = \sum_{j=0}^{\infty} \Phi^j \boldsymbol{B}_{t-1-j}\,, \quad t \in \mathbb{Z}\,, \tag{4.60}$$

is the only strict-sense stationary solution of (4.6).

Similar to the scalar case, an obvious question to ask is whether the top-Lyapunov condition is minimal. To do this, we assume that, as in the scalar case, $\{(A_t, \boldsymbol{B}_t),\ t \in \mathbb{Z}\}$ is i.i.d. Note that the situation is more complicated than in the scalar setting. We will state the main result and refer to Bougerol and Picard (1992) for the proof.

An affine subspace H of $\mathbb{R}^d$ is a translation, $z + V$, of a linear subspace V. An affine subspace H of $\mathbb{R}^d$ is said to be *invariant* under the model (4.43) if $\{A_0 \boldsymbol{x} + \boldsymbol{B}_0; \boldsymbol{x} \in H\}$ is included in H almost surely.

Theorem 4.29. *Consider the random coefficient autoregressive model* (4.43) *with i.i.d. coefficients and that $\mathbb{R}^d$ is the only affine invariant subspace.*

Assume that the process has a nonanticipative strictly stationary solution $\{\widetilde{\boldsymbol{Y}}_t,\ t \in \mathbb{Z}\}$ (i.e., for all $t \in \mathbb{Z}$, X_t is independent of $(A_s, \boldsymbol{B}_s)$ for $s \geq t$). Then, the following holds:

(i) $A_0 A_{-1} \cdots A_{-h}$ converges to 0 $\mathbb{P}$-a.s. as $h \to \infty$.

(ii) For any $t \in \mathbb{Z}$,

$$\widetilde{\boldsymbol{Y}}_t = \sum_{j=0}^{\infty} A_t A_{t-1} \cdots A_{t-j+1} \boldsymbol{B}_{t-j}\,, \tag{4.61}$$

where the series on the right-hand side converges absolutely almost surely.

(iii) This solution is the unique strictly stationary solution of (4.43).

Assume that $\mathbb{E}[\ln^+ \|A_0\|] < \infty$ and $\mathbb{E}[\ln^+ \|B_0\|] < \infty$. Then (4.43) *has a nonanticipative strictly stationary solution if and only if the top Lyapunov exponent γ, given in* (4.53) *is strictly negative.*

Proof. See Bougerol and Picard (1992, Theorem 2.4). ∎

4.2.2 Weak stationarity

We now consider conditions upon which (4.43) admits moments of order $p \geq 1$. In this case, it is necessary to strengthen the condition on the spectral radius (or equivalently, the top Lyapunov coefficient). For simplicity, we assume that $\{(A_t, \boldsymbol{B}_t),\ t \in \mathbb{Z}\}$ is an i.i.d. sequence. For $p \geq 0$, we define the p-th norm Lyapunov coefficient as

$$\gamma_p := \lim_{k \to \infty} \frac{1}{k} \ln \mathbb{E}^{1/p}[|||A_k \ldots A_1|||^p]\,. \tag{4.62}$$

Theorem 4.30. *Let $p \geq 1$ and assume that $\{(A_t, \boldsymbol{B}_t),\ t \in \mathbb{Z}\}$ is an i.i.d. sequence such that $\gamma_p < 0$ and $\|\boldsymbol{B}_0\|_p < \infty$. Then, the following hold.*

- *(i) The stochastic difference equation* (4.43) *has a unique strict-sense stationary solution $\{\widetilde{\boldsymbol{Y}}_t,\ t \in \mathbb{Z}\}$, given by* (4.56)*, and this solution satisfies $\|\widetilde{\boldsymbol{Y}}_0\|_p < \infty$. In addition, the series in the right-hand-side of* (4.56) *converges in p-th norm.*
- *(ii) Let $\boldsymbol{U}$ be a random variable such that $\|\boldsymbol{U}\|_p < \infty$. For any $t \in \mathbb{N}$, $\|\boldsymbol{Y}_t(\boldsymbol{U})\|_p < \infty$ where $\{\boldsymbol{Y}_t(\boldsymbol{U}), t \in \mathbb{N}\}$ is the solution of the stochastic difference equation* (4.43) *started at $t = 0$ from $\boldsymbol{U}$ defined by* (4.12)*. In addition, the sequence $\{\boldsymbol{Y}_t(\boldsymbol{U}), t \in \mathbb{N}\}$ converges in p-th norm to the stationary solution $\{\tilde{\boldsymbol{Y}}_t,\ t \in \mathbb{Z}\}$, i.e., $\lim_{t\to\infty} \|\boldsymbol{Y}_t(\boldsymbol{U}) - \widetilde{\boldsymbol{Y}}_t\|_p = 0$. In particular, $\lim_{t\to\infty} \mathbb{E}[\boldsymbol{Y}_t^{\otimes p}(\boldsymbol{U})] = \mathbb{E}[\widetilde{\boldsymbol{Y}}_0^{\otimes p}]$.*

Proof. Since the p-th norm top Lyapunov coefficient γ_p is larger than γ, we obtain that γ is strictly negative. Theorem 4.27 thus applies. Therefore, there exists a unique strict-sense stationary solution $\{\widetilde{\boldsymbol{Y}}_t, t \in \mathbb{Z}\}$. Also, there exists a constant C (depending only upon the choice of the norms) such that, for any $t \in \mathbb{Z}$,

$$
\begin{aligned}
\|A_t A_{t-1} \ldots A_{t-j+1} \boldsymbol{B}_{t-j}\|_p &\leq C \|\ |||A_t A_{t-1} \ldots A_{t-j+1}|||\ \|_p \|\boldsymbol{B}_{t-j}\|_p \\
&= C \|\ |||A_j A_{j-1} \ldots A_1|||\ \|_p \|\boldsymbol{B}_0\|_p \,.
\end{aligned}
$$

Under (4.62), $\lim_{j\to\infty} \left(\|\ |||A_j A_{j-1} \ldots A_1|||\ \|_p\right)^{1/j} < 1$, so that

$$
\sum_{j=0}^{\infty} \| \prod_{i=t-j+1}^{t} A_i \boldsymbol{B}_{t-j} \|_p < \infty \,.
$$

Hence, the series $\sum_{j=0}^{\infty} \prod_{i=t-j+1}^{t} A_i \boldsymbol{B}_{t-j}$ converges $\mathbb{P}$-a.s., and in the p-th norm to the same limit. The proof is concluded as in the scalar case. ■

Using Theorem 4.30, we can establish the existence of second-order stationary solutions.

Theorem 4.31. *Let $\{(A_t, \boldsymbol{B}_t), t \in \mathbb{Z}\}$ be an i.i.d. sequence such that $\rho(\mathbb{E}[A_0 \otimes A_0]) < 1$ and $\mathbb{E}[|\boldsymbol{B}_0|^2] < \infty$. Then,*

- *(i) the stochastic difference equation* (4.11) *has a unique nonanticipative weak-sense stationary solution $\{\widetilde{\boldsymbol{Y}}_t,\ t \in \mathbb{Z}\}$, given by* (4.56)*;*
- *(ii) for an arbitrary random variable $\boldsymbol{U}$ satisfying $\mathbb{E}[|\boldsymbol{U}|^2] < \infty$,*

$$
\lim_{t\to\infty} \|\boldsymbol{Y}_t(\boldsymbol{U}) - \widetilde{\boldsymbol{Y}}_t\|_2 = 0 \,,
$$

where $\boldsymbol{Y}_t(\boldsymbol{U})$ is defined by (4.12)*. In particular, $\lim_{t\to\infty} \mathbb{E}[\boldsymbol{Y}_t(\boldsymbol{U}) \boldsymbol{Y}_t'(\boldsymbol{U})] = \mathbb{E}[\widetilde{\boldsymbol{Y}}_0 \widetilde{\boldsymbol{Y}}_0']$.*

In addition $\rho(\mathbb{E}[A_0]) < 1$.

Proof. It suffices to show that $\rho(\mathbb{E}(A_0 \otimes A_0)) < 1$ implies that

$$
\gamma_2 = \lim_{k\to\infty} k^{-1} \ln \mathbb{E}(|||A_k A_{k-1} \ldots A_1|||^2) < 0 \,. \tag{4.63}
$$

Note that γ_2 does not depend upon the choice of the operator norm. For $M \in \mathbb{M}_d(\mathbb{R})$, we consider $|||M||| = \sqrt{\mathrm{Tr}(MM')}$. For $k \in \mathbb{N}$, we set $S_k = A_k A_{k-1} \ldots A_1$ and $F_k = \mathrm{Vec}(\mathbb{E}[S_k S_k'])$. We get

$$\begin{aligned}
F_{k+1} &= \mathrm{Vec}(\mathbb{E}[S_{k+1} S_{k+1}']) \\
&= \mathbb{E}\left[\mathrm{Vec}(A_{k+1} S_k S_k' A_{k+1})'\right] && \text{(since } S_{k+1} = A_{k+1} S_k) \\
&= \mathbb{E}\left[(A_{k+1} \otimes A_{k+1}) \mathrm{Vec}(S_k S_k')\right] && \text{(since } \mathrm{Vec}(ABC) = C' \otimes A \mathrm{Vec}(B)) \\
&= \mathbb{E}\left[A_0 \otimes A_0\right] F_k\,. && \text{(since } A_{k+1} \text{ and } S_k \text{ are independent)}
\end{aligned}$$

By Lemma 4.25, we may find $\gamma \in [\rho(\mathbb{E}[A_0 \otimes A_0]), 1)$ and a constant $C < \infty$, such that $\|F_k\| \le C\gamma^k$. Therefore, we get

$$\begin{aligned}
\mathbb{E}(|||A_k A_{k-1} \ldots A_1|||^2) &= \mathrm{Tr}(\mathbb{E}(S_k S_k')) \\
&= F_k' \mathrm{Vec}(\mathrm{I}[d \times d]) && \text{(since } \mathrm{Tr}(AB) = \mathrm{Vec}(B)' \mathrm{Vec}(A)) \\
&\le C d \gamma^k\,, && \text{(since } \|\mathrm{Vec}(\mathrm{I}[d \times d])\| \le d)\,,
\end{aligned}$$

which shows (4.63).

Then, since the sequence $\{A_t,\, t \in \mathbb{Z}\}$ is i.i.d. $\mathbb{E}(A_k A_{k-1} \ldots A_1) = [\mathbb{E}(A_0)]^k$. Therefore, we get

$$|||\,(\mathbb{E}A_0)^k\,||| = |||\mathbb{E}(A_k A_{k-1} \ldots A_1)||| \le \mathbb{E}|||A_k A_{k-1} \ldots A_1|||\,,$$

which implies that

$$\ln \rho(\mathbb{E}(A_0)) = \lim_{k \to \infty} k^{-1} \ln |||\mathbb{E}(A_0)^k||| \le \lim_{k \to \infty} k^{-1} \ln \mathbb{E}\,|||A_k A_{k-1} \ldots A_1||| < 0\,.$$

The condition (4.62) implies that the spectral radius $\rho(\mathbb{E}(A_0))$ is strictly less than one. ■

Assume that $\{(A_t, \boldsymbol{B}_t),\, t \in \mathbb{Z}\}$ is an i.i.d. sequence such that $\rho(\mathbb{E}[A_0 \otimes A_0]) < 1$ and $\mathbb{E}[|\boldsymbol{B}_0|^2] < \infty$. By taking the expectation on the both sides of (4.43), we obtain that, for any nonanticipative solution $\{\boldsymbol{Y}_t,\, t \in \mathbb{Z}\}$ of (4.43),

$$\mathbb{E}[\boldsymbol{Y}_t] = \mathbb{E}[A_0]\mathbb{E}[\boldsymbol{Y}_{t-1}] + \mathbb{E}[\boldsymbol{B}_0]\,.$$

Since $\rho(\mathbb{E}[A_0]) < 1$, the matrix $\mathrm{I} - \mathbb{E}[A_0]$ is invertible and the previous equation implies that, for the strict-sense stationary solution $\{\widetilde{\boldsymbol{Y}}_t,\, t \in \mathbb{Z}\}$,

$$\mathbb{E}[\widetilde{\boldsymbol{Y}}_0] = (\mathrm{I} - \mathbb{E}[A_0])^{-1} \mathbb{E}[\boldsymbol{B}_0]\,.$$

For any nonanticipative solution of (4.43), using $\mathrm{Vec}(ABC) = (C' \otimes A)\mathrm{Vec}(B)$, it follows that

$$\begin{aligned}
\mathrm{Vec}(\mathbb{E}[\boldsymbol{Y}_t \boldsymbol{Y}_t']) &= \mathbb{E}[A_0 \otimes A_0] \mathrm{Vec}(\mathbb{E}[\boldsymbol{Y}_{t-1} \boldsymbol{Y}_{t-1}']) \\
&\quad + (\mathbb{E}[A_0 \otimes \boldsymbol{B}_0] + \mathbb{E}[\boldsymbol{B}_0 \otimes A_0])\, \mathbb{E}[Y_t] + \mathrm{Vec}(\mathbb{E}[\boldsymbol{B}_0 \boldsymbol{B}_0'])\,,
\end{aligned}$$

from which we may deduce that the expression of the second order moment of the strict-sense stationary solution (4.43) is

$$\begin{aligned}
\mathrm{Vec}(\mathbb{E}[\tilde{\boldsymbol{Y}}_t \tilde{\boldsymbol{Y}}_t']) &= (\mathrm{I} - \mathbb{E}[A_0 \otimes A_0])^{-1} \\
&\quad \times \left[(\mathbb{E}[A_0 \otimes \boldsymbol{B}_0] + \mathbb{E}[\boldsymbol{B}_0 \otimes A_0])\, (\mathrm{I} - \mathbb{E}[A_0])^{-1} \mathbb{E}[\boldsymbol{B}_0] + \mathrm{Vec}(\mathbb{E}[\boldsymbol{B}_0 \boldsymbol{B}_0'])\right].
\end{aligned}$$

4.2.3 GARCH(p,q)

Definition 4.32 (GARCH(p,q)). *A process $\{X_t,\ t \in \mathbb{Z}\}$ is GARCH(p,q) if it is a solution of the following recursive equations*

$$X_t = \sigma_t \varepsilon_t \tag{4.64}$$

$$\sigma_t^2 = \alpha_0 + \alpha_1 X_{t-1}^2 + \cdots + \alpha_p X_{t-p}^2 + \beta_1 \sigma_{t-1}^2 + \cdots + \beta_q \sigma_{t-q}^2\,, \tag{4.65}$$

where $\{\varepsilon_t,\ t \in \mathbb{Z}\}$ is an i.i.d. sequence with zero-mean and unit-variance, and the coefficients α_j, $j \in \{0,\dots,p\}$ and β_j, $j \in \{1,\dots,q\}$ are nonnegative.

As in (3.29), we write a GARCH(p,q) model as a GARCH(m,m) model where $m = \max(p,q)$ and setting any additional coefficients to zero, i.e., $\alpha_{p+1} = \cdots = \alpha_q = 0$ if $p < q$ or $\beta_{q+1} = \cdots = \beta_p = 0$ if $p > q$. Set, for $t \in \mathbb{Z}$, $Z_t = X_t^2 - \sigma_t^2 = \sigma_t^2(\varepsilon_t^2 - 1)$. Since $\sigma_t^2 = X_t^2 - Z_t$, we may rewrite the recursion (4.64)–(4.65) as

$$X_t^2 - Z_t = \alpha_0 + \sum_{i=1}^m \alpha_i X_{t-i}^2 + \sum_{i=1}^m \beta_i (X_{t-i}^2 - Z_{t-i}) = \alpha_0 + \sum_{i=1}^m (\alpha_i + \beta_i) X_{t-i}^2 - \sum_{i=1}^m \beta_i Z_{t-i},$$

which implies that

$$X_t^2 - \sum_{i=1}^m (\alpha_i + \beta_i) X_{t-i}^2 = \alpha_0 + Z_t - \sum_{i=1}^m \beta_i Z_{t-i}\,. \tag{4.66}$$

Assume first that there exists a second-order stationary nonanticipative solution $\{(X_t, \sigma_t),\ t \in \mathbb{Z}\}$ to the equations (4.64)–(4.65), then $\mathbb{E}\left[Z_t \mid \mathcal{F}_{t-1}^X\right] = 0$ is martingale difference; this also implies that $\{X_t,\ t \in \mathbb{Z}\}$ is a martingale difference, $\mathbb{E}\left[X_t \mid \mathcal{F}_{t-1}^X\right] = 0$, and consequently it is white noise. If we further assume that the variance of process $\{Z_t,\ t \in \mathbb{Z}\}$ is finite, then $\{X_t^2,\ t \in \mathbb{Z}\}$ is a non-Gaussian ARMA(m,m) process. Taking the expectation in (4.66), we get

$$\mathbb{E}[X_0^2]\left(1 - \sum_{i=1}^m (\alpha_i + \beta_i)\right) = \alpha_0\,.$$

Therefore, a necessary condition for the existence of such solution is

$$1 - \sum_{i=1}^m (\alpha_i + \beta_i) > 0\,, \tag{4.67}$$

in which case,

$$0 < \mathbb{E}[X_0^2] = \alpha_0 (1 - \sum_{i=1}^p \alpha_i - \sum_{j=1}^q \beta_j)^{-1} < \infty\,. \tag{4.68}$$

We apply Theorem 4.27 to the GARCH(p,q) model. Contrary to the GARCH($1,1$) model, it is not straightforward to represent the recursion (4.64)–(4.65) as a first order random coefficient autoregressive process. There are several possibilities for doing this; here, we use the most straightforward approach. Consider the following state vectors:

$$\boldsymbol{Y}_t = \begin{pmatrix} X_t^2 \\ X_{t-1}^2 \\ \vdots \\ X_{t-p+1}^2 \\ \sigma_t^2 \\ \sigma_{t-1}^2 \\ \vdots \\ \sigma_{t-q+1}^2 \end{pmatrix}, \quad \boldsymbol{B}_t = \begin{pmatrix} \alpha_0 \varepsilon_t^2 \\ 0 \\ \vdots \\ 0 \\ \alpha_0 \\ 0 \\ \vdots \\ 0 \end{pmatrix}. \tag{4.69}$$

and the matrix

$$A_t = \left(\begin{array}{ccccc|ccccc} \alpha_1 \varepsilon_t^2 & & \cdots & & \alpha_p \varepsilon_t^2 & \beta_1 \varepsilon_t^2 & & \cdots & & \beta_q \varepsilon_t^2 \\ 1 & 0 & \cdots & & 0 & 0 & & \cdots & & 0 \\ 0 & 1 & \cdots & & 0 & 0 & & \cdots & & 0 \\ \vdots & \ddots & \ddots & \ddots & \vdots & \vdots & \ddots & \ddots & \ddots & \vdots \\ 0 & 0 & \cdots & 1 & 0 & 0 & 0 & \cdots & 0 & 0 \\ \hline \alpha_1 & & \cdots & & \alpha_p & \beta_1 & & \cdots & & \beta_q \\ 0 & & \cdots & & 0 & 1 & 0 & \cdots & & 0 \\ 0 & & \cdots & & 0 & 0 & 1 & \cdots & & 0 \\ \vdots & \ddots & \ddots & \ddots & \vdots & \vdots & \ddots & \ddots & \ddots & \vdots \\ 0 & & \cdots & 0 & 0 & 0 & & \cdots & 1 & 0 \end{array}\right). \tag{4.70}$$

It is straightforward to see that the recursion (4.64)–(4.65) can be equivalently written as

$$\boldsymbol{Y}_t = A_t \boldsymbol{Y}_{t-1} + \boldsymbol{B}_t \ .$$

The state vectors $\boldsymbol{Y}_t$ and $\boldsymbol{B}_t$ are $(p+q)$-dimensional and the matrix A_t belongs to $\mathbb{M}_{p+q}(\mathbb{R})$. In the ARCH($p$) case, the vector $\boldsymbol{Y}_t$ reduces to the p lagged values $[X_t^2, X_{t-1}^2, \ldots, X_{t-p+1}^2]$, and A_t to the upper-left $(p \times p)$ diagonal minor of the matrix displayed in (4.70).

Theorem 4.33 (Strict stationarity of GARCH(p,q)). *Consider the GARCH(p,q) model given by* (4.64)–(4.65). *Assume* $\alpha_0 > 0$.

(i) There exists a strictly stationary solution to the GARCH(p,q) recursion (4.64)–(4.65) *if and only if the top Lyapunov exponent* γ *of the sequence of matrices* $\{A_t,\ t \in \mathbb{Z}\}$ *given in* (4.70) *is strictly negative.*

(ii) In addition, if $\gamma < 0$*, the (unique) stationary solution is nonanticipative.*

Proof. If $\gamma < 0$, then the existence and the uniqueness of the stationary solution follows directly from Theorem 4.27, which shows that

$$\widetilde{\boldsymbol{Y}}_t = \sum_{n=0}^{\infty} A_t A_{t-1} \ldots A_{t-n+1} \boldsymbol{B}_{t-n} \ ,$$

is the unique strict-sense stationary solution of $\boldsymbol{Y}_t = A_t\boldsymbol{Y}_{t-1} + \boldsymbol{B}_t$. Note that the series in the right-hand-side of the previous equation converges normally with probability 1.

Next we establish that if $\gamma \geq 0$, then there is no stationary solution. We will use the following result from (Bougerol and Picard, 1992, Lemma 3.4), which we state without proof.

Lemma 4.34. *Let $\{M_t,\ t \in \mathbb{Z}\}$ be an ergodic strictly stationary sequence of matrices in $\mathbb{M}_d(\mathbb{R})$. We suppose that $\mathbb{E}[\ln^+ |||M_0|||]$ is finite and that, almost surely,*

$$\lim_{n\to\infty} |||M_n M_{n-1}\cdots M_1||| = 0\,.$$

Then the top Lyapunov exponent γ associated with this sequence is strictly negative.

We shall show that, if a strict-sense stationary solution exists, then for any $1 \leq i \leq p+q$,

$$\lim_{t\to\infty} A_0 \ldots A_{-t}\boldsymbol{e}_j = 0\,, \quad \mathbb{P}\text{-a.s.}, \tag{4.71}$$

where $\boldsymbol{e}_i$ is the i-th element of the canonical base of $\mathbb{R}^{p+q}$. Let $\{\boldsymbol{Y}_t,\ t \in \mathbb{Z}\}$ be a strictly stationary solution of $\boldsymbol{Y}_t = A_t\boldsymbol{Y}_{t-1} + \boldsymbol{B}_t$. For two $\mathbb{R}^{p+q}$ vectors $\boldsymbol{U}$ and $\boldsymbol{V}$, we write $\boldsymbol{U} \geq \boldsymbol{V}$ if each component of $\boldsymbol{U}$ is larger than the corresponding component of $\boldsymbol{V}$. Since the coefficients of the matrices $\{A_t,\ t \in \mathbb{Z}\}$, the vectors $\{\boldsymbol{B}_t,\ t \in \mathbb{Z}\}$ and the process $\{\boldsymbol{Y}_t,\ t \in \mathbb{Z}\}$ are nonnegative, for any $t > 0$, we get

$$\begin{aligned}\boldsymbol{Y}_0 &= \boldsymbol{B}_0 + A_0\boldsymbol{Y}_{-1} \\ &= \boldsymbol{B}_0 + \sum_{k=0}^{t-1} A_0 \ldots\ A_{-k}\boldsymbol{B}_{-k-1} + A_0 \ldots\ A_{-t}\boldsymbol{Y}_{-t-1} \\ &\geq \sum_{k=0}^{t-1} A_0 \ldots A_{-k}\boldsymbol{B}_{-k-1} \geq \boldsymbol{0}\,.\end{aligned}$$

It follows that the series $\sum_{k=0}^{t-1} A_0 \ldots A_{-k}\boldsymbol{B}_{-k-1}$ converges with probability 1 and thus $A_0 \ldots A_{-k}\boldsymbol{B}_{-k-1}$ converges with probability 1 to 0 as $k \to \infty$. But since $\boldsymbol{B}_{-k-1} = \alpha_0\varepsilon^2_{-k-1}\boldsymbol{e}_1 + \alpha_0\boldsymbol{e}_{p+1}$, it follows that $A_0 \ldots A_{-k}\boldsymbol{B}_{-k-1}$ can be decomposed into two positive terms, and we get

$$\lim_{k\to\infty} A_0 \ldots A_{-k}\varepsilon^2_{-k-1}\boldsymbol{e}_1 = 0, \quad \lim_{k\to\infty} A_0 \ldots A_{-k}\boldsymbol{e}_{p+1} = 0, \quad \mathbb{P}\text{-a.s.} \tag{4.72}$$

Since $\alpha_0 \neq 0$, (4.71) holds for $i = q+1$. It follows from the definition that

$$A_{-k}\boldsymbol{e}_{q+i} = \beta_i\varepsilon^2_{-k}\boldsymbol{e}_1 + \beta_i\boldsymbol{e}_{p+1} + \boldsymbol{e}_{p+i+1}, \quad i = 1,\ \ldots,\ q\,, \tag{4.73}$$

where we have set, by convention, $\boldsymbol{e}_{p+q+1} = 0$. Applying this relation with $i = 1$ yields

$$\boldsymbol{0} = \lim_{k\to\infty} A_0 \ldots A_{-k}\boldsymbol{e}_{p+1} \geq \lim_{k\to\infty} A_0 \ldots A_{-k+1}\boldsymbol{e}_{p+2} \geq \boldsymbol{0}\,,$$

which implies that (4.71) is true for $i = q+2$. Proceeding by induction, (4.72) shows that (4.71) is satisfied for $i = p+j,\ j = 1,\ \ldots,\ q$. Note in addition that $A_{-k}\boldsymbol{e}_p =$

$\alpha_p \varepsilon_{-k}^2 \boldsymbol{e}_1 + \alpha_p \boldsymbol{e}_{p+1}$. Using (4.72), this implies that (4.71) holds for $i = q$. For the remaining values of i the conclusion follows from

$$A_{-k}\boldsymbol{e}_i = \alpha_i \varepsilon_{-k}^2 \boldsymbol{e}_1 + \alpha_i \boldsymbol{e}_{p+1} + \boldsymbol{e}_{i+1}, \ i = 1, \ldots, p-1 .$$

The proof is completed by induction. ■

We conclude this section by giving conditions upon which there exists a weak-sense stationary condition to the GARCH(p,q) process.

Theorem 4.35. *Assume that $\alpha_0 > 0$ and that $\{\varepsilon_t,\ t \in \mathbb{Z}\}$ is an i.i.d. sequence with zero-mean $\mathbb{E}[\varepsilon_0] = 0$ and variance $\mathbb{E}[\varepsilon_0^2] = 1$. Then, the following hold.*

(i) If $\sum_{i=1}^p \alpha_i + \sum_{j=1}^q \beta_j < 1$, there exists a unique second-order stationary solution of (4.64)-(4.65). *This solution, denoted $\{\tilde{X}_t,\ t \in \mathbb{Z}\}$, is a martingale increment sequence.*

(ii) If $\sum_{i=1}^p \alpha_i + \sum_{j=1}^q \beta_j \geq 1$, there does not exist a nonanticipative and second-order stationary solution to the GARCH(p,q), (4.33)-(4.34), *recursions.*

Proof. Assume first $\sum_{i=1}^p \alpha_i + \sum_{j=1}^q \beta_j < 1$. Since $\mathbb{E}[X_t^2] = \mathbb{E}[\sigma_t^2]$, it suffices to show that, $\mathbb{E}[\bar{\sigma}_t^2] < \infty$, where $\{\bar{\sigma}_t^2,\ t \in \mathbb{Z}\}$ is the unique strict-sense stationary solution of (4.64)-(4.65). To establish the result, we use Theorem 4.30. Let $M \in \mathbb{M}_d(\mathbb{R})$ be a random $(d \times d)$-matrix with nonnegative entries. Consider $|||M||| = \sum_{i,j=1}^d |M_{i,j}|$. Then,

$$\mathbb{E}\,|||M||| = \sum_{i,j=1}^d \mathbb{E}|M_{i,j}| = \sum_{i,j=1}^d \mathbb{E}(M_{i,j}) = \sum_{i,j=1}^d |\mathbb{E}(M_{i,j})| = |||\mathbb{E}(M)||| .$$

Therefore, for any $k \in \mathbb{N}$, we have,

$$\mathbb{E}[|||A_k \ldots A_1|||] = |||\mathbb{E}(A_k \ldots A_1)||| = |||\, (\mathbb{E}(A_0))^k\, ||| ,$$

which implies that

$$\lim_{k \to \infty} \left(\mathbb{E}[|||A_k \ldots A_1|||]\right)^{1/k} = \rho\left(\mathbb{E}(A_0)\right) .$$

The characteristic polynomial of $\mathbb{E}(A_0)$ is given by

$$\det(\lambda \mathrm{I}_{p+q} - \mathbb{E}(A_0)) = \lambda^{p+q}\left(1 - \sum_{i=1}^p \alpha_i \lambda^{-i} - \sum_{j=1}^q \beta_j \lambda^{-j}\right) .$$

Thus, if $|\lambda| > 1$, we get

$$\left|\det(\lambda \mathrm{I}_{p+q} - \mathbb{E}(A_0))\right| \geq \left|1 - \sum_{i=1}^p \alpha_i - \sum_{j=1}^q \beta_j\right| \geq 1 - \sum_{i=1}^p \alpha_i - \sum_{j=1}^q \beta_j > 0 ,$$

showing that all the roots of the characteristic polynomial belong to the interior of the unit disk: $\rho(\,\mathbb{E}(A_0)\,) < 1$. The proof follows from Theorem 4.30.

When $\sum_{i=1}^p \alpha_i + \sum_{j=1}^q \beta_j \geq 1$, the conclusion follows directly from (4.67). ■

4.3 Iterated random function

4.3.1 Strict stationarity

In this section, we generalize Theorem 4.1 to functional autoregressive processes satisfying some "contraction in average" condition (as defined in Diaconis and Freedman, 1999). Let (X, d) be a Polish space and denote by $\mathcal{X}$ the associated Borel σ-field. Consider now $\{X_t,\, t \in \mathbb{Z}\}$ a stochastic process on $(\mathsf{X}, \mathcal{X})$ satisfying the following recurrence equation: for all $t \in \mathbb{Z}$,

$$X_t = f_{Z_t}(X_{t-1}) \,, \tag{4.74}$$

where $\{Z_t,\, t \in \mathbb{Z}\}$ is a stochastic process on $(\mathsf{Z}^{\mathbb{Z}}, \mathcal{Z}^{\otimes \mathbb{Z}})$ where $(\mathsf{Z}, \mathcal{Z})$ a measurable space. Moreover, throughout this section, the function $(z,x) \mapsto f_z(x)$ is assumed to be $(\mathcal{Z} \otimes \mathcal{X}, \mathcal{X})$-measurable. We will need the following additive assumptions.

Assumption A4.36. The sequence $\{Z_t,\, t \in \mathbb{Z}\}$ is strict-sense stationary and ergodic.

Assumption A4.37. There exists a measurable function $z \mapsto K_z$ such that

$$d(f_z(x), f_z(y)) \le K_z d(x,y) \,, \tag{4.75}$$

for all $(x,y,z) \in \mathsf{X} \times \mathsf{X} \times \mathsf{Z}$ and

$$\mathbb{E}\left[\ln^+(K_{Z_0})\right] < \infty \quad \text{and} \quad \mathbb{E}\left[\ln(K_{Z_0})\right] < 0 \,. \tag{4.76}$$

Assumption A4.38. There exists $x_0 \in \mathsf{X}$ such that

$$\mathbb{E}\left[\ln^+ d(x_0, f_{Z_0}(x_0))\right] < \infty \,. \tag{4.77}$$

Remark 4.39. Assume A4.37: for all $(x_0, x_0') \in \mathsf{X} \times \mathsf{X}$ and $z \in \mathsf{Z}$, we have

$$d(x_0', f_z(x_0')) \le (1 + K_z) d(x_0, x_0') + d(x_0, f_z(x_0)) \,,$$

so that

$$1 \wedge d(x_0', f_z(x_0')) \le 1 \wedge [(1 + K_z) d(x_0, x_0')] + 1 \wedge d(x_0, f_z(x_0)) \,.$$

Taking the logarithm function on both sides of the inequality and using that $\ln(x+y) \le \ln(x) + \ln(y)$ for all $x, y \ge 1$, we obtain that if A4.38 holds for some $x_0 \in \mathsf{X}$, then it also holds for all $x_0 \in \mathsf{X}$.

Theorem 4.40. *Assume A4.36, A4.37 and A4.38. Then, for all $x_0 \in \mathsf{X}$,*

$$X_{k,t}(x_0) = f_{Z_t} \circ \cdots \circ f_{Z_{t-k+1}}(x_0) \,, \quad t \in \mathbb{Z} \,, \tag{4.78}$$

converges $\mathbb{P}$-a.s. to a random variable $\tilde{X}_t$ which does not depend on x_0. Moreover, $\{\tilde{X}_t\,, t \in \mathbb{Z}\}$ is the only strict-sense stationary solution of (4.74). Furthermore, for an arbitrary random variable U,

$$\lim_{t \to \infty} d(X_t(U), \tilde{X}_t) = 0 \,, \quad \mathbb{P}\text{-a.s.} \,,$$

where $X_t(U)$ satisfies (4.74) for all $t \in \mathbb{N}$ and $X_0(U) = U$. In particular, the random variable $X_t(U)$ converges in distribution to $\tilde{X}_0$, i.e., as $t \to \infty$, $X_t(U) \stackrel{\mathbb{P}}{\Longrightarrow} \tilde{X}_0$.

Proof. For all $x \in \mathsf{X}$ and all $(k,t) \in \mathbb{N}^* \times \mathbb{Z}$, define

$$X_{k,t}(x) := f_{Z_t} \circ \cdots \circ f_{Z_{t-k+1}}(x) ,$$

with the convention $X_{0,t}(x) = x$. According to (4.75), we have for all $x, x' \in \mathsf{X}$ and all $(k,t) \in \mathbb{N} \times \mathbb{Z}$,

$$d(X_{k,t}(x), X_{k,t}(x')) \leq K_{Z_t} d(X_{k-1,t-1}(x), X_{k-1,t-1}(x')) ,$$

A straightforward induction yields

$$d(X_{k,t}(x), X_{k,t}(x')) \leq d(x,x') \prod_{s=t-k+1}^{t} K_{Z_s} . \tag{4.79}$$

Let $x_0 \in \mathsf{X}$ such that $\mathbb{E}[\ln^+ d(x_0, f_{Z_0}(x_0))] < \infty$. By setting $x = x_0$ and $x' = f_{Z_{t-k}}(x_0)$ in (4.79), we obtain

$$d(X_{k,t}(x_0), X_{k+1,t}(x_0)) \leq d(x_0, f_{Z_{t-k}}(x_0)) \prod_{s=t-k+1}^{t} K_{Z_s} . \tag{4.80}$$

Assumptions A 4.36, A 4.37 and A 4.38 allow to apply Lemma 4.2. Thus, the series $\sum_{k\geq 0} d(X_{k,t}(x_0), X_{k+1,t}(x_0))$ converges $\mathbb{P}$-a.s. and since (X, d) is complete, there exists a $\mathbb{P}$-a.s. finite random variable $\tilde{X}_t(x_0)$ such that for all $t \in \mathbb{N}$,

$$\lim_{k\to\infty} X_{k,t}(x_0) = \tilde{X}_t(x_0), \quad \mathbb{P}\text{-a.s.} \tag{4.81}$$

Moreover, using again (4.79) and applying again Lemma 4.2, we obtain that $\tilde{X}_t(x_0)$ does not depend on x_0 provided that $\mathbb{E}[\ln^+ d(x_0, f_{Z_0}(x_0))] < \infty$ and thus, we can write $\tilde{X}_t$ instead of $\tilde{X}_t(x_0)$. Since $X_{k,t}(x_0) = f_{Z_t}(X_{k-1,t-1}(x_0))$, we have

$$\begin{aligned} d(\tilde{X}_t, f_{Z_t}(\tilde{X}_{t-1})) &\leq d(\tilde{X}_t, X_{k,t}(x_0)) + d(X_{k,t}(x_0), f_{Z_t}(\tilde{X}_{t-1})) \\ &\leq d(\tilde{X}_t, X_{k,t}(x_0)) + K_{Z_t} d(X_{k-1,t-1}(x_0), \tilde{X}_{t-1}) . \end{aligned}$$

Using (4.81), we obtain that the right-hand side converges $\mathbb{P}$-a.s. to 0 as k goes to infinity. Thus, for all $t \in \mathbb{Z}$,

$$\tilde{X}_t = f_{Z_t}(\tilde{X}_{t-1}), \quad \mathbb{P}\text{-a.s.}$$

Moreover, let $h, t_1, \ldots, t_p \in \mathbb{N}$. Then, by stationarity of the process $\{A_t,\ t \in \mathbb{Z}\}$, we have that for all $k \in \mathbb{N}$, $(X_{k,t_1}(x_0), \ldots, X_{k,t_p}(x_0))$ has the same distribution as $(X_{k,t_1+h}(x_0), \ldots, X_{k,t_p+h}(x_0))$ so that by letting k go to infinity, $(\tilde{X}_{t_1}, \ldots, \tilde{X}_{t_p})$ has the same distribution as $(\tilde{X}_{t_1+h}, \ldots, \tilde{X}_{t_p+h})$, showing that the process $\{\tilde{X}_t, t \in \mathbb{N}\}$ is strict-sense stationary.

Let $\{\check{X}_t, t \in \mathbb{Z}\}$ be another strict-sense stationary solution of (4.74). Then, applying again (4.79) ,

$$d(\tilde{X}_t, \check{X}_t) \leq d(\tilde{X}_{t-k}, \check{X}_{t-k}) \prod_{\ell=t-k+1}^{t} K_{Z_\ell} ,$$

so that for all $M > 0$ and all $x_0 \in \mathsf{X}$,

$$\mathbb{P}(d(\tilde{X}_t, \check{X}_t) > \varepsilon) \le \mathbb{P}(d(\tilde{X}_{t-k}, \check{X}_{t-k}) > M) + \mathbb{P}\left(M \prod_{\ell=t-k+1}^{t} K_{Z_\ell} > \varepsilon\right)$$
$$\le \mathbb{P}(d(\tilde{X}_0, x_0) > M/2) + \mathbb{P}(d(x_0, \check{X}_0) > M/2) + \mathbb{P}\left(M \prod_{\ell=t-k+1}^{t} K_{Z_\ell} > \varepsilon\right),$$

where we have used the triangle inequality and the stationarity of the processes $\{\tilde{X}_t, t \in \mathbb{N}\}$ and $\{\check{X}_t, t \in \mathbb{Z}\}$. Now again, using Lemma 4.2, we let k go to infinity to obtain

$$\mathbb{P}(d(\tilde{X}_t, \check{X}_t) > \varepsilon) \le \mathbb{P}(d(\tilde{X}_0, x_0) > M/2) + \mathbb{P}(d(x_0, \check{X}_0) > M/2).$$

Finally, since M and ε are arbitrary, $\tilde{X}_t$ is equal to $\check{X}_t$ $\mathbb{P}$-a.s., so that $\{\tilde{X}_t, t \in \mathbb{Z}\}$ is the only strict-sense stationary solution of (4.74). Now, if $X_t(U)$ satisfies (4.74) for all $t \in \mathbb{N}$ and $X_0(U) = U$, then,

$$d(X_t(U), \tilde{X}_t) \le d(U, \tilde{X}_0) \prod_{\ell=0}^{t} K_{Z_\ell},$$

and thus $d(X_t(U), \tilde{X}_t) \xrightarrow{\mathbb{P}\text{-a.s.}} 0$ as t goes to infinity. The proof is completed. ■

We now generalize Theorem 4.40. The proof closely follows the lines of the proof of Theorem 4.40 and is left as an exercise for the reader.

Theorem 4.41. *Let $(z_1, \ldots, z_p) \mapsto K_{z_{1:p}}$ be a $(\mathcal{Z}^{\otimes p}, \mathcal{B}(\mathbb{R}))$-measurable function from Z^p to $\mathbb{R}$ and let $z \mapsto L_z$ be a $(\mathcal{Z}, \mathcal{B}(\mathbb{R}))$-measurable function from Z to $\mathbb{R}$. Assume that for all $x, x' \in \mathsf{X}$ and all $z_{1:p} = (z_1, \ldots, z_p) \in \mathsf{Z}^p$,*

$$d(f_{z_p} \circ \cdots \circ f_{z_1}(x), f_{z_p} \circ \cdots \circ f_{z_1}(x')) \le K_{z_{1:p}}\, d(x, x') \tag{4.82}$$
$$d(f_z(x), f_z(x')) \le L_z\, d(x, x'). \tag{4.83}$$

In addition, assume that $\mathbb{E}[\ln^+ L_{Z_0}] < \infty$, $\mathbb{E}[\ln^+(K_{Z_{0:p-1}})] < \infty$, $\mathbb{E}[\ln(K_{Z_{0:p-1}})] < 0$, and $\mathbb{E}[\ln^+ d(x_0, f_{Z_0}(x_0))] < \infty$ for some $x_0 \in \mathsf{X}$. Then, the conclusions of Theorem 4.40 hold.

Example 4.42 (Random coefficient autoregression: the vector case). Set $\mathsf{X} = \mathbb{R}^d$ and define the sequence $\{\boldsymbol{Y}_t,\ t \in \mathbb{Z}\}$ by the following linear recursion:

$$\boldsymbol{Y}_{t+1} = A_t \boldsymbol{Y}_t + \boldsymbol{B}_t, \quad t \in \mathbb{Z}, \tag{4.84}$$

where $\{(A_t, \boldsymbol{B}_t),\ t \in \mathbb{Z}\}$ is a sequence of i.i.d. random elements defined on $(\Omega, \mathcal{F}, \mathbb{P})$ taking values in $\mathbb{M}_d(\mathbb{R}) \times \mathbb{R}^d$, where $\mathbb{M}_d(\mathbb{R})$ is the set of $d \times d$ matrices with real coefficients. Let $|||\cdot|||$ be a matrix norm on $\mathbb{R}^d$ associated to a vector norm on $\mathbb{R}^d$ denoted by $\|\cdot\|$. Suppose that

$$\mathbb{E}[\ln^+ |||A_t|||] < \infty, \quad \mathbb{E}[\ln^+ \|\boldsymbol{B}_t\|] < \infty. \tag{4.85}$$

The existence and uniqueness of a stationary solution of (4.84) is obtained under the condition that γ, the top-Lyapunov exponent given in (4.53) satisfies

$$\gamma = \inf_{n>0} \frac{1}{n} \mathbb{E}[\ln |||A_1 \ldots A_n|||] < 0 . \tag{4.86}$$

This result may also be obtained from Theorem 4.41. Write (4.84) as $\boldsymbol{Y}_{t+1} = f_{Z_t}(\boldsymbol{Y}_t)$ with $Z_t = (A_t, \boldsymbol{B}_t)$ and $f_{(A,\boldsymbol{B})}(y) = Ay + \boldsymbol{B}$. Then,

$$f_{Z_p} \circ \cdots \circ f_{Z_1}(y) = A_p \ldots A_1 y + \sum_{j=1}^{p} (A_{j-1} \cdots A_1) \boldsymbol{B}_j ,$$

showing that

$$\|f_{Z_p} \circ \cdots \circ f_{Z_1}(y) - f_{Z_p} \circ \cdots \circ f_{Z_1}(y')\| \leq |||A_p \ldots A_1||| \, \|y - y'\| .$$

Then, Theorem 4.41 applies on the condition that for some $p > 0$,

$$\mathbb{E}[\ln |||A_p \ldots A_1|||] < 0 ,$$

which is exactly equivalent to the fact that the top-Lyapunov exponent γ is strictly less that 0. ◇

4.3.2 *Weak stationarity*

We now consider conditions upon which (4.74) admits moments of order $p \geq 1$.

Assumption A4.43. $\{Z_t,\, t \in \mathbb{Z}\}$ is a sequence of i.i.d. random elements.

Assumption A4.44. There exists $\alpha \in (0,1)$, $p \geq 1$ such that

$$\|d(f_{Z_0}(x), f_{Z_0}(y))\|_p \leq \alpha d(x,y) , \quad \text{for all } (x,y) \in \mathsf{X} \times \mathsf{X} . \tag{4.87}$$

Assumption A4.45. There exists $x_0 \in \mathsf{X}$ such that

$$\|d(x_0, f_{Z_0}(x_0))\|_p < \infty . \tag{4.88}$$

Remark 4.46. Note that, under A4.44, if (4.88) holds for one $x_0 \in \mathsf{X}$, then it holds for all $x_0' \in \mathsf{X}$. Indeed, by Minkowski's inequality, we have

$$\begin{aligned}\|d(x_0', f_{Z_0}(x_0'))\|_p &\leq d(x_0, x_0') + \|d(f_{Z_0}(x_0), f_{Z_0}(x_0'))\|_p + \|d(x_0, f_{Z_0}(x_0))\|_p \\ &\leq (1+\alpha) d(x_0, x_0') + \|d(x_0, f_{Z_0}(x_0))\|_p < \infty .\end{aligned}$$

Theorem 4.47. *Assume A4.43, A4.44 and A4.45. Then, for all* $x_0 \in \mathsf{X}$,

$$X_{k,t}(x_0) = f_{Z_t} \circ \cdots \circ f_{Z_{t-k+1}}(x_0) , \quad t \in \mathbb{Z} , \tag{4.89}$$

converges to a random variable $\tilde{X}_t$, $\mathbb{P}$-a.s., *which does not depend on* x_0 *and*

$\lim_{k\to\infty}\|d(X_{k,t}(x_0),\tilde{X}_t)\|_p = 0$. *Moreover,* $\{\tilde{X}_t, t\in\mathbb{Z}\}$ *is the only strict-sense stationary solution of* (4.74) *satisfying* $\mathbb{E}\left[d^p(x_0,\tilde{X}_0)\right] < \infty$ *for all* $x_0 \in \mathsf{X}$. *Furthermore, for an arbitrary random variable* U *such that* $\mathbb{E}\left[d^p(x_0,U)\right] < \infty$ *for all* $x_0 \in \mathsf{X}$,

$$\lim_{t\to\infty} d(X_t(U),\tilde{X}_t) = 0, \quad \mathbb{P}\text{-a.s.},$$

where $X_t(U)$ *satisfies* (4.74) *for all* $t\in\mathbb{N}$ *and* $X_0(U) = U$. *In particular, the random variable* $X_t(U)$ *converges in distribution to* $\tilde{X}_0$, *i.e., as* $t\to\infty$, $X_t(U) \stackrel{\mathbb{P}}{\Longrightarrow} \tilde{X}_0$.

Proof. For all $x\in\mathsf{X}$ and all $(k,t)\in\mathbb{N}\times\mathbb{Z}$, define

$$X_{k,t}(x) := f_{Z_t}\circ\cdots\circ f_{Z_{t-k+1}}(x),$$

with the convention $X_{0,t}(x) = x$. According to A4.43 and A4.44, we have for all $x,x'\in\mathsf{X}$ and all $(k,t)\in\mathbb{N}\times\mathbb{Z}$,

$$\|d(X_{k,t}(x),X_{k,t}(x'))\|_p \le \alpha\|d(X_{k-1,t-1}(x),X_{k-1,t-1}(x'))\|_p,$$

A straightforward induction yields

$$\|d(X_{k,t}(x),X_{k,t}(x'))\|_p \le \|d(x,x')\|_p\alpha^k. \tag{4.90}$$

By setting $x = x_0$ and $x' = f_{Z_{t-k}}(x_0)$ in (4.90), we obtain

$$\|d(X_{k,t}(x_0),X_{k+1,t}(x_0))\|_p \le \|d(x_0,f_{Z_{t-k}}(x_0))\|_p\alpha^k. \tag{4.91}$$

Using A4.45, the series $\sum_{k\ge0}\|d(X_{k,t}(x_0),X_{k+1,t}(x_0))\|_p$ is converging and thus, there exists a random variable $\tilde{X}_t(x_0)$ in L^p such that for all $t\in\mathbb{N}$,

$$\lim_{k\to\infty}\|d(X_{k,t}(x_0),\tilde{X}_t(x_0))\|_p = 0, \quad \mathbb{P}\text{-a.s.} \tag{4.92}$$

Moreover, using again (4.90), we obtain that $\tilde{X}_t(x_0)$ does not depend on x_0 and thus, we can write $\tilde{X}_t$ instead of $\tilde{X}_t(x_0)$. Moreover, since

$$\mathbb{E}^{1/p}\left[\left(\sum_{k=0}^{\infty} d(X_{k,t}(x_0),X_{k+1,t}(x_0))\right)^p\right] \le \sum_{k=0}^{\infty}\|d(X_{k,t}(x_0),X_{k+1,t}(x_0))\|_p < \infty,$$

the series $\sum_{k\ge0} d(X_{k,t}(x_0),X_{k+1,t}(x_0))$ converges $\mathbb{P}$-a.s.and thus, there exists a $\mathbb{P}$-a.s. random variable $X_{\infty,t}(x_0)$ such that $\lim_{k\to\infty} d(X_{k,t}(x_0),X_{\infty,t}(x_0)) = 0$. Now, by Fatou's Lemma,

$$\mathbb{E}\left[d(X_{\infty,t}(x_0),\tilde{X}_t)\right] = \mathbb{E}\left[\liminf_{k\to\infty} d(X_{k,t}(x_0),\tilde{X}_t)^p\right] \le \liminf_{k\to\infty}\mathbb{E}\left[d(X_{k,t}(x_0),\tilde{X}_t)^p\right] = 0,$$

which implies that $X_{\infty,t}(x_0) = \tilde{X}_t$.

Since $X_{k,t}(x_0) = f_{Z_t}(X_{k-1,t-1}(x_0))$, we have

$$\begin{aligned}\|d(\tilde{X}_t,f_{Z_t}(\tilde{X}_{t-1}))\|_p &\le \|d(\tilde{X}_t,X_{k,t}(x_0))\|_p + \|d(X_{k,t}(x_0),f_{Z_t}(\tilde{X}_{t-1}))\|_p \\ &\le \|d(\tilde{X}_t,X_{k,t}(x_0))\|_p + \alpha\|d(X_{k-1,t-1}(x_0),\tilde{X}_{t-1})\|_p.\end{aligned}$$

Using (4.92), we obtain that the right-hand side converges to 0 as k goes to infinity. Thus, for all $t \in \mathbb{Z}$,

$$\tilde{X}_t = f_{Z_t}(\tilde{X}_{t-1}), \quad \mathbb{P}\text{-a.s.}$$

Moreover, let $h, t_1, \ldots, t_p \in \mathbb{N}$. Then, by stationarity of the process $\{Z_t,\ t \in \mathbb{Z}\}$, we have that for all $k \in \mathbb{N}$, $(X_{k,t_1}(x_0), \ldots, X_{k,t_p}(x_0))$ has the same distribution as $(X_{k,t_1+h}(x_0), \ldots, X_{k,t_p+h}(x_0))$ so that by letting k go to infinity, $(\tilde{X}_{t_1}, \ldots, \tilde{X}_{t_p})$ has the same distribution as $(\tilde{X}_{t_1+h}, \ldots, \tilde{X}_{t_p+h})$ and the process $\{\tilde{X}_t, t \in \mathbb{N}\}$ is thus strict-sense stationary.

Let $\{\check{X}_t, t \in \mathbb{Z}\}$ be another strict-sense stationary solution of (4.74) such that $\mathbb{E}\left[d^p(x_0, \check{X}_0)\right] < \infty$ for all $x_0 \in \mathsf{X}$. Then, applying again (4.79) ,

$$\begin{aligned}
&\|d(\tilde{X}_t, \check{X}_t)\|_p \leq \|d(\tilde{X}_{t-k}, \check{X}_{t-k})\|_p \alpha^k \\
&\leq \alpha^k \left(\|d(\tilde{X}_{t-k}, x_0)\|_p + \|d(x_0, \check{X}_{t-k})\|_p\right) = \alpha^k \left(\|d(\tilde{X}_0, x_0)\|_p + \|d(x_0, \check{X}_0)\|_p\right) ,
\end{aligned}$$

By letting k go to infinity, we obtain that $\tilde{X}_t$ is $\mathbb{P}$-a.s. equal to X_t so that $\{\tilde{X}_t\,, t \in \mathbb{Z}\}$ is the only strict-sense stationary solution of (4.74) such that $\mathbb{E}\left[d^p(x_0, \check{X}_0)\right] < \infty$ for all $x_0 \in \mathsf{X}$. Now, if $X_t(U)$ satisfies (4.74) for all $t \in \mathbb{N}$ and $X_0(U) = U$, then,

$$d(X_t(U), \tilde{X}_t) \leq d(U, \tilde{X}_0) \prod_{\ell=0}^{t} K_{Z_\ell} .$$

so that $d(X_t(U), \tilde{X}_t) \xrightarrow{\mathbb{P}\text{-a.s.}} 0$ as t goes to infinity. The proof is completed. ∎

Exercises

4.1 (Proof of Theorem 4.6). We consider the case $t = 0$. For $k \geq 0$, let E_k be the event defined by

$$E_k = \left\{ \left(\prod_{i=0}^{k-1} |A_{-i}|,\ |B_{-k}| \right) \in [1,\ \infty) \times [b,\ \infty) \right\} .$$

Consider the increasing sequence of σ-algebras given by $\mathcal{G}_0 = \{\emptyset,\ \Omega\}$ and $\mathcal{G}_k = \sigma((A_i, B_i),\ -k \leq i \leq 0)$. Note that $E_k \in \mathcal{G}_k$.

(a) There is a constant $b > 0$ such that $\mathbb{P}(|B_0| \geq b) > 0$.

(b) Show that

$$\sum_{k=1}^{\infty} \mathbb{P}\left[E_k \mid \mathcal{G}_{k-1}\right] = \mathbb{P}(|B_0| \geq b) \sum_{k=1}^{\infty} \mathbb{1}\left\{ \sum_{i=0}^{k-1} \ln|A_{-i}| \geq 0 \right\} .$$

(c) Show that $\sum_{k=0}^{\infty} \mathbb{1}_{E_k} = \infty$, $\mathbb{P}$-a.s. . Deduce that

$$\mathbb{P}\left(\lim_{k\to\infty} B_k \prod_{j=0}^{k-1} A_{-j} = 0 \right) = 0 .$$

4.2 (The Fekete lemma). Let $\{u_n,\ n \in \mathbb{N}\}$ be a subadditive sequence in $[-\infty, \infty)$, i.e., for all $n, p \geq 1$, $u_{n+p} \leq u_n + u_p$. Show that the sequence $\{u_n/n, n \in \mathbb{N}\}$ either converges to its lower bound $\lim_{n\to\infty} u_n/n = \inf_{n \in \mathbb{N}^\star} u_n/n$ or diverges properly to $-\infty$.

4.3. Let $A \in \mathbb{M}_d(\mathbb{R})$.

(a) Show that, if $|||A||| < 1$, then $|||A^k||| \to 0$ as $k \to \infty$.

(b) Conclude that if there is a matrix norm $|||\cdot|||$ such that $|||A||| < 1$, then $\lim_{k\to\infty} A^k = 0$, that is, all the entries of A^k tend to zero as $k \to \infty$.

4.4. Let $A \in \mathbb{M}_d(\mathbb{R})$.

(a) Assume that $A^k \to 0$ as $k \to \infty$. Show that $\rho(A) < 1$.

(b) Conversely, assume that $\rho(A) < 1$. Show that $A^k \to 0$. Thus, $A^k \to 0$ as $k \to \infty$; then apply Exercise 4.3.]

4.5. Let $A \in \mathbb{M}_d(\mathbb{R})$ and $\varepsilon > 0$.

(a) Set $\tilde{A} \equiv [\rho(A) + \varepsilon]^{-1} A$. Show that $\tilde{A}^k \to 0$ as $k \to \infty$.

(b) Show that there is a constant $C = C(A,\ \varepsilon)$ such that $|(A^k)_{i,j}| \leq C(\rho(A) + \varepsilon)^k$ for all $k \in \mathbb{N}$ and all $(i, j) \in \{1, \ldots, d\}^2$.

4.6. Consider an ARCH(1) process $X_t = \sigma_t \varepsilon_t$, with $\sigma_t^2 = \alpha_0 + \alpha_1 X_{t-1}^2$, and $\{\varepsilon_t,\ t \in \mathbb{Z}\}$ is an i.i.d. sequence of standard gaussian random variables and $\alpha_0 \geq 0$ and $\alpha_1 \geq 0$.

(a) Under which conditions on α_0 and α_1 does the ARCH(1) process admit a strict-sense stationary solution?

(b) Under which conditions on α_0 and α_1 does the ARCH(1) process admit a weak sense stationary solution? In this case, compute the stationary variance of the process.

(c) Show that $\{X_t^2,\ t \in \mathbb{Z}\}$ is an AR(1) process. Compute the variance of the linear innovation.

(d) Use the R `fGarch` package to simulate $n = 2000$ observations from an ARCH(1) process with parameters $\alpha_0 = 0.3$ and $\alpha_1 = 0.9$.

(e) Compare the distribution of $\{X_t,\ t \in \mathbb{Z}\}$ and of a Gaussian white noise with same variance.

(f) Compare the distribution of $\{X_t^2,\ t \in \mathbb{Z}\}$ and of a Gaussian AR(1) process with the same parameters.

4.7. Consider a GARCH$(1, 1)$ process $X_t = \sigma_t \varepsilon_t$ with $\sigma_t^2 = \alpha_0 + \alpha_1 X_{t-1}^2 + \beta_1 \sigma_{t-1}^2$, where $\{\varepsilon_t,\ t \in \mathbb{Z}\}$ is an i.i.d. sequence with zero mean and finite 4th-order moment.

(a) Under which conditions does $\{X_t,\ t \in \mathbb{Z}\}$ have a nonanticipative second-order stationary solution?

(b) Under which conditions does $\{X_t^2,\ t \in \mathbb{Z}\}$ have a nonanticipative second-order stationary solution?

(c) Compute the autocovariance sequence of $\{X_t^2,\ t \in \mathbb{Z}\}$.

(d) For any integer h, compute the h-order forecast of X_{t+h}^2 and σ_{t+h}^2 given the history up to time t, $\mathcal{F}_t^X = \sigma(X_s, s \leq t)$.

(e) Using the R examples in Section 4.1.3 and Example 3.12 as a guide, complete the following.
   (i) Simulate $n = 2000$ observations of a GARCH$(1,1)$ process with parameters $\alpha_0 = 0.3$, $\alpha_1 = 0.1$ and $\beta_1 = 0.85$.
   (ii) Obtain and plot the forecasts and prediction intervals of σ_{2000+h} for $h = 1,\ldots,50$.

4.8. Consider the GARCH$(1,1)$ model given by (4.33)–(4.34).

(a) If ε_0 is a Rademacher random variable, $\mathbb{P}(\varepsilon = 1) = 1/2$ and $\mathbb{P}(\varepsilon = -1) = 1/2$, find necessary and sufficient conditions for a strict-sense stationary solution.

(b) Repeat part (a) when ε_0 is a uniform random variable.

4.9. Consider the following GARCH$(1,1)$ process: $X_t = \sigma_t \varepsilon_t$, with $\sigma_t^2 = \alpha_0(\varepsilon_{t-1}) + \alpha_1 X_{t-1}^2 + \beta_1 \sigma_{t-1}^2$, where $\{\varepsilon_t,\ t \in \mathbb{Z}\}$ is an i.i.d. sequence with zero mean and $\alpha_0 : \mathbb{R} \to \mathbb{R}_+$ is a function.

(a) State conditions upon which there exists a unique strict-sense stationary solution.

(b) State conditions upon which there exists a weak-sense stationary solution.

4.10. Consider the representation (4.32) of the bilinear process BL$(1,0,1,1)$

$$X_t = aX_{t-1} + bZ_t X_{t-1} + Z_t \ , \tag{4.93}$$

where $\{Z_t,\ t \in \mathbb{Z}\}$ is an i.i.d. sequence with zero mean, variance σ^2 and finite fourth order moment.

(a) Find a sufficient condition under which (4.93) admits a strict-sense stationary solution.

(b) Assume that $|a| < 1$ and $a^2 + b^2\sigma^2 < 1$. Show that (4.93) admits a unique nonanticipative second-order stationary solution.

(c) Show that if $|a| < 1$ and $a^2 + b^2\sigma^2 = 1$, then (4.93) and the distribution of Z_0 are not concentrated at two points, then $\{X_t,\ t \in \mathbb{Z}\}$ has a unique strict-sense stationary solution but that the second-order moment of this solution is infinite.

4.11. Consider the bilinear process BL$(0,0,2,1)$ given by

$$X_t = Z_t + bX_{t-2}Z_{t-1} \ , \tag{4.94}$$

where $\{Z_t,\ t \in \mathbb{Z}\}$ is an i.i.d. sequence with zero-mean, finite variance σ^2 and finite fourth-order moments; see Guégan (1981).

(a) Show that, if there exists a nonanticipative weak-sense stationary solution to (4.94), then $b^2\sigma^2 < 1$.

(b) Assume that $b^2\sigma^2 < 1$. Show that

$$\tilde{X}_t = Z_t + \sum_{j=0}^{\infty} b^j Z_{t-2j} \prod_{k=1}^{j} Z_{t-2k+1}$$

is a strict-sense stationary solution of (4.94) (show that the series on the right-hand side of the previous equation converges absolutely with probability 1).

(c) Show that $\mathbb{E}\bar{X}_t^2 < \infty$. Determine $\mathbb{E}(\bar{X}_0)$ and $\mathrm{Var}(\bar{X}_0)$.

(d) Assume that $\{X_t,\, t \in \mathbb{Z}\}$ is a nonanticipative strict-sense stationary solution of (4.94). For $t \in \mathbb{N}$, set $\gamma_t = \mathbb{E}(X_t\bar{X}_t)$. Show that $\gamma_t = b^2\sigma^2\gamma_{t-2} + \sigma^2$. Deduce that $\gamma_t = \mathrm{Var}(\bar{X}_0)$.

4.12. Consider the bilinear process (4.94). Assume throughout that $\|Z\|_\infty < \infty$.

(a) Show that if $|b|\|Z\|_\infty < 1$, then $\|X\|_\infty < \infty$.

(b) Assume that $b\|Z\|_\infty \geq 1$. Construct a sequence $\{Z_t,\, t \in \mathbb{Z}\}$ such that (4.94) admits a strict-sense stationary solution $\{\bar{X}_t,\, t \in \mathbb{Z}\}$, but $\|\bar{X}\|_\infty = \infty$.

(c) Assume that $\{Z_t,\, t \in \mathbb{Z}\}$ is standard Gaussian. For which values of m can we ensure that $\|X\|_m < \infty$?

(d) Assume that $\{Z_t,\, t \in \mathbb{Z}\}$ is standard Gaussian. Find a necessary and sufficient condition for the existence of a strict-sense stationary solution satisfying $\mathbb{E}(\bar{X}_t^4) < \infty$.

4.13. Consider the bilinear process (cf. Quinn, 1982)

$$X_t = Z_t + b_{j,k}X_{t-k}Z_{t-j}\,, \tag{4.95}$$

where $\{Z_t,\, t \in \mathbb{Z}\}$ is an i.i.d. sequence of random variables satisfying $\mathbb{E}|\ln|Z_0|| < \infty$ and $\mathbb{E}\ln^2|Z_0| < \infty$. Assume in addition that $k > j$.

(a) Show that a necessary and sufficient condition for the existence of a strict-sense stationary to (4.95) is

$$\ln|b_{j,k}| + \mathbb{E}\ln|Z_t| < 0\,.$$

(b) State a condition under which there exists a weak-sense stationary solution to (4.95).

4.14. Let X_t be a GARCH(1,1) process and assume there exists a nonanticipative weak-sense stationary solution.

(a) Show that $\{(X_t, \mathcal{F}_t^X),\, t \in \mathbb{Z}\}$ is a martingale increment process.

(b) Set $Z_t = X_t^2 - \sigma_t^2$. Show that $\{(Z_t, \mathcal{F}_t^X),\, t \in \mathbb{Z}\}$ is a martingale increment process and hence is white noise.

(c) Propose a representation of the squared process $\{X_t^2,\, t \in \mathbb{Z}\}$.

4.15. Give an example of a matrix A and two matrix norms $|||\cdot|||_\alpha$ and $|||\cdot|||_\beta$ such that $|||A|||_\alpha < 1$ and $|||A|||_\beta > 1$.

4.16. Consider, a stationary AR(1) process, whose innovation is a GARCH(1, 1) process:

$$\begin{cases} X_t = \phi X_{t-1} + \varepsilon_t \\ \varepsilon_t = \sigma_t \eta_t \\ \sigma_t^2 = \alpha_0 \ + \alpha_1\varepsilon_{t-1}^2 + \beta_1\sigma_{t-1}^2, \end{cases}$$

where

- $\alpha_0 > 0$, $\alpha_1 \geq 0$, $\beta_1 \geq 0$, $\alpha_1 + \beta_1 \leq 1$ and $|\phi| < 1$;

- $\{\eta_t,\ t\in\mathbb{Z}\}$ is a sequence of i.i.d. random variables with zero mean and unit variance.

(a) Show that, if $\phi^2 \neq \alpha_1+\beta_1$,

$$\operatorname{Var}\left(X_{t+h}\mid \mathcal{F}_{t-1}^X\right) = \frac{\alpha_0(1-\phi^{2(h+1)})}{\{1-(\alpha_1+\beta_1)\}(1-\phi^2)} + \left\{\sigma_t^2 - \frac{\alpha_0}{1-(\alpha_1+\beta_1)}\right\}\frac{\phi^{2(h+1)}-(\alpha_1+\beta_1)^{(h+1)}}{\phi^2-(\alpha_1+\beta_1)}$$

and that, if $\phi^2 = \alpha_1+\beta_1$,

$$\operatorname{Var}\left(X_{t+h}\mid \mathcal{F}_{t-1}^X\right) = \frac{\alpha_0(1-\phi^{2(h+1)})}{(1-\phi^2)^2} + \left\{\sigma_t^2 - \frac{\alpha_0}{1-(\alpha_1+\beta_1)}\right\}(h+1)\phi^{2h}\,.$$

(b) Discuss the behavior of $\operatorname{Var}\left(X_{t+h}\mid \mathcal{F}_{t-1}^X\right)$ as h increases for different values of the parameters α_0,α_1 and β_1.

4.17. The RiskMetrics model relies on the following equations: $X_t=\sigma_t\varepsilon_t$, $\sigma_t^2=(1-\lambda)\lambda X_t^2+\lambda\sigma_{t-1}^2$, where $\lambda\in(0,1)$ and $\{\varepsilon_t,\ t\in\mathbb{Z}\}$ is an i.i.d. sequence of normal random variables.

(a) Show that this model does not admit a strict-sense stationary non-degenerated solution.

(b) Show that, for all $h>0$,

$$\sigma_t^2 = \lambda\left(X_t^2+\cdots+\lambda^h X_{t-h}^2\right)+\lambda^{h+1}\sigma_{t-(h+1)}^2\,.$$

(Riskmetrics recommends the use of $\lambda=0.94$ for daily returns and $\lambda=0.97$ for monthly return.)

4.18. Consider a GARCH(p,q) model with coefficients $\{\alpha_i,\beta_j, i=0,\ldots,p, j=1\ldots,q\}$. Assume that this model admits a strict-sense stationary solution. Consider a GARCH(p,q) model with coefficients $\{\alpha_i',\beta_j', i=0,\ldots,p, j=1,\ldots,q\}$ satisfying $0\le\alpha_i'\le\alpha_i$, $i=1,\ldots,p$ and $0\le\beta_j'\le\beta_j$, $j=1,\ldots,q$.

(a) Show that the top Lyapunov coefficient of the sequence of matrices $\{A_t',\ t\in\mathbb{Z}\}$ with α_i' and β_j' (instead of α_i and β_j) is smaller than the matrix top Lyapunov exponent of the sequence $\{A_t,\ t\in\mathbb{Z}\}$.

(b) Show that the GARCH(p,q) model with coefficients $\{\alpha_i',\beta_j', i=0,\ldots,p, j=1,\ldots,q\}$ also admits a strict-sense stationary solution.

4.19. (a) Assume that the top Lyapunov coefficient γ of the sequence $\{A_t,\ t\in\mathbb{Z}\}$, defined in (4.70), is strictly negative. Show the spectral radius of the sequence $\{A_t,\ t\in\mathbb{Z}\}$ is larger than the spectral radius of the deterministic obtained by zeroing the coefficients of the first p rows and of the first p columns of the matrix A_t.

(b) Show that this matrix has the same nonzero eigenvalues as B (defined in (4.96)), and thus the same spectral radius. Deduce that $\gamma<0$, then $\rho(B)<1$.

(c) Show that the three following conditions are equivalent:

(i) $\sum_{j=1}^{p} \beta_j < 1$,

(ii) $1 - \beta_1 z - \cdots - \beta_p z^p \neq 0$ for $|z| \leq 1$,

(iii) The spectral radius $\rho(B) < 1$, where B is defined as

$$B = \begin{pmatrix} \beta_1 & \beta_2 & \dots & \beta_q \\ 1 & 0 & \dots & 0 \\ \vdots & \ddots & \ddots & 0 \\ 0 & \dots & 1 & 0 \end{pmatrix} . \tag{4.96}$$

(d) Conclude.

4.20. Let $\{\varepsilon_t,\ t \in \mathbb{Z}\}$ be an i.i.d. sequence with zero mean and unit variance. A process $\{X_t,\ t \in \mathbb{Z}\}$ is said to be an exponential GARCH (or EGARCH[1]) process if it is a solution of an equation of the form

$$X_t = \sigma_t \varepsilon_t \tag{4.97}$$

$$\ln \sigma_t^2 = \alpha_0 + \sum_{i=1}^{p} \alpha_i g\,(\varepsilon_{t-i}) + \sum_{j=1}^{q} \beta_j \ln \sigma_{t-j}^2, \tag{4.98}$$

where

$$g(\varepsilon) = \theta\varepsilon + \zeta(|\varepsilon| - \mathbb{E}|\varepsilon|)\,, \tag{4.99}$$

and α_0, α_j, β_j, θ and ζ are real numbers. The relation

$$\sigma_t^2 = \mathrm{e}^{\alpha_0} \prod_{i=1}^{p} \exp\{\alpha_i g(\varepsilon_{t-i})\} \prod_{j=1}^{q} (\sigma_{t-j}^2)^{\beta_j}$$

shows that, in contrast to the classical GARCH, the volatility has multiplicative dynamics. The positivity constraints on the coefficients can be avoided, because the logarithm can be of any sign Assume that $g(\varepsilon_0)$ is not almost surely equal to zero and that the polynomials $\alpha(z) = \sum_{i=1}^{p} \alpha_i z^i$ and $\beta(z) = 1 - \sum_{j=1}^{q} \beta_j z^j$ have no common root, and that $\alpha(z)$ is not identically null.

(a) Suppose that $\beta(z) \neq 0$ for $|z| \leq 1$. Show that $\ln \sigma_t^2$ can be expressed as

$$\ln \sigma_t^2 = \alpha_0^* + \sum_{i=1}^{\infty} \lambda_i g(\varepsilon_{t-i}), \quad \mathbb{P}\text{-a.s.}$$

where $\alpha_0^* = \alpha_0/\beta(1)$ and $\{\lambda_i,\ i \in i \geq 1\}$ are the coefficients of the expansion of the rational function $\alpha(z)/\beta(z)$ converging in the unit disk.

[1]This model is used to reproduce asymmetry in the return of a stock. When the price of a stock falls, the debt-equity ratio of the company increases. This entails an increase of the risk and hence of the volatility of the stock. When the price rises, the volatility also increases but by a smaller amount. Another model used for asymmetric behavior is the *Asymmetric Power ARCH* model presented in Example 3.12.

(b) Show that the log-volatility recursion (4.98) admits a strictly stationary and nonanticipative solution $\{\ln \sigma_t^2, t \in \mathbb{Z}\}$ if and only if $\beta(z) \neq 0$ for $|z| \leq 1$.

(c) Show that the equations (4.97)–(4.98) admit a strictly stationary and nonanticipative solution if and only if $\beta(z) \neq 0$ for $|z| \leq 1$. In the sequel, we assume that $\beta(z) \neq 0$ for $|z| \leq 1$.

(d) Assume that $\mathbb{E}(\ln \varepsilon_t^2)^2 < \infty$ and $\mathbb{E}g^2(\varepsilon_t) < \infty$. Show that $\mathbb{E}(\ln X_t^2)^2 < \infty$.

(e) Show that if

$$\prod_{i=1}^{\infty} \mathbb{E} \exp\{|\lambda_i g(\varepsilon_0)|\} < \infty , \tag{4.100}$$

then $\{X_t,\ t \in \mathbb{Z}\}$ is a white noise with variance

$$\mathbb{E}(X_t^2) = \mathbb{E}(\sigma_t^2) = \mathrm{e}^{\alpha_0^*} \prod_{i=1}^{\infty} \mathbb{E}(\exp\{\lambda_i g(\varepsilon_0)\})$$

(f) Assume that $\{\varepsilon_t,\ t \in \mathbb{Z}\}$ is an i.i.d. sequence of standard Gaussian variables. Show that

$$\ln \mathbb{E} \exp\{|\lambda_i g(\varepsilon_t)| = O(\lambda_i) , \quad i \geq 1 ,$$

and that (4.100) is satisfied.

(g) Assume that, for some integer m,

$$\mu_{2m} = \mathbb{E}(\varepsilon_0^{2m}) < \infty, \ \prod_{i=1}^{\infty} \mathbb{E} \exp\{|m\lambda_i g(\varepsilon_0)|\} < \infty,$$

Show that $\{X_t^2,\ t \in \mathbb{Z}\}$ admits a moment of order m given by

$$\mathbb{E}(X_t^{2m}) = \mu_{2m} \mathrm{e}^{m\alpha_0^*} \prod_{i=1}^{\infty} \mathbb{E}(\exp\{m\lambda_i g(\varepsilon_0)\}).$$

4.21. Consider the model (4.97)–(4.98). Assume that $\mathbb{E}(\ln \varepsilon_0^2)^2 < \infty$

(a) Show that

$$\ln X_t^2 - \sum_{i=1}^{p} \beta_j \ln X_{t-j}^2 = \alpha_0 + \ln \varepsilon_t^2 + \sum_{i=1}^{p} \alpha_i g(\varepsilon_{t-i}) - \sum_{j=1}^{p} \beta_j \ln \varepsilon_{t-j}^2 .$$

(b) Put $v_t = \ln X_t^2 - \sum_{i=1}^{p} \beta_j \ln X_{t-j}^2$. Show that $\{v_t,\ t \in \mathbb{Z}\}$ is an MA process of order $r = \max(p, q)$.

(c) Show that $\{\ln X_t^2, t \in \mathbb{Z}\}$ is an ARMA process.

4.22. Consider the following observation-driven[2] time series model:

$$\begin{aligned} Y_{k+1} | \mathcal{F}_k &\sim \mathrm{Exp}(X_k^{-1}) \\ X_{k+1} &= d + aX_k + bY_{k+1} \end{aligned} \tag{4.101}$$

where $\mathcal{F}_k = \sigma(X_0, \ldots, X_k, Y_1, \ldots, Y_k)$. Moreover, we assume that $d > 0$ and $a, b \geq 0$.

[2] See Definition 5.21 for an explicit definition of *observation-driven*.

(a) Let $\{X_k,\ k \in \mathbb{N}\}$ be a sequence of random variables satisfying

$$X_{k+1} = d + aX_k + bX_k \ln(U_{k+1}) \qquad (4.102)$$

where $\{U_k,\ k \in \mathbb{N}\}$ are such that $U_k \sim$iid $\mathrm{U}[0,1]$. Moreover, it is assumed that X_0 is independent of $\{U_k,\ k \in \mathbb{N}\}$ and for simplicity, it is also assumed that $X_0 \geq d$ almost surely. Show that $\{X_k,\ k \in \mathbb{N}\}$ satisfies (4.101).

(b) Show that the random variable

$$\bar{X}_k = \sum_{\ell=0}^{\infty} d \prod_{s=0}^{\ell-1} (a - b\ln(U_{k-s}))$$

is $\mathbb{P}$-a.s. finite if and only if a and b satisfy

$$\mathbb{E}\left[\ln(a - b\ln(U))\right] < 0\,, \quad \text{where} \quad U \sim \mathrm{U}[0,1]\,. \qquad (4.103)$$

(c) Show that $\{\bar{X}_t,\ t \in \mathbb{Z}\}k \in \mathbb{N}$ is a stationary solution of (4.102).

(d) Let $\{X_k,\ k \in \mathbb{N}\}$ be another solution of (4.102). Show that $|\bar{X}_k - X_k|$ goes $\mathbb{P}$-a.s. to 0 as k tends to infinity. Deduce that $\{\bar{X}_k,\ k \in \mathbb{N}\}$ is the unique stationary solution of (4.102).

(e) We now assume that $a + b < 1$. Justify that there exists a unique stationary solution of (4.102), which moreover, has a finite expectation.

(f) Conversely, assume now that there exists a stationary solution of (4.102) with finite expectation. Denoting by $\{\bar{X}_k,\ k \in \mathbb{N}\}$ this solution, compute $\mathbb{E}\left[\bar{X}_k\right]$. Deduce that $a + b < 1$.

(g) Does there exist $a, b > 0$ such that $a + b \geq 1$ and such that there exists a unique stationary solution of (4.102)?

(h) Now, assume that $d = 0$ and $a, b > 0$. Find a necessary and sufficient condition on a, b such that there exists a unique stationary solution of (4.102). Give the explicit expression of this solution.

Part II

Markovian Models

Chapter 5

Markov Models: Construction and Definitions

There are two rather different approaches to constructing a Markov chain. The first is through the transition laws of the chain. The second is through the use of iterating functions. This is particularly suited to the use of Markov chains in a time series context, where we wish to construct sample paths of the chain in an explicit manner rather than work with the distributions of the chain. This can also be carried out in a great degree of generality.

For both of these approaches we will discuss the existence of invariant measures (sometimes called stationary distributions) for the chain. This is the key property for most models because it implies (and is in some sense equivalent to) the existence of strict-sense stationary solutions. More precisely, when a Markov chain admits an invariant distribution, then we may construct strictly stationary versions of Markov chains; these stationary versions play a key role when considering the inference of Markov chains.

We will illustrate these constructions with several important time series models. It is worthwhile to note that, while many models in the nonlinear time series context might be viewed as (most of the time straightforward) extensions of ARMA models, finding necessary and sufficient conditions for the existence of an invariant distribution is a difficult task that will be covered in some generality in subsequent chapters.

Some of the material in this chapter will seem, to those interested in applications, to be both technical and exotic. However, we use the formal structures to define, in a rigorous manner, different time-series models. We encourage the reader to go through this material, perhaps skipping the proofs in a first pass.

5.1 Markov chains: Past, future, and forgetfulness

A Markov chain is a discrete time stochastic process $X = \{X_t,\ t \in \mathbb{N}\}$, i.e., a countable collection of random variables defined on a probability space $(\Omega, \mathcal{F}, \mathbb{P})$. In this definition, t is thought of as a time index and the set of times is taken by convention as $\mathbb{N}$.

The chain X evolves on a *state space* X. Although we let X be a general set (not assuming a priori any topological structure), in practice, it is most likely that we will be considering the set of real numbers or, more generally, the d-dimensional

Euclidean space, so that $\mathsf{X} = \mathbb{R}^d$, or some countable or uncountable subset of $\mathbb{R}^d$. In order to define probabilities, we use $\mathcal{X}$ to denote a countably generated σ-field on X. When $\mathsf{X} = \mathbb{R}^d$, then $\mathcal{X}$ will be taken as the Borel σ-field $\mathcal{X} = \mathcal{B}(\mathsf{X})$. When X is countable, $\mathcal{X}$ contains all the subsets, that is $\mathcal{X} = \mathcal{P}(\mathsf{X})$, the set of all parts of X. All random variables are assumed measurable individually with respect to $\mathcal{B}(\mathsf{X})$, and we shall in general denote elements of X by letters x, y, z, ... and elements of $\mathcal{X}$ by A, B, C.

The distinguishing feature of a Markov chain as opposed to an arbitrary stochastic process is that the future, given the present, is forgetful of the past. This is reflected in the conditional distributions of the future random variable X_{t+1} given its present, X_t, and its past $\{X_0, \ldots, X_{t-1}\}$.

Definition 5.1 (Markov chain). *Let $(\Omega, \mathcal{F}, \{\mathcal{F}_t,\ t \in \mathbb{N}\}, \mathbb{P})$ be a filtered probability space (see Definition A.13). An adapted stochastic process (see Definition A.14) $\{(X_t, \mathcal{F}_t),\ t \in \mathbb{N}\}$ is called a Markov chain of* order m *(or with* memory m*) if, for all $t \geq m-1$ and $A \in \mathcal{X}$, $\mathbb{P}\left[X_{t+1} \in A \mid \mathcal{F}_t\right] = \mathbb{P}\left[X_{t+1} \in A \mid X_t, \ldots, X_{t-m+1}\right]$,* $\mathbb{P}$-a.s.

When the order is one, we simply say that $\{(X_t, \mathcal{F}_t),\ t \in \mathbb{N}\}$ is a Markov chain. This definition is equivalent to assuming that for all $f \in \mathbb{F}_+(\mathsf{X}, \mathcal{X})$,

$$\mathbb{E}\left[f(X_t) \mid \mathcal{F}_{t-1}\right] = \mathbb{E}\left[f(X_t) \mid X_{t-1}\right], \quad \mathbb{P}\text{-a.s.}$$

Note that a Markov chain $\{(X_t, \mathcal{F}_t),\ t \in \mathbb{N}\}$ is always a Markov chain with respect to the *natural filtration* $\{\mathcal{F}_t^X,\ t \in \mathbb{N}\}$, where $\mathcal{F}_t^X = \sigma(X_0, \ldots, X_t)$.

5.2 Kernels

The construction of Markov chains is based on the definition of kernels.

Definition 5.2. *Let $(\mathsf{X}, \mathcal{X})$ and $(\mathsf{Y}, \mathcal{Y})$ be two measurable spaces.*

- *A* kernel *is a mapping N from $\mathsf{X} \times \mathcal{Y}$ into $[0, \infty]$ satisfying the following conditions:*
  (i) for every $x \in \mathsf{X}$, the mapping $N(x, \cdot) : A \mapsto N(x, A)$ is a measure on $\mathcal{Y}$,
  (ii) for every $A \in \mathcal{Y}$ the mapping $N(\cdot, A) : x \to N(x, A)$ is a measurable function from $(\mathsf{X}, \mathcal{X})$ to $[0, \infty]$.
- *The kernel N is said to be* finite *if $N(x, \mathsf{Y}) < \infty$ for all $x \in \mathsf{X}$.*
- *The kernel N is said to be* bounded *if $\sup_{x \in \mathsf{X}} N(x, \mathsf{Y}) < \infty$.*
- *A kernel N is said to be* Markovian *if $N(x, \mathsf{Y}) = 1$, for all $x \in \mathsf{X}$.*

Example 5.3 (Discrete state-space kernel). Assume that X and Y are countable sets. Each element $x \in \mathsf{X}$ is then called a state. A kernel N on $\mathsf{X} \times \mathcal{P}(\mathsf{Y})$, where $\mathcal{P}(\mathsf{Y})$ is the set of all parts of Y, is specified by a (possibly infinite) matrix $N = (N(x,y) : x, y \in \mathsf{X} \times \mathsf{Y})$ with nonnegative entries. Each row $(N(x,y) : y \in \mathsf{Y})$ is a measure on $(\mathsf{Y}, \mathcal{P}(\mathsf{Y}))$ defined by

$$N(x, A) = \sum_{y \in A} N(x, y),$$

for $A \subset \mathsf{Y}$. The matrix N is said to be *Markovian* if every row $(N(x,y) : y \in \mathsf{Y})$ is a probability measure on $(\mathsf{Y}, \mathcal{P}(\mathsf{Y}))$, i.e. $\sum_{y \in \mathsf{Y}} N(x,y) = 1$ for all $x \in \mathsf{X}$. ◇

Example 5.4 (Kernel density). Let $\lambda \in \mathbb{M}_1(\mathcal{Y})$ and $n : \mathsf{X} \times \mathsf{Y} \to \mathbb{R}_+$ be a nonnegative function, measurable with respect to the product σ-algebra $\mathcal{X} \otimes \mathcal{Y}$. Then, the application N defined on $\mathsf{X} \times \mathcal{Y}$ by

$$N(x,A) = \int_A n(x,y)\lambda(\mathrm{d}y) ,$$

is a kernel. The function n is called the density of the kernel N with respect to the measure λ. The kernel N is Markovian if and only if $\int_{\mathsf{Y}} n(x,y)\lambda(\mathrm{d}y) = 1$ for all $x \in \mathsf{X}$. ◇

Let N be a kernel and $f \in \mathbb{F}_+(\mathsf{Y},\mathcal{Y})$. A function $Nf : \mathsf{X} \to \mathbb{R}^+$ is defined by setting, for any $x \in \mathsf{X}$,

$$Nf : x \mapsto \int_{\mathsf{Y}} N(x,\mathrm{d}y) f(y) .$$

Proposition 5.5. *Let N be a kernel on $\mathsf{X} \times \mathcal{Y}$. For any $f \in \mathbb{F}_+(\mathsf{Y},\mathcal{Y})$, $Nf \in \mathbb{F}_+(\mathsf{X},\mathcal{X})$.*

Proof. See Exercise 5.5. ■

With a slight abuse of notation, we will use the same symbol N for the kernel and the associated operator $N : \mathbb{F}_+(\mathsf{Y},\mathcal{Y}) \to \mathbb{F}_+(\mathsf{X},\mathcal{X})$, $f \mapsto Nf$. By defining $Nf = Nf^+ - Nf^-$, we may extend the application $f \mapsto Nf$ to all functions f of $\mathbb{F}(\mathsf{Y},\mathcal{Y})$ such that Nf^+ and Nf^- are not both infinite. We will sometimes write $N(x,f)$ for $Nf(x)$ and make use of the notations $N(x,\mathbb{1}_A)$ and $N\mathbb{1}_A(x)$ for $N(x,A)$.

Let μ be a positive measure on $(\mathsf{X},\mathcal{X})$ and for $A \in \mathcal{Y}$, define

$$\mu N(A) = \int_{\mathsf{X}} \mu(\mathrm{d}x)\, N(x,\, A) .$$

Proposition 5.6. *Let N be a kernel on $\mathsf{X} \times \mathcal{Y}$ and $\mu \in \mathbb{M}_+(\mathcal{X})$. Then $\mu N \in \mathbb{M}_+(\mathcal{Y})$.*

Proof. See Exercise 5.6. ■

Let $(\mathsf{X},\mathcal{X})$, $(\mathsf{Y},\mathcal{Y})$ and $(\mathsf{Z},\mathcal{Z})$ be measurable spaces and let M and N be two kernels on $\mathsf{X} \times \mathcal{Y}$ and $\mathsf{Y} \times \mathcal{Z}$. For any $A \in \mathcal{Z}$, $y \mapsto N(y,A)$ is a measurable function, and by Proposition 5.5, $x \mapsto \int_{\mathsf{Y}} M(x,\mathrm{d}y)N(y,A)$ is a measurable function. For any $x \in \mathsf{X}$, $M(x,\cdot)$ is a measure on $(\mathsf{Y},\mathcal{Y})$ and by Proposition 5.6, $A \mapsto \int M(x,\mathrm{d}y)N(y,A)$ is a measure on $(\mathsf{Z},\mathcal{Z})$. Hence, the function $MN : (x,A) \mapsto \int_{\mathsf{X}} M(x,\mathrm{d}y)N(y,A)$ is a kernel on $\mathsf{X} \times \mathcal{Z}$.

The *composition* or *product* of the kernels M on $\mathsf{X} \times \mathcal{Y}$ and N on $\mathsf{Y} \times \mathcal{Z}$ is the kernel MN defined for $x \in \mathsf{X}$ and $A \in \mathcal{Z}$ by

$$MN(x,A) = \int_{\mathsf{Y}} M(x,\mathrm{d}y)N(y,A) . \tag{5.1}$$

Since MN is a kernel on $\mathsf{X} \times \mathcal{Z}$, for any $f \in \mathbb{F}_+(\mathsf{Z},\mathcal{Z})$, we may define the function $MNf : x \mapsto MNf(x)$, which by Proposition 5.5 belongs to $\mathbb{F}_+(\mathsf{X},\mathcal{X})$. On the other hand, $Nf : y \mapsto Nf(y)$ is a function belonging to $\mathbb{F}_+(\mathsf{Y},\mathcal{Y})$, and since M is a kernel on $\mathsf{X} \times \mathcal{Y}$, we may consider the function $x \mapsto M[Nf](x)$. As shown in the following proposition, these two quantities coincide.

Proposition 5.7. *Let M be a kernel on $\mathsf{X} \times \mathcal{Y}$ and N be a kernel on $\mathsf{Y} \times \mathcal{Z}$. Then, for each $x \in \mathsf{X}$, and $f \in \mathbb{F}_+(\mathsf{Z}, \mathcal{Z})$,*

$$MNf(x) = M[Nf](x) . \tag{5.2}$$

Proof. See Exercise 5.7. ∎

For $x \in \mathsf{X}$ and $A \in \mathcal{X}$, we set $N^0(x, A) = \delta_x(A)$ and for $n \geq 1$, we define inductively N^n by

$$N^n(x, A) = \int_{\mathsf{X}} N(x, \mathrm{d}y) N^{n-1}(y, A) . \tag{5.3}$$

We finally define the tensor products of kernels.

Proposition 5.8. *Let M be a kernel on $\mathsf{X} \times \mathcal{Y}$ and N be a kernel on $\mathsf{Y} \times \mathcal{Z}$. Then, there exists a kernel $M \otimes N$ on $\mathsf{X} \times (\mathcal{Y} \otimes \mathcal{Z})$, called the tensor product of M and N, such that, for all $f \in \mathbb{F}_+(\mathsf{Y} \times \mathsf{Z}, \mathcal{Y} \otimes \mathcal{Z})$,*

$$M \otimes Nf(x) = \int_{\mathsf{Y}} M(x, \mathrm{d}y) \int_{\mathsf{Z}} f(y,z) N(y, \mathrm{d}z) . \tag{5.4}$$

Proof. See Exercise 5.8. ∎

For $n \geq 1$, the n-th tensorial power $P^{\otimes n}$ of a kernel P on $X \times \mathcal{Y}$ is the kernel on $(\mathsf{X}, \mathcal{X}^{\otimes n})$ defined by

$$P^{\otimes n} f(x) = \int_{\mathsf{X}^n} f(x_1, \ldots, x_n) P(x, \mathrm{d}x_1) P(x_1, \mathrm{d}x_2) \cdots P(x_{n-1}, \mathrm{d}x_n) .$$

5.3 Homogeneous Markov chain

Definition 5.9 (Homogeneous Markov chain). *Let $(\Omega, \mathcal{F}, \{\mathcal{F}_t, t \in \mathbb{N}\}, \mathbb{P})$ be a filtered probability space (see Definition A.13). Let $m \geq 1$ be an integer and P a Markov kernel on $\mathsf{X}^m \times \mathcal{X}$. An adapted stochastic process $\{(X_t, \mathcal{F}_t), t \in \mathbb{N}\}$ is called an* homogeneous Markov chain of order *m with Markov kernel P and initial distribution $\nu \in \mathbb{M}_1(X^m, \mathcal{X}^{\otimes m})$ if*

(i) for all $A \in \mathcal{X}^{\otimes m}$, $\mathbb{P}[(X_0, \ldots, X_{m-1}) \in A] = \nu(A)$,

(ii) for all $t \geq m-1$ $\mathbb{P}\left[X_{t+1} \in A \mid \mathcal{F}_t\right] = P(X_t, \ldots, X_{t-m+1}, A)$, $\mathbb{P}$-a.s.

By construction, for any function $f_0 \in \mathbb{F}_+(\mathsf{X}, \mathcal{X})$, $\mathbb{E}[f_0(X_0)] = \int \nu(\mathrm{d}x_0) f_0(x_0) = \nu(f_0)$. For all functions $f_0, f_1 \in \mathbb{F}_+(\mathsf{X}, \mathcal{X})$, we may write similarly

$$\begin{aligned} \mathbb{E}[f_0(X_0) f_1(X_1)] &= \mathbb{E}[f_0(X_0) \mathbb{E}\left[f_1(X_1) \mid \mathcal{F}_0\right]] = \mathbb{E}[f_0(X_0) P f_1(X_0)] \\ &= \iint \nu(\mathrm{d}x_0) P(x_0, \mathrm{d}x_1) f_0(x_0) f_1(x_1) = \nu \otimes P(f_0 \otimes f_1) , \end{aligned}$$

where $f_0 \otimes f_1(x, x') = f_0(x) f_1(x')$. Assume that for some $t \geq 1$, and for all functions $f_0, \ldots, f_t \in \mathbb{F}_+(\mathsf{X}, \mathcal{X})$,

$$\begin{aligned} \mathbb{E}[f_0(X_0) f_1(X_1) \ldots f_t(X_t)] &= \int \nu(\mathrm{d}x_0) \prod_{s=1}^{t} P(x_{s-1}, \mathrm{d}x_s) \prod_{s=0}^{t} f_s(x_s) \\ &= \nu \otimes P^{\otimes t}(f_0 \otimes f_1 \otimes \cdots \otimes f_t) . \end{aligned} \tag{5.5}$$

Then, for any $f_{t+1} \in \mathbb{F}_+(\mathsf{X}, \mathcal{X})$,

$$\begin{aligned}\mathbb{E}[f_0(X_0)\dots f_t(X_t)f_{t+1}(X_{t+1})] &= \mathbb{E}[f_0(X_0)\dots f_t(X_t)\mathbb{E}\left[f_{t+1}(X_{t+1}) \mid \mathcal{F}_t\right]] \\ &= \mathbb{E}[f_0(X_0)\dots f_t(X_t)Pf_{t+1}(X_{t+1})] = \nu \otimes P^{\otimes(t+1)}(f_0 \otimes f_1 \otimes \dots \otimes f_{t+1}),\end{aligned}$$

showing that (5.5) is still true with t replaced by $t+1$, and hence is satisfied for all $t \in \mathbb{N}$. Since the cylinder sets generate the product σ-algebra $\mathcal{X}^{\otimes\mathbb{N}}$, the initial distribution ν on $(\mathsf{X}, \mathcal{X})$ and the Markov kernel P allows to define a unique distribution on $(\mathsf{X}^{\mathbb{N}}, \mathcal{X}^{\otimes\mathbb{N}})$.

Conversely, assume that for all $t \in \mathbb{N}$, (5.5) is satisfied. Then, using the Tonelli-Fubini theorem, we may write

$$\begin{aligned}&\mathbb{E}[f_0(X_0)\dots f_{t-1}(X_{t-1})f_t(X_t)] \\ &= \int \nu(\mathrm{d}x_0)\prod_{s=1}^{t-1} P(x_{s-1}, \mathrm{d}x_s)\prod_{s=0}^{t-1} f_s(x_s)\int P(x_{t-1}, \mathrm{d}x_t)f_t(x_t) \\ &= \mathbb{E}[f_0(X_0)\dots f_{t-1}(X_{t-1})Pf_t(X_{t-1})],\end{aligned}$$

showing that for any $\mathcal{F}_{t-1}$-measurable random variable $Y = f_0(X_0)\dots f_{t-1}(X_{t-1})$, $\mathbb{E}[Yf_t(X_t)] = \mathbb{E}[YPf_t(X_{t-1})]$, which implies that $\mathbb{E}\left[f_t(X_t) \mid \mathcal{F}_{t-1}\right] = Pf_t(X_{t-1})$. Hence, an adapted stochastic process satisfying (5.5) is an homogeneous Markov chain with initial distribution ν and Markov kernel P. We summarize these conclusions in the following theorem.

Theorem 5.10. *Let $\{(X_t, \mathcal{F}_t),\ t \in \mathbb{N}\}$ be a Markov chain on $(\mathsf{X}, \mathcal{X})$ with initial distribution ν and Markov kernel P. For any $f \in \mathbb{F}_b(\mathsf{X}^{t+1}, \mathcal{X}^{\otimes(t+1)})$, and $t \in \mathbb{N}$, we have*

$$\mathbb{E}[f(X_0, \dots, X_t)] = \nu \otimes P^{\otimes t}(f). \tag{5.6}$$

Conversely, let $\{X_t,\ t \in \mathbb{N}\}$ be a stochastic process on $(\mathsf{X}, \mathcal{X})$ satisfying (5.6) for some probability ν and a Markov kernel P on $\mathsf{X} \times \mathcal{X}$. Then, $\{X_t,\ t \in \mathbb{N}\}$ is a Markov chain with initial distribution ν and transition probability P.

Proof. The essence of the proof is presented above; technical details are filled in Exercise 5.10 and Exercise 5.11. ∎

Example 5.11 (Autoregressive process of order 1). Consider the AR(1) process

$$X_{t+1} = \phi X_t + Z_{t+1}, \quad t \in \mathbb{N}, \tag{5.7}$$

where $\{Z_t,\ t \in \mathbb{N}^*\}$ is a sequence of zero-mean i.i.d. random variables independent from X_0. For $t \geq 0$, define $\mathcal{F}_t = \sigma(X_0, Z_s, s \leq t)$. The process $\{(X_t, \mathcal{F}_t),\ t \in \mathbb{N}\}$ given by (5.7) is a Markov chain. For any $x \in \mathsf{X}$ and $A \in \mathcal{B}(\mathbb{R})$, the Markov kernel of this chain is given by $P(x, A) = \mathbb{E}[\mathbb{1}_A(\phi x + Z_0)] = \mu(A - \phi x)$, where μ is the distribution of Z_0. If the law of Z_0 has a density q, with respect to Lebesgue's measure on $\mathbb{R}$, then the Markov kernel also has a density $p(x, y) = q(y - \phi x)$, i.e.

$$P(x, A) = \int_A q(y - \phi x)\,\mathrm{Leb}(\mathrm{d}y).$$

◇

Example 5.12 (ARCH(1) process). An autoregressive conditional heteroscedastic (ARCH) model of order 1 is defined as

$$X_t = \sigma_t Z_t \,, \quad \sigma_t^2 = \alpha_0 + \alpha_1 X_{t-1}^2 \,, \quad t \geq 1 \,, \tag{5.8}$$

where the coefficients α_0, α_1 are positive and $\{Z_t,\, t \in \mathbb{N}\}$ is an i.i.d. sequence such that $\mathbb{E}[Z_0] = 0$, $\mathbb{E}\left[Z_0^2\right] = 1$, and $\{Z_t,\, t \in \mathbb{N}\}$ is independent of X_0. Equation (5.8) may be equivalently rewritten as

$$X_t = \sqrt{\alpha_0 + \alpha_1 X_{t-1}^2}\, Z_t \,,$$

which shows that $\{X_t,\, t \in \mathbb{N}\}$ is an homogeneous Markov chain. The Markov kernel of this chain is given by $P(x,A) = \mathbb{P}(\sqrt{\alpha_0 + \alpha_1 x^2} Z \in A)$, for $x \in \mathbb{R}$ and $A \in \mathcal{B}(\mathbb{R})$. If the distribution of Z_0 has a density q with respect to the Lebesgue measure on $(\mathbb{R}, \mathcal{B}(\mathbb{R}))$, then the Markov kernel P has a density p given by

$$p(x,x') = \frac{1}{\sqrt{\alpha_0 + \alpha_1 x^2}}\, q\left(\frac{x'}{\sqrt{\alpha_0 + \alpha_1 x^2}}\right) .$$

The square volatility $\{\sigma_t^2, t \geq 1_t,\, t \in \mathbb{Z}\}$ satisfies the recursion: for $t \geq 2$,

$$\sigma_t^2 = \alpha_0 + \alpha_1 Z_{t-1}^2 \sigma_{t-1}^2$$

and $\sigma_1^2 = \alpha_0 + \alpha_1 X_0^2$. This is again a Markov chain with state space $\mathbb{R}^+$ and Markov kernel Q defined by $Q(s,A) = \mathbb{P}(\alpha_0 + \alpha_1 s Z^2 \in A)$ for all $s \in \mathbb{R}^+$ and $A \in \mathcal{B}(\mathbb{R}^+)$. ◇

Example 5.13 (A simple Markov bilinear process). Consider the following bilinear process

$$X_t = aX_{t-1} + bZ_t X_{t-1} + Z_t \,,$$

where $\{Z_t,\, t \in \mathbb{N}\}$ is an i.i.d. sequence independent from X_0. The Markov kernel of the chain $\{X_t,\, t \in \mathbb{N}\}$ is given by $P(x,A) = \mathbb{P}(ax + (1+bx)Z \in A)$ for $x \in \mathbb{R}$ and $A \in \mathcal{B}(\mathbb{R})$. If the distribution of Z_0 has a density q with respect to the Lebesgue measure and if $x \neq -1/b$, then the Markov kernel P has a density, denoted p, given by

$$p(x,x') = \frac{1}{|1+bx|}\, q\left(\frac{x'-ax}{1+bx}\right) .$$

If $x = -1/b$, then $P(-1/b, \cdot) = \delta_{-a/b}$. ◇

5.4 Canonical representation

In this section, we show that, given an initial distribution $\nu \in \mathbb{M}_1(\mathcal{X})$ and a Markov kernel P on $\mathsf{X} \times \mathcal{X}$, we can construct a filtered probability space and a Markov chain with initial distribution ν and transition kernel P. A first construction has been given by Theorem 5.24, under some topological restriction. The following construction imposes no additional assumption on $(\mathsf{X}, \mathcal{X})$. Let $\mathsf{X}^{\mathbb{N}}$ be the set of X-valued sequences

$w = (w_0, \dots, w_n, \dots)$ endowed with the product σ-field $\mathcal{X}^{\otimes \mathbb{N}}$. We define the coordinate process $\{X_t,\ t \in \mathbb{N}\}$ by

$$X_t(w) = w_t \,. \tag{5.9}$$

The natural filtration of the coordinate process is denoted by $\{\mathcal{F}_t^X,\ t \in \mathbb{N}\}$; by construction, $\bigvee_{n \geq 0} \mathcal{F}_n^X = \mathcal{X}^{\otimes \mathbb{N}}$.

Theorem 5.14. *Let P be a Markov kernel P on $\mathsf{X} \times \mathcal{X}$ and $\nu \in \mathbb{M}_1(\mathcal{X})$. Then, there exists a unique probability measure $\mathbb{P}_\nu$ on $(\mathsf{X}^{\mathbb{N}}, \mathcal{X}^{\otimes \mathbb{N}})$ such that the coordinate process is a Markov chain with initial distribution ν and transition probability P.*

Proof. Set for $t \in \mathbb{N}$ and $f \in \mathbb{F}_+(\mathsf{X}^{t+1}, \mathcal{X}^{\otimes(t+1)})$,

$$\mu_t(f) = \nu \otimes P^{\otimes t}(f) = \int \nu(\mathrm{d}x_0) \int P(x_0, \mathrm{d}x_1) \dots \int P(x_{t-1}, \mathrm{d}x_t) f(x_0, x_1, \dots, x_t) \,.$$

By Proposition 5.8, $\mu_t \in \mathbb{M}_1(\mathcal{X}^{\otimes(t+1)})$ and the family of finite dimensional distributions $\{\mu_t,\ t \in \mathbb{N}\}$ satisfies the usual consistency conditions. We conclude by applying Theorem 1.5. ■

For $x \in \mathsf{X}$, we use the shorthand notation $\mathbb{P}_x = \mathbb{P}_{\delta_x}$. We denote by $\mathbb{E}_\nu$ the expectation associated to $\mathbb{P}_\nu$. For all $A \in \mathcal{F}$, the function $x \mapsto \mathbb{P}_x(A)$ is $\mathcal{X}$-measurable. For all $\nu \in \mathbb{M}_1(\mathcal{X})$ and $A \in \mathcal{F}$, $\mathbb{P}_\nu(A) = \int_{\mathsf{X}} \mathbb{P}_x(A) \nu(\mathrm{d}x)$.

5.5 Invariant measures

Stochastic processes are *strict-sense stationary* if, for any integer k, the distribution of the random vector $(X_t, \dots, X_{t+k})$ does not depend on the time-shift t. In general, a Markov chain will not be stationary. For example, suppose we define an AR(1) model $\{X_t,\ t \in \mathbb{N}\}$, as $X_t = \phi X_{t-1} + Z_t$ where $|\phi| < 1$ and $Z_t \sim \text{iid N}(0, \sigma^2)$. To initialize the process, set for example $X_0 = Z_0$. In this case, $X_0 \sim \mathrm{N}(0, \sigma^2)$, but $X_1 = \phi X_0 + Z_1 \sim \mathrm{N}(0, \sigma^2(1 + \phi^2))$, so that X_0 and X_1 do not have the same distribution. A simple fix is to put $X_0 = Z_0 / \sqrt{1 - \phi^2}$, in which case

$$X_t = \phi^t X_0 + \sum_{s=0}^{t-1} \phi^s Z_{t-s} \sim \mathrm{N}(0, \sigma^2/(1 - \phi^2)) \,,$$

for all $t \geq 0$.

Under appropriate conditions on the Markov kernel P, it is possible to produce a stationary process with a proper choice of the initial distribution π. Assuming that such a distribution exists, the stationarity of the marginal distribution implies that $\mathbb{E}_\pi[\mathbb{1}_A(X_0)] = \mathbb{E}_\pi[\mathbb{1}_A(X_1)]$ for any $A \in \mathcal{X}$. This can equivalently be written as $\pi(A) = \pi P(A)$, or $\pi = \pi P$. We can iterate to give, for any $h \in \mathbb{N}$, $\pi P^h = \pi$. For all integers h and n, and all $A_0, \dots, A_n \in \mathcal{X}$,

$$\mathbb{P}_\pi(X_h \in A_0, X_{h+1} \in A_1, \dots, X_{h+n} \in A_n) = \int \cdots \int \pi P^h(\mathrm{d}x_0) \prod_{i=1}^n P(x_{i-1},\, \mathrm{d}x_i) \mathbb{1}_{A_i}(x_i)$$

$$= \int \cdots \int \pi(\mathrm{d}x_0) \prod_{i=1}^n P(x_{i-1},\, \mathrm{d}x_i) \mathbb{1}_{A_i}(x_i) = \mathbb{P}_\pi(X_0 \in A_0,\, X_1 \in A_1,\, \dots,\, X_n \in A_n) \,,$$

which shows that, for any integers h and n, the random vectors $(X_h, X_{h+1}, \ldots, X_{h+n})$ and $(X_0, X_1, \ldots, X_n)$ have the same distributions. Therefore, the Markov property implies that all finite-dimensional distributions of $\{X_t,\ t \in \mathbb{N}\}$ are also invariant under translation in time. These considerations lead to the definition of the *invariant measure*.

Definition 5.15 (Invariant measure). *Given a Markov kernel P, a σ-finite measure π on $(\mathsf{X}, \mathcal{X})$ with the property*

$$\pi(A) = \pi P(A)\,, \quad A \in \mathcal{X}\,,$$

will be called invariant (with respect to P).

If an invariant measure is finite, it may be normalized to an *invariant probability measure*. In general, there may exist more than one invariant measure, and if X is not finite, an invariant measure may not exist. As a trivial example, consider $\mathsf{X} = \mathbb{N}$ and $P(x, x+1) = 1$.

It turns out that we can extend a stationary Markov chain to have time t take on negative values, as well.

Proposition 5.16. *Let $(\mathsf{X}, \mathcal{X})$ be a measurable space. Let P a Markov kernel on $(\mathsf{X}, \mathcal{X})$ admitting π has an invariant distribution. Then, there exists on some probability space $(\Omega, \mathcal{F}, \mathbb{P})$ a stochastic process $\{X_t,\ t \in \mathbb{Z}\}$ such that, for any integer p, any p-tuple $t_1 < t_2 < \cdots < t_p$, and any $A \in \mathcal{X}^{\otimes p}$,*

$$\begin{aligned}\mathbb{P}((X_{t_1}, X_{t_2}, \ldots, X_{t_p}) \in A) \\ = \int \cdots \int \pi(\mathrm{d}x_1) P^{t_2 - t_1}(x_1, \mathrm{d}x_2) \ldots P^{t_p - t_{p-1}}(x_{p-1}, \mathrm{d}x_p) \mathbb{1}_A(x_1, \ldots, x_p)\,. \quad (5.10)\end{aligned}$$

Proof. This is a direct consequence of the Kolmogorov extension theorem; see Theorem 1.5. ∎

Remark 5.17. We may take $\Omega = \prod_{t=-\infty}^{\infty} \mathsf{X}_t$ and $\mathcal{F} = \bigvee_{t=-\infty}^{\infty} \mathcal{X}_t$, where $(\mathsf{X}_t, \mathcal{X}_t)$ is a copy of $(\mathsf{X}, \mathcal{X})$. For $t \in \mathbb{Z}$, X_t is the t-th coordinate mapping, i.e. for $\{\omega_t,\ t \in \mathbb{Z}\}$ an element of Ω, $X_t(\omega) = \omega_t$. This is the canonical two-sided extension of a stationary Markov chain.

Example 5.18 (Markov chain over a finite state-space). Consider a Markov chain on a finite state space $\mathsf{X} = \{1, \ldots, n\}$ with transition kernel P. A probability measure is a vector ξ with nonnegative entries summing to 1, $\sum_x \xi(x) = 1$. If ξ is the initial distribution, then after one step, the distribution of the chain is ξP; after t steps, the distribution is ξP^t. The probability π is stationary if and only if $\pi P = \pi$. This means that 1 should be an eigenvalue of P and in this case π is the left-eigenvector of P associated to the eigenvalue 1. It may be shown that, provided there exists an integer m such that $P^m(x, x') > 0$ for all $(x, x') \in \mathsf{X} \times \mathsf{X}$ (the Markov kernel P is said to be irreducible in such a case), such distribution exists and is unique. ◇

Example 5.19 (Gaussian AR(1) processes). Consider a Gaussian AR(1) process, $X_t = \mu + \phi X_{t-1} + \sigma Z_t$, where $\{Z_t,\ t \in \mathbb{N}\}$ is an i.i.d. sequence of standard Gaussian random variables, independent of X_0. Assume that $|\phi| < 1$ and that X_0 is Gaussian with mean μ_0 and variance γ_0^2. Then X_1 is Gaussian with mean $\mu + \phi\mu_0$, and variance $\phi^2\gamma_0^2 + \sigma^2$. If we choose

$$\begin{cases} \mu + \phi\mu_0 = \mu_0 \\ \phi^2\gamma_0^2 + \sigma^2 = \gamma_0^2 \end{cases} \quad \Rightarrow \quad \begin{cases} \mu_0 = \mu/(1-\phi) \\ \gamma_0^2 = \sigma^2/(1-\phi^2) \end{cases}$$

then X_1 and X_0 have the same distribution. Therefore, the Gaussian distribution with mean $\mu/(1-\phi)$ and variance $\sigma^2/(1-\phi^2)$ is a stationary distribution. We will show later that this distribution is unique. ◇

Example 5.20. Consider a Markov chain whose state space $\mathsf{X} = (0,1)$ is the open unit interval. If the chain is at x, it picks one of the two intervals $(0, x)$ or $(x, 1)$ with equal probability $1/2$, and then moves to a point y which is uniformly distributed in the chosen interval. This Markov chain has a transition density with respect to Lebesgue measure on the interval $(0,1)$, which is given by

$$k(x,y) = \frac{1}{2}\frac{1}{x}\mathbb{1}_{(0,x)}(y) + \frac{1}{2}\frac{1}{1-x}\mathbb{1}_{(x,1)}(y) . \tag{5.11}$$

The first term in the sum corresponds to a move from x to the interval $(0,x)$; the second, to a move from x to the interval $(x,1)$.

This Markov chain can be equivalently represented as an iterated random sequence. Let $\{U_t,\ t \in \mathbb{N}\}$ be a sequence of i.i.d. random variable uniformly distributed on the interval $(0,1)$. Let $\{\varepsilon_t,\ t \in \mathbb{N}\}$ be a sequence of i.i.d. random Bernoulli variables distributed with probability of success $1/2$, independent of $\{U_t,\ t \in \mathbb{N}\}$. Let X_0, the initial state, be distributed according to some initial distribution ξ on $(0,1)$, and be independent of $\{U_t,\ t \in \mathbb{N}\}$ and $\{\varepsilon_t,\ t \in \mathbb{N}\}$. Define the sequence $\{X_t,\ t \in \mathbb{N}\}$ for $t \geq 1$ as follows

$$X_t = \varepsilon_t\left[X_{t-1}U_t\right] + (1-\varepsilon_t)\left[X_{t-1} + U_t(1-X_{t-1})\right] , \tag{5.12}$$

Begin by assuming that the stationary distribution has a density p with respect to Lebesgue measure. From (5.11),

$$p(y) = \int_0^1 k(x,\ y)p(x)\mathrm{d}x = \frac{1}{2}\int_y^1 \frac{p(x)}{x}\mathrm{d}x + \frac{1}{2}\int_0^y \frac{p(x)}{1-x}\mathrm{d}x . \tag{5.13}$$

Differentiation gives

$$p'(x) = -\frac{1}{2}\frac{p(x)}{x} + \frac{1}{2}\frac{p(x)}{1-x} \quad \text{or} \quad \frac{p'(x)}{p(x)} = \frac{1}{2}\left(-\frac{1}{x} + \frac{1}{1-x}\right).$$

The solutions of this first-order non-linear differential equation are given by

$$f_C(x) = \frac{C}{\sqrt{x(1-x)}} , \tag{5.14}$$

where $C \in \mathbb{R}$ is a constant. Note that for $x \in (0,1)$ and $C \geq 0$, $f_C(x) \geq 0$. For all $z \in (0,1)$,

$$\int_0^z f_C(y)\mathrm{d}y = 2C\arcsin(\sqrt{z}) . \tag{5.15}$$

Choosing $C = 1/\pi$ makes $\int_0^1 f_{1/\pi}(y)\mathrm{d}y = 1$. Therefore, the *arcsine density* $p = f_{1/\pi}$ is a stationary distribution. It is the unique stationary distribution admitting a density with respect to Lebesgue measure. We will show later that p is indeed the unique stationary distribution. ◇

5.6 Observation-driven models

Of course there are many time series models that are not Markov chains. However, in many useful situations, Markov chains are not very far from the surface.

Definition 5.21 (Observation-driven model). *Let* $(\mathsf{X},\mathcal{X})$ *and* $(\mathsf{Y},\mathcal{Y})$ *be a measurable space,* Q *be Markov kernel on* $\mathsf{X}\times\mathcal{Y}$ *and* $(x,y)\mapsto f_y(x)$ *a measurable function from* $(\mathsf{X}\times\mathsf{Y},\mathcal{X}\otimes\mathcal{Y})$ *to* $(\mathsf{X},\mathcal{X})$. *Let* $(\Omega,\mathcal{F},\{\mathcal{F}_t,\ t\in\mathbb{Z}\},\mathbb{P})$ *be a filtered probability space (see Definition A.13).*

An observation-driven time series model *is a stochastic process* $\{(X_t,Y_t),\ t\in\mathbb{N}\}$ *adapted to* $(\Omega,\mathcal{F},\{\mathcal{F}_t,\ t\in\mathbb{Z}\},\mathbb{P})$ *taking values in* $\mathsf{X}\times\mathsf{Y}$ *satisfying the following recursions: for all* $t\in\mathbb{N}^*$,

$$\mathbb{P}\left[Y_t\in A \mid \mathcal{F}_{t-1}\right] = \mathbb{P}\left[Y_t\in A \mid X_{t-1}\right] = Q(X_{t-1},A)\,,\quad \textit{for any } A\in\mathcal{Y}\,, \tag{5.16}$$

$$X_t = f_{Y_t}(X_{t-1})\,. \tag{5.17}$$

The name *observation-driven* models was introduced in Cox (1981), but the definition used here is slightly different from the one used in this original contribution. For any integer $t>1$, $X_{t-1} = f_{Y_{t-1}}\circ f_{Y_{t-2}}\circ\cdots\circ f_{Y_1}(X_0)$: therefore X_{t-1} is a function of the trajectory, up to time $t-1$ and the initial condition X_0. The state X_{t-1} summarizes the information available on the conditional distribution of Y_t given $X_0,Y_1,\ldots,Y_{t-1}$. For any $C\in\mathcal{X}\otimes\mathcal{Y}$, and $t\geq 1$, we get

$$\mathbb{E}\left[\mathbb{1}_C(X_t,Y_t) \mid \mathcal{F}_{t-1}\right] = \mathbb{E}\left[\mathbb{1}_C(f_{Y_t}(X_{t-1})) \mid \mathcal{F}_{t-1}\right] = \int Q(X_{t-1},\mathrm{d}y)\mathbb{1}_C(f_y(X_{t-1}))\,,$$

showing that $\{((X_t,Y_t),\mathcal{F}_t),t\geq 0\}$ is a Markov chain on the product space $\mathsf{X}\times\mathsf{Y}$, with transition kernel $P((x,y),C) = \int Q(x,\mathrm{d}y')\mathbb{1}_C(f_{y'}(x),y')$, for any $C\in\mathcal{X}\otimes\mathcal{Y}$. Note that this transition kernel depends only upon $x\in\mathsf{X}$. For any $A\in\mathcal{X}$, we may write similarly,

$$\mathbb{E}\left[\mathbb{1}_A(X_t) \mid \mathcal{F}_{t-1}\right] = \mathbb{E}\left[\mathbb{1}_A(f_{Y_t}(X_{t-1})) \mid \mathcal{F}_{t-1}\right] = \int Q(X_{t-1},\mathrm{d}y)\mathbb{1}_A(f_y(X_{t-1}))\,,$$

showing that $\{(X_t,\mathcal{F}_t),\ t\in\mathbb{N}\}$ is also a Markov chain on X, with transition kernel $H(x,A) = \int Q(x,\mathrm{d}y)\mathbb{1}_A(f_y(x))$, for any $x\in\mathsf{X}$ and $A\in\mathcal{X}$.

Example 5.22 (ARMA(1,1) model). Consider for example an ARMA(1,1) model,

$$Y_t - \phi_1 Y_{t-1} = Z_t + \theta_1 Z_{t-1}\,,\quad t\geq 1\,, \tag{5.18}$$

where $\{Z_t,\ t\in\mathbb{N}\}$ is a sequence of i.i.d. random variables with density q with respect to the Lebesgue measure on $\mathbb{R}$, and $\{Z_t,\ t\in\mathbb{N}\}$ is independent of Y_0, which has distribution ξ. The process of interest, $\{Y_t,\ t\in\mathbb{N}\}$ is referred to as the *observations*. Consider $X_t = [Y_t\,, Z_t]' \in \mathsf{X} = \mathbb{R}^2$. Since $Y_t = [\phi_1\,,\theta_1]X_{t-1} + Z_t$, the conditional distribution of Y_t given $\mathcal{F}_{t-1} = \sigma((X_s, Y_s), 0 \leq s \leq t-1)$ is given for any $A \in \mathcal{B}(\mathbb{R})$ by

$$\mathbb{P}\left[Y_t \in A \mid \mathcal{F}_{t-1}\right] = \int_A q(y - [\phi_1, \theta_1] X_{t-1}) \lambda(\mathrm{d}y) ,$$

and

$$X_t = \begin{bmatrix} Y_t \\ Z_t \end{bmatrix} = \begin{bmatrix} Y_t \\ Y_t \end{bmatrix} - \begin{pmatrix} 0 & 0 \\ \phi_1 & \theta_1 \end{pmatrix} X_{t-1} .$$

◇

Example 5.23 (GARCH(1,1) model). The GARCH(1,1) model is defined as

$$Y_t = \sigma_t \varepsilon_t , \quad \sigma_t^2 = \alpha_0 + \beta_1 \sigma_{t-1}^2 + \alpha_1 Y_{t-1}^2 , \quad t \geq 1 , \tag{5.19}$$

where the driving noise $\{\varepsilon_t,\, t \in \mathbb{N}\}$ is an i.i.d. sequence, σ_0^2 is a random variable distributed according to some initial distribution ξ on $(\mathbb{R}^+, \mathcal{B}(\mathbb{R}^+))$ assumed to be independent of $\{\varepsilon_t,\, t \in \mathbb{N}\}$. The case $\beta_1 = 0$ corresponds to an ARCH(1) model.

The GARCH(1,1) model can be cast into the framework of observation-driven models by setting $X_t = \sigma_{t+1}^2$ and $f_y(x) = \alpha_0 + \alpha_1 y^2 + \beta_1 x$.

The square volatility $\{\sigma_t^2,\, t \in \mathbb{N}\}$ satisfies the following stochastic recurrence equation

$$\sigma_t^2 = \alpha_0 + \beta_1 \sigma_{t-1}^2 + \alpha_1 \sigma_{t-1}^2 \varepsilon_{t-1}^2 = \alpha_0 + (\beta_1 + \alpha_1 \varepsilon_{t-1}^2) \sigma_{t-1}^2 , \quad t \geq 1 , \tag{5.20}$$

By construction, $\{\sigma_t^2,\, t \in \mathbb{N}\}$ is a Markov chain. By iterating the relation $\sigma_t^2 = \alpha_0 + \beta_1 \sigma_{t-1}^2 + \alpha_1 Y_{t-1}^2$ backwards in time, we get

$$\sigma_t^2 = \sum_{j=0}^{t-1} \beta_1^j (\alpha_0 + \alpha_1 Y_{t-j-1}^2) + \beta_1^t \sigma_0^2 , \tag{5.21}$$

showing that the state σ_t^2 at time t is a deterministic function of the lagged observations $Y_0, \ldots, Y_{t-1}$ and of the initial state σ_0^2. ◇

5.7 Iterated random functions

Under weak conditions on the structure of the state space, every homogeneous Markov chain $\{X_t,\, t \in \mathbb{N}\}$ may be represented as a functional autoregressive process, i.e., $X_{t+1} = f_{Z_{t+1}}(X_t)$ where $\{Z_t,\, t \in \mathbb{N}\}$ is a strong white noise and X_0 is independent of $\{Z_t,\, t \in \mathbb{N}\}$ and is distributed according to some initial probability ν. For simplicity, consider the case of a real valued Markov chain $\{X_t,\, t \in \mathbb{N}\}$ with initial distribution ν and Markov kernel P. Let X be a real-valued random variable and let $F(x) = \mathbb{P}(X \leq x)$ be the cumulative distribution function of X. Let F^{-1} be the quantile function, defined as the generalized inverse of F by

$$F^{-1}(u) = \inf\{x \in \mathbb{R} : F(x) \geq u\} . \tag{5.22}$$

The right continuity of F implies that $u \leq F(x) \Leftrightarrow F^{-1}(u) \leq x$. Therefore, if U is uniformly distributed on $[0,1]$, $F^{-1}(U)$ has the same distribution as X, since $\mathbb{P}(F^{-1}(U) \leq t) = \mathbb{P}(U \leq F(t)) = F(t) = \mathbb{P}(X \leq t)$.

Define $F_0(t) = \nu((-\infty, t])$ and $g = F_0^{-1}$. Consider the function F from $\mathbb{R} \times \mathbb{R}$ to $[0,1]$ defined by $F(x, x') = P(x, (-\infty, x'])$. Then, for all $x \in \mathbb{R}$, $F(x, \cdot)$ is a cumulative distribution function; the associated quantile function $f(x, \cdot)$ is

$$f(x,u) = f_u(x) = \inf\left\{x' \in \mathbb{R} : F(x,x') \geq u\right\} . \tag{5.23}$$

The function $(x,u) \mapsto f_u(x)$ is a Borel function since $(x,x') \mapsto F(x,x')$ is itself a Borel function; see Exercise 5.25. If U is uniformly distributed on $[0,1]$, then, for all $x \in \mathbb{R}$ and $A \in \mathcal{B}(\mathbb{R})$, we have

$$\mathbb{P}\left[f_U(x) \in A\right] = P(x,A) .$$

Let $\{U_t,\ t \in \mathbb{N}\}$ be a sequence of i.i.d. random variables, uniformly distributed on $[0,1]$. Define a sequence of random variables $\{X_t,\ t \in \mathbb{N}\}$ by $X_0 = g(U_0)$ and for $t \geq 0$,

$$X_{t+1} = f_{U_{t+1}}(X_t) .$$

Then, $\{X_t,\ t \in \mathbb{N}\}$ is a Markov chain with Markov kernel P and initial distribution ν. The general case is proved in (Borovkov and Foss, 1992, Theorem 8).

Theorem 5.24. *Assume that $(\mathsf{X}, \mathcal{X})$ is a measurable space and that $\mathcal{X}$ is countably generated. Let P be a Markov kernel and ν be a probability on $(\mathsf{X}, \mathcal{X})$. Let $\{U_t,\ t \in \mathbb{N}\}$ be a sequence of i.i.d. random variables uniformly distributed on $[0,1]$. There exists a measurable application g from $([0,1], \mathcal{B}([0,1]))$ to $(\mathsf{X}, \mathcal{X})$ and a measurable application f from $(\mathsf{X} \times [0,1], \mathcal{X} \otimes \mathcal{B}([0,1]))$ to $(\mathsf{X}, \mathcal{X})$ such that the sequence $\{X_t,\ t \in \mathbb{N}\}$ defined by $X_0 = g(U_0)$ and $X_{t+1} = f_{U_{t+1}}(X_t))$ for $t \geq 0$, is a Markov chain with initial distribution ν and transition probability P defined on $\mathsf{X} \times \mathcal{X}$ by*

$$P(x,A) = \mathbb{P}\left[f_U(x) \in A\right] .$$

We consider the Markov chain $\{X_t,\ t \in \mathbb{N}\}$ defined by the following recurrence equation

$$X_t = f(X_{t-1}, Z_t) = f_{Z_t}(X_{t-1}) ,\ t \geq 1 , \tag{5.24}$$

where $\{Z_t,\ t \in \mathbb{N}\}$ is a strong white noise independent of the initial condition X_0. As shown in Theorem 5.24, most Markov chains can be represented in this way, but the function f will in general not display any useful property. We assume in this section that the function f has some contraction properties. We will establish, under these conditions, the existence and uniqueness of the invariant measure.

For $x_0 \in \mathsf{X}$, define the *forward iteration* starting from $X_0^{x_0} = x_0$ by

$$X_t^{x_0} = f_{Z_t}(X_{t-1}^{x_0}) = f_{Z_t} \circ \cdots \circ f_{Z_1}(x_0) ;$$

this is just a rewrite of equation (5.24). Now, define the *backward iteration* as

$$Y_t^{x_0} = f_{Z_1} \circ \cdots \circ f_{Z_t}(x_0) . \tag{5.25}$$

Of course, $Y_t^{x_0}$ has the same distribution as X_t for each $t \in \mathbb{N}$. We will show that, under

A4.37-A4.38, the forward process $\{X_t,\ t \in \mathbb{N}\}$ has markedly different behavior from the backward process $\{Y_t^{x_0},\ t \in \mathbb{N}\}$: the forward process moves *ergodically* in X, while the backward process converges to a limit $\mathbb{P}$-a.s. The distribution of this limit, which does not depend on the initial state x_0, is the unique stationary distribution.

Lemma 5.25. *Assume that there exists a $\mathbb{P}$-a.s. finite random variable Y_∞ such that for all $x_0 \in \mathsf{X}$, the backward process $\{Y_t^{x_0},\ t \in \mathbb{N}\}$ defined in* (5.25) *satisfies*

$$\lim_{t\to\infty} Y_t^{x_0} = Y_\infty\,, \quad \mathbb{P}\text{-a.s.} \tag{5.26}$$

Then, the Markov chain $\{X_t,\ t \in \mathbb{N}\}$ defined in (5.24) *with Markov kernel P implicitly defined by* (5.24)*, admits a unique invariant distribution π. In addition, π is the distribution of Y_∞ and for any $\xi \in \mathbb{M}_1(\mathcal{X})$, the sequence of probability measures $\{\xi P^t, t \in \mathbb{N}\}$ converges weakly to π.*

Proof. Denote by π the distribution of Y_∞. Since $\{Z_t,\ t \in \mathbb{Z}\}$ is an i.i.d. sequence, it holds that

$$Y_{t+1}^{x_0} = f_{Z_1} \circ \cdots \circ f_{Z_{t+1}}(x_0) \stackrel{\mathrm{d}}{=} f_{Z_0} \circ \cdots \circ f_{Z_t}(x_0) = f_{Z_0}(Y_t^{x_0})\,.$$

Since f_{Z_0} is continuous, passing to the limit implies that $Y_\infty \stackrel{\mathrm{d}}{=} f_{Z_0}(Y_\infty)$, hence π is an invariant probability measure for P.

We now prove that π is the unique invariant probability measure of P. For any $x \in \mathsf{X}$, the distribution of $X_t^x = f_{Z_t} \circ \cdots \circ f_{Z_1}(x)$ is $\delta_x P^t$. Since $\{Z_t,\ t \in \mathbb{N}\}$ is a strong white noise, for each $t \in \mathbb{N}$, $X_t^x = f_{Z_t} \circ \cdots \circ f_{Z_1}(x)$ has the same distribution as $Y_t^x = f_{Z_1} \circ \cdots \circ f_{Z_t}(x)$. Thus the sequence $\{X_t^x,\ t \in \mathbb{N}\}$ converges weakly to π, i.e., for all $h \in \mathbb{F}_{bc}(\mathsf{X}, \mathcal{X})$ and $x \in \mathsf{X}$, we get

$$\lim_{t\to\infty} \delta_x P^t h = \lim_{t\to\infty} \mathbb{E}\left[h(X_t^x)\right] = \lim_{n\to\infty} \mathbb{E}\left[h(Y_t^x)\right] = \pi(h)\,.$$

By dominated convergence, this yields, for any probability measure ξ and $h \in \mathbb{F}_{bc}(\mathsf{X}, \mathcal{X})$,

$$\lim_{t\to\infty} \xi P^t h = \lim_{t\to\infty} \int_{\mathsf{X}} P^t h(x) \xi(\mathrm{d}x) = \pi(h)\,. \tag{5.27}$$

If π' is an invariant distribution for P, then $\pi' P^t = \pi'$ for all $t \in \mathbb{N}$, so that (5.27) with $\xi = \pi'$ yields $\pi'(h) = \lim_{t\to\infty} \pi' P^t(h) = \pi(h)$ for any $h \in \mathbb{F}_{bc}(\mathsf{X}, \mathcal{X})$, showing that $\pi' = \pi$. ■

Example 5.26 (Example 5.20 cont.). Consider the model in Example 5.20,

$$X_t = \varepsilon_t U_t X_{t-1} + (1-\varepsilon_t)[X_{t-1} + U_t(1-X_{t-1})]\,, \tag{5.28}$$

where $\{U_t,\ t \in \mathbb{N}\}$ is i.i.d. uniform on $[0,1]$, $\{\varepsilon_t,\ t \in \mathbb{N}\}$ is i.i.d. Bernoulli with probability of success 1/2, $\{U_t,\ t \in \mathbb{Z}\}$, $\{\varepsilon_t,\ t \in \mathbb{Z}\}$ and X_0 are independent. Set $\mathbb{T} = [0,1] \times \{0,1\}$ and $\mathcal{T} = \mathcal{B}([0,1]) \otimes \mathcal{P}\{0,1\}$. Then, $X_t = f(X_{t-1}, Z_t)$ with $Z_t = (U_t, \varepsilon_t)$ and $f_{u,\varepsilon}(x)(u) = xu\varepsilon + (1-\varepsilon)[x + u(1-x)]$. For any $(x,y) \in [0,1] \times [0,1]$, $|f_{u,\varepsilon}(x) - f_{u,\varepsilon}(y)| \le K_{(u,\varepsilon)}|x-y|$ with

$$K_{u,\varepsilon} = \varepsilon u + (1-\varepsilon)(1-u)\,. \tag{5.29}$$

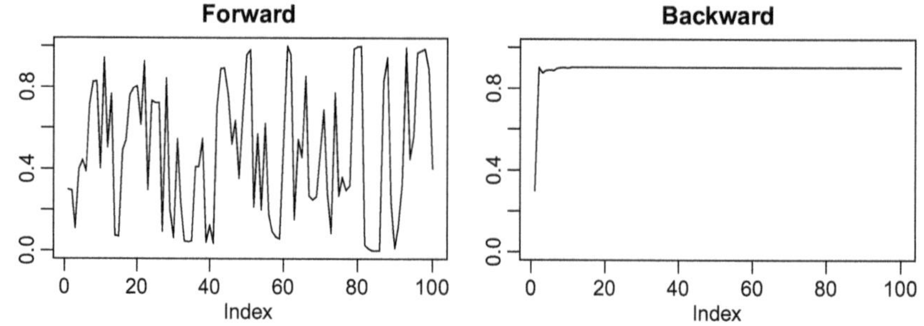

Figure 5.1 *Display for Example 5.26. The left hand panel shows ergodic behavior by the forward process; the right hand side panel shows convergence of the backward process.*

Figure 5.1 illustrates the difference between the backward process (right hand panel, convergence) and the forward process (left hand panel, ergodic behavior). Both processes start from $x_0 = 0.3$, and use the same random functions to move. The order in which the functions are composed is the only difference. In the right panel, the limit 0.2257 is random because it depends on the functions being iterated; we will see in Theorem 5.27 that this limiting value does not depend on the initial condition x_0. $\diamond$

We now state the main result of this section, i.e., conditions upon which the Markov chain (5.24) admits a unique stationary distribution.

Theorem 5.27. *Assume A4.37 and A4.38. Then, the Markov chain $\{X_t,\ t \in \mathbb{N}\}$ defined in* (5.24) *admits a unique invariant probability π, which is the distribution of the almost-sure limit of the sequence of random variables $\{Y_t^{x_0},\ t \in \mathbb{N}\}$, where $Y_t^{x_0} = f_{Z_1} \circ f_{Z_2} \circ \cdots \circ f_{Z_t}(x_0)$. In addition, for any $\xi \in \mathbb{M}_1(\mathcal{X})$, the sequence of probability measures $\{\xi P^t, t \in \mathbb{N}\}$ converges weakly to π.*

Proof. According to Lemma 5.25, we only need to show that the backward process $\{Y_t^{x_0},\ t \in \mathbb{N}\}$ converges $\mathbb{P}$-a.s. to a random variable Y_∞, which is $\mathbb{P}$-a.s. finite and does not depend on x_0. First note that for all $x_0, x \in \mathsf{X}$,

$$\begin{aligned} d(Y_t^{x_0}, Y_t^x) &= d(f_{Z_1}[f_{Z_2} \circ \cdots \circ f_{Z_t}(x_0)], f_{Z_1}[f_{Z_2} \circ \cdots \circ f_{Z_t}(x)]) \\ &\le K_{Z_1} d(f_{Z_2} \circ \cdots \circ f_{Z_t}(x_0), f_{Z_2} \circ \cdots \circ f_{Z_t}(x)) \,. \end{aligned}$$

Thus, by induction,

$$d(Y_t^{x_0}, Y_t^x) \le d(x_0, x) \prod_{i=1}^{t} K_{Z_i} \,. \tag{5.30}$$

Noting that $Y_{t+1}^{x_0} = Y_t^x$ with $x = f_{Z_{t+1}}(x_0)$, we obtain

$$d(Y_t^{x_0}, Y_{t+1}^{x_0}) \le d(x_0, f_{Z_{t+1}}(x_0)) \prod_{i=1}^{t} K_{Z_i} .$$

Since $\mathbb{E}\left[\ln K_{Z_0}\right] < 0$ (possibly $\mathbb{E}\left[\ln K_{Z_0}\right] = -\infty$), the strong law of large numbers implies that $\limsup_{t\to\infty} t^{-1}\sum_{i=1}^{t}\ln(K_{Z_t}) < 0$, $\mathbb{P}$-a.s. Thus,

$$\limsup_{t\to\infty}\left\{\prod_{i=1}^{t} K_{Z_i}\right\}^{1/t} = \exp\left\{\limsup_{t\to\infty} t^{-1}\sum_{i=1}^{t}\ln(K_{Z_i})\right\} < 1 \quad \mathbb{P}\text{-a.s.} \tag{5.31}$$

Next, since $\mathbb{E}\left[\ln_+(d(x_0, f_{Z_0}(x_0)))\right] < \infty$, by the strong law of large numbers, it holds that $\lim_{t\to\infty} t^{-1}\sum_{k=1}^{t}\ln_+ d(x_0, f_{Z_k}(x_0)) < \infty$ $\mathbb{P}$-a.s. which implies that $\lim_{t\to\infty} t^{-1}\ln_+ d(x_0, f_{Z_t}(x_0)) = 0$, $\mathbb{P}$-a.s.[1] Thus $\limsup_{t\to\infty} t^{-1}\ln d(x_0, f_{Z_t}(x_0)) \le 0$ $\mathbb{P}$-a.s. which, with (5.31), yields

$$\limsup_{t\to\infty}\left\{d(Y_t^{x_0}, Y_{t+1}^{x_0})\right\}^{1/t} = \limsup_{t\to\infty}\left\{d(x_0, f(x_0, Z_{t+1}))\prod_{i=1}^{t} K_{Z_i}\right\}^{1/t} < 1 \quad \mathbb{P}\text{-a.s.}$$

By the Cauchy root test, this implies that the series $\sum_{t=1}^{\infty} d(Y_t^{x_0}, Y_{t+1}^{x_0})$ is almost surely absolutely convergent. Since (X, d) is complete, this in turn implies that $\{Y_t^{x_0},\, t\in\mathbb{Z}\}$ is almost surely convergent to a $\mathbb{P}$-a.s. finite random variable. Denote by $Y_\infty^{x_0}$ the almost sure limit.

Using again (5.30) and $\lim_{t\to\infty}\prod_{i=1}^{t} K_{Z_i} = 0$ $\mathbb{P}$-a.s., we get that for any $x_0, x\in\mathsf{X}$, $\lim_{t\to\infty} d(Y_t^{x_0}, Y_t^{x}) = 0$ $\mathbb{P}$-a.s. so that $Y_\infty^{x_0} = Y_\infty^{x}$. The proof then follows by applying Lemma 5.25. ■

Example 5.28 (GARCH(1,1) model). $\{\sigma_t^2,\, t\in\mathbb{Z}\}$ is a Markov chain and, for $t \ge 1$,

$$\sigma_t^2 = \alpha_0 + \alpha_1\sigma_{t-1}^2 Z_{t-1}^2 + \beta_1\sigma_{t-1}^2 = f(\sigma_{t-1}^2, Z_{t-1}^2)\,,$$

where $f(x,z) = \alpha_0 + (\alpha_1 z^2 + \beta_1)x$ and $\{Z_t,\, t\in\mathbb{N}\}$ is a sequence of i.i.d. random variables. Thus the GARCH(1,1) model satisfies (4.75) with $K_z = \alpha_1 z^2 + \beta_1$. Consequently, a sufficient condition for the existence and uniqueness of an invariant distribution is

$$\mathbb{E}\left[\ln(\alpha_1 Z_1^2 + \beta_1)\right] < 0\,. \tag{5.32}$$

◇

Example 5.29 (Bilinear model). Consider the bilinear process $\{X_t,\, t\in\mathbb{N}\}$ defined by

$$X_t = aX_{t-1} + bZ_tX_{t-1} + Z_t\,, \tag{5.33}$$

where $\{Z_t,\, t\in\mathbb{N}\}$ is a sequence of integrable i.i.d. random variables and a and b are constants.

Tong (1981) studies this process and concludes that if $\{Z_t,\, t\in\mathbb{Z}\}$ is a Gaussian white noise, a sufficient condition for the existence of a stationary distribution is $\mathbb{E}[|a+bZ_0|] < 1$. Cline and Pu (2002) relax the Gaussian assumption and simply assume that the law Z_0 is absolutely continuous with respect to Lebesgue's measure

[1] Recall that if $\{u_k,\, k\in\mathbb{N}\}$ is such that $U_n/n = \sum_{k=1}^{n} u_k/n$ converges as n goes to infinity, then $u_n/n = U_n/n - [(n-1)/n]U_{n-1}/(n-1) \to 0$.

on $\mathbb{R}$, and its density is positive and lower semicontinuous. We will show that none of these assumptions on the distribution of Z_0 are necessary.

Define $f(x,z) = (a+bz)x+z$. The bilinear model (5.33) may be written as $X_t = f(X_{t-1}, Z_t)$. For any $(x,y,z) \in \mathbb{R}$, $|f(x,z) - f(y,z)| \leq |a+bz||x-y|$, which implies that A4.37 is satisfied with $K_z = |a+bz|$ as soon as $\mathbb{E}[\ln(|a+bZ_0|)] < 0$. If in addition, $\mathbb{E}[\ln_+(|Z_0|)] < \infty$, then A4.38 is also satisfied. Theorem 5.27 shows that the bilinear model (5.33) has a unique stationary distribution π, and that, starting from any initial distribution, the distribution of the iterates of the chain converge to π. ◇

Example 5.30 (Example 5.20 cont.). Using (5.29), we get

$$\mathbb{E}[\ln(K_Z)] = (1/2)\mathbb{E}[\ln(U_0)] + (1/2)\mathbb{E}[\ln(1-U_0)] = \mathbb{E}[\ln(U_0)] = -1 .$$

Therefore, A4.37 is satisfied and the Markov chain $\{X_t,\, t \in \mathbb{N}\}$ defined by (5.28) has a unique stationary distribution. ◇

Instead of assuming that the contraction condition is satisfied pathwise, we may alternatively try to consider a contraction condition in the p-th norm.

Theorem 5.31. *Assume A4.44 and A4.45. Then, the Markov chain $\{X_t,\, t \in \mathbb{N}\}$ defined in* (5.24) *admits a unique invariant probability π, which is the distribution of the almost-sure limit of the sequence of random variables $Y_t^{x_0} = f_{Z_1} \circ f_{Z_2} \circ \cdots \circ f_{Z_t}(x_0)$. In addition,*

$$\int_{\mathsf{X}} d^p(x_0, x)\pi(\mathrm{d}x) < \infty .$$

For any $\xi \in \mathbb{M}_1(\mathcal{X})$, the sequence of probability measures $\{\xi P^n, n \in \mathbb{N}\}$ converges weakly to π.

Proof. Using Lemma 5.25, we will show that the backward process $\{Y_t^{x_0},\, t \in \mathbb{N}\}$ converges $\mathbb{P}$-a.s. to a random variable Y_∞ which is $\mathbb{P}$-a.s. finite and does not depend on x_0. Since $\{Z_t,\, t \in \mathbb{N}\}$ is a strong white noise, we have

$$\begin{aligned}
(Y_t^{x_0}, Y_{t+1}^{x_0}) &= (f_{Z_1}[f_{Z_2} \circ \cdots \circ f_{Z_t}(x_0)], f_{Z_1}[f_{Z_2} \circ \cdots \circ f_{Z_{t+1}}(x_0)]) \\
&\stackrel{\mathrm{d}}{=} (f_{Z_0}[f_{Z_1} \circ \cdots \circ f_{Z_{t-1}}(x_0)], f_{Z_0}[f_{Z_1} \circ \cdots \circ f_{Z_t}(x_0)]) \\
&= (f_{Z_0}(Y_{t-1}^{x_0}), f_{Z_0}(Y_t^{x_0})) .
\end{aligned}$$

Applying (4.87), we obtain

$$\|d(Y_{t+1}^{x_0}, Y_t^{x_0})\|_p = \|d(f_{Z_0}(Y_t^{x_0}), f_{Z_0}(Y_{t-1}^{x_0}))\|_p \leq \alpha \|d(Y_t^{x_0}, Y_{t-1}^{x_0})\|_p .$$

Thus, by induction, $\|d(Y_{t+1}^{x_0}, Y_t^{x_0})\|_p \leq \alpha^t \|d(x_0, f_{Z_0}(x_0))\|_p$. By the Markov inequality, for all $\varepsilon > 0$,

$$\mathbb{P}(\beta^t d(Y_{t+1}^{x_0}, Y_t^{x_0}) > \varepsilon) \leq \frac{\beta^{tp}}{\varepsilon^p}(\|d(Y_{t+1}^{x_0}, Y_t^{x_0})\|_p)^p \leq \frac{(\alpha\beta)^{tp}}{\varepsilon^p}(\|d(x_0, f_{Z_0}(x_0))\|_p)^p ,$$

where β is chosen in $(1, 1/\alpha)$. The Borel-Cantelli Lemma then implies that $\lim_{t\to\infty} \beta^t d(Y_{t+1}^{x_0}, Y_t^{x_0}) = 0$, $\mathbb{P}$-a.s. and since $\beta > 1$, this in its turn yields that the sequence $\{Y_t^{x_0},\, t \in \mathbb{N}\}$ is $\mathbb{P}$-a.s. a Cauchy sequence in (X, d). Since (X, d) is complete, the sequence $\{Y_t^{x_0},\, t \in \mathbb{N}\}$ converges $\mathbb{P}$-a.s. to a random variable $Y_\infty^{x_0}$, which is $\mathbb{P}$-a.s. finite. We will now prove that $Y_\infty^{x_0}$ does not depend on x_0. For $x \in \mathsf{X}$ and $t \in \mathbb{N}$, define $Y_t^x = f_{Z_1} \circ \cdots \circ f_{Z_t}(x)$. Again, applying (4.87), we obtain

$$\|d(Y_t^{x_0}, Y_t^x)\|_p = \|d(f_{Z_0}(Y_{t-1}^{x_0}), f_{Z_0}(Y_{t-1}^x))\|_p \leq \alpha \|d(Y_{t-1}^{x_0}, Y_{t-1}^x)\|_p\,,$$

which implies $\|d(Y_t^{x_0}, Y_t^x)\|_p \leq \alpha^t \|d(x_0, x)\|_p$. As above, this implies that $d(Y_\infty^{x_0}, Y_\infty^x) = 0$, $\mathbb{P}$-a.s. Thus, $Y_\infty^{x_0}$ does not depend on x_0 and we can thus set $Y_\infty = Y_\infty^{x_0}$. Moreover, using $Y_\infty = \lim_{t\to\infty} Y_t^{x_0}$, $\mathbb{P}$-a.s., we obtain, by Fatou's lemma,

$$\begin{aligned}\|d(x_0, Y_\infty)\|_p &= \|\liminf_{t\to\infty} d(x_0, Y_t^{x_0})\|_p \leq \liminf_{t\to\infty} \|d(x_0, Y_t^{x_0})\|_p \\ &\leq \liminf_{t\to\infty} \sum_{n=0}^{t-1} \|d(Y_n^{x_0}, Y_{n+1}^{x_0})\|_p \leq (1-\alpha)^{-1} \|d(x_0, f_{Z_0}(x_0))\|_p < \infty\,.\end{aligned}$$

Thus,

$$\int_{\mathsf{X}} d^p(x_0, x)\pi(\mathrm{d}x) = \left(\|d(x_0, Y_\infty)\|_p\right)^p < \infty\,.$$

■

Example 5.32 (GARCH(1,1); Example 5.28 cont.). By Theorem 5.31, a sufficient condition for $\mathbb{E}_\pi[\sigma_1^2] < \infty$ is $\mathbb{E}\left[\alpha_1 Z_1^2 + \beta_1\right] < 1$. Since we have assumed that $\mathbb{E}\left[Z_1^2\right] = 1$, this condition boils down to $\alpha_1 + \beta_1 < 1$. It is easily seen that this is also a necessary condition, since the condition $\mathbb{E}_\pi[\sigma_1^2] < \infty$ implies that

$$\mathbb{E}_\pi[\sigma_t^2] = \alpha_0 + (\alpha_1 + \beta_1)\mathbb{E}_\pi[\sigma_{t-1}^2]\,, \quad \mathbb{E}_\pi[\sigma_1^2] = \mathbb{E}_\pi[\sigma_t^2] = \mathbb{E}_\pi[\sigma_{t-1}^2]\,,$$

which admits a finite solution only if $\alpha_1 + \beta_1 < \infty$. If, moreover,

$$\mathbb{E}_\pi[(\alpha_1 Z_1^2 + \beta_1)^p] < 1\,, \tag{5.34}$$

for some $p \geq 1$, then the stationary distribution of σ_t^2 has a finite moment of order p. For $p = 2$, this condition becomes $\alpha_1^2 \mathbb{E}\left[Z_1^4\right] + 2\alpha_1\beta_1 + \beta_1^2 < 1$. ◇

Example 5.33 (Bilinear model; Example 5.29 cont.). Assume that there exists $p \geq 1$ such that $\mathbb{E}\left[Z_0^p\right] < \infty$, then,

$$\|f(x, Z_0) - f(y, Z_0)\|_p = \|(a + bZ_0)\|_p |x - y|\,.$$

Thus, Assumption (4.87) holds if $\mathbb{E}\left[|a + bZ_0|^p\right] < 1$. In that case, (4.88) also holds and thus there exists an unique invariant probability measure π such that $\mathbb{E}\left[|X_0|^p\right] < \infty$ if the distribution of X_0 is π. ◇

Example 5.34 (Example 5.20 cont.). Using (5.29), for any $p \geq 1$, we have $\mathbb{E}\left[K_{U,\varepsilon}^p\right] \leq 2\mathbb{E}\left[U^p\right] = 2/(p+1)$, therefore, (4.87) is satisfied for any $p \geq 1$. (5.31) is satisfied showing that Theorem 5.31 is satisfied for any $p \geq 1$. We have already shown that the probability density $p(x) = \sqrt{x(1-x)}/\pi$ is a stationary distribution; Theorem 5.31 shows that this distribution is unique. In Figure 5.2, we have sampled

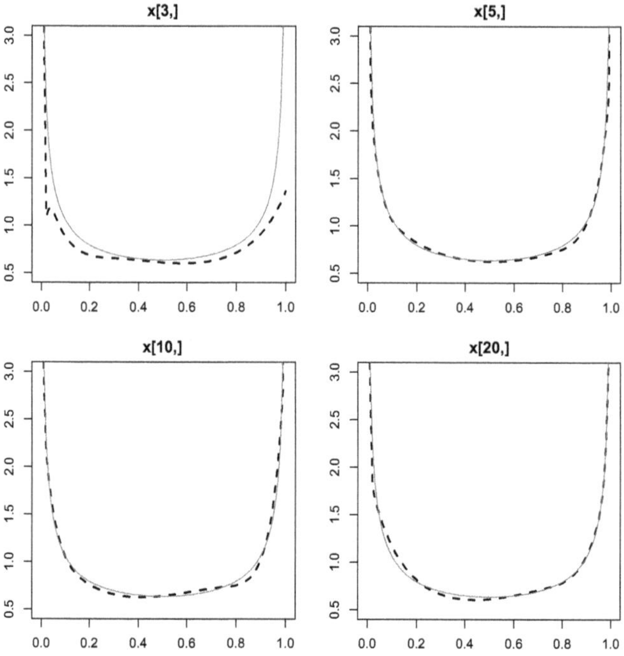

Figure 5.2 *Convergence of the stationary distribution for the iterative model* (5.28). *We have compared the stationary distribution (solid line) with the marginal distribution (dashed line) estimated using the log-spline estimator computed from* 10^4 *independent replications of the process.*

10^4 independent trajectories of the process started from $X_0 = 0.01$. We have displayed the marginal stationary distribution after 3, 5, 10 and 20 iterations, together with the limiting distribution. In this case, the convergence to the stationary distribution is very quick. ◇

Example 5.35 (Log-linear Poisson autoregression). Consider the following log-linear autoregressive models introduced by Fokianos and Tjøstheim (2011), defined as follows:

$$\mathcal{L}(Y_t|\mathcal{F}^Y_{t-1}) = \text{Poisson}(\exp(U_t)) \tag{5.35a}$$

$$U_t = a + bU_{t-1} + c\ln(1+Y_{t-1}),\ t \geq 1\,, \tag{5.35b}$$

where $\mathcal{F}^Y_t = \sigma(U_0, Y_s, s \leq t)$ and d, a, b are real-valued parameters. Let $\{N_k, k \geq 1\}$ be a sequence of independent unit rate homogeneous Poisson process on the real line, independent of U_0. Then $\{U_n,\ n \in \mathbb{N}\}$ can be expressed as $U_k = F(U_{k-1}, N_k)$, where F is the function defined on $\mathbb{R} \times \mathbb{N}^{\mathbb{R}}$ by

$$F(u,N) = a + bu + c\ln\{1 + N(\mathrm{e}^u)\}\,.$$

The transition kernel P of the Markov chain $\{U_t,\ t \in \mathbb{N}\}$ can be expressed as

$$Pf(v) = \mathbb{E}\left[f(a + bv + c\ln\{1 + N(\mathrm{e}^v)\})\right]\,,$$

where N is a unit rate homogeneous Poisson process.

Given the coefficients (a,b,c) and the initial intensity U_0, the log-intensity U_t can be expressed explicitly from the lagged responses by expanding (5.35b):

$$U_t = a\frac{1-b^t}{1-b} + b^t U_0 + c\sum_{i=0}^{t-1} b^i \ln(1+Y_{t-i-1}) \,.$$

Hence this model belongs to the class of observation-driven models. See Cox (1981). In this log-linear model, the lagged observation Y_t is fed into the autoregressive equation for U_t via the term $\ln(Y_{t-1}+1)$. Adding one to the integer valued observation is a standard way to avoid potential problems with zero counts. In addition, both the intensity and the counts Y_t are transformed onto the same logarithmic scale. Covariates can be included in the right-hand side of (5.35b).

Let us check that A4.44 holds for this model. By Minkowski's inequality, we have, for $v \geq u \geq 0$,

$$\|F(u,N)-F(v,N)\|_1 \leq |b||u-v| + |c| \left\| \ln\left(\frac{1+N(\mathrm{e}^v)}{1+N(\mathrm{e}^u)}\right)\right\|_1 \,.$$

We will prove below that for $v \geq u \geq 0$,

$$\mathbb{E}\left[\ln\left(\frac{1+N(\mathrm{e}^v)}{1+N(\mathrm{e}^u)}\right)\right] \leq v-u \,. \tag{5.36}$$

This yields, for $u,v \geq 0$,

$$\|F(v,N)-F(u,N)\|_1 \leq (|b|+|c|)|v-u| \,.$$

The contraction property A4.44 holds if $|b|+|c|<1$.

We now prove (5.36). Since N has independent increments, we can write $(1+N(e^v))/(1+N(e^u)) = 1+W/(1+U)$, where W and U are independent Poisson random variable with respective means $\mathrm{e}^v - \mathrm{e}^u$ and e^u. The function $x \mapsto \ln(1+x)$ is concave, thus, by Jensen's inequality, we obtain

$$\begin{aligned}\mathbb{E}\left[\ln\left(\frac{1+N(\mathrm{e}^v)}{1+N(\mathrm{e}^u)}\right)\right] &= \mathbb{E}\left[\ln\left(1+\frac{W}{1+U}\right)\right] \\ &\leq \ln\left(1+\mathbb{E}\left[\frac{W}{1+U}\right]\right) \\ &= \ln\left(1+(\mathrm{e}^v-\mathrm{e}^u)\mathbb{E}\left[\frac{1}{1+U}\right]\right) \,.\end{aligned}$$

Note now that

$$\mathbb{E}\left[\frac{1}{1+U}\right] = \mathrm{e}^{-\mathrm{e}^u}\sum_{k=0}^{+\infty}\frac{1}{1+k}\frac{\mathrm{e}^{ku}}{k!} = \mathrm{e}^{-u}\mathrm{e}^{-\mathrm{e}^u}\sum_{k=1}^{\infty}\frac{\mathrm{e}^{ku}}{k!} \leq \mathrm{e}^{-u} \,.$$

This yields

$$\mathbb{E}\left[\ln\left(\frac{1+N(\mathrm{e}^v)}{1+N(\mathrm{e}^u)}\right)\right] \leq \ln\left(1+(\mathrm{e}^v-\mathrm{e}^u)\mathrm{e}^{-u}\right) = v-u \,. \qquad \diamond$$

5.8 Markov chain Monte Carlo methods

5.8.1 Metropolis-Hastings algorithm

In this section, we introduce the *Metropolis-Hastings* algorithm, which is a generic example of Monte-Carlo Markov Chains (MCMC) algorithms. Assume that π is a probability density of interest on $\mathbb{R}^d$. Let $q(x,\,\cdot)$ be a *proposal* density that is easy to sample from. Note that, for simplicity, all these densities are considered with respect to the Lebesgue measure.

The Metropolis-Hastings algorithm proceeds in the following way. An initial starting value X_0 is chosen for the algorithm. Given X_t, a *candidate* Y_{t+1} is sampled from $q(X_t,\,\cdot)$, and is then accepted with probability $\alpha(X_t,\,Y_{t+1})$, given by

$$\alpha(x,\,y) = \begin{cases} \min\left(\frac{\pi(y)}{\pi(x)}\frac{q(y,x)}{q(x,y)},\,1\right) & \text{if } \pi(x)q(x,\,y) > 0 \\ 1 & \text{if } \pi(x)q(x,\,y) = 0\,. \end{cases} \tag{5.37}$$

If accepted, we set $X_{t+1} = Y_{t+1}$. Otherwise, Y_{t+1} is not accepted and we set $X_{t+1} = X_t$. This procedure produces a Markov chain, $\{X_t,\, t \in \mathbb{N}\}$, with transition kernel P given by

$$P(x,A) = \int \alpha(x,y)q(x,y)\mathbb{1}_A(y)\mathrm{Leb}(\mathrm{d}y) + \mathbb{1}_A(x)\int q(x,\,y)[1-\alpha(x,\,y)]\mathrm{Leb}(\mathrm{d}y)\,.$$

The Markov kernel P is *reversible* with respect to a probability measure π if for any functions f and $g \in \mathbb{F}_+(\mathsf{X},\mathcal{X})$,

$$\iint \pi(\mathrm{d}x)P(x,\mathrm{d}y)f(x)g(y) = \iint \pi(\mathrm{d}x)P(x,\mathrm{d}y)g(x)f(y)\,. \tag{5.38}$$

Lemma 5.36. *If P is reversible with respect to π, then π is a stationary distribution for P.*

Proof. Setting $f \equiv 1$, we indeed obtain for any function $g \in \mathbb{F}_+(\mathsf{X},\mathcal{X})$,

$$\iint \pi(\mathrm{d}x)P(x,\mathrm{d}y)g(y) = \iint \pi(\mathrm{d}x)P(x,\mathrm{d}y)g(x) = \int \pi(\mathrm{d}x)g(x)\,.$$

∎

For any $(x,y) \in \mathbb{R}^d \times \mathbb{R}^d$, we get

$$\pi(x)\alpha(x,y)q(x,y) = \pi(x)q(x,y) \wedge \pi(y)q(y,x) = \pi(y)\alpha(y,x)q(y,x)\,, \tag{5.39}$$

which implies that (5.38) is satisfied. Hence π is the invariant distribution of this Markov chain (we will find later conditions upon which the stationary distribution is unique). Note that the acceptance ratio $\alpha(x,y)$ only depends on the ratio $\pi(y)/\pi(x)$; therefore, we only need to know π up to a normalizing constant. In Bayesian inference, this property plays a crucial role.

Example 5.37 (Metropolis algorithm). The symmetric random walk Metropolis algorithm was introduced in the seminal paper by Metropolis et al. (1953). In this algorithm, the proposal transition density is given by $q(x,y) = \bar{q}(y-x)$ where $\bar{q}$ is an arbitrary density on $\mathbb{R}^d$. In practice, this means that if the current state is X_t, an increment Z_{t+1} is drawn from $\bar{q}$, and the candidate $Y_{t+1} = X_t + Z_{t+1}$ is proposed. We shall also assume that $\bar{q}$ is symmetric with respect to 0, so that $\bar{q}(-y) = \bar{q}(y)$ for all $y \in \mathbb{R}^d$. Since $\bar{q}$ is symmetric with respect to 0, for all $(x,y) \in \mathbb{R}^d \times \mathbb{R}^d$, we get $q(x,y) = q(y,x)$. In this case, (5.37) is given by

$$\alpha(x,\ y) = 1 \wedge \frac{\pi(y)}{\pi(x)} . \tag{5.40}$$

With probability $\alpha(X_t, Y_{t+1})$, the move is accepted and $X_{t+1} = Y_{t+1}$; otherwise the move is rejected and $X_{t+1} = X_t$. Note that if $\pi(Y_{t+1}) \geq \pi(X_t)$, then the move is always accepted. When $\pi(Y_{t+1}) < \pi(X_t)$, then the move is accepted, but with a probability strictly less than one. The choice of the incremental distribution is crucial here. The transition kernel P of the Metropolis algorithm is then given for $x \in \mathsf{X}$ and $A \in \mathcal{X}$ by

$$P(x,A) = \int_{A-x} \left(1 \wedge \frac{\pi(x+z)}{\pi(x)}\right) q(z)\,\mathrm{Leb}(\mathrm{d}z) \\ + \mathbb{1}_A(x) \int_{\mathsf{X}-x} \left(1 - 1 \wedge \frac{\pi(x+z)}{\pi(x)}\right) q(z)\,\mathrm{Leb}(\mathrm{d}z) . \tag{5.41}$$

A classical choice for q is the multivariate normal distribution with zero-mean and covariance matrix Γ, $\mathcal{N}(0,\Gamma)$. It is well known that either too small or too large a covariance matrix will result in highly positively correlated Markov chains. When the scale is very small, almost all the moves are accepted but the chain mixes slowly. When the scale is too large, most of the moves get rejected and the chain remains stuck for long intervals before a move is accepted again. Gelman et al. (1996) have shown that the "optimal" covariance matrix (under restrictive technical conditions not given here) is $(2.38^2/d)\Gamma_\pi$, where Γ_π is the covariance matrix of the target distribution. In practice this covariance matrix Γ is determined by trial and error, using several realisations of the Markov chain.

We used our script `metronorm` to consider a centered bivariate normal ($d = 2$) with variance-covariance matrix given by $\Gamma_\pi = \left[\begin{smallmatrix} 10 & -5 \\ -5 & 5 \end{smallmatrix}\right]$. We then proposed three different scales, one which is too small (.1), one which is approximately optimal ($3 \approx 2.38^2/2$) and one which is too large (100). Figure 5.3 shows the results of the first component for the final 1000 draws. ◇

Example 5.38 (Independent sampler). Another possibility is to set the transition density to be $q(x,y) = \bar{q}(y)$, where $\bar{q}$ is again a density on $\mathbb{R}^d$. In this case, the next candidate is drawn independently of the current state of the chain. In this case, the acceptance ratio (5.37) is given by

$$\alpha(x,\ y) = 1 \wedge \frac{\bar{q}(x)\pi(y)}{\bar{q}(y)\pi(x)} . \tag{5.42}$$

This method is closely related to the Acceptance-Rejection method for i.i.d. simulation (see Exercise 5.23).

Suppose for example that $\pi \sim \mathrm{N}(0,1)$, and $q \sim \mathrm{N}(0,\ \sigma^2)$. Assume that $\sigma^2 > 1$

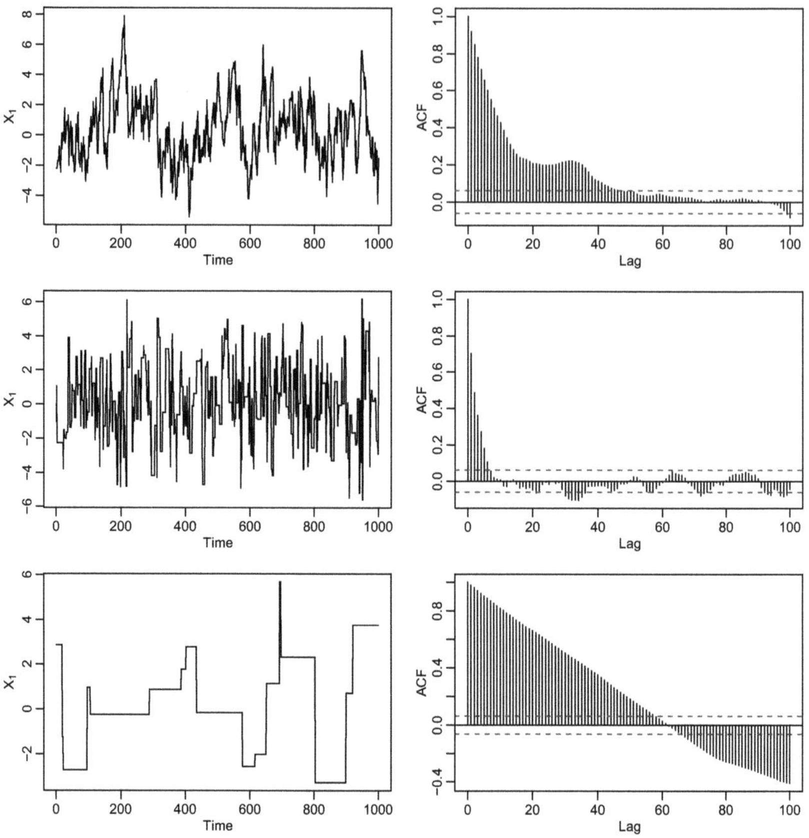

Figure 5.3 *Output for Example 5.37. The top row shows the sampled values and the sample ACF of a component of a bivariate normal when the scale is too small (.1). The middle row shows the values when the scale is nearly optimal (3). The bottom row shows the values when the scale is too large (100).*

so that the values being proposed are sampled from a distribution with heavier tails than π. Then given X_t, the algorithm proposes Y_{t+1} from $\mathrm{N}(0,\sigma^2)$, accepting it with probability $\alpha(X_t, Y_{t+1})$ given by (from (5.42))

$$\alpha(x,y) = \begin{cases} 1 & |y| \leq |x| \\ \exp(-(y^2 - x^2)(1-\sigma^{-2})/2) & |y| > |x| \end{cases}$$

Thus the algorithm compensates for the heavy tailed proposal by rejecting some moves that increase $|X_t|$, while accepting all moves that decrease $|X_t|$.

On the other hand, if $\sigma^2 < 1$, the values being proposed are sampled from a lighter tailed distribution than π and the accept/reject mechanism compensates for this:

$$\alpha(x,y) = \begin{cases} \exp(-(x^2 - y^2)(1 - \sigma^{-2})/2) & |y| \le |x| \\ 1 & |y| > |x| \end{cases} \qquad \diamond$$

5.8.2 *Gibbs sampling*

When the distribution of interest is multivariate, it may be the case that for each particular variable, its conditional distribution given all remaining variables has a simple form. In this case, a natural algorithm is the *Gibbs sampler*, which is now described. Its name somehow inappropriately stems from its use for the simulation of Gibbs Markov random fields by Geman and Geman (1984).

Assume that $\mathsf{X} = \mathsf{X}_1 \times \cdots \times \mathsf{X}_m$ is a product space equipped with the product σ-algebra $\mathcal{X} = \mathcal{X}_1 \otimes \cdots \otimes \mathcal{X}_m$. Let λ_k, $k \in \{1, \ldots, m\}$, be σ-finite measures on $(\mathsf{X}_k, \mathcal{X}_k)$, and let $\lambda = \lambda_1 \otimes \lambda_2 \otimes \cdots \otimes \lambda_m$ be the product measure. Suppose we are given a joint distribution with probability density function π with respect to the product measure λ. For simplicity, we assume that π is everywhere positive. An element $x \in \mathsf{X}$ may be decomposed into m components $x = (x^{[1]}, \ldots, x^{[m]})$, where $x^{[k]} \in \mathsf{X}_k$. If k is an index in $\{1, \ldots, m\}$, we shall denote by $x^{[k]}$ the kth component of x and by $x^{[-k]} = \{x^{[l]}\}_{l \neq k}$ the collection of remaining components. We further denote by $\pi_k(\cdot \mid x^{[-k]})$ the conditional probability density function, defined as

$$\pi_k(x^{[k]} \mid x^{[-k]}) = \frac{\pi(x^{[1]}, x^{[2]}, \ldots, x^{[m]})}{\int \pi(x^{[1]}, x^{[2]} \ldots, x^{[m]}) \lambda_k(\mathrm{d}x^{[k]})} .$$

We assume that sampling from this conditional distribution is feasible (for $k = 1, \ldots, m$). Note that $x^{[k]}$ is not necessarily scalar but may be itself vector-valued.

The *deterministic scan Gibbs* sampler is an MCMC algorithm which, starting from an initial arbitrary state X_0, updates the current state $X_i = (X_i^{[1]}, \ldots, X_i^{[m]})$ to a new state $X^{[i+1]}$ as follows.

For $k = 1, 2, \ldots, m$: Simulate $X_{i+1}^{[k]}$ from $\pi_k(\cdot \mid X_{i+1}^{[1]}, \ldots, X_{i+1}^{[k-1]}, X_i^{[k+1]}, \ldots, X_i^{[m]})$.

In other words, in the kth round of the cycle needed to simulate X_{i+1}, the kth component is updated by simulation from its conditional distribution given all other components, which remain fixed. This new value then supersedes the old one and is used in the subsequent simulation steps. A complete cycle of m conditional simulations is usually referred to as a *sweep* of the algorithm. We denote by K_k the corresponding Markov kernel:

$$\begin{aligned} K_k(x, A_1 \times \cdots \times A_m) &= K_k(x^{[1]}, \ldots, x^{[m]}; A) \\ &= \int \cdots \int \prod_{j \neq k} \delta_{x^{[j]}}(\mathrm{d}x'^{[j]}) \pi_k(x'^{[k]} \mid x^{[-k]}) \, \mathbb{1}_A(x') \, \lambda_k(\mathrm{d}x'^{[k]}) . \end{aligned} \tag{5.43}$$

Proposition 5.39 (Reversibility of individual Gibbs steps). *Each of the m individual kernels K_k, $k \in \{1, \ldots, m\}$ is reversible with respect to π and thus admits π as a stationary probability density function.*

Proof. See Exercise 5.26. ∎

Each step corresponds to a very special type of Metropolis-Hastings move where the acceptance probability is equal to 1, due to choice of π_k as the proposal distribution. However, Proposition 5.39 does not suffice to establish the convergence of the Gibbs sampler. Only the combination of the m moves in the complete cycle has a chance of producing a chain with the ability to visit the whole space X from any starting point.

Example 5.40 (Scalar normal–inverse gamma). In a statistical problem one may be presented with a set of independent observations $\mathbf{y} := \{y_1, \ldots, y_n\}$, which we assume to be normally distributed, but with unknown mean μ and variance τ^{-1} (τ is often referred to as the *precision*). The Bayesian approach to this problem is to assume that μ and τ are themselves random variables, with a given prior distribution. For example, we might assume that

$$\mu \sim \mathrm{N}(\mu_0, \tau_0^{-1}), \quad \tau \sim \Gamma(a_0, b_0),$$

i.e., μ is normal with mean μ_0 and variance τ_0^{-1} and τ has gamma distribution with parameters a_0 and b_0. [2]

The parameters μ_0, τ_0, a_0 and b_0 are assumed to be known. Then the prior density for (μ, τ) is given by

$$\pi(\mu, \tau) \propto \exp\{-\tau_0(\mu-\mu_0)^2/2\}\tau^{a_0-1}\exp\{-b_0\tau\}.$$

The posterior density for (μ, τ) defined as the conditional density given the observations, is then given, using the Bayes formula, by

$$\pi(\mu, \tau \mid \mathbf{y}) \propto \pi(\mu, \tau) f(\mathbf{y} \mid \mu, \tau) \propto \exp(-\tau_0(\mu-\mu_0)^2/2) \\ \times \exp\left(-\tau\sum_{i=1}^{n}(y_i-\mu)^2/2\right)\tau^{a_0-1+n/2}\exp(-b_0\tau).$$

Conditioning with respect to the observations has introduced a dependence between μ and τ. Nevertheless, the *full conditional distributions* still have a simple form

$$\pi(\mu \mid \mathbf{y}, \tau) = \mathrm{N}(m_n(\tau), t_n^{-1}(\tau)),$$
$$\pi(\tau \mid \mathbf{y}, \mu) = \Gamma(a_n, b_n(\mu)),$$

where, denoting $\bar{y} = n^{-1}\sum_{i=1}^{n} y_i$

$$m_n(\tau) := (\tau_0\mu_0 + n\tau\bar{y})/(\tau_0 + n\tau), \quad t_n(\tau) := \tau_0 + n\tau,$$
$$a_n := a_0 + n/2, \quad b_n(\mu) := b_0 + 1/2\sum_{i=1}^{n}(y_i-\mu)^2.$$

The *Gibbs sampler* provides a particularly simple approach to sample from $\pi(\mu, \tau \mid \mathbf{y})$. We set

$$\mathsf{X} = \mathsf{X}_1 \times \mathsf{X}_2 = \mathbb{R} \times [0, \infty).$$

[2] Z has an inverse gamma distribution if $1/Z$ has a gamma distribution; general properties can be found, for example, in Box and Tiao (1973, Section 8.5)

We wish to simulate $X = (\mu, \tau)$ with density $\pi(\mu, \tau \mid \mathbf{y})$. First we simulate X_0, say from the product form density $\pi(\mu, \tau)$. At the t-th stage, given $X_{t-1} = (\mu_{t-1}, \tau_{t-1})$, we first simulate $N_t \sim \mathrm{N}(0,1)$ and $G_t \sim \Gamma(a_n, 1)$ and we put

$$\begin{aligned} \mu_t &= m_n(\tau_{t-1}) + t_n^{-1/2}(\tau_{t-1})N_t \\ \tau_t &= b_n(\mu_t)G_t , \end{aligned}$$

and $X_t = (\mu_t, \tau_t)$. In the simple case where $\mu_0 = 0$ and $\tau_0 = 0$, which corresponds to a *flat prior* for μ (an improper distribution with a "constant" density on $\mathbb{R}$), the above equation can be rewritten as

$$\begin{aligned} \mu_t &= \bar{y} + (n\tau_{t-1})^{-1/2}N_t \\ \tau_t &= \left(b_0 + 1/2\sum_{i=1}^{n}(y_i - \mu_t)^2 \right) G_t . \end{aligned}$$

By alternating and iterating these two updates, we see that τ_t^{-1} is a Markov chain that iterates following a random coefficient autoregression (see Chapter 4),

$$\tau_t^{-1} = A_t \tau_{t-1}^{-1} + B_t , \tag{5.44}$$

where $A_t = N_t^2/(2G_t)$, and $B_t = (b_0 + nS^2/2)/G_t)$. It is easily checked that $(\mu_t - \bar{y})^2$ is also a random coefficient autoregressive process. The Gibbs sampler can be implemented in R using the script provided for this example. ◇

Exercises

5.1. Let $\{Z_t,\, t \in \mathbb{N}^*\}$ be an i.i.d. Bernoulli sequence such that $\mathbb{P}(Z_t = 1) = p = 1 - \mathbb{P}(Z_t = 0)$, and $p \in (0,1)$. Define $\{X_t,\, t \in \mathbb{N}\}$ by $X_0 = 0$ and $X_t = X_{t-1} + Z_t = \sum_{s=1}^{t} Z_s$, for $t \geq 1$. Denote by $\mathcal{F}_t^X = \sigma(X_s, 0 \leq s \leq t)$ the σ-algebra generated by the random variables $\{X_0, \dots, X_t\}$.

(a) Show that

$$\mathbb{P}(X_{t+1} = x_{t+1} \mid X_t = x_t, \dots, X_1 = x_1) = \mathbb{P}(X_{t+1} = x_{t+1} \mid X_t = x_t) = \begin{cases} p & \text{if } x_{t+1} - x_t = 1, \\ 1-p & \text{if } x_{t+1} - x_t = 0, \\ 0 & \text{otherwise.} \end{cases}$$

(b) Consider $A = \{X_1 = 1, X_2 = 1\}$ and $B = \{X_2 = 1, X_3 = 1\}$. Show that $A = \{Z_1 = 1, Z_2 = 0\}$ and $B = \{(Z_1 = 0, Z_2 = 1, Z_3 = 0) \cup (Z_1 = 1, Z_2 = Z_3 = 0)\}$. Show that A and B are not independent.

(c) Show that A and B are conditionally independent given X_1. $\mathbb{P}(A \cap B \mid X_2 = 1) = (1-p)/2 = \mathbb{P}(A \mid X_2 = 1)\mathbb{P}(B \mid X_2 = 1)$.

5.2. Let $\{(X_t, \mathcal{F}_t),\, t \in \mathbb{N}\}$ be a Markov chain.

(a) Show that for all $0 \leq r \leq t$ and $f \in \mathbb{F}_b(\mathsf{X}, \mathcal{X})$

$$\mathbb{E}\left[f(X_{t+1}) \mid X_t\right] = \mathbb{E}\left[f(X_{t+1}) \mid \sigma(X_r, \ldots, X_t)\right] = \mathbb{E}\left[f(X_{t+1}) \mid \mathcal{F}_t\right],$$

(b) Consider the property $(\mathcal{P}_n)$: for all $Y = \prod_{k=0}^{n} g_k(X_{t+k})$ where $g_k \in \mathbb{F}_b(\mathsf{X}, \mathcal{X})$, $\mathbb{E}\left[Y \mid \mathcal{F}_t\right] = \mathbb{E}\left[Y \mid X_t\right]$, $\mathbb{P}$-a.s. Show that $\mathcal{P}_0$ is satisfied.

(c) Assume that $(\mathcal{P}_n)$ is satisfied. Then, for any $g_k \in \mathbb{F}_b(\mathsf{X}, \mathcal{X})$, prove that

$$\mathbb{E}\left[g_0(X_t)\ldots g_n(X_{t+n})g_{n+1}(X_{t+n+1}) \mid \mathcal{F}_t\right] \\ = \mathbb{E}\left[g_0(X_t)\ldots g_n(X_{t+n})g_{n+1}(X_{t+n+1}) \mid X_t\right],$$

showing that $(\mathcal{P}_{n+1})$ is true.

(d) Consider the vector space

$$\mathcal{H} = \{Y \in \sigma(X_s, s \geq t), \mathbb{E}\left[Y \mid \mathcal{F}_t\right] = \mathbb{E}\left[Y \mid X_t\right], \mathbb{P}\text{-a.s.}\}.$$

Let $\{Y_n,\ n \in \mathbb{N}\}$ be an increasing sequence of nonnegative random variables in $\mathcal{H}$ such that $Y = \lim_{n\to\infty} Y_n$ is bounded. Show that

$$\mathbb{E}\left[Y \mid \mathcal{F}_t\right] = \lim_{n\to\infty} \mathbb{E}\left[Y_n \mid \mathcal{F}_t\right] = \lim_{n\to\infty} \mathbb{E}\left[Y_n \mid X_t\right] = \mathbb{E}\left[Y \mid X_t\right], \quad \mathbb{P}\text{-a.s.}$$

(e) Show that, for all $t \in \mathbb{N}$ and all bounded $\sigma(X_s, s \geq t)$-measurable random variables Y,

$$\mathbb{E}\left[Y \mid \mathcal{F}_t\right] = \mathbb{E}\left[Y \mid X_t\right], \quad \mathbb{P}\text{-a.s.} \tag{5.45}$$

5.3. Let $\{(X_t, \mathcal{F}_t),\ t \in \mathbb{N}\}$ be an adapted stochastic process such that for all $t \in \mathbb{N}$ and all bounded $\sigma(X_s, s \geq t)$-measurable random variables Y, $\mathbb{E}\left[Y \mid \mathcal{F}_t\right] = \mathbb{E}\left[Y \mid X_t\right]$, $\mathbb{P}$-a.s.. For $A \in \mathcal{F}_t$ and $B \in \sigma(X_s, s \geq t)$, denote $Z = \mathbb{1}_A$ and $Y = \mathbb{1}_B$. Show that $\mathbb{P}\left[A \cap B \mid \mathcal{F}_t\right] = Z\mathbb{E}\left[Y \mid X_t\right]$ $\mathbb{P}$-a.s.and that $\mathbb{P}\left[A \cap B \mid X_t\right] = \mathbb{P}\left[A \mid X_t\right]\mathbb{P}\left[B \mid X_t\right]$ $\mathbb{P}$-a.s.

5.4. Let $\{(X_t, \mathcal{F}_t),\ t \in \mathbb{N}\}$ be an adapted stochastic process such that for all $t \in \mathbb{N}$, $A \in \mathcal{F}_t$ and $B \in \sigma(X_s, s \geq t)$, $\mathbb{P}\left[A \cap B \mid X_t\right] = \mathbb{P}\left[A \mid X_t\right]\mathbb{P}\left[B \mid X_t\right]$, $\mathbb{P}$-a.s.

(a) Show that $\mathbb{E}\left[YZ \mid X_t\right] = \mathbb{E}\left[Y \mid X_t\right]\mathbb{E}\left[Z \mid X_t\right]$ for all bounded $\sigma(X_s, s \geq t)$-measurable random variables Y and $\mathcal{F}_t$-measurable random variables Z.

(b) Show that $\{(X_t, \mathcal{F}_t),\ t \in \mathbb{N}\}$ is a Markov chain.

5.5. Prove Proposition 5.5.

5.6. Prove Proposition 5.6.

5.7. Prove Proposition 5.7.

5.8. Prove Proposition 5.8.

5.9. Let $(\mathsf{X}, \mathcal{X})$, $(\mathsf{Y}, \mathcal{Y})$, and $(\mathsf{Z}, \mathcal{Z})$ be measurable spaces. Let M be a kernel on $\mathsf{X} \times \mathcal{Y}$ and N be a kernel on $\mathsf{Y} \times \mathcal{Z}$.

(a) Show that, if M and N are both finite (resp. bounded) kernels, then $M \otimes N$ is finite (resp. bounded) kernel.

(b) Show that, if M and N are both Markov kernels, then $M \otimes N$ is a Markov kernel.

(c) Show that, if $(\mathsf{U},\mathcal{U})$ is a measurable space and P is a kernel on $\mathsf{Z} \times \mathcal{U}$, then $(M \otimes N) \otimes P = M \otimes (N \otimes P)$, the tensor product of kernels is associative.

5.10 (Proof of Theorem 5.10: direct implication). Denote by $\mathcal{H}$ the set of measurable functions $f \in \mathbb{F}_b(\mathsf{X}^{t+1}, \mathcal{X}^{\otimes(t+1)})$ such that (5.6) is satisfied.

(a) Show that $\mathcal{H}$ is a vector space.

(b) Let $\{f_n,\ n \in \mathbb{N}\}$ be an nondecreasing sequence of nonnegative functions in $\mathcal{H}$ such that $\lim_{n\to\infty} f_n := f$ is bounded. Show that f belongs to $\mathcal{H}$.

(c) For $t \geq 1$, assume that (5.6) holds for $t-1$ and f of the form

$$f_0 \otimes \cdots \otimes f_t(x_0,\ldots,x_t) = f_0(x_0)\cdots f_t(x_t)\,, \quad f_s \in \mathbb{F}_b(\mathsf{X},\mathcal{X})\,. \tag{5.46}$$

Show that

$$\begin{aligned}\mathbb{E}\left[\prod_{s=0}^{t} f_s(X_s)\right] &= \mathbb{E}\left[\prod_{s=0}^{t-1} f_s(X_s)\mathbb{E}\left[f_t(X_t) \mid \mathcal{F}_{t-1}\right]\right] = \mathbb{E}\left[\prod_{s=0}^{t-1} f_s(X_s) P f_t(X_{t-1})\right] \\ &= \nu \otimes P^{\otimes(t-1)}(f_0 \otimes \cdots \otimes f_{t-1} P f_t) = \nu \otimes P^{\otimes t}(f_0 \otimes \cdots \otimes f_t)\,.\end{aligned}$$

(d) Show that $\mathcal{H}$ contains the functions of the form (5.46).

(e) Conclude.

5.11 (Proof of Theorem 5.10: converse). For $t = 0$, (5.6) implies that ν is the law of X_0. We have to prove that, for all $t \geq 1$, $f \in \mathbb{F}_+(\mathsf{X},\mathcal{X})$ and $\mathcal{F}^X_{t-1}$-measurable random variable Y:

$$\mathbb{E}\left[f(X_t)Y\right] = \mathbb{E}\left[Pf(X_{t-1})Y\right]\,. \tag{5.47}$$

Denote by $\mathcal{H}$ the set of $\mathcal{F}^X_{t-1}$-measurable random variables Y satisfying (5.47).

(a) Show that $\mathcal{H}$ is a vector space.

(b) Show that if $\{Y_t,\ t \in \mathbb{N}\}$ is an increasing sequence of nonnegative random variables such that $Y = \lim_{n\to\infty} Y_n$ is bounded, then $Y \in \mathcal{H}$.

(c) Show that (5.47) holds for $Y = f_0(X_0)f_1(X_1)\ldots f_{t-1}(X_{t-1})$ where $f_s \in \mathbb{F}_b(\mathsf{X},\mathcal{X})$ and conclude.

5.12 (Functional autoregressive process). Consider the following recursion

$$X_t = g(X_{t-1}) + \sigma(X_{t-1})Z_t\,,$$

where $\{Z_t,\ t \in \mathbb{N}\}$ is an i.i.d. sequence, $g : \mathbb{R} \to \mathbb{R}$ and $\sigma : \mathbb{R} \to \mathbb{R}^*_+$ are measurable functions.

(a) Determine the Markov kernel of this chain.

(b) Assume that the distribution of Z_0 has a density q with respect to Lebesgue's measure on $\mathbb{R}$; show that the Markov kernel P has a density. Give the expression of this density.

5.13 (Setwise convergence / existence of a stationary distribution). Assume that for some initial probability ξ, the sequence of probabilities $\{\xi P^t, t \in \mathbb{N}\}$ converges setwise, i.e., for any $A \in \mathcal{X}$, the sequence $\{\xi P^t(A), t \in \mathbb{N}\}$ has a limit, denoted $\gamma_\xi(A)$, i.e., $\lim_{t\to\infty} \xi P^t(A) = \gamma_\xi(A)$.

(a) Show that, $\gamma_\xi(A) = \gamma_\xi P(A)$.

(b) Assume that there exists a unique invariant probability measure. Show that the limit γ_ξ is independent of the initial distribution ξ.

(c) Conversely, assume that for any initial distribution ξ on $(\mathsf{X}, \mathcal{X})$, the sequence $\{\xi P^t, t \in \mathbb{N}\}$ has the same setwise limit. Show that there exists a unique stationary distribution.

5.14 (Weak convergence / existence of a stationary distribution). Let (X, d) be a Polish space. We assume that for some initial measure ξ, the sequence of probability measures $\{\xi P^t, t \in \mathbb{N}\}$ converges weakly to γ_ξ. In addition, for any $f \in \mathrm{C}_b(\mathsf{X})$ (the set of bounded continuous functions), $Pf \in \mathrm{C}_b(\mathsf{X})$.

(a) Show that $\gamma_\xi = \gamma_\xi P$.

(b) Show that if the kernel P has a unique invariant distribution, then the weak limit of the sequence $\{\xi P^t, t \in \mathbb{N}\}$ does not depend on the initial distribution ξ.

(c) Conversely, if for any initial distribution ξ the sequence of probability $\{\xi P^t, t \in \mathbb{N}\}$ converges weakly to the same limiting distribution, then the invariant distribution is unique.

5.15 (Existence and uniqueness of the stationary measure of an AR(1)). Consider a first order autoregressive process $X_t = \phi X_{t-1} + Z_t$, $t \geq 1$, where $\{Z_t,\ t \in \mathbb{N}\}$ is a zero mean i.i.d. sequence and X_0 is independent of $\{Z_t,\ t \in \mathbb{N}\}$ and is distributed according to some probability ξ on $(\mathbb{R}, \mathcal{B}(\mathbb{R}))$. We assume that $|\phi| < 1$.

(a) Show that $X_t = \phi^t X_0 + \sum_{i=0}^{t-1} \phi^i Z_{t-i}$

(b) Show that, for each integer t, the random variable X_t has the same distribution as $\phi^t X_0 + \sum_{i=0}^{t-1} \phi^i Z_{i+1}$.

(c) Define $\mathcal{F}_t^Z = \sigma(\{Z_s, 0 \leq s \leq t\})$ and

$$Y_t = \sum_{i=0}^{t-1} \phi^i Z_{i+1} = Y_{t-1} + \phi^{t-1} Z_t \,.$$

Show that $\{(Y_t, \mathcal{F}_t^Z),\ t \in \mathbb{N}\}$ is an L^1-bounded martingale.

(d) Show that the sequence $\{Y_t,\ t \in \mathbb{N}\}$ converges $\mathbb{P}$-a.s. to the integrable random variable Y_∞. We denote by π the law of Y_∞.

(e) Show that for any function f and any initial distribution ξ, $\mathbb{E}_\xi[f(X_t)] \to \pi(f)$.

(f) Using (5.14), show that π is the unique stationary distribution.

5.16. The DAR(1) (see Example 4.5) is given by $X_t = V_t X_{t-1} + (1 - V_t) Z_t$ where $\{V_t,\ t \in \mathbb{Z}\}$ is an i.i.d. sequence of binary random variables with $\mathbb{P}(V_t = 1) = \alpha$, $\{Z_t,\ t \in \mathbb{Z}\}$ is an i.i.d. sequence of random variables with distribution given by π on $(\mathsf{X}, \mathcal{X})$ and $\{V_t,\ t \in \mathbb{Z}\}$ and $\{Z_t,\ t \in \mathbb{Z}\}$ are independent. The initial random variable

X_0 is assumed to be independent of $\{V_t,\ t \in \mathbb{N}\}$ and $\{Z_t,\ t \in \mathbb{N}\}$ and is distributed according to ξ.

(a) Show that, for all $t \geq 1$ and any bounded measurable function,

$$\mathbb{E}_\xi\left[f(X_t)\right] = \alpha \mathbb{E}_\xi\left[f(X_{t-1})\right] + (1-\alpha)\pi(f)\ .$$

(b) Show that $\mathbb{E}_\xi\left[f(X_{t-1})\right] = \pi(f)$, then $\mathbb{E}_\xi\left[f(X_t)\right] = \pi(f)$ and that π is the unique stationary distribution.

(c) Assume that $\mathsf{X} = \mathbb{N}$ and that $\sum_{k=0}^{\infty} k^2 \pi(k) < \infty$. For any positive integer h, show that $\mathrm{Cov}_\pi(X_h, X_0) = \alpha \mathrm{Cov}_\pi(X_{h-1}, X_0)$ and $\mathrm{Cov}_\pi(X_h, X_0) = \alpha^h \mathrm{Var}_\pi(X_0)$.

(d) Show that the DAR(1) process has exactly the same autocovariance structure as an AR(1) process.

(e) Explain why the DAR(1) process cannot exhibit negative dependence.

5.17. Consider the self-excited threshold autoregressive model defined by

$$X_t = \phi_0^{(j)} + \sum_{i=1}^{p_j} \phi_i^{(j)} X_{t-i} + Z_t^{(j)} \quad \text{if} \quad X_{t-d} \in \left(r_{j-1}, r_j\right] \tag{5.48}$$

where $\{Z_t^{(j)},\ t \in \mathbb{N}\} \sim \text{iid N}(0, \sigma_j^2)$, for $j = 1, \ldots, k$, the positive integer d is a specified delay and $r_0 = -\infty < r_1 < \cdots < r_{p-1} < r_p = +\infty$ is a partition of $\mathbb{R}$.

(a) Show that $\{X_t,\ t \in \mathbb{N}\}$ is a m-th order Markov chain.

(b) Determine the Markov kernel of this chain.

5.18. Let $\{Z_t,\ t \in \mathbb{N}\}$ be Gaussian white noise with variance σ^2 and let $|\phi| < 1$ be a constant. Consider the process $X_0 = Z_0$, and $X_t = \phi X_{t-1} + Z_t$, for $t = 1, 2, \ldots$.

(a) Find the mean and the variance of X_t. Is $\{X_t,\ t \in \mathbb{N}\}$ stationary?

(b) Show that for all $h \in \{0, \ldots, t\}$,

$$\mathrm{Cor}(X_t, X_{t-h}) = \phi^h \left[\frac{\mathrm{Var}(X_{t-h})}{\mathrm{Var}(X_t)}\right]^{1/2} .$$

(c) Show that $\lim_{t\to\infty} \mathrm{Var}(X_t) = \sigma^2/(1-\phi^2)$ and that $\lim_{t\to\infty} \mathrm{Cor}(X_t, X_{t-h}) = \phi^h,\ h \geq 0$.

(d) Comment on how you could use these results to simulate n observations of a stationary Gaussian AR(1) model from simulated i.i.d. $\mathrm{N}(0,1)$ values.

(e) Now suppose $X_0 = Z_0/\sqrt{1-\phi^2}$. Is $\{X_t,\ t \in \mathbb{N}\}$ stationary?

5.19 (Simple Markovian bilinear model). Consider the simple Markovian bilinear model

$$X_t = aX_{t-1} + bZ_t X_{t-1} + Z_t = (a + bZ_t)X_{t-1} + Z_t\ . \tag{5.49}$$

where $\{Z_t,\ t \in \mathbb{N}\}$ is an i.i.d. sequence of standard Gaussian variables and is independent of $X_0 \sim \xi$. We assume that $a^2 + b^2 < 1$.

(a) Show that, for all $t \in \mathbb{N}$,

$$X_t = \prod_{i=1}^{t}(a+bZ_i)X_0 + Z_t + \sum_{j=1}^{t-1} Z_{t-j} \prod_{i=t-j+1}^{t} (a+bZ_i) .$$

(b) Show that $\lim_{t\to\infty} \left(\prod_{i=1}^{t} |a+bZ_i|\right)^{1/t}$ exists $\mathbb{P}$-a.s. and that this limit is strictly less than 1.

(c) Set $Y_t = Z_0 + \sum_{j=1}^{t-1} Z_j \prod_{i=0}^{j-1}(a+bZ_i)$. Show that the random variables Y_t and $Z_t + \sum_{j=1}^{t-1} Z_{t-j} \prod_{i=t-j+1}^{t}(a+bZ_i)$ have the same distributions.

(d) Show that $\{(Y_t, \mathcal{F}_t^Z),\ t \in \mathbb{N}\}$ is an L^2-bounded martingale converging $\mathbb{P}$-a.s. and in $L^2(\mathbb{P})$ to $Y_\infty = Z_0 + \sum_{j=1}^{\infty} Z_j \prod_{i=0}^{j-1}(a+bZ_i)$.

(e) Show that for any $f \in C_b(\mathbb{R})$, $\lim_{t\to\infty} \mathbb{E}_\xi[f(X_t)] = \pi(f)$, where $\pi_{a,b}$ is the distribution of Y_∞ (the subscripts (a,b) stress the dependence of the stationary distributions in a and b).

(f) Using Exercise 5.14, show that $\pi_{a,b}$ is the unique stationary distribution.

(g) Show that $\int x \pi_{a,b}(\mathrm{d}x) = 0$ and $\int x^2 \pi_{a,b}(\mathrm{d}x) = 1/(1-a^2-b^2)$.

5.20 (Simple Markovian bilinear model cont.). Consider again the simple bilinear model

$$X_t^{(a,b)} = aX_{t-1}^{(a,b)} + bZ_t X_{t-1}^{(a,b)} + Z_t = (a+bZ_t)X_{t-1}^{(a,b)} + Z_t , \tag{5.50}$$

where $\{Z_t,\ t \in \mathbb{N}\}$ is an i.i.d. sequence of standard Gaussian variables and is independent of $X_0 \sim \pi_{a,b}$ where $\pi_{a,b}$ is the stationary distribution. The superscript (a,b) is used to stress the dependence of the process in these coefficients.

(a) Show that

$$\mathbb{E}|X_t^{(a,b)} - X_t^{(a,0)}|^2 = \frac{b^2}{(1-a^2-b^2)(1-a^2)} \to 0 , \quad \text{as } b \to 0 .$$

(b) Assume that $b > 0$. Show that the cumulative distribution function $x \mapsto F_{a,b}(x)$ of $\pi_{a,b}$ is the solution of an integral equation.

(c) Show that, if $x \neq a/b$, the cumulative dist is differentiable at x.

(d) Show that $F_{a,b}$ does not have a discrete component at $x = -a/b$.

(e) Conclude that the stationary distribution $\pi_{a,b}$ has a density satisfying the following integral equation

$$f_{a,b}(x) = \frac{1}{\sqrt{2\pi}} \int_{-\infty}^{\infty} \frac{f_{a,b}(s)}{|1+bs|} \exp\left[-\frac{1}{2}\left(\frac{x-as}{1+bs}\right)^2\right] \mathrm{Leb}(\mathrm{d}s) .$$

(f) Show that $f_{a,b}(-1/b) > 0$ and that $\lim_{x\to -a/b} f_b(x) = \infty$.

5.21. Consider a GARCH(p,q) model

$$X_t = \sigma_t \varepsilon_t \tag{5.51}$$

$$\sigma_t^2 = \alpha_0 + \alpha_1 X_{t-1}^2 + \cdots + \alpha_p X_{t-p}^2 + \beta_1 \sigma_{t-1}^2 + \cdots + \beta_q \sigma_{t-q}^2 , \tag{5.52}$$

where $\{\varepsilon_t,\ t \in \mathbb{N}\}$ is an i.i.d. sequence with zero-mean and unit-variance, and the coefficients $\alpha_j,\ j \in \{0,\ldots,p\}$ and $\beta_j,\ j \in \{1,\ldots,q\}$ are nonnegative. State conditions upon which the GARCH(p,q) model admits a unique invariant stationary distribution.

5.22 (Basic Gibbs). Suppose we wish to obtain samples from a bivariate normal distribution,

$$\begin{pmatrix} X \\ Y \end{pmatrix} \sim \mathrm{N}\left[\begin{pmatrix} 0 \\ 0 \end{pmatrix}, \begin{pmatrix} 1 & \theta \\ \theta & 1 \end{pmatrix}\right],$$

where $|\theta| < 1$.

(a) Determine the univariate conditionals $\mathrm{p}_1^\theta(X \mid Y)$ and $\mathrm{p}_2^\theta(Y \mid X)$; see (1.32)–(1.33).

(b) Consider the Markov chain: Pick $X^{(0)} = x_0$, and then iterate the process $X^{(0)} \mapsto Y^{(0)} \mapsto X^{(1)} \mapsto Y^{(1)} \mapsto \cdots \mapsto X^{(t)} \mapsto Y^{(t)} \mapsto \cdots$, where $Y^{(t)}$ is a sample from $\mathrm{p}_2^\theta(\cdot \mid X^{(t)})$, and $X^{(t)}$ is a sample from $\mathrm{p}_1^\theta(\cdot \mid Y^{(t-1)})$. Write the joint distribution of $(X^{(t)}, Y^{(t)})$ in terms of the starting value x_0 and the correlation θ.

(c) What is the asymptotic distribution of $(X^{(t)}, Y^{(t)})$ as $t \to \infty$?

(d) How can you use the results of part (c) to obtain a pseudo random sample of n bivariate normals using only pseudo samples from a univariate normal distribution? How does the value of θ affect the procedure?

5.23 (Accept-reject algorithm). We wish to sample the target density π (with respect to a dominating measure λ), which is known up to a multiplicative constant. Let q be a proposal density (assumed to be easier to sample.) We assume that there exists a constant $M < \infty$ such that $\pi(x)/q(x) \le M$ for all $x \in \mathsf{X}$. The Accept-Reject goes as follows: first, we generate $Y \sim g$ and, independently, we generate $U \sim \mathrm{U}([0,1])$. If $Mq(Y)U \le \pi(Y)$, then we set $X = Y$. If the inequality is not satisfied, we then discard Y and U and start again.

(a) Show that X is distributed according to π.

(b) What is the distribution of the number of trials required per sample?

(c) What is the average number of samples needed for one simulation?

(d) Propose a method to estimate the normalizing constant of π.

(e) Compare with the Metropolis-Hastings algorithm with Independent proposal.

5.24 (The independent sampler MCMC). We wish to sample a target distribution $\Gamma(\alpha,\beta)$ using the independent MCMC algorithm. We will use as the proposal distribution $\gamma(\lfloor\alpha\rfloor, b)$.

(a) Derive the corresponding Accept-Reject method. Explain why, when $\beta = 1$, the choice of b minimizing the average number of simulations per sample is $b = \lfloor\alpha\rfloor/\alpha$.

(b) Generate 10000 $\Gamma(5, 5/5.25)$ random variables to derive a $\Gamma(5.25, 1)$ sample.

(c) Use the same sample in the corresponding Metropolis-Hastings algorithm to generate 10000 $\Gamma(5.25, 1)$ random variables.

(d) Compare the algorithms using (i) their acceptance rates and (ii) the estimates of the mean and variance of the $\Gamma(5.25, 1)$ along with their errors.

5.25 (Functional autoregressive). Let P be a Markov kernel on $(\mathbb{R}, \mathcal{B}(\mathbb{R}))$. Show that $(x,u) \mapsto G(x,u) = \inf\{x' \in \mathbb{R} : F(x,x') \geq u\}$ is a Borel function where $F : (x,x') \mapsto P(x, (\infty, x'])$ is a Borel function.

5.26 (Reversibility of the Gibbs sampler). We use the notations introduced in Section 5.8.2.

(a) Show that for any functions $f, g \in \mathbb{F}_+(\mathsf{X}, \mathcal{X})$,

$$\iint f(x) g(x') \pi(x) \lambda(\mathrm{d}x) K_k(x, \mathrm{d}x') =$$
$$\int \left\{ f(x) \pi(x) \lambda_k(\mathrm{d}x^{[k]}) \int g(x'^{[k]}, x^{[-k]}) \pi_k(x'^{[k]} \mid x^{[-k]}) \lambda_k(\mathrm{d}x'^{[k]}) \right\} \lambda_{-k}(\mathrm{d}x^{[-k]}),$$

where $(x'^{[k]}, x^{[-k]})$ refers to the element u of X such that $u_k = x'^{[k]}$ and $(u_{\ell \neq k}) = x^{[-k]}$

(b) Show that

$$\pi(x^{[k]}, x^{[-k]}) \pi_k(x'^{[k]} \mid x^{[-k]}) = \pi_k(x^{[k]} \mid x^{[-k]}) \pi(x'^{[k]}, x^{[-k]}),$$

(c) Conclude.

5.27 (A simple time series of counts). Consider the following iterated random function:

$$X_{k+1} = d + aX_k + b\lfloor -X_k \ln(U_{k+1}) \rfloor, \quad k \geq 0 \tag{5.53}$$

where $\{U_k,\ k \in \mathbb{N}^\star\}$ is a sequence of i.i.d. random variables such that $U_k \sim \mathrm{U}[0,1]$. We assume that X_0 is independent of $\{U_k,\ k \in \mathbb{N}^\star\}$ and that $d, a, b > 0$ and $a + b < 1$. By simplification, we assume $X_0 \geq d$, so that $\{X_k,\ k \in \mathbb{N}\}$ is a Markov chain taking values on $\mathsf{X} = [d, \infty)$.

(a) Show that for some function f, the recursion on the $\{X_k,\ k \in \mathbb{N}\}$ may be written as: for all $k \geq 0$,

$$\begin{aligned} Y_{k+1} | \mathcal{F}_k &\sim \mathrm{Geom}(f(X_k)), \\ X_{k+1} &= d + aX_k + b(Y_{k+1} - 1). \end{aligned} \tag{5.54}$$

where $\mathcal{F}_k = \sigma(X_0, \ldots, X_k, Y_1, \ldots, Y_k)$.

(b) Denote $X_1(x) = d + ax + b\lfloor -x \ln(U) \rfloor$ where $U \sim \mathrm{U}[0,1]$. Compute $\mathbb{E}[X_1(x)]$ explicitly.

(c) Deduce that $\varphi(x) : x \mapsto \mathbb{E}[X_1(x)]$ can be differentiated and show that $|\varphi'(x)| < 1$ for all $x \in \mathsf{X}$.

(d) By noting that $X_1(x) \geq X_1(x')$ for all $x \geq x'$, deduce that there exists $\rho < 1$ such that $\mathbb{E}[|X_1(x) - X_1(x')|] \leq \rho |x - x'|$ for all $x, x' \in \mathsf{X}$. Conclude.

Chapter 6

Stability and Convergence

Invariant measures are important because they define the stationary versions of Markov chains. In addition, they turn out to be the measures that define the long-term (or ergodic) behavior of a Markov chain.

In this chapter, we consider convergence results of the type $\lim_{t\to\infty} \xi P^t(A) = \pi(A)$. We examine different types of ergodicity for Markov chains. The first is the *uniform ergodicity*, the oldest and strongest form of convergence in the study of Markov chains. This concept dates back to some of the earliest studies of Markov chains on general state space presented in the seminal papers by Doeblin (1938) and Doob (1953). Uniform ergodicity ensures that the convergence of the sequence of iterates $\{\xi P^t, t \in \mathbb{N}\}$ with respect to the total variation distance does not depend on the initial value; hence the name *uniform*; more precisely we will show that, if the Markov kernel P is uniformly ergodic,

$$d_{\mathrm{TV}}(\xi P^t, \pi) := \sup_{|f|\le 1} \left| \int \xi(\mathrm{d}x) P^t(x, \mathrm{d}y) f(y) - \int \pi(\mathrm{d}y) f(y) \right| \tag{6.1}$$

goes to zero for any initial distribution ξ. It turns out that uniform ergodicity implies that the rate of convergence is necessarily geometric. To establish uniform ergodicity of a Markov chain, we check the uniform Doeblin condition, which basically states that there exists an $\varepsilon > 0$ and an integer m such that, starting from any $x \in \mathsf{X}$, the probability of entering in any set $A \in \mathcal{X}$ after m steps is lower bounded by $\varepsilon\pi(A)$, i.e., $P^m(x, A) \ge \varepsilon\pi(A)$.

Many time-series models turn out not to be uniformly ergodic. For example, as we will see later, the AR(1) model is generally not uniformly ergodic. The type of conditions typically required for this process to be uniformly ergodic are stringent. We therefore need to generalize ergodicity results beyond uniform ergodicity. We will show that there exist, under conditions that are satisfied in many practical settings, a function $V : \mathsf{X} \to [1, \infty)$ (not necessarily bounded), and constants $\rho \in [0, 1)$ and $C < \infty$, satisfying

$$d_V(\xi P^t, \pi) := \sup_{|f|\le V} \left| \int \xi(\mathrm{d}x) P^t(x, \mathrm{d}y) f(y) - \int \pi(\mathrm{d}y) f(y) \right| \le C\rho^t \xi(V) , \tag{6.2}$$

for any initial distribution $\xi \in \mathbb{M}_1(\mathcal{X})$. Because of the power of the final form of these results, and the wide range of Markov chains for which they hold, it is not too strong a statement that this *geometrically ergodic* context constitutes the most useful framework for nonlinear time series.

6.1 Uniform ergodicity

6.1.1 Total variation distance

We preface this section with a brief introduction to the total variation distance. Let $(\mathsf{X}, \mathcal{X})$ be a measurable space and ξ be a finite signed measure on $(\mathsf{X}, \mathcal{X})$. Then, by the Jordan decomposition theorem (Theorem A.3), there exists a unique pair of positive finite measures ξ_+, ξ_- on $(\mathsf{X}, \mathcal{X})$ such that ξ_+ and ξ_- are singular and $\xi = \xi_+ - \xi_-$. The couple (ξ_+, ξ_-) is referred to as the Jordan decomposition of the signed measure ξ. The finite measure $|\xi| = \xi_+ + \xi_-$ is called the *total variation* of ξ. Any set S such that $\xi_+(S^c) = \xi_-(S) = 0$ is called a *Jordan set* for ξ. The *total variation norm* of ξ is defined by

$$\|\xi\|_{\mathrm{TV}} = |\xi|(\mathsf{X}) .$$

The *total variation distance* between $\xi, \xi' \in \mathbb{M}_1(\mathcal{X})$ is defined by

$$d_{\mathrm{TV}}(\xi, \xi') = \frac{1}{2} \|\xi - \xi'\|_{\mathrm{TV}} .$$

Let $\mathbb{M}_0(\mathcal{X})$ be the set of finite signed measures ξ such that $\xi(\mathsf{X}) = 0$. We now give equivalent characterizations of the total variation norm for signed measures. Let the oscillation $\operatorname{osc}(f)$ of a bounded function f be defined by

$$\operatorname{osc}(f) = \sup_{x, x' \in \mathsf{X}} |f(x) - f(x')| .$$

It is easily seen that $\operatorname{osc}(f) = 2 \inf_{c \in \mathbb{R}} |f - c|_\infty$.

Theorem 6.1. *For any $\xi \in \mathbb{M}(\mathcal{X})$,*

$$\|\xi\|_{\mathrm{TV}} = \sup\{\xi(f) : f \in \mathbb{F}_b(\mathsf{X}, \mathcal{X}), |f|_\infty \le 1\} . \tag{6.3}$$

Let ξ be a finite signed measure such that $\xi(\mathsf{X}) = 0$. Then,

$$\|\xi\|_{\mathrm{TV}} = 2 \sup\{\xi(f) : f \in \mathbb{F}_b(\mathsf{X}, \mathcal{X}),\ \operatorname{osc}(f) \le 1\} . \tag{6.4}$$

Proof. Since ξ_+ and ξ_- are finite positive measures, for any $f \in \mathbb{F}_b(\mathsf{X}, \mathcal{X})$ and $\xi \in \mathbb{M}(\mathcal{X})$, we have

$$\xi(f) = \xi_+(f) - \xi_-(f) \le \xi_+(|f|) + \xi_-(|f|) = |\xi|(|f|) \le |\xi|(\mathsf{X}) |f|_\infty .$$

Thus $\sup\{|\xi(f)| : f \in \mathbb{F}_b(\mathsf{X}, \mathcal{X}), |f|_\infty \le 1\} \le \|\xi\|_{\mathrm{TV}}$. On the other hand, let S be a Jordan set for ξ and set $f := \mathbb{1}_S - \mathbb{1}_{S^c}$, then $\xi(f) = \|\xi\|_{\mathrm{TV}}$ and $|f|_\infty = 1$, proving that (6.3) holds. Consider now $\xi \in \mathbb{M}_0(\mathcal{X})$. Then, $\xi(f) = \xi(f + c)$ for all $c \in \mathbb{R}$ and thus, for all $c \in \mathbb{R}$,

$$|\xi(f)| = |\xi(f - c)| \le \|\xi\|_{\mathrm{TV}} |f - c|_\infty .$$

Since this inequality if valid for all $c \in \mathbb{R}$, this yields

$$|\xi(f)| \leq \frac{1}{2} \|\xi\|_{\mathrm{TV}} \operatorname{osc}(f) . \tag{6.5}$$

Conversely, if we set $f = (1/2)(\mathbb{1}_S - \mathbb{1}_{S^c})$ where S is a Jordan set for ξ, then $\operatorname{osc}(f) = 1$ and

$$\xi(f) = \frac{1}{2}(\xi_+(S) + \xi_-(S^c)) = \frac{1}{2}(\xi_+(\mathsf{X}) + \xi_-(\mathsf{X})) = \frac{1}{2}\|\xi\|_{\mathrm{TV}} .$$

Together with (6.5), this shows (6.4). ∎

Corollary 6.2. *If $\xi, \xi' \in \mathbb{M}_1(\mathcal{X})$, then $\xi - \xi' \in \mathbb{M}_0(\mathcal{X})$ and for any $f \in \mathbb{F}_b(\mathsf{X}, \mathcal{X})$,*

$$|\xi(f) - \xi'(f)| \leq d_{\mathrm{TV}}(\xi, \xi') \operatorname{osc}(f) . \tag{6.6}$$

We now state a very important completeness result:

Proposition 6.3. *The normed vector space $(\mathbb{M}(\mathcal{X}), \|\cdot\|_{\mathrm{TV}})$ is complete. The space $(\mathbb{M}_1(\mathcal{X}), d_{\mathrm{TV}})$ is a Polish space.*

The proof is a special case of Proposition 6.16 which will be proven later.

6.1.2 Dobrushin coefficient

Definition 6.4 (Dobrushin coefficient). *Let P be a Markov kernel on a measurable space $(\mathsf{X}, \mathcal{X})$. The* Dobrushin coefficient *$\Delta_{\mathrm{TV}}(P)$ of P is given by*

$$\Delta_{\mathrm{TV}}(P) := \sup_{(\xi,\xi') \in \mathbb{M}_1(\mathcal{X}) \times \mathbb{M}_1(\mathcal{X}), \xi \neq \xi'} \frac{d_{\mathrm{TV}}(\xi P, \xi' P)}{d_{\mathrm{TV}}(\xi, \xi')} . \tag{6.7}$$

For any $(\xi, \xi') \in \mathbb{M}_1(\mathcal{X})$ and P a Markov kernel on $\mathsf{X} \times \mathcal{X}$, $\xi P, \xi' P \in \mathbb{M}_1(\mathcal{X})$ and $d_{\mathrm{TV}}(\xi P, \xi' P) \leq \Delta_{\mathrm{TV}}(P) d_{\mathrm{TV}}(\xi, \xi')$. If R and Q are two Markov kernels on $\mathsf{X} \times \mathcal{X}$, then

$$d_{\mathrm{TV}}(\xi PR, \xi' PR) \leq \Delta_{\mathrm{TV}}(R) d_{\mathrm{TV}}(\xi P, \xi' P) \leq \Delta_{\mathrm{TV}}(R) \Delta_{\mathrm{TV}}(P) d_{\mathrm{TV}}(\xi, \xi') ,$$

showing that the Dobrushin coefficient is submultiplicative,

$$\Delta_{\mathrm{TV}}(PR) \leq \Delta_{\mathrm{TV}}(P) \Delta_{\mathrm{TV}}(R) . \tag{6.8}$$

The following Lemma shows that the maximum of the ratio $d_{\mathrm{TV}}(\xi P, \xi' P)/d_{\mathrm{TV}}(\xi, \xi')$ over all probability measures $\xi \neq \xi' \in \mathbb{M}_1(\mathcal{X})$, is the same as the maximum of $d_{\mathrm{TV}}(\delta_x P, \delta_{x'} P)/d_{\mathrm{TV}}(\delta_x, \delta_{x'})$, $x \neq x' \in \mathbb{M}_1(\mathcal{X})$: it suffices to consider probability measures whose mass is concentrated at one point.

Lemma 6.5. *Let P be a Markov kernel on $(\mathsf{X}, \mathcal{X})$. Then,*

$$\Delta_{\mathrm{TV}}(P) = \sup_{(x,x') \in \mathsf{X} \times \mathsf{X}} d_{\mathrm{TV}}(\delta_x P, \delta_{x'} P) . \tag{6.9}$$

Moreover, $0 \leq \Delta_{\mathrm{TV}}(P) \leq 1$.

Proof. The right-hand side of (6.9) is less than or equal to $\Delta_{\mathrm{TV}}(P)$. We now prove the converse inequality. Using (6.9), Theorem 6.1 and the bound (6.5), we have, for all probability measures ξ, ξ',

$$\begin{aligned} d_{\mathrm{TV}}(\xi P, \xi' P) &= \frac{1}{2} \left\| \xi P - \xi' P \right\|_{\mathrm{TV}} = \sup_{f:\operatorname{osc}(f)\leq 1} |\xi P(f) - \xi' P(f)| \\ &= \sup_{f:\operatorname{osc}(f)\leq 1} |\xi(Pf) - \xi'(Pf)| \leq d_{\mathrm{TV}}(\xi,\xi') \sup_{f:\operatorname{osc}(f)\leq 1} \operatorname{osc}(Pf) . \end{aligned}$$

Note now that

$$\begin{aligned} \sup_{f:\operatorname{osc}(f)\leq 1} \operatorname{osc}(Pf) &= \sup_{f:\operatorname{osc}(f)\leq 1} \sup_{x,x'} |Pf(x) - Pf(x')| \\ &= \sup_{x,x'} \sup_{f:\operatorname{osc}(f)\leq 1} \left| \{P(x,\cdot) - P(x',\cdot)\} f \right| \\ &= \frac{1}{2} \sup_{x,x'} \left\| P(x,\cdot) - P(x',\cdot) \right\|_{\mathrm{TV}} = \sup_{x,x'} d_{\mathrm{TV}}(\delta_x P, \delta_x' P) . \end{aligned}$$

This proves the converse inequality. The fact that $\Delta_{\mathrm{TV}}(P) \leq 1$ follows immediately from (6.9). ■

A kernel P on $\mathsf{X} \times \mathcal{X}$ defines a mapping on the Polish space $(\mathbb{M}_1(\mathcal{X}), d_{\mathrm{TV}})$. If $\Delta_{\mathrm{TV}}(P) < 1$, then P is a *contraction mapping*. The *Banach fixed point theorem*, also known as the contraction mapping theorem or contraction mapping principle (see Theorem 6.39), guarantees the existence and uniqueness of fixed points of contraction mapping and provides a constructive method to find those fixed points: given any $\xi \in \mathbb{M}_1(\mathcal{X})$, the sequence ξ, ξP, ξP^2, ... converges to the fixed point. Clearly, a fixed point π of P is an invariant probability measure, $\pi P = \pi$. Therefore, if $\Delta_{\mathrm{TV}}(P) < 1$, then P admits a unique invariant probability measure. The Banach fixed point theorem also provides an estimate of the rate of convergence to those fixed points. A slight improvement of this argument leads to the following elementary but very powerful theorem, proved by Dobrushin (1956).

Theorem 6.6. *Let $(\mathsf{X}, \mathcal{X})$ be a measurable space. Let P be a Markov kernel such that, for some integer m, $\Delta_{\mathrm{TV}}(P^m) \leq \rho < 1$. Then, P admits a unique invariant probability measure π. In addition, for all $\xi \in \mathbb{M}_1(\mathcal{X})$,*

$$d_{\mathrm{TV}}(\xi P^n, \pi) \leq \rho^{\lfloor n/m \rfloor} d_{\mathrm{TV}}(\xi, \pi) ,$$

where $\lfloor u \rfloor$ is the integer part of u.

Proof. This is a direct application of the fixed point theorem for contraction mapping (Theorem 6.40). ■

Using this result, it is natural to define some ergodicity concepts.

Definition 6.7 (Ergodicity, uniform ergodicity). *Let P be a Markov kernel and π be a probability distribution on $(\mathsf{X}, \mathcal{X})$.*

- *The Markov kernel P is* uniformly ergodic *if* $\lim_{n\to\infty} \sup_{x\in\mathsf{X}} \left\| P^n(x,\cdot) - \pi \right\|_{\mathrm{TV}} = 0$.

- *The Markov kernel P is* uniformly geometrically ergodic *if there exist constants $C < \infty$ and $\rho \in [0,1)$ such that* $\sup_{x \in \mathsf{X}} \|P^n(x,\cdot) - \pi\|_{\mathrm{TV}} \le C\rho^n$.

As a consequence of Theorem 6.6, we obtain:

Proposition 6.8. *The Markov kernel P on $(\mathsf{X},\mathcal{X})$ is uniformly ergodic if and only if P is uniformly geometrically ergodic. In this case, there exists an integer m such that $\Delta_{\mathrm{TV}}(P^m) < 1$.*

Proof. Assume that P is uniformly ergodic. Since $\lim_{n\to\infty} \sup_{x \in \mathsf{X}} \|P^n(x,\cdot) - \pi\|_{\mathrm{TV}} = 0$, there exists an integer m such that $\sup_{x \in \mathsf{X}} \|P^m(x,\cdot) - \pi\|_{\mathrm{TV}} < 1/2$, showing that

$$\tfrac{1}{2} \sup_{(x,x') \in \mathsf{X}\times\mathsf{X}} \left\|P^m(x,\cdot) - P^m(x',\cdot)\right\|_{\mathrm{TV}} \le \sup_{x \in \mathsf{X}} \|P^m(x,\cdot) - \pi\|_{\mathrm{TV}} < 1 .$$

Theorem 6.6 shows that there exist constants $C < \infty$ and $\rho \in [0,1)$ such that $\sup_{x \in \mathsf{X}} \|P^n(x,\cdot) - \pi\|_{\mathrm{TV}} \le C\rho^n$ for all n. ∎

6.1.3 The Doeblin condition

Definition 6.9 (Doeblin condition). *A Markov kernel P on $(\mathsf{X},\mathcal{X})$ satisfies the* Doeblin condition *if there exists an integer $m \ge 1$, $\varepsilon > 0$ and a probability measure ν on $(\mathsf{X},\mathcal{X})$ such that for all $x \in \mathsf{X}$ and $A \in \mathcal{X}$,*

$$P^m(x,A) \ge \varepsilon\nu(A) . \tag{6.10}$$

Note that if $\varepsilon > 0$ satisfies (6.10), then, necessarily, $\varepsilon \in (0,\, 1]$ since by taking $A = \mathsf{X}$,

$$1 = P^m(x,\mathsf{X}) \ge \varepsilon\nu(\mathsf{X}) = \varepsilon .$$

Lemma 6.10. *Assume that the kernel P satisfies the Doeblin condition* (6.10). *Then, $\Delta_{\mathrm{TV}}(P^m) \le 1 - \varepsilon$.*

Proof. Let Q be defined on $\mathsf{X} \times \mathcal{X}$ by $Q(x,A) = (1-\varepsilon)^{-1}(P^m(x,A) - \varepsilon\nu(A))$. The Doeblin condition implies that Q is a Markov kernel and $P^m(x,A) = \varepsilon\nu(A) + (1-\varepsilon)Q(x,A)$. Note now that, for $x,x' \in \mathsf{X}$

$$P^m(x,\cdot) - P^m(x',\cdot) = (1-\varepsilon)\{Q(x,\cdot) - Q(x',\cdot)\} .$$

This implies that

$$\frac{1}{2}\left\|P^m(x,\cdot) - P^m(x',\cdot)\right\|_{\mathrm{TV}} = \frac{1}{2}(1-\varepsilon)\left\|Q(x,\cdot) - Q(x',\cdot)\right\|_{\mathrm{TV}} \le 1 - \varepsilon .$$

Since this inequality is valid for any $(x,x') \in \mathsf{X}\times\mathsf{X}$, we obtain $\Delta_{\mathrm{TV}}(P^m) \le 1 - \varepsilon$. ∎

6.1.4 Examples

Example 6.11 (Functional autoregressive model of order 1). Consider the functional autoregressive model $X_t = f(X_{t-1}) + \sigma(X_{t-1})Z_t$, where $\{Z_t,\, t \in \mathbb{N}\}$ are i.i.d. standard Gaussian random variables, $f : \mathbb{R} \to \mathbb{R}$ is a bounded function, $-\infty < \inf f \le$

$f(x) \leq \sup f < \infty$, $x \in \mathbb{R}$ and $\sigma : \mathbb{R} \to \mathbb{R}^+$ is a bounded function that is also bounded away from zero, $0 < \inf \sigma \leq \sigma(x) \leq \sup \sigma < \infty$, $x \in \mathbb{R}$. The kernel of this Markov chain has a density with respect to Lebesgue measure given by

$$p(x,x') = \frac{1}{\sqrt{2\pi\sigma^2(x)}} \exp\left(-\frac{1}{2\sigma^2(x)}(x' - f(x))^2\right) .$$

For $x' \in [-1,1]$, we have $\inf_{x\in\mathbb{R}} p(x,x') \geq q(x') > 0$, where

$$q(x') := \frac{1}{\sqrt{2\pi \sup^2 \sigma}} \exp\left(-\frac{1}{2\inf^2 \sigma}(x' - \inf f)^2 \vee (x' - \sup f)^2\right) > 0 .$$

Hence, this Markov chain is uniformly ergodic; see Exercise 6.3. Setting $\Delta_{\mathrm{TV}}(P) = 1 - \varepsilon$, with $\varepsilon = \int_C q(x')\mathrm{d}x'$ and $C = [-1,1]$, we obtain an explicit convergence bound. ◇

Example 6.12 (Functional autoregressive model of order m). Consider the functional autoregressive model

$$X_t = f(X_{t-1},\ldots,X_{t-m}) + \sigma(X_{t-1},\ldots,X_{t-m})Z_t ,$$

where $\{Z_t,\ t \in \mathbb{N}\}$ are i.i.d. standard Gaussian random variables, $f : \mathbb{R}^m \to \mathbb{R}$ is a bounded function,

$$-\infty < \inf f \leq f(x_1,\ldots,x_m) \leq \sup f < \infty , \quad \text{for all } (x_1,\ldots,x_m) \in \mathbb{R}^m$$

and $\sigma : \mathbb{R}^m \to \mathbb{R}^+$ is a bounded function that is bounded away from zero,

$$0 < \inf \sigma \leq \sigma(x_1,\ldots,x_m) \leq \sup \sigma < \infty , \quad \text{for all } (x_1,\ldots,x_m) \in \mathbb{R}^m.$$

Define $\boldsymbol{X}_t$ the state vector

$$\boldsymbol{X}_t = [X_t, X_{t-1},\ldots,X_{t-m+1}]' .$$

The vector process $\{\boldsymbol{X}_t,\ t \in \mathbb{N}\}$ is a first-order Markov chain. Denote by P the transition kernel of this Markov chain. The m-th iterate of this Markov kernel has a density with respect to the Lebesgue measure on $\mathbb{R}^m$ given by

$$p_m(x_1,\ldots,x_m;x'_1,\ldots,x'_m) = p(x_1,\ldots,x_m;x'_1)\, p(x'_1,x_2,\ldots,x_{m-1};x'_2) \\ \ldots p(x'_{m-1},\ldots,x'_1,x_1;x'_m) ,$$

where

$$p(x_1,\ldots,x_m;x') = \\ \frac{1}{\sqrt{2\pi\sigma^2(x_1,\ldots,x_m)}} \exp\left(-\frac{1}{2\sigma^2(x_1,\ldots,x_m)}(x' - f(x_1,\ldots,x_m))^2\right) .$$

We set $C = [-1, 1]$. For $x' \in C$, we get

$$q(x') := \inf_{(x_1,\ldots,x_m)\in\mathbb{R}^m} p(x_1,\ldots,x_m;x') \geq \frac{1}{\sqrt{2\pi \sup^2 \sigma}} \exp\left(-\frac{1}{2\inf^2 \sigma}(x' - \inf f)^2 \vee (x' - \sup f)^2\right) > 0\,.$$

Then, for all $(x'_1,\ldots,x'_m) \in C^m$,

$$p_m(x_1,\ldots,x_m;x'_1,\ldots,x'_m) \geq q(x'_1)q(x'_2)\ldots q(x'_p)\,.$$

Indeed, set $\varepsilon := \int_C q(x')\mathrm{Leb}(\mathrm{d}x') > 0$ and define for $A \in \mathcal{X}^m$,

$$\nu_m(A) = \frac{\int\cdots\int \mathbb{1}_{A\cap C^m}(x'_1,\ldots,x'_m)q(x'_1)\ldots q(x'_m)\mathrm{Leb}(\mathrm{d}x'_1)\ldots\mathrm{Leb}(\mathrm{d}x'_m)}{\int \mathbb{1}_{C^m}(x'_1,\ldots,x'_m)q(x'_1)\ldots q(x'_m)\mathrm{Leb}(\mathrm{d}x'_1)\ldots\mathrm{Leb}(\mathrm{d}x'_m)}\,.$$

By construction, for any $(x_1,\ldots,x_m) \in \mathsf{X}^m$ and $A \in \mathcal{X}^{\otimes m}$,

$$P^m(x_1,\ldots,x_m;A) \geq \varepsilon^m \nu_m(A)\,.$$

The kernel P^m satisfies the Doeblin condition and is therefore uniformly ergodic. ◇

Example 6.13 (Count data). The following model has been proposed in Davis et al. (2003) to model count data.

$$X_t = \beta + \gamma(N_t - \mathrm{e}^{X_{t-1}})\mathrm{e}^{-X_{t-1}}\,, \tag{6.11}$$

where, conditionally on $(X_0,\ldots,X_t)$, N_t has a Poisson distribution with intensity e^{X_t}. We now show that the chain is uniformly ergodic. By definition, for $t \geq 1$, we have,

$$X_t \leq \beta - \gamma \quad \text{if} \quad \gamma < 0\,, \tag{6.12}$$

$$X_t \geq \beta - \gamma \quad \text{if} \quad \gamma \geq 0\,. \tag{6.13}$$

In order to establish the Doeblin condition we consider two cases separately.

Case $\gamma < 0$. In that case, (6.12) shows that the state space is $(-\infty,\ \beta-\gamma]$. Let B be a Borel set such that $\beta - \gamma \in B$. Then, for all $x \leq \beta - \gamma$,

$$\begin{aligned} P(x,B) &= \mathbb{P}\left[X_1 \in B \,\middle|\, X_0 = x\right] \geq \mathbb{P}\left[X_1 = \beta - \gamma \,\middle|\, X_0 = x\right] \\ &= \mathbb{P}\left[N_1 = 0 \,\middle|\, X_0 = x\right] = \mathrm{e}^{-\mathrm{e}^x} \geq \mathrm{e}^{-\mathrm{e}^{\beta-\gamma}} = \mathrm{e}^{-\mathrm{e}^{\beta-\gamma}}\delta_{\beta-\gamma}(B)\,, \end{aligned}$$

which yields Doeblin's condition with $m = 1$.

Case $\gamma > 0$. Now, (6.13) shows that the state space is $[\beta-\gamma,\ \infty)$. Let $C = [\beta-\gamma,\ \beta+\gamma]$. Then, for $x \in C$ and any Borel set B containing $\beta - \gamma$,

$$\begin{aligned} P(x,B) &= \mathbb{P}\left[X_1 \in B \,\middle|\, X_0 = x\right] \geq \mathbb{P}\left[X_1 = \beta - \gamma \,\middle|\, X_0 = x\right] \\ &= \mathbb{P}\left[N_1 = 0 \,\middle|\, X_0 = x\right] = \mathrm{e}^{-\mathrm{e}^x} \geq \mathrm{e}^{-\mathrm{e}^{\beta+\gamma}} \\ P^2(x,B) &\geq \mathbb{P}\left[X_{t+1} = \beta-\gamma,\ X_t = \beta-\gamma \,\middle|\, X_{t-1} = x\right] \geq \mathrm{e}^{-2\mathrm{e}^{\beta+\gamma}}\,. \end{aligned} \tag{6.14}$$

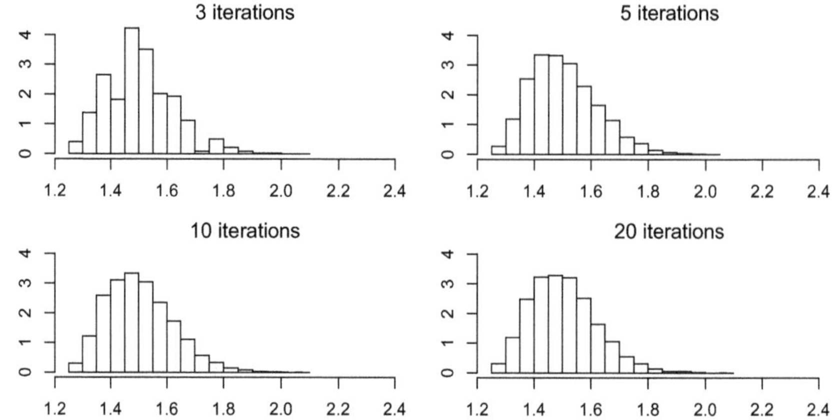

Figure 6.1 *Convergence of the stationary distribution for the iterative model* (5.28). *The marginal distribution is estimated using a histogram computed from* 10^4 *independent replications of the process.*

On the other hand, if $x > \beta + \gamma$, then, noting that $\mathbb{E}\left[X_1 \mid X_0 = x\right] = \beta$, we have

$$\begin{aligned}
P(x,C) &= \mathbb{P}\left[\beta - \gamma \le X_1 \le \beta + \gamma \mid X_0 = x\right] \\
&= \mathbb{P}\left[|X_1 - \beta| \le \gamma \mid X_0 = x\right] \\
&\ge 1 - \gamma^{-2}\mathrm{Var}\,(X_1 \mid X_0 = x) \\
&= 1 - \gamma^{-2}\gamma^2 \mathrm{e}^{-x} \ge 1 - \mathrm{e}^{-(\beta+\gamma)} .
\end{aligned}$$

Then, for $x \ge \beta + \gamma$, we have

$$\begin{aligned}
P^2(x,B) &= \mathbb{P}\left[X_2 \in B \mid X_0 = x\right] \ge \mathbb{P}\left[X_2 \in B,\, X_1 \in C \mid X_0 = x\right] \\
&= \mathbb{E}\left[\mathbb{1}_B(X_2)\mathbb{1}_C(X_1) \mid X_0 = x\right] \\
&= \mathbb{E}\left[\mathbb{1}_C(X_1)P(X_1,B) \mid X_0 = x\right] \\
&\ge \mathrm{e}^{-\mathrm{e}^{\beta+\gamma}} P(x,C) \ge \mathrm{e}^{-\mathrm{e}^{\beta+\gamma}}(1 - \mathrm{e}^{-(\beta+\gamma)}) \\
&= \mathrm{e}^{-\mathrm{e}^{\beta+\gamma}}(1 - \mathrm{e}^{-(\beta+\gamma)})\delta_{\beta-\gamma}(B) . \qquad (6.15)
\end{aligned}$$

(6.14) and (6.15) show that Doeblin's condition (6.10) holds with $m = 2$, $\nu = \delta_{\beta-\gamma}$, and $\varepsilon = \min(\mathrm{e}^{-\mathrm{e}^{\beta+\gamma}}(1 - \mathrm{e}^{-(\beta+\gamma)}, \mathrm{e}^{-2\mathrm{e}^{\beta+\gamma}})$.

The stationary distribution has a component that has a density with respect to the Lebesgue measure and atoms located at $\beta - \gamma(k - \exp(\beta - \gamma))\exp(-(\beta - \gamma))$, $k \in \mathbb{N}$. We have sampled 10^4 independent trajectories of the process with parameters $\beta = 1.5$ and $\gamma = 0.25$, started from $X_0 = 0$. We have displayed the marginal stationary distribution after 3, 5, 10 and 20 iterations in Figure 6.1. ◇

6.2 V-geometric ergodicity

6.2.1 V-total variation distance

We first generalized the total variation distance to be able to consider *unbounded functions*.

Definition 6.14 (V-norm). *Let $V \in \mathbb{F}(\mathsf{X}, \mathcal{X})$ taking values in $[1, \infty)$. The space of finite signed measures ξ such that $|\xi|(V) < \infty$ is denoted by $\mathbb{M}_V(\mathcal{X})$. The V-norm of a measure $\xi \in \mathbb{M}_V(\mathcal{X})$ is defined by*

$$\|\xi\|_{\mathrm{V}} = |\xi|(V) .$$

Of course, when $V = \mathbb{1}_{\mathsf{X}}$, then $\|\xi\|_{\mathbb{1}_{\mathsf{X}}} = \|\xi\|_{\mathrm{TV}}$ by (6.3). It also holds that $\|\xi\|_{\mathrm{V}} = \|V \cdot \xi\|_{\mathrm{TV}}$ where we denote by $V \cdot \xi$ the measure defined by $(V \cdot \xi)(A) = \xi(V \mathbb{1}_A)$. We now provide different characterizations of the V-norm similar to the characterizations of the TV-norm provided in Theorem 6.1. Define

$$\operatorname{osc}_V(f) := \sup_{x \neq y} \frac{|f(x) - f(y)|}{V(x) + V(y)} . \tag{6.16}$$

Theorem 6.15. *For $\xi \in \mathbb{M}_V(\mathcal{X})$,*

$$\|\xi\|_{\mathrm{V}} = \sup\{\xi(f) : f \in \mathbb{F}_b(\mathsf{X}, \mathcal{X}), |f/V|_\infty \leq 1\} . \tag{6.17}$$

Let $\xi \in \mathbb{M}_0(\mathcal{X}) \cap \mathbb{M}_V(\mathcal{X})$. Then,

$$\|\xi\|_{\mathrm{V}} = \sup\{\xi(f) : \operatorname{osc}_V(f) \leq 1\} . \tag{6.18}$$

Proof. First, let S be a Jordan set for ξ. Because $\xi(V \mathbb{1}_S - V \mathbb{1}_{S^c}) = |\xi|(V)$ and $\operatorname{osc}_V(V \mathbb{1}_S - V \mathbb{1}_{S^c}) \leq 1$, we get

$$\|\xi\|_{\mathrm{V}} = |\xi|(V) \leq \sup\{|\xi(f)| : \operatorname{osc}_V(f) \leq 1\} .$$

Then, using the fact that $\operatorname{osc}_V(f) \leq \inf_{c \in \mathbb{R}} |f - c|_{\mathrm{V}}$ and $\xi(f) = \xi(f - c)$ for all $\xi \in \mathbb{M}_0(\mathcal{X})$, we get $\|\xi\|_{\mathrm{V}} \geq \sup\{|\xi(f)| : \operatorname{osc}_V(f) \leq 1\}$. ■

Note that when $V = \mathbb{1}_{\mathsf{X}}$, then $\operatorname{osc}_V(f) = \operatorname{osc}(f)/2$ and thus Theorem 6.1 is a consequence of Theorem 6.15. Define $\mathbb{M}_{1,V}(\mathcal{X})$ the subset of probability measures such that $\xi(V) < \infty$:

$$\mathbb{M}_{1,V}(\mathcal{X}) := \{\xi \in \mathbb{M}_1(\mathcal{X}), \xi(V) < \infty\} , \tag{6.19}$$

the set of probability measures such that $\xi(V) < \infty$. Define a distance d_V on $\mathbb{M}_{1,V}(\mathcal{X})$ by

$$d_V(\xi, \xi') := \|\xi - \xi'\|_{\mathrm{V}} . \tag{6.20}$$

Proposition 6.16. *The space $(\mathbb{M}_V(\mathcal{X}), \|\cdot\|_{\mathrm{V}})$ is complete. The space $(\mathbb{M}_{1,V}(\mathcal{X}), d_V)$ is a Polish space.*

Proof. Let $\{\xi_n,\, n \in \mathbb{N}\}$ be a Cauchy sequence in $\mathbb{M}_V(\mathcal{X})$. Define

$$\lambda = \sum_{n=0}^{\infty} \frac{1}{2^n |\xi_n|(V)} |\xi_n| \, ,$$

which is a measure, as a limit of an increasing sequence of measures. By construction, $\lambda(V) < \infty$ and $|\xi_n| \ll \lambda$ for any $n \in \mathbb{N}$. Therefore, there exist functions $f_n \in \mathrm{L}^1(V \cdot \lambda)$ such that $\xi_n = f_n . \lambda$ and $\|\xi_n - \xi_m\|_{\mathrm{V}} = \int |f_n - f_m| V \mathrm{d}\lambda$. This implies that $\{f_n,\, n \in \mathbb{N}\}$ is a Cauchy sequence in $\mathrm{L}^1(V \cdot \lambda)$ that is complete. Thus, there exists $f \in \mathrm{L}^1(V \cdot \lambda)$ such that $f_n \to f$ in $\mathrm{L}^1(V \cdot \lambda)$. Setting $\xi = f.\lambda$, we obtain that $\xi \in \mathbb{M}_V(\mathcal{X})$ and $\lim_{n\to\infty} \|\xi_n - \xi\|_{\mathrm{V}} = \lim_{n\to\infty} \int |f_n - f| V \mathrm{d}\lambda = 0$. ■

6.2.2 V-Dobrushin coefficient

The uniform ergodicity condition is restrictive and we will consider in this section different conditions that may be shown to hold for a wider class of Markov chains.

Definition 6.17 (V-Dobrushin coefficient). *Let P be a Markov kernel on $(\mathsf{X}, \mathcal{X})$. Assume that, for any $\xi \in \mathbb{M}_{1,V}(\mathcal{X})$, $\xi P \in \mathbb{M}_{1,V}(\mathcal{X})$, where $\mathbb{M}_{1,V}(\mathcal{X})$ is defined in* (6.19). *The V*-Dobrushin coefficient *$\Delta_V(P)$ is defined by*

$$\Delta_V(P) = \sup_{\xi \neq \xi' \in \mathbb{M}_{1,V}(\mathcal{X})} \frac{d_V(\xi P, \xi' P)}{d_V(\xi, \xi')} \, . \tag{6.21}$$

Contrary to the Dobrushin coefficient, the V-Dobrushin coefficient is not necessarily finite, unless V is bounded.

Lemma 6.18. *Let P be a Markov kernel on $(\mathsf{X}, \mathcal{X})$. Then,*

$$\Delta_V(P) = \sup_{\mathrm{osc}_V(f) \le 1} \mathrm{osc}_V(Pf) = \sup_{x \neq x' \in \mathsf{X}} \frac{d_V(P(x,\cdot), P(x',\cdot))}{d_V(\delta_x, \delta_{x'})} \, . \tag{6.22}$$

Proof. The right-hand side of (6.22) is obviously less than or equal to $\Delta_V(P)$. By Theorem 6.15 and (6.18), we have, for all probability measures ξ, ξ',

$$\begin{aligned} d_V(\xi P, \xi' P) = \left\|\xi P - \xi' P\right\|_{\mathrm{V}} &= \sup_{f : \mathrm{osc}_V(f) \le 1} |\xi P(f) - \xi' P(f)| \\ &= \sup_{f : \mathrm{osc}_V(f) \le 1} |\xi(Pf) - \xi'(Pf)| \le d_V(\xi, \xi') \sup_{f : \mathrm{osc}_V(f) \le 1} \mathrm{osc}_V(Pf) \, . \end{aligned}$$

To conclude, we apply again Theorem 6.15 to obtain

$$\begin{aligned} \sup_{\mathrm{osc}_V(f) \le 1} \mathrm{osc}_V(Pf) &= \sup_{\mathrm{osc}_V(f) \le 1} \sup_{x \neq x'} \frac{|Pf(x) - Pf(x')|}{V(x) + V(x')} \\ &= \sup_{x \neq x'} \sup_{\mathrm{osc}_V(f) \le 1} \frac{|[P(x,\cdot) - P(x',\cdot)] f|}{V(x) + V(x')} \\ &= \sup_{x \neq x'} \frac{\|P(x,\cdot) - P(x',\cdot)\|_{\mathrm{V}}}{V(x) + V(x')} \, . \end{aligned}$$

■

Assume that P defines a mapping on $\mathbb{M}_{1,V}(\mathcal{X})$, i.e., for any $\xi \in \mathbb{M}_{1,V}(\mathcal{X})$, $\xi P(V) < \infty$. If $\Delta_V(P) < 1$, then P defines a contraction mapping on $(\mathbb{M}_{1,V}(\mathcal{X}), d_V)$. Since $(\mathbb{M}_{1,V}(\mathcal{X}), d_V)$ is a Polish space, we may again apply the Banach fixed point theorem to show the existence and uniqueness of a fixed point π, which is an invariant probability measure satisfying $\pi(V) < \infty$ (since $\pi \in \mathbb{M}_{1,V}(\mathcal{X})$). The Banach fixed point theorem also provides an estimate for the rate at which the sequence of iterates $\{\xi P^k, k \in \mathbb{N}\}$ converges to the fixed point π, in d_V. If we assume that $\Delta_V(P^m) < 1$ (but not $\Delta_V(P)$), then the argument is slightly more difficult.

Theorem 6.19. *Let $(\mathsf{X}, \mathcal{X})$ be a measurable space. Let P be a Markov kernel satisfying*

(a) for any $\xi \in \mathbb{M}_{1,V}(\mathcal{X})$, $\xi P \in \mathbb{M}_{1,V}(\mathcal{X})$, where $\mathbb{M}_{1,V}(\mathcal{X})$ is defined in (6.19),

(b) $\Delta_V(P) < \infty$ and for some integer m, $\Delta_V(P^m) \le \rho < 1$.

Then, P admits a unique invariant probability measure π, satisfying $\pi(V) < \infty$. In addition, for all $\xi \in \mathbb{M}_1(\mathcal{X})$,

$$d_V(\xi P^n, \pi) \le \max_{0 \le r \le m-1} \Delta_V(P^r) \rho^{\lfloor n/m \rfloor} d_V(\xi, \pi) .$$

Proof. This is a direct application of the fixed point theorem for contraction mapping in Polish space (see Theorem 6.40). ■

6.2.3 *Drift and minorization conditions*

Definition 6.20 (m-small set). *A set $C \in \mathcal{X}$ is a m-small set for the Markov kernel P if there exist $\varepsilon > 0$ and a probability measure ν on $(\mathsf{X}, \mathcal{X})$ such that*

$$P^m(x, A) \ge \varepsilon \nu(A) , \quad \text{for all } A \in \mathcal{X} \text{ and all } x \in C. \tag{6.23}$$

In the following, it will sometimes be convenient to specify the constant ε and the probability measure ν in the notations. In that case, we will say that C is a (m, ε, ν)-small set. It is perhaps easiest to develop an intuition about what a small set is by considering examples.

Example 6.21 (Functional autoregressive model). The first-order functional autoregressive model on $\mathbb{R}^d$ is defined iteratively by $X_t = m(X_{t-1}) + Z_t$, where $\{Z_t, t \in \mathbb{N}\}$ is an i.i.d. sequence of random vectors independent of X_0 and $m : \mathbb{R}^d \to \mathbb{R}^d$ is a measurable function, bounded on compact sets.

Assume that the noise distribution has an everywhere positive density q with respect to Lebesgue measure on $\mathbb{R}^d$, that q is lower bounded on compact sets and that $\mathbb{E}[|Z_0|] < \infty$. Let K be a compact set. Then, for all $x \in K$,

$$\begin{aligned} P(x, A) &= \int_A q(x' - m(x)) \mathrm{Leb}(\mathrm{d}x') \\ &\ge \int_{A \cap K} q(x' - m(x)) \mathrm{Leb}(\mathrm{d}x') \ge \underline{q}_K \mathrm{Leb}(K) \mathrm{Leb}_K(A \cap K) , \end{aligned}$$

where $\underline{q}_K = \min_{(x,x') \in K \times K} q(x' - m(x))$ and $\mathrm{Leb}_K(A) = \mathrm{Leb}(A \cap K)/\mathrm{Leb}(K)$. Therefore, every compact subset of $\mathbb{R}$ is a $(1, \underline{q}_K \mathrm{Leb}(K), \mathrm{Leb}_K)$-small set. ◇

Example 6.22 (ARCH(1) model). Consider an ARCH(1) model given by $X_t = (\alpha_0 + \alpha_1 X_{t-1}^2)^{1/2} Z_t$, where $\{Z_t,\ t \in \mathbb{N}\}$ is a strong white noise. Assume that $\alpha_0 > 0$, $\alpha_1 > 0$ and that the random variable Z_0 has a density, denoted g, which is bounded away from zero on a neighborhood $[-\tau, \tau]$ of zero, i.e., $g(z) \geq g_{\min} \mathbb{1}_{[-\tau,\tau]}(z)$.

We will now show that any interval $[-a,a]$ with $a > 0$ is a 1-small set. For any $A \in \mathcal{B}(\mathbb{R})$ and $x \in [-a,a]$, we get

$$\begin{aligned} P(x,A) &= \int_{-\infty}^{\infty} \mathbb{1}_A \left[(\alpha_0 + \alpha_1 x^2)^{1/2} z \right] g(z) \mathrm{Leb}(\mathrm{d}z) \\ &= (\alpha_0 + \alpha_1 x^2)^{-1/2} \int_{-\infty}^{\infty} \mathbb{1}_A(v) g((\alpha_0 + \alpha_1 x^2)^{-1/2} v) \mathrm{Leb}(\mathrm{d}v) \\ &\geq (\alpha_0 + \alpha_1 a^2)^{-1/2} g_{\min} \int_{-\infty}^{\infty} \mathbb{1}_A(v) \mathbb{1}_{[-\tau,\tau]}(\alpha_0^{-1/2} v) \mathrm{Leb}(\mathrm{d}v) \, . \end{aligned}$$

If we set

$$\nu(A) = \frac{\int_{-\infty}^{\infty} \mathbb{1}_A(v) \mathbb{1}_{[-\tau,\tau]}(\alpha_0^{-1/2} v) \mathrm{Leb}(\mathrm{d}v)}{\int_{-\infty}^{\infty} \mathbb{1}_{[-\tau,\tau]}(\alpha_0^{-1/2} v) \mathrm{Leb}(\mathrm{d}v)} = \frac{\int_{-\infty}^{\infty} \mathbb{1}_A(v) \mathbb{1}_{[-\tau,\tau]}(\alpha_0^{-1/2} v) \mathrm{Leb}(\mathrm{d}v)}{2\tau \alpha_0^{1/2}} ,$$

and $\varepsilon = (\alpha_0 + \alpha_1 a^2)^{-1/2} g_{\min} 2\tau \alpha_0^{1/2}$ then for all $x \in [-a,a]$, $P(x,A) \geq \varepsilon \nu(A)$. Every compact interval (and similarly every compact set) is a 1-small set. ◇

We next consider the following drift condition:

Definition 6.23 (Geometric drift condition). *A Markov kernel P satisfies a geometric drift (or a Foster-Lyapunov drift) condition if there exists a measurable function $V : \mathsf{X} \to [1, \infty)$ and constants $(\lambda, b) \in [0,1) \times \mathbb{R}^+$ such that*

$$PV \leq \lambda V + b \, . \tag{6.24}$$

The function V is sometimes called an *energy* function. The suggested image is that V is a potential energy surface and (6.24) implies that the chain goes *downhill* to states of lower energy in expectation and so drifts toward states of low energy. This condition is also closely related to the Lyapunov functions, which were developed by A. Lyapunov at the end of the nineteenth century for the study of stability of dynamical systems described by systems of ordinary differential equations (motivated by the study of the stability of mechanical systems). For this reason, the function V is sometimes referred to as a Lyapunov function. Again, it is easiest to develop an intuition about what a drift condition is by considering examples.

Example 6.24 (Functional AR model; Example 6.21 cont.). Assume that

$$\limsup_{|x| \to \infty} \frac{|m(x)|}{|x|} < 1 \, . \tag{6.25}$$

Set $V(x) = 1 + |x|$. We get

$$PV(x) = 1 + \mathbb{E}\left[|m(x) + Z_1| \right] \leq 1 + |m(x)| + \mathbb{E}\left[|Z_1| \right] \, .$$

Using (6.25), there exist $\lambda \in [0,1)$ and $r \in \mathbb{R}_+$ such that, for all $|x| \geq r$, $|m(x)|/|x| \leq \lambda$. For $|x| \geq r$, this implies

$$PV(x) \leq 1 + \lambda |x| + \mathbb{E}[|Z_1|] = \lambda V(x) + 1 - \lambda + \mathbb{E}[|Z_1|] .$$

For $|x| \leq r$, since m is bounded on compact sets, we get $PV(x) \leq b_r$. Setting $b = b_r \vee (1 - \lambda + \mathbb{E}[|Z_1|])$, we obtain $PV(x) \leq \lambda V(x) + b$. Thus, P satisfies a geometric drift condition. ◇

Example 6.25 (ARCH(1) model; Example 6.22 cont.). Assume that for some $s \in (0,1]$, $\alpha_1^s \mu_{2s} < 1$, where $\mu_{2s} = \mathbb{E}\left[Z_0^{2s}\right]$. Set $V(x) = 1 + x^{2s}$. Then, since $(a+b)^s \leq a^s + b^s$, we get

$$\begin{aligned} PV(x) = \mathbb{E}_x[V(X_1)] &= 1 + (\alpha_0 + \alpha_1 x^2)^s \mu_{2s} \\ &\leq 1 + \alpha_0^s \mu_{2s} + \alpha_1^s \mu_{2s} x^{2s} \leq \lambda V(x) + b , \end{aligned}$$

with $\lambda = \alpha_1^s \mu_{2s}$ and $b = 1 - \alpha_1^s \mu_{2s} + \alpha_0^s \mu_{2s}$. ◇

Lemma 6.26. *Let P be a Markov kernel on $(\mathsf{X}, \mathcal{X})$. Assume that P satisfies a geometric drift condition associated to a drift function V and constants $(\lambda, b) \in [0,1) \times \mathbb{R}^+$. Then, for all $\xi \in \mathbb{M}_1(\mathcal{X})$ such that $\xi(V) < \infty$,*

$$\xi(P^n V) \leq \lambda^n \xi(V) + b/(1-\lambda) < \infty . \tag{6.26}$$

Proof. Let $k \geq 0$. Applying ξP^k to (6.24) yields $\xi P^{k+1} V \leq \lambda \xi P^k V + b$. By straightforward induction, we obtain for all $n \geq 0$,

$$\xi P^n V \leq \lambda^n \xi(V) + b \sum_{k=0}^{n-1} \lambda^k \leq \lambda^n \xi(V) + b/(1-\lambda) ,$$

so that (6.26) holds. ■

It is sometimes easier to check the geometric drift condition on an iterate P^m of the kernel rather than directly on the kernel P. The following proposition shows that checking the drift condition for P^m or P is essentially equivalent.

Proposition 6.27. *Assume that the Markov kernel P^m satisfies a geometric drift condition, i.e., there exist $V_m : \mathsf{X} \to [1, \infty)$, $\lambda_m \in [0,1)$ and $b_m \in \mathbb{R}_+$ such that $P^m V_m \leq \lambda_m V_m + b_m$. Then the Markov kernel P satisfies a geometric drift condition with*

$$V := V_m + \lambda_m^{-1/m} P V_m + \cdots + \lambda_m^{-(m-1)/m} P^{m-1} V_m ,$$

and constants $\lambda := \lambda_m^{1/m}$ and $b_m := \lambda_m^{-(m-1)/m} b_m$.

Proof.

$$\begin{aligned} PV &= PV_m + \lambda_m^{-1/m} P^2 V_m + \cdots + \lambda_m^{-(m-1)/m} P^m V_m , \\ &\leq PV_m + \lambda_m^{-1/m} P^2 V_m + \cdots + \lambda_m^{-(m-1)/m} (\lambda_m V_m + b_m) , \\ &= \lambda_m^{1/m} (V_m + \lambda_m^{-1/m} P V_m + \cdots + \lambda_m^{-(m-1)/m} P^{m-1} V_m) + \lambda_m^{-(m-1)/m} b_m . \end{aligned}$$

■

The main result of this section is that the minorization and the drift condition imply the existence and uniqueness of the stationary distribution, and also provides the existence of an explicit rate of convergence to stationarity.

Assumption A6.28. The Markov kernel P satisfies a geometric drift condition associated to a drift function V and constants $(\lambda, b) \in [0,1) \times \mathbb{R}^+$. In addition, for some $R > 2b/(1-\lambda)^2$, the level set $\{x \in \mathsf{X} : V(x) \leq R\}$ is a m-small set for P.

Lemma 6.29. *Under assumption A6.28, there exists a pair* $(\alpha, \beta) \in (0,1)^2$ *such that* $\Delta_{V_\beta}(P^m) \leq \alpha$, *where*

$$V_\beta(x) := 1 - \beta + \beta V(x), \quad x \in \mathsf{X}. \tag{6.27}$$

Proof. Under A6.28, there exist $(R, \varepsilon, m) \in (2b/(1-\lambda)^2, \infty) \times \mathbb{R}_+^* \times \mathbb{N}^*$ such that for all $x \in \mathsf{X}$ such that $V(x) \leq R$, $P^m(x, \cdot) \geq \varepsilon \nu(\cdot)$. Then, for $x \in \mathsf{X}$ such that $V(x) \leq R$, define the probability measure

$$Q(x, \cdot) := \frac{P^m(x, \cdot) - \varepsilon \nu(\cdot)}{1 - \varepsilon}.$$

Note that, if $V(x) \leq R$, then $(1-\varepsilon)Q(x, \cdot) \leq P^m(x, \cdot)$. Now, set

$$\alpha_\beta := \left(1 - \varepsilon + \frac{\beta b}{(1-\beta)(1-\lambda)}\right) \vee \lambda^m. \tag{6.28}$$

Choose $\beta \in (0,1)$ sufficiently small so that $\alpha_\beta < 1$. Let $x, y \in \mathsf{X}$ be such that $V(x) + V(y) < R$ and f such that $\operatorname{osc}_{V_\beta}(f) \leq 1$. Then, since $V(x) \leq R$, $V(y) \leq R$ and $|f(u) - f(v)| \leq 2(1-\beta) + \beta V(u) + \beta V(v)$, we have

$$\begin{aligned}
|P^m f(x) - P^m f(y)| &= (1-\varepsilon)|Qf(x) - Qf(y)| \\
&\leq (1-\varepsilon) \iint Q(x, du) Q(y, dv) |f(u) - f(v)| \\
&\leq 2(1-\beta)(1-\varepsilon) + \beta(1-\varepsilon)[QV(x) + QV(y)] \\
&\leq 2(1-\beta)(1-\varepsilon) + \beta[P^m V(x) + P^m V(y)].
\end{aligned}$$

Using Lemma 6.26, the previous inequality implies

$$\begin{aligned}
&|P^m f(x) - P^m f(y)| \leq 2(1-\beta)(1-\varepsilon) + \beta[\lambda^m[V(x) + V(y)] + 2b/(1-\lambda)] \\
&\leq 2(1-\beta)\left(1 - \varepsilon + \frac{b\beta}{(1-\beta)(1-\lambda)}\right) + \lambda^m[\beta V(x) + \beta V(y)] \\
&\leq \alpha_\beta[2(1-\beta) + \beta V(x) + \beta V(y)] = \alpha_\beta(V_\beta(x) + V_\beta(y)),
\end{aligned}$$

where α_β is defined in (6.28).

Assume now that x, y are such that $V(x) + V(y) \geq R$. Using $R > 2b/(1-\lambda)^2$, we have $\lambda_0 := \lambda^m + 2b/[(1-\lambda)R] < 1$. Now, since $R = 2b/[(1-\lambda)(\lambda_0 - \lambda^m)]$, the inequality $V(x) + V(y) \geq R$ implies

$$\begin{aligned}
\lambda^m[V(x) + V(y)] + \frac{2b}{1-\lambda} &= \lambda^m[V(x) + V(y)] + R(\lambda_0 - \lambda^m) \\
&\leq \lambda_0[V(x) + V(y)].
\end{aligned} \tag{6.29}$$

Then, using again $|f(u) - f(v)| \leq 2(1-\beta) + \beta[V(u) + V(v)]$, we get

$$\begin{aligned} |P^m f(x) - P^m f(y)| &\leq 2(1-\beta) + \beta[P^m V(x) + P^m V(y)] \\ &\leq 2(1-\beta) + \beta(\lambda^m(V(x) + V(y)) + 2b/(1-\lambda)) \\ &\leq 2(1-\beta) + \beta\lambda_0[V(x) + V(y)] . \end{aligned} \tag{6.30}$$

where the last inequality follows from (6.29). Now, since $\lambda_0 < 1$, we have

$$\lambda_\beta := \frac{2(1-\beta) + R\beta\lambda_0}{2(1-\beta) + R\beta} < 1 .$$

Moreover, by definition of λ_β, the inequality $V(x) + V(y) \geq R$ implies

$$2(1-\beta) + \beta\lambda_0[V(x) + V(y)] \leq \lambda_\beta[2(1-\beta) + \beta V(x) + \beta V(y)] . \tag{6.31}$$

Combining this inequality with (6.30), we obtain that, if $V(x) + V(y) \geq R$,

$$|P^m f(x) - P^m f(y)| \leq \lambda_\beta [V_\beta(x) + V_\beta(y)] .$$

Finally, for all $x, y \in \mathsf{X}$ and for all f such that $\operatorname{osc}_{V_\beta}(f) \leq 1$,

$$|P^m f(x) - P^m f(y)| \leq (\lambda_\beta \vee \alpha_\beta)[V_\beta(x) + V_\beta(y)] ,$$

which concludes the proof. ■

Lemma 6.29 almost directly implies the following theorem, which is the most important result of this chapter.

Theorem 6.30. *Assume A6.28. Then, there exists $n \in \mathbb{N}$ such that $\Delta_V(P^n) < 1$.*

Proof. According to Lemma 6.29, there exists $(\alpha, \beta) \in (0,1)^2$ such that $\Delta_{V_\beta}(P^m) \leq \alpha$ so that for all $n \geq 0$,

$$\Delta_{V_\beta}(P^n) \leq \alpha^{\lfloor n/m \rfloor} \sup_{r=0,\ldots,m-1} \Delta_{V_\beta}(P^r) .$$

The drift condition implies that $PV_\beta \leq \lambda V_\beta + b_\beta$, with $b_\beta := (1-\beta)(1-\lambda) + b\beta$. Lemma 6.26 therefore implies that $\max_{0 \leq r \leq m-1} \Delta_{V_\beta}(P^r) \leq 2\lambda + 2b_\beta/(1-\lambda)$. Theorem 6.19 shows that there exists π such that $\pi(V_\beta) < \infty$ and for any $n \in \mathbb{N}^\star$ and $\xi, \xi' \in \mathbb{M}_{1,V_\beta}(\mathcal{X})$,

$$d_{V_\beta}(\xi P^n, \xi' P^n) \leq 2\left(\lambda + b_\beta/(1-\lambda)\right) \alpha^{\lfloor n/m \rfloor} d_{V_\beta}(\xi, \xi') .$$

Since $\|\xi\|_{\mathrm{V}_\beta} = (1-\beta)\|\xi\|_{\mathrm{TV}} + \beta\|\xi\|_{\mathrm{V}}$ and $\|\xi\|_{\mathrm{TV}} \leq \|\xi\|_{\mathrm{V}}$, we have: $\beta\|\xi\|_{\mathrm{V}} \leq \|\xi\|_{\mathrm{V}_\beta} \leq \|\xi\|_{\mathrm{V}}$, i.e., the norm $\|\cdot\|_{\mathrm{V}_\beta}$ and $\|\cdot\|_{\mathrm{V}}$ are equivalent on $\mathbb{M}_V(\mathcal{X})$. Therefore, we get

$$d_V(\xi P^n, \xi' P^n) \leq 2\beta^{-1}\left(\lambda + b_\beta/(1-\lambda)\right) \alpha^{\lfloor n/m \rfloor} d_V(\xi, \xi') ,$$

which implies that

$$\Delta_V(P^n) \leq 2\beta^{-1}\left(\lambda + b_\beta/(1-\lambda)\right) \alpha^{\lfloor n/m \rfloor} .$$

The proof is concluded by taking n large enough so that the right hand side of the previous equation is strictly less than 1. ■

Definition 6.31 (Harris V-geometric ergodicity). *Let $V : \mathsf{X} \to [1,\infty)$ be a measurable function. The Markov kernel P is said to be* Harris V-geometrically ergodic *if there exist constants $C < \infty$ and $\rho \in [0,1)$, such that for all $n \in \mathbb{N}$ and $x \in \mathsf{X}$,*

$$\|P^n(x,\cdot) - \pi\|_{\mathrm{V}} \le C\rho^n V(x) . \tag{6.32}$$

So long as x is not some bad starting value (i.e., $V(x)$ is not large), geometric ergodicity guarantees quick convergence for the Markov chain. Since $\|\xi\|_{\mathrm{TV}} = 2\|\xi\|_{\mathbb{1}_{\mathsf{X}}}$, the concept of V-geometric ergodicity generalizes the notion of uniform ergodicity. The upper-bound for the V-norm of the difference between the n-th iterate of the Markov chain $P^n(x,\cdot)$ and the invariant probability measure π is allowed to depend on the starting point x through the function $V(x)$.

Geometric ergodicity (in fact, uniform ergodicity) holds for every irreducible and aperiodic Markov chain on a finite state space. However, this is not true for Markov chains on general state spaces.

6.2.4 Examples

In this section, we consider the general functional autoregressive model

$$X_t = F(X_{t-1},\ldots,X_{t-p};Z_t) , \tag{6.33}$$

where $\{Z_t,\ t \in \mathbb{N}\}$ is an i.i.d. sequence of random vectors taking values in some subset $\mathsf{Z} \subset \mathbb{R}^m$, independent of the initial state $\boldsymbol{X}_0 = (X_0,X_{-1},\ldots,X_{-p+1})$ and $F : \mathsf{X}^p \times \mathsf{Z} \to \mathsf{X}$ is a measurable function. Consider the following assumptions:

Assumption A6.32.

(a) X and Z are open subsets of $\mathbb{R}^m$.

(b) The distribution of the excitation Z_0 has a nontrivial Lebesgue component with density q.

(c) For any $\boldsymbol{x} := (x_{-1},\ldots,x_{-p}) \in \mathsf{X}^p$, $F(\boldsymbol{x};\cdot) : \mathsf{Z} \to \mathsf{X}$ is a diffeomorphism from Z to X.

The Jacobian of the inverse $F^{-1}(\boldsymbol{x};\cdot)$ of $F(\boldsymbol{x},\cdot)$ is denoted by $J_{F^{-1}(\boldsymbol{x};\cdot)}$. Set $\boldsymbol{X}_t := (X_t,\ldots,X_{t-p+1})$; $\{\boldsymbol{X}_t,\ t \in \mathbb{N}\}$ is a Markov chain on X^p. Denote by P its Markov kernel. For $\boldsymbol{x} = (x_{-1},\ldots,x_{-p}) \in \mathsf{X}^p$ and $A \in \mathcal{X}^{\otimes p}$, we set

$$R(\boldsymbol{x},A) = \int \mathbb{1}_A(F(\boldsymbol{x};z),\ x_{-1},\ldots,x_{-p+1})q(z)\mathrm{Leb}(\mathrm{d}z) . \tag{6.34}$$

Clearly, under A6.32, for all $\boldsymbol{x} \in \mathsf{X}^p$ and $A \in \mathcal{X}^{\otimes p}$, $P(\boldsymbol{x},A) \ge R(\boldsymbol{x},A)$ and for any integer k, $P^k(\boldsymbol{x},A) \ge R^k(\boldsymbol{x},A)$. Consider the function $G = (G_0,\ldots,G_{p-1}) : \mathsf{X}^p \times \mathsf{Z}^p \to \mathsf{X}$, defined recursively as follows

$$\begin{aligned}
G_0(\boldsymbol{x},\boldsymbol{z}) &:= F(x_{-1},\ldots,x_{-p};z_0) , \\
G_1(\boldsymbol{x},\boldsymbol{z}) &:= F(G_0(\boldsymbol{x},\boldsymbol{z}),\ x_{-1},\ldots,x_{-p+1};z_1) , \\
&\ \vdots \\
G_{p-1}(\boldsymbol{x},\boldsymbol{z}) &:= F(G_{p-2}(\boldsymbol{x},\boldsymbol{z}),\ \ldots,\ G_0(\boldsymbol{x},\boldsymbol{z}),\ x_{-1};z_{p-1}) ,
\end{aligned}$$

where $\boldsymbol{x} = (x_{-1}, \ldots, x_{-p})$ and $\boldsymbol{z} = (z_0, \ldots, z_{p-1})$. For any given $\boldsymbol{x} = (x_{-1}, \ldots, x_{-p})$, the function $\boldsymbol{z} = (z_0, \ldots, z_{p-1}) \mapsto G(\boldsymbol{x}, \boldsymbol{z})$ is invertible. Denote by $G^{-1}(\boldsymbol{x}, \cdot) = H(\boldsymbol{x}, \cdot) = [H_0(\boldsymbol{x}, \cdot), \ldots, H_{p-1}(\boldsymbol{x}, \cdot)]$ its inverse. If $G(\boldsymbol{x}; \boldsymbol{z}) = \boldsymbol{x}' = (x'_0, \ldots, x'_{p-1}) \in \mathsf{X}^p$, this inverse is given by

$$
\begin{aligned}
z_0 &:= F^{-1}(x_{-1}, \ldots, x_{-p}; x'_0) = H_0(\boldsymbol{x}, \boldsymbol{x}') \,, \\
z_1 &:= F^{-1}(x'_0,\, x_{-1}, \ldots, x_{-p+1}; x'_1) = H_1(\boldsymbol{x}, \boldsymbol{x}') \,, \\
&\vdots \\
z_{p-1} &:= F^{-1}(x'_{p-2}, \ldots, x'_0,\, x_{-1}; x'_{p-1}) = H_{p-1}(\boldsymbol{x}, \boldsymbol{x}') \,.
\end{aligned}
$$

The Jacobian matrix of $G^{-1}(\boldsymbol{x}, \cdot) = H(\boldsymbol{x}, \cdot)$

$$
J_{G^{-1}(\boldsymbol{x},\cdot)}(\boldsymbol{x}') = \begin{pmatrix} \frac{\partial H_0(\boldsymbol{x},\boldsymbol{x}')}{\partial x'_0} & \cdots & \frac{\partial H_0(\boldsymbol{x},\boldsymbol{x}')}{\partial x'_{p-1}} \\ \vdots & \ddots & \vdots \\ \frac{\partial H_{p-1}(\boldsymbol{x},\boldsymbol{x}')}{\partial x'_0} & \cdots & \frac{\partial H_{p-1}(\boldsymbol{x},\boldsymbol{x}')}{\partial x'_{p-1}} \end{pmatrix} \tag{6.35}
$$

is block-triangular since $\partial H_i(\boldsymbol{x}, \cdot)/\partial x'_j = 0$ for $0 \le i < j \le p-1$. The Jacobian determinant $|J_{G^{-1}(\boldsymbol{x},\cdot)}|$ is therefore the product of the block-diagonal Jacobians, i.e.,

$$
|J_{G^{-1}(\boldsymbol{x},\cdot)}(\boldsymbol{x}')| = \left| J_{F^{-1}(x_{-1},\ldots,x_{-p},\cdot)}(x'_0) \right| \cdots \left| J_{F^{-1}(x'_{p-2},\ldots,x'_1,x_{-1},\cdot)}(x'_{p-1}) \right| \,,
$$

where $J_{F^{-1}(x_{-1},\cdots,x_{-p},\cdot)}$ is the Jacobian matrix of the function $F^{-1}(x_{-1}, \cdots, x_{-p}, \cdot)$. Denoting $\bar{q}(\boldsymbol{z}) = \bar{q}(z_0, \ldots, z_{p-1}) = q(z_0) \ldots q(z_{p-1})$, we get for $A \in \mathcal{X}^{\otimes p}$,

$$
R^p(\boldsymbol{x}, A) = \int \mathbb{1}_A(G(\boldsymbol{x}, \boldsymbol{z}))\bar{q}(\boldsymbol{z})\mathrm{Leb}(\mathrm{d}\boldsymbol{z}) = \int_A g_p(\boldsymbol{x}, \boldsymbol{x}')\mathrm{Leb}(\mathrm{d}\boldsymbol{x}') \,, \tag{6.36}
$$

where

$$
g_p(\boldsymbol{x}, \boldsymbol{x}') := \prod_{k=0}^{p-1} q \circ F^{-1}(x'_{k-1}, \ldots, x'_0,\, x_{-1}, \ldots, x_{-p+k}; x'_k) \\
\times \prod_{k=0}^{p-1} J_{F^{-1}(x'_{k-1},\ldots,x'_0,\, x_{-1},\ldots,x_{-p+k},\cdot)}(x'_k) \,. \tag{6.37}
$$

Theorem 6.33. *Assume A 6.32, that* $(x_{-1}, \ldots, x_{-p}, x'_0) \mapsto q \circ F^{-1}(x_{-1}, \ldots, x_{-p}; x'_0)$ *and* $(x_{-1}, \ldots, x_{-p}, x'_0) \mapsto |J_{F^{-1}(x_{-1},\ldots,x_{-p},\cdot)}(x'_0)|$ *are bounded away from* 0 *on the compact sets of* X^{p+1}. *Then, any compact set* C *such that* $\mathrm{Leb}(C) > 0$ *is* $(p, \varepsilon, \mathrm{Leb}_C)$*-small set with* $\mathrm{Leb}_C(\cdot) = [\mathrm{Leb}(C)]^{-1}\mathrm{Leb}(C \cap \cdot)$ *and* $\varepsilon > 0$.

Proof. Let C be a compact set. Since $q \circ F^{-1}$ and $|J_{F^{-1}}|$ are bounded away from zero on C, it follows from (6.37) that $\inf_{(\boldsymbol{x},\boldsymbol{x}') \in C \times C} g_p(\boldsymbol{x}, \boldsymbol{x}') \ge c > 0$. Therefore, for any $\boldsymbol{x} \in C$ and $A \in \mathcal{X}$, we get

$$R^p(\boldsymbol{x},A) = \int_A g_p(\boldsymbol{x},\boldsymbol{x}')\mathrm{Leb}(\mathrm{d}\boldsymbol{x}') \geq \int_{A\cap C} g_p(\boldsymbol{x},\boldsymbol{x}')\mathrm{Leb}(\mathrm{d}\boldsymbol{x}')$$
$$\geq c\mathrm{Leb}(C)\mathrm{Leb}(C \cap A)/\mathrm{Leb}(C) ,$$

showing that C is a small set. ■

Example 6.34. Consider the functional autoregressive model

$$X_t = a(X_{t-1},\ldots,X_{t-p}) + b(X_{t-1},\ldots,X_{t-p})Z_t \tag{6.38}$$

where $\{Z_t,\ t \in \mathbb{N}\}$ is an i.i.d. sequence of random vectors taking velues in $\mathbb{R}^m$, and $a : \mathbb{R}^{pm} \to \mathbb{R}^m$ and $b : \mathbb{R}^{pm} \to \mathbb{R}^{m\times m}$ are measurable functions. Set

$$F(\boldsymbol{x};z) = a(\boldsymbol{x}) + b(\boldsymbol{x})z, \ , \quad \boldsymbol{x} := (x_{-1},\ldots,x_{-p}) \in \mathbb{R}^{pm}\ , \ z \in \mathbb{R}^m\ .$$

Assume that for each $\boldsymbol{x} \in \mathbb{R}^{pm}$, $b(\boldsymbol{x})$ is an $m \times m$ invertible matrix, the functions $\boldsymbol{x} \mapsto b^{-1}(\boldsymbol{x})$ and $\boldsymbol{x} \mapsto a(\boldsymbol{x})$ are bounded on compact sets and $\boldsymbol{x} \mapsto \det(b^{-1}(\boldsymbol{x}))$ is bounded away from zero on compact sets. Under these assumptions, $q \circ F^{-1}(\boldsymbol{x};\boldsymbol{x}') = b^{-1}(\boldsymbol{x})(\boldsymbol{x}' - a(\boldsymbol{x}))$ and $J_{F^{-1}(\boldsymbol{x};\cdot)}(\boldsymbol{x}') = \det(b^{-1}(\boldsymbol{x}))$.

The model (6.38) includes some "regular" nonlinear autoregressive models such as the ARCH(p) model, wherein $a(\boldsymbol{x}) = 0$ and $b^2(\boldsymbol{x}) = \alpha_0 + \alpha_1 x_{-1}^2 + \ldots \alpha_p x_{-p}^2$, with $\alpha_i \geq 0\ i \in \{0,\cdots,p\}$, and the amplitude dependent exponential autoregressive models (EXPAR), wherein $a(\boldsymbol{x}) = (\phi_1 + \pi_1 \mathrm{e}^{-\gamma x_{-1}^2})x_{-1} + \cdots + (\phi_p + \pi_p \mathrm{e}^{-\gamma x_{-p}^2})x_{-p}$ and $b(\boldsymbol{x}) = 1$.

These assumptions also cover models for which the regression function is not regular, like the k regimes scalar ($m = 1$) SETAR model (see Section 3.6),

$$X_t = \begin{cases} \sum_{i=1}^{q_1} \phi_i^{(1)} X_{t-i} + \sigma_1 Z_t & \text{if } X_{t-d} < r_1\,, \\ \sum_{i=1}^{q_2} \phi_i^{(2)} X_{t-i} + \sigma_2 Z_t & \text{if } r_1 \leq X_{t-d} < r_2\,, \\ \vdots & \vdots \\ \sum_{i=1}^{q_k} \phi_i^{(k)} X_{t-i} + \sigma_k Z_t & \text{if } r_{k-1} \leq X_{t-d}\,, \end{cases} \tag{6.39}$$

where $-\infty = r_0 < r_1 < r_2 < \cdots < r_{k-1} < r_k = +\infty$, $\sigma_i > 0$, for $i \in \{1,\ldots,k\}$ and $d \leq \max(q_1,\ldots,q_k)$. In such a case, the regression functions are piecewise linear functions

$$a(\boldsymbol{x}) = \sum_{j=1}^{k} \mathbb{1}_{R_j}(x_{-d}) \sum_{i=1}^{q_i} \phi_i^{(j)} x_{-i}\,, \quad \text{and} \quad b(\boldsymbol{x}) = \sum_{j=1}^{k} \mathbb{1}_{R_j}(x_{-d})\sigma_j\,,$$

where $p = \max(q_1,q_2,\ldots,q_k)$ and $R_j = [r_{j-1},r_j)$, $j \in \{1,\ldots,k\}$. ◇

We now consider the Foster-Lyapunov condition given in Definition 6.23. Checking this condition is complicated and it is difficult to state general conditions under which it is satisfied.

Example 6.35 (Autoregressive model of order p). Consider a scalar autoregressive process of order p, $X_t = \sum_{j=1}^{p} \phi_j X_{t-j} + Z_t$ for $t \geq 0$, where $\{Z_t,\, t \in \mathbb{N}\}$ is a strong white noise independent of the initial condition $\boldsymbol{X}_0 = (X_{-1}, \ldots, X_{-p})$. Assume A6.32, $\mathbb{E}[|Z_0|] < \infty$ and that the roots of the prediction polynomial $\phi(z) = 1 - \sum_{j=1}^{p} \phi_j z^j$ are outside the unit-circle, i.e., $\phi(z) \neq 0$ for $|z| \leq 1$.

In this case $F(x_{-1}, \ldots, x_{-p}; z) = \sum_{j=1}^{p} \phi_j x_{-j} + z$ and $F^{-1}(x_{-1}, \ldots, x_{-p}; y) = y - \sum_{j=1}^{p} \phi_j x_{-j}$. The function $(x_{-1}, \ldots, x_{-p}; y) \mapsto F^{-1}(x_{-1}, \ldots, x_{-p}; y)$ is continuous and $J_{F^{-1}(x_{-1}, \ldots, x_{-p}, \cdot)}$ is the identity matrix. Theorem 6.33 implies that every compact set of $\mathbb{R}^p$ is p-small.

We now establish the Foster-Lyapunov drift condition. Note that

$$\mathbb{E}_{\boldsymbol{x}}[\|\boldsymbol{X}_1\|] \leqq |||\boldsymbol{\Phi}|||\, \|\boldsymbol{x}\| + \mathbb{E}[\|\boldsymbol{Z}_1\|] ,$$

where $\boldsymbol{\Phi}$ is the companion matrix (4.7) and $|||\cdot|||$ is the matrix norm associated to the vector norm $\|\cdot\|$ and $\boldsymbol{Z}_1 = [Z_1, 0, \ldots, 0]'$. Therefore, (6.24) will be fulfilled with $V(\boldsymbol{x}) = 1 + \|\boldsymbol{x}\|$ provided that $|||\boldsymbol{\Phi}||| < 1$.

As shown in Chapter 4, the Markov chain $\{\boldsymbol{X}_t,\, t \in \mathbb{N}\}$ has a unique invariant stationary distribution when $\rho(\boldsymbol{\Phi}) < 1$, where $\rho(\boldsymbol{\Phi})$ is the spectral radius of $\boldsymbol{\Phi}$ (see Definition 4.23). We know from Theorem 4.28 that this condition is equivalent to assume that $\phi(z) \neq 0$ for $|z| \leq 1$. For a given norm, the Foster-Lyapunov drift condition in general fails to establish this property, since it is easy to find examples, where $|||\boldsymbol{\Phi}||| > 1$ and $\rho(\boldsymbol{\Phi}) < 1$. On the other hand, if $\rho(\boldsymbol{\Phi}) < 1$ there exists a matrix norm $|||\cdot|||_{\boldsymbol{\Phi}}$ such that $|||\boldsymbol{\Phi}|||_{\boldsymbol{\Phi}} < 1$, since there exists a norm such that $|||\boldsymbol{\Phi}|||_{\boldsymbol{\Phi}} - \rho(\boldsymbol{\Phi})$ is arbitrarily small (see Proposition 4.24). We will now construct such a norm. By the Schur triangularization theorem, there is an unitary matrix U and an upper triangular matrix Δ, such that $\boldsymbol{\Phi} = U \Delta U'$. The diagonal elements of Δ are the (possibly complex) eigenvalues of $\boldsymbol{\Phi}$, denoted $\lambda_1, \ldots, \lambda_p \in \mathbb{C}^p$. For $\gamma > 0$, let $D_\gamma = \operatorname{diag}(\gamma, \gamma^2, \ldots, \gamma^p)$ and compute

$$D_\gamma \Delta D_\gamma^{-1} = \begin{pmatrix} \lambda_1 & \gamma^{-1} d_{1,2} & \gamma^{-2} d_{1,3} & \cdots & \gamma^{-p+1} d_{1,p} \\ 0 & \lambda_2 & \gamma^{-1} d_{2,3} & \cdots & \gamma^{-p+2} d_{2,p} \\ \cdot & \cdot & \cdot & \cdots & \cdot \\ 0 & 0 & 0 & \ddots & \gamma^{-1} d_{p-1,p} \\ 0 & 0 & 0 & 0 & \lambda_p \end{pmatrix} .$$

We can choose γ large enough so that the sum of all absolute values of the off-diagonal entries of $D_\gamma \Delta D_\gamma^{-1}$ are less than ε, where $0 < \varepsilon < 1 - \rho(\boldsymbol{\Phi})$. For a $p \times p$ matrix $A = (a_{i,j})$, denote by $|||A|||_1 = \max_{1 \leq j \leq p} \sum_{i=1}^{p} |a_{i,j}|$ the maximum column sum matrix norm. If we set $|||A|||_{\boldsymbol{\Phi}} = |||D_\gamma U' A U D_\gamma^{-1}|||_1$, then we have constructed a matrix norm such that $|||\boldsymbol{\Phi}|||_{\boldsymbol{\Phi}} < \rho(\boldsymbol{\Phi}) + \varepsilon < 1$. For $\boldsymbol{x} = (x_{-1}, \ldots, x_{-p}) \in \mathbb{R}^p$, denote $\|\boldsymbol{x}\|_1 = \sum_{j=1}^{p} |x_{-j}|$. Note that $\|A\boldsymbol{x}\|_1 \leq |||A|||_1 \|\boldsymbol{x}\|_1$. Finally, define the vector norm $\|\boldsymbol{x}\|_{\boldsymbol{\Phi}} = \|D_\gamma U' \boldsymbol{x}\|_1$. We get

$$\begin{aligned} \|\boldsymbol{\Phi}\boldsymbol{x}\|_{\boldsymbol{\Phi}} &= \|D_\gamma U' \boldsymbol{\Phi} \boldsymbol{x}\|_1 = \|D_\gamma U' \boldsymbol{\Phi} U D_\gamma^{-1} D_\gamma U' \boldsymbol{x}\|_1 \\ &= \|D_\gamma \Delta D_\gamma^{-1} D_\gamma U' \boldsymbol{x}\|_1 \leq |||D_\gamma \Delta D_\gamma^{-1}|||_1 \|\boldsymbol{x}\|_{\boldsymbol{\Phi}} \leq (\rho(\boldsymbol{\Phi}) + \varepsilon) \|\boldsymbol{x}\|_{\boldsymbol{\Phi}} . \end{aligned}$$

Taking $V(\boldsymbol{x}) = 1 + \|\boldsymbol{x}\|_{\Phi}$, we obtain

$$PV(\boldsymbol{x}) = 1 + \mathbb{E}_{\boldsymbol{x}}[\|X_1\|_{\Phi}] \le 1 + \|\boldsymbol{\Phi x}\|_{\Phi} + \mathbb{E}[\|\boldsymbol{Z}_1\|_{\Phi}]$$
$$\le (\rho(\Phi)+\varepsilon)V(\boldsymbol{x}) + 1 - (\rho(\Phi)+\varepsilon) + |||D_\gamma U'|||_1 \mathbb{E}[|\boldsymbol{Z}_1|] ,$$

showing that the Foster-Lyapunov condition is satisfied. ◇

Extensions of these types of techniques to more general functional autoregressive models should be considered on a case-by-case basis.

Example 6.36 (Nonlinear autoregressive models). Consider the p-th order nonlinear autoregressive process (NLAR(p)),

$$X_t = a(X_{t-1},\ldots,X_{t-p}) + Z_t , \tag{6.40}$$

where $\{Z_t,\ t \in \mathbb{N}\}$ is a strong white noise independent of $\boldsymbol{X}_0 = [X_{-1},\ldots,X_{-p}]'$. Assume A6.32 and $\mathbb{E}[|Z_0|] < \infty$. In vector form, the recursion (6.40) may be written as

$$\boldsymbol{X}_t = \boldsymbol{a}(\boldsymbol{X}_{t-1}) + \boldsymbol{Z}_t , \tag{6.41}$$

where $\boldsymbol{X}_t = [X_t, X_{t-1},\ldots,X_{t-p+1}]'$, and

$$\boldsymbol{a}(x_{-1},\ldots,x_{-p}) = [a(x_{-1},\ldots,x_{-p}), x_{-1},\ldots,x_{-p+1}]'$$

with $\boldsymbol{Z}_t = [Z_t, 0,\ldots,0]'$. Assume that a is bounded on compact sets and

$$\lim_{\|\boldsymbol{x}\|\to\infty} \frac{|a(\boldsymbol{x}) - \phi'\boldsymbol{x}|}{\|\boldsymbol{x}\|} = 0 , \quad \boldsymbol{x} = (x_{-1},\ldots,x_{-p}) , \tag{6.42}$$

where $\|\cdot\|$ is the euclidean norm and $\phi := [\phi_1, \phi_2, \ldots, \phi_p]'$ is such that $\phi(z) = 1 - \phi_1 z - \cdots - \phi_p z^p \neq 0$ for $|z| \le 1$. In other words, this condition implies that the model is asymptotically linear and that the limiting linear model is geometrically ergodic. Define as in the previous example $V(\boldsymbol{x}) = 1 + \|\boldsymbol{x}\|_{\Phi}$ where $\boldsymbol{\Phi}$ is the companion matrix of the polynomial $\phi(z)$. Note that (6.42) implies that

$$\lim_{\|\boldsymbol{x}\|_{\Phi}\to\infty} \|\boldsymbol{a}(\boldsymbol{x}) - \boldsymbol{\Phi x}\|_{\Phi} / \|\boldsymbol{x}\|_{\Phi} = 0 .$$

We get

$$PV(\boldsymbol{x}) \le 1 + \|\boldsymbol{a}(\boldsymbol{x})\|_{\Phi} + \mathbb{E}[\|\boldsymbol{Z}_1\|_{\Phi}] \le 1 + \|\boldsymbol{\Phi x}\|_{\Phi} + \|\boldsymbol{a}(\boldsymbol{x}) - \boldsymbol{\Phi x}\|_{\Phi} + \mathbb{E}[\|\boldsymbol{Z}_1\|_{\Phi}]$$
$$\le \left(\rho(\Phi) + \frac{\|\boldsymbol{a}(\boldsymbol{x}) - \boldsymbol{\Phi x}\|_{\Phi}}{\|\boldsymbol{x}\|_{\Phi}}\right) \|\boldsymbol{x}\|_{\Phi} + \mathbb{E}[\|\boldsymbol{Z}_1\|_{\Phi}] .$$

Since, $\limsup_{\|\boldsymbol{x}\|\to\infty} (\rho(\Phi) + \|\boldsymbol{a}(\boldsymbol{x}) - \boldsymbol{\Phi x}\|_{\Phi}| / \|\boldsymbol{x}\|_{\Phi}) < 1$, the geometric drift condition is satisfied. ◇

Example 6.37 (Nonlinear AR models). Consider again the NLAR model (6.40) and assume that there exist $\lambda < 1$ and a constant c such that

$$|a(x_{-1},\ \ldots,\ x_{-p})| \le \lambda \max\{|x_{-1}|,\ \ldots,\ |x_{-p}|\} + c . \tag{6.43}$$

We check that, under this condition, the model is geometrically ergodic. Put $\boldsymbol{X}_t = (X_t, \ldots, X_{t-p+1})$ and define the vector norm $\|\boldsymbol{x}\|_0 = \|(x_0,\ldots,x_{-p+1})'\|_0 = \max(|x_0|, \ldots, |x_{-p+1}|)$. This recursion may be rewritten as $\boldsymbol{X}_0 = \boldsymbol{x}$, $\boldsymbol{X}_1 = \boldsymbol{Z}_1 + \boldsymbol{a}(\boldsymbol{X}_0) = \boldsymbol{Z}_1 + \boldsymbol{a}(\boldsymbol{x})$, $\boldsymbol{X}_2 = \boldsymbol{Z}_2 + \boldsymbol{a}(\boldsymbol{X}_1) = \boldsymbol{Z}_2 + \boldsymbol{a}(\boldsymbol{Z}_1 + \boldsymbol{a}(\boldsymbol{x}))$ and, by induction $\boldsymbol{X}_p = \boldsymbol{Z}_p + \boldsymbol{a}(\boldsymbol{X}_{p-1}) = \boldsymbol{Z}_p + \boldsymbol{a}(\boldsymbol{Z}_{p-1} + \cdots + \boldsymbol{a}(\boldsymbol{x}))$. Using (6.43), we obtain

$$
\begin{aligned}
|x_1| &\le |Z_1| + c + \lambda \max\{|x_0|, \ldots, \||x_{-p+1}|\} \le |Z_1| + c + \lambda \|\boldsymbol{x}\|_0 , \\
|x_2| &\le |Z_2| + c + \lambda \max\{|x_1|, \ldots, |x_{-p+2}|\} \\
&\le |Z_2| + c + \lambda \max\{|Z_1| + c + \lambda \max\{|x_0|, \ldots, |x_{-p+1}|\}, |x_0|, \ldots, |x_{-p+2}|\} \\
&\le |Z_2| + c + \lambda(|Z_1| + c) + \lambda \max\{\lambda \max\{|x_0|, \ldots, |x_{-p+1}|\}, |x_0|, \ldots, |x_{-p+2}|\} \\
&\le |Z_2| + c + \lambda(|Z_1| + c) + \lambda \max\{|x_0|, \ldots, |x_{-p+2}|, \lambda|x_{-p+1}|\} \\
&\le |Z_2| + c + \lambda(|Z_1| + c) + \lambda \|X\|_0 \le |Z_2| + c + (|Z_1| + c) + \lambda \|\boldsymbol{x}\|_0 ,
\end{aligned}
$$

and similarly,

$$
\begin{aligned}
|x_p| \le |Z_p| &+ c + \lambda(|Z_{p-1}| + c) + \lambda^2(|Z_{p-2}| + c) + \cdots + \lambda^{p-1}(|Z_1| + c) \\
&+ \lambda \max\{|x_0|, \ldots, |x_{-p+2}|, \lambda|x_{-p+1}|\} \\
&\le |Z_p| + c + \lambda(|Z_{p-1}| + c) + \cdots + \lambda^{p-1}(|Z_1| + c) + \lambda \|\boldsymbol{x}\|_0 \\
&\le |Z_p| + c + (|Z_{p-1}| + c) + \cdots + (|Z_1| + c) + \lambda \|\boldsymbol{x}\|_0 .
\end{aligned}
$$

From the above inequalities and $\mathbb{E}\,[|Z_t|] < \infty$ we can easily get

$$
\begin{aligned}
\mathbb{E}\,[\|\boldsymbol{X}_p\|_0 \mid \boldsymbol{X}_0 = \boldsymbol{x}] &= \mathbb{E}\,[\max(|X_p|, |X_{p-1}|, \ldots, |X_1|)] \\
&\le \{(\mathbb{E}\,[|Z_p|] + c) + (\mathbb{E}\,[|Z_{p-1}|] + c) + \cdots + (\mathbb{E}\,[|Z_1|] + c)\} + \lambda \|\boldsymbol{x}\|_0 \\
&\le \lambda \|\boldsymbol{x}\|_0 + c',
\end{aligned}
$$

where c' is a constant number. Thus, by taking the test function V to be the norm $V(\boldsymbol{x}) = 1 + \|\boldsymbol{x}\|_0$, the Foster-Lyapunov condition is satisfied by the p-th iterate of the Markov kernel. As shown by Proposition 6.27, this implies that the drift condition is also satisfied in one-step. ◇

Example 6.38. We consider the NLAR(p) model (6.40) of Example 6.36 with

$$
a(\boldsymbol{x}) = f_1(\boldsymbol{x})x_{-1} + \cdots + f_p(\boldsymbol{x})x_{-p} , \quad \boldsymbol{x} := (x_{-1}, x_{-2}, \ldots, x_{-p}) \in \mathbb{R}^p , \tag{6.44}
$$

where p is an integer and $f_i : \mathbb{R}^p \mapsto \mathbb{R}$ are Borel functions. With $f_i(\boldsymbol{x}) = a_i + b_i \exp(-\lambda_i x_{-d})$, where the $d \in \{1, \ldots, p\}$ is a delay, this model reduces to an EXPAR model (see Section 3.7); with $f_i(\boldsymbol{x}) = \phi_{i,1} \mathbb{1}_{\{x_{-d} \le c\}} + \phi_{i,2} \mathbb{1}_{\{x_{-d} > c\}}$, the model reduces to a TAR model. Set

$$
A(\boldsymbol{x}) = \begin{pmatrix}
f_1(\boldsymbol{x}) & f_2(\boldsymbol{x}) & \cdots & f_{p-1}(\boldsymbol{x}) & f_p(\boldsymbol{x}) \\
1 & 0 & \cdots & 0 & 0 \\
0 & 1 & \cdots & 0 & 0 \\
\vdots & \vdots & \ddots & \vdots & \vdots \\
0 & 0 & \cdots & 1 & 0
\end{pmatrix} . \tag{6.45}
$$

We can rewrite the NLAR(p) model (6.44) as

$$\boldsymbol{X}_t = A(\boldsymbol{X}_{t-1}) + \boldsymbol{Z}_t \,. \tag{6.46}$$

Denote by $\|\cdot\|$ the euclidean norm. Assume that $|f_i|_\infty = c_i < \infty$ for all $i \in \{1,\ldots,p\}$. Then, for all $\boldsymbol{y} = [y_{-1}, y_{-2}, \ldots, y_{-p}]'$, we have

$$\|A(\boldsymbol{x})\boldsymbol{y}\| \le \|C|\boldsymbol{y}|\| \,, \quad \text{where } |\boldsymbol{y}| = [|y_{-1}|, |y_{-2}|, \cdots, |y_{-p}|]' \,,$$

where the matrix C is the companion matrix of the polynomial $c(z) = 1 - c_1 z - \cdots - c_p z^p$. Assume that $c(z) \neq 0$ for $|z| \le 1$ which implies that the spectral radius of C is strictly less than 1: $\rho(C) < 1$. Because $\rho(C) < 1$ and $\lim_{n\to\infty} |||C^n|||^{1/n} = \rho(C)$ (see Lemma 4.25), we can find $0 < \delta < 1$ and $h \in \mathbb{N}_*$ such that $|||C^h||| \le 1 - \delta$. Assume that $\mathbb{E}\left[|Z_0|\right] < \infty$. We have

$$\begin{aligned}
\mathbb{E}_x[\|\boldsymbol{X}_h\|] &\le \mathbb{E}\left[\|\prod_{i=0}^{h-1} A(\boldsymbol{X}_i)\boldsymbol{x} + \sum_{i=1}^{h}\left\{\prod_{j=i}^{h-1} A(\boldsymbol{X}_j)\right\}\boldsymbol{Z}_i\|\right] \\
&\le \|C^h|\boldsymbol{x}|\| + \mathbb{E}\left[\sum_{i=1}^{h} C^{h-i}|\boldsymbol{Z}_i|\right] \\
&\le (1-\delta)\|\boldsymbol{x}\| + \mathbb{E}\left[\sum_{i=1}^{h} C^{h-i}|\boldsymbol{Z}_i|\right] .
\end{aligned}$$

Therefore the geometric drift condition is satisfied in h-steps (and thus also in one step using Proposition 6.27). Suppose that $\{X_t,\ t \in \mathbb{N}\}$ satisfies the EXPAR model

$$X_t = \{a_1 + b_1 \exp(-\lambda_1 X_{t-d}^2)\}X_{t-1} + \cdots + \{a_p + b_q \exp(-\lambda_p X_{t-d}^2)\}X_{t-p} + Z_t \,, \tag{6.47}$$

where $\lambda_i \ge 0$ for $i \in \{1,\ldots,p\}$ where $\{Z_t,\ t \in \mathbb{N}\}$ is a strong white noise satisfying A 6.32. Assume that $c(z) = 1 - c_1 z - \cdots - c_p z^p \neq 0$ for $|z| \le 1$, where $c_i = \max(|a_i|, |a_i + b_i|)$. In particular, if $p = 1$, then the condition becomes $|a_1| < 1$ and $|a_1 + b_1| < 1$ (see Chen and Tsay, 1993). Then, by (6.47), $\{X_t,\ t \in \mathbb{N}\}$ is V-geometrically ergodic with $V(\boldsymbol{x}) = 1 + \|\boldsymbol{x}\|$. ◇

6.3 Some proofs

To establish a fixed point theorem for a contraction mapping, we will use the Banach fixed-point Theorem. Although it is well known, we state and prove it here in order to exhibit a rate of convergence and precise constants.

Theorem 6.39 (Banach fixed-point theorem). *Consider* $(\mathbb{F}, \mathrm{d})$, *a complete metric space. Let* $T : \mathbb{F} \to \mathbb{F}$ *be a continuous operator such that, for some positive integer* m, $\alpha \in (0,1)$ *and all* $u, v \in \mathbb{F}$,

$$\mathrm{d}(T^m u, T^m v) \le \alpha \mathrm{d}(u,v) \,. \tag{6.48}$$

Then there exists a unique fixed point $a \in \mathbb{F}$ and for all $u \in \mathbb{F}$,

$$\mathrm{d}(T^n u, a) \le (1-\alpha^{1/m})^{-1} \max_{0\le i<m} \alpha^{-i/m} \mathrm{d}(T^i u, T^{i+1} u)\, \alpha^{n/m} . \tag{6.49}$$

Moreover, if there exists $A \ge 1$ such that $\mathrm{d}(Tu, Tv) \le A\mathrm{d}(u,v)$ for all $u, v \in \mathbb{F}$, then

$$\mathrm{d}(T^n u, a) \le (\alpha^{-1/m} A)^{m-1} \mathrm{d}(u,a)\, \alpha^{n/m} . \tag{6.50}$$

Proof. Let us first prove the uniqueness. If $Ta = a$ and $Tb = b$, we have $T^m a = a$ and $T^m b = b$, thus

$$\mathrm{d}(a,b) = \mathrm{d}(T^m a, T^m b) \le \alpha \mathrm{d}(a,b) < \mathrm{d}(a,b)$$

since $\alpha \in (0,1)$. This yields $a = b$. To prove the existence, consider $u, v \in \mathbb{F}$ and an integer n. Write $n = km + r$ with $r \in \{0, \ldots, m-1\}$. Then

$$\mathrm{d}(T^n u, T^n v) = \mathrm{d}(T^{km+r} u, T^{km+r} v) \le \alpha^k \mathrm{d}(T^r u, T^r v) .$$

Taking $v = Tu$ we obtain

$$\begin{aligned}\mathrm{d}(T^n u, T^{n+1} u) &= \alpha^k \mathrm{d}(T^r u, T^{r+1} u) = \alpha^{n/m} \alpha^{-r/m} \mathrm{d}(T^r u, T^{r+1} u) \\ &\le \alpha^{n/m} \max_{0\le r<m} \alpha^{-r/m} \mathrm{d}(T^r u, T^{r+1} u) .\end{aligned}$$

This implies that $\{T^n u\}$ is a Cauchy sequence and denoting its limit by a, we obtain

$$\begin{aligned}\mathrm{d}(T^n u, a) &\le \max_{0\le r<m} \alpha^{-r/m} \mathrm{d}(T^r u, T^{r+1} u) \sum_{q=n}^{\infty} \alpha^{q/m} \\ &= (1-\alpha^{1/m})^{-1} \max_{0\le r<m} \alpha^{-r/m} \mathrm{d}(T^r u, T^{r+1} u)\, \alpha^{n/m} .\end{aligned}$$

Since T is continuous, $Ta = T \lim_{n\to\infty} T^n u = \lim_{n\to\infty} T^{n+1} u = a$, hence a is a fixed point. Assume now that $\mathrm{d}(Tu, Tv) \le A\mathrm{d}(u,v)$ for all u, v. Then $\mathrm{d}(T^r u, T^r v) \le A^r \mathrm{d}(u,v)$ and, denoting the integer part of $x \in \mathbb{R}$ by $[x]$, we have, for all $u \in \mathbb{F}$,

$$\begin{aligned}\mathrm{d}(T^n u, a) &= \mathrm{d}(T^n u, T^n a) \le \alpha^{[n/m]} \mathrm{d}(T^{n-m[n/m]} u, T^{n-m[n/m]} a) \\ &\le \alpha^{[n/m]} A^{n-m[n/m]} \mathrm{d}(u,a) \le \alpha^{n/m} \{\alpha^{-1/m} A\}^{m-1} \mathrm{d}(u,a) .\end{aligned}$$

■

Note that (6.49) yields

$$\mathrm{d}(T^n u, a) \le (1-\alpha^{1/m})^{-1} (\alpha^{-1/m} A)^r \mathrm{d}(u, Tu)\, \alpha^{n/m}. \tag{6.51}$$

By an application of the triangle inequality, we can show that

$$\mathrm{d}(u,a) \le (1-\alpha^{1/m})^{-1} \{\alpha^{-1/m} A\}^{m-1} \mathrm{d}(u, Tu) .$$

Thus (6.50) and (6.51) cannot be easily compared, except if $m = 1$, in which case the bound in (6.50) is sharper than the one in (6.51). In practice, however, it might be easier to obtain a bound for $\mathrm{d}(u, Tu)$ than for $\mathrm{d}(u,a)$, since the fixed point may not be explicitly known.

We now apply the fixed point theorem to a Markov kernel P, considered as an operator on a subset $\mathbb{F}$ of $\mathbb{M}_1(\mathcal{X})$.

Theorem 6.40. *Let $\mathbb{F}$ be a subset of $\mathbb{M}_1(\mathcal{X})$ and d be a metric on $\mathbb{F}$ such that $\delta_x \in \mathbb{F}$ for all $x \in \mathsf{X}$ and $(\mathbb{F}, \mathrm{d})$ is complete. Let P be a Markov kernel such that $\xi P \in \mathbb{F}$ if $\xi \in \mathbb{F}$. Assume that there exists a positive integer m, a constant $A > 0$ and $\alpha \in (0,1)$ such that, for all $\xi, \xi' \in \mathbb{F}$,*

$$\mathrm{d}(\xi P, \xi' P) \leq A\mathrm{d}(\xi, \xi'), \quad \mathrm{d}(\xi P^m, \xi' P^m) \leq \alpha \mathrm{d}(\xi, \xi') . \tag{6.52}$$

Then there exists a unique invariant probability measure $\pi \in \mathbb{F}$ and for all $\xi \in \mathbb{F}$,

$$\mathrm{d}(\xi P^n, \pi) \leq (1-\alpha^{1/m})^{-1}(\alpha^{-1/m}A)^{m-1}\mathrm{d}(\xi, \xi P)\alpha^{n/m} , \tag{6.53}$$

$$\mathrm{d}(\xi P^n, \pi) \leq (\alpha^{-1/m}A)^{m-1}\mathrm{d}(\xi, \pi)\alpha^{n/m} . \tag{6.54}$$

Assume moreover that the convergence of a sequence of probability measures in $\mathbb{F}$ with respect to d implies its weak convergence. Then π is the unique P-invariant probability measure in $\mathbb{M}_1(\mathcal{X})$.

Proof. We only need to prove the last part of the theorem, since the first part is a rephrasing of (6.39). Let $\pi \in \mathbb{F}$ be the unique invariant probability in $\mathbb{F}$ and let $\tilde{\pi}$ be an invariant probability in $\mathbb{M}_1(\mathcal{X})$. Then, for all continuous bounded functions f, we have

$$\tilde{\pi}(f) = \tilde{\pi}P^n(f) = \int P^n f(x) \tilde{\pi}(\mathrm{d}x) .$$

By the first part of theorem, the sequence $\{\delta_x P^n, n \in \mathbb{N}\}$ converges weakly to the probability measure π, which implies that $\lim_{n\to\infty} P^n f(x) = \pi(f)$ for all $x \in \mathsf{X}$ and all bounded continuous functions f. Since, in addition, $|P^n f(x)| \leq |f|_\infty$, the dominated convergence theorem implies that $\lim_{n\to\infty} \int P^n f(x)\tilde{\pi}(\mathrm{d}x) = \pi(f)$, which yields $\tilde{\pi}(f) = \pi(f)$ for all bounded continuous functions f. Therefore, $\tilde{\pi} = \pi$, which concludes the proof. ■

The second part of the theorem is very important. It states that if convergence with respect to d implies weak convergence (i.e., the topology induced by d is finer than the topology of weak convergence), then the invariant probability is not only unique in $\mathbb{F}$, but also in $\mathbb{M}_1(\mathcal{X})$. If $\mathbb{F} = \mathbb{M}_1(\mathcal{X})$, then this condition is superfluous to obtain the uniqueness of the invariant probability measure in $\mathbb{M}_1(\mathcal{X})$.

6.4 Endnotes

Classical uniform convergence to equilibrium for Markov processes was studied during the first half of the 20th century by Doeblin, Kolmogorov, and Doob, under various conditions. Doob (1953) gave a unifying form to these conditions, which he named *Doeblin type conditions*. Uniform convergence can also be tackled (as it is done in this chapter) by a straightforward application of the fixed point theorem in complete spaces; this line of research was opened by Dobrushin (1956).

Starting in the 1970s, an increasing interest in non-uniform convergence of

Markov processes arose. An explanation for this interest is that many useful processes do not converge uniformly to equilibrium, while they do satisfy weaker properties such as a geometric convergence. It later became clear that non-uniform convergence is related to *local* Doeblin-type conditions, i.e., small sets (see Definition 6.20) and Foster-Lyapunov type conditions (see Definition 6.23). Under such conditions, the transition probabilities converge exponentially fast to the unique invariant measure, with the constant controlled by the Foster-Lyapunov function and the minorization constant. Many different proofs of this result exist. Traditional proofs rely on the decomposition of the Markov chain into excursions outside the small set and a careful analysis of the exponential tail of the length of these excursions; see Nummelin (1984) and Baxendale (2005). The book Meyn and Tweedie (2009) provides an outstanding survey of these techniques. Another approach, based on coupling, produces explicit ergodicity constants; see Rosenthal (1995), Roberts and Tweedie (1999), Douc et al. (2004a), Roberts and Rosenthal (2004).

Alternately, in this chapter, we used the fact that the existence of a small set and Foster-Lyapunov conditions implies the existence of a spectral gap for the transition kernel seen as an operator on the Banach space of measurable functions growing no faster than V equipped with weighted supremum norm. The proof is due to Hairer and Mattingly (2008) and Hairer and Mattingly (2011). The proof relies on introducing a family of equivalent weighted norms indexed by a parameter, and on making an appropriate choice of this parameter that allows us to combine, in a very elementary way, the existence of a Foster-Lyapunov function and of a small set. This approach is not standard for establishing the ergodicity of Markov chains. The main advantage of this approach stems from the fact that it is elementary and does not rely on advanced concepts in probability theory (unlike the regenerative or the coupling approach).

The analysis of functional autoregressive processes was considered in the early work by Doukhan and Ghindès (1983) and Tjostheim (1990); see Tjøstheim (1994) for a survey of this early efforts. Several conditions implying geometric ergodicity have been worked out in Chen and Tsay (1993), Bhattacharya and Lee (1995) and An and Huang (1996), from which we have borrowed several conditions and examples.

Exercises

6.1 (Convergence: Examples and counterexamples). A sequence $\{\xi_n,\ n \in \mathbb{N}\}$ converges setwise to ξ, if for any $A \in \mathcal{X}$, $\lim_{n\to\infty} \xi_n(A) = \xi(A)$.

(a) Show that total variation convergence implies setwise convergence.

(b) Set $\mathsf{X} = \mathbb{R}$. Show that the sequence $\{\delta_{1/n}, n \in \mathbb{N}^*\}$ converges weakly to δ_0 but does not converge setwise.

(c) Let $\mathsf{X} = [0,1]$ and let ξ_n denote the probability measure with density $x \mapsto 1 + \sin(2\pi n x)$ with respect to Lebesgue measure on $[0,1]$. Show that the sequence $\{\xi_n,\ n \in \mathbb{N}\}$ converges setwise to the uniform distribution on $[0,1]$ but does not converge in the total variation distance.

6.2 (Uniform ergodicity of a finite Markov Chain). Assume that X is a finite set. Assume that for some integer m, $P^m(x,x') > 0$ for all $(x,x') \in \mathsf{X} \times \mathsf{X}$.

(a) Show that the Markov kernel P is uniformly ergodic.

(b) Let π be the unique invariant distribution. Find a value $\rho \in [0,1)$ for which $d_{\mathrm{TV}}(\xi P^n, \pi) \leq \rho^n d_{\mathrm{TV}}(\xi, \pi)$.

(c) Show that, for all $x \in \mathsf{X}$, $\pi(x) > 0$.

6.3. Assume that P^m is a kernel on $(\mathsf{X}, \mathcal{X})$ having a density with respect to some dominating measure λ, i.e., $P^m(x,A) = \int_A p_m(x,x')\lambda(\mathrm{d}x')$, for $A \in \mathcal{X}$. Assume that there exists a set C such that $\lambda(C) > 0$ and $q(x') := \inf_{x \in \mathsf{X}} p_m(x,x') > 0$, for all $x' \in C$. Set $\varepsilon := \int_C q(x')\lambda(\mathrm{d}x') > 0$. Define for $A \in \mathcal{X}$,

$$\nu(A) = \frac{\int \mathbb{1}_{A \cap C}(x')q(x')\lambda(\mathrm{d}x')}{\int \mathbb{1}_C(x')q(x')\lambda(\mathrm{d}x')} .$$

Show that for any $x \in \mathsf{X}$ and $A \in \mathcal{X}$, $P^m(x,A) \geq \varepsilon\nu(A)$ and that P^m is uniformly ergodic.

6.4 (Bounded functional autoregressive model). Consider the scalar FAR model: $X_t = \phi(X_{t-1}) + Z_t$, where $\{Z_t,\ t \in \mathbb{N}\}$ is a sequence of i.i.d. random variables with a lower semi-continuous positive density and finite first order moment, independent of X_0, where $\mathbb{E}[|X_0|] < \infty$. Assume that ϕ is bounded on $\mathbb{R}$.

(a) Show that the Markov chain is uniformly ergodic.

(b) Show that the Markov chain is V-uniformly ergodic, for an appropriately defined function V.

6.5 (Functional autoregressive model). Consider the scalar model given by $X_t = \phi(X_t) + Z_t$, where $\{Z_t,\ t \in \mathbb{N}\}$ is a sequence of i.i.d. random variables with a lower semi-continuous positive density and $\mathbb{E}[\exp(sZ_0)] < \infty$ for some $s > 0$. Assume that $X_0 = 0$ and that, for $|x| \geq R$, $|\phi(x)| \leq |x| - a$ for some $a > 0$. Show that the Markov chain is V-geometrically ergodic.

6.6. Let $\{X_t,\ t \in \mathbb{N}\}$ be a uniformly ergodic Markov chain on $(\mathsf{X}, \mathcal{X})$, with transition kernel P. Let $\ell, m \in \mathbb{N}_*$ be positive integers. Finally, let Y be a bounded random variable, measurable with respect to $\sigma(X_s, s \geq \ell + m)$. Show that $\left|\mathbb{E}\left[Y \mid \mathcal{F}_\ell^X\right] - \mathbb{E}\left[Y \mid \mathcal{F}_{\ell-1}^X\right]\right| \leq \Delta_{\mathrm{TV}}(P^m)\, |Y|_\infty$, where $\mathcal{F}_\ell^X = \sigma(X_0, \ldots, X_\ell)$.

6.7 (A first order scalar RCA model). Consider the first order RCA model

$$X_t = (\phi + \lambda B_t)X_{t-1} + \sigma Z_t \,, \tag{6.55}$$

where ϕ is a (deterministic) regression coefficient and $\{B_t,\ t \in \mathbb{Z}\}$ and $\{Z_t,\ t \in \mathbb{Z}\}$ are independent strong white noises with zero mean and unit-variance. Assume that for any integer k, $\mathbb{E}\left[|B_0|^k\right] < \infty$ and $\mathbb{E}\left[|Z_0|^k\right] < \infty$.

(a) Assume that $\mathbb{E}[|\phi + \lambda B_0|^p] < 1$. Show that the Markov chain defined by (6.55) admits a unique invariant distribution π satisfying $\int \pi(\mathrm{d}x)|x|^p < \infty$ for $p \geq 1$.

(b) Show that for any integer p,

$$\mathbb{E}\left[(\phi + \lambda B_0)^{2p}\right] = \sum_{k=0}^{2p} \binom{2p}{j} \phi^j \lambda^{2p-j} m_{2p-j}$$

where for any integer k, $m_k := \mathbb{E}\left[B_0^k\right]$.

(c) Let P be the Markov kernel associated to (6.55). Show that if $\mathbb{E}\left[(\phi+\lambda B_0)^{2p}\right] < 1$, then, for all $\lambda \in \left(\mathbb{E}\left[(\phi+\lambda B_0)^{2p}\right], 1\right)$ there exists $b_p < \infty$ such that

$$PV_p(x) \le \lambda_p V_p(x) + b_p \,,$$

with $V_p(x) = 1 + x^{2p}$.

(d) State conditions upon which P is V_p geometrically ergodic.

6.8 (β-ARCH model). The β-ARCH model may be written as

$$X_t = Z_t(a_0 + a_1 X_{t-1}^{2\beta} + \cdots + a_p X_{t-p}^{2\beta})^{1/2}, \quad t \ge p \,, \tag{6.56}$$

where β, a_0, a_1, ..., a_p are non-negative constant numbers, $\{Z_t,\ t \in \mathbb{N}\} \sim$iid $\mathrm{N}(0,1)$. Define $Y_t = \ln X_t^2$ and $\varepsilon_t = \ln Z_t^2$.

(a) Show that $Y_t = \phi(Y_{t-1}, \ldots, Y_{t-p}) + \varepsilon_t$ where

$$|\phi(y_1, \ldots, y_p)| \le \beta \max\{|y_1|, \ldots, |y_p|\} + \ln\left(\sum_{i=0}^{p} a_i\right) .$$

(b) Show that, if $\beta < 1$, the β-ARCH model is geometrically ergodic.

6.9 (Generalized linear AR model). Consider the model

$$X_t = \phi(\theta_1 x_{t-1} + \theta_2 x_{t-2} + \cdots + \theta_p x_{t-p}) + Z_t \,, \tag{6.57}$$

where $\phi(\cdot)$ is a nonlinear function, $\theta = (\theta_1, \ldots, \theta_p)'$ is a parameter vector satisfying $\sum_{i=1}^{p} \theta_i^2 = 1$. Assume that $\{Z_t,\ t \in \mathbb{N}\}$ is a strong white noise and that Z_0 has a lower-semi-continuous positive density, $\mathbb{E}[|Z_0|] < \infty$ and $\mathbb{E}[Z_0] = 0$ and that the function $\phi(\cdot)$ satisfies $|\phi(x)| \le \rho|x|/\sqrt{p} + c$, for any $x \in \mathbb{R}$ where c and ρ are some positive constants and $\rho < 1$. Show that $|\phi(\theta_1 x_{t-1} + \cdots + \theta_p x_{t-p})| \le \rho \max\{|x_{t-1}|, \ldots, |x_{t-p}|\} + c$ and that the model is V-geometrically ergodic.

6.10 (Exercise 6.9, continued). Assume now that $\phi(\cdot)$ satisfies $|\phi(x) - cx|/|x| \to 0$, as $|x| \to \infty$, where c is a constant number and $\alpha \equiv c\theta$ satisfies condition $1 - \alpha_1 z - \cdots - \alpha_p z^p \ne 0$ for $|z| \le 1$, and that $\phi(\cdot)$ is bounded over a bounded set. Show that the model is V-geometrically ergodic.

6.11 (Stationary distribution of the absolute autoregression). The following exercise is adapted from Tong (1990, Example 4.7). Consider the functional autoregressive model

$$X_t = \alpha|X_{t-1}| + Z_t \,, \tag{6.58}$$

where $\{Z_t,\ t \in \mathbb{N}\}$ are i.i.d. random variables whose distributions are absolutely continuous with symmetric density q with respect to Lebesgue measure. Assume in addition that X_0 is independent of $\{Z_t,\ t \in \mathbb{N}\}$.

(a) Assume that $\mathbb{E}[|Z_0|^p] < \infty$ and $\mathbb{E}[|X_0|^p] < \infty$ for some $p > 0$. Set $V(x) = 1 + |x|^p$. Show that V is a Lyapunov function for P. Show that P admits a unique stationary distribution.

(b) Show that the chain converges exponentially fast to its limiting distribution.

(c) Show that π satisfies the integral equation

$$\pi(y) = \int_0^\infty \pi(x)q(y-\alpha x)\text{Leb}(dx) + \int_{-\infty}^0 \pi(x)q(y+\alpha x)\text{Leb}(dx) .$$

(d) Show that

$$\pi(-y) = \int_0^\infty \pi(x)q(y+\alpha)\text{Leb}(dx) + \int_{-\infty}^0 \pi(x)q(y-\alpha x)\text{Leb}(dx) .$$

(e) Set $\tilde{\pi}(y) = \pi(y) + \pi(-y)$. Show that $\tilde{\pi}$ is the stationary distribution of the Markov chain $\tilde{X}_t = \alpha\tilde{X}_{t-1} + Z_t$

(f) Assume that $\{Z_t,\ t \in \mathbb{N}\} \sim \text{iid}\, \mathrm{N}(0,1)$. Compute $\tilde{\pi}$.

(g) Assume that $\{Z_t,\ t \in \mathbb{N}\} \sim \text{iid}\,\text{Cauchy}(0,1)$. Compute $\tilde{\pi}$.

(h) Show that $\pi(y) = \int_0^\infty \tilde{\pi}(x)q(y-\alpha x)\text{Leb}(dx)$. Determine π in the Gaussian case and in the Cauchy case.

6.12 (Deterministic scan Gibbs sampler). For $m \geq 5$, let $Y_1, \ldots, Y_m$ be i.i.d. $\mathrm{N}(\mu, \theta)$ where the joint prior density for (μ, θ) is $f(\mu, \theta) \propto 1/\sqrt{\theta}$. Also, let $y = (y_1, \ldots, y_m)$ denote the sample data with mean $\bar{y}$ and (scaled) variance $s^2 = \sum(y_i - \bar{y})^2$. This model yields posterior density

$$\pi(\mu, \theta|y) \propto \theta^{-(m+1)/2} \exp\left\{ -\frac{1}{2\theta} \sum_{j=1}^m (y_j - \mu)^2 \right\}$$

and full conditional distributions

$$\theta|\mu, y \sim \text{IG}\left(\frac{m-1}{2}, \frac{s^2 + m(\mu-\bar{y})^2}{2}\right)$$
$$\mu|\theta, y \sim N(\bar{y}, \theta/m)$$

where we say $X \sim \text{IG}(a, b)$ if X has a density proportional to $x^{-(a+1)}e^{-b/x}\mathbb{1}(x > 0)$.

Consider exploring $\pi(\mu, \theta|y)$ using a deterministic-scan Gibbs sampler (DUGS) with update scheme $(\theta', \mu') \to (\theta, \mu') \to (\theta, \mu)$. In each iteration, this sampler consecutively updates θ' and μ' by drawing from full conditional distributions $\theta|\mu', y$ and $\mu|\theta, y$, respectively.

(a) Show that the corresponding transition density can be written as

$$p((\mu', \theta'), (\mu, \theta)) = \pi(\theta|\mu', y)\pi(\mu|\theta, y) ,$$

where $\pi(\theta|\mu, y)$ and $\pi(\mu|\theta, y)$ represent the probability densities corresponding to the full conditional distributions $\theta|\mu, y$ and $\mu|\theta, y$, respectively. Let P denote the transition kernel corresponding to p.

(b) Define $V(\mu, \theta) = (\mu - \bar{y})^2$. Show that

$$\mathbb{E}\left[V(\mu, \theta) \mid \mu', \theta'\right] = \mathbb{E}\left[V(\mu, \theta) \mid \mu'\right] = \mathbb{E}\left[\mathbb{E}\left[V(\mu, \theta) \mid \theta\right] \mid \mu'\right] .$$

(c) Show that $\mathbb{E}\left[V(\mu, \theta) \mid \theta\right] = \theta/m$ and $\mathbb{E}\left[\theta \mid \mu'\right] = (s^2 + m(\mu' - \bar{y}))/(m-3)$.

(d) Show that

$$\mathbb{E}\left[V(\mu, \theta) \mid \mu', \theta'\right] = \frac{1}{m-3} V(\mu', \theta') + \frac{s^2}{m(m-3)} .$$

(e) Denote $C = \{(\mu, \theta) : V(\mu, \theta) \le d\}$ for some $d > 0$. Show that for any $(\mu', \theta') \in C$ and $(\mu, \theta) \in \mathbb{R} \times \mathbb{R}_+$,

$$k((\mu', \theta'), (\mu, \theta)) \ge \pi(\mu|\theta, y) \inf_{(\mu', \theta') \in C} \pi(\theta|\mu', y) .$$

(f) Show that

$$g(\theta) = \inf_{(\mu', \theta') \in C} \pi(\theta|\mu', y) = \begin{cases} \mathrm{IG}\left(\frac{m-1}{2}, \frac{s^2}{2} + \frac{md}{2}; \theta\right) & \text{if } \theta < \theta^* \\ \mathrm{IG}\left(\frac{m-1}{2}, \frac{s^2}{2}; \theta\right) & \text{if } \theta \ge \theta^* \end{cases} ,$$

for some θ^*. Show that C is a small set.

6.13. Consider Example 6.38. Assume that, for some integer $q \ge 1$, $\mathbb{E}\left[|Z_0|^q\right] < \infty$. Show that (6.46) is V-geometrically ergodic with $V(\boldsymbol{x}) = 1 + \|\boldsymbol{x}\|^q$.

6.14. Consider the EXPAR(p) process

$$X_t = \{a_1 + b_1 \sin(\phi_1 X_{t-d})\} X_{t-1} + \cdots + \{a_p + b_p \sin(\phi_p X_{t-d})\} X_{t-p} + Z_t ,$$

where $\{Z_t, t \in \mathbb{N}\}$ is a strong white noise satisfying A6.32. Find sufficient conditions upon which $\{X_t, t \in \mathbb{N}\}$ is geometrically ergodic. See Chen and Tsay (1993).

6.15. Consider Example 6.38, equation (6.44). Assume that the functions f_i can be written as $f_i(\boldsymbol{x}) = g_i(\boldsymbol{x}) + h_i(\boldsymbol{x})$, $i \in \{1, \ldots, p\}$, with $\sup_{\boldsymbol{x} \in \mathbb{R}^p} |g_i(\boldsymbol{x})| = c_i$ and $\sup_{\boldsymbol{x} \in \mathbb{R}^p} |x_{-i} h_i(\boldsymbol{x})| < \infty$, with $\boldsymbol{x} = [x_{-1}, x_{-2}, \ldots, x_{-p}]'$. Assume that $1 - c_1 z - \cdots - c_p z^p \ne 0$ for $|z| \le 1$.

(a) Show that the geometric drift condition is satisfied in h-step, with $V(\boldsymbol{x}) = 1 + \|\boldsymbol{x}\|$.

(b) Consider the EXPAR(1) model (6.47) with $p = 1$ and $d = 1$. Show that the geometric drift condition is satisfied as soon as $|a_1| < 1$.

See Chen and Tsay (1993).

6.16. Consider the linear TAR process (6.39). Using Example 6.38, show that this model satisfies the geometric drift condition if the polynomial $c(z) = 1 - c_1 z - \cdots - c_p z^p \ne 0$ for $|z| \le 1$, where $c_j = \max\{|\phi_j^{(1)}|, \ldots, |\phi_j^{(k)}|\}$. See Chan et al. (1985).

6.17 (Independent Metropolis-Hastings sampler). We consider the Independent Metropolis-Hastings algorithm (see Example 5.38) Let μ be a σ-finite measure on $(\mathsf{X}, \mathcal{X})$. Let π denote the density with respect to μ of the target distribution and q the *proposal density*. Assume that $\sup_{x \in \mathsf{X}} \pi(x)/q(x) < \infty$.

(a) Show that the transition kernel P is reversible with respect to π and that π is a stationary distribution for P.

(b) Assume now that there exists $\eta > 0$ such that for all $x \in \mathsf{X}$, $q(x) \ge \eta \pi(x)$. Show that the kernel is uniformly ergodic.

Chapter 7

Sample Paths and Limit Theorems

Let X be a real-valued random variable on a measurable space (X,σ), and let $\{X_t,\ t \in \mathbb{N}\}$ be an infinite sequence of i.i.d. copies of X. Let f be a Borel function and define $\bar{f}_n = n^{-1}\big(f(X_0) + \cdots + f(X_{n-1})\big)$ to be the cumulative averages of the sequence $\{f(X_t), t \in \mathbb{N}\}$. A fundamental theorem in probability is the law of large numbers: If the first moment of $f(X)$ is finite, then $\bar{f}_n$ converges almost surely to $\mathbb{E}[f(X)]$, i.e., $\mathbb{P}(\lim_{n\to\infty} \bar{f}_n = \mathbb{E}[f(X)]) = 1$.

If one strengthens the first moment assumption to that of finiteness of the second moment, then additionally, we have the Lindeberg–Lévy central limit theorem: $\sqrt{n}(\bar{f}_n - \mathbb{E}[f(X)]) \Rightarrow_{\mathbb{P}} \mathrm{N}(0, \mathrm{Var}[f(X)])$. With even more conditions on X, one similarly has more precise versions of the strong law of large numbers, such as the Hoeffding or the Bernstein inequality.

A question that arises very naturally is whether these results extend to Markov chains or observation-driven models. The answer is generally yes, but some care must be exercised and some additional assumptions beyond the existence and uniqueness of the stationary distribution π and the finiteness of the first and the second moment with respect to the stationary distributions are typically required.

Let $\{X_t,\ t \in \mathbb{N}\}$ be a Markov chain on $(\mathsf{X},\mathcal{X})$. For the law of large numbers, the basic result we will establish is that, under the weak condition that a Markov chain has a *unique* invariant distribution, π, and if the function f is integrable with respect to π, $\pi(|f|) < \infty$, then $\mathbb{P}_\pi(\lim_{n\to\infty} \bar{f}_n = \pi(f)) = 1$, where $\mathbb{P}_\pi$ is the law of the Markov chain started from π (the stationary version of the Markov chain $\{X_t,\ t \in \mathbb{N}\}$).

This result extends to the nonstationary case: we will show that there exists a set X_0 of π-measure one, such that for all $x \in \mathsf{X}_0$,

$$\mathbb{P}_x\Big(\lim_{n\to\infty} \bar{f}_n = \pi(f)\Big) = 1 , \tag{7.1}$$

where $\mathbb{P}_x$ is the distribution of the Markov chain started from x. The condition on the existence and uniqueness of the stationary distribution may be strengthened to ensure that (7.1) is satisfied for all $x \in \mathsf{X}$ (and not only on a subset of X). Markov chains satisfying (7.1) for all $x \in \mathsf{X}$ are referred to as *Harris ergodic*.

For a central limit theorem (CLT), the situation is more complicated. We will provide conditions upon which

$$\sqrt{n}(\bar{f}_n - \pi(f)) \overset{\mathbb{P}_x}{\Longrightarrow} \mathrm{N}(0, \sigma^2(f)) , \tag{7.2}$$

for any $x \in \mathsf{X}$, where the asymptotic variance is given by

$$\sigma^2(f) = \mathrm{Var}_\pi[f(X_0)] + 2\sum_{k=1}^{\infty} \mathrm{Cov}_\pi[f(X_0), f(X_k)] ;$$

the subscript π indicates that the expectations are computed under the stationary distribution π. The conditions we will require in this chapter are sufficient to guarantee that the convergence occurs under a wide-class of functions f (not necessarily bounded) and any initial distribution ξ. We focus on the use of drift and minorization conditions; for simplicity, we restrict our discussion to V-geometrically ergodic Markov chains (Definition 6.31), but many of the arguments used in this chapter can be adapted to deal with more general drift. The key argument of the proof is the *martingale decomposition*, which allows us to transform the sum $\sum_{t=0}^{n-1}\{f(X_t) - \pi(f)\}$ into a sum of martingale increments. Coupled with the Lindeberg–Levy martingale central limit theorem, this approach quickly yields to the desired results under general conditions.

We conclude this chapter by introducing some concentration type inequalities. Concentration of measure is a fairly general phenomenon which, roughly speaking, asserts that a Borel function $f(X_0, X_1, \ldots, X_{n-1})$ with suitably small local oscillations almost always takes values that are close to the average (or median) value $\mathbb{E}_x[f(X_0, X_1, \ldots, X_{n-1})]$. Under various assumptions on the function f, this phenomenon has been quite extensively studied when $X_0, X_1, \ldots, X_n$ are i.i.d. The situation is naturally far more complex for nonproduct measures, where one can trivially construct examples where the concentration property fails. For functions of dependent random variables $\{X_t,\ t \in \mathbb{N}\}$, the crux of the problem is often to quantify and bound the dependence among the random variables in terms of various types of mixing coefficients. For simplicity, we focus on uniformly ergodic Markov chains (Definition 6.7); some recent results will allow us to extend such bounds to more general V-geometric Markov chains, but these results are technically much more involved. We prove a form of the McDiarmid bounded difference inequality [see (7.62)] for uniformly ergodic Markov chains, starting from any initial distribution. These results are illustrated using examples from nonparametric statistics.

7.1 Law of large numbers

7.1.1 Dynamical system and ergodicity

We preface this section with a brief account of dynamical systems and ergodic theory.

Definition 7.1 (Dynamical system). *A* dynamical system $(\Omega, \mathcal{B}, \mathbb{Q}, \mathsf{T})$ *consists of a probability space* $(\Omega, \mathcal{B}, \mathbb{Q})$ *together with a measurable transformation* $\mathsf{T} : \Omega \to \Omega$ *of* Ω *into itself.*

The name *dynamical system* comes from the focus of the theory on the dynamical behavior of repeated applications of the transformation T on the underlying space Ω. An alternative to modeling a stochastic process as a sequence of random variables defined on a common probability space is to consider a dynamical system $(\Omega, \mathcal{B}, \mathbb{Q}, \mathsf{T})$

and a single random variable. The stochastic process will then be defined as the successive values of the random variable taken on transformed points in the original space. More precisely, let $(\Omega, \mathcal{B}, \mathbb{Q}, \mathsf{T})$ be a dynamical system. Since T is measurable and the composition of measurable functions is also measurable, the transformation T^t, defined recursively as $\mathsf{T}^t = \mathsf{T} \circ \mathsf{T}^{t-1}$ (with the convention that T^0 is the identity on Ω) is a measurable transformation on Ω for all positive integers. The sequence $\{\mathsf{T}^t \omega, t \in \mathbb{N}\}$ is referred to as the *orbit* of $\omega \in \Omega$.

Let $Y : \Omega \mapsto \mathbb{R}$ be a Borel-measurable random variable. The function $Y \circ T^t$ is also a random variable for all nonnegative integers. Thus a dynamical system together with a random variable Y defines a *one-sided* stochastic process $\{Y_t,\ t \in \mathbb{N}\}$, with $Y_t = Y \circ \mathsf{T}^t$ for all $t \in \mathbb{N}$. The transformation T is said to be *invertible* if T is one-to-one and its inverse, denoted T^{-1} is invertible. In such cases, we may define a *two-sided* stochastic process $\{Y_t,\ t \in \mathbb{Z}\}$, with $Y_t = Y \circ T^t$, for all $t \in \mathbb{Z}$.

The most useful dynamical systems for time-series is that consisting of all one-sided or two-sided X-valued sequences, where $(\mathsf{X}, \mathcal{X})$ is a measurable space together with the *shift* transformation. Let $\Omega = \mathsf{X}^{\mathbb{N}}$ be the set of X-valued sequences indexed by $\mathbb{N}$, i.e., an $\omega \in \Omega$ is a sequence $\{\omega_t,\ t \in \mathbb{N}\} = (\omega_0, \omega_1, \ldots)$, with $\omega_t \in \mathsf{X}$ for all $t \in \mathbb{N}$. Define by $\mathsf{S} : \Omega \to \Omega$ the shift operator: for $\omega = \{\omega_t,\ t \in \mathbb{N}\}$, $\mathsf{S}(\omega)$ is the sequence with coordinates $[\mathsf{S}(\omega)]_t = \omega_{t+1}$, for all $t \in \mathbb{N}$, or, more explicitly,

$$\mathsf{S}(\omega_0, \omega_1, \ldots) = (\omega_1, \omega_2, \ldots) .$$

The shift-transformation maps a sequence $\omega = (\omega_0, \omega_1, \ldots)$ into a sequence $\mathsf{S}(\omega) = (\omega_1, \omega_2, \ldots)$ wherein each coordinate has been shifted to the left by one time-unit. Note that in this shift operation, the first element of the sequence is lost, which implies that the shift is not invertible for one-sided sequences. The shift for two-sided sequences is defined similarly. In such a case, we have $\Omega = \mathsf{X}^{\mathbb{Z}}$ the set of two-sided X-valued sequences, i.e., $\omega \in \Omega$ is a sequence $\omega = \{\omega_t,\ t \in \mathbb{Z}\}$ indexed by $\mathbb{Z}$. We sometimes use the notations $\omega = (\ldots, \omega_{-1}, \dot{\omega}_0, \omega_1, \ldots)$ for an element of $\mathsf{X}^{\mathbb{Z}}$ where the point $\cdot$ indicates the 0-th position) and then, the shift $\tilde{\mathsf{S}}$ can be written as $\tilde{\mathsf{S}}(\ldots, \omega_{-1}, \dot{\omega}_0, \omega_1, \ldots) = (\ldots, \omega_{-1}, \omega_0, \dot{\omega}_1, \omega_2, \ldots)$. For two-sided sequences, the shift operator $\tilde{S}$ is invertible.

The transformation T is measure-preserving if the probability of any event is unchanged by the transformation. The fact that transforming by T preserves the probability $\mathbb{Q}$ can also be interpreted by stating that the probability $\mathbb{Q}$ is *invariant* or *stationary* with respect to T.

Definition 7.2 (Measure preserving transformation). *The transformation denoted by* $\mathsf{T} : \Omega \to \Omega$ *is* measure-preserving *if* T *is measurable and for all* $A \in \mathcal{B}$,

$$\mathbb{Q}[\mathsf{T}^{-1}(A)] = \mathbb{Q}[A] .$$

In such cases, the probability $\mathbb{Q}$ *is said to be invariant under the transformation* T.

We say that T *is an* invertible measure-preserving transformation *if* T *is invertible and measure-preserving and* T^{-1} *is also measure-preserving.*

A stationary stochastic process is a collection $\tilde{X} = \{\tilde{X}_t,\ t \in \mathbb{N}\}$ of random variables, defined on a probability space denoted $(\tilde{\Omega}, \mathcal{F}, \mathbb{P})$ with values in some space

$(\mathsf{X},\mathcal{X})$ such that the joint distribution of $(\tilde{X}_{t_1}, \cdots, \tilde{X}_{t_k})$ is the same as that of $(\tilde{X}_{t_1+t}, \cdots, \tilde{X}_{t_k+t})$ for any $k \geq 1$, and $t, t_1, \cdots, t_k \in \mathbb{N}$ (see Definition 1.6). By using the Kolmogorov consistency theorem (Theorem 1.5), we can build a measure $\mathbb{Q}$ on the countable product space Ω of sequences $\{\omega_t, t \in \mathbb{N}\} = (\omega_0, \omega_1, \ldots)$, with $\omega_t \in \mathsf{X}$ for all $t \in \mathbb{N}$, equipped with the product σ-field $\mathcal{B}$ satisfying

$$\mathbb{Q}(A) = \mathbb{P}((\tilde{X}_0, \tilde{X}_1, \ldots) \in A) .$$

The probability $\mathbb{Q}$ might be seen as the image probability of $\mathbb{P}$ under the transformation $\tilde{X} : \tilde{\Omega} \to \Omega$, defined by each $\tilde{\omega} \in \tilde{\Omega}$ by:

$$\tilde{X}(\tilde{\omega}) = (\tilde{X}_0(\tilde{\omega}), \tilde{X}_1(\tilde{\omega}), \ldots) \in \Omega = \mathsf{X}^{\mathbb{N}} .$$

Define the canonical random variables (or coordinate functions) by $X_t(\omega) = \omega_t$, for $t \in \mathbb{N}$ and $\omega \in \Omega$. The canonical random variables are essentially equivalent to $\tilde{X}_t$, in the sense that, for any $A \in \mathcal{B}$,

$$\mathbb{Q}(A) = \mathbb{Q}((X_0, X_1, \cdots) \in A) = \mathbb{P}((\tilde{X}_0, \tilde{X}_1, \cdots) \in A) .$$

The process $X = \{X_t, t \in \mathbb{N}\}$ is the *canonical version* of the process $\{\tilde{X}_t, t \in \mathbb{N}\}$, defined on the product space $\Omega = \mathsf{X}^{\mathbb{N}}$, equipped with the product σ-algebra $\mathcal{B} = \mathcal{X}^{\otimes \mathbb{N}}$.

The stationarity of the process is reflected in the invariance of $\mathbb{Q}$ with respect to the shift S, i.e., $\mathbb{Q} \circ \mathsf{S}^{-1} = \mathbb{Q}$, since for any $A \in \mathcal{X}^{\otimes \mathbb{N}}$,

$$\begin{aligned}\mathbb{Q}(\mathsf{S}^{-1}(A)) = \mathbb{Q}((X_1, X_2, \ldots) \in A) &= \mathbb{P}((\tilde{X}_1, \tilde{X}_2, \ldots) \in A) \\ &= \mathbb{P}((\tilde{X}_0, \tilde{X}_1, \ldots,) \in A) = \mathbb{Q}((X_0, X_1, \ldots) \in A) = \mathbb{Q}(A) .\end{aligned}$$

This construction associates each strict-sense stationary stochastic process to a dynamical system $(\Omega, \mathcal{B}, \mathbb{Q}, \mathsf{S})$, referred to as the *canonical dynamical system.*

Conversely, let $(\Omega, \mathcal{B}, \mathbb{Q}, \mathsf{T})$ be a dynamical system and assume that T is a measure-preserving transformation. Let $Y : \Omega \mapsto \mathsf{X}$ be a random element. Consider the X-valued stochastic process defined as $\{Y \circ \mathsf{T}^t, t \in \mathbb{N}\}$. For any $k \in \mathbb{N}$ and $t, t_1, t_2, \ldots, t_k \in \mathbb{N}$, we get

$$\begin{aligned}\mathbb{Q}((Y \circ \mathsf{T}^{t_1+t}, \ldots, Y \circ \mathsf{T}^{t_k+t}) \in A) &= \mathbb{Q}((Y \circ \mathsf{T}^{t_1}, \ldots, Y \circ \mathsf{T}^{t_k}) \circ \mathsf{T}^t \in A) \\ &= \mathbb{Q}((Y \circ \mathsf{T}^{t_1}, \ldots, Y \circ \mathsf{T}^{t_k}) \in A)\end{aligned}$$

for any cylinder $A \in \mathcal{X}^{\otimes k}$. Therefore, the process $\{Y \circ \mathsf{T}^t, t \in \mathbb{N}\}$ is strict sense stationary. The study of the properties of strict stationary process is then more or less the same as the study of measure-preserving transformations.

In practice, it is difficult to check if a given mapping T is measure-preserving using Definition 7.2. However, we often do have an explicit knowledge of a class of sets $\mathcal{B}_0$, stable by finite intersection and generating $\mathcal{B}$ (e.g., if Ω is a direct product-space, then $\mathcal{B}$ may be the collection of measurable rectangles). In such a cases, the following result is useful in checking whether a given mapping is measure-preserving.

Lemma 7.3. *Let $(\Omega, \mathcal{B}, \mathbb{Q})$ be a probability space and $\mathsf{T} : \Omega \to \Omega$ be a measurable map. Let $\mathcal{B}_0$ be a family of sets, stable under finite intersection and generating $\mathcal{B}$. If for all $B \in \mathcal{B}_0$, $\mathbb{Q}[\mathsf{T}^{-1}(B)] = \mathbb{Q}[B]$, then T is measure-preserving.*

Proof. The probability measures $\mathbb{Q} \circ \mathsf{T}^{-1}$ and $\mathbb{Q}$ coincide on $\mathcal{B}_0$ which is stable by finite intersection. Set $\mathcal{C} = \{B \in \mathcal{B}, \mathbb{Q}[\mathsf{T}^{-1}(B)] = \mathbb{Q}(B)\}$. By construction, $\mathcal{B}_0 \subset \mathcal{C}$. We will show that $\mathcal{B} = \sigma(\mathcal{B}_0) \subset \mathcal{C}$ by applying the monotone class Theorem (Theorem A.5). Note first that $\Omega \in \mathcal{C}$, since $\mathsf{T}^{-1}(\Omega) = \Omega$. Let $A \in \mathcal{C}$ and $B \in \mathcal{C}$ such that $A \subset B$; since $\mathbb{Q} \circ \mathsf{T}^{-1}$ and $\mathbb{Q}$ are probability measures,

$$\mathbb{Q}[\mathsf{T}^{-1}(B \setminus A)] = \mathbb{Q}[\mathsf{T}^{-1}(B)] - \mathbb{Q}[\mathsf{T}^{-1}(A)] = \mathbb{Q}[B] - \mathbb{Q}[A] = \mathbb{Q}[B \setminus A] .$$

Finally, let $\{A_n,\ n \in \mathbb{N}\}$ be an increasing sequence of elements of $\mathcal{C}$. We get

$$\mathbb{Q} \circ \mathsf{T}^{-1}\left(\bigcup_{n \in \mathbb{N}} A_n\right) = \lim_{n \to \infty} \mathbb{Q}[\mathsf{T}^{-1}(A_n)] = \lim_{n \to \infty} \mathbb{Q}[A_n] = \mathbb{Q}\left[\bigcup_{n \in \mathbb{N}} A_n\right] .$$

Therefore, $\mathcal{C}$ is a monotone class containing $\mathcal{B}_0$. Thus, by Theorem A.5, $\mathcal{B} = \sigma(\mathcal{B}_0) \subset \mathcal{C}$. ■

Example 7.4 (One-sided Markov shift). Let $\Omega = \mathsf{X}^{\mathbb{N}}$ the set of one-sided X-valued sequences, equipped with the product σ-algebra $\mathcal{X}^{\otimes \mathbb{N}}$. Let A be a cylinder set

$$A = \{\omega \in \Omega : (\omega_{t_1}, \ldots, \omega_{t_n}) \in H,\ (t_1, \ldots, t_n) \in \mathbb{N}^n,\ 0 \le t_1 \le t_2 \le \cdots \le t_n\} \quad (7.3)$$

for some $H \in \mathcal{X}^{\otimes n}$. Then,

$$\mathsf{S}^{-1}(A) = \{\omega \in \Omega : (\omega_{t_1+1}, \ldots, \omega_{t_n+1}) \in H\} , \quad (7.4)$$

which is another cylinder. Because the cylinders generate the product σ-algebra $\mathcal{X}^{\otimes \mathbb{Z}} = \sigma(\mathcal{C}_0)$, where $\mathcal{C}_0$ is the semialgebra of cylinders, the one-sided S is measurable.

Let P be a kernel on $(\mathsf{X}, \mathcal{X})$ and $\mu \in \mathbb{M}_1(\mathcal{X})$ be a probability measure. Define the probability of the cylinder A by

$$\mathbb{Q}(A) = \int \cdots \int \mu(\mathrm{d}x_0) \prod_{i=1}^{n} P^{t_i - t_{i-1}}(x_{i-1}, \mathrm{d}x_i) \mathbb{1}_H(x_1, \ldots, x_n) , \quad (7.5)$$

where by convention we have set $t_0 = 0$. Then $\mathbb{Q}$ is consistently defined and countably additive on the semialgebra $\mathcal{C}_0$ and hence extends to a probability measure on $\mathcal{X}^{\otimes \mathbb{N}} = \sigma(\mathcal{C}_0)$. For all $t \in \mathbb{N}$, define by $X_t : \Omega \to \mathsf{X}$ the coordinate functions. The coordinate functions $\{X_t,\ t \in \mathbb{N}\}$ form a Markov chain with kernel P and initial distribution μ (see Theorem 5.10).

Assume that $\mu = \pi$ is an invariant probability for P, i.e., $\pi P = \pi$. Then, for the cylinder A given by (7.3),

$$\mathbb{Q}(\mathsf{S}^{-1} A) = \int \cdots \int \pi(\mathrm{d}x_1) \prod_{i=2}^{n} P^{t_i - t_{i-1}}(x_{i-1}, \mathrm{d}x_i) \mathbb{1}_A(x_1, \ldots, x_n) = \mathbb{Q}(A) ,$$

from which it follows that S is measure-preserving for $\mathbb{Q}$ (see Lemma 7.3). ◇

Example 7.5 (Two-sided Markov shift). Let $\Omega = \mathsf{X}^{\mathbb{Z}}$ be the set of two-sided X-valued sequences, equipped with the product σ-algebra $\mathcal{X}^{\otimes\mathbb{Z}}$. Let A be a cylinder set

$$A = \{\omega \in \Omega : (\omega_{t_1}, \ldots, \omega_{t_n}) \in H\,,\ (t_1, \ldots, t_n) \in \mathbb{Z}^n\,,\ t_1 \leq t_2 \leq \cdots \leq t_n\} \tag{7.6}$$

for some $H \in \mathcal{X}^{\otimes n}$. Because the cylinders generate the basic σ-algebra, $\mathcal{X}^{\otimes\mathbb{Z}} = \sigma(\mathcal{C}_0)$, where $\mathcal{C}_0$ is the semialgebra of cylinders, the two-sided shift $\tilde{\mathsf{S}}$ and its inverse $\tilde{\mathsf{S}}^{-1}$ are both measurable. Let P be a kernel on $(\mathsf{X}, \mathcal{X})$ and assume that P has an invariant probability denoted π. Define the probability of the cylinder A by

$$\mathbb{Q}(A) = \int \cdots \int \pi(\mathrm{d}x_1) \prod_{i=2}^{n} P^{t_i - t_{i-1}}(x_{i-1}, \mathrm{d}x_i) \mathbb{1}_H(x_1, \ldots, x_n)\,. \tag{7.7}$$

Then $\mathbb{Q}$ is consistently defined and countably additive on the semialgebra $\mathcal{C}_0$ and hence extends to a probability measure on $\mathcal{X}^{\otimes\mathbb{Z}} = \sigma(\mathcal{C}_0)$. It is easily seen that both $\tilde{\mathsf{S}}$ and $\tilde{\mathsf{S}}^{-1}$ are invariant for $\mathbb{Q}$ given by (7.7); see Exercise 7.4. The two-sided shift transformation $\tilde{\mathsf{S}}$ is an invertible measure-preserving transformation for $\mathbb{Q}$. For all $t \in \mathbb{Z}$, define by $X_t : \Omega \to \mathsf{X}$ the coordinate functions. Note that, by construction, $X_t \circ \tilde{\mathsf{S}} = X_{t+1}$ and $X_t \circ \tilde{\mathsf{S}}^{-1} = X_{t-1}$. The double-sided sequence $\{X_t,\ t \in \mathbb{Z}\}$ is a strict-sense stationary (two-sided) Markov chain with Markov kernel P and marginal distribution π. ◇

Let $(\Omega, \mathcal{B}, \mathbb{Q}, \mathsf{T})$ be a dynamical system and consider a random variable $Y : \Omega \to \mathbb{R}$. In ergodic theory, we are interested in events of the form

$$F = \left\{ \omega \in \Omega : \lim_{n\to\infty} n^{-1} \sum_{t=0}^{n-1} Y \circ \mathsf{T}^t(\omega) \text{ exists} \right\}\,. \tag{7.8}$$

Consider the set $\mathsf{T}^{-1}(F)$. Then $\omega \in \mathsf{T}^{-1}(F)$ or equivalently $\mathsf{T}(\omega) \in F$ if and only if the limit

$$\lim_{n\to\infty} n^{-1} \sum_{t=0}^{n-1} Y \circ \mathsf{T}^t(\mathsf{T}\,\omega) \text{ exists}\,. \tag{7.9}$$

Provided that Y is real-valued (and hence cannot take on an infinite as a value), the limit in (7.9) is exactly the same as the limit $n^{-1} \sum_{t=0}^{n-1} Y \circ \mathsf{T}^t(\omega)$: either limit exists if the other one does and hence $\omega \in \mathsf{T}^{-1}(F)$ if and only if $\omega \in F$. More generally, an event F such that $\mathsf{T}^{-1}(F) = F$ is referred to as an *invariant* event. Invariants events are closed to transforming: if $\omega \in F$ where F is an invariant event, then $\mathsf{T}^t(\omega) \in F$ for all $t \in \mathbb{N}$ and thus, the whole orbit of ω also belongs to F.

Denote by $\hat{Y}(\omega) = \lim_{n\to\infty} n^{-1} \sum_{t=0}^{n-1} Y \circ \mathsf{T}^t(\omega)$ when this limit exists and set $\hat{Y}(\omega) = 0$ if the limit does not exist: in mathematical terms,

$$\hat{Y}(\omega) = \mathbb{1}_F(\omega) \lim_{n\to\infty} n^{-1} \sum_{t=0}^{n-1} Y \circ \mathsf{T}^t(\omega)\,,$$

where F is defined in (7.8). The above discussion implies that

$$\begin{aligned}\hat{Y}(\mathsf{T}(\omega)) &= \mathbb{1}_F(\mathsf{T}(\omega)) \lim_{n\to\infty} n^{-1} \sum_{t=0}^{n-1} Y \circ \mathsf{T}^t(\mathsf{T}(\omega)) \\ &= \mathbb{1}_F(\mathsf{T}(\omega)) \lim_{n\to\infty} n^{-1} \sum_{t=0}^{n-1} Y \circ \mathsf{T}^t(\mathsf{T}(\omega)) = \hat{Y}(\omega)\,.\end{aligned}$$

Therefore the limit $\hat{Y} = \hat{Y} \circ \mathsf{T}$ is unaffected by the transformation T: such random variables are referred to as *invariant*. Limiting sample averages are not the only invariant random variables; other examples are $\limsup_{n\to\infty} n^{-1} \sum_{t=0}^{\infty} Y \circ \mathsf{T}^t$ or $\liminf_{n\to\infty} n^{-1} \sum_{t=0}^{\infty} Y \circ \mathsf{T}^t$. Invariant random variables are random variables that yield the same value for all points in the orbit $\{\mathsf{T}^t\,\omega, t \in \mathbb{N}\}$.

Definition 7.6 (Invariant random variable and invariant event). *A random variable Y is invariant for the dynamical system $(\Omega, \mathcal{B}, \mathbb{Q}, \mathsf{T})$ if $Y \circ \mathsf{T} = Y$. An event A is* invariant *for $(\Omega, \mathcal{B}, \mathbb{Q}, \mathsf{T})$ if $A = \mathsf{T}^{-1}(A)$, or equivalently, if its indicator function $\mathbb{1}_A$ is invariant for $(\Omega, \mathcal{B}, \mathbb{Q}, \mathsf{T})$.*

It is easily shown that the family of invariant sets for $(\Omega, \mathcal{B}, \mathbb{Q}, \mathsf{T})$ is a sub-σ-algebra of $\mathcal{B}$, which is denoted $\mathcal{I}$; see Exercise 7.2. Note that if Y is invariant, then for all $t \in \mathbb{N}$, $Y = Y \circ \mathsf{T}^t$. The following proposition gives another way of expressing that a random variable is invariant.

Proposition 7.7. *A $\mathcal{B}$-measurable random variable $Y : \Omega \to \mathbb{R}$ is invariant for $(\Omega, \mathcal{B}, \mathbb{Q}, \mathsf{T})$ if and only if the random variable Y is $\mathcal{I}$-measurable.*

Proof. If Y is invariant, then for $a \in \mathbb{R}$, $\mathbb{1}_{\{Y>a\}} \circ \mathsf{T} = \mathbb{1}_{\{Y\circ\mathsf{T}>a\}} = \mathbb{1}_{\{Y>a\}}$, showing that $\{Y > a\} \in \mathcal{I}$, and that Y is $\mathcal{I}$-measurable. Conversely, assume that Y is simple and $\mathcal{I}$-measurable, $Y = \sum_{i\in I} \alpha_i \mathbb{1}_{A_i}$ where I is finite, $\{A_i\}_{i\in I} \subset \mathcal{I}$ and $\{\alpha_i\}_{i\in I} \subset \mathbb{R}$. Then, $Y \circ \mathsf{T} = \sum_{i\in I} \alpha_i \mathbb{1}_{A_i} \circ \mathsf{T} = \sum_{i\in I} \alpha_i \mathbb{1}_{A_i}$, showing that Y is invariant. The proof follows by the usual arguments. ■

In some applications, we consider events or random variables that are $\mathbb{Q}$-a.s. invariant in the sense that they are not modified after the transformation T with $\mathbb{Q}$-probability one instead of being invariant in the strict sense. A random variable $Y : \Omega \to \mathbb{R}$ is $\mathbb{Q}$-a.s. *invariant* for the dynamical system $(\Omega, \mathcal{B}, \mathbb{Q}, \mathsf{T})$ if $Y \circ \mathsf{T} = Y$, $\mathbb{Q}$-a.s. An event A is $\mathbb{Q}$-a.s. *invariant* for $(\Omega, \mathcal{B}, \mathbb{Q}, \mathsf{T})$ if $\mathbb{Q}(A \ominus \mathsf{T}^{-1}(A)) = 0$ ($A = \mathsf{T}^{-1}(A)$, $\mathbb{Q}$-a.s.) or, equivalently, if its indicator function $\mathbb{1}_A$ is $\mathbb{Q}$-a.s. invariant.

The following lemma shows that we may approximate any $\mathbb{Q}$-a.s. invariant event, resp. random variable, by an invariant event, resp. random variable, up to a $\mathbb{Q}$-negligible set.

Lemma 7.8. *If Y is $\mathbb{Q}$-invariant, there exists a $\mathcal{I}$-measurable random variable Z such that $Y = Z$, $\mathbb{Q}$-a.s. If A is T-invariant, there exists $B \in \mathcal{I}$ such that $\mathbb{1}_A = \mathbb{1}_B$, $\mathbb{Q}$-a.s.*

Proof. If $Y = Y \circ \mathsf{T}$, $\mathbb{Q}$-a.s., $Z := \limsup_{n\to\infty} Y \circ T^n = Y$, $\mathbb{Q}$-a.s. and Z is invariant. If $Y = \mathbb{1}_A$, then Z only takes the values 0 and 1. Therefore $Z = \mathbb{1}_B$ for some $B \in \mathcal{I}$. ■

Proposition 7.9. *Let $(\Omega,\mathcal{B},\mathbb{Q},\mathsf{T})$ be a dynamical system. If Y is a random variable such that $\mathbb{E}[Y^+] < \infty$ then, for all nonnegative integer t,*

$$\mathbb{E}\left[Y \circ \mathsf{T}^t \,\middle|\, \mathcal{I}\right] = \mathbb{E}[Y \mid \mathcal{I}], \quad \mathbb{Q}\text{-a.s.}$$

Proof. Let $A \in \mathcal{I}$. Then, using that $1_A \circ \mathsf{T}^t = \mathbb{1}_A$ and the fact that T is measure-preserving,

$$\mathbb{E}\left[\mathbb{1}_A\, Y \circ \mathsf{T}^t\right] = \mathbb{E}\left[\mathbb{1}_A \circ \mathsf{T}^t\; Y \circ \mathsf{T}^t\right] = \mathbb{E}[\mathbb{1}_A Y] = \mathbb{E}[\mathbb{1}_A \mathbb{E}[Y \mid \mathcal{I}]]$$

which thus implies that $\mathbb{E}\left[Y \circ \mathsf{T}^t \,\middle|\, \mathcal{I}\right] = \mathbb{E}[Y \mid \mathcal{I}]$, $\mathbb{Q}$-a.s. ∎

Let $(\Omega,\mathcal{B},\mathbb{Q},\mathsf{T})$ be a dynamical system and $Y : \Omega \mapsto \mathbb{R}$ be a random variable. In many cases, it is possible to observe realizations at successive time instants, $\{Y \circ \mathsf{T}^t, t \in \mathbb{N}\}$ and to compute sample average $n^{-1}\sum_{t=0}^{n-1} Y \circ \mathsf{T}^t$. On the other hand, the *ensemble average* of the random variable Y with respect to the invariant measure is simply $\mathbb{E}[Y]$, provided it exists. In this case, one hopes that for n sufficiently large, the sample average will be close to the ensemble average, in some sense. Unfortunately, it turns out that this is not true for any measure-preserving transformation T. Nevertheless, the *pointwise* or *almost everywhere* ergodic theorem, due to Birkhoff, shows that if $\mathbb{Q}$ is invariant under the transformation T, then the sample averages of a measure-invariant dynamical system can be interpreted as the conditional expectation of Y given the invariant σ-field $\mathcal{I}$.

Theorem 7.10 (Birkhoff ergodic theorem). *Let $(\Omega,\mathcal{B},\mathbb{Q},\mathsf{T})$ be a dynamical system and Y be a random variable such that $\mathbb{E}[Y^+] < \infty$ or $\mathbb{E}[Y^-] < \infty$. Then,*

$$n^{-1}\sum_{t=0}^{n-1} Y \circ \mathsf{T}^t \overset{\mathbb{Q}\text{-a.s.}}{\longrightarrow} \mathbb{E}[Y \mid \mathcal{I}] \,.$$

Proof. See Section 7.4. ∎

Of particular interest are the cases when the limiting sample average converges to a constant instead of random variables. This will occur if the dynamical system is such that all invariant random variables equal constants with $\mathbb{Q}$-probability 1. If this property is true for all random variables, then it is true for the indicator functions of invariant events. Since such functions can only take two values 0 or 1, for such a dynamical system, $\mathbb{Q}(F) = 0$ or 1, for all $F \in \mathcal{I}$. Conversely, if this property holds for every indicator function of an invariant event, every simple invariant function is equal to a constant with $\mathbb{Q}$-probability 1. Since every invariant function is the pointwise limit of a sequence of simple invariant functions, every invariant random variable is constant with $\mathbb{Q}$-probability 1. This yields to the following definition.

Definition 7.11 (Ergodicity). *A dynamical system $(\Omega,\mathcal{B},\mathbb{Q},\mathsf{T})$ is* ergodic *if the invariant σ-algebra $\mathcal{I}$ is trivial for $\mathbb{Q}$, i.e., for all $A \in \mathcal{I}$, $\mathbb{Q}(A) \in \{0,1\}$.*

A strict sense stationary stochastic process $\{\tilde{X}_t,\, t \in \mathbb{N}\}$ is ergodic *if its canonical dynamical system is ergodic.*

For an ergodic system, the conditional expectation $\mathbb{E}\left[Y \mid \mathcal{I}\right] = \mathbb{E}[Y]$, $\mathbb{Q}$-a.s. and Theorem 7.10 therefore implies

Corollary 7.12. *Let $(\Omega, \mathcal{B}, \mathbb{Q}, \mathsf{T})$ be an ergodic dynamical system and Y be a random variable such that $\mathbb{E}[Y^+] < \infty$ or $\mathbb{E}[Y^-] < \infty$. Then,*

$$n^{-1} \sum_{t=0}^{n-1} Y \circ \mathsf{T}^t \overset{\mathbb{Q}\text{-a.s.}}{\longrightarrow} \mathbb{E}[Y] .$$

Example 7.13. Let $\{\tilde{X}_t,\ t \in \mathbb{N}\}$ be an ergodic strict sense stationary process defined on $(\tilde{\Omega}, \mathcal{F}, \mathbb{P})$ with values in some space $(\mathsf{X}, \mathcal{X})$. Let $f : \mathsf{X}^{\mathbb{N}} \to \mathbb{R}$ be Borel function defined on the one-sided sequence space $\mathsf{X}^{\mathbb{N}}$ such that $\mathbb{E}\left[|f(\tilde{X}_0, \tilde{X}_1, \ldots)|\right] < \infty$. Let $\{X_t,\ t \in \mathbb{N}\}$ be the canonical version of $\{\tilde{X}_t,\ t \in \mathbb{N}\}$ defined on the sequence space $\Omega = \mathsf{X}^{\mathbb{N}}$ equipped with the product σ-field $\mathcal{B} = \mathcal{X}^{\otimes \mathbb{N}}$. Denoting by S the natural shift operator on the sequence space and using $X_t = X_0 \circ \mathsf{S}^t$, Theorem 7.10 implies that

$$n^{-1} \sum_{t=0}^{n-1} f(X_t, X_{t+1}, \ldots) = n^{-1} \sum_{t=0}^{n-1} f(X_0, X_1, \ldots) \circ \mathsf{S}^t \overset{\mathbb{Q}\text{-a.s.}}{\longrightarrow} \mathbb{E}[f(X_0, X_1, \ldots)] .$$

Since, by construction

$$\mathbb{P}\left(\lim_{n \to \infty} n^{-1} \sum_{t=0}^{n-1} f(\tilde{X}_t, \tilde{X}_{t+1}, \ldots) = \mathbb{E}\left[f(\tilde{X}_0, \tilde{X}_1, \ldots)\right] \right)$$
$$= \mathbb{Q}\left(\lim_{n \to \infty} n^{-1} \sum_{t=0}^{n-1} f(X_t, X_{t+1}, \ldots) = \mathbb{E}[f(X_0, X_1, \ldots)] \right) = 1 ,$$

the Birkhoff ergodic theorem shows that

$$\lim_{n \to \infty} n^{-1} \sum_{t=0}^{n-1} f(\tilde{X}_t, \tilde{X}_{t+1}, \ldots) \overset{\mathbb{P}\text{-a.s.}}{\longrightarrow} \mathbb{E}\left[f(\tilde{X}_0, \tilde{X}_1, \ldots)\right] . \tag{7.10}$$

Similarly, let $\{\tilde{X}_t,\ t \in \mathbb{Z}\}$ be an ergodic strict sense stationary process defined on $(\tilde{\Omega}, \mathcal{F}, \mathbb{P})$ with values in some space $(\mathsf{X}, \mathcal{X})$. Let $f : \mathsf{X}^{\mathbb{Z}} \to \mathbb{R}$ be Borel function defined on the two-sided sequence space $\mathsf{X}^{\mathbb{Z}}$ such that $\mathbb{E}\left[|f(\ldots, \tilde{X}_{-1}, \dot{\tilde{X}}_0, \tilde{X}_1, \ldots)|\right] < \infty$. Then, the Birkhoff theorem shows

$$\lim_{n \to \infty} n^{-1} \sum_{t=0}^{n-1} f(\ldots, \tilde{X}_{t-1}, \dot{\tilde{X}}_t, \tilde{X}_{t+1}, \ldots) \overset{\mathbb{P}\text{-a.s.}}{\longrightarrow} \mathbb{E}\left[f(\ldots, \tilde{X}_{-1}, \dot{\tilde{X}}_0, \tilde{X}_1, \ldots)\right] .$$

◇

7.1.2 Markov chain ergodicity

We now specialize the results of the previous section to Markov chains. Let $(\mathsf{X}, \mathcal{X})$ be a measurable space and P be a Markov kernel on $\mathsf{X} \times \mathcal{X}$. We consider the canonical Markov chain introduced in Section 5.4. Denote by $\Omega = \mathsf{X}^{\mathbb{N}}$ the set of one-sided and $\mathcal{F} = \mathcal{X}^{\otimes \mathbb{N}}$ the associated product σ-field. Denote by $\{X_t,\ t \in \mathbb{N}\}$ the canonical process $X_t(\omega) = \omega_t$ for all $t \in \mathbb{N}$ and $\omega = (\omega_0, \omega_1, \ldots) \in \Omega$ and by S the one-sided shift operator: $\mathsf{S}(\omega_0, \omega_1, \omega_2, \dots) = (\omega_1, \omega_2, \omega_3, \dots)$. The shift operator S is measurable with respect to $\mathcal{X}^{\otimes \mathbb{N}}$. With these notations, for $(t,h) \in \mathbb{N}^2$, we get

$$X_t \circ \mathsf{S}^h = X_{t+h} \,. \tag{7.11}$$

Moreover, for all $p,t \in \mathbb{N}$ and $A_0, \dots, A_p \in \mathcal{X}$,

$$\begin{aligned}\mathsf{S}^{-t}\{X_0 \in A_0, \dots, X_p \in A_p\} &= \{X_0 \circ \mathsf{S}^t \in A_0, \dots, X_p \circ \mathsf{S}^t \in A_p\} \\ &= \{X_t \in A_0, \dots, X_{t+p} \in A_p\} \,,\end{aligned}$$

thus S^t is $\sigma(X_s, s \geq t)$-measurable.

Proposition 7.14 (Markov property). *For all $\mathcal{X}^{\mathbb{N}}$-measurable bounded random variables Y, initial distribution $\nu \in \mathbb{M}_1(\mathcal{X})$ and $t \in \mathbb{N}$, it holds that*

$$\mathbb{E}_\nu\left[Y \circ \mathsf{S}^t \mid \mathcal{F}_t^X\right] = \mathbb{E}_{X_t}[Y] \quad \mathbb{P}_\nu\text{-a.s.} \tag{7.12}$$

Proof. We use the monotone class theorem Theorem A.6. Let $\mathcal{H}$ be the vector space of bounded random variables Y satisfying (7.12) and $\mathcal{C}$ the set of cylinders. By the monotone convergence theorem, if $\{Y_n,\ n \in \mathbb{N}\}$ is a nondecreasing sequence of non-negative random variables in $\mathcal{H}$ such that $\lim_{n\to\infty} Y_n = Y$ is bounded, then Y satisfies (7.12). We need to show that for any integer t and any $A \in \mathcal{X}^{t+1}$,

$$\mathbb{E}_\nu[\mathbb{1}_A(X_0, \dots, X_t) Y \circ \mathsf{S}^t] = \mathbb{E}_\nu[\mathbb{1}_A(X_0, \dots, X_t) \mathbb{E}_{X_t}(Y)] \,. \tag{7.13}$$

Again by Theorem A.7, it suffices to check (7.13) for $Y = \mathbb{1}_B(X_0, \dots, X_s)$, for any integer $s \geq 0$ and any $B \in \mathcal{X}^{s+1}$, i.e.,

$$\mathbb{E}_\nu[\mathbb{1}_A(X_0, \dots, X_t)\mathbb{1}_B(X_t, \dots, X_{t+s})] = \mathbb{E}_\nu[\mathbb{1}_A(X_0, \dots, X_t)\mathbb{E}_{X_t}[\mathbb{1}_B(X_0, \dots, X_s)]]$$

which follows easily from (5.6). ∎

In the sequel, it is assumed that P admits at least an invariant distribution π. Let $\mathbb{P}_\pi$ be the probability distribution on Ω under which the canonical process is a Markov chain with initial distribution π and transition kernel P; recall that under this distribution, the process $\{X_t,\ t \in \mathbb{N}\}$ is strict-sense stationary.

Lemma 7.15. *Let P be a Markov kernel on $(\mathsf{X}, \mathcal{X})$ with an invariant probability π. Let $(\Omega, \mathcal{F}, \mathbb{P}_\pi, \mathsf{S})$ be the associated dynamical system.*

(i) Let $Y \in \mathrm{L}^1(\mathbb{P}_\pi)$ be an invariant random variable. Then, $Y = \mathbb{E}_{X_0}[Y]$, $\mathbb{P}_\pi$-a.s.

(ii) If $I \in \mathcal{I}$ is an invariant set, there exists $B \in \mathcal{X}$ such that $\mathbb{1}_I = \mathbb{1}_B(X_0)$, $\mathbb{P}_\pi$-a.s.

(iii) If $Z \in \mathrm{L}^1(\mathbb{P}_\pi)$, then $\mathbb{E}_\pi\left[Z \mid \mathcal{I}\right] = \mathbb{E}_{X_0}[\mathbb{E}_\pi\left[Z \mid \mathcal{I}\right]]$, $\mathbb{P}_\pi$-a.s.

Proof. Note that since $Y \in \mathrm{L}^1(\mathbb{P}_\pi)$, $Y \in \mathrm{L}^1(\mathbb{P}_x)$ π-a.e. By the Markov property Proposition 7.14, using that Y is invariant (i.e., $Y \circ \mathsf{S}^t = Y$), we get

$$\mathbb{E}_{X_t}[Y] = \mathbb{E}_\pi\left[Y \circ \mathsf{S}^t \mid \mathcal{F}_t\right] = \mathbb{E}_\pi\left[Y \mid \mathcal{F}_t\right] , \quad \mathbb{P}_\pi\text{-a.s.}$$

Therefore, $\{(\mathbb{E}_{X_t}[Y], \mathcal{F}_t^X), t \in \mathbb{N}\}$ is a uniformly integrable martingale. By Corollary B.13, $\lim_{t\to\infty} \mathbb{E}_{X_t}[Y] = Y$, $\mathbb{P}_\pi$-a.s. and in $\mathrm{L}^1(\mathbb{P}_\pi)$. Then

$$\mathbb{E}_\pi[|Y - \mathbb{E}_{X_0}(Y)|] = \mathbb{E}_\pi[|Y - \mathbb{E}_{X_0}(Y)| \circ \mathrm{S}^t] \\ = \mathbb{E}_\pi[|Y - \mathbb{E}_{X_t}(Y)|] = \lim_{t\to\infty} \mathbb{E}_\pi[|Y - \mathbb{E}_{X_t}(Y)|] = 0 \,.$$

Let $I \in \mathcal{I}$ be an invariant set. The random variable $Y = \mathbb{1}_I$ is invariant; therefore, $\mathbb{1}_I = \mathbb{P}_{X_0}(I)$, $\mathbb{P}_\pi$-a.s. and, setting $B = \{x \in \mathsf{X},\ \mathbb{P}_x(I) = 1\}$, we get $\mathbb{1}_I = \mathbb{1}_B(X_0)$, $\mathbb{P}_\pi$-a.s.

Finally, since $\mathbb{E}_\pi\left[Z \mid \mathcal{I}\right]$ is $\mathcal{I}$-measurable and is thus invariant by Proposition 7.7, $\mathbb{E}_\pi\left[Z \mid \mathcal{I}\right] = \mathbb{E}_{X_0}[\mathbb{E}_\pi\left[Z \mid \mathcal{I}\right]]$, $\mathbb{P}_\pi$-a.s.. ■

Proposition 7.16. *Let $\{X_t,\ t \in \mathbb{N}\}$ be a canonical Markov chain on $(\mathsf{X}, \mathcal{X})$ with an invariant probability π and $(\Omega, \mathcal{F}, \mathbb{P}_\pi, \mathrm{S})$ the associated dynamical system. We assume that there exists an invariant set $I \in \mathcal{I}$ such that $\alpha = \mathbb{P}_\pi(I) \in (0,1)$. Then, there exists $B \in \mathcal{X}$ with $\pi(B) = \alpha$ such that $\mathbb{1}_I = \mathbb{1}_B(X_0)$, $\mathbb{P}_\pi$-a.s. Furthermore, setting $\pi_B(\cdot) = \alpha^{-1}\pi(B \cap \cdot)$ and $\pi_{B^c}(\cdot) = \alpha^{-1}\pi(B^c \cap \cdot)$, π_B and π_{B^c} are invariant probabilities and*

$$\mathbb{P}_{\pi_B}(X_t \in B, \textit{for all } t \geq 0) = \mathbb{P}_{\pi_{B^c}}(X_t \in B^c, \textit{for all } t \geq 0) = 1 \,.$$

Proof. Since $I \in \mathcal{I}$, Lemma 7.15-(i) implies that $\mathbb{1}_I = \mathbb{1}_B(X_0)$, $\mathbb{P}_\pi$-a.s. for some $B \in \mathcal{X}$. In addition, $\mathbb{P}_\pi$-a.s., we have $\mathbb{1}_B(X_0) = \cdots = \mathbb{1}_B(X_t) = \cdots = \prod_{t=0}^\infty \mathbb{1}_B(X_t)$. For all $A \in \mathcal{X}$, we get

$$\mathbb{P}_{\pi_B}(X_1 \in A) = \int \pi_B(\mathrm{d}x_0) P(x_0, \mathrm{d}x_1) \mathbb{1}_A(x_1) \\ = \alpha^{-1} \int \pi(\mathrm{d}x_0) \mathbb{1}_B(x_0) P(x_0, \mathrm{d}x_1) \mathbb{1}_A(x_1) = \alpha^{-1} \mathbb{P}_\pi(X_0 \in B, X_1 \in A) \,.$$

Using $\mathbb{1}_B(X_0) = \mathbb{1}_B(X_1)$, $\mathbb{P}_\pi$-a.s., the previous identity implies that

$$\mathbb{P}_{\pi_B}(X_1 \in A) = \alpha^{-1} \mathbb{P}_\pi(\{X_1 \in A\} \cap \{X_1 \in B\}) \\ = \alpha^{-1} \mathbb{P}_\pi(X_1 \in A \cap B) = \alpha^{-1} \pi(A \cap B) = \pi_B(A) \,,$$

showing that π_B is an invariant probability. Finally

$$\mathbb{P}_{\pi_B}(X_t \in B,\ \forall t \geq 0) = \alpha^{-1} \mathbb{P}_\pi(X_t \in B,\ \forall t \geq 0) = \alpha^{-1} \mathbb{P}_\pi(X_0 \in B) = 1 \,.$$

The same result holds for π_{B^c} by replacing I by I^c. ■

This proposition shows that if the Markov chain $\{X_t,\ t \in \mathbb{N}\}$ is not ergodic, then it necessarily has several distinct invariant distributions. Therefore, if we know that the Markov chain has a *unique* invariant distribution, then it is necessarily ergodic. This is the essence of the following important corollary:

Corollary 7.17. *Let $\{X_t,\ t \in \mathbb{N}\}$ be a canonical Markov chain on $(\mathsf{X}, \mathcal{X})$ with a unique invariant probability π. Then the dynamical system $(\Omega, \mathcal{F}, \mathbb{P}_\pi, \mathrm{S})$ is ergodic.*

Using the Birkhoff ergodic theorem, it is straightforward to formulate the following law of large numbers:

Theorem 7.18. *Let $\{X_t,\ t \in \mathbb{N}\}$ be a canonical Markov chain on $(\mathsf{X},\mathcal{X})$ with an invariant probability π, and let $(\Omega,\mathcal{F},\mathbb{P}_\pi,\mathsf{S})$ be the associated dynamical system. For any $Y \in \mathrm{L}^1(\mathbb{P}_\pi)$,*

$$\lim_{n\to\infty} \frac{1}{n}\sum_{t=0}^{n-1} Y \circ \mathsf{S}^t = \mathbb{E}_\pi\left[Y \,\middle|\, \mathcal{I}\right], \quad \mathbb{P}_\pi\text{-a.s.} \tag{7.14}$$

and there exists $B_Y \in \mathcal{X}$ with $\pi(B_Y) = 1$ such that, for all $x \in B_Y$,

$$\lim_{n\to\infty} \frac{1}{n}\sum_{t=0}^{n-1} Y \circ \mathsf{S}^t = \mathbb{E}_\pi\left[Y \,\middle|\, \mathcal{I}\right], \quad \mathbb{P}_x\text{-a.s.} \tag{7.15}$$

Proof. (7.14) follows directly from the Birkhoff ergodic theorem (Theorem 7.10). For any event $I \in \mathcal{F}$, we get

$$\int \mathbb{P}_x(I)\,\pi(\mathrm{d}x) = \mathbb{P}_\pi(I)\,. \tag{7.16}$$

■

If we apply this identity to the invariant event $I = \{\lim_{n\to\infty} n^{-1}\sum_{t=0}^{n-1} Y \circ \mathsf{S}^t = \mathbb{E}_\pi\left[Y \,\middle|\, \mathcal{I}\right]\}$, using that $\mathbb{P}_\pi(I) = 1$ and $\mathbb{P}_x(I) \leq 1$, (7.16) implies that $\mathbb{P}_x(I) = 1$ π-a.e.

If the Markov chain has a unique stationary distribution, then it is ergodic (see Corollary 7.17) and the Birkhoff theorem for ergodic processes applies.

Theorem 7.19 (Birkhoff's theorem for ergodic Markov chains). *Let $\{X_t,\ t \in \mathbb{N}\}$ be a canonical Markov chain on $(\mathsf{X},\mathcal{X})$ with a unique invariant probability π. Then, the dynamical system $(\Omega,\mathcal{F},\mathbb{P}_\pi,\mathsf{S})$ is ergodic. Let Y be a $\sigma(X_{0:\infty})$-measurable random variable satisfying $\mathbb{E}_\pi[Y^+] < \infty$ or $\mathbb{E}_\pi[Y^-] < \infty$. Then,*

$$\lim_{n\to\infty} \frac{1}{n}\sum_{t=0}^{n-1} Y \circ \mathsf{S}^t = \mathbb{E}_\pi[Y]\,, \quad \mathbb{P}_\pi\text{-a.s.}$$

In addition, for some set $A_Y \in \mathcal{X}$ such that $\pi(A_Y) = 1$, we have, for all $x \in A_Y$,

$$\lim_{n\to\infty} \frac{1}{n}\sum_{t=0}^{n-1} Y \circ \mathsf{S}^t = \mathbb{E}_\pi[Y]\,, \quad \mathbb{P}_x\text{-a.s.} \tag{7.17}$$

Proof. The proof follows immediately from Corollary 7.12 and Theorem 7.18. ■

This theorem admits a converse: if there exists a probability measure $\pi \in \mathbb{M}_1(\mathcal{X})$ such that, for any bounded measurable function f, the Markov chain $\{X_t,\ t \in \mathbb{N}\}$ started from π converges to the ensemble average, then π is the unique stationary distribution and the Markov chain $\{X_t,\ t \in \mathbb{N}\}$ is ergodic. In mathematical terms, this reads:

Theorem 7.20. *Let $\{X_t,\ t \in \mathbb{N}\}$ be a canonical Markov chain on $(\mathsf{X},\mathcal{X})$ and $\pi \in \mathbb{M}_1(\mathcal{X})$. We assume that, for all $f \in \mathbb{F}_b(\mathsf{X},\mathcal{X})$,*

$$\frac{1}{n}\sum_{t=0}^{n-1} f(X_t) \overset{\mathbb{P}_\pi\text{-a.s.}}{\longrightarrow} \pi(f)\,. \tag{7.18}$$

Then π is an invariant probability and the dynamical system $(\Omega,\mathcal{F},\mathbb{P}_\pi,\mathsf{S})$ is ergodic.

Proof. Since $f \in \mathbb{F}_b(\mathsf{X}, \mathcal{X})$, (7.18) implies that

$$\lim_{n\to\infty} \mathbb{E}_\pi \left[\left| n^{-1} \sum_{t=0}^{n-1} f(X_t) - \pi(f) \right| \right] = 0 .$$

Then, since $\mathbb{E}_\pi[Pf(X_t)] = \mathbb{E}_\pi[f(X_{t+1})]$, we get

$$\pi(Pf) = \mathbb{E}_\pi \left[\lim_{n\to\infty} \frac{1}{n} \sum_{t=0}^{n-1} Pf(X_t) \right]$$
$$= \lim_{n\to\infty} \frac{1}{n} \sum_{t=0}^{n-1} \mathbb{E}_\pi[Pf(X_t)] = \mathbb{E}_\pi \left[\lim_{n\to\infty} \frac{1}{n} \sum_{t=1}^{n} f(X_t) \right] = \pi(f) ,$$

showing that π is an invariant probability. Let I be an invariant event, $I \in \mathcal{I}$. Lemma 7.15 shows that there exists $B \in \mathcal{X}$ satisfying $\mathbb{1}_I = \mathbb{1}_B(X_0)$, $\mathbb{P}_\pi$-a.s. and, since $\mathbb{1}_B(X_t) = \mathbb{1}_B(X_0) \circ \mathsf{S}^t = \mathbb{1}_I \circ \mathsf{S}^t = \mathbb{1}_I = \mathbb{1}_B(X_0)$ $\mathbb{P}_\pi$-a.s., we get

$$\mathbb{1}_B(X_0) = \frac{1}{n} \sum_{t=0}^{n-1} \mathbb{1}_B(X_t) \to_{n\to\infty} \pi(B), \quad \mathbb{P}_\pi\text{-a.s.}$$

This implies that $\pi(B) = 0$ or 1 and therefore, that $\mathbb{P}_\pi(I) = 0$ or 1, i.e., the invariant σ-field is trivial. Hence the dynamical system $(\Omega, \mathcal{F}, \mathbb{P}_\pi, \mathsf{S})$ is ergodic. ∎

It is possible to extend Theorem 7.19 to obtain (7.17) for any $x \in \mathsf{X}$ and not only for all x in a set of π-measure 1. To obtain such a result, we need to strengthen the condition on the Markov chain: instead of assuming that the Markov chain has a unique stationary distribution, we will assume that the chain is Harris ergodic.

Definition 7.21 (Harris ergodic Markov chain). *Let P be a Markov kernel and π be a probability measure on $(\mathsf{X}, \mathcal{X})$. Assume that P admits a unique invariant probability π. The Markov kernel P is said to be* Harris ergodic *if, for all $x \in \mathsf{X}$,*

$$\|P^n(x, \cdot) - \pi\|_{\mathrm{TV}} \to_{n\to\infty} 0 . \tag{7.19}$$

A uniformly ergodic Markov chain (see Definition 6.7) is of course Harris ergodic. A Harris V-geometrically ergodic Markov chain (see Definition 6.31) is also Harris ergodic. A function $h \in \mathbb{F}_b(\mathsf{X}, \mathcal{X})$ satisfying $Ph = h$ is said to be *harmonic*. For such a function, the Markov property shows that $\{(M_t, \mathcal{F}_t^X),\, t \in \mathbb{N}\}$ with $M_t = h(X_t)$ is a martingale. Note indeed that $\mathbb{E}\left[M_{t+1} \mid \mathcal{F}_t\right] = \mathbb{E}\left[h(X_{t+1}) \mid X_t\right] = Ph(X_t) = M_t$. Lemma 7.22 shows that, if $\{X_t,\, t \in \mathbb{N}\}$ is a Harris ergodic Markov chain, $M_t = \pi(h)$ for all $t \in \mathbb{N}$.

Lemma 7.22. *Let P be a Harris ergodic Markov chain. Let $h \in \mathbb{F}_b(\mathsf{X}, \mathcal{X})$ be such that $Ph(x) = h(x)$ for all $x \in \mathsf{X}$. Then, h is constant on X.*

Proof. The condition $Ph = h$ implies that for all $n \in \mathbb{N}$, $P^n h = h$. The condition $\lim_{n\to\infty} \|P^n(x, \cdot) - \pi\|_{\mathrm{TV}}$ for all $x \in \mathsf{X}$, implies that $\lim_{n\to\infty} P^n h(x) = \pi(h)$. ∎

This Lemma implies that, if I is an invariant event such that $\mathbb{P}_\pi(I) = 1$, then for all $x \in \mathsf{X}$, $\mathbb{P}_x(I) = 1$.

Lemma 7.23. *Assume that $\{X_t,\ t \in \mathbb{N}\}$ is a canonical Harris ergodic Markov chain. Let $I \in \mathcal{I}$ be an invariant event. If $\mathbb{P}_\pi(I) = 1$, then for all $x \in \mathsf{X}$, $\mathbb{P}_x(I) = 1$.*

Proof. For $x \in \mathsf{X}$, we set $h(x) = \mathbb{P}_x(I)$. By the Markov property, we get

$$Ph(x) = \mathbb{E}_x[\mathbb{P}_{X_1}(I)] = \mathbb{E}_x[\mathbb{E}\left[\mathbb{1}_I \circ \mathrm{S} \mid \mathcal{F}_1\right]] = \mathbb{E}_x[\mathbb{E}\left[\mathbb{1}_I \mid \mathcal{F}_1\right]] = h(x) .$$

Therefore, $Ph = h$ and by Lemma 7.22, $h(x) = \pi(h)$ for all $x \in \mathsf{X}$. This concludes the proof since $\pi(h) = \mathbb{P}_\pi(I) = 1$. ■

Using this property, we may extend Theorem 7.19 as follows:

Theorem 7.24 (Birkhoff's theorem for Harris ergodic chain). *Let $\{X_t,\ t \in \mathbb{N}\}$ be a Harris ergodic canonical Markov chain on $(\mathsf{X}, \mathcal{X})$ with a unique invariant probability π. Let Y be $\sigma(X_{0:\infty})$-measurable random variable satisfying $\mathbb{E}_\pi[Y^+] < \infty$ or $\mathbb{E}_\pi[Y^-] < \infty$. Then, for any initial distribution $\xi \in \mathbb{M}_1(\mathcal{X})$,*

$$\lim_{n\to\infty} \frac{1}{n} \sum_{t=0}^{n-1} Y \circ \mathrm{S}^t = \mathbb{E}_\pi(Y) , \quad \mathbb{P}_\xi\text{-a.s.}$$

Proof. The event $I = \{\lim_{n\to\infty} n^{-1} \sum_{k=0}^{n-1} Y \circ \mathrm{S}^k = \pi(h)\}$ is invariant and by Theorem 7.24, $\mathbb{P}_\pi(I) = 1$. The proof follows from Lemma 7.23. ■

If $Y = f(X_0)$, where $f \in \mathbb{F}(\mathsf{X}, \mathcal{X})$, then for all $t \in \mathbb{N}$, $Y \circ \mathrm{S}^t = f(X_t)$. We therefore obtain the classical form of the law of large numbers as stated below.

Corollary 7.25. *Let $f \in \mathbb{F}(\mathsf{X}, \mathcal{X})$ be a function such that $\pi(f^+) < \infty$ or $\pi(f^-) < \infty$. Under the assumptions of Theorem 7.24, for any initial distribution $\xi \in \mathbb{M}_1(\mathcal{X})$,*

$$\lim_{n\to\infty} \frac{1}{n} \sum_{t=0}^{n-1} f(X_t) = \pi(f) , \quad \mathbb{P}_\xi\text{-a.s.}$$

Let $\{X_t,\ t \in \mathbb{N}\}$ be a Harris ergodic Markov chain; it then admits with a unique invariant distribution π; from all $x \in \mathsf{X}$ and for any set $A \in \mathcal{X}$ with positive π-measure, $\pi(A) > 0$, the Markov chain reaches A with probability 1 infinitely often. In other words, regardless of the starting value, a Harris ergodic Markov chain is guaranteed to explore the entire state space without, at least asymptotically, getting "stuck." This property is sometime taken as a definition for Harris ergodicity.

Example 7.26 (ARCH – least square estimator). We consider the ordinary least squares (OLS) estimator of the parameters of an ARCH model given by

$$X_t = \sigma_t \varepsilon_t , \quad \sigma_t^2 = \alpha_{*,0} + \sum_{i=1}^{p} \alpha_{*,i} X_{t-i}^2 , \tag{7.20}$$

with $\alpha_{*,0} > 0$, $\alpha_{*,i} \geq 0$, $i \in \{1, \ldots, p\}$, and $\{\varepsilon_t,\ t \in \mathbb{N}\}$ is a strong white noise with $\mathbb{E}[\varepsilon_0] = 0$ and $\mathrm{Var}(\varepsilon_0) = 1$. The OLS method uses the AR representation of the squares $\{X_t^2,\ t \in \mathbb{N}\}$. The distribution of $\{\varepsilon_t,\ t \in \mathbb{N}\}$ does not need to be specified (beyond the existence of certain moments; see below).

The true value of the parameters is denoted by $\theta_* = (\alpha_{*,0},\ \alpha_{*,1},\ \ldots,\ \alpha_{*,p})'$; θ

denotes a generic value of the parameter. From (7.20), we may derive the AR(p) representation of the ARCH(q) model,

$$X_t^2 = \alpha_{*,0} + \sum_{i=1}^{p} \alpha_{*,i} X_{t-i}^2 + Z_t , \tag{7.21}$$

where $Z_t = X_t^2 - \sigma_t^2 = (\varepsilon_t^2 - 1)\sigma_t^2$. The sequence $\{(Z_t, \mathcal{F}_t),\ t \in \mathbb{N}\}$ is a martingale difference sequence as soon as $\mathbb{E}\left[X_t^2\right] = \mathbb{E}\left[\sigma_t^2\right] < \infty$, where $\mathcal{F}_t$ denotes the σ-field generated by $\{X_s : s \le t\}$. Assume that we observe $X_0, \ldots, X_n$, a realization of length $n+1$ of the process $\{X_t,\ t \in \mathbb{Z}\}$. Introducing the vector

$$\varphi_t = (1, X_{t-1}^2, \ldots, X_{t-p}^2)' , \quad t \ge p , \tag{7.22}$$

(7.21) may be rewritten, for $t \in \{p, \ldots, n\}$,

$$X_t^2 = \varphi_t' \theta_* + Z_t . \tag{7.23}$$

In matrix form, this system of equations may be expressed as

$$\boldsymbol{Y}_n = \Phi_n \theta_* + \boldsymbol{Z}_n ,$$

where the $(n-p+1) \times p$ regression matrix Φ_n, and the $(n-p+1) \times 1$ vectors $\boldsymbol{Y}_n$ and $\boldsymbol{Z}_n$ are given by

$$\Phi_n = \begin{pmatrix} \varphi_p' \\ \varphi_{p+1}' \\ \vdots \\ \varphi_n' \end{pmatrix} , \quad \boldsymbol{Y}_n = \begin{pmatrix} X_p^2 \\ X_2^2 \\ \vdots \\ X_n^2 \end{pmatrix} , \quad \boldsymbol{Z}_n = \begin{pmatrix} Z_p \\ Z_2 \\ \vdots \\ Z_n \end{pmatrix} .$$

If the regression matrix Φ_n has full column rank, the ordinary least-squares (OLS) estimator of θ_* is given by

$$\hat{\theta}_n = (\Phi_n' \Phi_n)^{-1} \Phi_n' \boldsymbol{Y}_n . \tag{7.24}$$

Assume that (7.21) admits a unique invariant stationary distribution (see Theorem 5.27 and Exercise 5.21; denote by π the marginal distribution of the lag vector φ_p. Assume that $\mathbb{E}_\pi[\|\varphi_p\|^2] < +\infty$ (see Theorem 4.30).

$$\mathbb{P}\left[\varepsilon_0^2 = 1\right] \neq 1 . \tag{7.25}$$

The non-degeneracy assumption $\mathbb{P}\left[\varepsilon_0^2 = 1\right] \neq 1$ also guarantees that the regression matrix $\Phi_n' \Phi_n$ is invertible when n is large enough.

Since $\mathbb{E}_\pi[\|\varphi_p\|^2] < \infty$, the Birkhoff ergodic theorem for Markov chains shows that

$$\frac{1}{n} \Phi_n' \Phi_n = \frac{1}{n} \sum_{t=p}^{n} \varphi_t \varphi_t' \stackrel{\mathbb{P}\text{-a.s.}}{\longrightarrow} \mathbb{E}_\pi[\varphi_p \varphi_p'] . \tag{7.26}$$

Similarly, we have

$$\frac{1}{n}\Phi_n'\mathbf{Z}_n = \frac{1}{n}\sum_{t=p}^{n}\varphi_t Z_t \overset{\mathbb{P}\text{-a.s.}}{\longrightarrow} \mathbb{E}_\pi[\varphi_p Z_p]\,. \tag{7.27}$$

The invertibility of the matrix $\mathbb{E}_\pi[\varphi_p\varphi_p']$ can be shown by contradiction. Assume that there exists a nonzero vector $c \in \mathbb{R}^{p+1}$ such that $c'\mathbb{E}_\pi[\varphi_p\varphi_p']c = 0$. Thus $\mathbb{E}_\pi[(c'\varphi_p)^2] = 0$, which implies that $c'\varphi_p = 0$ $\mathbb{P}_\pi$-a.s.. Therefore, there exists a linear combination of the variables $1, X_0^2, \ldots, X_{p-1}^2$ which is $\mathbb{P}_\pi$-a.s. equal to a constant. Without loss of generality, one can assume that, in this linear combination, the coefficient of X_{p-1}^2 is 1. Thus ε_{p-1}^2 is $\mathbb{P}_\pi$-a.s. a measurable function of the variables $X_0, \ldots, X_{p-1}$. However, by construction, ε_{p-1}^2 is independent of these variables. This implies that ε_{p-1}^2 is $\mathbb{P}_\pi$-a.s. equal to a constant. This constant is necessarily equal to 1, which contradicts (7.25).

As mentioned above, $\{Z_t,\ t \in \mathbb{N}\}$ is a martingale increment sequence, which implies $\mathbb{E}_\pi[Z_t] = \mathbb{E}_\pi[Z_t X_{t-1}^2] = \ldots = \mathbb{E}_\pi[Z_t X_{t-p}^2] = 0$, showing that $\mathbb{E}_\pi[Z_t\varphi_{t-1}] = 0$. Combining the results above, we get that the OLS estimator of the parameters θ_* is strongly consistent:

$$\begin{aligned}\hat{\theta}_n - \theta_* &= (n^{-1}\Phi_n'\Phi_n)^{-1}(n^{-1}\Phi_n'\mathbf{Z}_n)\\ &\overset{\mathbb{P}_\pi\text{-a.s.}}{\longrightarrow} \{\mathbb{E}_\pi[\varphi_{p-1}\varphi_{p-1}'))\}^{-1}\mathbb{E}_\pi[\varphi_{p-1}Z_p] = 0\,. \end{aligned}\tag{7.28}$$

For the asymptotic normality of the OLS estimator, we need to assume that

$$\mathbb{E}_\pi[\|\varphi_p\|^4] < \infty\,. \tag{7.29}$$

Consider the $(p+1)\times(p+1)$ matrices

$$A = \mathbb{E}_\pi[\varphi_p\varphi_p'], \quad B = \mathbb{E}\left[\sigma_p^4\varphi_p\varphi_p'\right]\,. \tag{7.30}$$

We have shown in Example 7.26 that A is invertible; the invertibility of B follows along the same lines, noting that $c'\sigma_p^2\varphi_p = 0$ if and only if $c'\varphi_p = 0$ because $\sigma_p^2 > 0$ ($\alpha_{*,0} > 0$). Using (7.28), we get

$$\sqrt{n}(\hat{\theta}_n - \theta_*) = \left(\frac{1}{n}\sum_{t=p}^{n}\varphi_t\varphi_t'\right)^{-1}\left(\frac{1}{\sqrt{n}}\sum_{t=p}^{n}\varphi_t Z_t\right)\,. \tag{7.31}$$

Let $\lambda \in \mathbb{R}^{p+1}$, $\lambda \neq 0$. The sequence $\{(\lambda'\varphi_t Z_t,\ \mathcal{F}_t), t \in \mathbb{N}\}$ is a square integrable ergodic stationary martingale difference, with variance given by

$$\mathrm{Var}_\pi(\lambda'\varphi_t Z_t) = \lambda'\mathbb{E}_\pi[\varphi_t\varphi_t' Z_t^2]\lambda = \lambda'\mathbb{E}_\pi[\varphi_t\varphi_t'(\varepsilon_t^2-1)^2\sigma_t^4]\lambda = (\kappa_\varepsilon - 1)\lambda' B\lambda\,,$$

where $\kappa_\varepsilon = \mathbb{E}\left[\varepsilon_0^4\right]$. By Theorem B.20, we obtain that, for all $\lambda \neq 0$,

$$\frac{1}{\sqrt{n}}\sum_{t=p}^{n}\lambda'\varphi_t Z_t \overset{\mathbb{P}_\pi}{\Longrightarrow} \mathrm{N}(0, (\kappa_\varepsilon - 1)\lambda' B\lambda).$$

The Cramér-Wold device implies that

$$\sqrt{n}(\hat{\theta}_n - \theta_*) \stackrel{\mathbb{P}_\pi}{\Longrightarrow} \mathrm{N}(0,\ (\kappa_\varepsilon - 1)A^{-1}BA^{-1}) , \tag{7.32}$$

The OLS procedure is numerically straightforward, but is not efficient and is outperformed by methods based on the likelihood or on the quasi-likelihood that will be presented in Chapter 8. Note that the least-squares methods are of interest in practice because they provide initial estimators for the optimization procedure that is used in the (quasi-)maximum likelihood method. ◇

7.2 Central limit theorem

Establishing sets of sufficient conditions under which

$$S_n(f) := n^{-1/2} \sum_{t=0}^{n-1} \{f(X_t) - \pi(f)\} , \tag{7.33}$$

where f is a function and $\{X_t,\, t \in \mathbb{N}\}$ is a Markov chain on $(\mathsf{X}, \mathcal{X})$ with transition kernel P and stationary distribution π is the main matter of this Section. Denote by $\xi \in \mathbb{M}_1(\mathcal{X})$ the initial distribution. The variance of the partial sum is given by

$$\begin{aligned} s_n^2(f) &= \mathrm{Var}_\xi \left(\sum_{t=0}^{n-1} f(X_t) \right) \\ &= \sum_{t=0}^{n-1} \mathrm{Var}_\xi (f(X_t)) + 2 \sum_{0 \le s < t \le n-1} \mathrm{Cov}_\xi (f(X_s), f(X_t)) . \end{aligned} \tag{7.34}$$

For $0 \le s \le t$ we get

$$\begin{aligned} \mathrm{Cov}_\xi (f(X_s), f(X_t)) &= \mathbb{E}_\xi [(f(X_s) - \pi(f))(f(X_t) - \pi(f))] \\ &= \mathbb{E}_\xi [(f(X_s) - \pi(f)) \mathbb{E} \left[(f(X_t) - \pi(f)) \mid \mathcal{F}_s \right]] \\ &= \mathbb{E}_\xi \left[(f(X_s) - \pi(f))(P^{t-s} f(X_s) - \pi(f)) \right] . \end{aligned} \tag{7.35}$$

Some conditions are required at this point to prove that the sequence $\{n^{-1} s_n^2(f), n \ge 1\}$ converges. Typically, the covariance $\mathrm{Cov}_\xi (f(X_s), f(X_t))$ should decrease to zero sufficiently fast so that the series appearing in the right hand side of (7.34) converges. We assume in the sequel that the kernel P is a V-geometric Markov chain.

Assumption A7.27 (V-geometric ergodicity). The Markov kernel given by P admits a unique invariant distribution π, and there exists a measurable function $V : \mathsf{X} \to [1, \infty)$ satisfying $\pi(V) < \infty$, and

(a) There exist some constants $(C, \rho) \in \mathbb{R}^+ \times [0, 1)$ such that for all $x \in \mathsf{X}$ and all $n \ge 0$,

$$\| P^n(x, \cdot) - \pi \|_V \le C \rho^n V(x) .$$

(b) There exists $\lambda \in [0, 1)$ and $b < \infty$, such that $PV \le \lambda V + b$.

As shown in the next lemma, if a Markov kernel P is V-geometrically ergodic, it is also V^α-geometrically ergodic, for any $\alpha \in (0,1]$.

Lemma 7.28. *Assume A7.27. Then, for any $\alpha \in (0,1]$, P is V^α-geometrically ergodic; more precisely,*

$$\|P^n(x,\cdot) - \pi\|_{V^\alpha} \leq 2^{1-\alpha} C^\alpha \rho^{\alpha n} V^\alpha(x)\,, \quad \textit{for all } n \in \mathbb{N} \textit{ and } x \in \mathsf{X}\,, \tag{7.36}$$

where C and ρ are defined in A7.27.

Proof. By Definition 6.14, for any $\xi \in \mathbb{M}_0(\mathcal{X}) \cap \mathbb{M}_V(\mathcal{X})$, we get $\|\xi\|_{V^\alpha} = |\xi|(V^\alpha)$. By the Jensen inequality, we get

$$\|\xi\|_{V^\alpha} = |\xi|(\mathsf{X})\frac{|\xi|(V^\alpha)}{|\xi|(\mathsf{X})} \leq |\xi|(\mathsf{X})\left(\frac{|\xi|(V)}{|\xi|(\mathsf{X})}\right)^\alpha = |\xi|^{1-\alpha}(\mathsf{X})\,\|\xi\|_V^\alpha\,.$$

The result follows by applying this identity to $\xi = P^n(x,\cdot) - \pi$ and noting that $|\xi|(\mathsf{X}) \leq 2$. Similarly, by the Jensen inequality, $PV^\alpha \leq (PV)^\alpha \leq (\lambda V + b)^\alpha \leq \lambda^\alpha V^\alpha + b^\alpha$. ■

Assume that $|f| \leq V^{1/2}$ and that P is V-geometrically ergodic. Then, for all $t \geq s$, $|P^{t-s} f(x) - \pi(f))| \leq \sqrt{2} C^{1/2} \rho^{(t-s)/2} V^{1/2}(x)$, which implies that

$$\begin{aligned}&\left|\mathbb{E}_\xi\left[(f(X_s) - \pi(f))(P^{t-s} f(X_s) - \pi(f))\right]\right| \\ &\qquad\qquad \leq \sqrt{2} C^{1/2} \rho^{(t-s)/2} (\mathbb{E}_\xi[V(X_s)] + \pi(V^{1/2})\mathbb{E}_\xi[V^{1/2}(X_s)])\,.\end{aligned}$$

On the other hand, since $\mathbb{E}_\xi[V(X_s)] = \xi P^s V$, Lemma 6.26 shows that

$$|\mathbb{E}_\xi[V(X_s)] \leq \lambda^s \xi(V) + b/(1-\lambda) < \infty\,,$$

where (λ, b) are defined in A7.27. Thus, for all $s \geq 0$, $\sum_{s<t\leq n-1} |\mathrm{Cov}_\xi(f(X_s), f(X_t))|$ is upper bounded independently of n and s. This is not enough to conclude, but hints at the asymptotic independence of $f(X_s)$ and $f(X_t)$ where $t-s$ is large, which is essential to establish the desired result.

Using the *Poisson equation*, the sum $\sum_{t=0}^{n-1}\{f(X_t) - \pi(f)\}$ can be rewritten as a martingale with some negligible remainder terms. By doing so, we will use the powerful martingale limit theory (see Appendix B) to establish the central limit theorems and deviation inequalities for $S_n(f)$. The key tool allowing us to connect Markov chains and martingales is the Poisson equation, which is defined below.

Definition 7.29. *Let $(\mathsf{X}, \mathcal{X})$ be a measurable space and consider P, a Markov kernel on $(\mathsf{X}, \mathcal{X})$ that admits a unique invariant measure π. For $f \in \mathbb{F}(\mathsf{X}, \mathcal{X})$ such that $\pi|f| < \infty$, the equation*

$$\hat{f} - P\hat{f} = f - \pi(f) \tag{7.37}$$

is called the Poisson equation *associated with the function f.*

Any $\hat{f} \in \mathbb{F}(\mathsf{X}, \mathcal{X})$ satisfying $P|\hat{f}|(x) < \infty$ for all $x \in \mathsf{X}$ and such that (7.37) holds, is then called a solution of the Poisson equation associated to f.

Provided that a solution of the Poisson equation exists, we can link the quantity $S_n(f)$ to a specific martingale in the following way:

Lemma 7.30. *Let $f \in \mathbb{F}(\mathsf{X},\mathcal{X})$ be a function satisfying $\pi|f| < \infty$. Assume that the Poisson equation* (7.37) *admits a solution $\hat{f}$. Then,*

$$S_n(f) = n^{-1/2} M_n(\hat{f}) + n^{-1/2}\left(\hat{f}(X_0) - \hat{f}(X_n)\right) ,$$

where

$$M_n(\hat{f}) = \sum_{t=1}^{n} \{\hat{f}(X_t) - \mathbb{E}\left[\hat{f}(X_t) \,|\, \mathcal{F}_{t-1}\right]\} = \sum_{t=1}^{n} \{\hat{f}(X_t) - P\hat{f}(X_{t-1})\} ,$$

and $\mathcal{F}_t := \sigma(X_0,\ldots,X_t)$. Moreover, $\left\{(M_n(\hat{f}),\mathcal{F}_n)\; n \geq 1\right\}$ is a martingale.

Proof. This follows immediately from

$$\mathbb{E}\left[\hat{f}(X_t) - P\hat{f}(X_{t-1}) \,|\, \mathcal{F}_{t-1}\right] = \mathbb{E}\left[\hat{f}(X_t) - \mathbb{E}\left[\hat{f}(X_t) \,|\, \mathcal{F}_{t-1}\right] \,|\, \mathcal{F}_{t-1}\right] = 0 .$$

■

We first show that a V-geometric Markov chain admits at least one solution to the Poisson equation provided that f belongs to some appropriately defined class of (possibly unbounded) functions.

Lemma 7.31. *Assume A7.27 holds. Then, for all functions f such that $|f|_{\mathrm{V}} < \infty$, $\sum_{n=0}^{\infty} |P^n f(x) - \pi(f)| < \infty$ for all $x \in \mathsf{X}$, and*

$$\hat{f}(x) := \sum_{n=0}^{\infty} [P^n f(x) - \pi(f)] ,$$

is a solution of the Poisson equation associated with f. Moreover,

$$|\hat{f}|_{\mathrm{V}} \leq C(1-\rho)^{-1} |f|_{\mathrm{V}} < \infty .$$

Proof. Let f be such that $|f|_{\mathrm{V}} < \infty$. Set $\hat{f}_n(x) := \sum_{t=0}^{n-1} [P^t f(x) - \pi(f)]$ and note that

$$\hat{f}_n(x) - P\hat{f}_n(x) = \sum_{t=0}^{n-1} \left[P^t f(x) - P^{t+1} f(x)\right] = f(x) - P^n f(x) .$$

For all $x \in \mathsf{X}$ and $n \in \mathbb{N}$, we get $|P^n f(x) - \pi(f)| \leq CV(x)\rho^n |f|_{\mathrm{V}}$. This shows that $\lim_{n\to\infty} \hat{f}_n(x)$ exists and we can set $\hat{f}(x) = \lim_{n\to\infty} \hat{f}_n(x) = \sum_{n=0}^{\infty} [P^n f(x) - \pi(f)]$. Moreover, since for all $x \in \mathsf{X}$,

$$\frac{|\hat{f}_n(x)|}{V(x)} \leq \sum_{n=0}^{\infty} \frac{|P^n f(x) - \pi(f)|}{V(x)} \leq C(1-\rho)^{-1} |f|_{\mathrm{V}} ,$$

the Lebesgue theorem implies that $\lim_{n\to\infty} P\hat{f}_n = P\hat{f}$, and thus,

$$\hat{f}(x) - P\hat{f}(x) = \lim_{n\to\infty} [\hat{f}_n(x) - P\hat{f}_n(x)] = \lim_{n\to\infty} [f(x) - P^n f(x)] = f(x) - \pi(f) .$$

The proof is completed. ■

Theorem 7.32. *Assume A7.27. Let $f \in \mathbb{F}(\mathsf{X}, \mathcal{X})$ be such that $|f|_{\mathrm{V}^{1/2}} < \infty$ such that $\sigma^2(f) := \pi(\hat{f}^2) - \pi(P\hat{f})^2 > 0$. Then, for all distributions $\xi \in \mathbb{M}_1(\mathcal{X})$ such that $\xi(V) < \infty$,*

$$n^{-1/2} \sum_{t=0}^{n-1} [f(X_t) - \pi(f)] \overset{\mathbb{P}_\xi}{\Longrightarrow} \mathrm{N}\left(0, \sigma^2(f)\right) . \tag{7.38}$$

Proof. By Lemma 7.28, the Markov kernel P is $V^{1/2}$-geometrically ergodic. Since $|f|_{\mathrm{V}^{1/2}} < \infty$, Lemma 7.31 shows that there exists a solution $\hat{f}$ to the Poisson equation such that $\left|\hat{f}\right|_{\mathrm{V}^{1/2}} < \infty$. Using the martingale decomposition (Lemma 7.30), we get

$$\sum_{t=0}^{n-1} \frac{f(X_t) - \pi(f)}{n^{1/2}} = \sum_{t=1}^{n} \frac{\hat{f}(X_t) - \mathbb{E}\left[\hat{f}(X_t) \,\middle|\, \mathcal{F}_{t-1}\right]}{n^{1/2}} + \frac{\hat{f}(X_0) - \hat{f}(X_n)}{n^{1/2}} . \tag{7.39}$$

Obviously, $\hat{f}(X_0)/n^{1/2} \overset{\mathbb{P}_\xi}{\longrightarrow} 0$. By the Markov inequality, we get

$$\begin{aligned} \mathbb{P}_\xi(|\hat{f}(X_n)|/n^{1/2} > \varepsilon) &\le \varepsilon^{-1} n^{-1/2} \mathbb{E}_\xi(|\hat{f}(X_n)|) \le \varepsilon^{-1} n^{-1/2} \xi P^n |\hat{f}| \\ &\le \varepsilon^{-1} n^{-1/2} [C \xi(V) \rho^n \left|\hat{f}\right|_{\mathrm{V}} + \pi(|\hat{f}|)] , \end{aligned}$$

where the last inequality follows from the fact that $\left|\hat{f}\right|_{\mathrm{V}^{1/2}} < \infty$ implies $\left|\hat{f}\right|_{\mathrm{V}} < \infty$. Finally, $\hat{f}(X_n)/n^{1/2} \to_{\mathbb{P}_\xi} 0$. It thus remains to show the weak convergence of the first term of (7.39). But this follows directly from Theorem B.20 by setting $U_{n,t} = \hat{f}(X_t)/n^{1/2}$, provided that we show

$$\sum_{t=1}^{n} \{\mathbb{E}\left[U_{n,t}^2 \,\middle|\, \mathcal{F}_{t-1}\right] - (\mathbb{E}\left[U_{n,t} \,\middle|\, \mathcal{F}_{t-1}\right])^2\} \overset{\mathbb{P}_\xi}{\longrightarrow} \sigma^2 \quad \text{for some } \sigma^2 > 0 , \tag{7.40}$$

$$\sum_{t=1}^{n} \mathbb{E}\left[U_{n,t}^2 \mathbb{1}_{\{|U_{n,t}| \ge \varepsilon\}} \,\middle|\, \mathcal{F}_{t-1}\right] \overset{\mathbb{P}_\xi}{\longrightarrow} 0 \quad \text{for any } \varepsilon > 0 , \tag{7.41}$$

where $\mathcal{F}_t = \sigma(X_0, \ldots, X_t)$. Consider first (7.41). Let $\varepsilon > 0$ and $M > 0$. For all $n \ge M$,

$$\mathbb{E}_\xi \left(\sum_{t=1}^{n} \mathbb{E}\left[U_{n,t}^2 \mathbb{1}_{\{|U_{n,t}| \ge \varepsilon\}} \,\middle|\, \mathcal{F}_{t-1}\right] \right) \le n^{-1} \sum_{t=1}^{n} \xi P^n g_M , \tag{7.42}$$

where $g_M = \hat{f}^2 \mathbb{1}\{|\hat{f}| > \sqrt{M}\varepsilon\}$. Since $|g_M| \le \hat{f}^2 \le \left|\hat{f}\right|_{\mathrm{V}^{1/2}}^2 V$, we have $\pi(g_M) < \infty$ and, since $\lim_{n\to\infty} \xi P^n g_M = \pi(g_M)$, we get

$$\lim_{n\to\infty} n^{-1} \sum_{t=1}^{n} \xi P^n g_M = \pi(g_M) ,$$

which can be taken arbitrarily small since by the Lebesgue dominated convergence theorem, $\lim_{M\to\infty} \pi(g_M) = 0$. This implies, using (7.42), that

$$\limsup_{n\to\infty} \mathbb{E}_\xi \left[\sum_{t=1}^{n} \mathbb{E}\left[U_{n,t}^2 \mathbb{1}_{\{|U_{n,t}| \ge \varepsilon\}} \,\middle|\, \mathcal{F}_{t-1}\right] \right] = 0 ,$$

and thus

$$\sum_{t=1}^{n} \mathbb{E}\left[U_{n,t}^2 \mathbb{1}_{\{|U_{n,t}|\geq \varepsilon\}} \,\middle|\, \mathcal{F}_{t-1}\right] \xrightarrow{\mathbb{P}_\xi} 0 .$$

We now turn to (7.40). Write

$$\sum_{t=1}^{n}\{\mathbb{E}\left[U_{n,t}^2 \,|\, \mathcal{F}_{t-1}\right] - (\mathbb{E}\left[U_{n,t} \,|\, \mathcal{F}_{t-1}\right])^2\} = n^{-1}\sum_{t=1}^{n} g(X_{t-1}) , \tag{7.43}$$

where $g := P(\hat{f}^2) - (P\hat{f})^2$. Note that

$$0 \leq g \leq P\hat{f}^2 \leq |\hat{f}|_{V^{1/2}}^2 PV \leq |\hat{f}|_{V^{1/2}}^2 (\lambda V + b) .$$

We can thus apply Theorem 7.24:

$$n^{-1}\sum_{t=1}^{n} g(X_{t-1}) \xrightarrow{\mathbb{P}_\xi} \pi(g) = \pi P(\hat{f}^2) - \pi(P\hat{f})^2 = \pi(\hat{f}^2) - \pi(P\hat{f})^2 .$$

Plugging this into (7.43) yields (7.40) with $\sigma^2 = \pi(\hat{f}^2) - \pi(P\hat{f})^2$. The proof is completed. ■

It is important to note that, due to the dependence structure of a Markov chain, $\sigma^2(f) \neq \mathrm{Var}_\pi(f)$ except in trivial cases. Suppose that we have a sequence of estimators such that $\hat{\sigma}_n^2(f) \to_{\mathbb{P}\text{-a.s.}} \sigma^2(f)$. Then, we can construct an asymptotically valid confidence interval of $\pi(f)$ with confidence level $1-\alpha$ given by $\left[\mu_n(f) - z_{\alpha/2}\hat{\sigma}_n(f)/\sqrt{n}, \mu_n(f) + z_{\alpha/2}\hat{\sigma}_n(f)/\sqrt{n}\right]$, where $\mu_n(f) = n^{-1}\sum_{t=1}^{n} f(X_t)$ and $z_{\alpha/2}$ is the $1-\alpha/2$ quantile of the standard Gaussian distribution.

The asymptotic variance $\sigma^2(f) = \pi(\hat{f}^2) - \pi(P\hat{f})^2$ in Theorem 7.32 is expressed in terms of $\hat{f}$, the solution of the Poisson equation, and thus estimation of this quantity is involved. We now provide another expression of the asymptotic variance that directly involves the function f instead of $\hat{f}$, which will be much easier to estimate.

Proposition 7.33. *Assume A 7.27. Then, the series* $\sum_{t=1}^{\infty} \mathrm{Cov}_\pi(f(X_0), P^t f(X_0))$ *is absolutely convergent and*

$$\mathrm{Var}_\pi(f(X_0)) + 2\sum_{t=1}^{\infty} \mathrm{Cov}_\pi\left(f(X_0), P^t f(X_0)\right) = \sigma^2(f) . \tag{7.44}$$

Moreover,

$$\sigma^2(f) \leq 2^{1/2} C^{1/2} \pi(V) \, |f|_{V^{1/2}}^2 \frac{1+\rho^{1/2}}{1-\rho^{1/2}} . \tag{7.45}$$

Proof. Let $\{X_t,\, t \in \mathbb{N}\}$ be a Markov chain of Markov kernel P and let $f \in \mathbb{F}(\mathsf{X}, \mathcal{X})$ be such that $|f|_{V^{1/2}} < \infty$. Using Lemma 7.28,

$$\begin{aligned} \left|\mathrm{Cov}_\pi\left(f(X_0), P^t f(X_0)\right)\right| &= \left|\mathbb{E}_\pi\left[f(X_0)\left(P^t f(X_0) - \pi(f)\right)\right]\right| \\ &\leq \mathbb{E}_\pi\left[|f(X_0)|\,|P^t f(X_0) - \pi(f)|\right] \\ &\leq 2^{1/2} C^{1/2} \rho^{t/2} \mathbb{E}_\pi[|f|(X_0) V^{1/2}(X_0)] \, |f|_{V^{1/2}} \\ &\leq 2^{1/2} C^{1/2} \rho^{t/2} \pi(V) \, |f|_{V^{1/2}}^2 . \end{aligned} \tag{7.46}$$

Thus, the series $\sum_{t=1}^{\infty} \mathrm{Cov}_\pi(f(X_0), P^t f(X_0))$ is absolutely convergent showing the first part of the Proposition 7.33. Now, since π is invariant for the kernel P,

$$n^{-1}\mathbb{E}_\pi\left[\left(\sum_{t=0}^{n-1}\{f(X_t)-\pi(f)\}\right)^2\right]$$
$$= \mathrm{Var}_\pi(f(X_0)) + 2\sum_{t=1}^{n-1}\left(1-\frac{t}{n}\right)\mathrm{Cov}_\pi\left(f(X_0), P^t f(X_0)\right),$$

and since $\sum_{t=1}^{\infty} \mathrm{Cov}_\pi(f(X_0), P^t f(X_0))$ is absolutely convergent, the Lebesgue dominated convergence theorem for series shows that

$$\lim_{n\to\infty} n^{-1}\mathbb{E}_\pi\left[\left(\sum_{t=0}^{n-1}\{f(X_t)-\pi(f)\}\right)^2\right] = \mathrm{Var}_\pi(f(X_0)) + 2\sum_{t=1}^{\infty}\mathrm{Cov}_\pi\left(f(X_0), P^t f(X_0)\right).$$

To complete the proof of (7.44), it thus remains to show that

$$\lim_{n\to\infty} n^{-1}\mathbb{E}_\pi\left[\left(\sum_{t=0}^{n-1}\{f(X_t)-\pi(f)\}\right)^2\right] = \sigma^2(f). \tag{7.47}$$

There exists a solution of the Poisson equation $\hat{f}$ such that $|\hat{f}|_{V^{1/2}} < \infty$. Then, denoting $M_n(\hat{f}) = \sum_{t=1}^{n}\{\hat{f}(X_t) - P\hat{f}(X_{t-1})\}$, we have by Lemma 7.30,

$$A_n := n^{-1/2}\sum_{t=0}^{n-1}\{f(X_t)-\pi(f)\} = B_n + C_n,$$

where $B_n := n^{-1/2}M_n(\hat{f})$ and $C_n := n^{-1/2}(\hat{f}(X_0) - \hat{f}(X_n))$. Then,

$$\begin{aligned}|\mathbb{E}_\pi(A_n^2) - \mathbb{E}_\pi(B_n^2)| &\le \mathbb{E}_\pi(C_n^2) + 2\mathbb{E}_\pi|B_n C_n| \\ &\le \mathbb{E}_\pi(C_n^2) + 2(\mathbb{E}_\pi B_n^2)^{1/2}(\mathbb{E}_\pi C_n^2)^{1/2}.\end{aligned} \tag{7.48}$$

Now, since $\{\hat{f}(X_t) - P\hat{f}(X_{t-1}), t \ge 1\}$ is a martingale difference sequence, we get

$$\begin{aligned}\mathbb{E}_\pi(B_n^2) = n^{-1}\mathbb{E}_\pi M_n^2(\hat{f}) &= n^{-1}\sum_{i,j=1}^{n}\mathbb{E}_\pi\left[(\hat{f}(X_i) - P\hat{f}(X_{i-1}))(\hat{f}(X_j) - P\hat{f}(X_{j-1}))\right] \\ &= n^{-1}\sum_{i=1}^{n}\mathbb{E}_\pi(\hat{f}(X_i) - P\hat{f}(X_{i-1}))^2 = \sigma^2(f).\end{aligned} \tag{7.49}$$

Moreover,

$$\begin{aligned}\mathbb{E}_\pi(C_n^2) = n^{-1}\mathbb{E}_\pi\left[(\hat{f}(X_0) - \hat{f}(X_n))^2\right] &\le 2n^{-1}\left\{\mathbb{E}_\pi\left[\hat{f}^2(X_0)\right] + \mathbb{E}_\pi\left[\hat{f}^2(X_n)\right]\right\} \\ &= 4n^{-1}\pi(\hat{f}^2) \le 4n^{-1}\pi(V)\left|\hat{f}\right|_{V^{1/2}}^2 \to_{n\to\infty} 0.\end{aligned}$$

Coimbining this with (7.49) and then plugging it into (7.48), we obtain that

$$\lim_{n\to\infty} n^{-1}\mathbb{E}_\pi\left[\left\{\sum_{t=0}^{n-1}(f(X_t)-\pi(f))\right\}^2\right] = \sigma^2(f)\,,$$

which completes the proof of (7.44). Now, (7.45) directly follows from (7.44) and (7.46). ■

Under the assumption of stationarity, the asymptotic variance $\sigma^2(f)$ is equal to $2\pi g(0)$, where $g(\lambda)$ is the discrete Fourier transform of the autocovariance sequence $\{\mathrm{Cov}_\pi(f(X_0),f(X_h))\,,\ h\in\mathbb{N}\}$, or the spectral density function of the stationary version of the Markov chain (see Proposition 5.16), given by

$$g(\lambda) = \frac{1}{2\pi}\left[\mathrm{Var}_\pi(f(X_0)) + 2\sum_{h=1}^{\infty}\mathrm{Cov}_\pi(f(X_0),f(X_h))\cos(\lambda h)\right]\,,$$

where the parameter $\lambda\in[0,\ \pi)$ is the normalized frequency.

The spectral variance estimator may be seen as a natural estimator of $\sigma^2(f)$. Such an estimator is called a *plug-in* estimator. In this case, the autocovariance coefficients, $\mathrm{Cov}_\pi(f(X_0),f(X_h))$, are replaced by their sample estimates. Define the lag h autocovariance $\gamma(h;f) := \mathrm{Cov}_\pi(Y_0,Y_h)$ where $Y_t = f(X_t)-\pi(f)$ for $t\in\mathbb{N}$. As discussed in Example 1.12, the preferred estimator of $\gamma(h)$ is

$$\gamma_n(h;f) := n^{-1}\sum_{t=1}^{n-h}(Y_t-\overline{Y}_n)(Y_{t+h}-\overline{Y}_n)\,, \tag{7.50}$$

where $\overline{Y}_n := n^{-1}\sum_{t=1}^{n} Y_t$ is the empirical mean. Provided that the Markov chain $\{X_t,\ t\in\mathbb{N}\}$ has a unique stationary distribution, then Theorem 7.24 shows that, for π-æ x,

$$\gamma_n(h;f)\overset{\mathbb{P}_x\text{-a.s.}}{\longrightarrow}\gamma(h;f)\,,\quad \text{as } n\to\infty\,.$$

The *spectral variance estimator* is obtained by truncating and weighting the empirical autocovariance sequence using

$$\hat{\sigma}_n^2(f) := \sum_{|h|\le b_n} w_n(h)\gamma_n(h;f)\,, \tag{7.51}$$

where $w_n(\cdot)$ is the *lag window* and b_n is the *truncation point*. Under appropriate conditions, $\hat{\sigma}_n^2(f)$ is shown to be a strongly consistent estimator of $\sigma^2(f)$ (see Flegal and Jones, 2010, Damerdji, 1994, 1995, Theorem 1).

Example 7.34. Consider the AR(1) model defined by

$$X_t = \mu + \phi(X_{t-1}-\mu) + Z_t\,,\quad t\in\mathbb{N}^*\,, \tag{7.52}$$

where $\phi\in(-1,1)$, X_0 satisfies $\mathbb{E}\left[X_0^2\right]<\infty$ and $\{Z_t,\ t\in\mathbb{Z}\}$ is an i.i.d. sequence of zero-mean random variables, independent of X_0, such that $\mathbb{E}\left[Z_0^2\right] = \eta^2$. Theorem 5.31 shows that the AR(1) process (7.52) has a unique stationary distribution, denoted π, which is the law of the random variable

$$X_\infty = \mu + \sum_{k=0}^{\infty} \phi^k Z_k .$$

We might be interested in the estimation of $\mathbb{E}_\pi[X_0] = \mu$ by the sample mean $\overline{X}_n = n^{-1}\sum_{t=1}^{n} X_t$. Theorem 7.32 and Proposition 7.33 show that the sample mean is asymptotically normal, i.e., $n^{1/2}(\overline{X}_n - \mu) \Rightarrow_{\mathbb{P}_\xi} \mathrm{N}\left(0, \sigma^2\right)$, where the asymptotic variance σ^2 is given by

$$\sigma^2 = \mathrm{Var}_\pi\left(X_0\right) + 2\sum_{t=1}^{\infty} \mathrm{Cov}_\pi\left(X_0, X_t\right) .$$

Now, for all $h \geq 0$,

$$\mathrm{Cov}_\pi\left(X_0, X_h\right) = \sum_{k=0}^{\infty} \phi^k \phi^{k+h} \mathbb{E}\left[Z_0^2\right] = \frac{\eta^2 \phi^h}{1-\phi^2} ,$$

so that

$$\sigma^2 = \frac{\eta^2}{1-\phi^2} + 2\eta^2 \sum_{h=1}^{\infty} \frac{\phi^h}{1-\phi^2} = \frac{\eta^2}{(1-\phi)^2} = \mathrm{Var}_\pi\left(X_0\right) \frac{1+\phi}{1-\phi} .$$

Given the data $\{X_1, \ldots, X_n\}$, we can thus estimate σ^2 by $\hat{\sigma}_n^2 = \hat{\eta}_n^2/(1-\hat{\phi}_n)^2$ where $\hat{\phi}_n$ and $\hat{\eta}_n^2$ are estimators of ϕ and η^2. Since $\phi = \mathrm{Cov}_\pi\left(X_0, X_1\right)/\mathrm{Var}\left(X_0\right)$, moment estimators of $\hat{\phi}_n$ and $\hat{\eta}_n^2$ are given by

$$\widehat{\phi}_n := \frac{\sum_{t=1}^{n-1}(X_t - \overline{X}_n)(X_{t+1} - \overline{X}_n)}{\sum_{t=1}^{n}(X_t - \overline{X}_n)^2} , \quad \widehat{\eta}_n^2 := n^{-1}\sum_{t=1}^{n-1}\left((X_{t+1} - \overline{X}_n) - \hat{\phi}_n(X_t - \overline{X}_{n}i)\right)^2 .$$

The coefficient ϕ plays a crucial role in the behavior of the chain. Figure 7.1 displays plots based on a realization of size $n = 100$ with $\phi = .95$ and $\phi = -.95$. In each figure, the top plot is a realization of the time series, and the middle plot is the sample ACF with the actual ACF superimposed on it. In this example, the mean of the stationary distribution is 0. The impact of the correlation is particularly apparent on the plot of the running estimate of the mean versus iterations of the Markov chain. Clearly, the more positively correlated sequence requires many more iterations to achieve a sensible estimate. ◇

7.3 Deviation inequalities for additive functionals

7.3.1 Rosenthal type inequality

Let $(\Omega, \mathcal{F}, \{\mathcal{F}_t,\, t \in \mathbb{N}\}, \mathbb{P})$ be a filtered probability space. Let $\{(Z_t, \mathcal{F}_t),\, t \in \mathbb{N}\}$ be a martingale difference sequence. Assume that for some $p \geq 2$, $\mathbb{E}(|Z_t|^p) < \infty$ for any

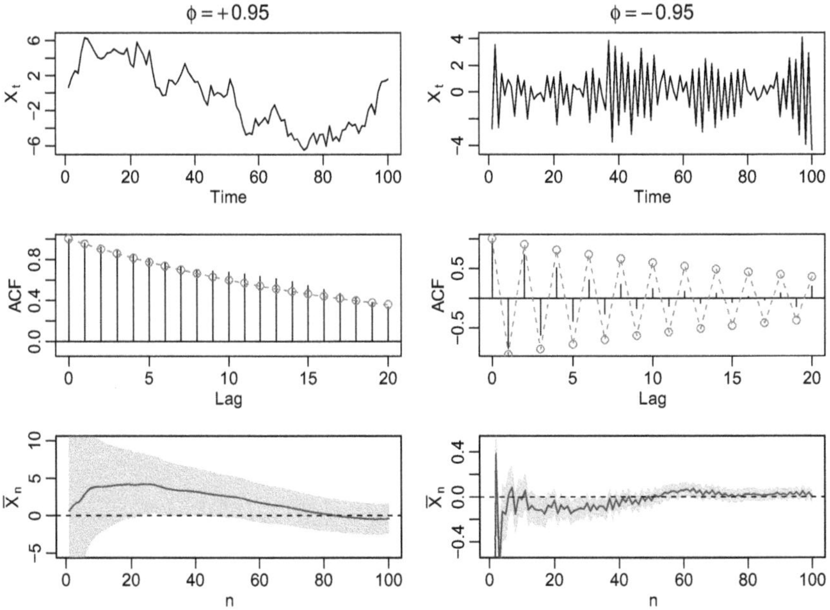

Figure 7.1 *Estimation of the mean of an AR(1) process. In the left panel, $\phi = .95$. In the right panel, $\phi = -.95$. The top plot is a realization of the time series with $n = 100$. The middle plot is the sample ACF with the actual ACF superimposed on it. The bottom plot is the running sample mean and the gray areas indicate 95% confidence intervals.*

$t \in \mathbb{N}$. Denote by $S_t = \sum_{j=1}^t Z_j$ the partial sum and by $\langle S \rangle_t = \sum_{j=1}^t \mathbb{E}[Z_j^2 \mid \mathcal{F}_{j-1}]$ the previsible quadratic variation (see Proposition B.5). The following inequality was established by Burkholder (1973): there exists a universal constant c_p, depending only on p, such that

$$\|\max_{1 \le t \le n} S_t\|_p \le c_p \left\{ \|\langle S \rangle_n^{1/2}\|_p + \|\max_{1 \le t \le n} |Z_t|\|_p \right\} . \tag{7.53}$$

For example, if $\{Z_j,\ j \in \mathbb{N}\}$ is an i.i.d. sequence of zero-mean p-th integrable random variables, then $\mathbb{E}[Z_j^2 \mid \mathcal{F}_{j-1}] = \|Z_1\|_2^2$ so that $\|\langle S \rangle_n^{1/2}\|_p = n^{1/2}\|Z_1\|_2$ whereas

$$\|\max_{t \le n} |Z_t|\|_p = \left(\mathbb{E} \max_{t \le n} |Z_t|^p \right)^{1/p} \le \left(\sum_{t=1}^n \mathbb{E}(|Z_t|^p) \right)^{1/p} = n^{1/p}\|Z_1\|_p ,$$

showing that

$$\|\max_{1 \le t \le n} S_t\|_p \le c_p (n^{1/2}\|Z_1\|_2 + n^{1/p}\|Z_1\|_p) .$$

This shows that the fluctuation of a partial sum around its mean is of order $n^{1/2}$, which is compatible with the central limit theorem. Using the martingale decomposition, we will show that similar bounds can be obtained for V-geometric Markov chains.

Theorem 7.35. *Assume A7.27. Let $p \geq 2$. Then, there exists a constant c_p such that for all $x \in \mathsf{X}$ and all $f \in \mathbb{F}(\mathsf{X}, \mathcal{X})$ such that $|f|_{\mathrm{V}^{1/p}} < \infty$, we get*

$$\left\{ \mathbb{E}_x \left[\left(\sum_{t=0}^{n-1} [f(X_t) - \pi(f)] \right)^p \right] \right\}^{1/p} \leq c_p |f|_{\mathrm{V}^{1/p}} V^{1/p}(x) n^{1/2} . \tag{7.54}$$

Proof. Set $x \in \mathsf{X}$. According to Lemma 7.28, P is $V^{1/p}$-geometrically ergodic and since $|f|_{\mathrm{V}^{1/p}} < \infty$, Lemma 7.31 shows that there exists $\hat{f}$ a solution of the Poisson equation associated to f satisfying

$$|\hat{f}|_{\mathrm{V}^{1/p}} \leq 2^{1-1/p} C^{1/p} (1 - \rho^{1/p})^{-1} |f|_{\mathrm{V}^{1/p}} < \infty . \tag{7.55}$$

Then, denoting $M_n = \sum_{t=1}^n Z_t$ with $Z_t := \hat{f}(X_t) - P\hat{f}(X_{t-1})$, Lemma 7.30 shows that

$$\sum_{t=0}^{n-1} [f(X_t) - \pi(f)] = M_n + \hat{f}(X_0) - \hat{f}(X_n) ,$$

so that

$$\begin{aligned} &\left| \left\{ \mathbb{E}_x \left[\left(\sum_{t=0}^{n-1} [f(X_t) - \pi(f)] \right)^p \right] \right\}^{1/p} - \{\mathbb{E}_x [M_n^p]\}^{1/p} \right| \leq |\hat{f}(x)| + (P^n \hat{f}^p(x))^{1/p} \\ &\leq |\hat{f}|_{\mathrm{V}^{1/p}} \left(V^{1/p}(x) + (P^n V(x))^{1/p} \right) \\ &\leq |\hat{f}|_{\mathrm{V}^{1/p}} \left(2V^{1/p}(x) + (b/(1-\lambda))^{1/p} \right) , \end{aligned} \tag{7.56}$$

where the last inequality follows from Lemma 6.26. Now, recalling that $Z_t := \hat{f}(X_t) - P\hat{f}(X_{t-1})$, we have $|Z_t|^p \leq 2^{p-1} |\hat{f}(X_t)|^p + 2^{p-1} P|\hat{f}|^p (X_{t-1})$, so that so that

$$\mathbb{E}_x[|Z_t|^p] \leq 2^p P^t |\hat{f}|^p (x) \leq 2^p P^t V(x) \, |\hat{f}|^p_{\mathrm{V}^{1/p}} , \tag{7.57}$$

which is finite by Lemma 6.26. This allows us to apply Rosenthal's inequality, Theorem B.7, to the martingale $\{M_t, \, t \in \mathbb{N}\}$: there exist a positive constant c_p (depending only on p) such that

$$\| \max_{1 \leq t \leq n} M_t \|_p \leq c_p \left\{ \| \langle M \rangle_n^{1/2} \|_p + \| \max_{1 \leq t \leq n} Z_t \|_p \right\} . \tag{7.58}$$

We now bound each term of the right-hand side of (7.58). Before doing this, we need some inequality that will be used repeatedly: using again Lemma 6.26,

$$\begin{aligned} \frac{\sum_{t=0}^n P^t |\hat{f}|^p (x)}{n} &\leq |\hat{f}|^p_{\mathrm{V}^{1/p}} \frac{\sum_{t=0}^n P^t V(x)}{n} \leq |\hat{f}|^p_{\mathrm{V}^{1/p}} \left(V(x) \frac{\sum_{t=0}^n \lambda^t}{n} + \frac{b}{1-\lambda} \right) \\ &\leq C(b, \lambda) \, |\hat{f}|^p_{\mathrm{V}^{1/p}} V(x) , \end{aligned} \tag{7.59}$$

where $C(b,\lambda) = 1 + b/(1-\lambda)$. Now, write

$$\langle M \rangle_n = \sum_{j=1}^{n} \mathbb{E}\left[Z_j^2 \mid \mathcal{F}_{j-1}\right] \leq n \frac{\sum_{j=0}^{n-1} P|\hat{f}|^2(X_j)}{n} ,$$

so that by Jensen's inequality and (7.59),

$$\begin{aligned}
\|\langle M \rangle_n^{1/2}\|_p &\leq n^{1/2} \left[\mathbb{E}_x \left(\frac{\sum_{j=0}^{n-1} P|\hat{f}|^2(X_j)}{n}\right)^{p/2}\right]^{1/p} \\
&\leq n^{1/2} \left[\mathbb{E}_x \left(\frac{\sum_{t=0}^{n-1} P|\hat{f}|^p(X_t)}{n}\right)\right]^{1/p} \\
&= n^{1/2} \left[\frac{\sum_{t=1}^{n} P^t |\hat{f}|^p(x)}{n}\right]^{1/p} \leq n^{1/2} C^{1/p}(b,\lambda) \left|\hat{f}\right|_{V^{1/p}} V^{1/p}(x) . \quad (7.60)
\end{aligned}$$

To bound the second term in the right-hand side of (7.58), we again use (7.57) and (7.59):

$$\begin{aligned}
\|\max_{1 \leq t \leq n} |Z_t|\|_p &\leq \left(\sum_{t=1}^{n} \mathbb{E}|Z_t|^p\right)^{1/p} \\
&\leq n^{1/p} \left(\frac{2^p \sum_{t=1}^{n} P^t |\hat{f}|^p(x)}{n}\right)^{1/p} \\
&\leq 2 n^{1/p} C^{1/p}(b,\lambda) \left|\hat{f}\right|_{V^{1/p}} V^{1/p}(x) .
\end{aligned}$$

Plugging this inequality and (7.60) into (7.58) and then combining with (7.56) and (7.55) complete the proof. ■

7.3.2 *Concentration inequality for uniformly ergodic Markov chains*

In this section, we briefly discuss some concentration inequalities. To keep the discussion simple, we consider only in this section uniformly ergodic Markov chains. The main result is a generalization of the McDiarmid inequality (McDiarmid, 1989), which is then used to prove the Hoeffding inequality. The McDiarmid inequality is a large deviation type inequality for functions of independent random variables satisfying the *bounded difference condition*; this condition states that there exist some nonnegative constants $\{\gamma_1, \ldots, \gamma_n\} \subset \mathbb{R}^+$ such that for all $(x_1, \ldots, x_n) \in \mathsf{X}^n$ and $(y_1, \ldots, y_n) \in \mathsf{X}^n$,

$$|f(x_1, \ldots, x_n) - f(y_1, \ldots, y_n)| \leq \sum_{i=1}^{n} \gamma_i \mathbb{1}_{\{x_i \neq y_i\}} . \quad (7.61)$$

The bounded difference condition specifies that by changing the value of the i-th variable, the value of the function cannot change by more than a constant γ_i. When the random variables $\{X_1,\dots,X_n\}$ are independent, the McDiarmid inequality states that, for all $\delta > 0$,

$$\mathbb{P}[|f(X_1,\dots,X_n) - \mathbb{E}[f(X_1,\dots,X_n)]| \geq \delta] \leq 2\exp\left(-\frac{2\delta^2}{\sum_{i=1}^n \gamma_i^2}\right) . \tag{7.62}$$

Taking $f_n(x_1,\dots,x_n) = n^{-1}\sum_{i=1}^n f(X_i)$ where $f : \mathsf{X} \to \mathsf{X}$ is a bounded function yields the Hoeffding inequality,

$$\mathbb{P}[|f_n(X_1,\dots,X_n) - \mathbb{E}[f_n(X_1,\dots,X_n)]| \geq \delta] \leq 2\exp\left(-\frac{2n\delta^2}{\operatorname{osc}^2(f)}\right) .$$

This inequality can be used to handle complicated functions. This result has found broad applicability in many different settings. See, for example, Serfling (1980) for various statistical contexts within which the inequality plays a central role. Devroye et al. (1996) illustrate the importance of this inequality in nonparametric statistics. The explicit nature of the constants in the bound makes it especially attractive in contexts within which one needs to establish that the error between the empirical estimate and the limiting value decays exponentially in n with explicit constants.

Theorem 7.36. *Let $f : \mathsf{X}^n \to \mathbb{R}$ be a measurable function satisfying the bounded difference condition* (7.61). *Let $\{X_n,\ n \in \mathbb{N}\}$ be a Markov chain with kernel P. Then, for all $\xi \in \mathbb{M}_1(\mathcal{X})$ and $\delta > 0$,*

$$\mathbb{P}_\xi\left(|f(X_1,\dots,X_n) - \mathbb{E}_\xi[f(X_1,\dots,X_n)]| > \delta\right) \leq 2\mathrm{e}^{-\delta^2/(2D_n)} , \tag{7.63}$$

where

$$D_n := \sum_{\ell=1}^n \left(\sum_{u=\ell}^n \gamma_u \Delta_{\mathrm{TV}}\left(P^{u-\ell}\right)\right)^2 ,$$

and $\Delta_{\mathrm{TV}}(P)$ is the Dobrushin coefficient of P (see Definition 6.4).

Proof. We only need to prove (7.63) under the assumption $\mathbb{E}_\xi[f(X_1,\dots,X_n)] = 0$. Actually, we show

$$\mathbb{P}_\xi\left(f(X_1,\dots,X_n) > \delta\right) \leq \mathrm{e}^{-\delta^2/(2D_n)} , \tag{7.64}$$

so that (7.63) follows by applying (7.64) to f and $-f$. We now focus on (7.64).

Denote for $\ell \geq 0$, $M_\ell := \mathbb{E}\left[f(X_1,\dots,X_n) \,|\, \mathcal{F}_\ell\right]$ where $\mathcal{F}_\ell = \sigma(X_i, 1 \leq i \leq \ell)$ with the convention $\mathcal{F}_0 := \{\emptyset, \Omega\}$ (with this choice, $M_0 = 0$). Note that $f(X_1,\dots,X_n) = M_n = \sum_{\ell=1}^n \{M_\ell - M_{\ell-1}\}$, and the sequence $\{(M_\ell, \mathcal{F}_\ell), \ell = 0,\dots,n\}$ is a martingale. We now bound the increment $M_\ell - M_{\ell-1}$. Denote for $\ell < s \leq n$,

$$F_{\ell,s} = \inf\{f(X_1,\dots,X_\ell,u_{\ell+1},\dots,u_s,X_{s+1},\dots,X_n) : \{u_{\ell+1},\dots,u_s\} \subset \mathsf{X}\} , \tag{7.65}$$

with the convention $F_{\ell,\ell} = f(X_1,\dots,X_n)$. Since $F_{\ell-1,n}$ is $\mathcal{F}_{\ell-1}$-measurable, it is also $\mathcal{F}_\ell$-measurable and we have for all $\ell \geq 1$,

$$\begin{aligned} &M_\ell - M_{\ell-1} \\ &= \mathbb{E}\left[f(X_1,\dots,X_n) - F_{\ell-1,n} \,\middle|\, \mathcal{F}_\ell\right] - \mathbb{E}\left[f(X_1,\dots,X_n) - F_{\ell-1,n} \,\middle|\, \mathcal{F}_{\ell-1}\right] . \end{aligned} \tag{7.66}$$

The inner term can be decomposed as

$$f(X_1,\ldots,X_n) - F_{\ell-1,n} = F_{\ell-1,\ell-1} - F_{\ell-1,n} = \sum_{s=\ell-1}^{n-1} \{F_{\ell-1,s} - F_{\ell-1,s+1}\} . \quad (7.67)$$

Now, by (7.65) and (7.61), $0 \le F_{\ell-1,s} - F_{\ell-1,s+1} \le \gamma_{s+1}$. Moreover, $F_{\ell-1,s}$ and $F_{\ell-1,s+1}$ are $\sigma(X_i, i \in \{1,\ldots,\ell-1\} \cup \{s+1,\ldots,n\})$-measurable. This implies (see Exercise 6.6)

$$\left|\mathbb{E}\left[F_{\ell-1,s} - F_{\ell-1,s+1} \,\middle|\, \mathcal{F}_\ell\right] - \mathbb{E}\left[F_{\ell-1,s} - F_{\ell-1,s+1} \,\middle|\, \mathcal{F}_{\ell-1}\right]\right| \le \gamma_{s+1} \Delta_{\mathrm{TV}}\left(P^{s+1-\ell}\right) .$$

Combining this with (7.67) and (7.66) yields

$$|M_\ell - M_{\ell-1}| \le \sum_{u=\ell}^{n} \gamma_u \Delta_{\mathrm{TV}}\left(P^{u-\ell}\right) ,$$

so that by Exercise 7.16,

$$\ln \mathbb{E}_\xi [\exp\{tf(X_1,\ldots,X_n)\}] = \ln \mathbb{E}_\xi [\exp\{tM_n\}] \le \frac{t^2}{8} \sum_{\ell=1}^{n} \left(2 \sum_{u=\ell}^{n} \gamma_u \Delta_{\mathrm{TV}}\left(P^{u-\ell}\right)\right)^2 .$$

The proof of (7.64) then follows by plugging this inequality into

$$\mathbb{P}_\xi (f(X_1,\ldots,X_n) > \delta) \le \exp\left\{-t\delta + \ln \mathbb{E}_\xi \exp\{tf(X_1,\ldots,X_n)\}\right\} , \quad t \ge 0 ,$$

and by minimizing the resulting right-hand side with respect to $t \in \mathbb{R}^+$. ■

As a byproduct of Theorem 7.36, the Hoeffding inequality for uniformly ergodic Markov chains is now stated and proved.

Corollary 7.37. *Let $\{X_t,\ t \in \mathbb{N}\}$ be a uniformly ergodic Markov chain with kernel P and stationary distribution π. Then, for all $\xi \in \mathbb{M}_1(\mathsf{X})$ and all $\delta \ge n^{-1}\Delta\,\mathrm{osc}\,(f)$,*

$$\mathbb{P}_\xi \left(\left|\sum_{i=1}^{n} f(X_i) - \pi(f)\right| > n\delta\right) \le 2\exp\left\{-\frac{n\left(\delta - n^{-1}\Delta\,\mathrm{osc}\,(f)\right)^2}{2\,\mathrm{osc}^2(f)\,\Delta^2}\right\} ,$$

where

$$\Delta := \sum_{\ell=0}^{\infty} \Delta_{\mathrm{TV}}\left(P^\ell\right) < \infty . \quad (7.68)$$

Proof. Set $f(x_1,\ldots,x_n) := \sum_{i=1}^n f(x_i)/n$ so that (7.61) holds with $\gamma_i = \mathrm{osc}\,(f)/n$. Note first that,

$$n^{-1} \sum_{\ell=1}^{n} \left|\mathbb{E}_\xi [f(X_i)] - \pi(f)\right| \le n^{-1}\Delta\,\mathrm{osc}\,(f) .$$

The proof then follows by applying Theorem 7.36 since

$$D_n = \sum_{\ell=1}^{n} \left(\sum_{u=\ell}^{n} \Delta_{\mathrm{TV}}\left(P^{u-\ell}\right) \gamma_u\right)^2 \le \frac{\mathrm{osc}^2(f)}{n} \left(\sum_{\ell=0}^{\infty} \Delta_{\mathrm{TV}}\left(P^\ell\right)\right)^2 .$$

■

In particular, if the Doeblin condition (Definition 6.9) holds for some $m \geq 1$ and constant $\varepsilon > 0$, then $\Delta_{\mathrm{TV}}(P^m) \leq 1 - \varepsilon$, so that

$$\sum_{\ell=0}^{\infty} \Delta_{\mathrm{TV}}\left(P^\ell\right) \leq \sum_{\ell=0}^{\infty} \Delta_{\mathrm{TV}}(P^m)^{\lfloor \ell/m \rfloor} = m/\varepsilon .$$

Thus, Corollary 7.37 provides the bound: for all $\delta \geq (n\varepsilon)^{-1} m \operatorname{osc}(f)$,

$$\mathbb{P}_\xi\left(\left|\sum_{i=1}^{n} f(X_i) - \mathbb{E} f(X_1)\right| > n\delta\right) \leq 2\exp\left\{-\frac{\varepsilon^2(n\delta - \varepsilon^{-1} m \operatorname{osc}(f))^2}{2\operatorname{osc}^2(f)\, m^2}\right\},$$

which is similar to the bound obtained by Glynn and Ormoneit (2002), who used a technique based on the Poisson equation.

Example 7.38 (Kolmogorov Smirnov). Let $\{X_t,\, t \in \mathbb{N}\}$ be a uniformly ergodic Markov chain on $\mathbb{R}$. Denote by π its stationary distribution and by $x \mapsto F_\pi(x) = \pi((-\infty, x])$ the associated distribution function. Denote by $x \mapsto F_n(x)$ the corresponding empirical distribution function, $\hat{F}_n(x) = n^{-1} \sum_{t=0}^{n-1} \mathbb{1}_{\{X_t \leq x\}}$. Denote the *Kolmogorov-Smirnov statistic* by

$$V_n = \sup_{x \in \mathbb{R}} |\hat{F}_n(x) - F_\pi(x)|$$

The Hoeffding inequality shows that, for any $n \geq 1$ and any initial distribution ξ, we have

$$\mathbb{P}_\xi(V_n \geq \delta) \leq 2\exp\left\{-\frac{n\left(\delta - n^{-1}\Delta\right)^2}{2\Delta^2}\right\},$$

where Δ is defined in (7.68) and $\delta \geq n^{-1}\Delta$. ◇

Theorem 7.36 may also be applied to more intricate functions than $f(x_1, \ldots, x_n) = \sum_{i=1}^{n} f(x_i)/n$. In particular, in the following example, we obtain deviation inequalities for nonparametric kernel density estimators in the case of uniformly ergodic Markov chains.

Example 7.39. Let $\{X_t,\, t \in \mathbb{N}\}$ be a uniformly ergodic Markov chain with kernel P, taking value in $\mathbb{R}^d$. Assume that the (unique) invariant distribution of P has a density π with respect to Lebesgue measure Leb on $\mathbb{R}^d$. Given $X_1, \ldots, X_n$, the nonparametric kernel density estimator π_n^* of the density π is defined by:

$$\pi_n^*(x) := \frac{1}{n} \sum_{i=1}^{n} \frac{1}{h_n^d} K\left(\frac{x - X_i}{h_n}\right), \quad x \in \mathbb{R}^d, \tag{7.69}$$

where $\{h_n,\, n \in \mathbb{N}^*\}$ is a sequence of positive numbers called *bandwiths*, which usually satisfy

$$\lim_{n\to\infty} h_n = 0, \quad \lim_{n\to\infty} n h_n^d = \infty, \tag{7.70}$$

and K is a measurable nonnegative function such that $\int_{\mathbb{R}^d} K(u)\mathrm{Leb}(du) = 1$.

Denote by $J_n := \int |\pi_n^*(x) - \pi(x)| \mathrm{Leb}(dx)$. Since the Markov chain $\{X_t,\, t \in \mathbb{N}\}$ is uniformly ergodic, we get $\sum_{\ell=0}^{\infty} \Delta_{\mathrm{TV}}\left(P^{\ell}\right) < \infty$. We will show that for all $n \geq 1$ and all $\delta > 0$,

$$\mathbb{P}_{\pi}(|J_n - \mathbb{E}_{\pi}[J_n]| > \delta) \leq 2\exp\left(-\frac{n\delta^2}{8\left[\sum_{\ell=0}^{\infty}\Delta_{\mathrm{TV}}(P^{\ell})\right]^2}\right) . \tag{7.71}$$

Let

$$f(x_1,\dots,x_n) := \int_{\mathbb{R}^d} \left| \frac{1}{n}\sum_{i=1}^{n} \frac{1}{h_n^d} K\left(\frac{x-x_i}{h_n}\right) - \pi(x) \right| \mathrm{Leb}(\mathrm{d}x) .$$

Then, for all $i \in \{1,\dots,n\}$ and all $y_1,\dots,y_n \in \mathbb{R}^d$ such that $x_j = y_j$ for $j \neq i$,

$$|f(x_1,\dots,x_n) - f(y_1,\dots,y_n)| \leq \frac{1}{n}\int \frac{1}{h_n^d}\left| K\left(\frac{x-x_i}{h_n}\right) - K\left(\frac{x-y_i}{h_n}\right)\right| \mathrm{Leb}(\mathrm{d}x) \leq \frac{2}{n} ,$$

so that (7.61) holds with $\gamma_i = 2/n$. The proof of (7.71) then follows by applying Theorem 7.36, since

$$D_n = \sum_{\ell=1}^{n}\left(\sum_{u=\ell}^{n}\Delta_{\mathrm{TV}}\left(P^{u-\ell}\right)\gamma_u\right)^2 \leq \frac{4}{n}\left(\sum_{\ell=0}^{\infty}\Delta_{\mathrm{TV}}\left(P^{\ell}\right)\right)^2 .$$

◇

7.4 Some proofs

Definition 7.40 (Subadditivity and additivity). *The sequence of random variables $\{Y_n,\ n \in \mathbb{N}^\star\}$ is said to be* subadditive *for the dynamical system $(\Omega,\mathcal{B},\mathbb{Q},\mathrm{T})$ if for all $(n,p) \in \mathbb{N}^\star$, $Y_{n+p} \leq Y_n + Y_p \circ \mathrm{T}^n$. The sequence is said to be* additive *if for all $(n,p) \in \mathbb{N}^\star$, $Y_{n+p} = Y_n + Y_p \circ \mathrm{T}^n$.*

Lemma 7.41 (Improved Fekete lemma). *Consider $\{a_n,\ n \in \mathbb{N}^\star\}$, a sequence in $[-\infty,\infty)$ such that, for all $(m,n) \in \mathbb{N}^\star \times \mathbb{N}^\star$, $a_{n+m} \leq a_n + a_m$. Then,*

$$\lim_{n\to\infty} \frac{a_n}{n} = \inf_{m\in\mathbb{N}^\star} \frac{a_m}{m} ;$$

in other words, the sequence $\{n^{-1}a_n, n \in \mathbb{N}^\star\}$ either converges to its lower bounds or diverges to $-\infty$.

Proof. Set $m \in \mathbb{N}^\star$. Any number n can be written in the form $n = q(n)m + r(n)$, where $r(n) \in \{0,\dots,m-1\}$. We define $a_0 = 0$. Then, we have $a_n = a_{q(n)m+r(n)} \leq q(n)a_m + a_{r(n)}$. Then, we have

$$\frac{a_n}{n} = \frac{a_{q(n)m+r(n)}}{q(n)m+r(n)} \leq \frac{q(n)m}{q(n)m+r(n)}\frac{a_m}{m} + \frac{a_{r(n)}}{n} ,$$

which implies that,

$$\inf_{n\in\mathbb{N}^\star}\frac{a_n}{n} \leq \liminf_{n\to\infty}\frac{a_n}{n} \leq \limsup_{n\to\infty}\frac{a_n}{n} \leq \frac{a_m}{m} .$$

Since this inequality is valid for all $m \in \mathbb{N}^\star$, the result follows. ■

Lemma 7.42. *Let* $(\Omega,\mathcal{B},\mathbb{Q},\mathsf{T})$ *be a dynamical system and consider* $\{Y_n,\ n\in\mathbb{N}^\star\}$ *a subadditive sequence of functions such that* $\mathbb{E}(Y_1^+)<\infty$*. Then, for any* $n\in\mathbb{N}^\star$, $\mathbb{E}(Y_n^+)\le n\mathbb{E}(Y_1^+)<\infty$ *and*

$$\lim_{n\to\infty} n^{-1}\mathbb{E}(Y_n)=\inf_{n\in\mathbb{N}^\star} n^{-1}\mathbb{E}(Y_n)\,, \tag{7.72}$$

$$\lim_{n\to\infty} n^{-1}\mathbb{E}\left[Y_n\,|\,\mathcal{I}\right]=\inf_{n\in\mathbb{N}^\star} n^{-1}\mathbb{E}\left[Y_n\,|\,\mathcal{I}\right]\,,\quad \mathbb{Q}\text{-a.s.} \tag{7.73}$$

Proof. By subadditivity of the sequence,

$$Y_n^+\le\left(\sum_{k=0}^{n-1} Y_1\circ T^k\right)^+\le\sum_{k=0}^{n-1} Y_1^+\circ T^k\,.$$

Now, take the expectation in both sides of the previous inequality and use that T is measure-preserving. We obtain that $\mathbb{E}(Y_n^+)\le n\mathbb{E}(Y_1^+)<\infty$. With a similar argument, $\mathbb{E}(Y_p^+\circ\mathsf{T}^n)<\infty$. This implies that $\mathbb{E}(Y_{n+p})$ or $\mathbb{E}(Y_p\circ\mathsf{T}^n)$ are well-defined and

$$\mathbb{E}(Y_{n+p})\le\mathbb{E}(Y_n+Y_p\circ\mathsf{T}^n)=\mathbb{E}(Y_n)+\mathbb{E}(Y_p\circ\mathsf{T}^n)=\mathbb{E}(Y_n)+\mathbb{E}(Y_p)\,.$$

The proof of (7.72) and (7.73) follows from the Fekete Lemma (see Lemma 7.41) applied to $u_n=\mathbb{E}(Y_n)$, resp. $u_n=\mathbb{E}\left[Y_n\,|\,\mathcal{I}\right]$ and Proposition 7.9. ∎

For subadditive or additive sequences of functions, $\{Y_n,\ n\in\mathbb{N}^\star\}$, we are interested in finding conditions under which the quantity Y_n/n has a limit as n tends to infinity. To obtain this, we bound from above $\limsup_{n\to\infty} Y_n/n$ by a quantity close to $\liminf_{n\to\infty} Y_n/n$. More precisely, we choose a subdivision of $\{0,\ldots,n\}$ denoted by $\{i_s,\ 0\le s\le t+1\}$, where $0=i_0<i_1<i_2<\ldots<i_{t+1}=n$, and apply the subadditivity of the sequence $\{Y_n,\ n\in\mathbb{N}^\star\}$ to obtain

$$Y_n\le\sum_{s=0}^{t} Y_{i_{s+1}-i_s}\circ\mathsf{T}^{i_s}\,. \tag{7.74}$$

A convenient choice of the subdivision will lead to the following technical lemma, which exhibits the upper-bound $\liminf_{n\to\infty} Y_n/n$. This is a key lemma that allows us to prove the convergence of additive (resp. subadditive) sequences of functions: see Birkhoff's (resp. Kingman's) theorem below.

Lemma 7.43. *Let* $(\Omega,\mathcal{B},\mathbb{Q},\mathsf{T})$ *be a dynamical system and denote by* $\mathcal{I}$ *the* σ*-algebra of invariant subsets with respect to* T*. Let* $\{Y_n,\ n\in\mathbb{N}^\star\}$ *be a sequence of subadditive functions such that* $\mathbb{E}(Y_1^+)<\infty$*. Denote by* $F_\infty:=\liminf_{n\to\infty} Y_n/n$ *and* $F_q:=F_\infty\vee(-q)$*. Then,* F_q *is an integrable and invariant random variable for the dynamical system* $(\Omega,\mathcal{B},\mathbb{Q},\mathsf{T})$ *and*

$$\lim_{n\to\infty} n^{-1}\mathbb{E}\left[Y_n\,|\,\mathcal{I}\right]=\inf_{n\in\mathbb{N}^\star} n^{-1}\mathbb{E}\left[Y_n\,|\,\mathcal{I}\right]\le\liminf_{n}(n^{-1}Y_n)\,,\quad \mathbb{Q}\text{-a.s.} \tag{7.75}$$

Proof. Note that $-q\le F_q\le F_\infty^+$. Moreover, by Fatou's lemma,

$$\begin{aligned}\mathbb{E}(F_\infty^+)&=\mathbb{E}(\liminf_{n\to\infty} Y_n^+/n)\le\liminf_{n\to\infty}\mathbb{E}(Y_n^+/n)\\&\le\mathbb{E}((Y_1^++Y_1^+\circ\mathsf{T}+\ldots+Y_1^+\circ\mathsf{T}^{n-1})/n)=\mathbb{E}(Y_1^+)<\infty\,,\end{aligned}$$

so that F_q is integrable, that is $\mathbb{E}(|F_q|) < \infty$. The subadditivity of the sequence $\{Y_n,\, n \in \mathbb{N}^\star\}$ implies

$$Y_{n+1}/n \leq (Y_1 + Y_n \circ \mathsf{T})/n\,.$$

By taking the lim inf as n goes to infinity, we obtain that $F_\infty \circ \mathsf{T} - F_\infty \geq 0$, $\mathbb{Q}$-a.s. This implies that F_q also satisfies $F_q \circ \mathsf{T} - F_q \geq 0$, $\mathbb{Q}$-a.s. But $0 = \mathbb{E}(F_q \circ T) - \mathbb{E}(F_q) = \mathbb{E}(F_q \circ T - F_q)$ where we have used that T is measure preserving and F_q is integrable. Finally, $\mathbb{Q}(F_q \circ T = F_q) = 1$, i.e., F_q is an invariant random variable for the dynamical system $(\Omega, \mathcal{B}, \mathbb{Q}, \mathsf{T})$.

The first equality of (7.75) follows from (7.73). Now, writing $G_n = Y_n - nF_0$, we have that $\{G_n,\, n \in \mathbb{N}^\star\}$ is a sequence of subadditive functions and since F_0 is integrable and $\mathcal{I}$-measurable, it is equivalent to show (7.75) for G_n or Y_n. But, $\liminf_{n\to\infty} G_n/n = F_\infty - F_0 \leq 0$, $\mathbb{Q}$-a.s. and thus it is sufficient to prove (7.75) under the assumption that $F_\infty = \liminf Y_n/n \leq 0$, $\mathbb{Q}$-a.s.

But (7.75) is a consequence of the following inequality: for all $p, q \in \mathbb{N}^\star$ and all $n \geq p$, $\mathbb{Q}$-a.s.

$$Y_n \leq n(F_q + 1/q) + \sum_{k=1}^{n-p} [(-F_q + Y_1^+)\mathbb{1}_{B(p,q)}] \circ \mathsf{T}^k + \sum_{k=n-p+1}^{n} (-F_q + Y_1^+) \circ \mathsf{T}^k\,, \tag{7.76}$$

where $B(p,q) := \{Y_\ell/\ell > F_q + 1/q : \text{for all } 1 \leq \ell \leq p\}$. To see why (7.76) implies (7.75), take the conditional distribution on both sides of (7.76) so that $\mathbb{Q}$-a.s.,

$$\mathbb{E}[Y_n \,|\, \mathcal{I}] \leq n(F_q + 1/q) + (n-p+1)\mathbb{E}\left[(Y_1^+ - F_q)\mathbb{1}_{B(p,q)} \,\middle|\, \mathcal{I}\right] + p\mathbb{E}\left[(Y_1^+ - F_q) \,\middle|\, \mathcal{I}\right]\,.$$

By dividing by n and letting n go to infinity,

$$\limsup_{n\to\infty} \mathbb{E}[Y_n/n \,|\, \mathcal{I}] \leq F_q + 1/q + \mathbb{E}\left[(Y_1^+ - F_q)\mathbb{1}_{B(p,q)} \,\middle|\, \mathcal{I}\right], \quad \mathbb{Q}\text{-a.s.}$$

Now, let p go to infinity. Since $B_{p+1,q} \subset B(p,q)$, we get $\lim_{p\to\infty} \mathbb{1}_{B(p,q)} = \mathbb{1}_{\cap_p B(p,q)}$ On the other hand, $\bigcap_{p=1}^\infty B(p,q) = \{\liminf_{\ell\to\infty} \ell^{-1} Y_\ell \geq F_q + 1/q\}$ so that $\lim_{p\to\infty} \mathbb{1}_{B(p,q)} = 0$, $\mathbb{Q}$-a.s. The conditional dominated convergence theorem finally implies that

$$\limsup_{n\to\infty} \mathbb{E}[Y_n/n \,|\, \mathcal{I}] \leq F_\infty \vee (-q) + 1/q, \quad \mathbb{Q}\text{-a.s.}$$

Since q is arbitrary in $\mathbb{N}^\star$,

$$\limsup_{n} \mathbb{E}[Y_n/n \,|\, \mathcal{I}] \leq F_\infty = \liminf_{n} (Y_n/n), \quad \mathbb{Q}\text{-a.s.} \tag{7.77}$$

The proof of (7.75) follows. It remains to prove (7.76). Since F_q is invariant and $\mathbb{Q}(F_\infty \leq 0) = 1$, we have $\mathbb{Q}$-a.s.,

$$F_q \circ \mathsf{T}^k = F_q \leq 0, \quad \text{for all } k \in \mathbb{N}^*. \tag{7.78}$$

We now use this property to conveniently choose a (random) subdivision of $\{1, \ldots, n\}$. Define $\{I_s,\, s \geq 0\}$ inductively in the following way:

(i) $I_0 = 0$.

(ii) If $\mathrm{T}^{I_s} \in B(n - I_s, q)$, set $I_{s+1} = I_s + 1$.

(iii) If $\mathrm{T}^{I_s} \notin B(n - I_s, q)$, then define

$$K := \inf\left\{k \in \{1, \ldots, n - I_s\} : Y_k \circ \mathrm{T}^{I_s}/k \leq F_q \circ \mathrm{T}^{I_s} + 1/q\right\}$$

and set $I_{s+1} = I_s + K$.

Let T such that $I_{T+1} = n$. Since $0 = I_0 < I_1 < \ldots < I_{T+1} = n$, (7.74) yields

$$\begin{aligned} Y_n &\leq \sum_{s=0}^{T} Y_{I_{s+1}-I_s} \circ \mathrm{T}^{I_s} \leq \sum_{s=0}^{T} Y_1^+ \circ \mathrm{T}^{I_s} \mathbb{1}\{\mathrm{T}^{I_s} \in B(n - I_s, q)\} \\ &+ (F_q + 1/q) \sum_{s=0}^{T} (I_{s+1} - I_s)(1 - \mathbb{1}\{\mathrm{T}^{I_s} \in B(n - I_s, q)\}) \\ &\leq n(F_q + 1/q) + \sum_{s=0}^{T} [(Y_1^+ - F_q) \circ \mathrm{T}^{I_s} \ \mathbb{1}\{\mathrm{T}^{I_s} \in B(n - I_s, q)\}] \,. \end{aligned}$$

Since $Y_1^+ \geq 0$ and $-F_q \geq 0$ and $B(p', q) \subset B(p, q)$ for $p \leq p'$,

$$\begin{aligned} &(Y_1^+ - F_q) \circ \mathrm{T}^{I_s} \ \mathbb{1}\{\mathrm{T}^{I_s} \in B(n - I_s, q)\} \\ &\leq (Y_1^+ - F_q) \circ \mathrm{T}^{I_s} \ \mathbb{1}\{\mathrm{T}^{I_s} \in B(p, q), I_s \leq n - p\} + (Y_1^+ - F_q) \circ \mathrm{T}^{I_s} \ \mathbb{1}\{I_s > n - p\} \end{aligned}$$

so that $\mathbb{Q}$-a.s.,

$$Y_n \leq n(F_q + 1/q) + \sum_{k=0}^{n-p} [(Y_1^+ - F_q)\mathbb{1}_{B(p,q)}] \circ \mathrm{T}^k + \sum_{k=n-p+1}^{n} (Y_1^+ - F_q) \circ \mathrm{T}^k$$

which completes the proof of (7.76). ■

Theorem 7.44 (Birkhoff's ergodic theorem). *Let $(\Omega, \mathcal{B}, \mathbb{Q}, \mathrm{T})$ be a dynamical system and denote by $\mathcal{I}$ the σ-algebra of invariant subsets with respect to T. Let $Y \in \mathrm{L}^p(\Omega, \mathcal{B}, \mathbb{Q})$, $p \geq 1$. Then,*

$$\frac{1}{n} \sum_{k=0}^{n-1} Y \circ \mathrm{T}^k \to_n \mathbb{E}\left[Y \mid \mathcal{I}\right], \quad \mathbb{Q}\text{-a.s. } \textit{and in } \mathrm{L}^p.$$

Proof. Assume that $\mathbb{E}|Y| < \infty$. Set $M_n(Y) = n^{-1} \sum_{k=0}^{n-1} Y \circ \mathrm{T}^k$. According to (7.75),

$$\mathbb{E}[Y \mid \mathcal{I}] = \lim_{n\to\infty} \mathbb{E}[M_n(Y) \mid \mathcal{I}] \leq \liminf_{n\to\infty} M_n(Y), \quad \mathbb{Q}\text{-a.s.} \tag{7.79}$$

Since $\{-Y_n, \, n \in \mathbb{N}^\star\}$ is also subadditive, one can replace Y by $-Y$ and $M_n(Y)$ by $-M_n(Y)$ in the above inequality. Then,

$$\limsup_{n\to\infty} M_n(Y) \leq \mathbb{E}[Y \mid \mathcal{I}], \quad \mathbb{Q}\text{-a.s.} \tag{7.80}$$

Therefore $M_n(Y) \to_n \mathbb{E}[Y \mid \mathcal{I}]$, $\mathbb{Q}$-a.s.

Consider now the L^p-convergence. If Y is bounded, $M_n(Y)$ is also bounded and by the dominated convergence theorem, $M_n(Y)$ converges to $\mathbb{E}[Y \mid \mathcal{I}]$ in L^p. Assume now that $\|Y\|_p < \infty$. For all bounded $\bar{Y}$, consider the decomposition

$$|M_n(Y) - \mathbb{E}[Y \mid \mathcal{I}]| \leq |M_n(Y) - M_n(\bar{Y})| + |M_n(\bar{Y}) - \mathbb{E}[\bar{Y} \mid \mathcal{I}]| + \mathbb{E}[|\bar{Y} - Y| \mid \mathcal{I}]$$

and note that $\|\mathbb{E}[|\bar{Y} - Y| \mid \mathcal{I}]\|_p \leq \|\bar{Y} - Y\|_p$ and

$$\||M_n(Y) - M_n(\bar{Y})\|_p \leq n^{-1} \sum_{k=0}^{n-1} \|(Y - \bar{Y}) \circ \mathrm{T}^k\|_p = \|Y - \bar{Y}\|_p \,.$$

Therefore, $\limsup_{n\to\infty} \|M_n(Y) - \mathbb{E}\left[Y \mid \mathcal{I}\right]\|_p \leq 2\|\bar{Y} - Y\|_p$. The proof follows by density of bounded functions in L^p. ■

Theorem 7.45 (Kingman's subadditive ergodic theorem). *Let $(\Omega, \mathcal{B}, \mathbb{Q}, \mathrm{T})$ be a dynamical system and $\{Y_n,\ n \in \mathbb{N}^\star\}$ a sequence of subadditive functions such that $\mathbb{E}(Y_1^+) < \infty$. Then,*

$$\lim_{n\to\infty} n^{-1} Y_n \text{ exists } \mathbb{Q}\text{-a.s. and } \lim_{n\to\infty} n^{-1} Y_n = \inf_{n\in\mathbb{N}^\star} n^{-1} \mathbb{E}[Y_n \mid \mathcal{I}],\ \mathbb{Q}\text{-a.s.} \tag{7.81}$$

$$\text{The convergence also holds in } L^1 \text{ if and only if } \inf_{n\in\mathbb{N}^\star} n^{-1}\mathbb{E}(Y_n) > -\infty. \tag{7.82}$$

Proof. We preface the proof by the following lemma, allowing us to restrict our attention to nonpositive subadditive sequences.

Lemma 7.46. *Assume that for any sequence $\{Y_n,\ n \in \mathbb{N}^\star\}$ of integrable nonpositive subadditive functions,* (7.81) *(resp.* (7.82)*) holds. Then* (7.81) *(resp.* (7.82)*) holds for any sequence of subadditive functions.*

Proof. First, let $\{Y_n,\ n \in \mathbb{N}^\star\}$ be a sequence of subadditive functions. If we set $Y_n' = Y_n - \sum_{k=0}^{n-1} Y_1 \circ \mathrm{T}^k$, then by subadditivity, $Y_n' \leq 0$, $\mathbb{Q}$-a.s., and by Proposition 7.9, $n^{-1}\mathbb{E}\left[Y_n \mid \mathcal{I}\right] = n^{-1}\mathbb{E}\left[Y_n' \mid \mathcal{I}\right] + \mathbb{E}\left[Y_1 \mid \mathcal{I}\right]$. By Theorem 7.44, denoting $M_n(Y_1) = n^{-1}\sum_{k=0}^{n-1} Y_1 \circ \mathrm{T}^k$, we get $\lim_{n\to\infty} M_n(Y_1) = \mathbb{E}\left[Y_1 \mid \mathcal{I}\right]$, $\mathbb{Q}$-a.s., which implies that $\lim_{n\infty} n^{-1} Y_n$ exists $\mathbb{Q}$-a.s. and

$$\begin{aligned}
\lim_{n\to\infty} n^{-1} Y_n &= \lim_{n\to\infty} n^{-1} Y_n' + \lim_{n\to\infty} M_n(Y_1) \\
&= \inf_{n\in\mathbb{N}^\star} n^{-1}\mathbb{E}\left[Y_n' \mid \mathcal{I}\right] + \mathbb{E}\left[Y_1 \mid \mathcal{I}\right] = \inf_{n\in\mathbb{N}^\star} n^{-1}\mathbb{E}\left[Y_n \mid \mathcal{I}\right], \quad \mathbb{Q}\text{-a.s.}
\end{aligned}$$

Second, we have $n^{-1}\mathbb{E}(Y_n) = n^{-1}\mathbb{E}(Y_n') + \mathbb{E}(Y_1)$ so $\inf_{n\in\mathbb{N}^\star} n^{-1}\mathbb{E}(Y_n) > -\infty$ if and only if $\inf_{n\in\mathbb{N}^\star} n^{-1}\mathbb{E}(Y_n') > -\infty$. Since $Y_n = Y_n' + M_n(Y_1)$ and, by Birkhoff's ergodic theorem (Theorem 7.44), $\{M_n(Y_1), n \in \mathbb{N}^\star\}$ converges in L^1, $\{Y_n, n \in \mathbb{N}^\star\}$ converges in L^1 if and only if $\{Y_n, n \in \mathbb{N}^\star\}$ converges in L^1 . ■

We first consider the case for which, for any $n \in \mathbb{N}^\star$, Y_n is integrable and nonpositive. The subadditivity property shows that the sequence $\{Y_n,\ n \in \mathbb{N}^*\}$ is nonincreasing. Thus, for all $m > 0$ and all $n > m$,

$$\sum_{k=0}^{n-1} Y_m \circ \mathrm{T}^k \geq \sum_{k=0}^{n-1} Y_{m+k} - \sum_{k=0}^{n-1} Y_k = \sum_{k=n}^{n+m-1} Y_k - \sum_{k=0}^{m-1} Y_k \geq m Y_{n+m-1} \,,$$

where the last inequality follows from the fact that $\{Y_n,\ n \in \mathbb{N}^*\}$ is nonincreasing and Y_k is nonpositive. Then, the Birkhoff's ergodic theorem (Theorem 7.44) applied to the function Y_m implies

$$\limsup_{n\to\infty} n^{-1}Y_n \le m^{-1}\mathbb{E}\left[Y_m \,|\, \mathcal{I}\right], \quad \mathbb{Q}\text{-a.s.} \tag{7.83}$$

for any $m \in \mathbb{N}^\star$. Combining with (7.75), we obtain that

$$\limsup_{n\to\infty} n^{-1}Y_n \le \inf_{m\in\mathbb{N}^\star} m^{-1}\mathbb{E}\left[Y_m \,|\, \mathcal{I}\right] \le \liminf_{n\to\infty} n^{-1}Y_n\,, \quad \mathbb{Q}\text{-a.s.}\,,$$

which completes the proof of (7.81).

We now consider (7.82). Obviously, if $\{n^{-1}Y_n, n \in \mathbb{N}^\star\}$ converges in L^1, $\{n^{-1}\mathbb{E}(Y_n), n \in \mathbb{N}^\star\}$ converges and $\inf_{n\in\mathbb{N}} n^{-1}\mathbb{E}(Y_n) > -\infty$. Conversely, assume that $\inf_{n\in\mathbb{N}^\star} n^{-1}\mathbb{E}(Y_n) \ge -M$. By (7.83) $L := \limsup_{n\to\infty} n^{-1}Y_n \le m^{-1}\mathbb{E}\left[Y_m \mid \mathcal{I}\right]$ for all $m \in \mathbb{N}$, which implies $\mathbb{E}(m^{-1}Y_m - L) \ge 0$. We have, by the Fatou Lemma,

$$\mathbb{E}\left[-L\right] = \mathbb{E}\left[\lim_{n\to\infty} n^{-1}(-Y_n)\right] \le \liminf_{n\to\infty} n^{-1}\mathbb{E}\left[-Y_n\right] \le M\,,$$

showing that $L \in \mathrm{L}^1$. Therefore, using $|a| = 2a^+ - a$, we get

$$\mathbb{E}(|n^{-1}Y_n - L|) \le 2\mathbb{E}\left[(n^{-1}Y_n - L)^+\right] - \mathbb{E}(n^{-1}Y_n - L) \le 2\mathbb{E}\left[(n^{-1}Y_n - L)^+\right].$$

Since $(n^{-1}Y_n - L)^+ \le -L = |L|$, the dominated convergence theorem shows that $\lim_{n\to\infty} \mathbb{E}|n^{-1}Y_n - L| = 0$, concluding the proof when Y_n is integrable for all $n \in \mathbb{N}^\star$. This concludes the proof of Kingman's subadditive theorem when Y_n is integrable for any $n \in \mathbb{N}^\star$.

We finally consider the case where $\mathbb{E}(Y_1^+) < \infty$. Lemma 7.42 shows that for any $n \in \mathbb{N}^\star$, $\mathbb{E}(Y_n^+) \le n\mathbb{E}(Y_1^+)$ and $\lim_{n\to\infty} n^{-1}\mathbb{E}\left[Y_n \mid \mathcal{I}\right] = \inf_{n\in\mathbb{N}^\star} n^{-1}\mathbb{E}\left[Y_n \mid \mathcal{I}\right] =: M$, $\mathbb{Q}$-a.s. Now we set, for $a > 0$, $Y_n^a := Y_n \mathbb{1}_{\{M \ge -a\}}$. Then,

$$n^{-1}\mathbb{E}\left[Y_n^a \mid \mathcal{I}\right] \ge \mathbb{1}_{\{M\ge -a\}} n^{-1}\mathbb{E}\left[Y_n \mid \mathcal{I}\right] \ge M\mathbb{1}_{\{M\ge -a\}} \ge -a\,,$$

showing that $\mathbb{E}(Y_n^a) \ge -na$. This implies that $Y_n^a \in \mathrm{L}^1$ for any $n \in \mathbb{N}^\star$. On the set $\{M \ge -a\} \in \mathcal{I}$, we have by applying (7.81) for integrable random variables

$$\begin{aligned}\lim_{n\to\infty} n^{-1}Y_n &= \lim_{n\to\infty} n^{-1}Y_n^a = \inf_{n\in\mathbb{N}^\star} n^{-1}\mathbb{E}\left[Y_n^a \mid \mathcal{I}\right] \\ &= \mathbb{1}_{\{M\ge -a\}} \inf_{n\in\mathbb{N}^\star} n^{-1}\mathbb{E}\left[Y_n \mid \mathcal{I}\right] = \inf_{n\in\mathbb{N}^\star} n^{-1}\mathbb{E}\left[Y_n \mid \mathcal{I}\right], \quad \mathbb{Q}\text{-a.s.}\end{aligned}$$

Therefore $\lim_{n\to\infty} n^{-1}Y_n = \inf_{n\in\mathbb{N}^\star} n^{-1}\mathbb{E}\left[Y_n \mid \mathcal{I}\right]$ $\mathbb{Q}$-a.s. on $\{M > -\infty\}$. Now, for any $a > 0$, define $\bar{Y}_n^a := Y_n \vee -na$. The sequence $\{\bar{Y}_n^a,\ n \in \mathbb{N}\}$ is subadditive for $(\Omega, \mathcal{B}, \mathbb{Q}, \mathrm{T})$ and for any $n \in \mathbb{N}^\star$, $\bar{Y}_n^a$ is integrable. By the Kingman's subadditive theorem for integrable sequences, $\lim_{n\to\infty} \bar{Y}_n^a$ exists $\mathbb{Q}$-a.s. and

$$\lim_{n\to\infty} n^{-1}\bar{Y}_n^a = \inf_{n\in\mathbb{N}^\star} n^{-1}\mathbb{E}\left[\bar{Y}_n^a \mid \mathcal{I}\right], \quad \mathbb{Q}\text{-a.s.}$$

Since $Y_n \le \bar{Y}_n^a$, we get

$$\limsup_{n\to\infty} n^{-1} Y_n \le \lim_{n\to\infty} n^{-1}\bar{Y}_n^a = \inf_{n\in\mathbb{N}^\star} n^{-1}\mathbb{E}\left[\bar{Y}_n^a \,\middle|\, \mathcal{I}\right] ,$$

showing that, for any $m \in \mathbb{N}^\star$, $\limsup_{n\to\infty} n^{-1}Y_n \le m^{-1}\mathbb{E}\left[Y_m \vee -ma \,\middle|\, \mathcal{I}\right]$. By the monotone convergence theorem we get $\lim_{a\to\infty}\mathbb{E}\left[Y_m \vee -ma \,\middle|\, \mathcal{I}\right] = \mathbb{E}\left[Y_m \,\middle|\, \mathcal{I}\right]$, $\mathbb{Q}$-a.s., which implies that,

$$\limsup_{n\to\infty} n^{-1}Y_n \le \inf_{m\in\mathbb{N}^\star} m^{-1}\mathbb{E}\left[Y_m \,\middle|\, \mathcal{I}\right] .$$

In particular, on the set $\{\inf_{m\in\mathbb{N}^\star} m^{-1}\mathbb{E}\left[Y_m \,\middle|\, \mathcal{I}\right] = -\infty\}$, $\limsup_{n\to\infty} n^{-1}Y_n = -\infty$. This concludes the proof. ■

Corollary 7.47 (Generalized Birkhoff's ergodic theorem). *Let $(\Omega, \mathcal{B}, \mathbb{Q}, \mathsf{T})$ be a dynamical system and Y be a random variable such that $\mathbb{E}(Y^+) < \infty$ or $\mathbb{E}(Y^-) < \infty$. Then,*

$$\frac{1}{n}\sum_{k=0}^{n-1} Y \circ \mathsf{T}^k \to_n \mathbb{E}\left[Y \,\middle|\, \mathcal{I}\right] , \quad \mathbb{Q}\text{-a.s.}$$

Proof. If $\mathbb{E}(Y^+) < \infty$, we apply Theorem 7.45 to $Y_n = \sum_{k=0}^{n-1} Y \circ \mathsf{T}^k$. If $E(Y^-) < \infty$, we replace Y by $-Y$. ■

Exercises

7.1. Suppose $(\Omega, \mathcal{B}, \mathbb{P})$ is a probability space and T and S are measure-preserving mappings. Show that $\mathsf{T} \circ \mathsf{S}$ is measure-preserving.

7.2. Show that the set $\mathcal{I}$ is a σ-field.

7.3. We use the notations and assumptions of Example 7.4. Show that $\limsup_{n\to\infty} X_n$, $\liminf_{n\to\infty} X_n$, $\limsup_{n\to\infty} n^{-1}(X_0 + \ldots + X_{n-1})$ or $\liminf_{n\to\infty} n^{-1}(X_0 + \ldots + X_{n-1})$ are invariant random variables.

7.4. We use the notations and assumptions of Example 7.5. Show that $\tilde{\mathsf{S}}$ and $\tilde{\mathsf{S}}^{-1}$ are both invariant for $\mathbb{Q}$.

7.5. Let $(\Omega, \mathcal{B}, \mathbb{Q}, \mathsf{T})$ be a dynamical system and T be a measure-preserving mapping.

(a) Let $\mathcal{B}_0$ be a semi-algebra generating $\mathcal{B}$. Show that for all $E \in \mathcal{B}$ and $\varepsilon > 0$, there exists $A \in \mathcal{B}_0$ such that $\mathbb{Q}(E \ominus A) \le \varepsilon$.

(b) Assume that E is an invariant set. Set $B = \mathsf{T}^{-n}A$. Prove that $\mathbb{Q}(E \ominus B) \le \varepsilon$.

(c) Prove that $\mathbb{Q}(E \ominus (A \cap B)) \le 2\varepsilon$ and that $|\mathbb{Q}(E) - \mathbb{Q}(A \cap B)| \le 2\varepsilon$.

(d) Assume that for some $n \in \mathbb{N}$, $\mathbb{Q}(A \cap B) = \mathbb{Q}(A)\mathbb{Q}(B)$. Prove that

$$|\mathbb{Q}(E) - \mathbb{Q}^2(E)| \le |\mathbb{Q}(E) - \mathbb{Q}(A \cap B)| - |\mathbb{Q}^2(A) - \mathbb{Q}^2(E)| \le 4\varepsilon .$$

(e) Assume that there exits a semi-algebra $\mathcal{B}_0$ generating $\mathcal{B}$ such that, for any $A \in \mathcal{B}_0$, there exists an integer n such that $\mathbb{Q}(A \cap \mathrm{T}^{-n} A) = \mathbb{Q}(A)\mathbb{Q}(\mathrm{T}^{-n} A)$. Show that T is ergodic.

7.6. Let $(\Omega, \mathcal{B}, \mathbb{Q}, \mathrm{T})$ be a dynamical system. Let $F \in \mathcal{B}$. Show that $\{\omega, \mathrm{T}^t(\omega) \in F \text{i.o.}\}$ (where i.o. means infinitely often) is an invariant event.

7.7. Let $\Omega = \mathsf{X}^{\mathbb{N}}$ be the set of X-valued sequences indexed by $\mathbb{N}$ (see Example 7.4) Define by $\mathrm{S} : \Omega \to \Omega$ the shift operator. For all $t \in \mathbb{N}$, define by $X_t : \Omega \to \mathsf{X}$ the coordinate functions (or natural projections): for $\omega = \{\omega_t,\ t \in \mathbb{N}\}$, $X_t(\omega) = \omega_t$. Let $\nu \in \mathbb{M}_1(\mathcal{X})$ be a probability measure. Define the probability of the cylinder A by

$$\mathbb{Q}(A) = \int \cdots \int \prod_{i=0}^{n} \nu(\mathrm{d}x_i) \mathbb{1}_A(x_0, \ldots, x_{n-1}) . \tag{7.84}$$

Then $\mathbb{Q}$ is consistently defined and countably additive on the semialgebra $\mathcal{C}_0$ and hence extends to a probability measure on $\mathcal{X}^{\otimes \mathbb{N}} = \sigma(\mathcal{C}_0)$.

(a) Show that S is measure-preserving for $\mathbb{Q}$.

(b) Show that S is ergodic.

7.8. Let $(\Omega, \mathcal{B}, \mathbb{Q}, \mathrm{T})$ be a dynamical system and T be a measure-preserving mapping. The system is said to be *mixing*, if for any pair of events A, $B \in \mathcal{B}$,

$$\lim_{n \to \infty} \mathbb{Q}(A \cap \mathrm{T}^{-k} B) = \mathbb{Q}(A)\mathbb{Q}(B) . \tag{7.85}$$

(a) Show that if a system is mixing, then it is ergodic.

(b) Assume that (7.85) is satisfied for any pair of events $A, B \in \mathcal{B}_0$, where $\mathcal{B}_0$ is a semi-algebra generating $\mathcal{B}$. Show that the system is mixing.

7.9. Let $(\Omega, \mathcal{B}, \mathbb{Q}, \mathrm{T})$ be a dynamical system and T be a measure-preserving mapping. We assume that for any pair of events $A, B \in \mathcal{B}$,

$$\lim_{n \to \infty} \frac{1}{n} \sum_{k=1}^{n} \mathbb{P}(\mathrm{T}^{-k} A \cap B) = \mathbb{P}(A)\mathbb{P}(B) . \tag{7.86}$$

(a) Show that if (7.86) is satisfied, then $(\Omega, \mathcal{B}, \mathbb{Q}, \mathrm{T})$ is ergodic.

(b) Show that, for any $B \in \mathcal{B}$,

$$\left| \frac{1}{n} \sum_{k=1}^{n} \mathbb{P}(\mathrm{T}^{-k} A \cap B) - \mathbb{P}(A)\mathbb{P}(B) \right| \leq \int \left| \frac{1}{n} \sum_{k=1}^{n} \mathbb{1}_A \circ \mathrm{T}^k - \mathbb{P}(A) \right| \mathrm{d}\mathbb{P} .$$

(c) Show that if $(\Omega, \mathcal{B}, \mathbb{Q}, \mathrm{T})$ is ergodic, then (7.86) holds.

7.10. We use the notations and assumptions of Example 7.4. Show that if $\{X_t,\ t \in \mathbb{N}\}$ is a stationary, ergodic and square integrable sequence, then

$$n^{-1} \sum_{k=0}^{n-1} \mathrm{Cov}(X_0, X_k) \to 0 .$$

7.11. Let $(\Omega, \mathcal{B}, \mathbb{Q}, \mathrm{T})$ be a dynamical system and $\mathcal{I}$ be the invariant σ-field.

(a) Show that if $Y = Y \circ T$, $\mathbb{P}$-a.s. there exists an $\mathcal{I}$-measurable random variable Z such that $Y = Z$, $\mathbb{P}$-a.s.

(b) Prove that if $\mathbb{1}_A = \mathbb{1}_A \circ T$, $\mathbb{P}$-a.s., there exists $B \in \mathcal{I}$ such that $\mathbb{1}_A = \mathbb{1}_B$, $\mathbb{P}$-a.s.

(c) Show that $\mathcal{J} = \{A \in \mathcal{B} : \mathbb{P}(A \ominus \mathrm{T}^{-1}A) = 0\}$ is a σ-field.

7.12. Let $(\mathsf{X}^{\mathbb{Z}}, \mathcal{X}^{\otimes\mathbb{Z}}, \mathbb{P}, \mathrm{S})$ be an ergodic dynamical system and let f be a measurable function from $(\mathsf{X}^{\mathbb{Z}}, \mathcal{X}^{\otimes\mathbb{Z}})$ to $(\tilde{\mathsf{X}}, \tilde{\mathcal{X}})$. We associate to f the measurable function F defined, for $\omega = \{\omega_t,\, t \in \mathbb{Z}\} \in \mathsf{X}^{\mathbb{Z}}$, by

$$F(\omega_{-\infty:\infty}) = \tilde{\omega}_{-\infty:\infty}, \quad \text{where for all } n \in \mathbb{Z}, \quad \tilde{\omega}_n = f \circ \mathrm{S}^n(\omega_{-\infty:\infty}).$$

Write $\tilde{\mathbb{P}} = \mathbb{P} \circ F^{-1}$ and denote by $\tilde{\mathrm{S}}$ the shift operator on $(\tilde{\mathsf{X}}^{\mathbb{Z}}, \tilde{\mathcal{X}}^{\otimes\mathbb{Z}})$, that is, for $\omega = \{\omega_t,\, t \in \mathbb{Z}\} \in \tilde{X}^{\mathbb{Z}}$, $\tilde{\mathrm{S}}(\omega)$ is the sequence with coordinates $\left[\tilde{\mathrm{S}}(\omega)\right]_t = \omega_{t+1}$, for all $t \in \mathbb{Z}$.

(a) Show that $\tilde{\mathrm{S}}$ is measure-preserving with respect to $(\tilde{\mathsf{X}}^{\mathbb{Z}}, \tilde{\mathcal{X}}^{\otimes\mathbb{Z}}, \tilde{\mathbb{P}})$.

(b) Show that $(\tilde{\mathsf{X}}^{\mathbb{Z}}, \tilde{\mathcal{X}}^{\otimes\mathbb{Z}}, \tilde{\mathbb{P}}, \tilde{\mathrm{S}})$ is an ergodic dynamical system.

7.13. Let $\{Z_t,\, t \in \mathbb{Z}\}$ be a strong white noise with zero mean and unit-variance and $\{\phi_k,\, k \in \mathbb{Z}\}$ be a deterministic sequence satisfying $\sum_{k=-\infty}^{\infty} \phi_k^2 < \infty$. Show that the process $Y_t = \sum_{k=-\infty}^{\infty} \phi_k Z_{t-k}$ is strict-sense stationary and ergodic.

7.14 (ARCH, consistent estimators of the asymptotic variance). We use the notations and the assumptions of Example 7.26.

(a) Show that

$$\hat{A} = \frac{1}{n} \sum_{t=1}^{n} \varphi_t \varphi_t', \quad \hat{B} = \frac{1}{n} \sum_{t=1}^{n} \hat{\sigma}_t^4 \varphi_1 \varphi_t',$$

where $\hat{\sigma}_t^2 = \varphi_t' \theta_n$ are consistent estimators of matrices A and B defined in (7.30).

(b) Show that the fourth order moment of the process $\varepsilon_t = X_t / \sigma_t$ is also consistently estimated by $\hat{\mu}_4 = n^{-1} \sum_{t=1}^{n} (X_t / \hat{\sigma}_t)^4$.

(c) Propose a consistent estimator of the asymptotic variance of the OLS estimator.

7.15 (ARCH(1)). We use the notations and the assumptions of Example 7.26, with $p = 1$.

(a) Show that the OLS estimator is consistent if $\kappa_\varepsilon \alpha_{*,1}^2 < 1$ and is asymptotically normal if $\mu_8 \alpha_{*,1}^4 < 1$, where $\mu_8 = \mathbb{E}\left[\varepsilon_0^8\right]$.

(b) Show that

$$\mathbb{E}_\pi[X_t^2] = \frac{\alpha_{*,0}}{1 - \alpha_{*,1}}, \quad \mathbb{E}_\pi[X_t^4] = \kappa_\varepsilon \mathbb{E}_\pi[\sigma_t^4] = \frac{\alpha_{*,0}^2 (1 + \alpha_{*1})}{(1 - \kappa_\varepsilon \alpha_{*,1}^2)(1 - \alpha_{*,1})} \kappa_\varepsilon.$$

(c) Compute the matrices A and B, defined in (7.30).

(d) Determine the restrictions for the parameter space when the distribution of ε_0 is Gaussian and Student-t (with different number of degrees-of-freedom).

7.16 (Generalized Hoeffding Lemma). Let V be a random variable, $\mathcal{G}$ be a σ-algebra such that

$$\mathbb{E}\left[V \mid \mathcal{G}\right] = 0 \quad \text{and} \quad A \le V \le B\,, \quad \mathbb{P}\text{-a.s.}$$

where A, B are two $\mathcal{G}$-measurable random variables.

(a) Show that $\mathbb{E}\left[e^{sV} \mid \mathcal{G}\right] \le e^{\phi(s(B-A))}$ where $P = -A/(B-A)$, and $\phi(u) = -Pu + \ln(1 - P + Pe^u)$.

(b) Show that $\phi(u) \le u^2/8$ and that, for any $s > 0$, $\mathbb{E}\left[e^{sV} \mid \mathcal{G}\right] \le e^{(s^2/8)(B-A)^2}$.

7.17. Let P be a Markov kernel satisfying the Doeblin condition (Definition 6.9) for some $m \ge 1$ and denote by π the unique stationary distribution of the Markov chain. Let $S_n = \sum_{i=0}^{n-1} f(X_i)$ where $f \in \mathbb{F}_b(\mathcal{X}, \mathcal{X})$. This exercise proves the Hoeffding inequality for uniformly ergodic Markov chains by following the original approach of Glynn and Ormoneit (2002). More specifically, we will show that for all $\delta > 0$ and all $n > 2\,\mathrm{osc}\,(f)\,m/(\varepsilon\delta)$,

$$\mathbb{P}_x(S_n - \mathbb{E}_\pi(S_n) \ge n\delta) \le \exp\left(-\frac{\varepsilon^2(n\delta - 2\,\mathrm{osc}\,(f)\,m/\varepsilon)^2}{2n\,\mathrm{osc}^2\,(f)\,m^2}\right).$$

(a) Show that $\hat{f}(x) = \sum_{n=0}^{\infty}\{P^n f(x) - \pi(f)\}$ is a solution of the Poisson equation associated to f and $|\hat{f}|_\infty \le \mathrm{osc}\,(f)\,m/\varepsilon$.

(b) Let $D_i := \hat{f}(X_i) - P\hat{f}(X_{i-1})$. Show that $\{D_n,\ n \in \mathbb{N}\}$ is a martingale and that

$$\mathbb{E}_x\left[\exp(s\{S_n - n\pi(f)\})\right] \le \exp(2s|\hat{f}|_\infty) \cdot \mathbb{E}_x\left[\exp\left(s\sum_{i=1}^{n} D_i\right)\right],$$

where $s > 0$.

(c) By applying the Generalized Hoeffding Lemma (Exercise 7.16), show that

$$\mathbb{E}\left[\exp(sD_n) \mid X_0,\ \ldots, X_{n-1}\right] \le \exp\left(s^2|\hat{f}|_\infty^2/2\right).$$

(d) Show that for all $s > 0$,

$$\mathbb{P}_x(S_n - n\pi(f) \ge n\delta) \le \exp(-sn\delta)\mathbb{E}_x\left[\exp(s\{S_n - n\pi(f)\})\right].$$

7.18. Let $\{X_t,\ t \in \mathbb{N}\}$ be a Markov chain on $\mathbb{R}^d$ with Markov kernel P and initial distribution ξ. We assume that P is uniformly geometrically ergodic (see Definition 6.7), i.e., there exists a probability distribution π and constants $C < \infty$ and $\rho \in [0,1)$ such that $\sup_{x \in \mathsf{X}} \|P^n(x,\cdot) - \pi\|_{\mathrm{TV}} \le C\rho^n$, for all $n \in \mathbb{N}$. We assume that π admits a twice continuous differentiable density denoted $f(\cdot)$ with respect to Lebesgue measure on $\mathbb{R}^d$. Let $K : \mathbb{R}^d \to \mathbb{R}$ be a bounded even positive function satisfying

$$\int_{\mathbb{R}^d} xK(x)\mathrm{d}x = 0 \quad \text{and} \quad \int_{\mathbb{R}^d} x^2 K(x)\mathrm{d}x < \infty\,.$$

Let $\{h_n,\ n \in \mathbb{N}\}$ be a sequence of nonincreasing positive numbers such that

$\lim_{n\to\infty} h_n = 0$ and $\lim_{n\to\infty} nh_n = \infty$. Consider the kernel estimator of the marginal density given by

$$\hat{f}_n(x) = (nh_n^d)^{-1} \sum_{k=0}^{n-1} K\left(h_n^{-1}(x - X_k)\right) = n^{-1} \sum_{k=1}^{n-1} K_{h_n}(x - X_k) , \tag{7.87}$$

where for $h > 0$, $K_h(x) = h^{-d} K(h^{-1}x)$.

(a) Show that for all $\xi \in \mathbb{M}_1(\mathcal{X})$, $\|\xi P^n - \pi\|_{\mathrm{TV}} \leq C\rho^n$.

(b) Show that for all $A \in \sigma(X_\ell, 0 \leq \ell \leq k)$ and $B \in \sigma(X_\ell, \ell \geq k+n)$,

$$|\mathbb{P}_\xi(A \cap B) - \mathbb{P}_\xi(A)\mathbb{P}_\xi(B)| \leq 2C\rho^n \mathbb{P}_\xi(A) .$$

(c) Show that for all $x \in \mathbb{R}^d$, $\lim_{h\to 0} |\int K_h(x-y) f(y) \mathrm{d}y - f(x)| = 0$.

(d) Show that for all $k \in \mathbb{N}$, $\mathbb{E}_\xi[K_{h_n}(x - X_k)] \leq C\rho^k |K|_\infty h_n^{-1}$.

(e) Show that, if $\lim_{n\to\infty} nh_n = \infty$, then the sequence of estimators $\{\hat{f}_n(x), n \in \mathbb{N}\}$ is asymptotically unbiased. Compute the variance of the estimator.

(f) Find conditions upon which the mean square error of the kernel density estimator converges to zero.

7.19 (Testing equality of the means). Let $\{X_t,\ t \in \mathbb{N}\}$ be a Markov chain on X satisfying A7.27. Let $f \in \mathbb{F}(\mathsf{X}, \mathcal{X})$ be a function such that $|f|_{V^{1/2}} < \infty$ and $\sigma^2(f) := \pi(\hat{f}^2) - \pi(P\hat{f})^2 > 0$, where $\hat{f}$ is a solution of the Poisson equation Definition 7.29.

(a) Show that for any initial distribution ξ such that $\xi(V) < \infty$, $(n^{-1/2}\sum_{t=0}^{n-1}\{f(X_t) - \pi(f)\}, n^{-1/2}\sum_{t=0}^{n-1}\{f(X_{t+n}) - \pi(f)\})$ are asymptotically normal with zero mean and covariance matrix $\sigma^2(f)\mathrm{I}_2$.

(b) Denote by $Y_t = f(X_t)$, $\bar{Y}_{n,1} = n^{-1}\sum_{t=0}^{n-1} Y_t$ and $\bar{Y}_{n,2} = n^{-1}\sum_{t=n}^{2n-1} Y_t$. Show that, for any initial distribution ξ such that $\xi(V) < \infty$, $\sqrt{n}(\bar{Y}_{n,1} - \bar{Y}_{n,2})/\sqrt{2\sigma^2(f)}$ is asymptotically Gaussian with zero-mean and unit-variance.

(c) Show that this result continues to hold if we replace $\sigma^2(f)$ by $\hat{\sigma}_n^2(f)$, where $\{\hat{\sigma}_n^2(f)\}$ is a consistent sequence of estimators of $\sigma^2(f)$.

(d) Use this result to construct a test of the equality of the mean.

This test has been proposed by Geweke (1992) and is implemented in the R package `coda`.

7.20 (Non-overlapping batch means). Let $\{X_t,\ t \in \mathbb{N}\}$ be a Markov chain on X. The non-overlapping batch mean is a classical estimator of the variance of a sample mean. This estimator is constructed by dividing the observations $\{X_0, \ldots, X_{nm-1}\}$ into m contiguous non-overlapping batches of size n (for simplicity, it is assumed that the number of observations is a multiple of the batch length). For a Borel function f, put $Y_t = f(X_t)$. Define by $\bar{Y}_{n,i} = n^{-1}\sum_{k=(i-1)n}^{in-1} Y_k$ the sample mean of the i-th batch, $i \in \{0, \ldots, m-1\}$ and $\bar{Y}_n = n^{-1}\sum_{j=0}^{n-1} Y_j$ the sample mean of the observations. We consider the following estimator of the variance

$$\hat{\sigma}_{m,n}^2 = \frac{n}{m(m-1)} \sum_{i=0}^{m-1} \{\bar{Y}_{n,i} - \bar{Y}_n\}^2 \tag{7.88}$$

which may be interpreted as the empirical variance of the blocks. In the sequel, we assume that the Markov chain satisfies A7.27. Assume that $|f|_{V^{1/2}} < \infty$ and $\sigma^2(f) := \pi(\hat{f}^2) - \pi(P\hat{f})^2 > 0$, where $\hat{f}$ is a solution of the Poisson equation Definition 7.29.

(a) Show that, for any initial distribution $\xi \in \mathbb{M}_1(\mathcal{X})$ such that $\xi(V) < \infty$, $\sqrt{n}\{\bar{X}_{n,0} - \pi(f), \dots, \bar{X}_{n,m-1} - \pi(f)\}$ is asymptotically normal with zero-mean and variance $\sigma^2(f)\mathrm{I}_m$.

(b) Show that $\hat{\sigma}^2_{m,n} \Rightarrow_{\mathbb{P}_\xi} \sigma^2(f)\chi^2_{m-1}/(m-1)$, where χ^2_{m-1} is a chi-square random variable with $(m-1)$ degrees of freedom.

(c) Show that, if $\{\chi^2_m,\ m \in \mathbb{N}\}$ is a sequence of chi-square random variables with m degrees of freedom, then $\sigma^2\chi^2_m/m \to_{\mathbb{P}} \sigma^2$ and $\sqrt{m}(\sigma^2\chi^2_m/ - \sigma^2) \Rightarrow_{\mathbb{P}} \mathrm{N}(0, 2\sigma^4)$.

(d) Argue that we can get (asymptotically) as close as we want to σ^2 by increasing the batch size.

7.21. Let $|||\cdot|||$ be any matrix norm. Let $\{C_k,\ k \in \mathbb{N}\}$ be a sequence of strict-sense stationary and ergodic matrices in $\mathbb{M}_d(\mathbb{R})$. Set $u_k = \ln |||C_1 C_2 \dots C_k|||$.

(a) Show that the sequence $\{u_n,\ n \in \mathbb{N}\}$ is subadditive, i.e., for all $n, p \geq 1$, $u_{n+p} \leq u_n + u_p$.

(b) Show that the sequence $\{u_n/n, n \in \mathbb{N}\}$ is convergent.

(c) Let $|||\cdot|||_a$ and $|||\cdot|||_b$ be two submultiplicative norms; assume that $\alpha|||A|||_a \leq |||A|||_b \leq \beta|||A|||_a$ for some $0 < \alpha < \beta$. For any product $C_1 C_2 \cdots C_k$, show that $\alpha^{1/k}|||C_1 C_2 \cdots C_k|||_a^{1/k} \leq |||C_1 C_2 \cdots C_k|||_b^{1/k} \leq \beta^{1/k}|||C_1 C_2 \cdots C_k|||_a^{1/k}$.

(d) Show that the sequence $\{|||C_1 C_2 \cdots C_k|||^{1/k}, k \in \mathbb{N}\}$ is convergent and that this limit is independent of the choice of the matrix norm.

7.22 (Stationary distribution estimator (after Lacour (2008))). Let P be a V-geometrically ergodic kernel on $\mathbb{R}$ having a unique invariant distribution with density π with respect to Lebesgue measure. Given $\{X_0, \dots, X_{n-1}\}$, we are willing to construct an estimator of π on a compact interval of $\mathbb{R}$. Without loss of generality, this compact set is assumed to be equal to $[0,1]$ and, from now on, f denotes the stationary density multiplied by the indicator function of $[0,1]$, $f = \pi\mathbb{1}_{[0,1]}$. For simplicity, it is assumed that P has a density p with respect to Lebesgue measure and that $|p|_\infty = \sup_{(x,y)\in\mathsf{X}\times\mathsf{X}} p(x,y) < \infty$. Assume in addition that $|f|_\infty < \infty$, which implies in particular that $f \in \mathrm{L}^2([0,1], \mathrm{Leb})$ (we simply put $\mathrm{L}^2 := \mathrm{L}^2([0,1], \mathrm{Leb})$).

Let $\{\phi_n,\ n \in \mathbb{N}\}$ be an orthonormal sequence of functions of L^2 and $\{D_m,\ m \in \mathbb{N}\}$ be a sequence of integers. We consider the sequence of subspaces $S_m = \mathrm{span}(\phi_0, \phi_1, \dots, \phi_{D_m})$ the subspace of L^2 spanned by the functions $(\phi_0, \phi_1, \dots, \phi_{D_m})$. We assume that there exists $r_0 \in \mathbb{R}$ such that for all m, $\phi_m \leq r_0$ where

$$\phi_m = \frac{1}{\sqrt{D_m}} \sup_{g \in S_m \setminus \{0\}} \frac{|g|_\infty}{|g|_{\mathrm{L}^2}},$$

where $|g|^2_{\mathrm{L}^2} = \int_0^1 g^2(x)\mathrm{Leb}(dx)$. Equivalently, for all $g \in S_m$,

$$|g|_\infty \leq r_0\sqrt{D_m}\,|g|_{\mathrm{L}^2}\ . \tag{7.89}$$

Define $\gamma_n(g) = n^{-1}\sum_{i=0}^{n-1}\{|g|^2_{\mathrm{L}^2} - 2g(X_i)\}$. Note that $\mathbb{E}_\pi[\gamma_n(g)] = |g-f|^2_{\mathrm{L}^2} - |f|^2_{\mathrm{L}^2}$. We consider the estimator $\hat{f}_m$ of f given by

$$\hat{f}_m = \arg\min_{g\in S_m} \gamma_n(g) . \tag{7.90}$$

The dependence of this estimator on n is not indicated for brevity.

(a) Show that (7.89) is satisfied by the models spanned by the *Histogram basis*: $S_m := \mathrm{span}(\varphi_1,\ldots,\varphi_{2^m})$ with $\varphi_j = 2^{m/2}\mathbb{1}_{\left[\frac{j-1}{2^m},\frac{j}{2^m}\right)}$ for $j=1,\ldots,2^m$. Here $D_m = 2^m$, $\mathcal{M}_n = \{1,\ldots,\lfloor \ln n/2\ln 2\rfloor\}$.

(b) Show that (7.89) is satisfied by the *Trigonometric basis*: $S_m = \mathrm{span}(\varphi_0,\ldots,\varphi_{m-1})$ with $\varphi_0(x) = \mathbb{1}_{[0,1]}(x)$, $\varphi_{2j} = \sqrt{2}\cos(2\pi jx)\mathbb{1}_{[0,1]}(x)$, $\varphi_{2j-1} = \sqrt{2}\sin(2\pi jx)\mathbb{1}_{[0,1]}(x)$ for $j\geq 1$.

(c) Show that $\hat{f}_m = \sum_{\ell=1}^{D_m}\hat{\beta}_\ell\varphi_\ell$ with $\hat{\beta}_\ell = n^{-1}\sum_{k=0}^{n-1}\varphi_\ell(X_k)$. Denote by ξ the initial distribution of the Markov chain. Show that $\mathbb{E}_\xi[\hat{\beta}_\ell] := \beta_\ell = \pi(\phi_\ell)$.

(d) Show that $f_m := \sum_{\ell=1}^{D_m}\beta_\ell\varphi_\ell$ is the orthogonal projection of f on S_m.

(e) Show that the mean integrated square error $\mathbb{E}_\xi\left[|f-\hat{f}_m|^2_{\mathrm{L}^2}\right]$ may be expressed as $\mathbb{E}_\xi[|f-\hat{f}_m|^2_{\mathrm{L}^2}] = |f-f_m|^2_{\mathrm{L}^2} + \sum_{\ell=1}^{D_m}\mathrm{Var}_\xi(\hat{\beta}_\ell)$.

(f) Assume that the initial distribution ξ has a density (also denoted ξ) with respect to the Lebesgue measure, and that $|\xi\mathbb{1}_{[0,1]}|_\infty < \infty$; show that there exists a constant $c<\infty$, such that for all $\ell\in\mathbb{N}^*$ and all $n\in\mathbb{N}^*$, $\mathrm{Var}_\xi(\hat{\beta}_\ell)\leq cn^{-1}$ and

$$\mathbb{E}_\xi[|f-\hat{f}_m||^2_{\mathrm{L}^2}] \leq |f-f_m|^2_{\mathrm{L}^2} + c\frac{D_m}{n} .$$

7.23. We use the notations and the assumptions of Exercise 7.22. By applying (7.54) with $p=2$, for any function $\phi\in\mathbb{F}(\mathsf{X},\mathcal{X})$,

$$\mathrm{Var}_\xi(n^{-1}\sum_{k=0}^{n-1}\{\phi(X_k)-\pi(\phi)\}) \leq c_2^2|\phi_\ell|^2_{\mathrm{V}^{1/2}}\xi(V)n^{-1} , \tag{7.91}$$

where c_2 is a universal constant. As shown below, this inequality can be improved: we may replace, up to constants, $|\phi|_{\mathrm{V}^{1/2}}$ by $|\phi|_{\mathrm{L}^2}$. Since $(\varphi_\ell)_{\ell=1}^{D_m}$ is an orthonormal basis of S_m, $|\phi_\ell|_{\mathrm{L}^2} = 1$ whereas $|\phi_\ell|_{\mathrm{V}^{1/2}} \leq |\phi_\ell|_\infty$ is bounded under (7.89) by $r_0\sqrt{D_m}$. In the sequel, ϕ is a function belonging to $\mathrm{L}^2([0,1],\mathrm{Leb})$.

(a) Show that

$$\mathbb{E}_\xi\left[\left(n^{-1}\sum_{k=0}^{n-1}\{\phi(X_k)-\pi(\phi)\}\right)^2\right]$$
$$\leq 2n^{-2}\sum_{k=1}^{n}\mathbb{E}_\xi[\{\hat{\phi}(X_k)-P\hat{\phi}(X_{k-1})\}^2] + 4n^{-2}\left\{\mathbb{E}_\xi[\hat{\phi}^2(X_0)] + \mathbb{E}_\xi[\hat{\phi}^2(X_n)]\right\} ,$$

where $\hat{\phi}$ is a solution of the Poisson equation $\hat{\phi} - P\hat{\phi} = \phi - \pi(\phi)$.

(b) Show that, for any $k \geq 1$, $|P^k\phi(x) - \pi(\phi)| \leq C\rho^{k-1} |P\phi|_\infty V(x)$, where the constant C and ρ are defined in A7.27. Show that $|P\phi|_\infty \leq |p|_\infty |\phi|_{L^2}$.

(c) Show that for all $x \in [0,1]$, $|\hat{\phi}(x)| \leq |\phi(x)| + C_1 |\phi|_{L^2}$, where $C_1 = C(1-\rho)^{-1}|p|_\infty + |f|_\infty$.

(d) Assume that the initial distribution ξ has a density (also denoted ξ) with respect to Lebesgue measure, bounded on $[0,1]$. Show that $\mathbb{E}_\xi[\hat{\phi}^2(X_0)] \leq C_2 |\phi|^2_{L^2}$ for some constant C_2 (give an explicit expression of $C-2$).

(e) Show that, for any integer $r \geq 1$, P^r has a density denoted p_r with respect to the Lebesgue measure and that $|p_r|_\infty \leq |p|_\infty$.

(f) Show that, for $r \geq 1$, $\mathbb{E}_\xi[\hat{\phi}^2(X_r)] \leq C_3 |\phi|^2_{L^2}$ (give an explicit expression of C_3).

(g) Conclude.

7.24 (The Glynn and Henderson density estimator). Let P be a Markov kernel on $\mathsf{X} \times \mathcal{X}$. Assume that P admits a stationary distribution π and that P is Harris ergodic (see Definition 7.21). Assume in addition that for some integer $m \geq 1$ and some reference measure $\mu \in \mathbb{M}_+(\mathcal{X})$, $P^m(x, \mathrm{d}y) = p(x,y)\mu(\mathrm{d}y)$.

(a) Show that π is absolutely continuous with respect to π.

(b) Show that $y \mapsto \int_{\mathsf{X}} \pi(\mathrm{d}x)p(x,y)$ is a version of the stationary density.

To avoid having to state pointwise convergence properties as holding for μ-almost every y, we will in the sequel define $\pi(y) = \int_{\mathsf{X}} \pi(\mathrm{d}x)p(x,y)$; this new definition changes π on a set of μ-measure 0 and is merely a theoretical convenience. The above result suggests to estimate $\pi(y)$ using the look-ahead estimator $\hat{\pi}_n(y) = n^{-1}\sum_{k=0}^{n-1} p(X_k, y)$.

(c) Show that, for all $x \in \mathsf{X}$ and $y \in \mathsf{X}$, $\hat{\pi}_n(y) \stackrel{\mathbb{P}_x\text{-a.s.}}{\longrightarrow} \pi(y)$.

(d) Show that $0 \leq |\hat{\pi}_n(y) - \pi(y)| \leq \hat{\pi}_n(y) + \pi(y)$ and that, for all $x \in \mathsf{X}$, $\mathbb{P}_x$-a.s.,

$$\lim_{n\to\infty} \int_{\mathsf{X}} \{\pi_n(y) + \pi(y)\}\mathrm{d}y = \int_{\mathsf{X}} \lim_{n\to\infty} \{\pi_n(y) + \pi(y)\}\mathrm{d}y\,.$$

(e) Show that, for all $x \in \mathsf{X}$, $\mathbb{P}_x$-a.s.,

$$\lim_{n\to\infty} \int_{\mathsf{X}} |\hat{\pi}_n(y) - \pi(y)|\mathrm{d}y = \int \lim_{n\to\infty} |\hat{\pi}_n(y) - \pi(y)|\mathrm{d}y = 0\,.$$

(f) Show that, for all $x \in \mathsf{X}$, $\lim_{n\to\infty} \mathbb{E}_x\left[\int_{\mathsf{X}} |\hat{\pi}_n(y) - \pi(y)|\mathrm{d}y\right] = 0$.

(g) State conditions upon which this estimator is asymptotically normal, and give an expression of the limiting variance.

Chapter 8

Inference for Markovian Models

8.1 Likelihood inference

Assume that $(X_1, \ldots, X_n)$ is an observation from a collection of distributions $(\mathbb{P}_\theta, \theta \in \Theta)$ that depends on a parameter θ ranging over a set Θ. A popular method for finding an estimator $\hat{\theta}_n = \hat{\theta}_n(X_1, \ldots, X_n)$ is to maximize a criterion of the type $\theta \mapsto M_n(\theta)$ over the parameter set Θ. For notational simplicity, the dependence of M_n in the observations is implicit. Such an estimator is often called an *M-estimator*. When $(X_1, \ldots, X_n)$ are i.i.d., the criterion $\theta \mapsto M_n(\theta)$ is often chosen to be the sample average of some known functions $m^\theta : \mathsf{X} \to \mathbb{R}$,

$$M_n(\theta) = n^{-1} \sum_{t=1}^{n} m^\theta(X_t) . \tag{8.1}$$

Sometimes, the maximizing value is computed by setting a derivative (or the set of partial derivatives in the multidimensional case) equal to zero. In such a case, the estimator $\hat{\theta}_n$ is defined as the solution of a system of equations of the type $\Psi_n(\theta) = 0$. For instance, if θ is d-dimensional, then Ψ_n typically has d coordinate functions $\Psi_n(\theta) = (\Psi_n^{(1)}(\theta), \ldots, \Psi_n^{(d)}(\theta))$. Such estimators are often referred to as Z-estimators. In the i.i.d. case, $\Psi_n(\theta)$ is often chosen to be $\Psi_n(\theta) = n^{-1} \sum_{t=1}^{n} \psi^\theta(X_t)$ where $\psi^\theta := (\psi_1^\theta, \ldots, \psi_d^\theta)$. In such a case, $\Psi_n(\theta) = 0$ is shorthand for the system of equations

$$n^{-1} \sum_{t=1}^{n} \psi_j^\theta(X_t) = 0, \; j = 1, 2, \ldots, d. \tag{8.2}$$

In many examples ψ_j^θ is taken to be the j-th partial derivative of the function m_θ, $\psi_j^\theta = \partial m^\theta / \partial \theta_j$.

In this section, we consider *maximum likelihood estimators*. Suppose first that $X_1, \ldots, X_n$ are i.i.d. with a common density $x \mapsto \mathrm{p}^\theta(x)$. Then the maximum likelihood estimator maximizes the likelihood or, equivalently, the log-likelihood, given by

$$\theta \mapsto n^{-1} \sum_{t=1}^{n} \ln \mathrm{p}^\theta(X_t) .$$

Thus, a maximum likelihood estimator is an M-estimator with $m^\theta = \ln p^\theta$. If the density is partially differentiable with respect to θ for each $x \in \mathsf{X}$, then the maximum likelihood estimator also solves (8.2), with

$$\psi^{\theta}(x) = \nabla \ln \mathrm{p}^{\theta}(x) = \left(\frac{\partial \ln \mathrm{p}^{\theta}(x)}{\partial \theta_1}, \ldots, \frac{\partial \ln \mathrm{p}^{\theta}(x)}{\partial \theta_d} \right)' .$$

This approach extends directly to the Markov chain context. Let (X, d) be a Polish space equipped with its Borel sigma-field $\mathcal{X}$ and p be a positive integer. Consider $\{Q^{\theta}, \theta \in \Theta\}$, a family of Markov kernels on $\mathsf{X}^p \times \mathcal{X}$ indexed by $\theta \in \Theta$ where (Θ, d) is a compact metric space. Assume that all $(\theta, x) \in \Theta \times \mathsf{X}^p$, $Q^{\theta}(x; \cdot)$ is dominated by some σ-finite measure μ on $(\mathsf{X}, \mathcal{X})$ and denote by $q^{\theta}(x; \cdot)$ its Radon-Nikodym derivative: $q^{\theta}(x; y) = \mathrm{d}Q^{\theta}(x; \cdot)/\mathrm{d}\mu(y)$. If $\{X_t,\ t \in \mathbb{N}\}$ is a Markov chain of order p associated to the Markov kernel Q^{θ}, then, the conditional distribution of the observations $(X_p, \ldots, X_n)$ given $X_0, \ldots, X_{p-1}$ has a density with respect to the product measure $\mu^{\otimes(n-p+1)}$ given by

$$x_{0:n} \mapsto \mathrm{p}^{\theta}(x_{p:n} | x_{0:p-1}) = \prod_{t=p}^{n} q^{\theta}(x_{t-p}, \ldots, x_{t-1}; x_t) . \tag{8.3}$$

The conditional Maximum Likelihood Estimator $\hat{\theta}_n$, based on the observations $X_{0:n}$ and on the family of likelihood functions (8.3) is defined by

$$\hat{\theta}_n \in \arg\max_{\theta \in \Theta} \ln \mathrm{p}^{\theta}(X_{p:n} | X_{0:p-1}) . \tag{8.4}$$

Starting at θ_0, optimization algorithms generate a sequence $\{\theta_k,\ k \in \mathbb{N}\}$ that terminate when either no progress can be made or when it is apparent that a solution point has been approximated with a sufficient accuracy. Prior knowledge of the model and the data set may be used to choose θ_0 to be a reasonable estimate of the solution; otherwise the starting point must be set by the algorithm, either by a systematic approach, or in some arbitrary manner. In deciding how to move from one iterate θ_k to the other, the algorithm will use information about the likelihood, and possibly information gathered by earlier iterates $\theta_0, \theta_1, \ldots, \theta_{k-1}$.

A classical strategy is to use a line search: at each iteration, the algorithm chooses a direction δ_k, and searches along this direction from the current iterate θ_k for a new iterate with a higher function value. The distance to move along the search direction δ_k can be found by solving approximately the following one-dimensional optimization problem

$$\max_{\alpha > 0} \ln \mathrm{p}^{\theta_k + \alpha \delta_k}(X_{p:n} | X_{0:p-1}) .$$

By solving exactly this optimization problem, we would derive the maximum benefit from moving along the direction δ_k, but an exact maximization may be quite expensive and is usually worthless. Instead, the line search algorithm will typically generate a limited number of trial step lengths, until it finds one that loosely approximates the maximum.

The most obvious choice for search direction is the *steepest ascent* function $\theta \mapsto \nabla \ln \mathrm{p}^{\theta}(X_{p:n} | X_{0:p-1})$:

$$\delta_k = \nabla \ln \mathrm{p}^{\theta_k}(X_{p:n} | X_{0:p-1}) .$$

The gradient $\nabla \ln \mathrm{p}^{\theta_k}(X_{p:n}|X_{0:p-1})$ is the *score* vector computed at $\theta = \theta_k$. Along all the directions we could move from θ_k, it is the one along which $\theta \mapsto \ln \mathrm{p}^{\theta}(X_{p:n}|X_{0:p-1})$ increases most rapidly. One of the advantage of the steepest ascent algorithm is that it requires only the calculation of the score function; however, it is known to be slow on difficult problems. Line search methods may use other search directions than the gradient. In general, any direction that makes a positive angle with $\nabla \ln \mathrm{p}^{\theta_k}(X_{p:n}|X_{0:p-1})$ is guaranteed to produce an increase in the likelihood, provided that the step-size is chosen appropriately. Another important search direction (and perhaps the most important one of all), is the *Newton direction*. This direction is derived from the second order Taylor series approximation of log-likelihood $\mathrm{p}^{\theta}(X_{p:n}|X_{0:p-1})$ in the neighborhood of the current fit θ_k. Assuming that $\nabla^2 \ln \mathrm{p}^{\theta_k}(X_{p:n}|X_{0:p-1})$ is positive definite, the Newton direction is

$$\delta_k^N = -\left[\nabla^2 \ln \mathrm{p}^{\theta_k}(X_{p:n}|X_{0:p-1})\right]^{-1} \nabla \ln \mathrm{p}^{\theta_k}(X_{p:n}|X_{0:p-1}) . \tag{8.5}$$

Unlike the steepest ascent algorithm, there is a "natural" step length of 1 associated with the Newton direction. Most line search implementations of Newton's method use the unit step $\alpha = 1$ where it is possible to adjust α only when it does not produce a satisfactory increase in the value of the likelihood.

When $\nabla^2 \ln \mathrm{p}^{\theta_k}(X_{p:n}|X_{0:p-1})$ is not positive definite, the Newton direction may not even be defined, since $\left(\nabla^2 \ln \mathrm{p}^{\theta_k}(X_{p:n}|X_{0:p-1})\right)^{-1}$ may not exist. Even when it is defined, it may not satisfy the descent property $\nabla \ln \mathrm{p}^{\theta_k}(X_{p:n}|X_{0:p-1})'\delta_k^N < 0$, in which case it is unsuitable as a search direction. In these situations, line search methods modify the definition of the search direction δ_k^N to make it satisfy the descent condition while retaining the benefit of the second-order information.

Methods that use the Newton direction have a fast rate of local convergence, typically quadratic (at least when the score and the Hessian can be estimated exactly). After a neighborhood of the solution is reached, convergence to a high accuracy solution occurs most often in a few iterations.

The main drawback of the Newton direction is the need for the Hessian (also called the *information matrix*). Explicit computation of this matrix of second derivatives can sometimes be cumbersome. Finite-difference and automatic differentiation techniques may be useful in avoiding the need to calculate second derivatives. *Quasi-Newton* search directions provide an attractive alternative to Newton's method in that they do not require computation of the Hessian and yet still attain a superlinear rate of convergence. In place of the true Hessian, Quasi-Newton strategies use an approximation that is updated after each step. This update uses the changes in the gradient to gain information about the second derivatives along the search direction.

Before going further, consider some examples:

Example 8.1 (Discrete-valued Markov chain). In this case, the set X is at most countable; the dominating measure ν on X is chosen to be the counting measure. If $\mathsf{X} = \{1, \ldots, K\}$ is finite and if the parameters $\theta = (\theta_{i,j})_{K \times K}$ are the transition probabilities, $\theta_{i,j} = q^{\theta}(i,j)$ for all $(i,j) \in \{1, \ldots, K\}^2$, then the maximum likelihood estimator is given by

$$\hat{\theta}_{n,i,j} = \frac{\sum_{t=1}^{n-1} \mathbb{1}_{i,j}(X_t, X_{t+1})}{\sum_{t=1}^{n} \mathbb{1}_i(X_t)} .$$

◇

Example 8.2 (AR models). Consider the AR(p) model, $X_t = \sum_{i=1}^{p} \phi_i X_{t-i} + \sigma Z_t$, where $\{Z_t,\, t \in \mathbb{N}\}$ is a strong white Gaussian noise. Here $\theta = (\phi_1, \ldots, \phi_p, \sigma^2)$ and Θ is a compact subset of $\mathbb{R}^p \times \mathbb{R}_+$. The conditional log-likelihood of the observations may be written as

$$\ln \mathrm{p}^{\theta}(X_{p:n} | X_{0:p-1}) = -\frac{n-p+1}{2} \ln(2\pi\sigma^2) - \frac{1}{2\sigma^2} \sum_{t=p}^{n} (X_t - \sum_{j=1}^{p} \phi_j X_{t-j})^2 .$$

The conditional likelihood function is therefore a quadratic function in the regression parameter and a convex function in the innovation variance. The maximum likelihood estimator can be computed in such a case explicitly as follows:

$$\begin{pmatrix} \hat{\phi}_{n,1} \\ \hat{\phi}_{n,2} \\ \vdots \\ \hat{\phi}_{n,p} \end{pmatrix} = \hat{\Gamma}_n^{-1} \begin{pmatrix} n^{-1} \sum_{t=p}^{n} X_t X_{t-1} \\ n^{-1} \sum_{t=p}^{n} X_t X_{t-2} \\ \vdots \\ n^{-1} \sum_{t=p}^{n} X_t X_{t-p} \end{pmatrix} \tag{8.6}$$

where $\hat{\Gamma}_n$ is the $(p \times p)$ empirical covariance matrix for which the i,j-th element is defined by $\hat{\Gamma}_n(i,j) = n^{-1} \sum_{t=p}^{n} X_{t-i} X_{t-j}$. The maximum likelihood estimator for the innovation variance is given by

$$\hat{\sigma}_n^2 = \frac{1}{n-p+1} \sum_{t=p}^{n} \left(X_t - \sum_{j=1}^{p} \hat{\phi}_{n,j} X_{t-j} \right)^2 . \tag{8.7}$$

◇

Example 8.3 (Threshold autoregressive models). Consider a two-regime threshold autoregressive TAR model

$$X_t = \left\{ \phi_{1,0} + \sum_{j=1}^{p_1} \phi_{1,j} X_{t-j} + \sigma_1 Z_t \right\} \mathbb{1}_{\{X_{t-d} \leq r\}} + \left\{ \phi_{2,0} + \sum_{j=1}^{p_2} \phi_{2,j} X_{t-j} + \sigma_2 Z_t \right\} \mathbb{1}_{\{X_{t-d} > r\}} , \tag{8.8}$$

where $\{Z_t,\, t \in \mathbb{N}\}$ is a strong Gaussian white noise with zero mean and variance 1. While the delay d may be theoretically larger than the maximum autoregressive order p, this is seldom the case in practice; hence we assume that $d \leq p$. The normal error assumption implies that for every $t \geq 0$, the conditional distribution of X_t given $X_{t-1}, \ldots, X_{t-p}$ is normal, where $p = \max(p_1, p_2)$. To estimate the parameters, we use the likelihood conditional to the p initial values. Assume first that the threshold parameter r and the delay parameter d are known. In such a case, the observations may be split into two parts according to whether or not $X_{t-d} \leq r$. Denote by $n_1(r,d)$ the number of observations in the lower regimes. With the observations in the lower regime, we can regress X_t on $X_{t-1}, \ldots, X_{t-p_1}$, to find estimates of the autoregressive coefficients $\{\hat{\phi}_{1,j}(r,d)\}_{j=1}^{p_1}$ and the noise variance estimate (see (8.6) and (8.7))

$$\sigma_1^2(r,d) = \frac{1}{n_1(r,d)} \sum_{t=p+1}^{n} \left(X_t - \sum_{j=1}^{p_1} \hat{\phi}_{1,j}(r,d) X_{t-j} \right)^2 \mathbb{1}\{X_{t-d} \leq r\} . \tag{8.9}$$

Similarly, using the $n_2(r,d)$ observations in the upper regime (note that $n - p = n_1(r,d) + n_2(r,d)$), we can find estimates of the autoregressive coefficients $\{\hat{\phi}_{2,j}(r,d)\}_{j=1}^{p_2}$ and the noise variance, $\sigma_2^2(r,d)$. To estimate (r,d), we consider the profile likelihood

$$\ell(r,d) = -\frac{n-p+1}{2}\{1+\ln(2\pi)\} - \sum_{i=1}^{2} \frac{n_i(r,d)}{2} \ln(\hat{\sigma}_i{}^2(r,d)) , \tag{8.10}$$

which is maximized over r and d. The optimization need only be performed with r over the observations $X_{p+1}, \ldots, X_n$ and $d \in \{1, \ldots, p\}$. Note indeed that, for a given d, the functions are constant if r lies in between two consecutive observations. To avoid inconsistent estimators, we typically restrict the search of the threshold to be between two predetermined quantiles of $X_{p+1}, \ldots, X_t$. ◇

Example 8.4 (Functional autoregressive models). Consider a functional autoregressive model (see Example 6.21)

$$X_t = a^\theta(X_{t-1}, \ldots, X_{t-p}) + b^\theta(X_{t-1}, \ldots, X_{t-q}) Z_t , \tag{8.11}$$

where $\{Z_t,\ t \in \mathbb{N}\}$ is a strong Gaussian noise with zero mean and unit variance, $\theta \in \Theta \subset \mathbb{R}^d$ and for each $\theta \in \Theta$. This model includes many of the popular non linear autoregressive models such as

(a) *ARCH(p)* $a^\theta(x_1, \ldots, x_p) = 0$ and

$$b^\theta(x_1, \ldots, x_q) = \sqrt{\alpha_0 + \alpha_1 x_1^2 + \cdots + \alpha_q x_q^2} , \tag{8.12}$$

with $\theta = (\alpha_0, \alpha_1, \ldots, \alpha_q) \in \Theta$ a compact subset of $\mathbb{R}_+^* \times \mathbb{R}_+^q$.

(b) *Autoregressive models with ARCH errors* $a^\theta(x_1, \ldots, x_p) = \phi_1 x_1 + \cdots + \phi_p x_p$ and

$$b^\theta(x_1, \ldots, x_q) = \sqrt{\alpha_0 + \sum_{j=1}^{\ell} \alpha_j (x_j - a^\theta(x_{j+1}, x_{j+2}, \ldots, x_{j+p}))^2} , \tag{8.13}$$

where $\theta = (\phi_1, \phi_2, \ldots, \phi_p, \alpha_0, \alpha_1, \ldots, \alpha_\ell) \in \mathbb{R}^p \times \mathbb{R}_+^{\ell+1}$, $\alpha_0 > 0$ and $q = p + \ell$.

(c) *Logistic Smooth Transition AR(p)* $b^\theta(x_1,\ldots,x_p) = 1$ and

$$a^\theta(x_1,\ldots,x_p) = \mu_1 + \sum_{j=1}^{p} \phi_{1,j}x_j + \left(\mu_2 + \sum_{j=1}^{p} \phi_{2,j}x_j\right)(1+\exp(-\gamma(x_d - c)))^{-1} \tag{8.14}$$

where $d \in \{1,\ldots,j\}$ and $\theta = (\mu_i, \phi_{i,j}, i \in \{1,2\}, j = \{1,\ldots,p\}, \gamma, c) \in \Theta$, a compact subset of $\mathbb{R}^{2(p+1)+2}$.

(d) *Exponential Smooth Transition AR(p)* $b^\theta(x_1,\ldots,x_p) = 1$ and

$$a^\theta(x_1,\ldots,x_p) = \mu_1 + \sum_{j=1}^{p} \phi_{2,j}x_j + \left(\mu_2 + \sum_{j=1}^{p} \phi_{2,j}x_j\right)(1-\exp(-\gamma(x_d - c)^2)) \tag{8.15}$$

where $d \in \{1,\ldots,j\}$ and $\theta = (\mu_i, \phi_{i,j}, i \in \{1,2\}, j = \{1,\ldots,p\}, \gamma, c) \in \Theta$, a compact subset of $\mathbb{R}^{2(p+1)+2}$.

The functional autoregressive models also cover situations for which the regression function is not regular, like the 2 regimes scalar threshold autoregressive model, $b^\theta(x_1,\ldots,x_p) = \sigma_1^2 \mathbb{1}_{\{x_d \le c\}} + \sigma_2^2 \mathbb{1}_{\{x_d > c\}}$, $d \in \{1,\ldots,p\}$ and

$$a^\theta(x_1,\ldots,x_p) = \mathbb{1}_{\{x_d \le c\}} \sum_{j=1}^{p} \phi_{1,j}x_j + \mathbb{1}_{\{x_d > c\}} \sum_{j=1}^{p} \phi_{2,j}x_j\,, \tag{8.16}$$

where $\theta = (\phi_{i,j}, \sigma_i^2, i \in \{1,2\}, j \in \{1,\ldots,p\}) \in \Theta \subset \mathbb{R}^{2(p+1)}$.

Assume that for any $\theta \in \Theta$ and $(x_1,\ldots,x_p) \in \mathbb{R}^p$, $b^\theta(x_1,\ldots,x_p) > 0$. In such a case, the conditional likelihood of the observations may be written as

$$\ln \mathrm{p}_\theta(X_{p:n}|X_{0:p-1}) = -\frac{1}{2}\sum_{t=p}^{n} \ln\left[2\pi(b^\theta(X_{t-1},\ldots,X_{t-p}))^2\right] - \frac{1}{2}\sum_{t=p}^{n} \frac{\{X_t - a^\theta(X_{t-1},X_{t-2},\ldots,X_{t-p})\}^2}{(b^\theta(X_{t-1},X_{t-2},\ldots,X_{t-p}))^2}\,.$$

The solution of this maximization problem cannot be in general obtained in closed form; a numerical optimization technique is therefore the only option. The choice of "good" initial points is in general mandatory for the optimization algorithm to converge to a sensible optimum. There are several ways to obtain preliminary estimators for these models, but these are in general model dependent. ◇

8.2 Consistency and asymptotic normality of the MLE

8.2.1 Consistency

We consider first the inference of the parameter θ for *misspecified models*. We postulate a model $\{\mathrm{p}^\theta(\cdot) : \theta \in \Theta\}$ for the observations $(X_0, \ldots, X_n)$. However, the model is misspecified in that the true underlying distribution does not belong to the model. We use the postulated model anyway, and obtain an estimate $\hat{\theta}_n$ from maximizing the log-likelihood (8.4) where

$$\mathrm{p}^\theta(x_{p:n}|x_{0:p-1}) = \prod_{t=p}^{n} q^\theta(x_{t-p:t-1}; x_t) ,$$

and $\{q^\theta : \theta \in \Theta\}$ is a set of Markov kernel densities (with respect to some dominating measure μ) associated to a parametric family of Markov chains of order p. We derive in this section the asymptotic behavior of $\hat{\theta}_n$. Perhaps surprisingly, $\hat{\theta}_n$ does not behave erratically despite the use of a wrong family of models. First, we show that $\hat{\theta}_n$ is asymptotically consistent, i.e., converges to a value $\theta_\star$ that maximizes $\theta \mapsto \mathbb{E}[\ln q^\theta(X_{0:p-1}; X_p)]$ (or a set of value if the solution of this problem is not unique), where the expectation is taken under the true underlying distribution. The density $q^{\theta_\star}$ can be viewed as the "projection" of the true underlying distribution on the model using the Kullback-Leibler divergence, which is defined as $\mathbb{E}[\ln q^\theta(X_{0:p-1}; X_p)]$, i.e., as a "distance" measure: $q^{\theta_\star}$ optimizes this quantity over all transition densities in the model. Consider the following assumptions:

Assumption A8.5. For any $n \in \mathbb{N}$, the vector of observations $(X_0, X_1, \ldots, X_n)$ is a realization of a (strict-sense) stationary and ergodic process $\{X_t,\, t \in \mathbb{Z}\}$.

We denote by $\mathbb{P}$ the probability induced on $(\mathsf{X}^{\mathbb{Z}}, \mathcal{X}^{\otimes \mathbb{Z}})$ by $\{X_t,\, t \in \mathbb{Z}\}$ and by $\mathbb{E}$ the associated expectation. In particular, $\{X_t,\, t \in \mathbb{Z}\}$ is a stationary process but it is not necessarily a stationary Markov chain.

Assumption A8.6.

(a) $\mathbb{P}$-a.s., the function $\theta \mapsto q^\theta(X_{0:p-1}; X_p)$ is continuous.

(b) $\mathbb{E}\left[\sup_{\theta \in \Theta} \ln^+ q^\theta(X_{0:p-1}; X_p)\right] < \infty$

Theorem 8.7. *Assume A8.5 and A8.6. Then, any estimator $\hat{\theta}_n$ belonging to the set $\arg\max_{\theta \in \Theta} \ln \mathrm{p}^\theta(X_{p:n}|X_{0:p-1})$ is strongly consistent in the sense that:*

$$\lim_{n \to \infty} \mathsf{d}(\hat{\theta}_n, \Theta_\star) = 0,\ \mathbb{P}\text{-a.s.} \tag{8.17}$$

where

$$\Theta_\star := \arg\max_{\theta \in \Theta} \mathbb{E}[\ln q^\theta(X_{0:p-1}; X_p)] . \tag{8.18}$$

Proof. We have for $n \geq p$,

$$\hat{\theta}_n \in \arg\max_{\theta \in \Theta} \mathrm{L}_n^\theta(X_{p:n}|X_{0:p-1}) ,$$

where

$$\mathsf{L}_n^{\theta}(X_{p:n}|X_{0:p-1}) := (n-p+1)^{-1}\left(\sum_{t=p}^{n} \ln q^{\theta}(X_{t-p:t-1};X_t)\right) .$$

The proof follows from Theorem 8.42 since

(a) $\mathbb{E}\left[\sup_{\theta\in\Theta} \ln^+ q^{\theta}(X_{0:p-1};X_p)\right] < \infty$,

(b) $\mathbb{P}$-a.s., the function $\theta \mapsto \ln q^{\theta}(X_{0:p-1};X_p)$ is continuous. ∎

Thus, in misspecified models, the MLE strongly converges to the set of parameters that maximize the *relative entropy* (or Kullback-Leibler divergence) between the true distribution and the family of postulated likelihoods. We now consider well-specified models, that is, we assume that $\{X_t,\ t \in \mathbb{N}\}$ is the observation process of a Markov chain of order p associated to the Markov kernel Q^{θ} with $\theta = \theta_\star \in \Theta$. In well-specified models, we stress the dependence in $\theta_\star$ by using the notations

$$\mathbb{P}^{\theta_\star} := \mathbb{P}, \quad \mathbb{E}^{\theta_\star} := \mathbb{E} . \tag{8.19}$$

In this situation, the consistency of the sequence of conditional MLE $\{\hat{\theta}_n\}_{n\geq 0}$ follows from Theorem 8.7 provided the set $\Theta_\star$ is reduced to the singleton $\{\theta_\star\}$, that is: $\theta_\star$ is the only parameter θ satisfying

$$\mathbb{E}^{\theta_\star}\left[\ln \frac{q^{\theta}(X_{0:p-1};X_p)}{q^{\theta_\star}(X_{0:p-1};X_p)}\right] = 0 .$$

Assumption A8.8. $Q^{\theta}(X_{0:p-1};\cdot) = Q^{\theta_\star}(X_{0:p-1};\cdot)$, $\mathbb{P}^{\theta_\star}$-a.s. if and only if $\theta = \theta_\star$.

The following Corollary is immediate.

Corollary 8.9. *Under A8.6 and A8.8, if $\{X_t,\ t \in \mathbb{N}\}$ is a strict-sense stationary and ergodic sequence under $\mathbb{P}^{\theta_\star}$, then*

$$\lim_{n\to\infty} \hat{\theta}_n = \theta_\star , \quad \mathbb{P}^{\theta_\star}\text{-a.s.}$$

Proof. According to Theorem 8.7, it is sufficient to show that $\Theta_\star = \{\theta_\star\}$. Now, by the tower property, we have for all $\theta \in \Theta$,

$$\mathbb{E}^{\theta_\star}\left[\ln \frac{q^{\theta_\star}(X_{0:p-1};X_p)}{q^{\theta}(X_{0:p-1};X_p)}\right] = \mathbb{E}^{\theta_\star}\left[\mathbb{E}^{\theta_\star}\left[\ln \frac{q^{\theta_\star}(X_{0:p-1};X_p)}{q^{\theta}(X_{0:p-1};X_p)} \,\middle|\, X_{0:p-1}\right]\right] . \tag{8.20}$$

The RHS is always nonnegative as the expectation of a conditional Kullback-Leibler divergence, which implies that $\theta_\star \in \Theta_\star$. Moreover, using that $\ln(u) = u - 1$ if and only if $u = 1$, (8.20) also shows that if $\theta \in \Theta_\star$, then

$$q^{\theta_\star}(X_{0:p-1};X_p) = q^{\theta}(X_{0:p-1};X_p) , \quad \mathbb{P}^{\theta_\star}\text{-a.s.}$$

which concludes the proof under A8.8. ∎

8.2.2 *Asymptotic normality*

Suppose a sequence of estimators $\{\hat{\theta}_n\}_{n\geq 0}$ is consistent for a parameter $\theta_\star$ which belongs to an open subset of a Euclidean space. A question of interest relates to the order at which the error $\hat{\theta}_n - \theta_\star$ converges to zero. The answer of course depends on the specificity of the model, but for i.i.d. samples and *regular* statistical models, the order for sensible estimators based on n observations is $n^{1/2}$. Multiplying the estimation error $\hat{\theta}_n - \theta_\star$ by $n^{1/2}$ produces a proper tradeoff so that, most often, the sequence $\sqrt{n}(\hat{\theta}_n - \theta_\star)$ converges in distribution to a Gaussian random variable with zero-mean and a covariance that can be computed explicitly. With the help of the results gathered in Chapter 7, we will see that this property extends to the Markov chain case without much technical trouble (but of course the expression of the covariance is more involved, except when the model is well-specified). The convergence of $\sqrt{n}(\hat{\theta}_n - \theta_\star)$ is interesting not only from a theoretical point of view but also in practice, since it makes it possible to construct asymptotic confidence regions.

Before going further, just recall briefly the outline of the standard proof in the i.i.d. case. The proof in the Markov case follows almost exactly along the same lines. Let $X_1, \ldots, X_n$ be a sample from some distribution $\mathbb{P}$, and let the random and the "true" criterion function be of the form:

$$M_n(\theta) \equiv \frac{1}{n}\sum_{i=1}^{n} m^\theta(X_i), \quad M(\theta) = \mathbb{E}[m^\theta(X_0)] .$$

Assume that the estimator $\hat{\theta}_n$ is a maximizer of $M_n(\theta)$ and converges in probability to an element $\theta_\star$ of $\arg\max_\theta M(\theta)$ such that $\theta_\star$ is in the interior of Θ. Assume for simplicity that θ is one-dimensional. Because $\hat{\theta}_n \to_{\mathbb{P}} \theta_\star$, it makes sense to expand $\dot{M}_n(\hat{\theta}_n)$ in a Taylor series around $\theta_\star$. Then,

$$0 = \dot{M}_n(\hat{\theta}_n) = \dot{M}_n(\theta_\star) + (\hat{\theta}_n - \theta_\star)\ddot{M}_n(\tilde{\theta}_n) ,$$

where $\tilde{\theta}_n$ is a point between $\hat{\theta}_n$ and $\theta_\star$. This can be rewritten as

$$\sqrt{n}(\hat{\theta}_n - \theta_\star) = \frac{-\sqrt{n}\dot{M}_n(\theta_\star)}{\ddot{M}_n(\tilde{\theta}_n)} . \tag{8.21}$$

If, in this case, $\mathbb{E}\left[(\dot{m}^{\theta_\star})^2(X_0)\right]$ is finite, then by the central limit theorem, the numerator $-\sqrt{n}\dot{M}_n(\theta_\star) = -n^{-1/2}\sum_{i=1}^n \dot{m}^{\theta_\star}(X_i)$ is asymptotically normal with zero-mean $\mathbb{E}\left[\dot{m}^{\theta_\star}(X_0)\right] = \dot{M}(\theta_\star) = 0$ and variance $\mathbb{E}\left[(\dot{m}^{\theta_\star})^2(X_0)\right]$.

Next consider the denominator of (8.21). The denominator $\ddot{M}_n(\tilde{\theta}_n)$ is an average and its limit can be found by using some "uniform" version of the law of large numbers: $\ddot{M}_n(\tilde{\theta}_n) \to_{\mathbb{P}} \mathbb{E}\left[\ddot{m}^{\theta_\star}(X_0)\right]$, provided the expectation exists. The difficulty here stems from the fact that $\tilde{\theta}_n$ is itself a random variable: this is why it is required to use a law of large number that is valid uniformly over at least a vanishing neighborhood of $\theta_\star$. Together with Slutsky's lemma, these two conclusions yield

$$\sqrt{n}(\hat{\theta}_n - \theta_\star) \overset{\mathbb{P}}{\Longrightarrow} \mathrm{N}\left(0, \frac{\mathbb{E}\left[(\dot{m}^{\theta_\star})^2(X_0)\right]}{(\mathbb{E}\left[\ddot{m}^{\theta_\star}(X_0)\right])^2}\right) . \tag{8.22}$$

The preceding derivation can be made rigorous by imposing appropriate technical conditions (on the regularity of the function $\theta \mapsto \dot{M}_n(\theta)$, the validity of interchanging derivation and expectation), often called *regularity conditions* in the i.i.d. case. Perhaps surprisingly, the most challenging part of the proof consists of showing that $\ddot{M}_n(\tilde{\theta}_n)$ is converging to $\mathbb{E}\left[\ddot{m}^{\theta_\star}(X_0)\right]$ (see above); all the other steps are direct consequences of classical limit theorems.

The derivation can be extended to higher-dimensional parameters. For a d-dimensional parameter, we use d estimating equations $\dot{M}_n(\theta) = 0$, where $\dot{M}_n : \mathbb{R}^d \mapsto \mathbb{R}^d$. The derivatives $\ddot{M}_n(\theta_\star)$ are $(d \times d)$-matrices that converge to the $(d \times d)$ matrix $\Gamma_\star := \mathbb{E}\left[\ddot{m}^{\theta_\star}(X_0)\right]$ with entries $\mathbb{E}\left[\partial^2 m^{\theta_\star}(X_0)/\partial\theta_i\partial\theta_j\right]$. The limiting distribution (see (8.22)) may be expressed as

$$\sqrt{n}(\hat{\theta}_n - \theta_\star) \stackrel{\mathbb{P}}{\Longrightarrow} \mathrm{N}_d(0, \Gamma_\star^{-1}\Sigma_\star\Gamma_\star^{-1}) , \tag{8.23}$$

where $\Sigma_\star = \mathbb{E}\left[\dot{m}^{\theta_\star}(X_0)(\dot{m}^{\theta_\star}(X_0))'\right]$ is the covariance matrix of the "pseudo-scores." Note that the matrix $\Gamma_\star$ should be nonsingular.

In this section we extend these results to Markov chains. We consider first the case of (possibly) misspecified models. Assume that $\Theta \subset \mathbb{R}^d$ and recall that Θ^o is the interior of Θ.

Assumption A8.10. There exists $\theta_\star \in \Theta^o$ such that $\hat{\theta}_n \stackrel{\mathbb{P}}{\longrightarrow} \theta_\star$.

Assumption A8.11.

(a) The function $\theta \mapsto q^\theta(X_{0:p-1}; X_p)$ is $\mathbb{P}$-a.s. twice continuously differentiable in an open neighborhood of $\theta_\star$.

(b) There exists $\rho \in (0,1)$ such that for all indexes $i, j \in \{1, \dots, d\}$,

$$\mathbb{E}\left[\sup_{\theta \in \mathrm{B}(\theta_\star, \rho)} \left| \frac{\partial^2 \ln q^\theta}{\partial\theta_i \partial\theta_j}(X_{0:p-1}; X_p) \right| \right] < \infty .$$

Assumption A8.12. There exists a $d \times d$ nonsingular matrix $\Sigma_\star$ such that

$$n^{-1/2} \sum_{t=p}^{n} (\nabla \ln q^{\theta_\star}(X_{t-p:t-1}; X_t) - \mathbb{E}[\nabla \ln q^{\theta_\star}(X_{0:p-1}; X_p)]) \stackrel{\mathbb{P}}{\Longrightarrow} \mathrm{N}_d(0, \Sigma_\star) . \tag{8.24}$$

Theorem 8.13. *Assume A8.10, A8.11 and A8.12. In addition, assume that* $\{X_t,\ t \in \mathbb{N}\}$ *is a strict-sense stationary and ergodic sequence under* $\mathbb{P}$ *and that* $\Gamma_\star := \mathbb{E}[\nabla^2 \ln q^{\theta_\star}(X_{0:p-1}; X_p)]$ *is nonsingular. Then,*

$$n^{1/2}(\hat{\theta}_n - \theta_\star) \stackrel{\mathbb{P}}{\Longrightarrow} \mathrm{N}(0, \Gamma_\star^{-1}\Sigma_\star\Gamma_\star^{-1})$$

where $\Sigma_\star$ *is defined in* (8.24).

We preface the proof by the following technical Lemma, which will be useful in misspecified and well-specified models as well.

Lemma 8.14. *Assume that A8.11 hold for some* $\theta_\star \in \Theta$. *Assume in addition that* $\{X_t,\ t \in \mathbb{N}\}$ *is a strict-sense stationary and ergodic sequence under* $\mathbb{P}$ *and let*

$\{\theta_n,\ n \in \mathbb{N}\}$ be a sequence of random vectors such that $\theta_n \to_{\mathbb{P}} \theta_\star$. Then, for all $i, j \in \{1, \ldots, d\}$,

$$n^{-1} \sum_{t=p}^{n} \frac{\partial^2 \ln q^{\theta_n}(X_{t-p:t-1}; X_t)}{\partial \theta_i \partial \theta_j} \xrightarrow{\mathbb{P}} \mathbb{E}\left[\frac{\partial^2 \ln q^{\theta_\star}(X_{0:p-1}; X_p)}{\partial \theta_i \partial \theta_j}\right] .$$

Proof. Denote $A_t(\theta) = \frac{\partial^2 \ln q^{\theta}(X_{t-p:t-1}; X_t)}{\partial \theta_i \partial \theta_j}$. Since by the Birkhoff ergodic theorem, $n^{-1} \sum_{t=p}^{n} A_t(\theta_\star) \to_{\mathbb{P}} \mathbb{E}\left[A(\theta_\star)\right]$, we only need to show that

$$n^{-1} \sum_{t=p}^{n} |A_t(\theta_\star) - A_t(\theta_n)| \xrightarrow{\mathbb{P}} 0 . \tag{8.25}$$

Let $\varepsilon > 0$ and choose $0 < \eta < \rho$ such that

$$\mathbb{E}\left(\sup_{\theta \in \mathrm{B}(\theta_\star, \eta)} |A_p(\theta_\star) - A_p(\theta)| \right) < \varepsilon . \tag{8.26}$$

The existence of such η follows from the $\mathbb{P}$-a.s. continuity of $\theta \mapsto A_t(\theta)$ under A8.11-(a) and by the Lebesgue convergence theorem under A8.11-(b). We then have

$$\limsup_{n \to \infty} \mathbb{P}\left(n^{-1} \sum_{t=p}^{n} |A_t(\theta_\star) - A_t(\theta_n)| \geq \varepsilon,\ \theta_n \in \mathrm{B}(\theta_\star, \eta) \right)$$
$$\leq \limsup_{n \to \infty} \mathbb{P}\left(n^{-1} \sum_{t=p}^{n} \sup_{\theta \in \mathrm{B}(\theta_\star, \eta)} |A_t(\theta_\star) - A_t(\theta)| \geq \varepsilon \right) = 0 ,$$

where the last equality follows from (8.26) and the Birkhoff ergodic theorem. Moreover, since $\hat{\theta}_n \xrightarrow{\mathbb{P}} \theta_\star$, $\lim_{n \to \infty} \mathbb{P}(\theta_n \notin \mathrm{B}(\theta_\star, \eta)) = 0$ so that finally,

$$\lim_{n \to \infty} \mathbb{P}\left(n^{-1} \sum_{t=p}^{n} |A_t(\theta_\star) - A_t(\theta_n)| \geq \varepsilon \right) = 0 .$$

Thus, (8.25) holds and the proof follows. ∎

Proof (of Theorem 8.13). Since $\hat{\theta}_n$ cancels the derivatives of the log-likelihood, a Taylor expansion at the point $\theta = \theta_\star$ with an integral form of the remainder yields

$$n^{-1/2} \sum_{t=p}^{n} \nabla \ln q^{\hat{\theta}_n}(X_{t-p:t-1}; X_t) = 0 = n^{-1/2} \sum_{t=p}^{n} \nabla \ln q^{\theta_\star}(X_{t-p:t-1}; X_t)$$
$$+ n^{-1} \sum_{t=p}^{n} \left(\int_0^1 \nabla^2 \ln q^{\theta_{n,s}}(X_{t-p:t-1}; X_t) \mathrm{d}s \right) \sqrt{n}(\hat{\theta}_n - \theta_\star) , \tag{8.27}$$

where $\theta_{n,s} = s\hat{\theta}_n + (1-s)\theta_\star$. According to A8.12, we have

$$n^{-1/2} \sum_{t=p}^{n} \nabla \ln q^{\theta_\star}(X_{t-p:t-1}; X_t) \overset{\mathbb{P}^{\theta_\star}}{\Longrightarrow} \mathrm{N}(0, \Sigma_\star) .$$

We now turn to the asymptotic properties of the second term of the right-hand side in (8.27). More precisely, since by A8.10, $\hat{\theta}_n \to_{\mathbb{P}^{\theta_\star}} \theta_\star$, Lemma 8.14 shows that for all $i, j \in \{1, \ldots, d\}$,

$$n^{-1} \sum_{t=p}^{n} \int_0^1 \frac{\partial^2 \ln q^{\theta_{n,s}}}{\partial \theta_i \partial \theta_j}(X_{t-p:t-1}; X_t) \mathrm{d}s \xrightarrow{\mathbb{P}^{\theta_\star}} \mathbb{E}^{\theta_\star}\left[\frac{\partial^2 \ln q^{\theta_\star}}{\partial \theta_i \partial \theta_j}(X_{0:p-1}; X_p)\right] .$$

This implies that

$$n^{-1} \sum_{t=p}^{n} \left(\int_0^1 \nabla^2 \ln q^{\theta_{n,s}}(X_{t-p:t-1}; X_t) \mathrm{d}s \right) \xrightarrow{\mathbb{P}^{\theta_\star}} \Gamma_\star := \mathbb{E}[\nabla^2 \ln q^{\theta_\star}(X_{0:p-1}; X_p)] .$$

The proof then follows by applying the Slutsky Lemma to (8.27). ■

We now turn to well-specified models, that is, the observation process $\{X_n,\ n \in \mathbb{N}\}$ is the realization of a Markov chain associated to an unknown parameter $\theta_\star$.

Before stating the result, we need some additional assumptions. The parameter set Θ is now a compact subset of $\mathbb{R}^d$ with a non-empty interior and we assume that $\theta_\star \in \Theta^o$, where Θ^o is the interior of Θ. Moreover,

Assumption A8.15.

(a) The function $\theta \mapsto q^\theta(X_{0:p-1}; X_p)$ is, $\mathbb{P}^{\theta_\star}$-a.s., twice continuously differentiable in an open neighborhood of $\theta_\star$,

(b) there exists $\rho > 0$ such that for all $i, j \in \{1, \ldots, d\}$,

$$\mathbb{E}^{\theta_\star}\left[\sup_{\theta \in \mathrm{B}(\theta_\star, \rho)} \left| \frac{\partial^2 \ln q^\theta}{\partial \theta_i \partial \theta_j}(X_{0:p-1}; X_p) \right| \right] < \infty .$$

A8.15 is similar to A8.11. In well-specified models, it is possible to obtain the weak convergence of the score function using a martingale argument (see Exercise 8.1). Some additional assumptions are needed.

Assumption A8.16. There exists $\rho > 0$ such that

$$\mathbb{E}^{\theta_\star}\left[\left| \frac{\partial \ln q^{\theta_\star}}{\partial \theta_i}(X_{0:p-1}; X_p) \right|^2\right] < \infty , \tag{8.28}$$

$$\int \sup_{\theta \in \mathrm{B}(\theta_\star, \rho)} \left| \frac{\partial q^\theta}{\partial \theta_i}(X_{0:p-1}; x_p) \right| \mathrm{d}\mu(x_p) < \infty, \quad \mathbb{P}^{\theta_\star}\text{-a.s.} , \tag{8.29}$$

$$\int \sup_{\theta \in \mathrm{B}(\theta_\star, \rho)} \left| \frac{\partial^2 q^\theta}{\partial \theta_i \partial \theta_j}(X_{0:p-1}; x_p) \right| \mathrm{d}\mu(x_p) < \infty, \quad \mathbb{P}^{\theta_\star}\text{-a.s.} , \tag{8.30}$$

and the Fisher information matrix $\mathcal{J}(\theta_\star)$ defined by

$$\begin{aligned} \mathcal{J}(\theta_\star) &:= -\mathbb{E}^{\theta_\star}\left[\nabla^2 \ln q^{\theta_\star}(X_{0:p-1}; X_p)\right] \\ &= \mathbb{E}^{\theta_\star}\left[\nabla \ln q^{\theta_\star}(X_{0:p-1}; X_p) \nabla \ln q^{\theta_\star}(X_{0:p-1}; X_p)^T\right] \end{aligned} \tag{8.31}$$

is nonsingular.

Note that the Fisher information matrix has two expressions given in (8.31). These two expressions are classically obtained under the assumption (8.30). We now state and prove the weak convergence of the score function using a martingale-type approach.

Lemma 8.17. *Under A8.16,*

$$n^{-1/2}\sum_{t=p}^{n}\nabla\ln q^{\theta_\star}(X_{t-p:t-1};X_t)\overset{\mathbb{P}^{\theta_\star}}{\Longrightarrow}\mathrm{N}(0,\mathcal{J}(\theta_\star))\,,$$

where $\mathcal{J}(\theta_\star)$ is defined in (8.31).

Proof. Let $\mathcal{F}$ be the filtration $\mathcal{F}=(\mathcal{F}_n)_{n\in\mathbb{N}}$ where $\mathcal{F}_n=\sigma(X_0,\ldots,X_n)$ and let

$$M_n:=\sum_{t=p}^{n}\nabla\ln q^{\theta_\star}(X_{t-p:t-1};X_t)\,.$$

Note that according to (8.28), $\mathbb{E}(|M_t|^2)<\infty$ and that

$$\begin{aligned}\mathbb{E}^{\theta_\star}\left[\nabla\ln q^{\theta_\star}(X_{t-p:t-1};X_t)\,\middle|\,\mathcal{F}_{t-1}\right]&=\int\nabla q^{\theta_\star}(X_{t-p:t-1};x_t)\mathrm{d}x_t\\&=\nabla\int q^{\theta}(X_{t-p:t-1};x_t)\mathrm{d}x_t\Big|_{\theta=\theta_\star}=0\,,\end{aligned}$$

where the assumption (8.29) allows us to interchange $\int$ and ∇. Finally, $\{M_t\,,t\geq p\}$ is a square integrable $\mathcal{F}$-martingale with stationary increments. The proof follows by applying Theorem B.21. ■

We now have all the ingredients for obtaining the central limit theorem of the MLE in well-specified models.

Theorem 8.18. *Assume that $\hat{\theta}_n\overset{\mathbb{P}^{\theta_\star}}{\longrightarrow}\theta_\star$ and that A8.15 and A8.16 hold. Then,*

$$\sqrt{n}(\hat{\theta}_n-\theta_\star)\overset{\mathbb{P}}{\Longrightarrow}\mathrm{N}(0,\mathcal{J}^{-1}(\theta_\star))\,,$$

where $\mathcal{J}(\theta_\star)$ is defined in (8.31).

Proof. See Exercise 8.6 ■

Note that $\mathcal{J}(\theta_\star)$ is the asymptotic Fisher information matrix in the sense that

$$\mathcal{J}(\theta_\star)=-\lim_{n\to\infty}n^{-1}\mathbb{E}^{\theta_\star}\left[\nabla^2\ln\left(\prod_{k=p}^{n}q^{\theta}(X_{k-p:k-1};X_k)\right)\Bigg|_{\theta=\theta_\star}\right]\,.$$

Example 8.19 (Autoregressive models). Let $(\phi_{*,1},\phi_{*,2},\ldots,\phi_{*,p})\in\mathbb{R}^p$ be coefficients such that $\phi_*(z)=1-\sum_{j=1}^{p}\phi_{*,j}z^j\neq 0$ for $|z|\leq 1$, where $p>0$ is an integer. Assume that $\{X_t,\,t\in\mathbb{Z}\}$ is the causal autoregressive process: $X_t=\sum_{j=1}^{p}\phi_{*,j}X_{t-j}+\sigma_*Z_t$

where $\{Z_t,\ t \in \mathbb{Z}\}$ is white Gaussian noise with unit-variance (see Definition 1.30 and Proposition 1.31). Set $\theta = (\phi_1, \dots, \phi_p, \sigma^2)$, $\theta_\star = (\phi_{*,1}, \dots, \phi_{*,p}, \sigma_*^2)$ and

$$\ln q^\theta(x_{0:p-1}, x_p) = -\frac{1}{2}\ln(2\pi\sigma^2) - \frac{1}{2\sigma^2}\left(x_p - \sum_{j=1}^{p} \phi_j x_{p-j}\right)^2 .$$

By differentiating twice with respect to σ^2 and ϕ_i, we get

$$\frac{\partial^2 \ln q^\theta(X_{0:p-1}; X_p)}{\partial\phi_i \partial\phi_j} = -\frac{1}{\sigma^2} X_{p-i} X_{p-j}\,, \qquad (i,j) \in \{1,\dots,p\}^2$$

$$\frac{\partial^2 \ln q^\theta(X_{0:p-1}; X_p)}{\partial\phi_j \partial\sigma^2} = -\frac{1}{\sigma^4}\left(X_p - \sum_{j=1}^{p} \phi_j X_{p-j}\right) X_{p-i}\,, \qquad i \in \{1,\dots,p\}$$

$$\frac{\partial^2 \ln q^\theta(X_{0:p-1}; X_p)}{\partial\sigma^2 \partial\sigma^2} = \frac{1}{2\sigma^4} - \frac{1}{\sigma^6}\left(X_p - \sum_{j=1}^{p} \phi_j X_{p-j}\right)^2 .$$

Since, for any $i \in \{1,\dots,p\}$,

$$\mathbb{E}^{\theta_\star}\left[\left(X_p - \sum_{j=1}^{p} \phi_{*,j} X_{p-j}\right) X_{p-i}\right] = \mathbb{E}^{\theta_\star}[Z_p X_{p-i}] = 0\,,$$

and, similarly

$$\mathbb{E}^{\theta_\star}\left[\left(X_p - \sum_{j=1}^{p} \phi_{*,j} X_{p-j}\right)^2\right] = \mathbb{E}^{\theta_\star}[Z_p^2] = \sigma_*^2\,,$$

we get that $\mathcal{J}(\theta_\star)$ is a block diagonal matrix, with block diagonal elements equal to $1/2\sigma_*^4$ and $\sigma_*^{-2}\Gamma_\star$, where $\Gamma_\star$ is the $p \times p$ covariance matrix, $[\Gamma_\star]_{i,j} = \mathrm{Cov}^{\theta_\star}(X_0, X_{i-j})$, $1 \le i,j \le p$. If $p = 1$, then $\Gamma_\star = \sigma_*^2/(1-\phi_{*,1}^2)$, and the asymptotic variance for the autoregressive parameter is equal to $1/(1-\phi_{*,1}^2)$. For $p = 2$, the reader can verify (Exercise 8.4) that

$$\sigma_*^{-2}\Gamma_\star = \begin{bmatrix} 1-\phi_{*,2}^2 & -\phi_{*,1}(1+\phi_{*,2}) \\ -\phi_{*,1}(1+\phi_{*,2}) & 1-\phi_{*,2}^2 \end{bmatrix} . \qquad (8.32)$$

Note that the asymptotic variance of the estimate of $\phi_{*,1}$ depends only on $\phi_{*,2}$. ◇

Example 8.20 (Threshold autoregressive models). We use the notations and definitions of Example 8.3. Denote by $\theta = [\eta_1, \eta_2, r, d]$ with $\eta_i = [(\phi_{i,j})_{j=0}^{p_1}, \sigma_i^2]$, $i = 1,2$, the unknown parameters. In this case the log-likelihood may be expressed as

$$
\begin{aligned}
&\ln q^{\theta}(x_{0:p-1};x_p) \\
&= \left\{ -\frac{1}{2}\ln(2\pi\sigma_1^2) - \frac{1}{2\sigma_1^2}\left(X_p - \phi_{1,0} - \sum_{j=1}^{p_1} \phi_{1,j} X_{p-j} \right)^2 \right\} \mathbb{1}_{\{X_{t-d} \leq r\}} \\
&\quad + \left\{ -\frac{1}{2}\ln(2\pi\sigma_2^2) - \frac{1}{2\sigma_2^2}\left(X_p - \phi_{2,0} - \sum_{j=1}^{p_2} \phi_{2,j} X_{p-j} \right)^2 \right\} \mathbb{1}_{\{X_{t-d} > r\}} ,
\end{aligned}
$$

where $p = \max(p_1, p_2)$. When the delay d and the threshold r are known, Theorem 8.7 shows that, provided that $\{X_t,\, t \in \mathbb{Z}\}$ is an ergodic sequence, then the regression parameters $\hat{\eta}_i(r,d) = [\hat{\phi}_{i,j}(r,d), \hat{\sigma}_i(r,d)]_{j=1}^{p_i}$, $i = 1,2$ are consistent in the sense of (8.17). It has been established by Chan (1993, Theorem 1) that the same remains true if the delay and the threshold are estimated as the maximum of the profile likelihood.

Assume now that $\{X_t,\, t \in \mathbb{Z}\}$ is a strict sense stationary geometrically ergodic SETAR process

$$
\begin{aligned}
X_t = &\left\{ \phi_{*,1,0} + \sum_{j=1}^{p_1} \phi_{*,1,j} X_{t-j} + \sigma_{*,1} Z_t \right\} \mathbb{1}_{\{X_{t-d_*} \leq r_*\}} \\
&\quad + \left\{ \phi_{*,2,0} + \sum_{j=1}^{p_2} \phi_{*,2,j} X_{t-j} + \sigma_{*,2} Z_t \right\} \mathbb{1}_{\{X_{t-d_*} > r_*\}} , \qquad (8.33)
\end{aligned}
$$

where $\{Z_t,\, t \in \mathbb{Z}\}$ is a strong white Gaussian noise with zero-mean and unit-variance. Such condition is satisfied, for example, if $\max_{i=1,2} \sum_{j=1}^{p_i} |\phi_{*,i,j}| < 1$ (see Example 6.37, condition (6.43)). Assume first that the delay d and the threshold r are both known: $r = r_*$ and $d = d_*$. We may then apply Theorem 8.18 to show that the distribution of the parameters in the two regimes are asymptotically normal,

$$
\sqrt{n}\left(\hat{\eta}_1(r_*, d_*) - \eta_{*,1}, \hat{\eta}_2(r_*, d_*) - \eta_{*,2}\right) \stackrel{\mathbb{P}}{\Longrightarrow} \mathrm{N}(0, \mathcal{J}^{-1}(\theta_\star))
$$

where $\theta_\star = [\eta_{*,1}, \eta_{*,2}, r_*, d_*]$, $\eta_{*,i} = [(\phi_{*,i,j})_{j=0}^{p_1}, \sigma_{*,i}^2]$, $i = 1,2$. Note that, for any $(i,j) \in \{1,2\}$, $i \neq j$, and any $(k,\ell) \in \{1,\ldots,p_i\}^2$, we get

$$
\frac{\partial^2 \ln q^{\theta}(x_{0:p-1};x_p)}{\partial \phi_{i,k} \partial \phi_{j,\ell}} = 0 , \qquad \frac{\partial^2 \ln q^{\theta}(x_{0:p-1};x_p)}{\partial \sigma_i^2 \partial \phi_{j,k}} = 0 ,
$$

showing that the matrix $\mathcal{J}(\theta_\star)$ is block diagonal and therefore that the estimators in the two regimes are asymptotically independent. The entries of the Fisher information matrix are given by

$$\mathbb{E}^{\theta_\star}\left[\frac{\partial^2 \ln q^{\theta_\star}(X_{0:p-1};X_p)}{\partial\phi_{1,k}\partial\phi_{1,\ell}}\right] = \frac{1}{\sigma_1^2}\mathbb{E}^{\theta_\star}\left[X_{p-k}X_{p-\ell}\mathbb{1}_{\{X_{p-d}\le r\}}\right], \quad (k,\ell)\in\{1,\ldots,p_1\}$$

$$\mathbb{E}^{\theta_\star}\left[\frac{\partial^2 \ln q^{\theta_\star}(X_{0:p-1};X_p)}{\partial\phi_{1,0}\partial\phi_{1,k}}\right] = \frac{1}{\sigma_1^2}\mathbb{E}^{\theta_\star}\left[X_{p-k}\mathbb{1}_{\{X_{p-d}\le r\}}\right], \quad k\in\{1,\ldots,p_1\}$$

$$\mathbb{E}^{\theta_\star}\left[\frac{\partial^2 \ln q^{\theta_\star}(X_{0:p-1};X_p)}{\partial\phi_{1,k}\partial\sigma_1^2}\right] = 0\,,$$

$$\mathbb{E}^{\theta_\star}\left[\frac{\partial^2 \ln q^{\theta_\star}(X_{0:p-1};X_p)}{\partial\sigma_1^2\partial\sigma_1^2}\right] = \frac{1}{2\sigma_1^2}\mathbb{P}^{\theta_\star}(X_0\le r)\,.$$

The Fisher information matrix for the parameters of the upper regime can be written similarly. When the threshold r and d are both unknown, the assumptions of Theorem 8.18 are no longer satisfied. It is shown in Chan (1993, Theorem 2) that

$$\left\{n(\hat{r}_n - r_*), \sqrt{n}(\hat{\eta}_1 - \eta_{*,1}, \hat{\eta}_2 - \eta_{*,2})\right\}$$

converges in distribution, that $n(\hat{r}_n - r_*)$ and $\sqrt{n}(\hat{\eta}_1 - \eta_{*,1}, \hat{\eta}_2 - \eta_{*,2})$ are asymptotically independent. Chan (1993) identified the distribution of $n(\hat{r}_n - r_*)$: this is an interval on which a compound Poisson process (whose parameters depend on $\theta_\star$) attains its global minimum.

In general, the problem of determining the set of parameters Θ_* for $\{X_t,\, t\in\mathbb{Z}\}$ to be geometrically ergodic is still an open problem. For $p_1 = p_2 = 1$ and $d = 1$, however, the problem is solved. The condition in this case is $\phi_{*,1,1} < 1$, $\phi_{*,2,1} < 1$ and $\phi_{*,1,1}\phi_{*,2,1} < 1$. For details, see Chan et al. (1985). ◇

8.3 Observation-driven models

Let (X, d) and (Y, δ) be Polish spaces equipped with their Borel sigma-field $\mathcal{X}$ and $\mathcal{Y}$. Consider $\{Q^\theta, \theta\in\Theta\}$ a family of Markov kernels on $\mathsf{X}\times\mathcal{Y}$ indexed by $\theta\in\Theta$ where (Θ, d) is a compact metric space. Assume that all $(\theta, x)\in\Theta\times\mathsf{X}$, $Q^\theta(x,\cdot)$ is dominated by some σ-finite measure μ on $(\mathsf{Y},\mathcal{Y})$ and denote by $q^\theta(x,\cdot)$ its Radon-Nikodym derivative: $q^\theta(x,y) = \mathrm{d}Q^\theta(x,\cdot)/\mathrm{d}\mu(y)$. Finally, let $\left\{f^\theta : \theta\in\Theta\right\}$ be a family of measurable functions from $(\mathsf{X}\times\mathsf{Y}, \mathcal{X}\otimes\mathcal{Y})$ to $(\mathsf{X},\mathcal{X})$. We consider the following observation-driven model (see Definition 5.21)

$$\mathbb{P}^\theta\left[Y_t\in A \,\middle|\, \mathcal{F}_{t-1}\right] = Q^\theta(X_{t-1}, A) = \int_A q^\theta(X_{t-1}, y)\mu(\mathrm{d}y)\,, \quad \text{for any } A\in\mathcal{Y}\,,$$

$$X_t = f^\theta_{Y_t}(X_{t-1})\,.$$

With these notations, the distribution of $(Y_1,\ldots,Y_n)$ conditionally on $X_0 = x$ has a density with respect to the product measure $\mu^{\otimes n}$ given by

$$y_{1:n}\mapsto\prod_{t=1}^n q^\theta(f^\theta_{y_{1:t-1}}(x), y_t)\,, \tag{8.34}$$

where we have set for all $s \leq t$ and all $y_{s:t} \in \mathsf{Y}^{t-s+1}$,

$$f^{\theta}_{y_{s:t}} = f^{\theta}_{y_t} \circ f^{\theta}_{y_{t-1}} \circ \cdots \circ f^{\theta}_{y_s} , \tag{8.35}$$

with the convention $f^{\theta}_{y_{1:0}}(x_0) = x_0$. Note that, for any $t \geq 0$, X_t is a deterministic function of $Y_{1:t}$ and X_0, i.e.,

$$X_t = f^{\theta}_{Y_{1:t}}(X_0) = f^{\theta}_{Y_t} \circ f^{\theta}_{Y_{t-1}} \circ \cdots \circ f^{\theta}_{Y_1}(X_0) . \tag{8.36}$$

In this section, we study the asymptotic properties of $\hat{\theta}_{n,x}$, the conditional Maximum Likelihood Estimator (MLE) of the parameter θ based on the observations $(Y_1, \ldots, Y_n)$ and associated to the parametric family of likelihood functions given in (8.34), that is, we consider

$$\hat{\theta}_{n,x} \in \operatorname*{arg\,max}_{\theta \in \Theta} \mathsf{L}^{\theta}_{n,x}(Y_{1:n}) , \tag{8.37}$$

where

$$\mathsf{L}^{\theta}_{n,x}(y_{1:n}) := n^{-1} \ln \left(\prod_{t=1}^{n} q^{\theta}(f^{\theta}_{y_{1:t-1}}(x), y_t) \right) . \tag{8.38}$$

We are especially interested here in inference for *misspecified models*, that is, we *do not assume* that the distribution of the observations belongs to the set of distributions where the maximization occurs. In particular, $\{Y_t,\ t \in \mathbb{Z}\}$ are not necessarily the observation process associated to the recursion (8.36).

Nevertheless, consider the following assumptions

Assumption A8.21. $\{Y_t,\ t \in \mathbb{Z}\}$ is a strict-sense stationary and ergodic stochastic process.

Under A8.21, denote by $\mathbb{P}$ the distribution of $\{Y_t,\ t \in \mathbb{Z}\}$ on $(\mathsf{Y}^{\mathbb{Z}}, \mathcal{Y}^{\mathbb{Z}})$. Write $\mathbb{E}$, the associated expectation.

Assumption A8.22. For all $(x,y) \in \mathsf{X} \times \mathsf{Y}$, the functions $\theta \mapsto f^{\theta}_y(x)$ and $(y,\theta) \mapsto q^{\theta}(x,y)$ are continuous.

Assumption A8.23. There exists a family of $\mathbb{P}$-a.s. finite random variables

$$\left\{ f^{\theta}_{Y_{-\infty:t}} : (\theta, t) \in \Theta \times \mathbb{Z} \right\}$$

such that for all $x \in \mathsf{X}$,

(a) $\lim_{m \to \infty} \sup_{\theta \in \Theta} d(f^{\theta}_{Y_{-m:0}}(x), f^{\theta}_{Y_{-\infty:0}}) = 0$, $\mathbb{P}$-a.s.,

(b) $\lim_{t \to \infty} \sup_{\theta \in \Theta} |\ln q^{\theta}(f^{\theta}_{Y_{1:t-1}}(x), Y_t) - \ln q^{\theta}(f^{\theta}_{Y_{-\infty:t-1}}, Y_t)| = 0$, $\mathbb{P}$-a.s.,

(c) $\mathbb{E}\left[\sup_{\theta \in \Theta} \left(\ln q^{\theta}(f^{\theta}_{Y_{-\infty:t-1}}, Y_t) \right)_+ \right] < \infty$.

In the following, we set for all $(\theta, t) \in \Theta \times \mathbb{N}$,

$$\bar{\ell}^{\theta}(Y_{-\infty:t}) := \ln q^{\theta}(f^{\theta}_{Y_{-\infty:t-1}}, Y_t) . \tag{8.39}$$

Remark 8.24. When checking A8.23, we usually introduce $f^{\theta}_{Y_{-\infty:0}}$ by showing that for all $(\theta,x) \in \Theta \times \mathsf{X}$, $f^{\theta}_{Y_{-m:0}}(x)$ converges, $\mathbb{P}$-a.s., as m goes to infinity to a limit that does not depend on x. We can therefore denote by $f^{\theta}_{Y_{-\infty:0}}$ this limit. With this definition, we then check A8.23-(a)-(b)-(c).

Note that under A8.22, $\hat{\theta}_{n,x}$ is well-defined. The following theorem establishes the consistency of the sequence of estimators $\{\hat{\theta}_{n,x}\,,\, n \in \mathbb{N}\}$.

Theorem 8.25. *Assume A8.21, A8.22 and A8.23. Then, for all $x \in \mathsf{X}$,*

$$\lim_{n\to\infty} \mathsf{d}(\hat{\theta}_{n,x}, \Theta_\star) = 0\,, \quad \mathbb{P}\text{-a.s.}$$

where $\Theta_\star := \arg\max_{\theta\in\Theta} \mathbb{E}[\bar{\ell}^{\theta}(Y_{-\infty:1})]$ and $\bar{\ell}^{\theta}(Y_{-\infty:1})$ is defined in (8.39).

Proof. The proof directly follows from Theorem 8.42 provided that

$$\mathbb{E}[\sup_{\theta\in\Theta}(\bar{\ell}^{\theta}(Y_{-\infty:1}))_+] < \infty, \tag{8.40}$$

$$\text{the function } \theta \mapsto \bar{\ell}^{\theta}(Y_{-\infty:1}) \text{ is upper-semicontinuous, } \mathbb{P}\text{-a.s.,} \tag{8.41}$$

$$\lim_{n\to\infty} \sup_{\theta\in\Theta} |\mathsf{L}^{\theta}_{n,x}(Y_{1:n}) - \bar{\mathsf{L}}^{\theta}_{n}(Y_{-\infty:n})| = 0, \ \mathbb{P}\text{-a.s.,} \tag{8.42}$$

where $\bar{\mathsf{L}}^{\theta}_{n}(Y_{-\infty:n}) = n^{-1}\sum_{t=1}^{n} \bar{\ell}^{\theta}(Y_{-\infty:t})$.

But, (8.40) follows from A8.23-(c), (8.41) follows by combining A8.23-(a) and A8.22 since a uniform limit of continuous functions is continuous and (8.42) is direct from A 8.23-(b) and the definitions of $\mathsf{L}^{\theta}_{n,x}(Y_{1:n})$ and $\bar{\mathsf{L}}^{\theta}_{n}(Y_{-\infty:n})$. The proof is completed. ■

Example 8.26 (GARCH(1,1)). Assume that the observations $\{Y_t,\, t \in \mathbb{Z}\}$ are a strict-sense stationary ergodic process. We fit to the observations a GARCH(1,1) model with Student t-innovations,

$$\begin{cases} Y_t = \sigma_t \left(\frac{\nu-2}{\nu}\right)^{1/2} \varepsilon_t \\ \sigma_t^2 = \alpha_0 + \alpha_1 Y_{t-1}^2 + \beta_1 \sigma_{t-1}^2, \quad t \in \mathbb{Z} \end{cases} \tag{8.43}$$

where $\{\varepsilon_t,\, t \in \mathbb{N}\}$ is an i.i.d. sequence of t_ν-distributed random variables (where ν is the number of degrees of freedom) and $\theta = (\nu, \alpha_0, \alpha_1, \beta_1) \in \Theta$ is a compact subset of

$$\{(\nu, \alpha_0, \alpha_1, \beta_1), \nu > 2, \alpha_0 > 0, \alpha_1 > 0, \beta_1 > 0, \alpha_1 + \beta_1 < 1\}\,.$$

The restriction on the degrees of freedom parameter ν ensures the conditional variance to be finite. The variance of ε_1 is equal to $\nu/(\nu-2)$, therefore, $((\nu-2)/\nu)^{1/2}\varepsilon_1$ has unit-variance. Recall that the density of a Student t-distribution with ν degrees of freedom is given by

$$t_\nu(z) = \frac{\Gamma(\frac{\nu+1}{2})}{\Gamma(\frac{1}{2})\Gamma(\frac{\nu}{2})} \nu^{-1/2} [1 + z^2/\nu]^{-(\nu+1)/2}\,.$$

Setting $X_t = \sigma_{t+1}^2$, (8.43) defines an observation-driven model with $q^{\theta}(x,y) =$

$t_\nu(y/\sqrt{x})$ and $f_y^\theta(x) = \alpha_0 + \alpha_1 y^2 + \beta_1 x$. Given initial values Y_0 and σ_0^2 to be specified below, the conditional log-likelihood may be expressed as

$$\mathsf{L}_{n,\sigma_0}^\theta(Y_{1:n}) = \sum_{t=1}^n \ln t_\nu(Y_t/\sigma_t) ,$$

where σ_t^2 are computed recursively using (8.43). For a given value of θ, the unconditional variance (corresponding to the stationary value of the variance) is a sensible choice for the unknown initial values σ_0^2

$$\sigma_0^2 = \frac{\alpha_0}{1 - \alpha_1 - \beta_1} . \tag{8.44}$$

In this case, for any integer m, we have (see (5.21))

$$f_{Y_{-m:0}}^\theta(x) = \beta_1^{m+1} x + \sum_{j=0}^m \beta_1^j (\alpha_0 + \alpha_1 Y_{-j}^2) .$$

Assume that $\mathbb{E}_*[Y_0^2] < \infty$. Since Θ is compact, there exist $b_1 < 1$, $a_0 > 0$ and $a_1 < 1$ such that, for all $\theta \in \Theta$, $0 \leq \beta_1 \leq b_1$, $\alpha_0 \leq a_0$ and $\alpha_1 \leq a_1$. The series $\sum b_1^j (a_0 + a_1 Y_{-j}^2)$ converges $\mathbb{P}_*$-a.s. Define

$$f_{Y_{-\infty:0}}^\theta = \sum_{j=0}^\infty \beta_1^j (\alpha_0 + \alpha_1 Y_{-j}^2) .$$

With these definitions, we get

$$\sup_{\theta \in \Theta} |f_{Y_{-m:0}}^\theta(x) - f_{Y_{-\infty:0}}^\theta| \leq b_1^{m+1} x + \sum_{j=m+1}^\infty b_1^j (a_0 + a_1 Y_{-j}^2) \stackrel{\mathbb{P}_*\text{-a.s.}}{\longrightarrow} 0 ,$$

showing A8.23-(a). Note similarly that

$$\begin{aligned} &\left| \ln q^\theta (f_{Y_{1:t-1}}^\theta(x), Y_t) - \ln q^\theta (f_{Y_{-\infty:t-1}}^\theta, Y_t) \right| \\ &\qquad = \frac{\nu+1}{2} \left| \ln \left(1 + \frac{Y_t^2}{\nu (f_{Y_{1:t-1}}^\theta(x))^2} \right) - \ln \left(1 + \frac{Y_t^2}{\nu (f_{Y_{-\infty:t-1}}^\theta)^2} \right) \right| . \end{aligned}$$

Since for any $a > 0$, $b > 0$ and $z > 0$, $|\ln(1 + z^2/a^2) - \ln(1 + z^2/b^2)| \leq 2|a - b|/(a \wedge b)$ and $f_{Y_{1:t-1}}^\theta(x) > \alpha_0$, $f_{Y_{-\infty:t-1}}^\theta > \alpha_0$, there exists a constant $K > 0$ such that

$$\begin{aligned} &\sup_{\theta \in \Theta} \left| \ln q^\theta (f_{Y_{1:t-1}}^\theta(x), Y_t) - \ln q^\theta (f_{Y_{-\infty:t-1}}^\theta, Y_t) \right| \\ &\qquad \leq K \left(b_1^{m+1} x + \sum_{j=m+1}^\infty b_1^j (a_0 + a_1 Y_{-j}^2) \right) \stackrel{\mathbb{P}_*\text{-a.s.}}{\longrightarrow} 0 , \end{aligned}$$

showing A8.23-(b). The proof of A8.23-(c) is along the same lines. Theorem 8.25 shows the consistency of the estimator. ◇

Of course, in well-specified models where the observations process is associated to a parameter $\theta_\star \in \Theta$, Theorem 8.25 allows us to obtain the convergence of the MLE to $\theta_\star$ on the condition that $\Theta_\star$ is reduced to the singleton $\{\theta_\star\}$.

An important subclass of misspecified models corresponds to the case where the kernel density $q^\theta = q$ does not depend on θ and where the observation process is assumed to follow the recursions:

$$\mathbb{P}\left[Y_t \in A \,\middle|\, \mathcal{F}_{t-1}\right] = Q^\star(X_{t-1}, A) = \int_A q^\star(X_{t-1}, y)\mu(\mathrm{d}y)\,, \quad \text{for any } A \in \mathcal{Y}\,,$$
$$X_t = f_{Y_t}^{\theta_\star}(X_{t-1})\,, \quad t \in \mathbb{Z}\,.$$

The MLE is based on the likelihood functions $y_{1:n} \mapsto \mathsf{L}_{n,x}^\theta(y_{1:n})$ defined (8.38) whose expression includes a kernel density $q \neq q^\star$ and a family of functions $\{f^\theta : \theta \in \Theta\}$. Since $\theta_\star$ is assumed to be in Θ^o, the interior of Θ but $q^\star \neq q$, this situation falls into the misspecified models framework. In such cases, Maximum Likelihood Estimators $\{\hat{\theta}_{n,x}\}$ defined in (8.37) are called Quasi Maximum Likelihood estimators (QMLE) and $\theta_\star$ is not the true value of the parameter in the sense that the distribution of the observation process is not characterized by $\theta_\star$ only. Nevertheless, perhaps surprisingly, it can be shown that, under some additional assumptions, the QMLE $\{\hat{\theta}_{n,x}\}$ are consistent and asymptotically normal with respect to the parameter $\theta_\star$. For simplicity, we assume here that $\mathsf{X} = \mathbb{R}$.

The main assumption that links $q^\star$ and q is the following:

Assumption A8.27. For all $x^\star \in \mathbb{R}$,

$$\operatorname*{arg\,max}_{x \in \mathbb{R}} \int Q^\star(x^\star, \mathrm{d}y) \ln q(x, y) = \{x^\star\}\,. \tag{8.45}$$

Theorem 8.28. *Assume A8.21, A8.22, A8.23 and A8.27. Moreover, assume that* $f_{Y_{-\infty:0}}^\theta = f_{Y_{-\infty:0}}^{\theta_\star}$, $\mathbb{P}$-a.s. *implies that* $\theta = \theta_\star$. *Then, for all* $x \in \mathsf{X}$,

$$\lim_{n \to \infty} \mathsf{d}(\hat{\theta}_{n,x}, \theta_\star) = 0\,, \quad \mathbb{P}\text{-a.s.}$$

Proof. See Exercise 8.10. ■

Example 8.29 (Normal innovations). Assume that the observations $\{Y_t,\, t \in \mathbb{Z}\}$ is a strict-sense stationary ergodic process associated to

$$\mathbb{P}\left[Y_t \in A \,\middle|\, \mathcal{F}_{t-1}\right] = Q^\star(\sigma_{t-1}^2, A)\,, \quad \text{for any } A \in \mathcal{Y}\,, \tag{8.46}$$
$$\sigma_t^2 = f_{Y_t}^{\theta_\star}(\sigma_{t-1}^2)\,, \quad t \in \mathbb{Z}\,.$$

We fit to the observations an observation-driven model with normal innovations,

$$Y_t = \sigma_t \varepsilon_t \tag{8.47}$$
$$\sigma_t^2 = f_{Y_t}^\theta(\sigma_{t-1}^2)\,, \quad (t, \theta) \in \mathbb{N} \times \Theta\,,$$

where $\{\varepsilon_t, t \in \mathbb{N}\} \sim$ iid $\mathrm{N}(0,1)$. This model typically covers the GARCH$(1,1)$ model with normal innovations. Now, the function

$$x \mapsto \int Q^\star(x^\star, \mathrm{d}y) \ln q(x,y) = \int Q^\star(x^\star, \mathrm{d}y) \left(-\frac{y^2}{2x} - \frac{1}{2}\ln(2\pi x)\right)$$
$$= \left(-\frac{\int Q^\star(x^\star, \mathrm{d}y) y^2}{2x} - \frac{1}{2}\ln(2\pi x)\right),$$

is maximized at $x = \int Q^\star(x^\star, \mathrm{d}y) y^2$ by straightforward algebra. Thus, A8.27 implies

$$\int Q^\star(x^\star, \mathrm{d}y) y^2 = x^\star .$$

Plugging this equality into (8.46),we obtain that the observations $\{Y_t, t \in \mathbb{Z}\}$ is a strict-sense stationary ergodic process associated to

$$Y_t \sim \sigma_t \varepsilon_t^\star$$
$$\sigma_t^2 = f_{Y_t}^{\theta_\star}(\sigma_{t-1}^2), \quad t \in \mathbb{Z},$$

where $\{\varepsilon_t^\star, t \in \mathbb{Z}\}$ is an i.i.d. sequence of random variables with potentially any unknown distribution provided that $\mathbb{E}[(\varepsilon_t^\star)^2] = 1$. ◇

Example 8.30 (Exponential families). Assume that the observations $\{Y_t, t \in \mathbb{Z}\}$ are a strict-sense stationary ergodic process associated to

$$\mathbb{P}\left[Y_t \in A \mid \mathcal{F}_{t-1}\right] = Q^\star(X_{t-1}, A) = \int_A q^\star(X_{t-1}, y) \mu(\mathrm{d}y), \quad \text{for any } A \in \mathcal{Y},$$
$$X_t = f_{Y_t}^{\theta_\star}(X_{t-1}), \quad t \in \mathbb{Z}.$$

We fit to the observations the following observation-driven model

$$\mathbb{P}\left[Y_t \in A \mid \mathcal{F}_{t-1}\right] = Q(X_{t-1}, A), \quad \text{for any } A \in \mathcal{Y},$$
$$X_t = f_{Y_t}^{\theta}(X_{t-1}), \quad (t, \theta) \in \mathbb{Z} \times \Theta .$$

where $Q(x, \cdot)$ is assumed to belong to the class of exponential family distributions. More precisely, we assume that for all $(x,y) \in \mathsf{X} \times \mathsf{Y}$, $q(x,y) = \exp(xy - A(x))h(y)$ for some twice differentiable function $A : \mathsf{X} \to \mathbb{R}$ and some measurable function $h : \mathsf{Y} \to \mathbb{R}^+$. Using that

$$\int Q(x, \mathrm{d}y) \frac{\partial^2 \ln q(x,y)}{\partial x^2} \leq 0,$$

it can be readily checked that $A'' \geq 0$ so that A is concave. Thus, the function

$$x \mapsto \int Q^\star(x^\star, \mathrm{d}y) \ln q(x,y) = \int Q^\star(x^\star, \mathrm{d}y) (xy - A(x) + \ln h(y))$$
$$= x \int Q^\star(x^\star, \mathrm{d}y) y - A(x) + \int Q^\star(x^\star, \mathrm{d}y) \ln h(y),$$

is convex. The maximum of this function can thus be obtained by cancelling the derivatives with respect to x, which yields $\int Q^\star(x^\star, \mathrm{d}y)y - A'(x) = 0$. Then, A8.27 implies that

$$\int Q^\star(x^\star, \mathrm{d}y)y = A'(x^\star) . \tag{8.48}$$

For example in the log-linear Poisson autoregression model (see Example 5.35),

$$q(x,y) = \exp(xy - \mathrm{e}^x)/y!$$

so that $A(x) = \exp(x)$. Thus, (8.48) yields that $\int Q^\star(x^\star, \mathrm{d}y)y = \exp(x^\star)$. ◇

Remark 8.31. In well-specified models, $Q^\star = Q$. Since the Kullback-Leibler divergence is nonnegative, we obtain

$$\int Q^\star(x^\star, \mathrm{d}y) \ln q^\star(x,y) \le \int Q^\star(x^\star, \mathrm{d}y) \ln q^\star(x^\star, y) ,$$

and, provided that $x \mapsto Q(x, \cdot)$ is one-to-one, the equality holds if and only if $x = x^\star$. Thus, (8.45) in A8.27 is satisfied.

If for all $y \in \mathsf{Y}$, $x \mapsto q(x,y)$ is twice differentiable, we may define for all vector $\boldsymbol{u}$ in $\mathbb{R}^d$, all matrix Γ of size $d \times d$ with real entries and all $(x,y) \in \mathbb{R} \times \mathsf{Y}$,

$$\varphi(\boldsymbol{u}, x, y) := \boldsymbol{u} \frac{\partial \ln q}{\partial x}(x,y) , \tag{8.49}$$

$$\psi(\Gamma, \boldsymbol{u}, x, y) := \Gamma \frac{\partial \ln q}{\partial x}(x,y) + \boldsymbol{u}\boldsymbol{u}' \frac{\partial^2 \ln q}{\partial x^2}(x,y) . \tag{8.50}$$

These functions appear naturally when differentiating $\theta \mapsto \ln q(f(\theta), y)$ where $\theta \mapsto f(\theta)$ is a twice differentiable function. More precisely, we have in such a case by straightforward algebra,

$$\nabla \ln q(f(\theta), y) = \varphi(\nabla f(\theta), f(\theta), y) ,$$
$$\nabla^2 \ln q(f(\theta), y) = \psi(\nabla^2 f(\theta), \nabla f(\theta), f(\theta), y) .$$

Assumption A8.32. For all $y \in \mathsf{Y}$, the function $x \mapsto q(x,y)$ is twice continuously differentiable. Moreover, there exist $\rho > 0$ and a family of $\mathbb{P}$-a.s. finite random variables

$$\left\{ f^\theta_{Y_{-\infty:t}} : (\theta, t) \in \Theta \times \mathbb{Z} \right\}$$

such that $\theta \mapsto f^\theta_{Y_{-\infty:0}}$ is, $\mathbb{P}$-a.s., twice continuously differentiable on some ball $\mathrm{B}(\theta_\star, \rho)$ and for all $x \in \mathsf{X}$,

(a) $\mathbb{P}$-a.s. ,

$$\lim_{t \to \infty} \| \varphi(\nabla f^{\theta_\star}_{Y_{1:t-1}}(x), f^{\theta_\star}_{Y_{1:t-1}}(x), Y_t) - \varphi(\nabla f^{\theta_\star}_{Y_{-\infty:t-1}}, f^{\theta_\star}_{Y_{-\infty:t-1}}, Y_t) \| = 0 ,$$

where $\| \cdot \|$ is any norm on $\mathbb{R}^d$,

(b) $\mathbb{P}$-a.s. ,

$$\lim_{t\to\infty} \sup_{\theta\in \mathrm{B}(\theta_\star,\rho)} \| \psi(\nabla^2 f^\theta_{Y_{1:t-1}}(x), \nabla f^\theta_{Y_{1:t-1}}(x), f^\theta_{Y_{1:t-1}}(x), Y_t) \\ - \psi(\nabla^2 f^\theta_{Y_{-\infty:t-1}}, \nabla f^\theta_{Y_{-\infty:t-1}}, f^\theta_{Y_{-\infty:t-1}}, Y_t)\| = 0 ,$$

where by abuse of notation, we use again $\|\cdot\|$ to denote any norm on the set of $d\times d$-matrices with real entries,

(c)

$$\mathbb{E}\left[\|\varphi(\nabla f^{\theta_\star}_{Y_{-\infty:0}}, f^{\theta_\star}_{Y_{-\infty:0}}, Y_1)\|^2\right] < \infty , \tag{8.51}$$

$$\mathbb{E}[\sup_{\theta\in \mathrm{B}(\theta_\star,\rho)} \|\psi(\nabla^2 f^\theta_{Y_{-\infty:0}}, \nabla f^\theta_{Y_{-\infty:0}}, f^\theta_{Y_{-\infty:0}}, Y_1)\|] < \infty . \tag{8.52}$$

Moreover, the matrix $\mathcal{J}(\theta_\star)$ defined by

$$\mathcal{J}(\theta_\star) := \mathbb{E}\left[(\nabla f^{\theta_\star}_{Y_{-\infty:0}})(\nabla f^{\theta_\star}_{Y_{-\infty:0}})'\left(\frac{\partial^2 \ln q}{\partial x^2}(f^{\theta_\star}_{Y_{-\infty:0}}, y)\right)\right] , \tag{8.53}$$

is nonsingular.

Theorem 8.33. *Assume A8.21, A8.27 and A8.32. Assume in addition that* $\hat\theta_{n,x} \to_{\mathbb{P}} \theta_\star$. *Then,*

$$\sqrt{n}(\hat\theta_{n,x} - \theta_\star) \overset{\mathbb{P}}{\Longrightarrow} \mathrm{N}(0, \mathcal{J}(\theta_\star)^{-1}\mathcal{I}(\theta_\star)\mathcal{J}(\theta_\star)^{-1}) ,$$

where $\mathcal{J}(\theta_\star)$ *is given by* (8.53) *and* $\mathcal{I}(\theta_\star)$ *is defined by*

$$\mathcal{I}(\theta_\star) := \mathbb{E}\left[(\nabla f^{\theta_\star}_{Y_{-\infty:0}})(\nabla f^{\theta_\star}_{Y_{-\infty:0}})'\left(\frac{\partial \ln q}{\partial x}(\nabla f^{\theta_\star}_{Y_{-\infty:0}}, y)\right)^2\right] .$$

We preface the proof by two Lemmas, whose proofs are given as exercises.

Lemma 8.34. *Assume A8.27 and A8.32. Then,*

$$n^{-1/2}\sum_{t=1}^n \nabla \ln q(f^{\theta_\star}_{Y_{-\infty:t-1}}, Y_t) \overset{\mathbb{P}}{\Longrightarrow} \mathrm{N}(0, \mathcal{I}(\theta_\star)) .$$

Proof. See Exercise 8.11. ■

Lemma 8.35. *Assume A8.27 and A8.32. Let* $\{\theta_n,\ n\in\mathbb{N}\}$ *be a sequence of random vectors such that* $\theta_n \to_{\mathbb{P}} \theta_\star$. *Then, for all* $i, j \in \{1,\dots,d\}$,

$$n^{-1}\sum_{t=1}^n \frac{\partial^2 \ln q(f^{\theta_n}_{Y_{-\infty:t-1}}, Y_t)}{\partial\theta_i\partial\theta_j} \xrightarrow{\mathbb{P}} \mathbb{E}\left[\frac{\partial^2 \ln q(f^{\theta_\star}_{Y_{-\infty:t-1}}, Y_t)}{\partial\theta_i\partial\theta_j}\right] .$$

Proof. See Exercise 8.12. ■

Proof (of Theorem 8.33). Since the MLE $\hat{\theta}_{n,x}$ cancels the derivatives of the log-likelihood, a Taylor expansion at $\theta = \theta_\star$ with an integral form of the remainder yields

$$n^{-1/2}\sum_{t=1}^{n} \nabla \ln q(f_{Y_{1:t-1}}^{\hat{\theta}_{n,x}}(x), Y_t) = 0 = n^{-1/2}\sum_{t=1}^{n} \nabla \ln q(f_{Y_{1:t-1}}^{\theta_\star}(x), Y_t)$$
$$+ n^{-1}\sum_{t=1}^{n}\left(\int_0^1 \nabla^2 \ln q(f_{Y_{1:t-1}}^{\theta_{n,x,s}}(x), Y_t)\mathrm{d}s\right)\sqrt{n}(\hat{\theta}_n - \theta_\star)\,, \quad (8.54)$$

where $\theta_{n,x,s} = s\hat{\theta}_{n,x} + (1-s)\theta_\star$. The proof of Theorem 8.33 then follows from (8.54) and the Slutsky Lemma, provided we show that for all $\theta_n \to_{\mathbb{P}} \theta_\star$,

$$n^{-1/2}\sum_{t=1}^{n} \nabla \ln q(f_{Y_{1:t-1}}^{\theta_\star}(x), Y_t) \overset{\mathbb{P}}{\Longrightarrow} \mathrm{N}(0, \mathcal{J}(\theta_\star))\,, \quad (8.55)$$

$$n^{-1}\sum_{t=1}^{n} \frac{\partial^2 \ln q(f_{Y_{1:t-1}}^{\theta_n}(x), Y_t)}{\partial\theta_i \partial\theta_j} \overset{\mathbb{P}}{\longrightarrow} \mathbb{E}\left[\frac{\partial^2 \ln q(f_{Y_{-\infty:0}}^{\theta_\star}, Y_1)}{\partial\theta_i \partial\theta_j}\right]\,, \quad (8.56)$$

$$\mathcal{J}(\theta_\star) = \mathbb{E}\left[\psi(\nabla^2 f_{Y_{-\infty:0}}^{\theta}, \nabla f_{Y_{-\infty:0}}^{\theta}, f_{Y_{-\infty:0}}^{\theta}, Y_1)\right]\,. \quad (8.57)$$

By noting that

$$\nabla \ln q(f_{Y_{1:t-1}}^{\theta_\star}(x), Y_t) = \varphi(\nabla f_{Y_{1:t-1}}^{\theta_\star}(x), f_{Y_{1:t-1}}^{\theta_\star}(x), Y_t)\,,$$
$$\nabla^2 \ln q(f_{Y_{1:t-1}}^{\theta}(x), Y_t) = \psi(\nabla^2 f_{Y_{1:t-1}}^{\theta}(x), \nabla f_{Y_{1:t-1}}^{\theta}(x), f_{Y_{1:t-1}}^{\theta}(x), Y_t)\,,$$

we obtain, according to A 8.32-(a)-(b), that showing (8.55)-(8.56) is equivalent to showing that for all $\theta_n \to_{\mathbb{P}} \theta_\star$,

$$n^{-1/2}\sum_{t=1}^{n} \nabla \ln q(f_{Y_{-\infty:t-1}}^{\theta_\star}, Y_t) \overset{\mathbb{P}}{\Longrightarrow} \mathrm{N}(0, \mathcal{I}(\theta_\star))\,, \quad (8.58)$$

$$n^{-1}\sum_{t=1}^{n} \frac{\partial^2 \ln q(f_{Y_{-\infty:t-1}}^{\theta_n}, Y_t)}{\partial\theta_i \partial\theta_j} \overset{\mathbb{P}}{\longrightarrow} \mathbb{E}\left[\frac{\partial^2 \ln q(f_{Y_{-\infty:t-1}}^{\theta_\star}, Y_t)}{\partial\theta_i \partial\theta_j}\right]\,. \quad (8.59)$$

But this follows from Lemma 8.34 and Lemma 8.35. It remains to show (8.57). Since under A8.27,

$$\mathbb{E}\left[\nabla^2 f_{Y_{-\infty:0}}^{\theta_\star} \frac{\partial \ln q}{\partial x}(\nabla f_{Y_{-\infty:0}}^{\theta_\star}, Y_1)\right] = 0\,,$$

we have, using (8.50),

$$\mathcal{J}(\theta_\star) = \mathbb{E}\left[\psi(\nabla^2 f_{Y_{-\infty:0}}^{\theta_\star}, \nabla f_{Y_{-\infty:0}}^{\theta_\star}, f_{Y_{-\infty:0}}^{\theta_\star}, Y_1)\right]\,.$$

The proof is completed. ∎

8.4 Bayesian inference

In addition to classical or likelihood inference, there is another approach termed Bayesian inference, wherein prior belief is combined with data to obtain posterior distributions on which statistical inference is based. Although we focus primarily on likelihood based inference in this text, it is also worthwhile to consider Bayesian inference for some problems. Except for some simple cases, Bayesian inference can be computationally intensive and may rely on computational techniques.

The basic idea in Bayesian analysis is that a parameter vector, say $\theta \in \Theta$, is unknown to a researcher, so a *prior* distribution, $\pi(\theta)$, is put on the parameter vector. The researcher also proposes a model or likelihood, $\mathrm{p}(x \mid \theta)$, that describes how the data X depend on the parameter vector. Inference about θ is then based on the *posterior* distribution, which is obtained via Bayes's theorem,

$$\pi(\theta \mid X = x) \propto \pi(\theta)\,\mathrm{p}(x \mid \theta).$$

In some simple cases, the prior and the likelihood are *conjugate* distributions that may be combined easily. For example, in n fixed repeated (i.i.d.) Bernoulli experiments with probability of success θ, a *Beta-Binomial* conjugate pair is taken. In this case the prior is Beta(a,b): $\pi(\theta) \propto \theta^a(1-\theta)^b$; the values $a,b > -1$ are called hyperparameters. The likelihood in this example is Binomial(n,θ): $\mathrm{p}(x \mid \theta) \propto \theta^x(1-\theta)^{n-x}$, from which we easily deduce that the posterior is also Beta, $\pi(\theta \mid X = x) \propto \theta^{x+a}(1-\theta)^{n+b-x}$, and from which inference may easily be achieved. As previously mentioned, in more complex experiments, the posterior distribution is often difficult to obtain by direct calculation, so MCMC techniques are employed. The main idea is that we may not be able to explicitly display the posterior, but we may be able to simulate from the posterior.

In the remaining part of this section, we give three examples of the Bayesian analysis of time series models. The first example is a simple linear AR(p) model. Then, we discuss two nonlinear examples, a threshold model and a GARCH model.

Example 8.36 (AR(p) Model). We now discuss the Bayesian approach for fitting a normal autoregressive model of order p. We write the model as $Y_t = \sum_{i=1}^{p} \phi_i Y_{t-i} + \sigma Z_t$ where $\{Z_t,\, t \in \mathbb{N}\} \sim$ iid $\mathrm{N}(0,1)$; replace Y_s by $Y_s - \mu$ if the mean is not zero. Define

$$\boldsymbol{Y} = \begin{pmatrix} Y_{p+1} \\ Y_{p+2} \\ \vdots \\ Y_n \end{pmatrix}, \quad \boldsymbol{\phi} = \begin{pmatrix} \phi_1 \\ \phi_2 \\ \vdots \\ \phi_p \end{pmatrix}, \quad \boldsymbol{Z} = \begin{pmatrix} Z_{p+1} \\ Z_{p+2} \\ \vdots \\ Z_n \end{pmatrix},$$

$$X = \begin{pmatrix} Y_p & Y_{p-1} & \dots & Y_1 \\ Y_{p+1} & Y_p & \dots & Y_2 \\ \vdots & \vdots & & \vdots \\ Y_n & Y_{n-1} & \dots & Y_{n-p+1} \end{pmatrix}.$$

The model can then be written as

$$\boldsymbol{Y} = X\boldsymbol{\phi} + \sigma \boldsymbol{Z},$$

which implies that $\boldsymbol{Y} \sim \mathrm{N}(X\phi,\ \sigma^2 \mathrm{I})$, with X being $(n-p)\times p$ matrix; for simplicity, X is assumed to be of full column rank p. It is easier to reparametrize the model in terms of a precision parameter, $\tau = 1/\sigma^2$. Then the likelihood function can be written as, up to a constant,

$$\mathrm{L}(\boldsymbol{\phi},\ \tau; \boldsymbol{Y}) = \mathrm{p}(\boldsymbol{Y} \mid \boldsymbol{\phi},\ \tau) \propto \tau^{(n-p)/2} \exp(-\tau(\boldsymbol{Y} - X\boldsymbol{\phi})'(\boldsymbol{Y} - X\boldsymbol{\phi})/2)\,. \tag{8.60}$$

In a Bayesian framework, there are two ways for considering the prior settings of the ARMA parameters. One is to constrain the prior to be non-zero only over the region defined by the stationarity conditions, such as in Nakatsuma (2000) and Philippe (2006). The second method is to simply ignore the stationary assumptions and proceed to use, e.g., the normal-gamma family as a conjugate prior distribution, such as in Chen (1999). In this example, we apply the latter method, i.e., no restriction for stationarity or invertibility is made in the prior distribution. We use the following specification

$$\begin{cases} \boldsymbol{Y} \mid (\boldsymbol{\phi},\ \tau) \sim \mathrm{N}(X\boldsymbol{\phi},\ \mathrm{I}/\tau) \\ \boldsymbol{\phi} \mid \tau \sim \mathrm{N}_p(\boldsymbol{\phi}_0,\ V_0/\tau) \\ \tau \sim \Gamma(a,\ b) \end{cases}$$

where $\boldsymbol{\phi}_0$ and V_0 are the prior mean and covariance matrix, and a and b are the shape and scale of the prior of the precision parameter. The matrix V_0 is assumed to be invertible (which is generally chosen to be $V_0 = \gamma_0^2 \mathrm{I}$, and we can take γ_0^2 large if the prior is meant to be noninformative). After straightforward simplification, the posterior joint density of $(\boldsymbol{\phi},\ \tau)$ becomes

$$\begin{aligned} \pi(\boldsymbol{\phi},\ \tau \mid \boldsymbol{Y}) &\propto \tau^{\tilde{a}-1} \exp(-\tau(b + (1/2)\boldsymbol{\phi}_0' V_0^{-1} \boldsymbol{\phi}_0 + (1/2)\|\boldsymbol{y} - X\boldsymbol{\phi}\|^2)) \\ &\propto \tau^{\tilde{a}-1} \exp(-\tau \tilde{b}) \exp\left(-(\tau/2)\cdot(\boldsymbol{\phi} - \overline{\boldsymbol{\phi}})'(X'X + V_0^{-1})(\boldsymbol{\phi} - \overline{\boldsymbol{\phi}})\right) \end{aligned}$$

where

$$\begin{aligned} \tilde{a} &= \frac{n}{2} + a \\ \tilde{b} &= b + \frac{\boldsymbol{Y}'\boldsymbol{Y} - (X'\boldsymbol{Y} + V_0^{-1}\boldsymbol{\phi}_0)'(X'X + V_0^{-1})^{-1}(X'\boldsymbol{Y} + V_0^{-1}\boldsymbol{\phi}_0)}{2} \\ \overline{\boldsymbol{\phi}} &= (X'X + V_0^{-1})^{-1}(X'\boldsymbol{Y} + V_0^{-1}\boldsymbol{\phi}_0). \end{aligned}$$

From the joint posterior density $f(\boldsymbol{\phi},\ \tau \mid \boldsymbol{Y})$, it can be seen that the conditional posterior density of $(\boldsymbol{\phi} \mid \tau,\ \boldsymbol{Y})$ is multivariate normal with mean $\overline{\boldsymbol{\phi}}$ and covariance matrix $\tau^{-1}(X'X + V_0^{-1})^{-1}$, i.e.,

$$\boldsymbol{\phi} \mid (\tau,\ \boldsymbol{Y}) \sim \mathrm{N}_p(\overline{\boldsymbol{\phi}},\ \tau^{-1}(X'X + V_0^{-1})^{-1})\,.$$

The conditional posterior density of $(\tau \mid \boldsymbol{\phi},\ \boldsymbol{Y})$ is $\Gamma(\tilde{a},\ \tilde{b}(\boldsymbol{\phi}))$,

$$\tilde{b}(\boldsymbol{\phi}) := b + (1/2)\boldsymbol{\phi}_0' V_0^{-1} \boldsymbol{\phi}_0 + (1/2)\|\boldsymbol{Y} - X\boldsymbol{\phi}\|^2\,.$$

As an example, we use our R script `arp.mcmc` to analyze a simulated AR(2) model with $\phi_1 = 1$, $\phi_2 = -.9$ and $\sigma^2 = 1$.

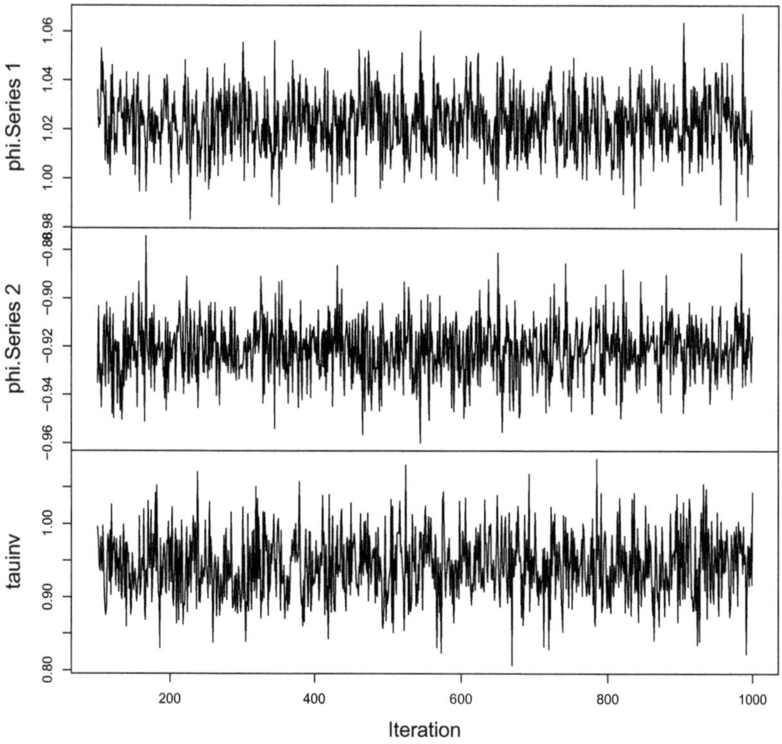

Figure 8.1 *Output of* `arp.mcmc` *for Example 8.36. Each plot shows the trace, after burnin, of the draws for each component.*

```
> x = arima.sim(list(order=c(2,0,0), ar=c(1,-.9)), n=1000)  # generate an ar2
> arp.mcmc(x, porder=2, n.iter=1000, n.warmup=100)    # sample from posterior
     phi.Series 1      phi.Series 2          tauinv
   Min.    :0.9831   Min.    :-0.9598   Min.    :0.8068
   1st Qu.:1.0142   1st Qu.:-0.9292   1st Qu.:0.9117
   Median :1.0228   Median :-0.9219   Median :0.9419
   Mean   :1.0226   Mean   :-0.9217   Mean   :0.9430
   3rd Qu.:1.0311   3rd Qu.:-0.9138   3rd Qu.:0.9739
   Max.   :1.0672   Max.   :-0.8744   Max.   :1.0886
```

A summary as well as trace plots are produced by the script; see Figure 8.1. ◇

Example 8.37 (SETAR model for the lynx series). A two-regime SETAR model is considered, and denoted by SETAR (2; p_1, p_2):

$$Y_t = \begin{cases} \phi_0^{(1)} + \sum_{i=1}^{p_1} \phi_i^{(1)} Y_{t-i} + Z_t^{(1)} & Y_{t-d} \le r \\ \phi_0^{(2)} + \sum_{i=1}^{p_2} \phi_i^{(2)} Y_{t-i} + Z_t^{(2)} & Y_{t-d} > r \end{cases} \tag{8.61}$$

where $Z_t^{(k)} = \sigma_k Z_t$, $k = 1, 2$, and $\{Z_t,\ t \in \mathbb{N}\} \sim$ iid N$(0,\ 1)$. This is a piecewise AR model, where regime switching is driven by lagged values of the series and a threshold value r. The unknown parameters in the model are $\boldsymbol{\phi}_k = (\phi_0^{(k)}, \phi_1^{(k)}, \ldots, \phi_{p_k}^{(k)})'$, $k = 1,2$, the threshold value r, the delay lag d and the regime error variances are σ_k^2, $k = 1,2$.

Likelihood approaches for estimating threshold models usually involve setting the threshold r and the delay d (e.g., $r = 0$ and $d = 1$), or choosing r, d via some information criterion. The likelihood inference for the parameters is then carried out conditionally upon these choices, but ignore the associated uncertainty in those parameters. The uncertainty in estimating r and d is explicitly captured via a Bayesian approach.

For SETAR models, the stationary and invertible conditions in the literature are rather involved. As in Example 8.36, we do not take these constraints into account in the Bayesian analysis. We use the following specification for the priors. The AR parameters are assumed to be normal with zero mean and covariance V_k, $k = 1,2$, i.e., $\boldsymbol{\phi}_k \sim \mathrm{N}(0, V_k)$, where V_k is the prior covariance matrix. We may for example set $V_k = \gamma_v^2 \mathrm{I}_{g_k}$, where γ_v^2 is the prior variance (generally taken to be large). An inverse-gamma prior is assumed for σ_k^2, i.e., $1/\sigma_k^2 \sim \Gamma(a_k, b_k)$, for $k = 1,2$, where the hyperparameters a_k and b_k are known constants. The prior for the threshold value r is uniformly distributed between $[r_{\min}, r_{\max}]$, where $r_{\min}$ and $r_{\max}$ are generally specified as percentiles of the observations (e.g., 10 and 90 percentiles), for identification purposes in each regime. A discrete uniform prior on $\{1, 2, \ldots, d_{\max}\}$ is employed for the delay d.

Let $\boldsymbol{Y} = (Y_{p+1}, \ldots, Y_n)$ be the observed values of the TAR model, where $p = \max(p_1, p_2)$. the conditional likelihood function of the model is:

$$L(\mathbf{y} \mid \boldsymbol{\phi}^{(1)}, \boldsymbol{\phi}^{(2)}\ \sigma_1^2,\ \sigma_2^2,\ r,\ d) \propto \sigma_1^{-n_1} \sigma_2^{-n_2}$$
$$\times \exp\left(\sum_{t=p+1}^{n} \sum_{k=1}^{2} -\frac{1}{2\sigma_k^2} \left\{ Y_t - \phi_0^{(k)} - \sum_{i=1}^{p_k} \phi_j^{(k)} Y_{t-i} \right\}^2 I_t^{(k)} \right)$$

where $n_1 := \sum_{t=p+1}^{n} \mathbb{1}\{Y_{t-d} \le r\}$, $n_2 = n - n_1$ and $I_t^{(k)}$ is the indicator variable $\mathbb{1}\{r_{k-1} \le Y_{t-d} < r_k\}$ with $r_0 = -\infty$, $r_1 = r$ and $r_2 = \infty$. For notational convenience, let the full parameter vector be $\boldsymbol{\theta} = \{\boldsymbol{\phi}^{(1)}, \boldsymbol{\phi}^{(2)}, \sigma_1^2, \sigma_2^2, r, d\}$. Under the prior specification, the k-th regime AR parameter $\boldsymbol{\phi}^{(k)} \mid (\boldsymbol{Y}, \boldsymbol{\theta} \setminus \{\boldsymbol{\phi}^{(k)}\})$ is Gaussian with mean $\bar{\boldsymbol{\phi}}^{(k)}$ and covariance $\bar{V}_k$ given by

$$\bar{\boldsymbol{\phi}}^{(k)} = \sigma_k^{-2} \bar{V}_k X_k' \boldsymbol{Y}_k$$
$$\bar{V}_k = (\sigma_k^{-2} X_k' X_k + V_k^{-1})^{-1},$$

where the design matrices X_1 and X_2 are given by

$$X_1 = \begin{bmatrix} 1 & Y_{\ell_1 - 1} & \cdots & \cdots & Y_{\ell_1 - p_1} \\ 1 & Y_{\ell_2 - 1} & \cdots & \cdots & Y_{\ell_2 - p_1} \\ \vdots & \vdots & & & \vdots \\ 1 & Y_{\ell_{n_1} - 1} & \cdots & \cdots & Y_{\ell_{n_1} - p_1} \end{bmatrix}$$

and

$$X_2 = \begin{bmatrix} 1 & Y_{u_1-1} & \cdots & \cdots & Y_{u_1-p_2} \\ 1 & Y_{u_2-1} & \cdots & \cdots & Y_{u_2-p_2} \\ \vdots & \vdots & & & \vdots \\ 1 & Y_{u_{n_1}-1} & \cdots & \cdots & Y_{u_{n_2}-p_2} \end{bmatrix}$$

where ℓ_i and u_j are the time indices of the i-th and j-th observations in the "lower" and "upper" regimes: ℓ_i is the index for regime 1 and u_j is for regime 2. The vectors $\boldsymbol{Y}_1$ and $\boldsymbol{Y}_2$ are set as $\boldsymbol{Y}_1 = (Y_{\ell_1}, \ldots, Y_{\ell_{n_1}})'$ and $\boldsymbol{Y}_2 = (Y_{u_1}, \ldots, Y_{u_{n_2}})'$.

The posterior distribution of the inverse of the variance $\sigma_k^{-2} \mid (\boldsymbol{Y}, \boldsymbol{\theta} \setminus \{\sigma_k^{-2}\})$ is Gamma with shape and scale

$$\tilde{a}_k = a_k + n_k/2\,,\ \tilde{\boldsymbol{b}}_k = b_k + (1/2)\|\boldsymbol{Y}_k - X_k \boldsymbol{\phi}_k\|^2\,.$$

The posterior distribution for the threshold is a bit more involved. By simple multiplication of the likelihood and prior for r, the posterior density of $r \mid (\boldsymbol{Y}, \boldsymbol{\theta} \setminus \{r\})$ is given by (up to a normalization constant)

$$\sigma_1^{-n_1} \sigma_2^{-n_2} \exp\left(-\frac{1}{2} \sum_{k=1}^{2} \frac{\|\boldsymbol{Y}_k - X_k \boldsymbol{\phi}_k\|^2}{\sigma_k^2} \right) \mathbb{1}_{\{r \in [r_{\min}, r_{\max}]\}}$$

where $n_1, n_2, \boldsymbol{Y}_1, \boldsymbol{Y}_2, X_1, X_2$ depend on r. This posterior distribution is not a known or standard form and therefore we cannot use the elementary Gibbs sampler. To solve this problem, we use a hybrid of the ideas behind the Gibbs sampler and the Metropolis-Hastings algorithm. The algorithm merely replaces the step that samples from the full-posterior by a Metropolis-Hastings step, which leaves the full-posterior stationary. Now by simulating sufficiently many of these Metropolis-Hastings steps before proceeding to the next component, we will approximately mimic the elementary Gibbs sampler itself. However, there is often a far more efficient procedure that only updates the component once (or a small finite number of times) before moving on to other coordinates. Like the Gibbs sampler, this algorithm can be thought of as a special case of the Metropolis-Hastings algorithm. In this special case, Chen et al. (2011) use a random walk Metropolis algorithm with a Gaussian proposal density with zero mean and variance σ^2, to sample this parameter. Finally, the posterior distribution of the delay $d \mid (\boldsymbol{Y}, \boldsymbol{\theta} \setminus \{d\})$ is a multinomial trial with probabilities

$$\frac{L(\boldsymbol{Y} \mid \boldsymbol{\phi}^{(1)}, \boldsymbol{\phi}^{(2)}\ \sigma_1^2,\ \sigma_2^2,\ r, j)}{\sum_{i=1}^{d_{\max}} L(\boldsymbol{Y} \mid \boldsymbol{\phi}^{(1)}, \boldsymbol{\phi}^{(2)}\ \sigma_1^2,\ \sigma_2^2,\ r, i)}\,,\ j = 1, 2, \ldots, d_{\max}\,.$$

The sampling scheme amounts to updating parameters in turn, which leads to the algorithm implemented in the R package `BAYSTAR`. However, sampling the delay distribution this way is not optimal and may lead to slow mixing.

We analyzed the lynx data set presented in Chapter 3. We used `BAYSTAR` to perform a Bayesian inference of a two-regime threshold model, which was considered in Example 3.11. In particular, our goal was to fit the same model, an AR(2) for each regime, with the threshold delay being 2. In this example, we set the maximum delay,

$d_{\max}$, equal to 2. In this case, the highest posterior probability for the delay lag was indeed 2. We also note that the variance of the proposal density, σ^2, which is called `step.thv`, must be specified. If the scaling parameter is too large, the acceptance ratio of the algorithm will be small. If it is too small, then the acceptance ratio is high but the algorithm mixes slowly (see Example 5.37 for a discussion of this issue). This can be monitored by observing the output produced by the package.

The R code and a partial output is given below. In addition to the output shown below, BAYSTAR exhibits various diagnostic plots, which we do not display.

```
> require(BAYSTAR)
> lynx.out = BAYSTAR(log10(lynx), lagp1=1:2, lagp2=1:2, Iteration=10000,
    Burnin=2000, d0=2, step.thv=.5)
                  mean  median     s.d.   lower   upper
   phi1.0       0.5620  0.5701   0.1683  0.2039  0.8733
   phi1.1       1.2614  1.2607   0.0724  1.1251  1.4022
   phi1.2      -0.4143 -0.4174   0.0919 -0.5867 -0.2274
   phi2.0       1.4567  1.5353   1.0021 -0.6318  3.2400
   phi2.1       1.5694  1.5643   0.1317  1.3221  1.8346
   phi2.2      -1.0623 -1.0759   0.2859 -1.5988 -0.4590
   sigma^2 1    0.0363  0.0356   0.0067  0.0258  0.0513
   simga^2 2    0.0557  0.0535   0.0143  0.0349  0.0899
   r            3.2057  3.2810   0.1818  2.5784  3.3848
   acceptance rate of r =  23.76 %
   The highest posterior prob. of lag is at :  2
```

We note that the result of the MCMC analysis is similar to the frequentist analysis presented in Example 3.11. ◇

Example 8.38 (Bayesian GARCH). In this example, we consider a GARCH(1,1) model with Student-t innovations for log-returns, i.e.,

$$Y_t = \sigma_t \left(\frac{\nu - 2}{\nu}\right)^{1/2} \varepsilon_t \tag{8.62}$$

$$\sigma_t^2 = \alpha_0 + \alpha_1 Y_{t-1}^2 + \beta_1 \sigma_{t-1}^2 \tag{8.63}$$

where $\{Z_t,\ t \in \mathbb{N}\}$ are i.i.d. t-distribution with $\nu > 2$ degrees of freedom, $\alpha_0 > 0$, $\alpha_1, \beta_1 \geq 0$ and $\nu > 2$. To perform the Bayesian analysis of this model, it is easier to use data augmentation; see Tanner (1993) and Geweke (1993). The idea is to represent the t-distribution as a scale mixture of Gaussians as suggested by Andrews and Mallows (1974). A random variable Z is a Gaussian scale mixture if it can be expressed as the product of a standard Gaussian G and an independent positive scalar random variable $H^{1/2}$: $Z = H^{1/2} G$. The variable H is the *multiplier* or the *scale*. If H has finite support, then Z is a finite mixture of Gaussians, whereas if H has a density (with respect to Lebesgue measure) on $\mathbb{R}_+$, then Z is a continuous mixture of Gaussian variables. Gaussian scale mixtures are symmetric, zero mean, and have leptokurtic marginal densities (tails heavier than those of a Gaussian distribution). Assume that the distribution of H is inverse gamma (denoted IG), with shape and scale parameters both equal to $\nu/2$; then the distribution of H has the density

$$p_\nu(h) = \left(\frac{\nu}{2}\right)^{\nu/2} \frac{h^{-(\nu+2)/2}}{\Gamma(\nu/2)} \exp\left(-\frac{\nu}{2h}\right), h \geq 0.$$

If $G \sim \mathrm{N}(0,1)$, then the distribution of $Z = H^{1/2}G$ has a density given by

$$f(z) = \frac{\left(\frac{\nu}{2}\right)^{\nu/2}}{\Gamma(\nu/2)\sqrt{2\pi}} \int_0^\infty h^{-(\nu+3)/2} \exp\left(-\frac{z^2+\nu}{2h}\right) \mathrm{d}h \,.$$

Using the result,

$$\int_0^\infty x^{-a/2} \exp\left(-\frac{b}{2x}\right) \mathrm{d}x = \left(\frac{2}{b}\right)^{(a-2)/2} \Gamma\left(\frac{a-2}{2}\right) ,$$

which derives from a simple change variable in the definition of the complete gamma function, we obtain that

$$f(z) = \frac{\left(\frac{\nu}{2}\right)^{\nu/2} \Gamma((\nu+1)/2)}{\Gamma(\nu/2)\sqrt{2\pi}} \left(\frac{2}{z^2+\nu}\right)^{(\nu+1)/2} ,$$

which is the density of a Student-t distribution with ν degrees of freedom. Using this representation of the t-distribution, we may rewrite (8.62) as follows,

$$\begin{aligned}
&Y_t = \sigma_t \left(\frac{\nu-2}{\nu}\right)^{1/2} H_t^{1/2} G_t \\
&G_t \sim \text{iid } \mathrm{N}(0,1) \\
&H_t \sim \text{iid } \mathrm{IG}\left(\frac{\nu}{2}, \frac{\nu}{2}\right) \\
&\sigma_t^2 = \alpha_0 + \alpha_1 Y_{t-1}^2 + \beta_1 \sigma_{t-1}^2 \,.
\end{aligned}$$

Assume that we have n observations from this model. In order to write the likelihood function, we define the vectors, $\boldsymbol{H} = (H_1, \ldots, H_n)'$ and $\boldsymbol{\alpha} = (\alpha_0, \alpha_1)'$. The model parameters are collected into the vector $\boldsymbol{\theta} = (\boldsymbol{\alpha}, \beta, \nu)$. Define the $n \times n$ diagonal matrix

$$\Sigma = \Sigma(\boldsymbol{\theta}, \boldsymbol{H}) = \mathrm{diag}\left(\left\{H_t \frac{\nu-2}{\nu} \sigma_t^2(\boldsymbol{\alpha}, \beta_1)\right\}_{i=1}^n\right) ,$$

where $\sigma_t^2(\boldsymbol{\alpha}, \beta_1) = \alpha_0 + \alpha_1 Y_{t-1}^2 + \beta_1 \sigma_{t-1}^2(\boldsymbol{\alpha}, \beta_1)$, and $\sigma_0^2(\boldsymbol{\alpha}, \beta_1) = 0$. We can express the conditional likelihood

$$L(\boldsymbol{Y} \mid \boldsymbol{\theta}, \boldsymbol{H}) \propto (\det \Sigma)^{-1/2} \exp\left(-\frac{1}{2} \boldsymbol{Y}' \Sigma^{-1} \boldsymbol{Y}\right) .$$

The prior distribution for the GARCH parameters $\boldsymbol{\alpha}$ and β_1 are chosen to be independent and distributed according to the truncated normal distribution (to ensure the positivity of the coefficients)

$$p(\boldsymbol{\alpha}) \propto (\det \Sigma_{\boldsymbol{\alpha}})^{-1/2} \exp\left(-\frac{1}{2}(\boldsymbol{\alpha} - \boldsymbol{\mu}_\alpha)' \Sigma_\alpha^{-1} (\boldsymbol{\alpha} - \boldsymbol{\mu}_\alpha)\right) \mathbb{1}_{\mathbb{R}_+^2}(\boldsymbol{\alpha})$$

$$p(\beta_1) \propto \sigma_\beta^{-1} \exp\left(-\frac{1}{2\sigma_\beta^2} (\beta_1 - \mu_\beta)^2\right) \mathbb{1}_{\mathbb{R}^+}(\beta_1),$$

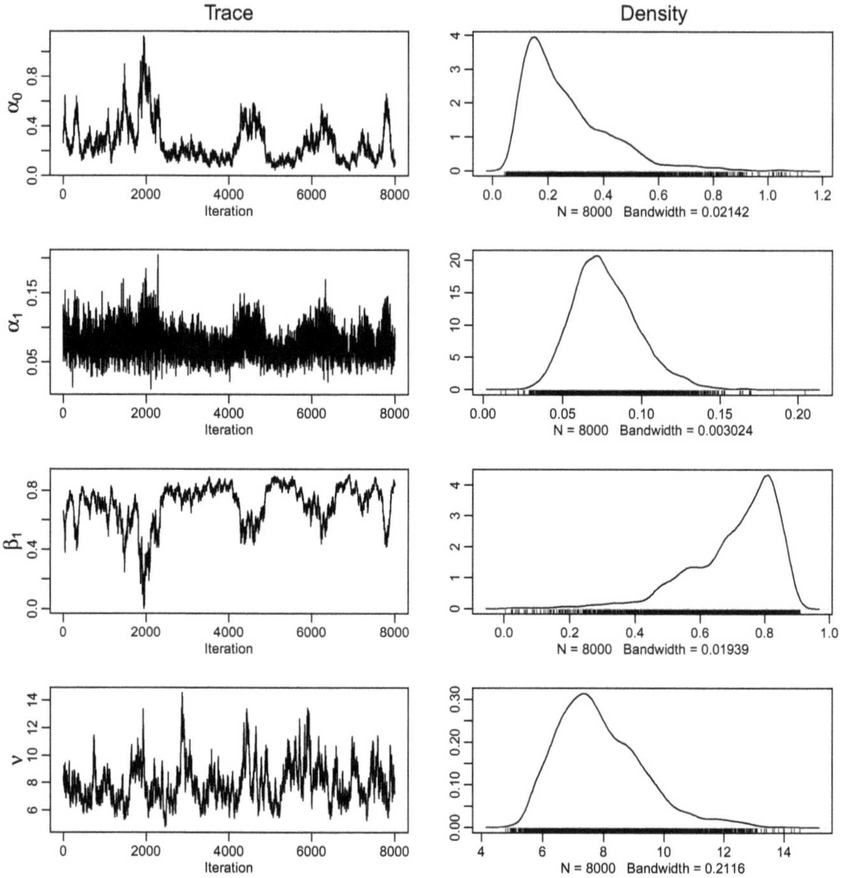

Figure 8.2: *Output of* `bayesGARCH` *for Example 8.38*

where μ_α, Σ_α, μ_β, σ_β are the hyperparameters. Only positivity constraints are taken into account in the prior specification; no stationarity conditions are imposed.

Since the components H_t are i.i.d. from the inverse gamma distribution, the conditional distribution of the vector $\boldsymbol{H}$ given ν has a density given by

$$p(\boldsymbol{h} \mid \nu) = \left(\frac{\nu}{2}\right)^{\frac{n\nu}{2}} \left[\Gamma(\frac{\nu}{2})\right]^{-n} \left(\prod_{t=1}^{n} h_t\right)^{-\frac{\nu}{2}-1} \exp\left(-\frac{1}{2}\sum_{t=1}^{n}\frac{\nu}{h_t}\right) .$$

The prior distribution on the degrees of freedom ν is chosen to be a translated exponential with parameters $\lambda > 0$ and $\delta \geq 2$

$$p(\nu) = \lambda \exp[-\lambda(\nu - \delta)] \mathbb{1}\{\nu \geq \delta\} .$$

For large values of λ, most of the mass of the prior is concentrated in the neighborhood of δ and a constraint on the degrees of freedom can be imposed in this

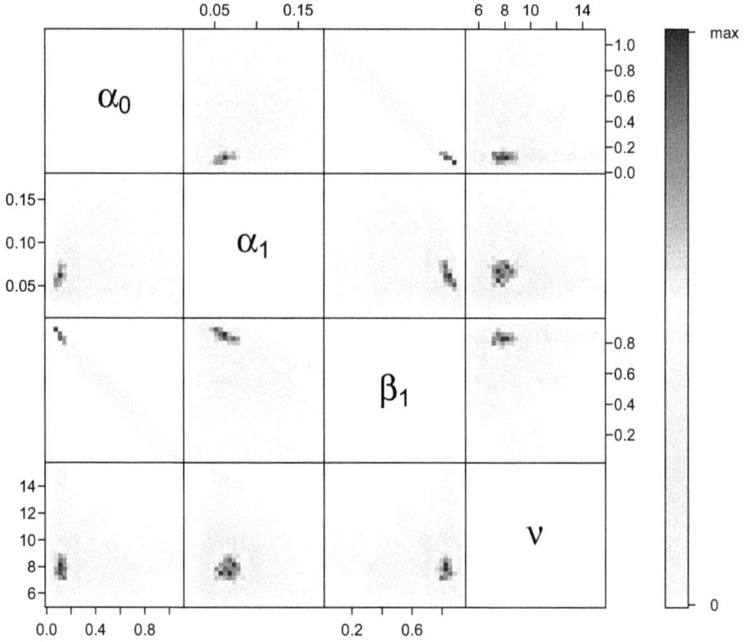

Figure 8.3: *Pairwise sampling distributions for Example 8.38.*

manner. A light tail prior may be imposed by taking both δ and λ large. The joint prior distribution is then obtained by assuming that the priors are independent, i.e., $p(\bar{\theta},\bar{h}) = p(\bar{\alpha})p(\beta_1)p(h \mid \nu)p(\nu)$. The joint posterior and the full conditional densities cannot be expressed in closed form. Therefore, we cannot use the elementary Gibbs sampler and need to rely on a Metropolis-within-Gibbs strategy to approximate the posterior density. In this case, there is an R package called `bayesGARCH` that can be used to perform the analysis. Implementations details are given in Ardia and Hoogerheide (2010). For this example, we use part of the `CAC` returns from the R dataset `EuStockMarkets`. The graphical output of the package is displayed in Figure 8.2, which shows the traces for each parameter (after burnin) and the corresponding posterior densities. We also use the R package `IDPmisc` to graph the pairwise results; see Figure 8.3. ◇

8.5 Some proofs

Let (X,d) be a Polish space equipped with its Borel sigma-field $\mathcal{X}$. Let $\mathsf{S} : \mathsf{X}^{\mathbb{N}} \to \mathsf{X}^{\mathbb{N}}$ and $\tilde{\mathsf{S}} : \mathsf{X}^{\mathbb{Z}} \to \mathsf{X}^{\mathbb{Z}}$ be the shift operators defined by: for all $\boldsymbol{x} = (x_t)_{t\in\mathbb{N}} \in \mathsf{X}^{\mathbb{N}}$ and all

$\tilde{\boldsymbol{x}} = (\tilde{x}_t)_{t\in\mathbb{Z}} \in \mathsf{X}^{\mathbb{Z}}$,

$$\mathsf{S}(\boldsymbol{x}) = (y_t)_{t\in\mathbb{N}}, \quad \text{where} \quad y_t = x_{t+1}, \quad \forall t \in \mathbb{N}, \tag{8.64}$$
$$\tilde{\mathsf{S}}(\tilde{\boldsymbol{x}}) = (\tilde{y}_t)_{t\in\mathbb{Z}}, \quad \text{where} \quad \tilde{y}_t = \tilde{x}_{t+1}, \quad \forall t \in \mathbb{Z}. \tag{8.65}$$

Assume that $(\mathsf{X}^{\mathbb{Z}}, \mathcal{X}^{\otimes\mathbb{Z}}, \mathbb{P}, \tilde{\mathsf{S}})$ is a measure-preserving ergodic dynamical system. Denote by $\mathbb{E}$ the expectation operator associated to $\mathbb{P}$.

Let $(\bar{\ell}^\theta, \theta \in \Theta)$ be a family of measurable functions, $\bar{\ell}^\theta : \mathsf{X}^{\mathbb{Z}} \to \mathbb{R}$, indexed by $\theta \in \Theta$ where (Θ, d) is a compact metric space and denote $\bar{\mathsf{L}}_n^\theta := n^{-1}\sum_{k=0}^{n-1} \bar{\ell}^\theta \circ \tilde{\mathsf{S}}^k$. Moreover, consider $(\mathsf{L}_n^\theta, n \in \mathbb{N}^*, \theta \in \Theta)$ a family of upper-semicontinuous functions $\mathsf{L}_n^\theta : \mathsf{X}^{\mathbb{Z}} \to \mathbb{R}$ indexed by $n \in \mathbb{N}^*$ and $\theta \in \Theta$. Consider the following assumptions:

Assumption A8.39. $\mathbb{E}[\sup_{\theta\in\Theta} \bar{\ell}_+^\theta] < \infty$,

Assumption A8.40. $\mathbb{P}$-a.s., the function $\theta \mapsto \bar{\ell}^\theta$ is upper-semicontinuous,

Assumption A8.41. $\lim_{n\to\infty} \sup_{\theta\in\Theta} |\mathsf{L}_n^\theta - \bar{\mathsf{L}}_n^\theta| = 0, \quad \mathbb{P}$-a.s.

Let $\{\bar{\theta}_n : n \in \mathbb{N}^*\} \subset \Theta$ and $\{\hat{\theta}_n : n \in \mathbb{N}^*\} \subset \Theta$ such that for all $n \geq 1$,

$$\bar{\theta}_n \in \operatorname*{arg\,max}_{\theta\in\Theta} \bar{\mathsf{L}}_n^\theta, \quad \hat{\theta}_n \in \operatorname*{arg\,max}_{\theta\in\Theta} \mathsf{L}_n^\theta.$$

The proofs are adapted from Pfanzagl (1969).

Theorem 8.42. *Assume A8.39–A8.40. Then*

$$\lim_{n\to\infty} \mathsf{d}(\bar{\theta}_n, \Theta_\star) = 0, \ \mathbb{P}\text{-a.s. } \textit{where } \Theta_\star := \operatorname*{arg\,max}_{\theta\in\Theta} \mathbb{E}[\bar{\ell}^\theta]. \tag{8.66}$$

If, in addition, A8.41 holds, then $\lim_{n\to\infty} \mathsf{d}(\hat{\theta}_n, \Theta_\star) = 0$, $\mathbb{P}$-a.s. *and,*

$$\lim_{n\to\infty} \mathsf{L}_n^{\hat{\theta}_n} = \sup_{\theta\in\Theta} \mathbb{E}[\bar{\ell}^\theta], \quad \mathbb{P}\text{-a.s.} \tag{8.67}$$
$$\textit{for all } \theta \in \Theta, \quad \lim_{n\to\infty} \mathsf{L}_n^\theta = \mathbb{E}[\bar{\ell}^\theta], \quad \mathbb{P}\text{-a.s.} \tag{8.68}$$

Proof. First note that according to the Birkhoff ergodic theorem (Theorem 7.44) and A8.39, for all $\theta \in \Theta$, $\lim_{n\to\infty} \bar{\mathsf{L}}_n^\theta$ exists $\mathbb{P}$-a.s., and

$$\lim_{n\to\infty} \bar{\mathsf{L}}_n^\theta = \lim_{n\to\infty} n^{-1} \sum_{k=0}^{n-1} \bar{\ell}^\theta \circ \tilde{\mathsf{S}}^k = \mathbb{E}[\bar{\ell}^\theta], \quad \mathbb{P}\text{-a.s.} \tag{8.69}$$

Let K be a compact subset of Θ. For all $\theta_0 \in K$, $\mathbb{P}$-a.s.,

$$\begin{aligned}
&\limsup_{\rho\to 0} \limsup_{n\to\infty} \sup_{\theta\in\mathrm{B}(\theta_0,\rho)} n^{-1} \sum_{k=0}^{n-1} \bar{\ell}^\theta \circ \tilde{\mathsf{S}}^k \\
&\leq \limsup_{\rho\to 0} \limsup_{n\to\infty} n^{-1} \sum_{k=0}^{n-1} \sup_{\theta\in\mathrm{B}(\theta_0,\rho)} \bar{\ell}^\theta \circ \tilde{\mathsf{S}}^k = \limsup_{\rho\to 0} \mathbb{E}[\sup_{\theta\in\mathrm{B}(\theta_0,\rho)} \bar{\ell}^\theta],
\end{aligned} \tag{8.70}$$

where the last equality follows from A8.39 and the Birkhoff ergodic theorem (Theorem 7.44). Moreover, by the monotone convergence theorem applied to the nondecreasing function $\rho \mapsto \sup_{\theta\in\mathrm{B}(\theta_0,\rho)} \bar{\ell}^{\theta}$, we have

$$\limsup_{\rho\to 0} \mathbb{E}[\sup_{\theta\in\mathrm{B}(\theta_0,\rho)} \bar{\ell}^{\theta}] = \mathbb{E}[\limsup_{\rho\to 0} \sup_{\theta\in\mathrm{B}(\theta_0,\rho)} \bar{\ell}^{\theta}] \le \mathbb{E}[\bar{\ell}^{\theta_0}] , \tag{8.71}$$

where the last inequality follows from A8.40. Combining (8.70) and (8.71), we obtain that for all $\eta > 0$ and $\theta_0 \in K$, there exists $\rho^{\theta_0} > 0$ satisfying

$$\limsup_{n\to\infty} \sup_{\theta\in\mathrm{B}(\theta_0,\rho^{\theta_0})} n^{-1} \sum_{k=0}^{n-1} \bar{\ell}^{\theta} \circ \tilde{\mathrm{S}}^k \le \mathbb{E}[\bar{\ell}^{\theta_0}] + \eta \le \sup_{\theta\in K} \mathbb{E}[\bar{\ell}^{\theta}] + \eta , \quad \mathbb{P}\text{-a.s.}$$

Since K is a compact subset of Θ, we can extract a finite subcover of K from $\bigcup_{\theta_0\in K} \mathrm{B}(\theta_0, \rho^{\theta_0})$, so that

$$\limsup_{n\to\infty} \sup_{\theta\in K} n^{-1} \sum_{k=0}^{n-1} \bar{\ell}^{\theta} \circ \tilde{\mathrm{S}}^k \le \sup_{\theta\in K} \mathbb{E}[\bar{\ell}^{\theta}] + \eta , \quad \mathbb{P}\text{-a.s.} \tag{8.72}$$

Since η is arbitrary, we obtain

$$\limsup_{n\to\infty} \sup_{\theta\in K} n^{-1} \sum_{k=0}^{n-1} \bar{\ell}^{\theta} \circ \tilde{\mathrm{S}}^k \le \sup_{\theta\in K} \mathbb{E}[\bar{\ell}^{\theta}] , \quad \mathbb{P}\text{-a.s.} \tag{8.73}$$

Moreover, (8.71) implies

$$\limsup_{\rho\to 0} \sup_{\theta\in\mathrm{B}(\theta_0,\rho)} \mathbb{E}[\bar{\ell}^{\theta}] \le \limsup_{\rho\to 0} \mathbb{E}[\sup_{\theta\in\mathrm{B}(\theta_0,\rho)} \bar{\ell}^{\theta}] \le \mathbb{E}[\bar{\ell}^{\theta_0}] ,$$

This shows that $\theta \mapsto \mathbb{E}[\bar{\ell}^{\theta}]$ is upper-semicontinuous. As a consequence, $\Theta_\star := \arg\max_{\theta\in\Theta} \mathbb{E}[\bar{\ell}^{\theta}]$ is a closed and nonempty subset of Θ and therefore, for all $\varepsilon > 0$, $K_\varepsilon := \{\theta \in \Theta; \mathrm{d}(\theta, \Theta_\star) \ge \varepsilon\}$ is a compact subset of Θ. Using again the upper-semicontinuity of $\theta \mapsto \mathbb{E}[\bar{\ell}^{\theta}]$, there exists $\theta_\varepsilon \in K_\varepsilon$ such that for all $\theta_\star \in \Theta_\star$,

$$\sup_{\theta\in K_\varepsilon} \mathbb{E}[\bar{\ell}^{\theta}] = \mathbb{E}[\bar{\ell}^{\theta_\varepsilon}] < \mathbb{E}[\bar{\ell}^{\theta_\star}] .$$

Finally, combining this inequality with (8.73), we obtain that $\mathbb{P}$-a.s.,

$$\begin{aligned}\limsup_{n\to\infty} \sup_{\theta\in K_\varepsilon} \bar{\mathrm{L}}_n^{\theta} = \limsup_{n\to\infty} \sup_{\theta\in K_\varepsilon} n^{-1} \sum_{k=0}^{n-1} \bar{\ell}^{\theta} \circ \tilde{\mathrm{S}}^k &\le \sup_{\theta\in K_\varepsilon} \mathbb{E}[\bar{\ell}^{\theta}] \\ &< \mathbb{E}[\bar{\ell}^{\theta_\star}] \overset{(1)}{=} \lim_{n\to\infty} \bar{\mathrm{L}}_n^{\theta_\star} \le \liminf_{n\to\infty} \bar{\mathrm{L}}_n^{\bar{\theta}_n} ,\end{aligned} \tag{8.74}$$

where (1) follows from (8.69). This inequality ensures that $\bar{\theta}_n \notin K_\varepsilon$ for all n larger to some $\mathbb{P}$-a.s. finite integer-valued random variable. This completes the proof of (8.66) since ε is arbitrary.

Eq. (8.68) follows from (8.69) and A8.41. Let $\theta_\star$ be any point in $\Theta_\star$. Then, $\mathbb{P}$-a.s.,

$$\mathbb{E}[\bar{\ell}^{\theta_\star}] \stackrel{(1)}{=} \liminf_{n\to\infty} \bar{\mathsf{L}}_n^{\theta_\star} \stackrel{(2)}{\leq} \liminf_{n\to\infty} \bar{\mathsf{L}}_n^{\bar{\theta}_n} \leq \limsup_{n\to\infty} \bar{\mathsf{L}}_n^{\bar{\theta}_n}$$
$$= \limsup_{n\to\infty} \sup_{\theta\in\Theta} \bar{\mathsf{L}}_n^{\theta} \stackrel{(3)}{\leq} \sup_{\theta\in\Theta} \mathbb{E}[\bar{\ell}^{\theta}] = \mathbb{E}[\bar{\ell}^{\theta_\star}] ,$$

where (1) follows from (8.69), (2) follows from the definition of $\bar{\theta}_n$ and (3) is obtained by applying (8.73) with $K = \Theta$. Thus,

$$\lim_{n\to\infty} \bar{\mathsf{L}}_n^{\bar{\theta}_n} = \mathbb{E}[\bar{\ell}^{\theta_\star}] , \quad \mathbb{P}\text{-a.s.} \tag{8.75}$$

Denote $\delta_n := \sup_{\theta\in\Theta} |\mathsf{L}_n^{\theta} - \bar{\mathsf{L}}_n^{\theta}|$. We get

$$\bar{\mathsf{L}}_n^{\bar{\theta}_n} - \delta_n \stackrel{(1)}{\leq} \mathsf{L}_n^{\bar{\theta}_n} \stackrel{(2)}{\leq} \mathsf{L}_n^{\hat{\theta}_n} \stackrel{(1)}{\leq} \bar{\mathsf{L}}_n^{\hat{\theta}_n} + \delta_n \stackrel{(3)}{\leq} \bar{\mathsf{L}}_n^{\bar{\theta}_n} + \delta_n . \tag{8.76}$$

where (1) follows from the definition of δ_n, (2) from the definition of $\hat{\theta}_n$ and (3) from the definition of $\bar{\theta}_n$. Combining the above inequalities with (8.75) and A8.41 yields (8.67). (8.76) also implies that

$$\lim_{n\to\infty} \bar{\mathsf{L}}_n^{\hat{\theta}_n} = \mathbb{E}[\bar{\ell}^{\theta_\star}] , \quad \mathbb{P}\text{-a.s.}$$

which yields, using (8.74),

$$\limsup_{n\to\infty} \sup_{\theta\in K_\varepsilon} \bar{\mathsf{L}}_n^{\theta} < \liminf_{n\to\infty} \bar{\mathsf{L}}_n^{\hat{\theta}_n} = \limsup_{n\to\infty} \bar{\mathsf{L}}_n^{\hat{\theta}_n} = \mathbb{E}[\bar{\ell}^{\theta_\star}], \quad \mathbb{P}\text{-a.s.}$$

where $K_\varepsilon := \{\theta \in \Theta; \mathrm{d}(\theta, \Theta_\star) \geq \varepsilon\}$. Therefore, $\hat{\theta}_n \notin K_\varepsilon$ for all n larger to some $\mathbb{P}$-a.s.-finite integer-valued random variable. The proof is completed since ε is arbitrary. ■

8.6 Endnotes

The consistency of the MLE for well and misspecified models is of course classical. The assumptions used here are minimal. There is a large amount of literature on the maximum likelihood estimator of parameters of stationary ergodic Markov chains. An early systematic treatment of maximum likelihood estimator for Markovian models can be found in Billingsley (1961). Hall and Heyde (1980, Chapter 6) gives a detailed presentation of these early works in the common unifying framework of martingale representation of the log-likelihood. Many refinements and improvements have been published so far, and it is impossible to give a fair survey of these results in only a few sentences.

The consistency of observation-driven models under model misspecification is less classical. This presentation is adapted from Douc et al. (2013). Closely related results are reported in Fokianos and Tjøstheim (2011) and Neumann (2011), for integer-valued time-series.

The approach taken in this chapter to study the asymptotic distribution of the maximum likelihood estimate is elementary and uses only the most standard tools of asymptotic statistics. A more elaborate account on this topic is given in Hwang and Basawa (1993), Hwang and Basawa (1997), which established the local asymptotic normality (LAN) of the log-likelihood ratio for a class of Markovian nonlinear time series models, using the approach of quadratic mean differentiability. As a consequence of the LAN property, asymptotically optimal estimators of the model parameters can be constructed; such results are surveyed by Basawa (2001).

Exercises

8.1. Let $\{X_t,\, t \in \mathbb{N}\}$ be an X-valued stochastic process, where $(\mathsf{X}, \mathcal{X})$ is a measurable space. Assume that for each positive integer n, the random vector $(X_0, \ldots, X_{n-1})$ has a joint density $x_{0:n-1} \mapsto \mathrm{p}^\theta(x_{0:n-1})$ with respect to the product measure $\lambda^{\otimes n}$ on the product space $(\mathsf{X}^n, \mathcal{X}^{\otimes(n-1)})$, where $\lambda \in \mathbb{M}_1(\mathcal{X})$. For simplicity, it is assumed that for all $n \in \mathbb{N}$ and all $x_{0:n-1}$, $\mathrm{p}^\theta(x_{0:n-1}) > 0$. Denote by $\mathrm{p}^\theta(x_n | x_{0:n-1})$ the conditional density of X_n given $X_{0:n-1}$, $\mathrm{p}^\theta(x_n | x_{0:n-1}) = \mathrm{p}^\theta(x_{0:n}) / \mathrm{p}^\theta(x_{0:n-1})$. Assume that, for all $n \in \mathbb{N}$, and $x_{0:n-1} \in \mathsf{X}^n$, $\theta \mapsto \mathrm{p}^\theta(x_{0:n-1})$ is differentiable and that

$$\nabla \mathrm{p}^\theta(x_{0:n-1}) = \int \nabla \mathrm{p}^\theta(x_{0:n}) \lambda(\mathrm{d}x_n)\,.$$

Prove that the score function (the gradient of the log-likelihood) $\sum_{k=1}^n \nabla \ln \mathrm{p}^\theta(X_k | X_{0:k-1})$ is a martingale adapted to the natural filtration of the process.

8.2 (Conditionally Gaussian MLE for an AR(1)). Consider a causal AR(1) model, $X_t = \phi X_{t-1} + \sigma Z_t$, where $\{Z_t,\, t \in \mathbb{N}\}$ is a strong white noise having a density f, which is symmetric around the origin and satisfies $\int_{-\infty}^{\infty} x^2 f(x) \mathrm{Leb}(\mathrm{d}x) < \infty$. We put $\theta = (\phi, \sigma) \in \Theta$, where Θ is a compact subset of $(-1,1) \times \mathbb{R}_*^+$. We denote by $\theta_\star = (\phi_*, \sigma_*^2) \in \Theta^o$ the true value of the parameters. We observe $(X_0, \ldots, X_n)$ and we are interested in estimating these parameters. We fit the model using the conditional *Gaussian* maximum likelihood estimator. We denote by $(\hat{\phi}_n, \hat{\sigma}_n^2)$ these estimators.

(a) Derive the expressions of $(\hat{\phi}_n, \hat{\sigma}_n^2)$.

(b) Show that this estimator is strongly consistent.

(c) Show that the estimator $\hat{\phi}_n$ is asymptotically normal. Compute the asymptotic variance of this estimator.

(d) Assume that f is t-distribution with $\nu > 2$ degrees of freedom. Compute the variance of $\hat{\phi}_n$ (as a function of ν) and construct confidence intervals. Discuss the result.

(e) Assume that $\int_{-\infty}^{\infty} x^4 f(x) \mathrm{Leb}(\mathrm{d}x) < \infty$. Show that $(\hat{\phi}_n, \hat{\sigma}_n^2)$ is asymptotically normal.

(f) Determine confidence regions for $(\hat{\phi}_n, \hat{\sigma}_n^2)$ when f is a Student t-distribution with $\nu > 4$ degrees of freedom.

8.3 (Conditionally Gaussian MLE for an AR(1)). We use the same notations and definitions as in Exercise 8.2. We now fit the model using the conditional maximum likelihood estimator. We assume in this exercise that the model is well specified. We denote by $(\tilde{\phi}_n, \tilde{\sigma}_n^2)$ these estimators.

(a) Derive the estimating equations for $(\tilde{\phi}_n, \tilde{\sigma}_n^2)$.

(b) Derive explicitly the likelihood equations when f is a t-distribution with ν-degrees of freedom.

(c) Show that these estimators (assuming that the model is well-specified) are consistent and asymptotically normal.

(d) Compute the asymptotic variance of the resulting estimator.

(e) Compare the asymptotic variance with Exercise 8.2.

8.4. Verify (8.32).

8.5 (AR process with nonnegative innovation). Consider the nonnegative autoregressive process $X_t = \phi_1 X_{t-1} + \cdots + \phi_p X_{t-p} + Z_t$, where $\{Z_t,\, t \in \mathbb{N}\}$ is a strong white noise with left endpoint of their common distributions being zero and $(\phi_1, \ldots, \phi_p) \in (0,1)^p$. This type of model is suitable for modelling applications of data that are inherently nonnegative, such as streams flow in hydrology, interarrival times, etc. Based on the observations, $(X_0, \ldots, X_n)$, we are interested in estimating the parameters. The parameters $(\phi_1, \ldots, \phi_p)$ are estimated using the maximum likelihood estimator associated to the assumption that the common distribution of the Z's is unit exponential, so that $\mathbb{P}(Z_1 > x) = \mathrm{e}^{-x}$, $x \geq 0$. See Feigin and Resnick (1994).

(a) Consider first the case $p = 1$. Show that the conditional likelihood function is given (up to a constant) by

$$\prod_{t=1}^{n} \mathbb{1}_{\{X_t - \phi_1 X_{t-1} \geq 0\}} \exp\left(\phi_1 \sum_{t=1}^{n} X_{t-1}\right) .$$

(b) Show that the conditional maximum likelihood estimator is given by

$$\hat{\phi}_{1,n} = \min_{1 \leq t \leq n} (X_t / X_{t-1}) .$$

(c) Consider now the general autoregressive case. Show that the likelihood function is given (up to a constant) by

$$\prod_{t=p}^{n} \mathbb{1}_{\{X_t - \phi X_{t-1} \geq 0\}} \exp\left(\phi_1 \sum_{t=p}^{n} X_{t-1} + \cdots + \phi_p \sum_{t=p}^{n} X_{t-p}\right) .$$

(d) Argue that the maximum likelihood estimator is approximately determined by solving the linear program

$$\max\left(\sum_{i=1}^{p} \phi_i\right) \quad \text{subject to} \quad X_t \geq \sum_{i=1}^{p} \phi_i X_{t-i} \text{ for } t \in \{1, \ldots, n\} .$$

Determining the asymptotic properties of these estimators is difficult (see Feigin and Resnick (1994)); the rate of convergence can be in such a case faster than $n^{1/2}$.

8.6 (Proof of Theorem 8.18).

(a) Show that

$$n^{-1/2}\sum_{t=p}^{n}\nabla\ln q^{\hat{\theta}_n}(X_{t-p:t-1};X_t)=0=n^{-1/2}\sum_{t=p}^{n}\nabla\ln q^{\theta_\star}(X_{t-p:t-1};X_t)$$
$$+n^{-1}\sum_{t=p}^{n}\left(\int_0^1\nabla^2\ln q^{\theta_{n,s}}(X_{t-p:t-1};X_t)\mathrm{d}s\right)\sqrt{n}(\hat{\theta}_n-\theta_\star)\,, \quad (8.77)$$

where $\theta_{n,s}=s\hat{\theta}_n+(1-s)\theta_\star$.

(b) Show that

$$n^{-1/2}\sum_{t=p}^{n}\nabla\ln q^{\theta_\star}(X_{t-p:t-1};X_t)\overset{\mathbb{P}^{\theta_\star}}{\Longrightarrow}\mathrm{N}(0,\mathcal{J}(\theta_\star))\,.$$

(c) Show that, for all $i,j\in\{1,\ldots,d\}$,

$$n^{-1}\sum_{t=p}^{n}\int_0^1\frac{\partial^2\ln q^{\theta_{n,s}}}{\partial\theta_i\partial\theta_j}(X_{t-p:t-1};X_t)\mathrm{d}s\overset{\mathbb{P}^{\theta_\star}}{\longrightarrow}\mathbb{E}^{\theta_\star}\left[\frac{\partial^2\ln q^{\theta_\star}}{\partial\theta_i\partial\theta_j}(X_{0:p-1};X_p)\right]\,.$$

(d) Show that

$$n^{-1}\sum_{t=p}^{n}\left(\int_0^1\nabla^2\ln q^{\theta_{n,s}}(X_{t-p:t-1};X_t)\mathrm{d}s\right)\overset{\mathbb{P}^{\theta_\star}}{\longrightarrow}\mathcal{J}(\theta_\star)\,.$$

(e) Conclude.

8.7 (Autoregressive model with ARCH error). Consider the pth order autoregressive model with ARCH(1) error

$$Y_t=\alpha_1Y_{t-1}+\alpha_2Y_{t-2}+\ldots+\alpha_pY_{t-p}+\varepsilon_t\,, \quad (8.78)$$

where $\varepsilon_t|\mathcal{F}_{t-1}\sim\mathrm{N}(0,\ \beta_0+\beta_1\varepsilon_{t-1}^2)$ and $\{\mathcal{F}_t,\ t\in\mathbb{N}\}$ is the natural filtration of the process $\{Y_t,\ t\in\mathbb{N}\}$, i.e., for any nonnegative integer t, $\mathcal{F}_t=\sigma(Y_s,\ s\le t)$. It is assumed that β_0 is strictly positive and $0\le\beta_1<1$.

(a) Show that the unique strict-sense stationary solution to the recursion $\varepsilon_t=\sqrt{\beta_0+\beta_1\varepsilon_{t-1}^2}Z_t$, where $\{Z_t,\ t\in\mathbb{Z}\}$ is a strong Gaussian noise with zero-mean and unit-variance is given by

$$\varepsilon_t=\beta_0\sum_{l=0}^{\infty}\beta_1^l\left(\prod_{i=0}^{l}Z_{t-i}^2\right)\,. \quad (8.79)$$

(b) Let r be a nonnegative integer. Show that $m_{2r}=\mathbb{E}\left[\varepsilon_t^{2r}\right]$ exists if and only if $\eta_r=\beta_1^r\prod_{j=1}^r(2j-1)<1$.

(c) Show that if $3\beta_1^2<1$, then $\mathrm{Var}(\varepsilon_t^2)=2\sigma^4(1-3\beta_1^2)^{-1}$, and $\mathrm{Cov}(\varepsilon_t^2,\ \varepsilon_{t-j}^2)=\beta_1^j\,\mathrm{Var}(\varepsilon_t^2)$ where $\sigma^2=\beta_0(1-\beta_1)^{-1}$ is the stationary unconditional variance.

(d) Show that, if $\eta_r < 1$, then

$$n^{-1}\sum_{t=1}^{n} \varepsilon_t^{2r} \xrightarrow{\mathbb{P}\text{-a.s.}} \mathbb{E}\left[\varepsilon_t^{2r}\right], \quad \text{and} \quad n^{-1}\sum_{t=1}^{n} \varepsilon_t^{2r-1} \xrightarrow{\mathbb{P}\text{-a.s.}} 0\,.$$

(e) Show that if $3\beta_1^2 < 1$, then for a fixed $j \neq 0$,

$$n^{-1}\sum_{t=1}^{n} \varepsilon_t \varepsilon_{t-j} \xrightarrow{\mathbb{P}\text{-a.s.}} 0 \quad \text{and} \quad n^{-1}\sum_{t=1}^{n} \varepsilon_t^2 \varepsilon_{t-j}^2 \xrightarrow{\mathbb{P}\text{-a.s.}} \mathbb{E}[\varepsilon_t^2 \varepsilon_{t-j}^2]\,.$$

(f) Given $(Y_0,\ldots,Y_n)$ we want to estimate $\alpha = (\alpha_1,\ldots,\alpha_p)$ and $\beta = (\beta_0,\beta_1)$. We write $\theta = (\alpha,\beta)$. Show that the log-likelihood conditional on $(Y_p,Y_{p-1},\ldots,Y_0)$ is given by

$$\bar{\mathrm{L}}_n^{\theta}(Y_{p:n}|Y_{0:p-1}) = (n-p)^{-1} \sum_{t=p+1}^{n} \ln \mathrm{p}^{\theta}(Y_t|Y_{t-1:t-p-1})$$

with

$$\mathrm{p}^{\theta}(Y_t|Y_{t-1:t-p-1}) = (2\pi h_t(\theta))^{-1/2} \exp[-(Y_t - \boldsymbol{X}_{t-1}'\alpha)^2/2h_t(\theta)]\,,$$
$$h_t(\theta) = \beta_0 + \beta_1 (Y_{t-1} - \boldsymbol{X}_{t-2}'\alpha)^2\,,$$

where $\boldsymbol{X}_t = (Y_t,\, Y_{t-1},\, \ldots,\, Y_{t-p+1})$.

(g) We assume in the sequel that the model is well-specified. State sufficient conditions upon which the MLE estimator is strongly consistent.

(h) State sufficient conditions upon which the MLE estimator is asymptotically normal, i.e., $n^{1/2}(\hat{\theta}_n - \theta_\star) \stackrel{\mathbb{P}}{\Longrightarrow} \mathrm{N}(0,\, H^{-1}(\theta_\star))$.

(i) Show that the asymptotic covariance matrix $H(\theta_\star) = (H_{i,j}(\theta_\star))_{1\le i,j\le 2}$ is the $(p+2)\times(p+2)$ block matrix with $H_{1,2}(\theta_\star) = H_{2,1}(\theta_\star) = 0$,

$$H_{1,1}(\theta_\star) = \mathbb{E}^{\theta_\star}[v_t^{-1}(\theta_\star)\boldsymbol{X}_{t-1}'\boldsymbol{X}_{t-1}] + 2\beta_{*,1}\mathbb{E}^{\theta_\star}[v_t^{-1}(\theta_\star)\boldsymbol{X}_{t-2}'\boldsymbol{X}_{t-2}]$$
$$- 2\beta_{*,1}\beta_{*,0}\mathbb{E}^{\theta_\star}[v_t^{-2}(\theta_\star)\boldsymbol{X}_{t-2}'\boldsymbol{X}_{t-2}]\,,$$

where $\{v_t(\theta_\star), t \in \mathbb{Z}\}$ is the unique strict sense stationary non anticipative solution of $v_t(\theta_\star) = \beta_{*,0} + \beta_{*,1} v_{t-1}(\theta_\star) Z_{t-1}^2$. Compute similarly $H_{2,2}(\theta_\star)$.

(j) Show that the matrix H may be estimated consistently by replacing the parameters by their estimates and expectations by sample averages in the above expression.

(k) Construct asymptotic confidence regions for the parameters.

8.8 (EXPAR(1) model). Consider the EXPAR(1) model (see Example 6.38)

$$X_t = \{\phi_0 + \phi_1 \exp(-\gamma X_{t-1}^2)\} X_{t-1} + \sigma Z_t\,, \tag{8.80}$$

where $\{Z_t,\, t \in \mathbb{N}\}$ is a strong white noise with zero mean, unit variance independent of X_0. Assume that the distribution of Z_0 has a continuous density f with respect to the Lebesgue measure, which is everywhere positive and $\mathbb{E}\left[Z_0^{2q}\right] < \infty$ for some $q \in \mathbb{N}_*$. Denote by $\theta = (\phi_0, \phi_1, \gamma, \sigma^2) \in \Theta$ the unknown parameters, where Θ is a compact subset of

$$\left\{(\phi_0, \phi_1, \gamma, \sigma) : \gamma > 0, \sigma^2 > 0, \max(|\phi_0|, |\phi_0 + \phi_1|) < 1\right\} .$$

For $\theta \in \Theta$, denote $a^\theta(x) = \phi_0 + \phi_1 \exp(-\gamma x^2)$ and $V_q(x) = 1 + |x|^{2q}$.

(a) Show that the Markov chain is V_q-geometrically ergodic.

(b) We observe $(X_0, X_1, \ldots, X_n)$ $(n+1)$ observations from the strict sense stationary solution of (8.80) associated to the "true" parameters $\theta_\star \in \Theta$. We fit the parameters of the model using the Gaussian likelihood. Show that for all $(x_0, x_1) \in \mathbb{R}^2$,

$$\ln q^\theta(x_0, x_1) = -\frac{1}{2}\ln(2\pi\sigma^2) - \frac{1}{2\sigma^2}(x_1 - a^\theta(x_0))^2 .$$

(c) Assume that $q = 1$. Show that the sequence of estimators $\hat{\theta}_n \to_{\mathbb{P}_{\theta_\star}\text{-a.s.}} \theta_\star$.

(d) State conditions upon which the sequence of estimators $\{\hat{\theta}_n, n \in \mathbb{N}\}$ is asymptotically normal and compute the asymptotic covariance matrix.

8.9 (Nonlinear autoregression). Consider the NLAR(1) process

$$X_t = f^\theta(X_{t-1}) + Z_t , \tag{8.81}$$

where $\{Z_t,\ t \in \mathbb{N}\}$ is a strong white noise independent of X_0 and $\theta \in \Theta$ a compact subset of $\mathbb{R}$. Assume that for all $\theta \in \Theta$, $\lim_{|x| \to \infty} |f^\theta(x)|/|x| < 1$, $f^\theta : \mathbb{R} \to \mathbb{R}$ is a twice continuously differentiable function, the distribution of Z_0 has a continuous density with respect to the Lebesgue measure, which is everywhere positive and that $\mathbb{E}\left[Z_0^4\right] < \infty$.

(a) Show that the conditional Gaussian likelihood is given by

$$\ln \mathrm{p}^\theta(X_{p:n}|X_{0:p-1}) = -\frac{n}{2}\ln(2\pi) - \frac{1}{2}\sum_{t=1}^{n}\{X_t - f^\theta(X_{t-1})\}^2 .$$

(b) Assume $\{X_t,\ t \in \mathbb{N}\}$ is strictly stationary such that $\mathbb{E}\left[\sup_{\theta \in \Theta}(X_1 - f^\theta(X_0))^2\right] < \infty$. Then, show that the sequence of MLE estimators $\hat{\theta}_n$ is strongly consistent.

(c) Assume that the model is well-specified and denote by $\theta_\star \in \Theta$ the "true" values of the unknown parameters. Show that the model is identifiable if the condition $f^\theta(X_0) = f^{\theta_\star}(X_0)$ $\mathbb{P}^{\theta_\star}$-a.s. implies that $\theta = \theta_\star$.

(d) Give conditions upon which the sequence of estimators is asymptotically normal, and give the expressions of the asymptotic covariance matrix both in the well-specified and in the misspecified cases.

8.10 (Proof of Theorem 8.28).

(a) Show that, under A8.27, for all $\theta \in \Theta$,

$$
\begin{aligned}
&\mathbb{E}[\bar{\ell}^{\theta}(Y_{-\infty:1})] \\
&= \mathbb{E}\left[\mathbb{E}\left[\ln q(f^{\theta}_{Y_{-\infty:0}}, Y_1) \,\middle|\, Y_s,\, s \le 0\right]\right] = \mathbb{E}\left[\int Q^{\star}(f^{\theta_\star}_{Y_{-\infty:0}}, \mathrm{d}y) \ln q(f^{\theta}_{Y_{-\infty:0}}, y)\right] \\
&\le \mathbb{E}\left[\int Q^{\star}(f^{\theta_\star}_{Y_{-\infty:0}}, \mathrm{d}y) \ln q(f^{\theta_\star}_{Y_{-\infty:0}}, y)\right] = \mathbb{E}[\bar{\ell}^{\theta_\star}(Y_{-\infty:1})] \,. \qquad (8.82)
\end{aligned}
$$

(b) Show that (8.45) implies that $\theta = \theta_\star$.

(c) Conclude.

8.11 (Proof of Lemma 8.34). Let $\mathcal{F}$ be the filtration $\mathcal{F} = (\mathcal{F}_n)_{n \in \mathbb{N}}$ where $\mathcal{F}_n = \sigma(Y_s,\, s \le n)$ and let

$$
M_n := \sum_{t=1}^{n} \nabla \ln q(f^{\theta_\star}_{Y_{-\infty:t-1}}, Y_t) = \sum_{t=1}^{n} \varphi(\nabla f^{\theta_\star}_{Y_{-\infty:t-1}}, f^{\theta_\star}_{Y_{1:t-1}}(x), Y_t) \,.
$$

(a) Show that $\mathbb{E}(\|M_n\|^2) < \infty$ and

$$
\begin{aligned}
&\mathbb{E}^{\theta_\star}\left[\nabla \ln q(f^{\theta_\star}_{Y_{-\infty:t-1}}, Y_t) \,\middle|\, \mathcal{F}_{t-1}\right] \\
&\qquad = \nabla f^{\theta_\star}_{Y_{-\infty:t-1}} \int Q^{\star}(f^{\theta_\star}_{Y_{-\infty:t-1}}, \mathrm{d}y) \frac{\partial \ln q(f^{\theta_\star}_{Y_{-\infty:t-1}}, y)}{\partial x} = 0 \,.
\end{aligned}
$$

(b) Show that $\{M_t\,, t \ge 1\}$ is a square integrable $\mathcal{F}$-martingale with stationary and ergodic increments.

(c) Conclude.

8.12 (Proof of Lemma 8.35). Denote by $A_t(\theta) = \frac{\partial^2 \ln q}{\partial \theta_i \partial \theta_j}(f^{\theta}_{Y_{-\infty:t-1}}, Y_t)$.

(a) Show that $n^{-1} \sum_{t=1}^{n} A_t(\theta_\star) \to_{\mathbb{P}} \mathbb{E}\,[A(\theta_\star)]$.

(b) Let $\varepsilon > 0$. Show that we can choose $0 < \eta < \rho$ such that

$$
\mathbb{E}\left(\sup_{\theta \in \mathrm{B}(\theta_\star, \eta)} |A_1(\theta_\star) - A_1(\theta)|\right) < \varepsilon \,. \qquad (8.83)
$$

(c) Show that

$$
\begin{aligned}
&\limsup_{n \to \infty} \mathbb{P}\left(n^{-1} \sum_{t=1}^{n} |A_t(\theta_\star) - A_t(\theta_n)| \ge \varepsilon,\ \theta_n \in \mathrm{B}(\theta_\star, \eta)\right) \\
&\qquad \le \limsup_{n \to \infty} \mathbb{P}\left(n^{-1} \sum_{t=1}^{n} \sup_{\theta \in \mathrm{B}(\theta_\star, \eta)} |A_t(\theta_\star) - A_t(\theta)| \ge \varepsilon\right) = 0 \,,
\end{aligned}
$$

(d) Show that

$$\lim_{n\to\infty} \mathbb{P}\left(n^{-1} \sum_{t=1}^{n} |A_t(\theta_\star) - A_t(\theta_n)| \geq \varepsilon \right) = 0 .$$

(e) Show that $n^{-1}\sum_{t=1}^{n} |A_t(\theta_\star) - A_t(\theta_n)| \xrightarrow{\mathbb{P}} 0$ and conclude.

8.13 (Conditional least-squares estimators). Let p be an integer and $\{P^\theta : \theta = (\varphi, \gamma) \in \Theta\}$ be a parametric family of Markov kernels on $\mathsf{X}^p \times \mathcal{X}$, where Θ is a compact subset of $\mathbb{R}^d$ (here, φ is the parameter of interest and γ is a nuisance). Assume that for any $\theta \in \Theta$, P^θ has a unique stationary distribution satisfying $\mathbb{E}^\theta[|X_0|] < \infty$. Let $\theta_\star \in \Theta^o$. Assume that the vector of observations $(X_0, X_1, \ldots, X_n)$ is, for each n a realization of the stationary ergodic Markov chain $\{X_t,\, t \in \mathbb{Z}\}$ with transition kernel $P^{\theta_\star}$. Finally, assume that

- there exists $a^\varphi(X_{p-1}, \ldots, X_0)$ a version of the conditional expectation $a^\varphi(X_{p-1}, \ldots, X_0) = \mathbb{E}^\theta\left[X_p \mid X_0, \ldots, X_{p-1}\right]$ with $\theta = (\varphi, \gamma)$ satisfying

$$\mathbb{E}^{\theta_\star}\left[\sup_{\varphi \in \Theta_\varphi} |a^\varphi(X_{p-1}, \ldots, X_0)|^2 \right] < \infty ,$$

  where $\Theta_\varphi = \{\varphi, (\varphi, \gamma) \in \Theta\}$ is the canonical projection of Θ.
- for any $(x_0, \ldots, x_{p-1}) \in \mathsf{X}^p$, the function $\varphi \mapsto a^\varphi(x_{p-1}, \ldots, x_0)$ is continuous.

We estimate the parameter $\varphi \in \Theta_\varphi$ by the *conditional least-squares*, which amounts to estimating the model by using the likelihood of

$$Y_t = a^\varphi(Y_{t-1}, \ldots, Y_{t-p}) + \sigma^2 Z_t \tag{8.84}$$

where $\{Z_t,\, t \in \mathbb{Z}\}$ is a strong white noise with zero-mean and unit-variance Assume in addition that A6.32 is satisfied and $\mathbb{E}\left[Z_0^{2q}\right] < \infty$ for some $q \in \mathbb{N}_*$.

(a) Show that the log-likelihood for the misspecified model (8.84) is given by

$$\ln q^\varphi(x_{p-1:0}; X_p) = -\frac{1}{2}\ln(2\pi\sigma^2) - \frac{1}{2\sigma^2}\left\{x_p - a^\varphi(x_{p-1}, \ldots, x_0)\right\}^2 .$$

(b) Show that if $a^{\varphi_*}(X_{p-1}, \ldots, X_0) = a^\varphi(X_{p-1}, \ldots, X_0)$ $\mathbb{P}_{\theta_\star}$-a.s. implies that $\varphi = \varphi_*$, then the conditional least square estimator $\{\hat{\varphi}(n), n \in \mathbb{N}\}$ is strongly consistent.

(c) State conditions upon which the conditional least-squares estimator is asymptotically normal and compute its asymptotic covariance matrix.

8.14. Consider an EXPAR(p) model

$$X_t = \{a_1 + b_1 \exp(-\lambda_1 X_{t-d}^2)\} X_{t-1} + \ldots \\ + \{a_p + b_p \exp(-\lambda_p X_{t-p}^2)\} X_{t-p} + \sigma Z_t , \tag{8.85}$$

where $\{Z_t,\, t \in \mathbb{Z}\}$ is a strong white noise with zero-mean and unit-variance Assume in addition that A 6.32 is satisfied and $\mathbb{E}\left[Z_0^{2q}\right] < \infty$ for some $q \in \mathbb{N}_*$. .

We denote by $\theta = \{(a_i, b_i, \lambda_i), i = 1, \ldots, p, d, \sigma^2\} \in \Theta$ the unknown parameters, where Θ is a compact subset of the set $\{(a_i, b_i, \lambda_i), i = 1, \ldots, p, d, \sigma^2\}$ satisfying $\sigma^2 > 0$, $d \in \{0, \ldots, p\}$, $\lambda_i > 0$, $i \in \{1, \ldots, p\}$ and $c(z) = 1 - c_1 z - \cdots - c_p z^p$ where $c_i = \max(|a_i|, |a_i + b_i|)$. For $\theta \in \Theta$, the Markov chain $\boldsymbol{X}_t = [X_t, \ldots, X_{t-p+1}]'$ given by (8.85) is V_q-geometrically ergodic with $V_q(\boldsymbol{x}) = 1 + \|\boldsymbol{x}\|^q$; see Example 6.38 and Exercise 6.13. Denote by $\theta_\star \in \Theta$ the true parameter vector. Given the observations $(X_0, \ldots, X_n)$, we estimate the parameters using the conditional least square (Gaussian likelihood)

$$-\frac{n-p}{2} \ln(2\pi\sigma^2) - \frac{1}{2\sigma^2} \sum_{t=p}^{n} (X_t - a^\theta(X_{t-1}, \ldots, X_{t-p}))^2 ,$$

where

$$a^\theta(x_{p-1}, \ldots, x_0) = \sum_{i=1}^{p} \{a_i + b_i \exp(-\lambda_i x_{p-i}^2)\} x_{p-i} .$$

(a) Compute the conditional least squares estimator of θ.

(b) Assume that $(X_0, X_1, \ldots, X_n)$ is a sample record of $\{X_t,\, t \in \mathbb{N}\}$ which is the ergodic solution of (8.85) associated to the true parameter vector $\theta_\star$. Show that if $q = 1$, then $\hat{\theta}_n$ is strongly consistent.

(c) Show that if $q = 2$, this estimator is asymptotically normal and compute the limiting covariance.

8.15 (First order RCA; Exercise 6.7 cont.). We use the notations of Exercise 6.7. We denote by $\theta = (\phi, \lambda, \sigma)$ the unknown parameters. We assume that the noise Z_t cannot take on only two values asymptotically.

(a) Show that $\mathbb{E}^\theta\left[X_t \mid \mathcal{F}_{t-1}^X\right] = \phi X_{t-1}$ and $\mathrm{Var}^\theta\left(X_t \mid \mathcal{F}_{t-1}^X\right) = \lambda^2 X_{t-1}^2 + \sigma^2$, $\mathbb{P}^\theta$-a.s..

(b) Show that if $\{B_t,\, t \in \mathbb{Z}\}$ and $\{Z_t,\, t \in \mathbb{Z}\}$ are jointly Gaussian, then the conditional distribution of X_t given $\mathcal{F}_{t-1}^X$ is Gaussian with mean $\mathbb{E}^\theta\left[X_t \mid \mathcal{F}_{t-1}^X\right]$ and variance $\mathrm{Var}^\theta\left(X_t \mid \mathcal{F}_{t-1}^X\right)$. Write the likelihood of this model (referred to as in the sequel as the Gaussian likelihood).

(c) Assume that $\theta_\star \in \Theta$, where Θ is a compact subset of

$$\{(\phi, \lambda, \sigma) \in \mathbb{R} \times \mathbb{R}_+^2 : \phi^2 + \lambda^2 < 1, \sigma > 0\} .$$

Assume that the observations $(X_0, \ldots, X_n)$ is a sample record from the strict-sense stationary solution $\{X_t,\, t \in \mathbb{N}\}$ of (6.55) associated to some values of the parameters $\theta_\star \in \Theta$. Prove that the Gaussian MLE $\{\hat{\phi}_n\}_{n \geq 1}$ is a strongly consistent sequence of estimator of ϕ_*

(d) Under which conditions is the Gaussian MLE $\{(\hat{\lambda}_n, \hat{\sigma}_n^2)\}_{n \geq 0}$ a strongly consistent sequence of estimators of (λ, σ^2)?

(e) Assume that $\theta_\star \in \Theta^o$, where Θ is a compact subset of

$$\{(\phi, \lambda, \sigma) \in \mathbb{R} \times \mathbb{R}_+^2 : \phi^4 + 6\lambda^2\phi^2 + m_4\lambda^4 < 1, \sigma > 0\} .$$

Prove that the Gaussian MLE $\{\hat{\phi}_n\}_{n \geq 0}$ is an asymptotically normal sequence of estimators of $\phi_\star$. Compute a confidence interval for this parameter.

(f) Under which conditions is the sequence $\{(\hat{\lambda}_n, \hat{\sigma}_n^2)\}_{n\geq 0}$ asymptotically normal?

8.16. Consider the stochastic process

$$X_t = A_t X_{t-1} + B_t \ , \tag{8.86}$$

where $\{(A_t, B_t),\ t \in \mathbb{N}\}$ is a Gaussian white noise, independent of X_0, with mean zero and covariance matrix

$$\begin{pmatrix} \alpha^2 & \rho\alpha\beta \\ \rho\alpha\beta & \beta^2 \end{pmatrix} .$$

We denote by $\theta = (\alpha^2, \beta^2, \rho) \in \Theta$ the unknown parameters, where Θ is a compact subset of $(0,1) \times \mathbb{R}_*^+ \times (-1,1)$.

(a) Show that for any $\theta \in \Theta$, (8.86) admits a unique strict sense stationary solution, satisfying $\mathbb{E}^\theta[X_0^2] < \infty$.

(b) Find a necessary and sufficient condition on θ under which (8.86) admits a unique strict sense stationary solution satisfying $\mathbb{E}^\theta[X_0^4] < \infty$.

(c) Write the conditional maximum likelihood estimator for the model (8.86).

(d) Write an algorithm in R to solve numerically this equation.

(e) Assume that the observation $(X_0, \ldots, X_n)$ is a realization of (8.86) for some parameter $\theta_\star \in \Theta$. Prove that $\hat{\theta}_n$ is a strongly consistent sequence of estimators of $\theta_\star$.

(f) State conditions upon which this estimator is asymptotically normal and compute the asymptotic covariance matrix.

8.17 (APARCH(1,1); Example 3.12 cont.). The APARCH(1,1) model of Ding et al. (1993) can be defined as follows:

$$Y_t = \sigma_t(\theta)\varepsilon_t \tag{8.87}$$

$$\sigma_t^\delta(\theta) = \alpha_0 + \alpha_1(Y_{t-1} - \gamma Y_{t-1})^\delta + \beta_1 \sigma_{t-1}^\delta(\theta) \ , \tag{8.88}$$

where $\{\varepsilon_t,\ t \in \mathbb{N}\}$ is a strong white Gaussian noise with zero-mean and unit variance and $\theta = (\alpha_0, \alpha_1, \beta_1, \gamma, \delta) \in \Theta$ a compact set of

$$\{(\alpha_0, \alpha_1, \beta_1, \gamma, \delta) \,:\, \alpha_0 > 0, \alpha_1 > 0, 0 < \gamma < 1, \delta > 0\} \ .$$

The parameter δ, ($\delta > 0$) parameterizes a Box-Cox transformation of the conditional standard deviation $\sigma_t(\theta)$, while the parameters γ reflect the leverage effect.

(a) Show that the APARCH model is an observation-driven model. Determine the kernel q^θ and the function f_y^θ.

(b) Show that the maximum likelihood estimator is consistent.

(c) Compute the confidence intervals for the parameters.

Part III

State Space and Hidden Markov Models

Chapter 9

Non-Gaussian and Nonlinear State Space Models

The state space model has become a powerful tool for time series modeling and forecasting. Such models, in conjunction with the Kalman filter, have been used in a wide range of applications (see Chapter 3). A *nonlinear state space model* (NLSS) or equivalently a *Hidden Markov Model* (HMM), keeps the hierarchical structure of the Gaussian linear state space model, but removes the limitations of linearity and Gaussianity. An HMM is a discrete time process $\{(X_t, Y_t),\ t \in \mathbb{N}\}$, where $\{X_t,\ t \in \mathbb{N}\}$ is a Markov chain and, conditional on $\{X_t,\ t \in \mathbb{N}\}$, $\{Y_t,\ t \in \mathbb{N}\}$ is a sequence of independent random variables such that the conditional distribution of Y_t only depends on X_t. We denote by $(\mathsf{X}, \mathcal{X})$ the state space of the hidden Markov chain $\{X_t,\ t \in \mathbb{N}\}$ and by $(\mathsf{Y}, \mathcal{Y})$ the state space of the observations.

Of the two processes $\{X_t,\ t \in \mathbb{N}\}$ and $\{Y_t,\ t \in \mathbb{N}\}$, only $\{Y_t,\ t \in \mathbb{N}\}$ is actually observed, so that inference on the parameters of the model must be achieved using $\{Y_t,\ t \in \mathbb{N}\}$. Inference on the latent or state process, $\{X_t,\ t \in \mathbb{N}\}$, is often also of interest. As we shall see, these two statistical objectives are strongly intertwined.

In this chapter, we consider a number of prototype HMMs (used in some of these applications) in order to illustrate the variety of situations; e.g., finite-valued state spaces, nonlinear Gaussian state-space models, conditionally Gaussian state-space models, and so on.

9.1 Definitions and basic properties

9.1.1 Discrete-valued state space HMM

If both X and Y are discrete-valued, the hidden Markov model is said to be *discrete*, which is the case originally considered by Baum and Petrie (1966). Let M be a Markov transition matrix on X, so that for any $x \in \mathsf{X}$, $x' \mapsto M(x, x')$ is a probability on X. Thus, for any $x' \in \mathsf{X}$, $M(x, x') \geq 0$ and $\sum_{x' \in \mathsf{X}} M(x,\ x') = 1$. In the discrete state-space setting, we identify any function $f \colon \mathsf{X} \to \mathbb{R}$, i.e., $f \colon x \mapsto f(x)$, with a column vector $f = (f(x))_{x \in \mathsf{X}}$ and any finite measure ξ on X, with a row vector $\xi = (\xi(x))_{x \in \mathsf{X}}$; ξ is a probability if $\sum_{x \in \mathsf{X}} \xi(x) = 1$. Let $\{X_t,\ t \in \mathbb{N}\}$ be a Markov chain with initial distribution ξ and Markov transition matrix M. For $f \in \mathbb{F}(\mathsf{X}, \mathcal{X})$ and any $x \in \mathsf{X}$, we

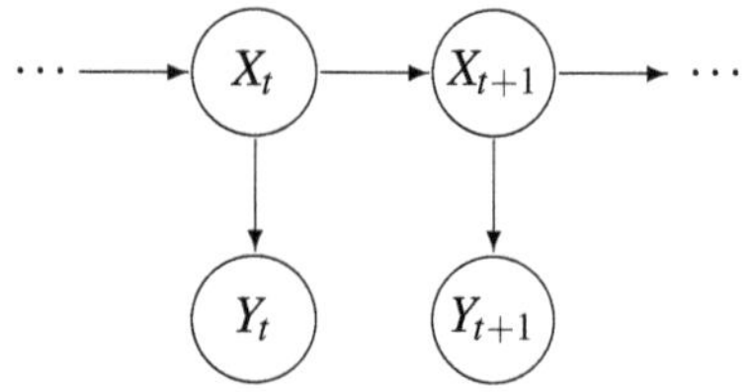

Figure 9.1 *Representation of the dependence structure of a hidden Markov model, where* $\{Y_t,\ t \in \mathbb{N}\}$ *are the observations and* $\{X_t,\ t \in \mathbb{N}\}$ *is the state sequence.*

get

$$\mathbb{E}\left[f(X_t) \mid X_{t-1} = x\right] = Mf(x) = \sum_{x' \in \mathsf{X}} M(x,x')f(x') .$$

For any integer t and any $f_{t+1} \in \mathbb{F}_b(\mathsf{X}^{t+1}, \mathcal{X}^{\otimes(t+1)})$, the joint distribution of the chain is given by

$$\mathbb{E}_\xi\left[f_{t+1}(X_0,X_1,\ldots,X_t)\right] = \sum_{x_0 \in \mathsf{X}} \cdots \sum_{x_t \in \mathsf{X}} \xi(x_0)M(x_0,x_1)\ldots M(x_{t-1},x_t)f_{t+1}(x_0,x_1,\ldots,x_t) .$$

The previous identity implies that $\xi_t(f) := \mathbb{E}_\xi[f(X_t)] = \xi M^t f$, where ξ_t denotes the marginal distribution of X_t.

Let G be a Markov transition matrix from X to Y, i.e., for any $x \in \mathsf{X}$, $G(x,\cdot)$ is a probability on Y, so that for any $y \in \mathsf{Y}$, $G(x,y) \geq 0$ and $\sum_{y \in \mathsf{Y}} G(x,y) = 1$. Consider K, the Markov kernel on $\mathsf{X} \times (\mathcal{X} \otimes \mathcal{Y})$ given by

$$K(x;x',y') = M(x,x')G(x',y'), \quad (x,x',y') \in \mathsf{X}^2 \times \mathsf{Y} .$$

For all $x \in \mathsf{X}$ and $(x',y') \in \mathsf{X} \times \mathsf{Y}$, $K(x;x',y') \geq 0$ and for any $x \in \mathsf{X}$,

$$\sum_{(x',y') \in \mathsf{X} \times \mathsf{Y}} K(x;x',y') = \sum_{x' \in \mathsf{X}} M(x,x') \sum_{y' \in \mathsf{Y}} G(x',y') = \sum_{x' \in \mathsf{X}} M(x,x') = 1 .$$

Let ξ be a probability on X. Consider the stochastic process $\{(X_t,Y_t),\ t \in \mathbb{N}\}$ with joint distribution given, for any $t \in \mathbb{N}$ and function $h_{t+1} \in \mathbb{F}_b((\mathsf{X} \times \mathsf{Y})^{t+1}, (\mathcal{X} \otimes \mathcal{Y})^{\otimes(t+1)})$ by

$$\mathbb{E}_\xi\left[h_{t+1}((X_0,Y_0),\ldots,(X_t,Y_t))\right] = \sum_{(x_0,y_0)} \cdots \sum_{(x_t,y_t)} h_{t+1}((x_0,y_0),\ldots,(x_t,y_t))\xi(x_0,y_0)G(x_0,y_0)\prod_{s=1}^{t} K(x_{s-1};x_s,y_s) . \tag{9.1}$$

The process $\{(X_t,Y_t),\ t \in \mathbb{N}\}$ is a Markov chain on $\mathsf{X} \times \mathsf{Y}$ with initial distribution ξ_G where $\xi_G(x,y) = \xi(x)G(x,y)$, $(x,y) \in \mathsf{X} \times \mathsf{Y}$ and transition kernel K. The marginal distribution of $\{X_t,\ t \in \mathbb{N}\}$ is obtained by marginalizing with respect to the observations:

$$\mathbb{E}_\xi\left[f_{t+1}(X_0,X_1,\ldots,X_t)\right] = \sum_{x_0 \in \mathsf{X}} \cdots \sum_{x_t \in \mathsf{X}} f_{t+1}(x_0,\ldots,x_t)\xi(x_0)\prod_{i=1}^{t} M(x_{i-1},x_i) , \tag{9.2}$$

where $f_{t+1} \in \mathbb{F}_b(\mathsf{X}^{t+1}, \mathcal{X}^{\otimes(t+1)})$, showing that $\{X_t,\ t \in \mathbb{N}\}$ is a Markov chain on X with initial distribution ξ and transition kernel M.

On the other hand, let $\{h_0, \ldots, h_t\}$ be a set of functions, $h_i \in \mathbb{F}_b(\mathsf{Y}, \mathcal{Y})$ and $f_{t+1} \in \mathbb{F}_b(\mathsf{X}^{t+1}, \mathcal{X}^{\otimes(t+1)})$. We get

$$\begin{aligned}\mathbb{E}_\xi\left[\prod_{s=0}^{t} h_s(Y_s) f_{t+1}(X_0, \ldots, X_t)\right] &= \mathbb{E}_\xi\left[f_{t+1}(X_0, \ldots, X_t)\mathbb{E}\left\{\prod_{s=0}^{t} h_s(Y_s) \mid X_0, \ldots, X_t\right\}\right] \\ &= \sum_{x_0 \in \mathsf{X}} \cdots \sum_{x_t \in \mathsf{X}} \xi(x_0) \prod_{s=1}^{t} M(x_{s-1}, x_s) \prod_{s=0}^{t} \sum_{y_s \in \mathsf{Y}} G(x_s, y_s) h_s(y_s) ,\end{aligned}$$

showing that the components of the vector of observations $(Y_0, \ldots, Y_t)$ are conditionally independent given the state sequence $X_0, \ldots, X_t$ and that the conditional distribution of Y_s given X_s is $G(X_s, \cdot)$:

$$\mathbb{E}\left[\prod_{s=0}^{t} h_s(Y_s) \mid X_0, \ldots, X_t\right] = \prod_{i=0}^{t} \mathbb{E}\left[h_s(Y_s) \mid X_s\right] = \prod_{s=0}^{t} G h_s(X_s) . \tag{9.3}$$

The joint distribution of the sequence of observations $Y_0, \ldots, Y_t$ may be deduced from (9.1) by marginalizing with respect to the state sequence:

$$\mathrm{p}_{\xi,t}(Y_{0:t}) = \sum_{x_0 \in \mathsf{X}} \cdots \sum_{x_t \in \mathsf{X}} \xi(x_0) \prod_{s=1}^{t} M(x_{s-1}, x_s) G(x_s, Y_s) . \tag{9.4}$$

This expression of the joint distribution of the observations might look a little daunting at first sight, because it involves evaluating the joint distribution of the state-sequence and the observations, and then marginalizing the state sequence. If the number of states is m, then the number of state sequences is m^{t+1}, so that the numerical complexity seems to grow exponentially with t. We will see later that the likelihood can be computed with an algorithm whose complexity grows quadratically in the number of states and linearly in the number of time steps t.

The marginal distribution of the t-th observation Y_t is obtained by marginalizing (9.4) with respect to $(Y_0, \ldots, Y_{t-1})$

$$\begin{aligned}\mathrm{p}_{\xi,t}(Y_t) &= \sum_{x_0 \in \mathsf{X}} \cdots \sum_{x_t \in \mathsf{X}} \xi(x_0) \prod_{s=1}^{t} M(x_{s-1}, x_s) G(x_t, Y_t) \\ &= \sum_{x \in \mathsf{X}} \mathbb{P}_\xi[X_t = x] G(x, Y_t) .\end{aligned}$$

The marginal distribution is a mixture of the distributions $\{G(x, Y_t), x \in \mathsf{X}\}$ with weights given by $\{\mathbb{P}_\xi[X_t = x], x \in \mathsf{X}\}$. If the Markov kernel M admits a stationary distribution π, then $\mathbb{P}_\pi[X_t = x] = \pi(x)$, and the weights of the mixture remain constants $\{\pi(x), x \in \mathsf{X}\}$. Such behavior is a key property of HMMs; their marginal distribution is a mixture of state-dependent distributions.

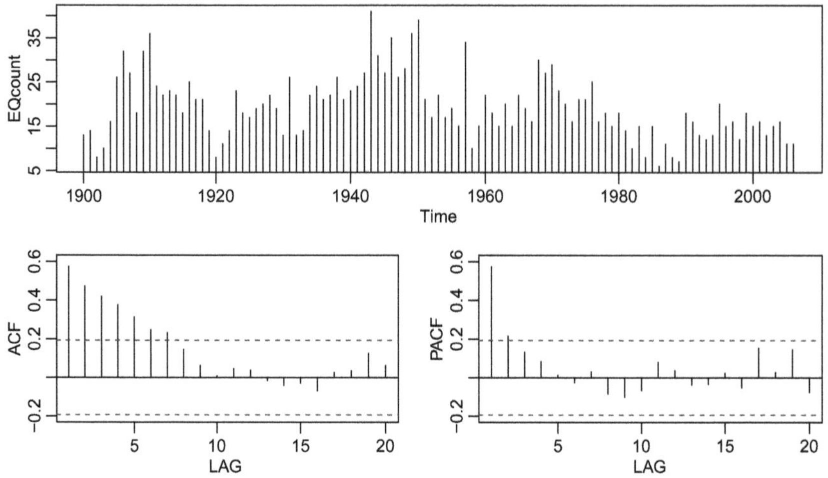

Figure 9.2 *Top: Series of annual counts of major earthquakes (magnitude 7 and above) in the world between 1900-2006. Bottom: Sample ACF and PACF of the square root of the counts.*

Example 9.1 (Number of major earthquakes). Consider the time series of annual counts of major earthquakes displayed in Figure 9.2; see MacDonald and Zucchini (2009, Chapter 1). As discussed in Example 3.3 and Example 3.13 an overdispersed Poisson or Negative Binomial distribution may be a satisfactory model for the marginal distribution; however, given the serial correlation, it is perhaps more important to model joint distributions. As suggested by MacDonald and Zucchini (2009), a simple and convenient way to capture both the marginal distribution and the serial dependence is to consider HMM model with a Poisson distribution. We denote the number of major earthquakes in year t as Y_t, whereas the state, or latent variable, is denoted by X_t. For simplicity, we consider the process $\{X_t,\, t \in \mathbb{N}\}$ to be a two-state Markov chain, $\mathsf{X} = \{1,2\}$, where M is a 2×2 matrix given by,

$$M = \begin{bmatrix} M(1,1) & M(1,2) \\ M(2,1) & M(2,2) \end{bmatrix}$$

with $M(1,1), M(2,2) \in (0,1)$, $M(1,2) = 1 - M(1,1)$, $M(2,1) = 1 - M(2,2)$. The stationary distribution of this Markov chain is given by

$$\pi(1) = \frac{M(2,1)}{2 - M(1,1) - M(2,2)}, \quad \text{and} \quad \pi(2) = \frac{M(1,2)}{2 - M(1,1) - M(2,2)}.$$

For $x \in \mathsf{X}$, denote λ_x as the parameter of the Poisson distribution:

$$G(x,y) = \frac{(\lambda_x)^y}{y!} \mathrm{e}^{-\lambda_x}, \quad y \in \mathbb{N}.$$

Assuming that the Markov chain is stationary ($\xi = \pi$), the marginal distribution is a mixture of Poisson distribution

$$p_{\pi,t}(Y_t) = \pi(1)G(1,Y_t) + \pi(2)G(2,Y_t) = \pi(1)\frac{(\lambda_1)^{Y_t}}{Y_t!}e^{-\lambda_1} + \pi(2)\frac{(\lambda_2)_t^{Y}}{Y_t!}e^{-\lambda_2} .$$

Let $f : \mathbb{R}_+ \to \mathbb{R}$ be a function. The mean and the variance of $f(Y_t)$ are given by (see Exercise 9.1)

$$\mathbb{E}_\pi[f(Y_t)] = \pi(1)\mu(1) + \pi(2)\mu(2) , \tag{9.5}$$

$$\mathrm{Var}_\pi[f(Y_t)] = \mathbb{E}_\pi[Y_0] + \pi(1)\pi(2)(\mu(2)-\mu(1))^2 \geq \mathbb{E}_\pi[f(Y_0)] , \tag{9.6}$$

where

$$\mu(x) = \sum_{y\in\mathbb{N}} G(x,y)f(y) = \sum_{y\in\mathbb{N}} \frac{\lambda_x^y}{y!} f(y) .$$

The marginal distribution of a two-state Poisson-HMM model is therefore overdispersed compared to the Poisson distribution. Using the conditional independence of the observations and the states (9.3),

$$\mathbb{E}_\pi[f(Y_t)f(Y_{t+k})] = \mathbb{E}_\pi\{\mathbb{E}_\pi\left[f(Y_t)f(Y_{t+k})\,|\,X_t,X_{t+k}\right]\} ,$$

$$\mathbb{E}_\pi[\mu(X_t)\mu(X_{t+k})] = \mathbb{E}_\pi[\mu(X_0)\mu(X_k)] = \sum_{x_0=0}^{1}\sum_{x_k=0}^{1} \pi(x_0)M^k(x_0,x_k)\mu(x_0)\mu(x_k) ,$$

where $M^k(x,x')$ denotes the (x,x')-element of the k-th iterate of the transition matrix M. Therefore, the process $\{f(Y_t), t\in\mathbb{N}\}$ is a covariance stationary process with autocovariance function

$$\begin{aligned}\mathrm{Cov}_\pi(f(Y_0),f(Y_k)) &= \sum_{x_0=0}^{1}\sum_{x_k=0}^{1} \pi(x_0)\{M^k(x_0,x_k)-\pi(x_k)\}\mu(x_0)\mu(x_k) ,\\ &= \pi(1)\pi(2)(\mu(2)-\mu(1))^2(1-M(1,2)-M(2,1))^k .\end{aligned}$$

Therefore, for any function $f : \mathbb{R}_+ \to \mathbb{R}$, a two-state Poisson-HMM has an exponentially decaying autocorrelation function (see Exercise 9.3). It is worthwhile to note that the rate of decay of the autocorrelation does not depend upon the choice of f. If we increase the number of states, more complex dependence structure may be obtained; see Exercise 9.3. ◇

A slightly more general example is when the state is discrete, but the observations take values in a general state space. Let $(\mathsf{Y},\mathcal{Y})$ be a measurable space, and let G be a kernel on $\mathsf{X}\times\mathcal{Y}$ (see Definition 5.2). Denote by K, the Markov kernel on $\mathsf{X}\times(\mathcal{X}\otimes\mathcal{Y})$ given for all $x,x'\in\mathsf{X}$, and $A\in\mathcal{Y}$ by

$$K(x;\{x'\}\times A) = M(x,x')G(x',A) . \tag{9.7}$$

Let ξ be a probability on X. Consider the stochastic process $\{(X_t,Y_t),\ t\in\mathbb{N}\}$ with joint distribution given, for any $t \in \mathbb{N}$ and function $h_{t+1} \in \mathbb{F}_b((\mathsf{X} \times \mathsf{Y})^{t+1}, (\mathcal{X} \otimes \mathcal{Y})^{\otimes(t+1)})$ by

$$\mathbb{E}_\xi\left[h_{t+1}(\{(X_s,Y_s)\}_{s=0}^t)\right] = \sum_{x_0} \cdots \sum_{x_t} \int \cdots \int h_{t+1}(\{(x_s,y_s)\}_{s=0}^t) \\ \times \xi(x_0, \mathrm{d}y_0) G(x_0, \mathrm{d}y_0) \prod_{s=1}^t K(x_{s-1}; x_s, \mathrm{d}y_s) \,. \quad (9.8)$$

The process $\{(X_t,Y_t),\ t \in \mathbb{N}\}$ is a Markov chain on $\mathsf{X} \times \mathsf{Y}$ with initial distribution ξ_G where $\xi_G(\{x\} \times A) = \xi(x) G(x,A)$, $x \in \mathsf{X}$ and $A \in \mathcal{Y}$ and transition kernel K. By marginalizing with respect to the observations, (9.8) implies that $\{X_t,\ t \in \mathbb{N}\}$ is a Markov chain with transition kernel M and initial distribution ξ; see (9.2). Similarly, proceeding as in (9.3), the sequence of observations $Y_0, \ldots, Y_t$ are independent conditional to the states.

If, for all $x \in \mathsf{X}$, $G(x,\cdot)$ is absolutely continuous with respect to μ, $G(x,\cdot) \ll \mu(\cdot)$, with transition density function $g(x,\cdot)$. Then, for $A \in \mathcal{Y}$, $G(x,A) = \int_A g(x,y)\, \mu(\mathrm{d}y)$ and the joint transition kernel K can be written as

$$K(x,C) = \iint_C M(x,dx') g(x',y')\, \mu(dy') \,, \quad C \in \mathcal{X} \otimes \mathcal{Y} \,. \quad (9.9)$$

In this case, the joint distribution of the sequence of observations $Y_0, \ldots, Y_t$ has a density with respect to the product measure $\mu^{\otimes(t+1)}$ given by

$$\mathrm{p}_{\xi,t}(Y_{0:t}) = \sum_{x_0 \in \mathsf{X}} \cdots \sum_{x_t \in \mathsf{X}} \xi(x_0) \prod_{s=1}^t M(x_{s-1}, x_s) g(x_s, Y_s) \,. \quad (9.10)$$

The marginal distribution of the t-th observation Y_t is obtained by marginalizing (9.10) with respect to the observations $(Y_0, \ldots, Y_{t-1})$ and is therefore a mixture of the densities $\{g(x,Y_t), x \in \mathsf{X}\}$.

$$\mathrm{p}_{\xi,t}(Y_t) = \sum_{x \in \mathsf{X}} \mathbb{P}_\xi\left[X_t = x\right] g(x, Y_t) \,.$$

If f is a function and ξ is chosen to be the stationary distribution of the Markov chain P (assuming that it exists) then $\mathbb{E}_\pi[f(Y_t)] = \sum_{x \in \mathsf{X}} \pi(x) \mu(f;x)$, where $\mu(f;x)$ is the conditional expectation of $f(Y_t)$ given state x, $\mu(f;x) = \mathbb{E}\left[f(Y_0) \,\middle|\, X_0 = x\right]$, $x \in \mathsf{X}$. For instance, if $f(y) = y^2$, then $\mu(f;x)$ equals the conditional second moment. Assume that the number of states, d, is finite. Defining $\Gamma(f) = \mathrm{diag}\{\mu(f;x), x \in \mathsf{X}\}$, the unconditional mean can be written more compactly as $\mathbb{E}_\pi[f(Y_t)] = \pi \Gamma(f) \mathbf{1}$. Furthermore, for $h > 0$,

$$\begin{aligned} \mathbb{E}_\pi[f(Y_t) f(Y_{t+h})] &= \sum_{x,x'} \mathbb{E}\left[f(Y_t) f(Y_{t+h}) \,\middle|\, (X_t, X_{t+h}) = (x,x')\right] \mathbb{P}_\pi[(X_t, X_{t+h}) = (x,x')] \\ &= \sum_{x,x'} \mu(f;x) \mu(f;x') \pi(x) P^h(x,x') \,, \end{aligned}$$

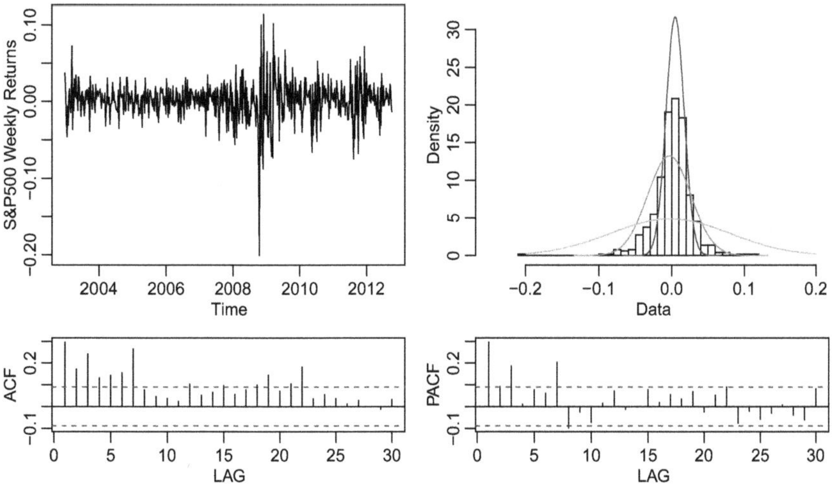

Figure 9.3 *Top: Weekly log-returns of S&P500 from January 2, 2003 to September 28, 2012. Histogram superimposed with a mixture of three Gaussian distributions. Bottom: Sample ACF and PACF of the squares of the log-returns.*

where we have used

$$\begin{aligned}\mathbb{E}\left[f(Y_t)f(Y_{t+h}) \mid (X_t,X_{t+h})=(x,x')\right] \\ = \mathbb{E}\left[f(Y_t) \mid X_t=x\right]\mathbb{E}\left[f(Y_{t+h}) \mid X_{t+h}=x'\right] = \mu(f;x)\mu(f;x') ,\end{aligned}$$

which follows from the conditional independence of the observations given the state and $\mathbb{P}_\pi[(X_t,X_{t+h})=(x,x')] = \mathbb{P}(X_{t+h}=x' \mid X_t=x)\,\mathbb{P}_\pi[X_t=x] = \pi(x)P^h(x,x')$. In matrix form, this can be written as

$$\mathbb{E}_\pi[f(Y_t)f(Y_{t+h})] = \pi\Gamma(f)P^h\Gamma(f)\mathbf{1} .$$

Finally, the covariance of $f(Y_t)$ and $f(Y_{t+h})$ is

$$\mathrm{Cov}_\pi(f(Y_t),f(Y_{t+h})) = \pi\Gamma(f)P^h\Gamma(f)\mathbf{1} - (\pi\Gamma(f)\mathbf{1})^2 .$$

Example 9.2 (S&P500 weekly returns). In this example, we consider the weekly S&P500 log-returns from January 3, 2003 until September 28, 2012. The time series is displayed in the upper left of Figure 9.3; the bottom row of the figure shows the sample ACF and PACF of the squared returns indicating that there is dependence among the returns.

Financial markets are usually characterized as *bullish* (most investors expect upward price movement), *neutral* or *bearish* (most investors expect downward price movement). It has also been reported that the equity market returns and volatility tend to move in opposite directions. To assess this assumption, we modeled the marginal distribution of the log-return by a mixture of three Gaussian distributions. We fitted the marginal distributions using the R package `mixtools`, which can be used to implement MLE to fit the normal mixtures via the EM algorithm. The procedure, however, assumes the data are independent, which is obviously not the case. We will consider a more appropriate analysis later in Example 9.12. Under the assumption of independence, we fitted the observations with three components yielding means $(\hat{\mu}_1, \hat{\mu}_2, \hat{\mu}_3) = (.005, -.003, -.002)$, standard deviations $(\hat{\sigma}_1, \hat{\sigma}_2, \hat{\sigma}_3) = (.013, .030, .082)$, and mixing probabilities $(\hat{\pi}_1, \hat{\pi}_2, \hat{\pi}_3) = (.55, .42, .03)$. A histogram of the data along with the three fitted normals are displayed in the upper right of Figure 9.3. Note that state 1 may be interpreted as the bullish state, with positive mean and a lower volatility. State 2 may be seen as the bearish state with negative mean and comparatively higher volatility. The meaning of state 3 should be interpreted more carefully, since it captures mostly the outliers occurring during the 2008 and 2011 crises.

As previously indicated, a Gaussian mixture model is not appropriate here because the log-returns are serially correlated. The market may stay in a bullish regime for some time before moving to another regime at a later date. As suggested by Rydén et al. (1998), a Hidden Markov model with Gaussian emission probability is a good candidate to capture these stylized facts. ◇

The previous example leads us to examine a more general model. Suppose that $\{X_t,\ t \in \mathbb{N}\}$ is a Markov chain with state space $\mathsf{X} := \{1, \ldots, m\}$ and that the observations $\{Y_t,\ t \in \mathbb{N}\}$, conditional on $\{X_t,\ t \in \mathbb{N}\}$, are independent Gaussian with means $\{\mu_{X_t},\ t \in \mathbb{N}\}$ and variance $\{\sigma^2_{X_t},\ t \in \mathbb{N}\}$. The distribution of the Markov chain is specified by a Markov transition matrix $M = \{M(x,x')\}_{(x,x') \in \mathsf{X}^2}$, which is assumed to have a unique invariant distribution denoted π. Assume for simplicity that the Markov chain is stationary. Suppose the marginal distribution of $\{Y_t,\ t \in \mathbb{N}\}$ is a mixture of m Gaussian distributions with mixing weights $(\pi(1), \ldots, \pi(m))$. The observations may be expressed as $Y_t = \mu_{X_t} + \sigma_{X_t} V_t$, where $\{V_t,\ t \in \mathbb{N}\}$ are i.i.d. $\mathrm{N}(0,1)$. The autocorrelation function of $\{Y_t,\ t \in \mathbb{N}\}$ is given by, for $h > 0$,

$$\frac{\operatorname{Cov}(Y_t, Y_{t+h})}{\operatorname{Var}(Y_t)} = \frac{\pi \Gamma_1 P^h \Gamma_1 \mathbf{1} - (\pi \Gamma_1 \mathbf{1})^2}{\pi \Gamma_2 \mathbf{1} - (\pi \Gamma_1 \mathbf{1})^2} ,$$

where $\Gamma_p = \operatorname{diag}\{\int y^p g(y; \mu_x, \sigma_x^2) dy,\ x \in \mathsf{X}\}$. For a two-state model, the autocorrelation is given by

$$\frac{\operatorname{Cov}(Y_t, Y_{t+h})}{\operatorname{Var}(Y_t)} = \frac{\pi(1)\pi(2)(\mu_1 - \mu_2)^2}{\pi(1)\sigma_1^2 + \pi(2)\sigma_2^2} \lambda^h ,$$

where $\lambda := 1 - M(1,2) - M(2,1)$. The process is not autocorrelated if $\mu_1 = \mu_2$. For the squared process, we have

$$\frac{\operatorname{Cov}(Y_t^2, Y_{t+h}^2)}{\operatorname{Var}(Y_t^2)} = \frac{\pi \Gamma_2 P^h \Gamma_2 \mathbf{1} - (\pi \Gamma_2 \mathbf{1})^2}{\pi \Gamma_4 \mathbf{1} - (\pi \Gamma_2 \mathbf{1})^2} .$$

For a two-state model, the autocovariance of the squared process is given by

$$\operatorname{Cov}(Y_t^2, Y_{t+h}^2) = \pi(1)\pi(2)(\mu_1^2 - \mu_2^2 + \sigma_1^2 - \sigma_2^2)\lambda^h, \quad h > 0 .$$

Note that if $\mu_1 = \mu_2$ and $\lambda \neq 0$, the process $\{Y_t, t \in \mathbb{N}\}$ is white noise, but $\{Y_t^2, t \in \mathbb{N}\}$

is autocorrelated. State dependent variances are neither necessary nor sufficient for autocorrelation in the squared process. Even if $\sigma_1 = \sigma_2$, the marginal process shows conditional heteroscedasticity provided that $M(1,1) \neq M(2,1)$. On the other hand, if $M(1,1) = M(2,1)$, the squared process is not autocorrelated even if $\sigma_1 \neq \sigma_2$. These results extend directly to more general families of distributions.

9.1.2 Continuous-valued state-space models

It is not necessary restrict the definition of HMM to discrete state-spaces.

Definition 9.3 (Hidden Markov model). *Let* $(\mathsf{X},\mathcal{X})$ *and* $(\mathsf{Y},\mathcal{Y})$ *be two measurable spaces and let* M *and* G *denote, respectively, a Markov kernel on* $(\mathsf{X},\mathcal{X})$ *and a Markov kernel from* $(\mathsf{X},\mathcal{X})$ *to* $(\mathsf{Y},\mathcal{Y})$*. Denote by* K *the Markov kernel on* $\mathsf{X} \times (\mathcal{X} \otimes \mathcal{Y})$ *by*

$$K(x;C) = \iint_C M(x,\mathrm{d}x')\,G(x',\mathrm{d}y')\,, \quad x \in \mathsf{X},\, C \in \mathcal{X} \otimes \mathcal{Y}\,. \tag{9.11}$$

The Markov chain $\{(X_t,Y_t),\, t \in \mathbb{N}\}$ *with Markov transition kernel* K *and initial distribution* $\xi \otimes G$*, where* ξ *is a probability measure on* $(\mathsf{X},\mathcal{X})$*, is called a* hidden Markov Model *(HMM).*

The definition above specifies the distribution of $\{(X_t,Y_t),\, t \in \mathbb{N}\}$; the term *hidden* is justified because $\{X_t,\, t \in \mathbb{N}\}$ is not observable. As before, we shall denote by $\mathbb{P}_\xi$ and $\mathbb{E}_\xi$ the probability measure and corresponding expectation associated with the process $\{(X_t,Y_t),\, t \in \mathbb{N}\}$, respectively.

An HMM is said to be *partially dominated* if there exists a probability measure μ on $(\mathsf{Y},\mathcal{Y})$ such that for all $x \in \mathsf{X}$, $G(x,\cdot)$ is absolutely continuous with respect to μ, $G(x,\cdot) \ll \mu(\cdot)$, with transition density function $g(x,\cdot)$. Then, for $A \in \mathcal{Y}$, $G(x,A) = \int_A g(x,y)\,\mu(\mathrm{d}y)$ and the joint transition kernel K can be written as

$$K(x;C) = \iint_C M(x,dx')g(x',y')\,\mu(dy')\,, \quad C \in \mathcal{X} \otimes \mathcal{Y}\,. \tag{9.12}$$

A partially dominated HMM is *fully dominated* if there exists a probability measure λ on $(\mathsf{X},\mathcal{X})$ such that $\xi \ll \lambda$ and, for all $x \in \mathsf{X}$, $M(x,\cdot) \ll \lambda(\cdot)$ with transition density function $m(x,\cdot)$. Then, for $A \in \mathcal{X}$, $M(x,A) = \int_A m(x,x')\,\lambda(\mathrm{d}x')$ and the joint Markov transition kernel K has a density k with respect to the product measure $\lambda \otimes \mu$

$$k\left(x;x',y'\right) := m(x,x')g(x',y')\,, \quad (x,x',y') \in \mathsf{X}^2 \times \mathsf{Y}\,. \tag{9.13}$$

Note that for the fully dominated model, we will generally use the notation ξ to denote the *probability density function* of the initial state X_0 (with respect to λ) rather than the distribution itself.

Proposition 9.4. *Let* $\{(X_t,Y_t),\, t \in \mathbb{N}\}$ *be a Markov chain over the product space* $\mathsf{X} \times \mathsf{Y}$ *with transition kernel* K *given by* (9.11)*. Then, for any integer* p *and any ordered set* $\{t_1 < \cdots < t_p\}$ *of indices the random variables* $Y_{t_1},\ldots,Y_{t_p}$ *are* $\mathbb{P}_\xi$*-conditionally independent given* $(X_{t_1},X_{t_2},\ldots,X_{t_p})$*, i.e., and all functions* $f_1,\ldots,f_p \in \mathbb{F}_b(\mathsf{Y},\mathcal{Y})$*,*

$$\mathbb{E}_\xi\left[\prod_{i=1}^{p} f_i(Y_{t_i}) \mid X_{t_1},\ldots,X_{t_p}\right] = \prod_{i=1}^{p} Gf_i(X_{t_i})\,, \tag{9.14}$$

where $Gf(x) = \int_{\mathsf{Y}} G(x, \mathrm{d}y) f(y)$.

Proof. See Exercise 9.10. ■

Assume that the HMM is partially dominated (see (9.12)). The joint probability of the unobservable states and observations up to index t is such that for any function $h_{t+1} \in \mathbb{F}_b((\mathsf{X}\times\mathsf{Y})^{t+1}, (\mathcal{X}\otimes\mathcal{Y})^{\otimes(t+1)})$,

$$\mathbb{E}_\xi\left[h_{t+1}(X_0, Y_0, \ldots, X_t, Y_t)\right] = \int \cdots \int h_{t+1}(x_0, y_0, \ldots, x_t, y_t) \\ \times \xi(\mathrm{d}x_0) g(x_0, y_0) \prod_{s=1}^{t} M(x_{s-1}, \mathrm{d}x_s) g(x_s, y_s) \prod_{s=0}^{t} \mu(\mathrm{d}y_s)\,, \quad (9.15)$$

Marginalizing with respect to the unobservable variables $X_0, \ldots, X_t$, one obtains the joint distribution of the observations $Y_{0:t}$

$$\mathrm{p}_{\xi,t}(Y_{0:t}) = \int \cdots \int \xi(\mathrm{d}x_0) g(x_0, Y_0) \prod_{s=1}^{t} M(x_{s-1}, \mathrm{d}x_s) g(x_s, Y_s)\,. \quad (9.16)$$

Example 9.5 (Stochastic volatility). Denote by Y_t the daily *log-returns* of some financial asset. Most models for return data that are used in practice are of a multiplicative form,

$$Y_t = \sigma_t V_t\,, \quad (9.17)$$

where $\{V_t,\ t \in \mathbb{N}\}$ is an i.i.d. sequence and the *volatility process* $\{\sigma_t,\ t \in \mathbb{N}\}$ is a non-negative stochastic process such that V_t is independent of σ_s for all $s \le t$. It is often assumed that V_t has zero mean and unit variance.

We have already discussed the ARCH/GARCH models in Section 3.5. An alternative to the ARCH/GARCH models is stochastic volatility (SV) models, in which the volatility is a non-linear transform of a hidden linear autoregressive process. The canonical model in SV for discrete-time data has been introduced by Taylor (1982) and worked out since then by many authors; see Hull and White (1987) and Jacquier et al. (1994) for early references and Shephard and Andersen (2009) for an up-to-date survey. In this model, the hidden volatility process, $\{X_t,\ t \in \mathbb{N}\}$, follows a first order autoregression,

$$X_{t+1} = \phi X_t + \sigma W_t\,, \quad (9.18a)$$

$$Y_t = \beta \exp(X_t/2) V_t\,. \quad (9.18b)$$

where $\{W_t,\ t \in \mathbb{N}\}$ is a white Gaussian noise with mean zero and unit variance and $\{V_t,\ t \in \mathbb{N}\}$ is a strong white noise. The error processes $\{W_t,\ t \in \mathbb{N}\}$ and $\{V_t,\ t \in \mathbb{N}\}$ are assumed to be mutually independent and $|\phi| < 1$. As W_t is normally distributed, X_t is also normally distributed. All moments of V_t exist, so that all moments of Y_t in (9.18) exist as well. Assuming that $X_0 \sim \mathrm{N}(0, \sigma^2/(1-\phi^2))$ (the stationary distribution of the Markov chain) the kurtosis[1] of Y_t is given by (see Exercise 9.12)

$$\kappa_4(Y) = \kappa_4(V) \exp(\sigma_X^2)\,, \quad (9.19)$$

[1] For an integer m and a random variable U, $\kappa_m(U) := \mathbb{E}\left[|U|^m\right] / (\mathbb{E}\left[|U|^2\right])^{m/2}$. Typically, κ_3 is called *skewness* and κ_4 is called *kurtosis*.

where $\sigma_X^2 = \sigma^2/(1-\phi^2)$ is the (stationary) variance of X_t. Thus $\kappa_4(Y_t) > \kappa_4(V_t)$, so that if $V_t \sim \mathrm{N}(0,1)$, the distribution of Y_t is leptokurtic. The autocorrelation function of $\{Y_t^{2m}, t \in \mathbb{N}\}$ for any integer m is given by

$$\mathrm{Cor}(Y_t, Y_{t+h}) = \frac{\exp(m^2\sigma_X^2\phi^h) - 1}{\kappa_{4m}(V)\exp(m^2\sigma_X^2) - 1}, \quad h \in \mathbb{N}. \tag{9.20}$$

The decay rate of the autocorrelation function is faster than exponential at small time lags and then stabilizes to ϕ for large lags. ◇

Example 9.6 (NGM model). We consider the univariate model introduced in Netto, Gimeo, and Mendes (1978)—hereafter referred to as the *NGM model*—discussed by Kitagawa (1987) and Carlin et al. (1992), given, in state-space form, by

$$X_t = F_t^\theta(X_{t-1}) + W_t \quad \text{and} \quad Y_t = H_t(X_t) + V_t\,, \tag{9.21}$$

with

$$F_t^\theta(X_{t-1}) = \alpha X_{t-1} + \beta X_{t-1}/(1 + X_{t-1}^2) + \gamma\cos[1.2(t-1)]\,, \tag{9.22a}$$

$$H_t(X_t) = X_t^2/20\,, \tag{9.22b}$$

where $X_0 \sim \mathrm{N}(\mu_0, \sigma_0^2)$, with $W_t \sim \mathrm{iid}\,\mathrm{N}(0, \sigma_w^2)$ independent of $V_t \sim \mathrm{iid}\,\mathrm{N}(0, \sigma_v^2)$ and each sequence independent of X_0. Figure 9.4 shows a typical data sequence Y_t and the corresponding state process X_t with all the variances equal to unity and, as in Kitagawa (1987) and Carlin et al. (1992), $\theta = (\alpha = .5, \beta = 25, \gamma = 8)$. Additionally, Figure 9.4 demonstrates the nonlinearity by exhibiting a scatterplot of the observations versus the states, and a phase space trajectory of the states that demonstrates that the states are bifurcating near ± 10.

Note that, in this case, there is no closed form for the covariance of the observations. However, because of the nonlinearity of the processes, the autocovariance function contains little information about the dynamics of the Y_t. In addition, the marginal distribution of the observations is highly complex and no longer known. This model has become a standard model for testing numerical procedures and is used throughout Chapter 12. ◇

9.1.3 *Conditionally Gaussian linear state-space models*

Conditionally Gaussian linear state-space models belong to a class of models that we will refer to as *hierarchical hidden Markov models*, whose dependence structure is depicted in Figure 9.5. In such models the variable I_t, which is the highest in the hierarchy, influences both the transition from W_{t-1} to W_t as well as the observation Y_t.

Conditionally Gaussian models related to the previous example are also commonly used to approximate non-Gaussian state-space models. Imagine that we are interested in the linear model given by (2.1)–(2.2) with both noise sequences still being i.i.d. but at least one of them with a non-Gaussian distribution. Assuming a very

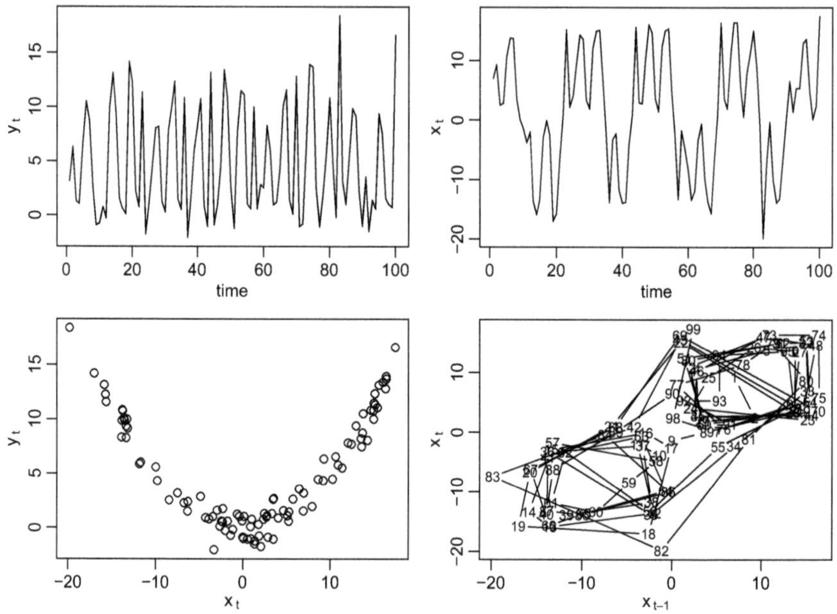

Figure 9.4 *Typical realization of the observations* (Y_t) *and state process* (X_t)*, for* $t = 1,\ldots,100$*, generated from the model* (9.21)*. The bottom row shows the quadratic relationship between the observations and states, and a phase space trajectory of the states indicating the bifurcating dynamics of the process.*

general form of the noise distribution would directly lead us into the world of (general) continuous state-space HMMs. As a middle ground, however, we can assume that the distribution of the noise is a mixture of Gaussians.

Let $\{I_t,\, t \in \mathbb{N}\}$ be a sequence of random variables taking values in a set $\mathbb{I}$, which can be finite or infinite. We often refer to these variables as the *indicator variables* when $\mathbb{I}$ is finite. To model non-Gaussian system dynamics, we will typically model the dynamic of the partial state sequence $\{W_t,\, t \in \mathbb{N}\}$ as follows

$$W_{t+1} = \mu_W(I_{t+1}) + A(I_{t+1})W_t + R(I_{t+1})W_t\,, \quad W_t \sim \mathrm{N}(0,I)\,,$$

where, μ_W, A and R are respectively vector-valued and matrix-valued functions of suitable dimensions on $\mathbb{I}$. When $\mathbb{I} = \{1,\ldots,r\}$ is finite, the distribution of the noise, $\mu_W(I_{t+1}) + R(I_{t+1})W_t$, driving the state equation is a finite mixture of multivariate Gaussian distributions. Similarly, the observation equation is modeled by

$$Y_t = \mu_Y(I_t) + B(I_t)W_t + S(I_t)V_t\,, \quad V_t \sim \mathrm{N}(0,I)\,,$$

where μ_Y, B and S are respectively vector-valued and matrix-valued functions. Here again, when $\mathbb{I} = \{1,\ldots,r\}$ is finite, then the distribution of the observation noise $\mu_Y(I_t) + S(I_t)V_t$ is a finite mixture of multivariate distribution, allowing us to model outliers, for example. Since B is also a function of I, this model may accommodate changes in the way the state is observed.

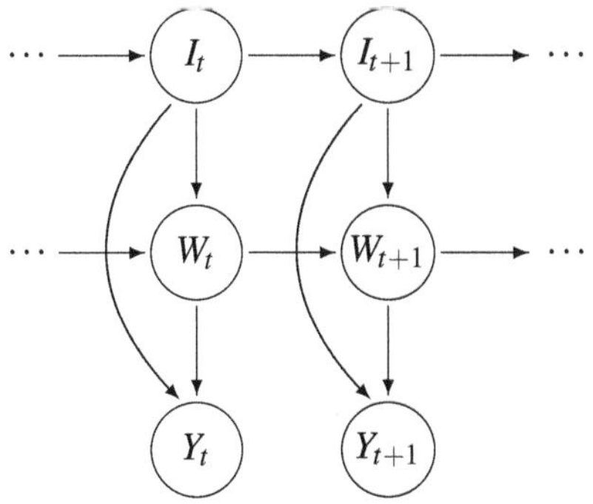

Figure 9.5: *Graphical representation of the dependence structure of a hierarchical HMM.*

Example 9.7 (Level shifts and outliers). Gerlach et al. (2000) have considered the following model to analyze data with level shifts and outliers in both the observations and innovations:

$$Y_t = W_t + \sigma_V I_{t,1} V_t \,, \tag{9.23}$$

$$W_t - \mu_t = \sum_{i=1}^{p} \phi_i (W_{t-i} - \mu_{t-i}) + \sigma_Z I_{t,2} Z_t \,, \tag{9.24}$$

$$\mu_t = \mu_{t-1} + \sigma_W I_{t,3} W_t \,, \tag{9.25}$$

where $\{(V_t, Z_t, W_t), t \in \mathbb{Z}\}$ is an i.i.d. sequence of Gaussian vectors with zero mean and identity covariance, and $\{(I_{t,1}, I_{t,2}, I_{t,3}), t \in \mathrm{Z}\}$ is i.i.d. taking values in $\mathbb{I}$, which is typically discrete. The time series $\{W_t,\ t \in \mathbb{Z}\}$ has mean level μ_t and is generated by an autoregressive model with coefficients $\phi = (\phi_1, \ldots, \phi_p)$.

If $I_{t,1}$ and $I_{t,2}$ are equal to 1, then the observations are a noisy version of an AR(p) process; recall Example 2.2. Observational outliers are modeled by assuming that $I_{t,1}$ takes some large values (like 10, 20). Similarly, innovation outliers are modeled by large values of $I_{t,2}$. If $I_{t,3} = 0$, then $\mu_t = \mu_{t-1}$. Level shifts occur at time points t for which $I_{t,3} \neq 0$. ◇

Example 9.8 (Stochastic volatility cont.). Another example of the use of mixtures is in the observational noise of the SVM, (9.18),

$$X_t = \phi X_{t-1} + \sigma W_t \,, \tag{9.26a}$$

$$\ln Y_t^2 = \beta + X_t + \ln V_t^2 \,, \tag{9.26b}$$

where $W_t \sim$ iid $\mathrm{N}(0,1)$, but where now, the observational noise, V_t, is not assumed to be normal. The assumption that the V_t are normal comes from the original ARCH model, which is an assumption that is typically violated empirically. Under the normal assumption, $\ln V_t^2$ is the log of a χ_1^2 random variable with density given by

$$f(x) = \frac{1}{\sqrt{2\pi}} \exp\left\{ -\frac{1}{2}(e^x - x) \right\}, \quad -\infty < x < \infty \,. \tag{9.27}$$

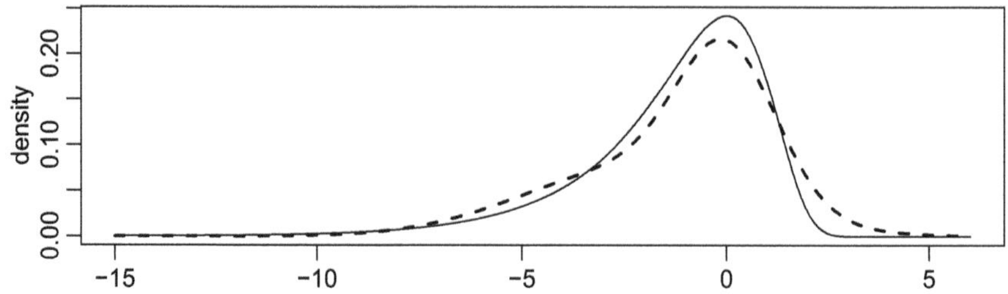

Figure 9.6 *Density of the log of a χ_1^2 as given by (9.27) (solid line) and a fitted normal mixture (dashed line) from Shumway and Stoffer (2011, Example 6.18).*

The mean of the distribution is $-(\gamma + \ln 2)$, where $\gamma \approx 0.5772$ is Euler's constant, and the variance of the distribution is $\pi^2/2$. It is a highly skewed density but it is, of course, not flexible because the distribution is fixed; i.e., there are no parameters to be estimated.

To avoid having a fixed observational noise distribution, Kim and Stoffer (2008) and Shumway and Stoffer (2011, Chapter 6) assumed that the observational noise in (9.26b) is a mixture of two normals with parameters to be estimated. That is,

$$\ln V_t^2 = I_t Z_{t,0} + (1 - I_t) Z_{t,1} , \tag{9.28}$$

where $I_t \sim$ iid Ber(π), with $\pi \in [0,1]$, $Z_{t,0} \sim$ iid N$(0, \sigma_0^2)$, and $Z_{t,1} \sim$ iid N(μ_1, σ_1^2). The advantage to this model is that it is easy to fit because it uses conditional normality and there are three additional parameters to provide flexibility in the analysis. Figure 9.6 compares the $\ln \chi_1^2$ density to a fitted mixture distribution taken from Shumway and Stoffer (2011, Example 6.18). Note that the mixture distribution is able to accommodate kurtosis when the volatility is large. ◇

9.1.4 *Switching processes with Markov regimes*

Markov-switching models perhaps constitute the most significant generalization of HMMs. In such models, the conditional distribution of Y_{t+1}, given all the past variables, depends not only on X_{t+1} but also on Y_t (and possibly more lagged Y-variables). Thus, conditional on the state sequence $\{X_t,\, t \in \mathbb{N}\}$, $\{Y_t,\, t \in \mathbb{N}\}$ forms a (non-homogeneous) Markov chain. Graphically, this is represented as in Figure 9.7. In state-space form, a Markov-switching model may be written as

$$X_{t+1} = a_t(X_t, W_t) , \tag{9.29}$$

$$Y_{t+1} = b_t(X_{t+1}, Y_t, V_{t+1}) . \tag{9.30}$$

We can even go a step further and assume that $\{(X_t, Y_t),\, t \in \mathbb{N}\}$ jointly forms a Markov chain, but that only $\{Y_t,\, t \in \mathbb{N}\}$ is actually observed.

A switching linear autoregression is a model of the form

$$Y_t = \mu(I_t) + \sum_{i=1}^{p} a_i(I_t)(Y_{t-i} - \mu(I_{t-i})) + \sigma(I_t) V_t , \qquad p \geq 1 , \tag{9.31}$$

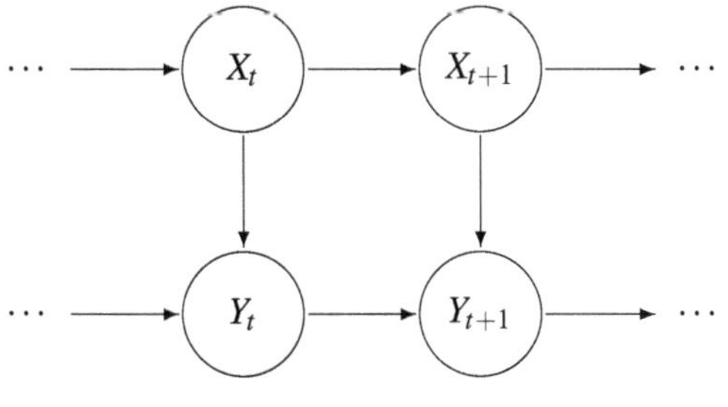

Figure 9.7 *Graphical representation of the dependence structure of a Markov-switching model, where* $\{Y_t,\ t \in \mathbb{N}\}$ *is the observable process and* $\{X_t,\ t \in \mathbb{N}\}$ *is the hidden chain.*

where $\{I_t,\ t \in \mathbb{N}\}$, called the *regime*, is a Markov chain on a finite state space $\mathbb{I} = \{1, 2, \ldots, r\}$, and $\{V_t,\ t \in \mathbb{N}\}$ is white noise independent of the regime; the functions $\mu : \mathbb{I} \to \mathbb{R}$, $a_i : \mathbb{I} \to \mathbb{R}$, $i = 1, \ldots, r$, and $\sigma : \mathbb{I} \to \mathbb{R}$ describe the dependence of the parameters on the realized regime.

This model can be rewritten in state-space form as follows. Let

$$\begin{aligned}
\boldsymbol{Y}_t &= [Y_t, Y_{t-1}, \ldots, Y_{t-p+1}]' , \\
\boldsymbol{I}_t &= [I_t, I_{t-1}, \ldots, I_{t-d+1}]' , \\
\mu(\boldsymbol{I}_t) &= [\mu(I_t), \ldots, \mu(I_{t-p+1})]' , \\
\boldsymbol{V}_t &= [V_t, 0, \ldots, 0]' ,
\end{aligned}$$

and denote by $A(i)$ the $p \times p$ companion matrix associated with the autoregressive coefficients of the state i,

$$A(i) = \begin{bmatrix} a_1(i) & a_2(i) & \ldots & \ldots & a_p(i) \\ 1 & 0 & & & 0 \\ 0 & 1 & 0 & & \vdots \\ \vdots & \ddots & \ddots & \ddots & \vdots \\ 0 & \ldots & 0 & 1 & 0 \end{bmatrix} . \tag{9.32}$$

The stacked observation vector $\boldsymbol{Y}_t$ then satisfies

$$\boldsymbol{Y}_t = \mu(I_t) + A(I_t)\left(\boldsymbol{Y}_{t-1} - \mu(\boldsymbol{I}_{t-1})\right) + \sigma(I_t)\boldsymbol{V}_t . \tag{9.33}$$

Note that the model is a random coefficient vector autoregression as discussed in Chapter 4.

Example 9.9 (Influenza mortality). In Example 3.1, we discussed the monthly pneumonia and influenza mortality series shown in Figure 3.1. We pointed out the non-reversibility of the series, which rules out the possibility that the data are generated by a linear Gaussian process. In addition, note that the series is irregular, and while mortality is highest during the winter, the peak does not occur in the same month each year. Moreover, some seasons have very large peaks, indicating flu epidemics, whereas other seasons are mild. In addition, it can be seen from Figure 3.1

that there is a slight negative trend in the data set (this is best seen by focusing on the troughs), indicating that flu prevention is getting better over the eleven year period.

Although it is not necessary, to ease the discussion, we focus on the differenced data, which will remove the trend. In this case, we denote $Y_t = \nabla \text{flu}_t$, where flu_t represents the data discussed in Example 3.1. Shumway and Stoffer (2011, Example 5.6) fit a threshold model to Y_t, but we might also consider a switching autoregessive model given in (9.31) or (9.33) where there are two hidden regimes, one for epidemic periods and one for more mild periods. In this case, the model is given by

$$Y_t = \begin{cases} \phi_0^{(1)} + \sum_{j=1}^{p} \phi_j^{(1)} Y_{t-j} + \sigma^{(1)} Z_t\,, & \text{for } I_t = 1\,, \\ \phi_0^{(2)} + \sum_{j=1}^{p} \phi_j^{(2)} Y_{t-j} + \sigma^{(2)} Z_t\,, & \text{for } I_t = 2\,, \end{cases} \tag{9.34}$$

where $Z_t \sim \text{iid N}(0,1)$, and I_t is a hidden, two-state Markov chain. ◇

9.2 Filtering and smoothing

Statistical inference for nonlinear state-space models involves computing the *posterior distribution* of a collection of state variables $X_{s:s'} := (X_s, \ldots, X_{s'})$, with $s < s'$ conditioned on a batch of observations, $Y_{0:t} = (Y_0, \ldots, Y_t)$, which we denote by $\phi_{\xi,s:s'|t}$ (where ξ is the initial distribution), the dependence on the observations being implicit for ease of notation. Specific problems include *filtering*, which corresponds to $s = s' = t$, *fixed lag smoothing*, when $s = s' = t - L$ and fixed interval smoothing, if $s = 0$ and $s' = t$ (see Section 2.2).

Definition 9.10 (Smoothing, filtering, prediction). *For non-negative indices s, t, and n with $t \geq s$, and any initial distribution ξ on $(\mathsf{X}, \mathcal{X})$, denote by $\phi_{\xi,s:t|n}$ (the dependence in the observations is implicit to avoid overloading the notation) the conditional distribution of $X_{s:t}$ given $Y_{0:n}$. Specific choices of s,t and n give rise to several particular cases of interest:*

Joint Smoothing: $\phi_{\xi,0:n|n}$, *for* $n \geq 0$;

(Marginal) Smoothing: $\phi_{\xi,t|n}$ *for* $0 \leq t \leq n$;

Prediction: $\phi_{\xi,t+1|t}$ *for* $t \geq 0$;

***p*-step Prediction:** $\phi_{\xi,t+p|t}$ *for* $t, p \geq 0$.

Filtering: $\phi_{\xi,t|t}$ *for* $t \geq 0$; *Because the use of filtering will be preeminent in the following, we shall most often abbreviate* $\phi_{\xi,t|t}$ *to* $\phi_{\xi,t}$.

Despite the apparent simplicity of the above problems, the smoothing distribution can be computed in closed form only in very specific cases, principally, the linear Gaussian model (see Section 2.2) and the discrete-valued Hidden Markov model (where the state $\{X_t,\ t \in \mathbb{N}\}$ takes its values in a finite alphabet).

9.2.1 *Discrete-valued state-space HMM*

We denote by $\phi_{\xi,t}$ the filtering distribution, i.e., the distribution of X_t given the observations up to time t, $Y_{0:t}$. To simplify the notations, the dependence of the filtering distribution with respect to the observations is implicit.

Denote by $\gamma_{\xi,t}$ the joint distribution of the state X_t and the observations $Y_{0:t}$:

$$\gamma_{\xi,t}(x_t) = \sum_{x_0} \cdots \sum_{x_{t-1}} \xi(x_0) g_0(x_0) \prod_{s=1}^{t} Q_s(x_{s-1}, x_s) ,$$

where we have set, for $s \in \mathbb{N}$

$$g_s(x_s) = g(x_s, Y_s) \quad \text{and} \quad Q_s(x_{s-1}, x_s) = M(x_{s-1}, x_s) g(x_s, Y_s) . \tag{9.35}$$

This equation may be rewritten in matrix form as follows

$$\gamma_{\xi,t} = \xi Q_0 Q_1 \cdots Q_t ,$$

where $Q_0 = \operatorname{diag}\{g(x, Y_0), x \in \mathsf{X}\}$ and

$$Q_s = M \operatorname{diag}\{g(x, Y_s), x \in \mathsf{X}\}, \quad \text{for } s \geq 1 . \tag{9.36}$$

This distribution may be computed recursively as follows

$$\gamma_{\xi,0}(x_0) = \xi(x_0) g_0(x_0) \quad \text{and} \quad \gamma_{\xi,t}(x_t) = \sum_{x_{t-1} \in \mathsf{X}} \gamma_{\xi,t-1}(x_{t-1}) Q_t(x_{t-1}, x_t) \tag{9.37}$$

or equivalently in matrix form $\gamma_{\xi,0} = \xi Q_0$ and for $t \geq 1$, $\gamma_{\xi,t} = \gamma_{\xi,t-1} Q_t$. The computational complexity grows like the square of the number of states. The joint distribution of $(Y_0, \ldots, Y_t)$ may be obtained by marginalizing the joint distribution $\gamma_{\xi,t}$ of $(Y_0, \ldots, Y_t, X_t)$ with respect to the state X_t, i.e., $\mathrm{p}_{\xi,t}(Y_{0:t}) = \sum_{x_t \in \mathsf{X}} \gamma_{\xi,t}(x_t)$ or, in matrix form, $\mathrm{p}_{\xi,t}(Y_{0:t}) = \gamma_{\xi,t} \mathbf{1}$. The filtering distribution is the conditional distribution of the state X_t given $(Y_0, \ldots, Y_t)$. It is obtained by dividing the joint distribution of $(X_t, Y_0, \ldots, Y_t)$ by $\mathrm{p}_{\xi,t}(Y_{0:t})$,

$$\phi_{\xi,t}(x_t) = \frac{\gamma_{\xi,t}(x_t)}{\sum_{x_t \in \mathsf{X}} \gamma_{\xi,t}(x_t)} , \tag{9.38}$$

or in matrix form $\phi_{\xi,t} = \gamma_{\xi,t} / \gamma_{\xi,t} \mathbf{1}$. By plugging the recursion (9.37), the filtering distribution can thus be updated recursively as follows

$$\phi_{\xi,t}(x_t) = \frac{\sum_{x_{t-1} \in \mathsf{X}} \phi_{\xi,t-1}(x_{t-1}) Q_t(x_{t-1}, x_t)}{\sum_{(x_{t-1}, x'_{t-1}) \in \mathsf{X}^2} \phi_{\xi,t-1}(x_{t-1}) Q_t(x_{t-1}, x'_t)} . \tag{9.39}$$

In matrix form, this recursion reads

$$\phi_{\xi,t} = \frac{\phi_{\xi,t-1} Q_t}{\phi_{\xi,t-1} Q_t \mathbf{1}} .$$

Algorithm 9.1 (Forward Filtering)

Initialization: For $x \in \mathsf{X}$,

$$\phi_{\xi,0|-1}(x) = \xi(x) .$$

Forward Recursion: For $t = 0, \ldots, n$,

$$c_{\xi,t} = \sum_{x \in \mathsf{X}} \phi_{\xi,t|t-1}(x) g_t(x) , \tag{9.40}$$

$$\phi_{\xi,t}(x) = \phi_{\xi,t|t-1}(x) g_t(x) / c_{\xi,t} , \tag{9.41}$$

$$\phi_{\xi,t+1|t}(x) = \sum_{x' \in \mathsf{X}} \phi_{\xi,t}(x') M(x, x') , \tag{9.42}$$

for each $x \in \mathsf{X}$.

The algorithm is summarized in Algorithm 9.1, which in the Rabiner (1989) terminology, corresponds to the normalized forward recursion. The computational cost of filtering is thus proportional to n, the number of observations, and scales like $|\mathsf{X}|^2$ (squared cardinality of the state space X) because of the $|\mathsf{X}|$ vector matrix products corresponding to (9.42).

The predictive distribution of the observation Y_t given $Y_{0:t-1}$ is equal to the ratio

$$\frac{\mathrm{p}_{\xi,t}(Y_{0:t})}{\mathrm{p}_{\xi,t-1}(Y_{0:t-1})} = \frac{\sum_{x_t \in \mathsf{X}} \gamma_{\xi,t}(x_t)}{\sum_{x_{t-1} \in \mathsf{X}} \gamma_{\xi,t-1}(x_{t-1})} = \sum_{(x_{t-1}, x_t) \in \mathsf{X}^2} \phi_{\xi,t-1}(x_{t-1}) Q_t(x_{t-1}, x_t) . \tag{9.43}$$

or in matrix form

$$\frac{\mathrm{p}_{\xi,t}(Y_{0:t})}{\mathrm{p}_{\xi,t-1}(Y_{0:t-1})} = \phi_{\xi,t-1} Q_t \mathbf{1} .$$

The likelihood of $n+1$ the observations may therefore be written as

$$\begin{aligned} \mathrm{p}_{\xi,n}(Y_{0:n}) &= \mathrm{p}_{\xi,0}(Y_0) \prod_{t=1}^{n} \frac{\mathrm{p}_{\xi,t}(Y_{0:t})}{\mathrm{p}_{\xi,t-1}(Y_{0:t-1})} \\ &= \mathrm{p}_{\xi,0}(Y_0) \prod_{t=1}^{n} \sum_{(x_{t-1}, x_t) \in \mathsf{X}^2} \phi_{\xi,t-1}(x_{t-1}) Q_t(x_{t-1}, x_t) . \end{aligned} \tag{9.44}$$

In matrix form, the likelihood may be expressed as

$$\mathrm{p}_{\xi,n}(Y_{0:n}) = \mathrm{p}_{\xi,0}(Y_0) \prod_{t=1}^{n} \phi_{\xi,t-1} Q_t \mathbf{1} .$$

The complexity to evaluate this joint distribution grows linearly with the number of observations n and quadratically with the number of states m, whereas the complexity of the direct evaluation of the likelihood (summing up on all the possible sequences of states) grows exponentially fast $O(m^n)$. The direct evaluation of the likelihood is therefore manageable even when the number of observations is large, which enables likelihood inference. We will discuss this issue in depth in Chapter 12.

The filtering recursion yields the probability distribution of the state X_t given the observations up to time t. When analyzing a time series by batch, the inference of the state X_t that incorporates all the observations $(Y_0,\dots,Y_n)$ is in general preferable. Such probability statements are given by the fixed interval smoothing probabilities. To simplify the derivations, we denote by $p_\xi(x_{s:t}, y_{s':t'})$ the density with respect to the counting measure of the vector $(X_{s:t}, Y_{s':t'})$. Note first that, for any $s \in \{0,\dots,n-1\}$,

$$\begin{aligned} p_\xi(x_s | x_{s+1:n}, y_{0:n}) &= \frac{p_\xi(y_{0:s}, x_s, x_{s+1}, y_{s+1:n}, x_{s+2:n})}{\sum_{x'_s \in \mathsf{X}} p_\xi(y_{0:s}, x'_s, x_{s+1}, y_{s+1:n}, x_{s+2:n})} \\ &\overset{(1)}{=} \frac{p_\xi(y_{0:s}, x_s, x_{s+1})}{\sum_{x'_s \in \mathsf{X}} p_\xi(y_{0:s}, x'_s, x_{s+1})} \\ &\overset{(2)}{=} \frac{\phi_{\xi,s}(x_s) M(x_s, x_{s+1})}{\sum_{x'_s \in \mathsf{X}} \phi_{\xi,s}(x'_s) M(x'_s, x_{s+1})} = p_\xi(x_s | x_{s+1}, y_{0:s}) \,, \end{aligned}$$

where (1) follows from

$$p_\xi(y_{s+1:t}, x_{s+2:n} | y_{0:s+1}, x_{0:s+1}) = p_\xi(y_{s+1:t}, x_{s+2:n} | x_{s+1})$$

which cancels in the numerator and the denominator, (2) from

$$p_\xi(x_{s+1} | x_s, y_{0:s}) = p(x_{s+1} | x_s) = M(x_s, x_{s+1}) \,,$$

where we have used (9.7) and the fact that $\{(X_t, Y_t),\ t \in \mathbb{N}\}$ is a Markov chain. This shows that $\{X_{n-s}, s \in \{0,1,\dots,n\}\}$ conditioned on the observations $Y_{0:n}$ is a Markov chain, with initial distribution $\phi_{\xi,n}$ and transition kernel $B_{\phi_{\xi,s}}$ where for any measure η on X, B_η is the Markov matrix given by

$$B_\eta(x, x') := \frac{\eta(x')\, m(x', x)}{\sum_{x'' \in \mathsf{X}} \eta(x'')\, m(x'', x)} \,. \tag{9.45}$$

In matrix form, the backward kernel may be written as

$$B_\eta = \operatorname{diag}(M' D_\eta \mathbf{1})^{-1} M' D_\eta \,, \quad D_\eta := \operatorname{diag}(\eta(x), x \in \mathsf{X}) \,.$$

Here, $\mathbf{1}$ denotes the matrix with all entries equal to one. For any integers $n > 0$, $s \in \{0,\dots,n-1\}$, the posterior distribution $\phi_{\xi,s:n|n}$ may be expressed as

$$\phi_{\xi,s:n|n}(x_{s:n}) = \phi_{\xi,n}(x_n)\, B_{\phi_{\xi,n-1}}(x_n, x_{n-1}) \dots B_{\phi_{\xi,s}}(x_{s+1}, x_s) \,. \tag{9.46}$$

In particular, the marginal smoothing distribution $\phi_{\xi,s|n}$ may be expressed in matrix form as

$$\phi_{\xi,s|n} = \phi_{\xi,n} B_{\phi_{\xi,n-1}} B_{\phi_{\xi,n-2}} \dots B_{\phi_{\xi,s}} \,.$$

Algorithm 9.2 (Backward marginal smoothing)

Given stored values of $\phi_{\xi,0}, \ldots, \phi_{\xi,n}$ and starting from n, backwards in time.

Initialization: For $x \in \mathsf{X}$,

$$\phi_{\xi,n|n}(x) = \phi_{\xi,n}(x) .$$

Backward Recursion: For $t = n-1, \ldots, 0$,

- Compute the backward transition kernel according to

$$B_{\phi_{\xi,t}}(x,x') = \frac{\phi_{\xi,t}(x')M(x',x)}{\sum_{x'' \in \mathsf{X}} \phi_{\xi,t}(x'')M(x'',x)}$$

  for $(x,x') \in \mathsf{X} \times \mathsf{X}$.
- Compute

$$\phi_{\xi,t|n}(x) = \sum_{x' \in \mathsf{X}} \phi_{\xi,t+1|n}(x') B_{\phi_{\xi,t}}(x',x) .$$

  for $(x,x') \in \mathsf{X} \times \mathsf{X}$.

The marginal smoothing distribution can be generated recursively, backwards in time as follows

$$\phi_{\xi,s|n} = \phi_{\xi,s+1|n} B_{\phi_{\xi,s}} . \tag{9.47}$$

This recursion, summarized in Algorithm 9.2, is the *forward-backward* algorithm or the *Baum-Welch* algorithm for discrete Hidden Markov Models. In the forward pass, the filtering distributions $\{\phi_{\xi,t}, t \in \{0, \ldots, n\}\}$ are computed and stored. In the backward pass, these filtering distributions are corrected by recursively applying the backward kernels.

When X is finite, it turns out that it is also possible to determine the path $\hat{X}_{0:n}$ which maximizes the joint smoothing probability

$$\hat{X}_{0:n} := \operatorname*{arg\,max}_{x_{0:n} \in \mathsf{X}^{n+1}} \mathbb{P}_\xi(X_{0:n} = x_{0:n} \mid Y_{0:n}) = \operatorname*{arg\,max}_{x_{0:n} \in \mathsf{X}^{n+1}} \phi_{\xi,0:n|n}(x_{0:n}) . \tag{9.48}$$

Solving the maximization problem (9.48) over all possible state sequences $x_{0:m}$ by brute force would involve m^{n+1} function evaluations, which is clearly not feasible except for small n. The algorithm that makes it possible to efficiently compute the *a posteriori most likely sequence of states* is known as the *Viterbi algorithm*, which is based on the well-known *dynamic programming* principle. The logarithm of the joint smoothing distribution may be written as

$$\ln \phi_{\xi,0:t|t}(x_{0:t}) = (\ell_{\xi,t-1} - \ell_{\xi,t}) \\ + \ln \phi_{\xi,0:t-1|t-1}(x_{0:t-1}) + \ln m(x_{t-1},x_t) + \ln g_t(x_t) , \tag{9.49}$$

where $\ell_{\xi,t}$ denotes the log-likelihood of the observations up to index t. The salient feature of (9.49) is that, except for a constant term that does not depend on the state

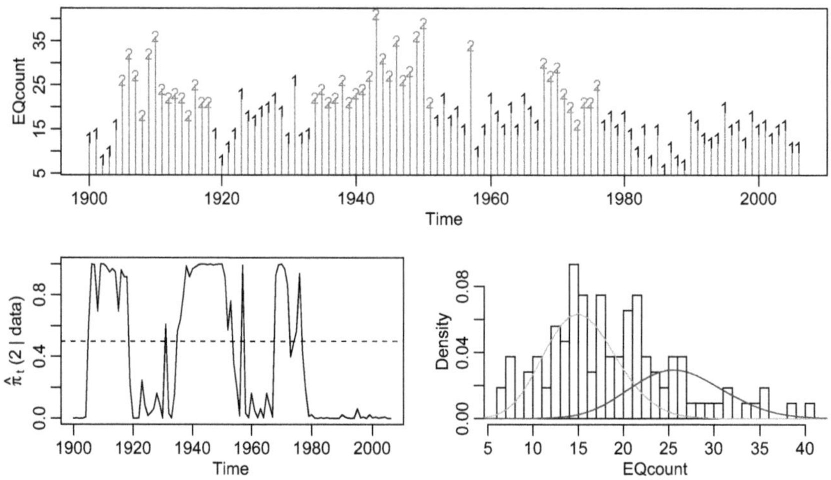

Figure 9.8 *Top: Earthquake count data and estimated states. Bottom left: Smoothing probabilities. Bottom right: Histogram of the data with the two estimated Poisson densities superimposed (solid lines).*

sequence (on the right-hand side of the first line), the *a posteriori* log-probability of the subsequence $x_{0:t}$ is equal to that of $x_{0:t-1}$ up to terms that only involve the pair (x_{t-1}, x_t). Define

$$\mu_t(x) = \max_{x_{0:t-1} \in \mathsf{X}^t} \ln \phi_{\xi,0:t|t}(x_{0:t-1}, x) + \ell_{\xi,t} , \tag{9.50}$$

that is, up to a number independent of the state sequence, the maximal conditional probability (on the log scale) of a sequence up to time t and ending with state $x \in \mathsf{X}$. Also define $b_t(x)$ to be that value in X of x_{t-1} for which the optimum is achieved in (9.50); in other words, $b_t(x)$ is the second final state in an optimal state sequence of length $t+1$ and ending with state x. Using (9.49), we then have the simple recursive relation

$$\mu_t(x') = \max_{x \in \mathsf{X}} \left[\mu_{t-1}(x) + \ln m(x,x')\right] + \ln g_t(x') , \tag{9.51}$$

and $b_t(x')$ equals the state for which the maximum is achieved. The backward recursion first identifies the final state of the optimal state sequence. Then, once the final state is known, the next to final one can be determined as the state that gives the optimal probability for sequences ending with the now known final state. After that, the second next to final state can be determined in the same manner, and so on.

Example 9.11 (Number of major earthquakes; Example 9.1, cont.). For a model with two states, we assume that the parameters of the Poisson distribution $(\lambda_1, \lambda_2) \in \mathbb{R}^+$ associated with each state and of the transition matrix $[M(x,x')]_{(x,x') \in \mathsf{X}^2}$ are unknown, where $\mathsf{X} = \{1, 2\}$. Denote by θ these parameters, which are assumed to be-

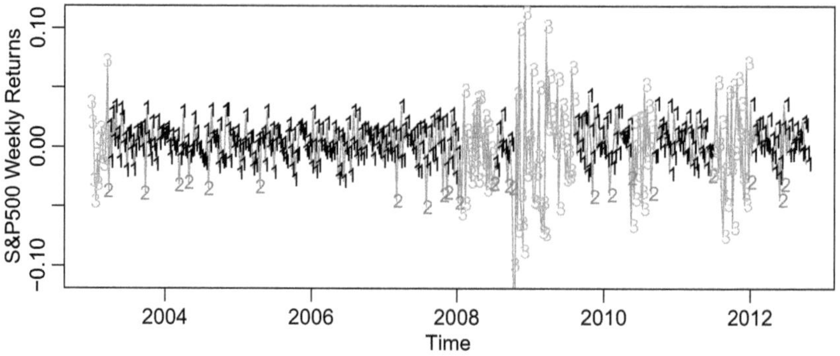

Figure 9.9 *S&P 500 weekly return from January 3, 2003 to September 30, 2012 and the estimated state based on the smoothing distributions. The states are indicated by points labeled 1,2 or 3. For display purposes, the vertical axis has been truncated; cf. Figure 9.3.*

long to a compact subset of

$$\Theta = \Big\{ \{\lambda_x\}_{x\in\mathsf{X}},\ [M(x,x')]_{(x,x')\in\mathsf{X}^2},\ M(x,x') \geq 0 \text{ and } \sum_{x'\in\mathsf{X}} M(x,x') = 1 \Big\}. \tag{9.52}$$

Given observations $Y_0, \dots, Y_n$, we may use (9.44) to write the log-likelihood as

$$\ln \mathrm{p}^{\theta}_{\xi,n}(Y_{0:n}) = \ln\left(\mathrm{p}^{\theta}_{\xi,0}(Y_0)\right) + \sum_{t=1}^{n} \ln\left(\sum_{(x_{t-1},x_t)\in\mathsf{X}^2} \phi^{\theta}_{\xi,t-1}(x_{t-1}) Q^{\theta}_t(x_{t-1},x_t) \right).$$

Consequently, MLE can be performed via numerical maximization.

We fit the model to the time series of earthquake counts using the R package `depmixS4`. The package does not provide standard errors, so we obtained them by a parametric bootstrap procedure; see Remillard (2011) for justification. We note, however, that the standard errors may be obtained as a by-product of the estimation procedure; see Chapter 12. The MLEs of the intensities, along with their standard errors, were $(\hat\lambda_1, \hat\lambda_2) = (15.4_{(.7)}, 26.0_{(1.1)})$. The MLE of the transition matrix was $[\hat M(1,1), \hat M(1,2), \hat M(2,1), \hat M(2,2)] = [.93_{(.04)}, .07_{(.04)}, .12_{(.09)}, .88_{(.09)}]$. Figure 9.8 displays the counts, the estimated state (displayed as points) and the smoothing distribution for the earthquakes data, modeled as a 2-state Poisson HMM model with parameters fitted using the MLEs. Finally, a histogram of the data is displayed along with the two estimated Poisson densities superimposed as solid lines. ◇

Example 9.12 (S&P500; Example 9.2, cont.). In Example 9.2, we fitted a mixture of three Gaussian distributions to the weekly S&P500 log-returns from January 3, 2003 until September 28, 2012. The results, which are displayed in Figure 9.3, are obtained under the unlikely assumption that the data are independent.

Here, we fit an HMM using the R package `depmixS4`, which takes into account that the data are dependent. As in Example 9.2, we chose a three-state model and we

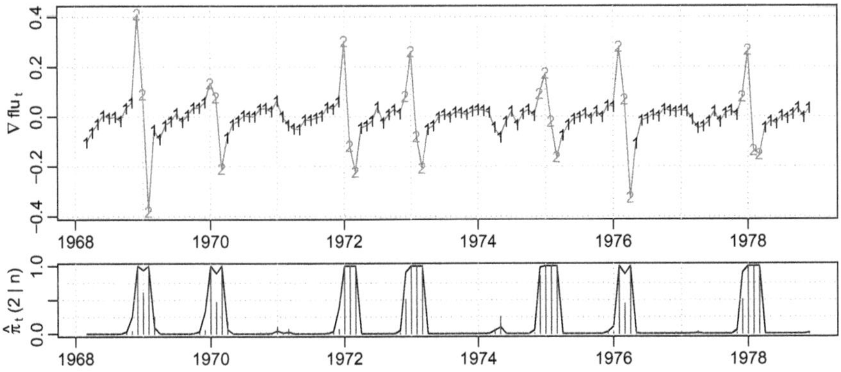

Figure 9.10 *The differenced flu mortality data of Figure 3.1 along with the estimated states (displayed as points). The smoothed state 2 probabilities are displayed in the bottom of the figure as a straight line. The filtered state 2 probabilities are displayed as vertical lines.*

leave it to the reader to investigate a two-state model (see Exercise 9.11). Standard errors (shown in parentheses below) were obtained via a parametric bootstrap based on a simulation script provided with the package.

The fitted transition matrix was

$$\widehat{M} = \begin{bmatrix} .945_{(.107)} & .055_{(.107)} & .000_{(.005)} \\ .739_{(.362)} & .000_{(.069)} & .261_{(.351)} \\ .031_{(.029)} & .027_{(.069)} & .942_{(.062)} \end{bmatrix},$$

and the three fitted normals were $\mathrm{N}(\hat{\mu}_1 = .004_{(.018)}, \hat{\sigma}_1 = .014_{(.020)})$, $\mathrm{N}(\hat{\mu}_2 = -.034_{(.020)}, \hat{\sigma}_2 = .009_{(.006)})$, and $\mathrm{N}(\hat{\mu}_3 = -.003_{(.006)}, \hat{\sigma}_3 = .044_{(.012)})$. The data, along with the predicted state (based on the smoothing distribution), are plotted in Figure 9.9.

The major differences between these results and the results from Example 9.2 are that regime 2 appears to represent a somewhat large-in-magnitude negative return, and may be a lone dip, or the start or end of a highly volatile period. States 1 and 3 represent clusters of regular or high volatility, respectively. Note that there is a large amount of uncertainty in the fitted normals, and in the transition matrix involving transitions from state 2 to states 1 or 3. ◇

Example 9.13 (Influenza mortality; Example 9.9, cont.). In Example 9.9, we considered fitting a two-state switching AR model given by (9.34). In particular, the idea was that data exhibit two different dynamics, one during an epidemic period, and another during a non-epidemic period.

We used the R package `MSwM` to fit the model specified in (9.34), with $p = 2$. The results were

$$\hat{Y}_t = \begin{cases} .006_{(.003)} + .293_{(.039)} Y_{t-1} + .097_{(.031)} Y_{t-2} + .024 Z_t\,, & \text{for } I_t = 1\,, \\ .199_{(.063)} - .313_{(.281)} Y_{t-1} - 1.604_{(.276)} Y_{t-2} + .112 Z_t\,, & \text{for } I_t = 2\,, \end{cases}$$

with estimated transition matrix

$$\widehat{M} = \begin{bmatrix} .927 & .073 \\ .300 & .700 \end{bmatrix}.$$

Figure 9.10 displays the data $Y_t = \nabla \text{flu}_t$ along with the estimated states (displayed as points labeled 1 or 2). The smoothed state 2 probabilities are displayed in the bottom of the figure as a straight line. The filtered state 2 probabilities are displayed in the same graph as vertical lines. ◇

9.2.2 *Continuous-valued state-space HMM*

The recursion developed for the discrete-valued state space extends directly to the general state-space setting. At time $t-1$, the filtering distribution $\phi_{\xi,t-1}$ summarizes all information the observations $Y_0, \ldots, Y_{t-1}$ contain about the state X_{t-1}.

Denote by $\gamma_{\xi,t}$

$$\gamma_{\xi,t}(f) = \int \cdots \int \xi(\mathrm{d}x_0) g_0(x_0) \prod_{s=1}^{t} Q_s(x_{s-1}, \mathrm{d}x_s) f(x_t) \tag{9.53}$$

where $f \in \mathbb{F}_+(\mathsf{X}, \mathcal{X})$ and

$$g_t(x_t) := g(x_t, Y_t) \text{ and } Q_t(x_{t-1}, A) := \int_A M(x_{t-1}, \mathrm{d}x_t) g_t(x_t) \text{ for all } A \in \mathcal{X}. \tag{9.54}$$

This distribution may be computed recursively as follows

$$\gamma_{\xi,0}(f) = \int \xi(\mathrm{d}x_0) g_0(x_0) f(x_0) = \xi(g_0 f)$$
$$\gamma_{\xi,t}(f) = \iint \gamma_{\xi,t-1}(\mathrm{d}x_{t-1}) Q_t(x_{t-1}, \mathrm{d}x_t) f(x_t) = \gamma_{\xi,t-1} Q_t(f).$$

The joint distribution of the observations $(Y_0, \ldots, Y_t)$ is obtained by marginalizing the joint distribution $\gamma_{\xi,t}$ with respect to the state X_t, i.e.,

$$\mathrm{p}_{\xi,t}(Y_{0:t}) = \int \gamma_{\xi,t}(\mathrm{d}x_t) = \gamma_{\xi,t}(\mathbf{1}). \tag{9.55}$$

The filtering distribution is the conditional distribution of the state X_t given $(Y_0, \ldots, Y_t)$. It is obtained by dividing the joint distribution of $(X_t, Y_0, \ldots, Y_t)$ by $\mathrm{p}_{\xi,t}(Y_{0:t})$,

$$\begin{aligned}
\phi_{\xi,t}(f) = \frac{\gamma_{\xi,t}(f)}{\gamma_{\xi,t}(\mathbf{1})} &= \frac{\int \cdots \int f(x_t) \xi(\mathrm{d}x_0) g_0(x_0) \prod_{s=1}^{t} M(x_{s-1}, \mathrm{d}x_s) g_s(x_s)}{\int \cdots \int \xi(\mathrm{d}x_0) g_0(x_0) \prod_{s=1}^{t} M(x_{s-1}, \mathrm{d}x_s) g_s(x_s)} \\
&= \frac{\int \cdots \int f(x_t) \phi_{\xi,t-1}(\mathrm{d}x_{t-1}) M(x_{t-1}, \mathrm{d}x_t) g_t(x_t)}{\int \cdots \int \phi_{\xi,t-1}(\mathrm{d}x_{t-1}) M(x_{t-1}, \mathrm{d}x_t) g_t(x_t)} \\
&= \frac{\phi_{\xi,t-1} Q_t f}{\phi_{\xi,t-1} Q_t \mathbf{1}}.
\end{aligned} \tag{9.56}$$

The forward recursion in (9.56) may be rewritten to highlight a two-step procedure involving both the predictive and filtering distributions. For $t \in \{0, 1, \ldots, n\}$ and $f \in \mathbb{F}_b(\mathsf{X}, \mathcal{X})$, with the convention that $\phi_{\xi,0|-1} = \xi$, (9.56) may be decomposed as

$$\phi_{\xi,t|t-1} = \phi_{\xi,t-1} M \,, \tag{9.57a}$$

$$\phi_{\xi,t}(f) = \frac{\phi_{\xi,t|t-1}(f g_t)}{\phi_{\xi,t|t-1}(g_t)} \,. \tag{9.57b}$$

***Filter to Predictor*:** The first equation in (9.57) means that the updated predictive distribution $\phi_{\xi,t|t-1}$ is obtained by applying the transition kernel M to the current filtering distribution $\phi_{\xi,t-1}$. The predictive distribution is the one-step distribution of the Markov chain with kernel M given its initial distribution.

***Predictor to Filter*:** The second equation in (9.57) is recognized as Bayes' rule to correct the predictive distribution through the information contained in the actual observation Y_t.

- X_t is distributed *a priori* according to the predictive distribution $\phi_{\xi,t|t-1}$,
- g_t is the conditional probability density function of Y_t given X_t.

Although the recursions (9.57) appear relatively simple, they in fact must be approximated by numerical methods. We discuss particle approximations in Chapter 10.

The joint smoothing distribution $\phi_{\xi,0:t|t}$ then satisfies, for $f_{t+1} \in \mathbb{F}_b(\mathcal{X}^{\otimes(t+1)})$,

$$\phi_{\xi,0:t|t}(f_{t+1}) = \left(\mathrm{p}_{\xi,t}(Y_{0:t})\right)^{-1} \int \cdots \int f_{t+1}(x_{0:t}) \xi(\mathrm{d}x_0) g(x_0, y_0) \prod_{s=1}^{t} Q_s(x_{s-1}, \mathrm{d}x_s) \tag{9.58}$$

assuming that $\mathrm{p}_{\xi,t}(Y_{0:t}) > 0$. Likewise, for indices $p \geq 0$,

$$\phi_{\xi,0:t+p|t}(f_{t+p+1}) = \int \cdots \int f_{t+p+1}(x_{0:t+p}) \times \phi_{\xi,0:t|t}(\mathrm{d}x_{0:t}) \prod_{s=t+1}^{t+p} M(x_{s-1}, \mathrm{d}x_s) \tag{9.59}$$

for all functions $f_{t+p+1} \in \mathbb{F}_b(\mathcal{X}^{\otimes(t+p+1)})$. Eq. (9.58) implicitly defines the filtering, the predictive and the smoothing distributions as these are obtained by marginalization of the joint smoothing distribution; see Exercise 9.13.

The expression of the joint smoothing distribution (9.58) implicitly defines all other particular cases of smoothing kernels as these are obtained by marginalization. For instance, the marginal smoothing kernel $\phi_{\xi,t|n}$ for $0 \leq t \leq n$ is such that for $f \in \mathbb{F}_+(\mathsf{X}, \mathcal{X})$,

$$\phi_{\xi,t|n}(f) := \int \cdots \int f(x_t)\, \phi_{\xi,0:n|n}(\mathrm{d}x_{0:n}) \,, \tag{9.60}$$

where $\phi_{\xi,0:n|n}$ is defined by (9.58).

Similarly, we note that the p-step predictive distribution $\phi_{\xi,n+p|n}$ may be obtained by marginalization of the joint distribution $\phi_{\xi,0:n+p|n}$ with respect to all variables x_t

except the last one (the one with index $t = n+p$). A closer examination of (9.59) directly shows that $\phi_{\xi,n+p|n} = \phi_{\xi,n} M^p$.

We now derive recursion for the smoothing distribution. For $\eta \in \mathbb{M}_1(\mathcal{X})$, assume that there exists a kernel B_η on $(\mathsf{X},\mathcal{X})$ that satisfies, for all $h \in \mathbb{F}_b(\mathsf{X}^2,\mathcal{X}^{\otimes 2})$,

$$\iint h(x,x')\eta(\mathrm{d}x)M(x,\mathrm{d}x') = \iint h(x,x')\eta M(\mathrm{d}x')B_\eta(x',\mathrm{d}x) . \tag{9.61}$$

This kernel is referred to as the *backward kernel* . When the HMM is fully dominated (see Definition 9.3), then the backward kernel may be explicitly written as

$$B_\eta(x,A) := \frac{\int \eta(\mathrm{d}x')\, m(x',x)\mathbb{1}_A(x')}{\int \eta(\mathrm{d}x')\, m(x',x)} , \quad A \in \mathcal{X} . \tag{9.62}$$

In other words, denote by η the distribution of X_0, and assume that the conditional distribution of X_1 given X_0 is $M(X_0,\cdot)$. Then the joint distribution of (X_0,X_1) is given by $\eta \otimes M$, i.e., for any $h \in \mathbb{F}_b(\mathsf{X}^2,\mathcal{X}^{\otimes 2})$, $\mathbb{E}_\eta[h(X_0,X_1)] = \iint \eta(\mathrm{d}x_0)M(x_0,\mathrm{d}x_1)h(x_0,x_1)$. The marginal distribution of X_1 is ηM, and the conditional distribution of X_0 given X_1 is specified by the kernel $B_\eta(X_1,\cdot)$.

Proposition 9.14. *Given a strictly positive index t, initial distribution ξ, and index $t \in \{0,\dots,n-1\}$,*

$$\mathbb{E}\left[f(X_t) \mid X_{t+1:n}, Y_{0:n}\right] = \mathbb{E}\left[f(X_t) \mid X_{t+1}, Y_{0:t}\right] = B_{\phi_{\xi,t}} f(X_{t+1})$$

for any $f \in \mathbb{F}_b(\mathsf{X},\mathcal{X})$. In addition,

$$\mathbb{E}_\xi\left[f(X_{0:n}) \mid Y_{0:n}\right] = \int \cdots \int f(x_{0:n})\, \phi_{\xi,n}(\mathrm{d}x_n) \prod_{s=0}^{n-1} B_{\phi_{\xi,s}}(x_{s+1},\mathrm{d}x_s) \tag{9.63}$$

for any $f \in \mathbb{F}_b(\mathsf{X}^{n+1},\mathcal{X}^{\otimes(n+1)})$.

Proof. See Exercise 9.14. ∎

It follows from Proposition 9.14 that, conditionally on $Y_{0:n}$, the joint distribution of the index-reversed sequence $\{X_n, X_{n-1},\dots,X_0\}$ is that of a non-homogeneous Markov chain with initial distribution $\phi_{\xi,n}$ and transition kernels $\{B_{\phi_{\xi,t}}\}_{n-1 \ge t \ge 0}$. The backward smoothing kernel depends neither on the future observations nor on the index n. Therefore, the sequence of backward transition kernels $\{B_{\phi_{\xi,t}}\}_{0 \le t \le n-1}$ may be computed by forward recurrence on t. This decomposition suggests Algorithm 9.3 to recursively compute the marginal smoothing decomposition. Although the algorithm is apparently simple, the smoother must be approximated numerically. We discuss particle methods in Chapter 11.

Example 9.15 (The Rauch-Tung-Striebel smoother). For a linear Gaussian state-space model, all the conditional distributions are Gaussian distributions. Therefore, in that case, only the mean vectors and the covariance matrices need to be evaluated, and correspondingly the filtering or smoothing equations become equivalent to the

Algorithm 9.3 (Forward Filtering/Backward Smoothing)

Forward Filtering: Compute, forward in time, the filtering distributions $\phi_{\xi,0}$ to $\phi_{\xi,n}$ using the recursion (9.56). At each index t, the backward transition kernel $B_{\phi_{\xi,t}}$ may be computed according to (9.61).

Backward Smoothing: From $\phi_{\xi,n}$, compute, for $t = n-1, n-2, \ldots, 0$,

$$\phi_{\xi,t|n} = \phi_{\xi,t+1|n} B_{\phi_{\xi,t}} .$$

---

ordinary Kalman filter / smoother. The smoothing algorithm introduced above leads to an alternative derivation of Proposition 2.7, which is referred to as the Rauch-Tung-Striebel smoother; see Rauch et al. (1965). Let $X_{t+1} = \Phi X_t + W_t$ and $Y_t = AX_t + V_t$, where $\{W_t,\, t \in \mathbb{N}\}$ is i.i.d. zero-mean Gaussian with covariance Q and $\{V_t,\, t \in \mathbb{N}\}$ is i.i.d. zero mean-mean Gaussian with covariance R, $\{V_t,\, t \in \mathbb{N}\}$ and $\{W_t,\, t \in \mathbb{N}\}$ are independent. The initial state X_0 has a Gaussian distribution and is independent of $\{V_t,\, t \in \mathbb{N}\}$ and $\{W_t,\, t \in \mathbb{N}\}$.

We first determine the backward kernel. Let η be a Gaussian distribution with mean μ_0 and covariance Γ_0, i.e., $\eta = \mathrm{N}(\mu_0, \Gamma_0)$. Assume that $X_0 \sim \eta$ and let $X_1 = \Phi X_0 + W_0$. Note that, under this model,

$$\begin{bmatrix} X_0 \\ X_1 \end{bmatrix} \sim \mathrm{N}\left(\begin{bmatrix} \mu_0 \\ \mu_1 \end{bmatrix}, \begin{bmatrix} \Gamma_0 & \Gamma_0 \Phi' \\ \Phi \Gamma_0 & Q + \Phi \Gamma_0 \Phi' \end{bmatrix} \right), \tag{9.64}$$

where $\mu_1 = \Phi \mu_0$. For any $x_1 \in \mathsf{X}$, the backward kernel, (9.61), is the conditional distribution of X_0 given $X_1 = x_1$, in the model (9.64). This conditional distribution is Gaussian, with mean and covariance

$$\mu_{0|1} = \mu_0 + J_0(x_1 - \mu_1) \tag{9.65}$$

$$\Gamma_{0|1} = \Gamma_0 - \Gamma_0 \Phi' (\Phi \Gamma_0 \Phi' + Q)^{-1} \Phi \Gamma_0 = \Gamma_0 - J_0 (\Phi \Gamma_0 \Phi' + Q) J_0' \tag{9.66}$$

where J_0 is the Kalman gain

$$J_0 = \Gamma_0 \Phi' (\Phi \Gamma_0 \Phi' + Q)^{-1} . \tag{9.67}$$

The action of this kernel is best understood by considering the Gaussian random vector

$$\tilde{X}_0 = \mu_0 + J_0(X_1 - \mu_1) + Z_0 \tag{9.68}$$

where Z_0 is a Gaussian random vector with zero-mean and covariance $\Gamma_{0|1}$ independent of X_1. Conditional to $X_1 = x_1$, $\tilde{X}_0$ is distributed according to $B_\eta(x_1, \cdot)$, provided that $\eta = \mathrm{N}(\mu_0, \Gamma_0)$. Assume now that $X_1 \sim \mathrm{N}(\mu_1, \Gamma_1)$. Then, the (unconditional) distribution of $\tilde{X}_0$ is Gaussian with mean and covariance given by

$$\tilde{\mu}_0 = \mu_0 + J_0(\mu_1 - \mu_0) , \tag{9.69}$$

$$\begin{aligned} \tilde{\Gamma}_0 &= J_0 \Gamma_1 J_0' + \Gamma_0 - J_0 (\Phi \Gamma_0 \Phi' + Q) J_0' \\ &= \Gamma_0 + J_0 \left(\Gamma_1 - (\Phi \Gamma_0 \Phi' + Q) \right) J_0' . \end{aligned} \tag{9.70}$$

To obtain the recursion for the smoother covariance, it suffices to replace (μ_0, Γ_0) in (9.65)-(9.66) by the filtering mean and covariance $(X_{t|t}, P_{t|t})$, defined in (2.11)-(2.12) and (μ_1, Γ_1) by $(X_{t+1|n}, P_{t+1|n})$, to obtain the *forward-filtering, backward smoothing* recursion

$$X_{t|n} = X_{t|t} + J_t(X_{t+1|n} - X_{t|t}) \tag{9.71}$$

$$P_{t|n} = P_{t|t} + J_t\left(P_{t+1|n} - P_{t+1|t}\right) J_t' \tag{9.72}$$

where $J_t = P_{t|t}\Phi'(\Phi P_{t|t}\Phi' + Q)^{-1}$ is the Kalman Gain. The filtering mean and covariance $(X_{t|t}, P_{t|t})$ are computed using the Kalman filter. The smoothing mean and covariance are obtained by running (9.71)-(9.72) backwards in time, starting from $(X_{n|n}, P_{n|n})$. ◇

9.3 Endnotes

Nonlinear state space models and their generalizations are nowadays used in many different areas. Several specialized books are available that largely cover applications of HMMs to some specific areas such as speech recognition (Rabiner and Juang, 1993, Jelinek, 1997), econometrics (Hamilton, 1989, Kim and Nelson, 1999), computational biology (Durbin et al., 1998, Koski, 2001), or computer vision (Bunke and Caelli, 2001). The elementary theory of HMM is covered in MacDonald and Zucchini (2009) and Fraser (2008), which discuss a lot of interesting examples of applications.

Most of the early references on filtering and smoothing, which date back to the 1960s, focused on the specific case of Gaussian linear state-space models, following the pioneering work by Kalman and Bucy (1961). The classic book by Anderson and Moore (1979) on *optimal filtering*, for instance, is fully devoted to linear state-space models; see also Kailath et al. (2000, Chapter 10) for a more exhaustive set of early references on the smoothing problem. Although some authors, for example, Ho and Lee (1964) considered more general state-space models, it is fair to say that the Gaussian linear state-space model was the dominant paradigm. Until the early 1980s, the works that *did not* focus on the linear state-space model were usually advertised by the use of the words "Bayes" or "Bayesian" in their title; see, e.g., Ho and Lee (1964) or Askar and Derin (1981).

Almost independently, the work by Baum and his colleagues on hidden Markov models (Baum et al., 1970) dealt with the case where the state space X of the hidden state is finite. These two streams of research (on Gaussian linear models and finite state space models) remained largely separated. The forward-backward algorithm is known to many, especially in the field of speech processing, as the *Baum-Welch algorithm*, although the first published description of the approach is due to Baum et al. (1970, p. 168).

The forward-backward algorithm was discovered several times in the early 1970s; see Fraser (2008) and MacDonald and Zucchini (2009). A salient example is the paper by Bahl et al. (1974) on the computation of posterior probabilities for a finite-state Markov channel encoder for transmission over a discrete memoryless channel. The algorithm described by Bahl et al. (1974) is fully equivalent to the forward-backward and is known in digital communication as the BCJR (for Bahl, Cocke, Jelinek, and Raviv) algorithm. Chang and Hancock (1966) is another less well-known reference, contemporary with the work of Baum and his colleagues, which also describes the forward-backward decomposition and its use for decoding in communication applications.

Approximately at the same time, in applied probability, the seminal work by Stratonovich (1960) stimulated a number of contributions that were to compose a body of work generally referred to as *filtering theory*. The object of filtering theory is to study inference about partially observable Markovian processes in *continuous time*. A number of early references in this domain indeed consider some specific form of discrete state space continuous-time equivalent of the HMM (Shiryaev 1966, Wonham 1965; see also Lipster and Shiryaev 2001, Chapter 9). Working in continuous time, however, implies the use of mathematical tools that are definitely more complex than those needed to tackle the discrete-time model of Baum et al. (1970). As a matter of fact, filtering theory and hidden Markov models evolved as two mostly independent fields of research. A poorly acknowledged fact is that the pioneering paper by Stratonovich (1960) (translated from an earlier Russian publication) describes, in its first section, an equivalent to the forward-backward smoothing approach of Baum et al. (1970). It turns out, however, that the formalism of Baum et al. (1970) generalizes well to models where the state space is *not* discrete anymore, in contrast to that of Stratonovich (1960).

Exercises

9.1. Let $(\Omega, \mathcal{F}, \mathbb{P})$ be a probability space and Y a random variable such that $\mathbb{E}\left[Y^2\right] < \infty$. Let $\mathcal{G} \subset \mathcal{F}$ be a σ-algebra. Show that

$$\mathbb{E}[(Y - \mathbb{E}[Y \mid \mathcal{G}])^2] = \mathbb{E}[(Y - \mathbb{E}[Y \mid \mathcal{G}])^2] + \mathbb{E}\left[(\mathbb{E}[Y \mid \mathcal{G}] - \mathbb{E}[Y])^2\right] ,$$

and check (9.6).

9.2. Consider a discrete state-space HMM. Denote by $\mathsf{X} = \{1, \ldots, m\}$ the state-space of the Markov chain, M, the $m \times m$ transition matrix and, for $y \in \mathsf{Y}$, by $\Gamma = \text{diag}(G(1,y), \ldots, G(m,Y))$. Assume that M admits a unique stationary distribution denoted by $\pi = [\pi(1), \ldots, \pi(m)]$.

(a) Show that the likelihood of the observations (9.4) may be expressed as

$$\mathrm{p}_{\xi,t}(Y_{0:t}) = \xi \Gamma(Y_0) M \Gamma(Y_1) M \ldots M \Gamma(Y_t) \mathbb{1}$$

where $\mathbb{1} = [1, 1, \ldots, 1]'$.

(b) Show that for any $h \in \mathbb{N}$,

$$\mathrm{p}_{\xi,t}(Y_{h:t+h}) = \xi M^h \Gamma(Y_0) M \Gamma(Y_1) M \ldots M \Gamma(Y_t) \mathbb{1}$$

and check that $\mathrm{p}_{\xi,t}(Y_{h:t+h}) = \mathrm{p}_{\xi,t}(Y_{0:t})$.

9.3. We use the notations of Exercise 9.2. Let $(\mu_x, x \in \mathsf{X})$ and $(\sigma_x^2, x \in \mathsf{X})$ denote the mean and variance of the distributions $(G(x,\cdot), x \in \mathsf{X})$.

(a) $\mathbb{E}_\pi[Y_t] = \sum_{x \in \mathsf{X}} \pi(x)\mu_x$.

(b) $\mathbb{E}_\pi[Y_t^2] = \sum_{x \in \mathsf{X}} \pi(x)(\sigma_x^2 + \mu_x^2)$.

(c) $\mathrm{Var}_\pi(Y_t) = \sum_{x \in \mathsf{X}} \pi(x)(\sigma_x^2 + \mu_x^2) - (\sum_{x \in \mathsf{X}} \pi(x)\mu_x)^2$.

(d) If $m = 2$, $\mathrm{Var}_\pi(Y_t) = \pi(1)\sigma(1)^2 + \pi(2)\sigma_2^2 + \pi(1)\pi(2)(\mu_1 - \mu_2)^2$.

(e) For $k \in \mathbb{N}$,

$$\mathbb{E}_\pi[Y_t Y_{t+k}] = \sum_{x_0 \in \mathsf{X}} \sum_{x_k \in \mathsf{X}} \pi(x_0)\mu_{x_0} M^k(x_0, x_k)\mu_{x_k} = \pi \mathrm{diag}(\mu) M^k \mu \;,$$

where $\mu = [\mu_1, \mu_2, \ldots, \mu_m]'$.

(f) Show that, if the eigenvalues of M are distinct, then $\mathrm{Cov}_\pi(X_0, X_k)$ may be expressed as a linear combination of the k-th powers of those eigenvalues.

9.4. Consider the state-space model with non-linear state evolution equation

$$X_t = A(X_{t-1}) + R(X_{t-1})W_t \;, \qquad W_t \sim \mathrm{N}(0, I) \;, \qquad (9.73)$$

$$Y_t = BX_t + SV_t \;, \qquad V_t \sim \mathrm{N}(0, I) \;, \qquad (9.74)$$

where A and R are matrix-valued functions of appropriate dimensions. Show that the conditional distribution of X_t given $X_{t-1} = x$ and Y_t is multivariate Gaussian with mean $m_t(x)$ and covariance matrix $\Sigma_t(x)$, given by

$$K_t(x) = R(x)R'(x)B' \left[BR(x)R'(x)B' + SS^t\right]^{-1} \;,$$
$$m_t(x) = A(x) + K_t(x)\left[Y_{t+1} - BA(x)\right] \;,$$
$$\Sigma_t(x) = \left[I - K_t(x)B\right] R(x)R'(x) \;.$$

9.5. Assume that $Y_t = \mu_{X_t} + \sigma_{X_t} Z_t$, where $\{Z_t,\, t \in \mathbb{N}\} \sim$ iid $\mathrm{N}(0,1)$ and $\{X_t,\, t \in \mathbb{N}\}$ is a two-state stationary Markov chain, independent of $\{Z_t,\, t \in \mathbb{N}\}$.

(a) Show that the unconditional distribution of Y_t is given by a mixture of two normal distributions: $p(y_t) = \pi_1 \mathfrak{g}(y_t; \mu_1,\, \sigma_1^2) + \pi_2 \mathfrak{g}(y_t; \mu_2,\, \sigma_2^2)$ where $\pi_x = \mathbb{P}(X_t = x)$, $x \in \{1, 2\}$.

(b) Show that the skewness is given by

$$\frac{\mathbb{E}\left[(Y_t - \mu)^3\right]}{(\mathbb{E}\left[(Y_t - \mu)^2\right])^{3/2}} = \pi_1\pi_2(\mu_1 - \mu_2)\frac{3(\sigma_2^2 - \sigma_1^2)^2 + (\pi_2 - \pi_1)(\mu_2 - \mu_1)^2}{\sigma^3} \;,$$

with $\mu = \mathbb{E}[Y_t]$ and $\sigma^2 = \mathrm{Var}(Y_t)$ being the mean and variance of the mixture distribution: $\mu = \pi_1\mu_1 + \pi_2\mu_2$ and $\sigma^2 = \pi_1\sigma_1^2 + \pi_2\sigma_2^2 + \pi_1\pi_2(\mu_2 - \mu_1)^2$.

(c) Show that the excess kurtosis is given by

$$\frac{\mathbb{E}\left[(Y_t - \mu)^4\right]}{(\mathbb{E}\left[(Y_t - \mu)^2\right])^2} - 3 = \pi_1\pi_2 \frac{3(\sigma_2^2 - \sigma_1^2)^2 + c(\mu_1, \mu_2)}{\sigma^4} \;,$$

where

$$c(\mu_1, \mu_2) = 6(\pi_1 - \pi_2)(\sigma_2^2 - \sigma_1^2)(\mu_2 - \mu_1)^2 + (\mu_2 - \mu_1)^4(1 - 6\pi_1\pi_2) \;.$$

Remark 9.16. Note that skewness in the marginal distribution will be present whenever both the means and the variances are different. For a model where the means are the same, no skewness is present. If the variances are the same and the means are different, skewness is possible only if $\pi_1 \neq \pi_2$. Thus, for a Markov mixture model with different means but equal variances, asymmetry is introduced into the marginal distribution only through asymmetry in the persistence probabilities, namely $M(1,1) \neq M(2,2)$. If $\mu_1 = \mu_2$, the marginal distribution has fatter tails than a normal distribution as long as $\sigma_1^2 \neq \sigma_2^2$; see Frühwirth-Schnatter (2006, p. 309).

9.6 (Filtering distribution for a 2-state HMM). Let $\{X_t,\, t \in \mathbb{N}\}$ be a two-state stationary Markov chain with transition kernel M and stationary distribution π. Let $\{(X_t, Y_t),\, t \in \mathbb{N}\}$ be a partially dominated HMM.

(a) Show that the predictive distribution of $X_t = 1$,

$$\phi_{t|t-1}(1) = \pi_1 + \lambda \pi_2 \phi_{t-1}(1) - \lambda \pi_1 \phi_{t-1}(2),$$

where $\lambda = M(1,1) - M(1,2)$ is equal to the second eigenvalue of M.

(b) Show that

$$\phi_{t|t-1}(1) = (1-\lambda)\pi_1 + \lambda \phi_{t-1}(1)\,.$$

Remark 9.17. When λ is close to 0 (the Markov chain is not very persistent), the predictive distribution for X_t is dominated by the stationary distribution of the chain. When λ is close to 1 (highly persistent Markov chains), the predictive distribution for X_t will be dominated by the filtered state probability ϕ_{t-1}.

9.7. Consider a two-state Gaussian HMM, $Y_t = \mu_{X_t} + \sigma_{X_t} V_t$, where $\{V_t,\, t \in \mathbb{N}\}$ is a strong white Gaussian noise and $\{X_t,\, t \in \mathbb{N}\}$ is a stationary two-state Markov chain with transition kernel M such that $M(1,2) \in (0,1)$ and $M(2,1) \in (0,1)$. Show that $\{Y_t,\, t \in \mathbb{N}\}$ is an ARMA(1,1) process.

9.8 (Spectral density of a discrete-valued Markov chain). Let $\mathsf{X} = \{x_1, \ldots, x_n\}$ be a finite set and M be a transition kernel on X. Assume that M admits a unique stationary distribution π. Let $\{X_t,\, t \in \mathbb{Z}\}$ be a stationary Markov chain on X with transition kernel M. Define by $M_\infty = \lim_{k \to \infty} M^k = \pi \mathbf{1}$.

(a) Show that $M_\infty = M M_\infty = M_\infty M$ and M_∞ is idempotent.

(b) Let $F = M - M_\infty$. Show that $(M^k - M_\infty) = F^k$, $k = 1, 2, \cdots$.

(c) Set $S = \operatorname{diag}(x_1, \ldots, x_n)$ $R = \operatorname{diag}(\pi_1, \cdots, \pi_n)$. Show that $\mu_X = \pi S \mathbf{1}$, $\gamma_X(0) = \pi S(\mathrm{I} - M_\infty)\mathbf{1}$ and $\gamma_X(k) = \pi S F^{|k|} S \mathbf{1}$, $k = \pm 1, \pm 2, \cdots$.

(d) Show that the eigenvalues of F are zero and λ_i, $|\lambda_i| < 1$, $i = 2, \cdots, n$, the subdominant eigenvalues of P counted with their algebraic multiplicity.

(e) Denote by $f_X(\omega)$ denote the spectral density of $\{X_t,\, t \in \mathbb{Z}\}$. Show that

$$2\pi f_X(\omega) = \pi S[(\mathrm{I} - M_\infty) + 2\mathrm{Re}(\mathrm{e}^{-\mathrm{i}\omega} F(\mathrm{I} - \mathrm{e}^{-\mathrm{i}\omega} F)^{-1})] S \mathbf{1}\,. \tag{9.75}$$

9.9 (Exercise 9.8, cont.). (a) Show that $z^{-1}F(\mathrm{I} - z^{-1}F)^{-1} = F(Iz - F)^{-1} = QJ(Iz - J)^{-1}Q^{-1}$ where $J = Q^{-1}FQ$ is the Jordan canonical form of F, where the Jordan matrix $J = \text{diag}(J_1 : \cdots : J_u)$ where each diagonal block J_i is a $\nu_i \times \nu_i$ matrix of the form

$$\begin{pmatrix} \varphi_i & 1 & 0 & \cdots & 0 \\ 0 & \varphi_i & 1 & \cdots & 0 \\ \vdots & \ddots & \ddots & \ddots & \vdots \\ 0 & & \ddots & \ddots & 1 \\ 0 & 0 & \cdots & 0 & \varphi_i \end{pmatrix},$$

$i = 1, \cdots, u$, the φ_j, $i = 1, \cdots, u$, being the unique eigenvalues of F and $\nu_i \geqq 1$, $\nu_1 + \cdots + \nu_u = n$, their respective algebraic multiplicities. Set $Q = [Q_1 : \cdots : Q_u]$ and $Q^{-1} = [Q^1 : \cdots : Q^u]'$ where Q_i and Q^i are $n \times \nu_i$ matrices, $i = 1, \cdots u$.

(b) Show that $J(\mathrm{I}z - J)^{-1} = \text{diag}(J_1(\mathrm{I}z - J_1)^{-1} : \cdots : J_n(\mathrm{I}z - J_u)^{-1})$ and hence that

$$QJ(\mathrm{I}z - J)^{-1}Q^{-1} = \sum_{i=1}^{u} Q_i J_i (\mathrm{I}z - J_i)^{-1}(Q^i)'. \tag{9.76}$$

(c) Let $q_{ij}, \cdots q_{i\nu_i}$, and $q^{i1}, \cdots, q^{i\nu_i}$ denote the columns of Q_i and Q^j, respectively. Show that $J_j(\mathrm{I}z - J_i)^{-1}$ equals an upper triangular Toeplitz matrix with first row

$$[\varphi_i(z-\varphi_i)^{-1},\ (z-2\varphi_i)(z-\varphi_i)^{-2},\ \cdots,\ (z-2\varphi_i)(z-\varphi_i)^{-\nu_i}(-1)^{\nu_i}].$$

(d) Deduce that each of the summands in (9.76) gives rise to an expansion of the form

$$R_{i1}\frac{\varphi_i}{(z-\varphi_i)} + \sum_{j=2}^{\nu_i} R_{ij}\frac{(z-2\varphi_i)(-1)^j}{(z-\varphi_i)^j}, \tag{9.77}$$

where the second and subsequent terms only appear if $\nu_i \geqq 2$ and

$$R_{ij} = \sum_{l=1}^{\nu_i - j+1} q_{il}(q^{il+j-1})'.$$

(e) Show that $\pi S Q_i J_i (\mathrm{I}z - J_i)^{-1}(Q^i)' S\mathbf{1} = b_i(z)/\beta_j(z)$ where

$$b_i(z) = \pi S[R_{i1}\varphi_i(z-\varphi_i)^{\nu_i-1} + \sum_{j=2}^{\nu_i} R_{ij}(z-2\varphi_i)(z-\varphi_i)^{\nu_i-j}(-1)^j]S\mathbf{1}$$

and $\beta_i(z) = (z-\varphi_i)^{\nu_i}$, $\quad i = 1, \cdots, u$.

(f) Show that

$$2\pi f_X(\omega) = \gamma_X(0) + \sum_{i=2}^{u} \frac{b_i(e^{i\omega})}{\beta_i(e^{i\omega})} + \frac{b_i(e^{-i\omega})}{\beta_i(e^{-i\omega})} = \gamma_X(0) + \frac{a(e^{i\omega})}{\alpha(e^{i\omega})} + \frac{a(e^{-i\omega})}{\alpha(e^{-i\omega})}$$

where $\alpha(z) = z^{1-n}\prod_{i=2}^{u}\beta_i(z) = \prod_{j=2}^{n}(1-\lambda_j z^{-1})$ and $a(z) = z^{1-n}\sum_{i=2}^{u}\{b_i(z)\prod_{j=2,j\neq i}^{u}$

(g) By combining the results above, show that the special density of a discrete-valued Markov chain can be expressed in the rational form

$$f_X(\omega) = \frac{1}{2\pi} \frac{|m(\mathrm{e}^{\mathrm{i}\omega})|^2}{|\alpha(\mathrm{e}^{\mathrm{i}\omega})|^2}, \quad -\pi < \omega < \pi,$$

where $m(z) = m_0 + m_1 z^{-1} + \cdots + m_{n-1} z^{-n+1}$ and $\alpha(z) = 1 + \alpha_1 z^{-1} + \cdots + \alpha_{n-1} z^{-n+1}$ are relatively prime and n is the state dimension. Furthermore, if λ_j, $j = 2, \cdots, n$, are the sub-dominant eigenvalues of P, then $\alpha(z) = \prod_{j=2}^{n}(1 - \lambda_j z^{-1})$.

9.10. (a) Show that

$$\mathbb{E}_\xi \left[\prod_{i=1}^{p} f_i(Y_{t_i}) h(X_{t_1}, \ldots, X_{t_p}) \right] = \mathbb{E}_\xi \left[h(X_{t_1}, \ldots, X_{t_p}) \prod_{i=1}^{p} G f_i(X_{t_i}) \right]$$

and check (9.14).

(b) Show that, for any integers t and p and any ordered t-tuple $\{t_1 < \cdots < t_p\}$ of indices such that $t \notin \{t_1, \ldots, t_p\}$, the random variables Y_t and $(X_{t_1}, \ldots, X_{t_p})$ are $\mathbb{P}_\xi$-conditionally independent given X_t.

9.11. Fit a two-state model to the S&P 500 weekly returns discussed in Example 9.12. Compare the AIC and BIC of the two-state model with the three-state model and state your conclusions. Note: For the 3-state model, `depmix` reports:

```
'log Lik.' 1236.996 (df=14),  AIC: -2445.992,  BIC: -2386.738.
```

9.12. We consider the autoregressive stochastic volatility model Example 9.5.

(a) Show that for any integer m,

$$\mathbb{E}\left[Y_t^{2m}\right] = \beta^{2m} \mathbb{E}\left[V_t^{2m}\right] \exp(m^2 \sigma_X^2 / 2),$$

where $\sigma_X^2 = \sigma^2/(1 - \phi^2)$.

(b) Show (9.19).

(c) Show that for any positive integer h, $\mathrm{Var}(X_t + X_{t+h}) = 2\sigma_X^2(1 + \phi^h)$.

(d) Show that

$$\mathrm{Cov}(Y_t^{2m}, Y_{t+h}^{2m}) = \beta^{4m} \left(\mathbb{E}\left[V_t^{2m}\right]\right)^2 \left(\exp(m^2 \sigma_X^2 (1 + \phi^h)) - \exp(m^2 \sigma_X^2)\right).$$

(e) Establish (9.20).

9.13. The purpose of this exercise is to prove (9.59). Consider two functions $f \in \mathbb{F}_b(\mathsf{X}^{n+p+1}, \mathcal{X}^{\otimes(n+p+1)})$ and $h \in \mathbb{F}_b(\mathsf{Y}^{n+1}, \mathcal{Y}^{\otimes(n+1)})$.

(a) Show that

$$\mathbb{E}_\xi[h(Y_{0:n}) f(X_{0:n+p})] = \int \cdots \int f(x_{0:n+p}) \xi(\mathrm{d}x_0) g(x_0, y_0) \times \left[\prod_{s=1}^{n} Q_s(x_{s-1}, \mathrm{d}x_s)\right] h(y_{0:n}) \left[\prod_{s=n+1}^{n+p} Q_s(x_{s-1}, \mathrm{d}x_s)\right] \mu_{n+p}(\mathrm{d}y_{0:n+p}),$$

where $Q_s(x_{s-1}, \mathrm{d}x_s) = M(x_{s-1}, \mathrm{d}x_s) g - x_s, y_s)$.

(b) Show that

$$\mathbb{E}_\xi[h(Y_{0:n})f(X_{0:n+p})] = \int \cdots \int h(y_{0:n})f(x_{0:n+p})$$
$$\phi_{\xi,0:n|n}(y_{0:n}, \mathrm{d}x_{0:n}) \left[\prod_{s=n+1}^{n+p} M(x_{s-1}, \mathrm{d}x_s) \right] \mathrm{p}_{\xi,n}(y_{0:n}) \mu_n(\mathrm{d}y_{0:n}) .$$

9.14. Let $t \in \{0, \ldots, n-1\}$ and $h \in \mathbb{F}_b(\mathsf{X}^{n-t}, \mathcal{X}^{\otimes(n-t)})$.

(a) Show that

$$\mathbb{E}_\xi \left[f(X_t)h(X_{t+1:n}) \mid Y_{0:n} \right] = \int \cdots \int f(x_t)h(x_{t+1:n}) \, \phi_{\xi,t:n|n}(\mathrm{d}x_{t:n}) .$$

(b) Show that

$$\mathbb{E}_\xi \left[f(X_t)h(X_{t+1:n}) \mid Y_{0:n} \right] = \frac{\mathrm{p}_{\xi,t}(Y_{0:t})}{\mathrm{p}_{\xi,n}(Y_{0:n})} \iint \phi_{\xi,t}(\mathrm{d}x_t) M(x_t, \mathrm{d}x_{t+1}) f(x_t) \tilde{h}(x_{t+1}) ,$$

where

$$\tilde{h}(x_{t+1}) := g(x_{t+1}, Y_{t+1}) \int \cdots \int \left[\prod_{s=t+2}^{n} M(x_{s-1}, \mathrm{d}x_s) g(x_s, Y_s) \right] h(x_{t+1:n}) .$$

(c) Show that

$$\iint \phi_{\xi,t}(\mathrm{d}x_t) M(x_t, \mathrm{d}x_{t+1}) f(x_t) \tilde{h}(x_{t+1})$$
$$= \int \phi_{\xi,t} M(\mathrm{d}x_{t+1}) B_{\phi_{\xi,t}} f(x_{t+1}) \tilde{h}(x_{t+1}) .$$

(d) Show that for any $\hat{h} \in \mathbb{F}_b(\mathsf{X}^{n-t}, \mathcal{X}^{\otimes(n-t)})$,

$$\phi_{\xi,t+1:n|n}(\hat{h}) = \frac{\int \cdots \int \phi_{\xi,t}(\mathrm{d}x_t) \prod_{s=t+1}^{n} M(x_{s-1}, \mathrm{d}x_s) g(x_s, Y_s) \hat{h}(x_{t+1:n})}{\int \cdots \int \phi_{\xi,t}(\mathrm{d}x_t) \prod_{s=t+1}^{n} M(x_{s-1}, \mathrm{d}x_s) g(x_s, Y_s)} .$$

(e) Deduce that

$$\mathbb{E}_\xi \left[f(X_t)h(X_{t+1:n}) \mid Y_{0:n} \right] = \int \cdots \int B_{\phi_{\xi,t}} f(x_{t+1}) h(x_{t+1:n}) \phi_{\xi,t+1|n}(\mathrm{d}x_{t+1:n}) .$$

(f) Show that

$$\mathbb{E}_\xi \left[f(X_t)h(X_{t+1:n}) \mid Y_{0:n} \right] = \mathbb{E}_\xi \left[h(X_{t+1:n}) B_{\phi_{\xi,t}} f(X_{t+1}) \mid Y_{0:n} \right] ,$$

and conclude.

9.15. Using Proposition 9.14, show the validity of Algorithm 9.3.

Chapter 10

Particle Filtering

Prior to the mid-1980s, a number of methods were developed to approximate the filtering/smoothing distribution for non-normal or nonlinear state-space models in an attempt to circumvent the computational complexity of inference for such models. With the advent of cheap and fast computing, a number of authors developed computer-intensive methods based on numerical integration. For example, Kitagawa (1987) proposed a numerical method based on piecewise linear approximations to the density functions for prediction, filtering, and smoothing for non-Gaussian and nonstationary state-space models. Pole and West (1989) used Gaussian quadrature techniques; see West and Harrison (1997, Chapter 13) and the references therein.

Sequential Monte Carlo (SMC) refers to a class of methods designed to approximate a *sequence of probability distributions* by a set of *particles* such that each have an assigned non-negative weight and are updated recursively. SMC methods are a combination of the sequential importance sampling method introduced in Handschin and Mayne (1969) and the sampling importance resampling algorithm proposed in Rubin (1987).

10.1 Importance sampling

Throughout this section, μ denotes a probability measure on a measurable space $(\mathsf{X},\mathcal{X})$, which is referred to as the *target distribution*. The aim of importance sampling is to approximate integrals of the form $\mu(f) = \int_{\mathsf{X}} f(x)\,\mu(\mathrm{d}x)$ for $f \in \mathbb{F}(\mathsf{X},\mathcal{X})$. The plain Monte Carlo approach consists in drawing an i.i.d. sample $\{X^i\}_{i=1}^N$, from the target distribution μ and then evaluating the sample mean $N^{-1}\sum_{i=1}^N f(X^i)$.

Importance sampling is based on the idea that in certain situations it is more appropriate to sample from a *proposal distribution* ν, and then to apply a change-of-measure formula. Assume that the target distribution μ is absolutely continuous with respect to ν and denote by $w = \mathrm{d}\mu/\mathrm{d}\nu$ the Radon-Nikodym derivative of μ with respect to ν, referred to in the sequel as the *weight function*. Then, for $f \in \mathrm{L}^1(\mu)$, the change of measure formula implies

$$\mu(f) = \int f(x)\,\mu(\mathrm{d}x) = \int f(x)\,w(x)\,\nu(\mathrm{d}x)\,. \tag{10.1}$$

If $\{X^i\}_{i=1}^N$ is an i.i.d. sample from ν, (10.1) suggests the following estimator of $\mu(f)$:

$$N^{-1}\sum_{i=1}^{N} f(X^i)w(X^i)\,. \tag{10.2}$$

Because $\{X^i\}_{i=1}^N$ is an i.i.d. sample from ν, the Strong Law of Large Numbers (SLLN) implies that $N^{-1}\sum_{i=1}^N f(X^i)w(X^i)$ converges to $\nu(fw) = \mu(f)$ almost surely as N tends to infinity; see Exercise 10.1. In addition, moments bounds, deviations inequalities, and central limit theorem for i.i.d. variables may be used to assess the fluctuations of this estimator around its mean.

In many situations, the target probability measure μ is known only up to a normalizing factor. This happens in particular when applying importance sampling ideas to solve filtering and smoothing problems in NLSS. The weight function w is then known up to a (constant) scaling factor only. It is however still possible to use the importance sampling paradigm, by adopting the self-normalized form of the importance sampling estimator,

$$\sum_{i=1}^{N} \frac{w(X^i)}{\sum_{j=1}^N w(X^j)} f(X^i)\,. \tag{10.3}$$

The self-normalized importance sampling estimator is defined as a ratio of the sample means $N^{-1}\sum_{i=1}^N f(X^i)w(X^i)$ and $N^{-1}\sum_{i=1}^N w(X^i)$. The SLLN implies that these two sample means converge, almost surely, to $\nu(fw) = \nu(w)\mu(f)$ and $\nu(w)$, respectively, showing that the self-normalized importance sampling estimator is a consistent estimator of $\mu(f)$; see Exercise 10.5. Importance sampling is of considerable generality and interest since it introduces very little restrictions on the choice of the proposal distribution. This choice is typically guided by two requirements: the proposal distribution should be easy to simulate and should lead to an efficient estimator. This is discussed in the very simple example below.

Example 10.1. Assume that the target distribution is a Gaussian mixture, with density $p(x) = \alpha \mathfrak{g}(x; m_1, \sigma_1^2) + (1-\alpha)\mathfrak{g}(x; m_2, \sigma_2^2)$. A natural choice for the proposal distribution is the Student t_κ-distribution,

$$q_\kappa(x) = \frac{\Gamma((\kappa+1)/2)}{\sqrt{\kappa\pi}\,\Gamma(\kappa/2)}\left(1+\frac{x^2}{\kappa}\right)^{-\frac{\kappa+1}{2}}\,,$$

where κ is the *number of degrees of freedom* and Γ is the Gamma function. The t_κ-distribution is symmetrical about $x = 0$ and has a single mode at $x = 0$. It is easy to show that $\lim_{\kappa\to\infty} q_\kappa(x) = (\sqrt{2\pi})^{-1}\mathrm{e}^{-x^2/2}$: As $\kappa \to \infty$, the t_κ distribution tends to the unit normal distribution. For small to moderate number of degrees of freedom, the fat-tailed behavior of the distribution is characterized by the kurtosis relative to that of a normal distribution that is equal to $6/(\kappa-4)$. When κ is an integer, a draw of a student t_κ may be obtained by sampling $Z_1, \ldots, Z_\kappa$ independent standard normal random variables and computing

$$T_\kappa = Z_0\left(\kappa^{-1}\sum_{i=1}^{\kappa} Z_i^2\right)^{-1/2}\,.$$

◇

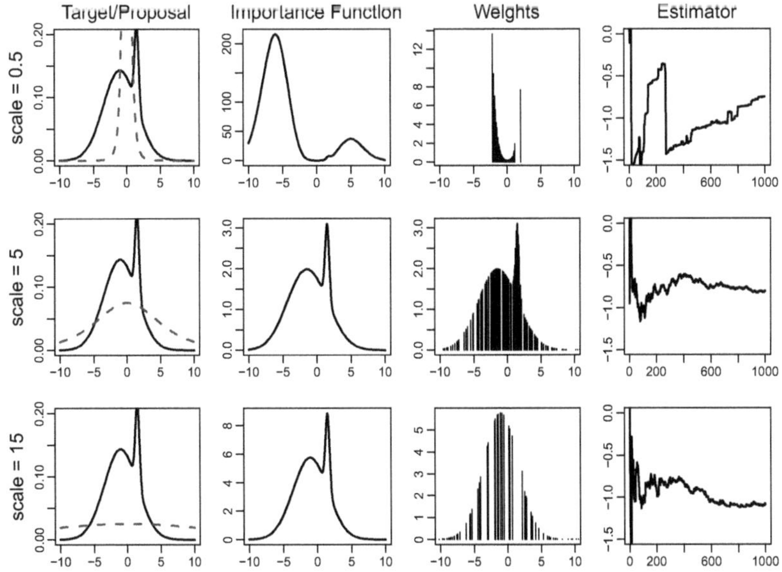

Figure 10.1 *The importance sampling estimator for different choices of proposal distributions. The proposal distribution is a Student t with 4 degrees of freedom. The scales are 0.5 (top row), 5 (middle row), and 15 (bottom row). The number of samples in each case is 1000. In the first column of each row, the target pdf is displayed as a solid line and the proposal is displayed as a dashed line.*

Another possibility consists in resorting to the inversion method. We use the importance sampling estimator to estimate the mean of the target distribution (here equal to $\alpha m_1 + (1-\alpha)m_2$) using t_κ distribution with $\kappa = 4$ and different scales (denoting s the scale, the proposal distribution is $x \mapsto s^{-1}q_\kappa(s^{-1}x)$). As illustrated in Figure 10.1, the choice of the scale plays a crucial role. When the scale is either too small or too large, then the importance sampling estimator becomes very poor.

We can take the idea further. Assume that instead of drawing an independent sample from ν, the distribution ν is already approximated by a set *particles*, each associated to a non-negative *weight*.

Definition 10.2. *A weighted sample $\{(X^{N,i}, \omega^{N,i})\}_{i=1}^N$ is said to be* adapted *to $\mathcal{F}^N \subset \mathcal{F}$ if $\sigma(\{(X^{N,i}, \omega^{N,i})\}_{i=1}^N) \subset \mathcal{F}^N$.*

A weighted sample $\{(X^{N,i}, \omega^{N,i})\}_{i=1}^N$ is consistent *for the probability measure $\mu \in \mathbb{M}_1(\mathcal{X})$ if, as $N \to \infty$,*

$$\sum_{i=1}^N \frac{\omega^{N,i}}{\Omega^N} f\left(X^{N,i}\right) \xrightarrow{\mathbb{P}} \mu(f)\,, \quad \textit{for any } f \in \mathbb{F}_b(\mathsf{X}, \mathcal{X})\,, \tag{10.4}$$

$$\max_{1 \le i \le N} \frac{\omega^{N,i}}{\Omega^N} \xrightarrow{\mathbb{P}} 0\,, \tag{10.5}$$

where Ω^N is the sum of the importance weights

$$\Omega^N := \sum_{i=1}^{N} \omega^{N,i} . \tag{10.6}$$

A weighted sample is a triangular array of random variables: For different values of N, say $N \neq M$, the $(\omega^{N,i}, X^{N,i})$ and $(\omega^{M,i}, X^{M,i})$ are not necessarily equal for any given $i \leq M \wedge N$. To simplify the notation, this dependence is not mentioned explicitly when it is obvious from the context.

Remark 10.3. It is not necessary for the weighted sample size to be equal to N. The weighted sample size may be equal to M_N, where M_N is a deterministic or even random sequence of integers satisfying $M_N \to \infty$ as $N \to \infty$. For simplicity, we assume in this chapter that $M_N = N$, for the weighted sample size. More general statements can be found in Douc and Moulines (2008).

We may transform a weighted sample $\{(X^{N,i}, \omega^{N,i})\}_{i=1}^N$ consistent for ν into a weighted sample $\{(X^{N,i}, \tilde{\omega}^{N,i})\}_{i=1}^N$ consistent for μ, simply by modifying the weights. Setting $\tilde{\omega}^{N,i} = w(X^{N,i})\omega^{N,i}$, then $\{X^{N,i}, \tilde{\omega}^{N,i}\}_{i=1}^N$ is a weighted sample consistent for μ; see Exercise 10.6. We can even consider a more complex transformation. Let Q be a finite kernel on $\mathsf{X} \times \mathcal{X}$ (not necessarily Markov) and assume that we are willing to construct a weighted sample consistent for μ, where μ is given by

$$\mu = \frac{\nu Q}{\nu Q(\mathbf{1})} . \tag{10.7}$$

This type of recursive update is ubiquitous in NLSS; see Section 10.2. Assume that we have already constructed a weighted sample $\{(X^{N,i}, \omega^{N,i})\}_{i=1}^N$ consistent for ν. We wish to apply a transformation to $\{(X^{N,i}, \omega^{N,i})\}_{i=1}^N$ to obtain a new weighted sample $\{(\tilde{X}^{N,i}, \tilde{\omega}^{N,i})\}_{i=1}^N$ consistent for μ. We will describe below one possible method to achieve this goal. Consider a Markov kernel denoted R on $\mathsf{X} \times \mathcal{X}$. Assume that there exists a function $w : \mathsf{X} \times \mathsf{X} \to \mathbb{R}_+$, such that, for each $x \in \mathsf{X}$ and $A \in \mathcal{X}$,

$$Q(x,A) = \int_{\mathsf{X}} w(x,x')R(x,\mathrm{d}x')\mathbb{1}_A(x') . \tag{10.8}$$

If the kernels Q and R have densities denoted by q and r with respect to the same dominating measure, then we simply have to set

$$w(x,x') = \begin{cases} q(x,x')/r(x,x') & r(x,x') \neq 0 \\ 0 & \text{otherwise.} \end{cases} \tag{10.9}$$

The new weighted sample $\{(\tilde{X}^{N,i}, \tilde{\omega}^{N,i})\}_{i=1}^N$ is constructed as follows. For $i = 1,\ldots,N$, we draw $\tilde{X}^{N,i}$ from the proposal kernel $R(X^{N,i},\cdot)$ conditionally independently given $\mathcal{F}^N$, where $\sigma(\{(X^{N,j}, \omega^{N,j})\}_{j=1}^N) \subset \mathcal{F}^N$. By construction, for any $f \in \mathbb{F}_+(\mathsf{X},\mathcal{X})$,

$$\mathbb{E}\left[f(\tilde{X}^{N,i}) \,\middle|\, \mathcal{F}^N\right] = Rf(X^{N,i}) . \tag{10.10}$$

Note that we can take $\mathcal{F}^N = \sigma(\{(X^{N,j}, \omega^{N,j})\}_{j=1}^N)$ if we are only performing a single step analysis, but the σ-algebra $\mathcal{F}^N$ can be chosen larger than that (we will see examples of this later, when we will apply these results sequentially). We then associate to each new particle positions the importance weight:

$$\tilde{\omega}^{N,i} = \omega^{N,i} w(X^{N,i}, \tilde{X}^{N,i}) , \quad \text{for } i = 1, \ldots, N. \tag{10.11}$$

We may now state the main consistency result for the importance sampling estimator.

Theorem 10.4. *Assume that the weighted sample* $\{(X^{N,i}, \omega^{N,i})\}_{i=1}^N$ *is adapted to* $\mathcal{F}^N$ *and consistent for* ν. *Then, the weighted sample* $\{(\tilde{X}^{N,i}, \tilde{\omega}^{N,i})\}_{i=1}^N$ *defined by* (10.10) *and* (10.11) *is consistent for* μ.

Proof. We show first that for any $f \in \mathbb{F}_b(\mathsf{X}^2, \mathcal{X}^{\otimes 2})$,

$$\frac{1}{\Omega^N} \sum_{j=1}^N \tilde{\omega}^{N,j} f(X^{N,j}, \tilde{X}^{N,j}) \xrightarrow{\mathbb{P}} \nu \otimes Q(f) , \tag{10.12}$$

where $\tilde{X}^{N,j}$ and $\tilde{\omega}^{N,j}$ are defined in (10.10) and (10.11), respectively. Here, $\nu \otimes Q$ is the tensor product of the probability ν and the kernel Q, which is the measure defined for any $C \in \mathcal{X}^{\otimes 2}$,

$$\nu \otimes Q(C) = \iint \nu(\mathrm{d}x) Q(x, \mathrm{d}x') \mathbb{1}_C(x, x') . \tag{10.13}$$

The definition (10.8) implies that,

$$\begin{aligned} \mathbb{E}\left[\tilde{\omega}^{N,i} f(X^{N,i}, \tilde{X}^{N,i}) \mid \mathcal{F}^N\right] &= \omega^{N,i} \int w(X^{N,i}, x') R(X^{N,i}, \mathrm{d}x') f(X^{N,i}, x') \\ &= \omega^{N,i} \int Q(X^{N,i}, \mathrm{d}x') f(X^{N,i}, x') = \omega^{N,i} \delta_{X^{N,i}} \otimes Q(f) . \end{aligned} \tag{10.14}$$

Therefore, we get

$$\sum_{i=1}^N \mathbb{E}\left[\frac{\tilde{\omega}^{N,i}}{\Omega^N} f(X^{N,i}, \tilde{X}^{N,i}) \middle| \mathcal{F}^N \right] = \sum_{i=1}^N \frac{\omega^{N,i}}{\Omega^N} \delta_{X^{N,i}} \otimes Q(f) .$$

Noting that the weighted sample $\{(X^{N,i}, \omega^{N,i})\}_{i=1}^N$ is consistent, and that the function $x \mapsto \delta_x \otimes Q(f)$ is bounded, using that $\int \nu(\mathrm{d}x) \delta_x \otimes Q(f) = \nu \otimes Q(f)$, $\sum_{i=1}^N (\omega^{N,i}/\Omega^N) \delta_{X^{N,i}} \otimes Q(f) \to_{\mathbb{P}} \nu \otimes Q(f)$ as N goes to infinity. We will now show that

$$\sum_{j=1}^N \left\{ \frac{\tilde{\omega}^{N,j}}{\Omega^N} f(X^{N,j}, \tilde{X}^{N,j}) - \mathbb{E}\left[\frac{\tilde{\omega}^{N,j}}{\Omega^N} f(X^{N,j}, \tilde{X}^{N,j}) \middle| \mathcal{F}^N \right] \right\} \xrightarrow{\mathbb{P}} 0 . \tag{10.15}$$

Put $U_{N,j} = (\tilde{\omega}^{N,j}/\Omega^N) f(X^{N,j}, \tilde{X}^{N,j})$ for $j = 1, \ldots, N$ and appeal to Theorem B.18 on the convergence of triangular array of random variables. There are two key conditions to check, the *tightness* (B.4) and the asymptotic negligibility (B.5). We first check (B.4). Note that

$$\sum_{j=1}^{N} \mathbb{E}\left[|U_{N,j}| \,\middle|\, \mathcal{F}^N\right] = \sum_{i=1}^{N} \frac{\omega^{N,i}}{\Omega^N} \delta_{X^{N,i}} \otimes Q(|f|) \xrightarrow{\mathbb{P}} \nu \otimes Q(|f|) \,,$$

showing that the sequence $\left\{\sum_{j=1}^N \mathbb{E}\left[|U_{N,j}| \,\middle|\, \mathcal{F}^N\right]\right\}_{N\geq 0}$ is tight (Theorem B.18-Eq.(B.4)). We now check the *negligibility* condition (B.5), i.e., for any $\varepsilon > 0$, put $A_N := \sum_{j=1}^N \mathbb{E}\left[|U_{N,j}|\mathbb{1}\{|U_{N,j}| \geq \varepsilon\} \,\middle|\, \mathcal{F}^N\right] \to_{\mathbb{P}} 0$. For all C, $\varepsilon > 0$,

$$A_N \mathbb{1}\left\{\max_{1\leq i\leq N} \omega^{N,i}/\Omega^N \leq \varepsilon/C\right\} \leq \sum_{j=1}^{N} (\omega^{N,j}/\Omega^N)\delta_{X^{N,j}} \otimes R\left([w|f|]_C\right)$$

$$\xrightarrow{\mathbb{P}} \nu \otimes R\left([w|f|]_C\right) \,, \quad (10.16)$$

where for $u \in \mathbb{R}^+$, $[u]_C = u\mathbb{1}_{\{u\geq C\}}$. By dominated convergence, the right-hand side can be made arbitrarily small by letting $C \to \infty$. Since $\max_{1\leq i\leq N} \omega^{N,i}/\Omega^N \to_{\mathbb{P}} 0$, A_N tends to zero in probability, showing (B.5). Thus Theorem B.18 applies and (10.12) holds; in addition, $\sum_{j=1}^N \tilde{\omega}^{N,j}/\Omega^N \to_{\mathbb{P}} \nu Q(1)$. Combined with (10.12) this shows that, for $f \in \mathbb{F}(\mathsf{X}, \mathcal{X})$,

$$\sum_{j=1}^{N} \frac{\tilde{\omega}^{N,j}}{\tilde{\Omega}^N} f(\tilde{X}^{N,j}) \xrightarrow{\mathbb{P}} \mu(f) \,.$$

It remains to prove that $\max_{1\leq j\leq N} \tilde{\omega}^{N,j}/\tilde{\Omega}^N \to_{\mathbb{P}} 0$. Because $\tilde{\Omega}^N/\Omega^N \to_{\mathbb{P}} \nu Q(1)$, it suffices to show that $\max_{1\leq j\leq N} \tilde{\omega}^{N,j}/\Omega^N \to_{\mathbb{P}} 0$. For any $C > 0$, by applying (10.12), we get

$$\max_{1\leq j\leq N} \frac{\tilde{\omega}^{N,j}}{\Omega^N} \mathbb{1}_{\{w(X^{N,j},\tilde{X}^{N,j})\leq C\}} \leq C \max_{1\leq i\leq N} \frac{\omega^{N,i}}{\Omega^N} \xrightarrow{\mathbb{P}} 0 \,,$$

$$\max_{1\leq j\leq N} \frac{\tilde{\omega}^{N,j}}{\Omega^N} \mathbb{1}_{\{w(\tilde{X}^{N,j})>C\}} \leq \sum_{j=1}^{N} \frac{\tilde{\omega}^{N,j}}{\Omega^N} \mathbb{1}_{\{w(X^{N,j},\tilde{X}^{N,j})>C\}} \xrightarrow{\mathbb{P}} \nu \otimes Q(\{w > C\}) \,.$$

The term in the RHS of the last equation goes to zero as $C \to \infty$, which concludes the proof. ■

Next, we discuss the asymptotic normality of the estimator. Asymptotic normality is crucial to assess the dispersion of the estimators and compute, in particular, confidence intervals. We first need to extend our definition of consistent weighted samples.

Definition 10.5 (Asymptotically normal weighted samples). *Let $\mu \in \mathbb{M}_1(\mathcal{X})$ and $\zeta \in \mathbb{M}_+(\mathcal{X})$. A weighted sample $\{(X^{N,i}, \omega^{N,i})\}_{i=1}^N$ on X is said to be* asymptotically normal *for* (μ,σ,ζ) *if, for any* $f \in \mathbb{F}_b(\mathsf{X},\mathcal{X})$,

$$N^{1/2}\sum_{i=1}^{N}\frac{\omega^{N,i}}{\Omega^N}\{f(X^{N,i})-\mu(f)\} \overset{\mathbb{P}}{\Longrightarrow} \mathrm{N}\{0,\sigma^2(f)\}, \tag{10.17}$$

$$N\sum_{i=1}^{N}\left(\frac{\omega^{N,i}}{\Omega^N}\right)^2 f(X^{N,i}) \overset{\mathbb{P}}{\longrightarrow} \zeta(f), \tag{10.18}$$

$$N^{1/2}\max_{1\le i\le N}\frac{\omega^{N,i}}{\Omega^N} \overset{\mathbb{P}}{\longrightarrow} 0, \tag{10.19}$$

where Ω^N *is defined in* (10.6).

We establish the asymptotic normality of the importance sampling estimator defined by (10.10) and (10.11) in the following theorem.

Theorem 10.6. *Suppose that the assumptions of Theorem 10.4 hold. Assume in addition that the weighted sample* $\{(X^{N,i},\omega^{N,i})\}_{i=1}^N$ *is asymptotically normal for* (ν,σ,ζ). *Then, the weighted sample* $\{(\tilde{X}^{N,i},\tilde{\omega}^{N,i})\}_{i=1}^N$ *is asymptotically normal for* $(\mu,\tilde{\sigma},\tilde{\zeta})$ *with*

$$\tilde{\zeta}(f) := \{\nu Q(1)\}^{-2}\iint \zeta(\mathrm{d}x)R(x,\mathrm{d}x')w^2(x,x')f(x')\,,$$

$$\tilde{\sigma}^2(f) := \frac{\sigma^2\{Q[f-\mu(f)]\}}{\{\nu Q(1)\}^2}+\tilde{\zeta}([f-\mu(f)]^2)-\frac{\zeta(\{Q[f-\mu(f)]\}^2)}{\{\nu Q(1)\}^2}\,.$$

Proof. Pick $f \in \mathbb{F}_b(\mathsf{X},\mathcal{X})$ and assume, without loss of generality, that $\mu(f)=0$. Write $\sum_{i=1}^N \tilde{\omega}^{N,i}/\widetilde{\Omega}^N f(\tilde{X}^{N,i}) = (\Omega^N/\widetilde{\Omega}^N)\,(A_N+B_N)$, with

$$A_N = \sum_{j=1}^{N}\mathbb{E}\left[\left.\frac{\tilde{\omega}^{N,j}}{\Omega^N}f(\tilde{X}^{N,j})\,\right|\,\mathcal{F}^N\right] = \sum_{j=1}^{N}\frac{\omega^{N,j}}{\Omega^N}Qf(X^{N,j})\,,$$

$$B_N = \sum_{j=1}^{N}\left\{\frac{\tilde{\omega}^{N,j}}{\Omega^N}f(\tilde{X}^{N,j})-\mathbb{E}\left[\left.\frac{\tilde{\omega}^{N,j}}{\Omega^N}f(\tilde{X}^{N,j})\,\right|\,\mathcal{F}^N\right]\right\}.$$

Because $\widetilde{\Omega}^N/\Omega^N \to_{\mathbb{P}} \nu Q1$ (see Theorem 10.4), the conclusion of the theorem follows from Slutsky's theorem if we prove that $N^{1/2}(A_N+B_N) \Rightarrow_{\mathbb{P}} \mathrm{N}(0,\sigma^2(Qf)+\eta^2(f))$ where

$$\eta^2(f) := \zeta\otimes R(w^2f^2)-\zeta(\{Qf\}^2)\,, \tag{10.20}$$

with w given in (10.8). Because $\{(X^{N,i},\omega^{N,i})\}_{i=1}^N$ is asymptotically normal for (ν,σ,ζ), $N^{1/2}A_N \Rightarrow_{\mathbb{P}} \mathrm{N}(0,\sigma^2(Qf))$. Next we prove that for any real u,

$$\mathbb{E}\left[\exp(\mathrm{i}uN^{1/2}B_N)\,\middle|\,\mathcal{F}^N\right] \overset{\mathbb{P}}{\longrightarrow} \exp\left(-(u^2/2)\eta^2(f)\right)\,,$$

where $\eta^2(f)$ is defined in (10.20). For that purpose we use Theorem B.20, and we thus need to check (B.10)-(B.11) with $U_{N,j} := N^{1/2}(\tilde{\omega}^{N,j}/\Omega^N) f(\tilde{X}^{N,j})$, $j = 1,\ldots,N$ Because $\{(X^{N,i}, \omega^{N,i})\}_{i=1}^N$ is asymptotically normal for (ν, σ, ζ), (10.18) implies

$$\sum_{j=1}^N \mathbb{E}\left[U_{N,j}^2 \,\middle|\, \mathcal{F}^N\right] \xrightarrow{\mathbb{P}} \zeta \otimes R(w^2 f^2)\,, \quad \sum_{j=1}^N (\mathbb{E}\left[U_{N,j} \,\middle|\, \mathcal{F}^N\right])^2 \xrightarrow{\mathbb{P}} \zeta\{Qf\}^2\,,$$

showing (B.10). It then remains for us to check (B.11). For $\varepsilon > 0$, denote $C_N := \sum_{j=1}^N \mathbb{E}\left[U_{N,j}^2 \mathbb{1}_{\{|U_{N,j}| \geq \varepsilon\}} \,\middle|\, \mathcal{F}^N\right]$. Proceeding like in (10.16), for all $C > 0$, it is easily shown that

$$C_N \leq N \sum_{i=1}^N \left(\frac{\omega^{N,i}}{\Omega^N}\right)^2 \delta_{X^{N,i}} \otimes R\left([wf]_C\right) \\ + \mathbb{1}\left\{\frac{N^{1/2} \max_{1 \leq i \leq N} \omega^{N,i}}{\Omega^N} \geq \frac{\varepsilon}{C}\right\} \sum_{j=1}^N \mathbb{E}\left[U_{N,j}^2 \,\middle|\, \mathcal{F}^N\right]\,,$$

where for $u \in \mathbb{R}^+$, $[u]_C = u^2 \mathbb{1}_{\{u \geq C\}}$. The RHS of the previous display converges in probability to $\zeta \otimes R\left([wf]_C\right)$, which can be made arbitrarily small by taking C sufficiently large. Therefore, condition (B.11) is satisfied and Theorem B.20 applies, showing that $N^{1/2}(A_N + B_N) \Rightarrow_{\mathbb{P}} \mathrm{N}\left(0, \sigma^2(Qf) + \eta^2(f)\right)$.

Consider now (10.18). Recalling that $\widetilde{\Omega}^N/\Omega^N \to_{\mathbb{P}} \nu Q(1)$, it is sufficient to show that for $f \in \mathbb{F}_b(\mathcal{X}^{\otimes 2}, \mathcal{X}^{\otimes \in})$,

$$\left(\frac{N^{1/2}}{\Omega^N}\right)^2 \sum_{j=1}^N (\tilde{\omega}^{N,j})^2 f(X^{N,j}, \tilde{X}^{N,j}) \xrightarrow{\mathbb{P}} \zeta \otimes R(w^2 f)\,, \tag{10.21}$$

Define $U_{N,j} = N(\tilde{\omega}^{N,j}/\Omega^N)^2 f(\tilde{X}^{N,j})$. Because $\{(X^{N,i}, \omega^{N,i})\}_{i=1}^N$ is asymptotically normal for (ν, σ, ζ),

$$\sum_{j=1}^N \mathbb{E}\left[U_{N,j} \,\middle|\, \mathcal{F}^N\right] = N \sum_{i=1}^N \left(\frac{\omega^{N,i}}{\Omega^N}\right)^2 \delta_{X^{N,i}} \otimes R(w^2 f) \xrightarrow{\mathbb{P}} \zeta \otimes R(w^2 f)\,.$$

The proof of (10.21) follows from Theorem B.18 along the same lines as above. Details are omitted. Thus Theorem B.18 applies and condition (10.18) is proved.

Consider finally (10.19). Combining with $\widetilde{\Omega}^N/\Omega^N \to_{\mathbb{P}} \nu Q(1)$ (see proof of Theorem 10.4) it is sufficient to show that $C_N := (N^{1/2}/\Omega^N)^2 \max_{1 \leq j \leq N} (\tilde{\omega}^{N,j})^2 \to_{\mathbb{P}} 0$. For any $C > 0$,

$$C_N \leq C^2 N \max_{1 \leq i \leq N} \left(\frac{\omega^{N,i}}{\Omega^N}\right)^2 + N \sum_{j=1}^N \left(\frac{\tilde{\omega}^{N,j}}{\Omega^N}\right)^2 \mathbb{1}_{\{w(X^{N,j}, \tilde{X}^{N,j}) \geq C\}}\,.$$

Applying (10.21) with $f \equiv \mathbb{1}_{\{w > C\}}$, the RHS of the previous display converges in probability to $\zeta \otimes R\left(w^2 \mathbb{1}\{w \geq C\}\right)$. The proof follows since this quantity can be made as small as we wish by taking C large enough. ∎

10.2 Sequential importance sampling

We now specialize the importance sampling to NLSS models. We use the notations introduced in Definition 9.3 where M denotes the Markov transition kernel of the hidden chain, ξ is the distribution of the initial state X_0, and g denotes the transition density function of the observation given the state with respect to the measure μ on $(\mathsf{Y},\mathcal{Y})$. We denote the filtering distribution by ϕ_t, omitting the dependence with respect to the initial distribution ξ and the observations for notational simplicity, and by Q_t the kernel on $\mathsf{X}\times\mathcal{X}$ defined, for all $x\in\mathsf{X}$ and $f\in\mathbb{F}_+(\mathsf{X},\mathcal{X})$, by

$$Q_t f(x) = \int_{\mathsf{X}} M(x,\mathrm{d}x')g(x',Y_t)f(x') \,. \tag{10.22}$$

According to (9.56), the filtering distribution ϕ_t is given by

$$\phi_t(f) = \frac{\gamma_t(f)}{\gamma_t(1)} \,, \quad \text{for all } f\in\mathbb{F}_+(\mathsf{X},\mathcal{X}) \tag{10.23}$$

where $\{\gamma_t, t\in\mathbb{N}\}$ are computed recursively as follows

$$\gamma_0(f) := \xi[g_0 f] \,, \quad \gamma_t(f) := \gamma_{t-1}Q_t(f) \,, \quad t\geq 1 \,,\ f\in\mathbb{F}_+(\mathsf{X},\mathcal{X}) \,. \tag{10.24}$$

Let $\{R_t\ , t\geq 1\}$ be a family of Markov kernels on $(\mathsf{X},\mathcal{X})$ and $r_0\in\mathbb{M}_1(\mathcal{X})$. The kernels R_t will be referred to as the *proposal kernels*. We assume that there exist weight functions $w_0:\mathsf{X}\to\mathbb{R}_+$ and $w_t:\mathsf{X}\times\mathsf{X}\to\mathbb{R}_+$ such that, for any $(x,x')\in\mathsf{X}$ and $f\in\mathbb{F}_+(\mathsf{X},\mathcal{X})$,

$$\xi[g_0 f] = r_0[w_0 f] \,, \tag{10.25}$$

$$Q_t f(x) = \int w_t(x,x')R_t(x,\mathrm{d}x')f(x') \,. \tag{10.26}$$

When the kernels Q_t and R_t have densities with respect to a common dominating measure, then

$$w_t(x,x') = \frac{q_t(x,x')}{r_t(x,x')} \,, \quad (x,x')\in\mathsf{X}\times\mathsf{X} \,. \tag{10.27}$$

Assume that the weighted sample $\{(X_{t-1}^{N,i},\omega_{t-1}^{N,i})\}_{i=1}^N$ is consistent for ϕ_{t-1}. We construct a weighted sample $\{(X_t^{N,i},\omega_t^{N,i})\}_{i=1}^N$ consistent for ϕ_t as follows. In the proposal step, each particle $X_{t-1}^{N,i}$ gives birth to a single offspring, $X_t^{N,i}$, $i\in\{1,\dots,N\}$ which is sampled conditionally independently from the past of the particles and weights, i.e., $\mathcal{F}_{t-1}^N = \sigma\{\{(X_s^{N,i},\omega_s^{N,i})\}_{i=1}^N,\ s\leq t-1\}$. The distribution of this offspring is specified by the proposal kernel $R_t(X_{t-1}^{N,i},\cdot)$. Next we assign to the new particle $X_t^{N,i}$, $i=1,\dots,N$, the importance weight

$$\omega_t^{N,i} = \omega_{t-1}^{N,i} w_t(X_{t-1}^{N,i},X_t^{N,i}) \,. \tag{10.28}$$

This construction yields to the *Sequential Importance Sampling (SIS)* algorithm (Algorithm 10.1) . The first obvious choice is that of setting $R_t = M$. The weight function

Algorithm 10.1 (SIS: Sequential Importance Sampling)

Initial State: Draw an i.i.d. sample $X_0^1, \ldots, X_0^N$ from r_0 and set

$$\omega_0^i = g_0(X_0^i) w_0(X_0^i) \qquad \text{for } i = 1, \ldots, N .$$

Recursion: For $t = 1, 2, \ldots,$

- Draw $(X_t^1, \ldots, X_t^N)$ conditionally independently given $\{X_s^j, j = 1, \ldots, N, s = 0, \ldots, t-1\}$ from the distribution $X_t^i \sim R_t(X_{t-1}^i, \cdot)$.
- Compute the updated importance weights

$$\omega_t^i = \omega_{t-1}^i w_t(X_{t-1}^i, X_t^i), \qquad i = 1, \ldots, N .$$

then simplifies to

$$w_t(x, x') = g_t(x') = g(x', Y_t) \qquad \text{for all } (x, x') \in \mathsf{X}^2 , \tag{10.29}$$

which *does not depend on* x. The prior kernel is often convenient: sampling from M is often straightforward, and computing the incremental weight amounts to evaluating the conditional likelihood of the new observation given the current particle. The *optimal kernel* is defined as

$$P_t^\star(x, A) = \frac{Q_t(x, A)}{Q_t(x, \mathsf{X})} , \tag{10.30}$$

where Q_t is given by (10.22). The kernel $P_t^\star$ may be interpreted as the conditional distribution of the hidden state X_t given X_{t-1} and the current observation Y_t. The optimal kernel was introduced in Zaritskii et al. (1975) and Akashi and Kumamoto (1977) and has been used since by many authors Liu and Chen (1995), Chen and Liu (2000), Doucet et al. (2000, 2001), Tanizaki (2003). The associated weight function

$$w_t(x, x') = Q_t(x, \mathsf{X}) \qquad \text{for } (x, x') \in \mathsf{X}^2, \tag{10.31}$$

is the conditional likelihood of the observation Y_t given the previous state $X_{t-1} = x$. Note that this weight does not depend on x'. The optimal kernel (10.30) incorporates information both on the state dynamics and on the current observation. There are however two problems with using $P_t^\star$. First, sampling from this kernel is most often computationally costly. Second, calculation of the incremental importance weight $Q_t(x, \mathsf{X})$ may be analytically intractable. However, when the observation equation is linear (Zaritskii et al., 1975), these difficulties can be overcome as illustrated in the following example:

Example 10.7 (Noisy ARCH(1)). We consider an ARCH(1) model observed in additive noise:

$$X_t = \sigma_w(X_{t-1}) W_t , \qquad W_t \sim \text{iid N}(0, 1) ,$$
$$Y_t = X_t + \sigma_v V_t , \qquad V_t \sim \text{iid N}(0, 1) ,$$

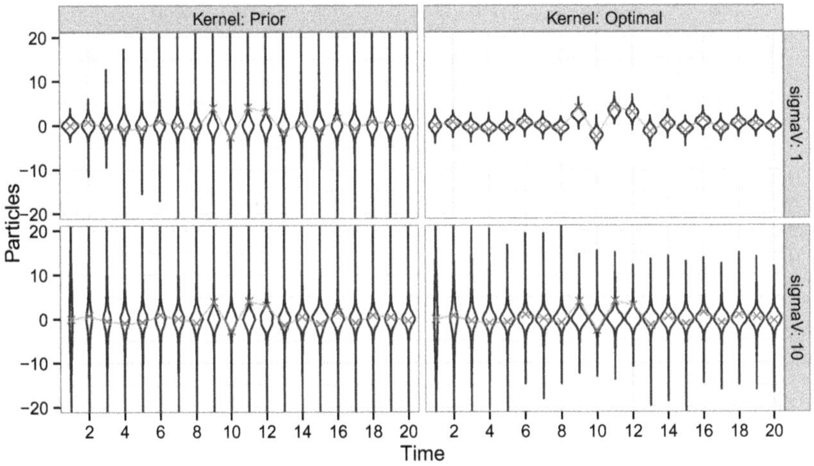

Figure 10.2 *Violin plots of the particle distributions, ignoring the importance weights, for* 5000 *particles, with prior kernel and with optimal kernel, for the ARCH(1) model of Example 10.7. The continuous line represents the actual hidden state trajectory, which is also the observation.*

with $\sigma_w^2(x) = \alpha_0 + \alpha_1 x^2$, where α_0 and α_1 are positive. The optimal kernel has density

$$p_t^\star(x, x') = \mathfrak{g}\left(x'; \tilde{m}_t(x, Y_t), \tilde{\sigma}_t^2(x)\right) , \tag{10.32}$$

where

$$\tilde{m}_t(x, Y_t) = \frac{\sigma_w^2(x) Y_t}{\sigma_w^2(x) + \sigma_v^2} , \quad \tilde{\sigma}_t^2(x) = \frac{\sigma_w^2(x)\sigma_v^2}{\sigma_w^2(x) + \sigma_v^2} . \tag{10.33}$$

The associated weight function is given by

$$Q_t(x, \mathsf{X}) = \mathfrak{g}\left(Y_t; 0, \sigma_w^2(x) + \sigma_v^2\right) . \tag{10.34}$$

When $\sigma_v^2 \gg \sigma_w^2(x)$ (the observation is non-informative), then $\tilde{m}_t(x, Y_t) \approx 0$ and $\tilde{\sigma}_t^2(x) \approx \sigma_w^2(x)$: the prior and the optimal kernel are almost identical. On the other hand, when $\sigma_v^2 \ll \sigma_w^2(x)$, then $\tilde{m}_t(x, Y_t) \approx Y_t$ and $\tilde{\sigma}_t^2(x) = \sigma_v^2$ (i.e., the observation is informative), the optimal kernel is markedly different from the prior kernel and it is expected that the optimal kernel would display better performance. This is illustrated in Figure 10.2, which compares the distribution of the posterior mean estimates achieved with both kernels, on one single simulated dataset of 10 timesteps, generated with parameters $(\alpha_0 = 1, \alpha_1 = 0.99)$. The observations Y_t were set to the simulated hidden state X_t, i.e., the mode of the local likelihood: this setting favors the prior kernel over the optimal kernel, by avoiding "extreme" observations, but makes the comparison meaningful across different values of σ_V.

SIS was then run in turn with 5000 particles for each of the two proposal kernels, for two distinct values of σ_V. The procedure was repeated independently 100 times, to obtain a sample of posterior mean estimates in the four cases. The results are

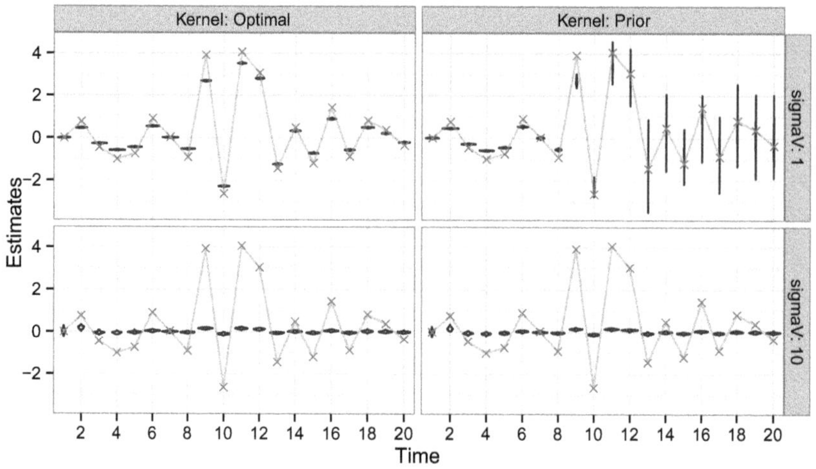

Figure 10.3 *Boxplots of the posterior mean estimates of* 100 *independent SIS runs, each using* 5000 *particles, with prior kernel and with optimal kernel, for the ARCH(1) model of Example 10.7. The continuous line represents the actual hidden state trajectory, which is also the observation.*

exhibited in Figure 10.3. With poorly informative observations, i.e., $\sigma_V = 10$, the distribution of the posterior mean estimates for both kernels are similar: the filtering distribution is mostly influenced by the dynamic equation. However, for informative observations, i.e., $\sigma_V = 1$, while the SIS for both kernels is centered around a same value, much closer to the actual hidden state, the variance of the point estimate is markedly smaller when using the optimal kernel. ◇

The weights ω_t^i measure the adequacy of the particle X_t^i to the target distribution ϕ_t. A particle such that the associated weight ω_t^i is orders of magnitude smaller than the sum Ω_t^N does not contribute to the estimator. If there are too many ineffective particles, the particle approximation is inefficient.

Unfortunately, this situation is the rule rather than the exception, as the importance weights will degenerate as the time index t increases, with most of the normalized importance weights ω_t^i/Ω_t^N close to 0 except for a few.

Example 10.8 (Example 10.7, cont.). Figure 10.4 displays the Lorenz curve of the normalized importance weights after 5, 10, 25, and 50 time steps for the noisy ARCH(1) with $\sigma_V = 1$ (see Example 10.7) for the prior kernel and the optimal kernel. The number of particles is set to 5000. The Lorenz curve is a graphical representation of the cumulative distribution function of the empirical probability distribution. Assume that X is a random variable with cumulative distribution function F and quantile function $F^{-1}(t) = \inf\{x : F(x) \geq t\}$. The Lorenz curve corresponding to any random variable X with cumulative distribution function F and finite mean $\mu = \int x \mathrm{d}F(x)$ is defined to be

$$L(p) = \mu^{-1} \int_0^p F^{-1}(t)\mathrm{d}t\,, \quad 0 \leq p \leq 1\,. \tag{10.35}$$

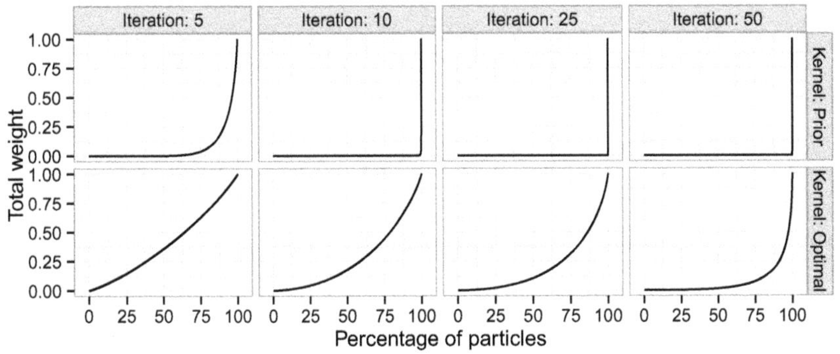

Figure 10.4 *Lorenz curves of the importance weights for the noisy ARCH model after 5, 10, 25, and 50 iterations with 5000 particles. Top panel: prior kernel; Bottom panel: optimal kernel.*

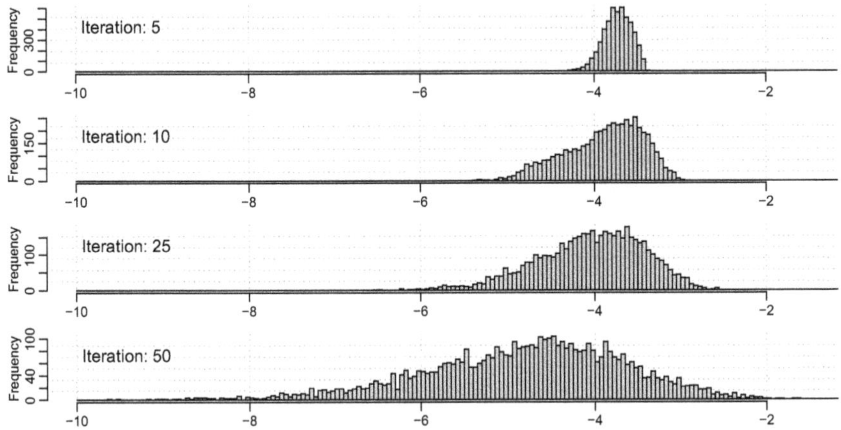

Figure 10.5 *Histograms of the base 10 logarithm of the normalized importance weights after (from top to bottom) 5, 10, 25, and 50 iterations for the noisy ARCH model with N=5000 particles.*

Applied in our setting, Figure 10.4 shows the fraction of the total sum of the importance weights that the particles of the lowest $x\%$ fraction possess. The Lorenz curves show that the normalized weights for the prior kernel quickly degenerate: the total mass is concentrated on a small fraction of particles. For the optimal kernel, the degeneracy is much slower, but after 50 steps, the bottom 75% of the importance weights accounts for less than 10% of the total weights.

Figure 10.5 displays the histogram of the base 10 logarithm of the normalized importance weights after 5, 10, 25, and 50 time steps for the same model using the optimal kernel. Figure 10.5 shows that the normalized importance weights degenerate as the number of iterations of the SIS algorithm increases. ◇

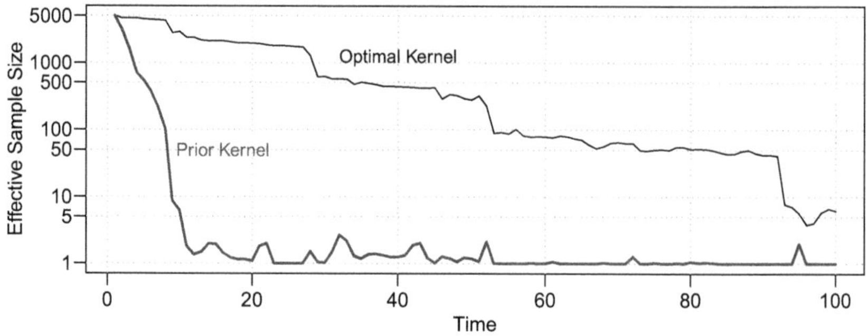

Figure 10.6 *Effective Sample Size curves of the importance weights for the noisy ARCH model, 5000 particles, 100 time points. Note how the optimal kernel improves over the prior kernel but eventually degenerates.*

A simple criterion to quantify the degeneracy of a set of importance weights $\{\omega^j\}_{i=1}^N$ is the *coefficient of variation* used by Kong et al. (1994), which is defined by

$$\mathrm{CV}\left(\{\omega^i\}_{i=1}^N\right) := \left[\frac{1}{N}\sum_{i=1}^{N}\left(N\frac{\omega^i}{\Omega^N}-1\right)^2\right]^{1/2} . \tag{10.36}$$

The coefficient of variation is minimal when the normalized weights are all equal to $1/N$, and then $\mathrm{CV}(\{\omega^i\}_{i=1}^N) = 0$. The maximal value of $\mathrm{CV}(\{\omega^i\}_{i=1}^N)$ is $\sqrt{N-1}$, which corresponds to one of the normalized weights being one and all others being null. A related criterion is the *effective sample size* ESS (Liu, 1996), defined as

$$\mathrm{ESS}(\{\omega^i\}_{i=1}^N) = \left[\sum_{i=1}^{N}\left(\frac{\omega^i}{\Omega^N}\right)^2\right]^{-1} = \frac{N}{1+\left[\mathrm{CV}(\{\omega^i\}_{i=1}^N)\right]^2} , \tag{10.37}$$

which varies between 1 and N (equal weights). The effective sample size may be understood as a proxy for the equivalent number of i.i.d. samples, but this interpretation can sometimes be rather misleading.

As an example, Figure 10.6 shows the ESS curves of the importance weights for a simulated noisy ARCH model as given in Example 10.7, with 100 time points. Using 5000 particles, note how the optimal kernel performs better than the prior kernel, but eventually degenerates.

10.3 Sampling importance resampling

The solution proposed by Gordon et al. (1993) to avoid the degeneracy of the importance weights is based on *resampling* using the normalized weights as probabilities of selection. Thus, particles with small importance weights are eliminated, whereas those with large importance weights are replicated. After resampling, all importance weights are reset to one.

Algorithm 10.2 (SIR: Sampling Importance Resampling)

Sampling: Draw an i.i.d. sample $X^1,\dots,X^N$ from the instrumental distribution ν.

Weighting: Compute the (normalized) importance weights

$$\omega^i = \frac{w(X^i)}{\sum_{j=1}^N w(X^j)} \quad \text{for } i = 1,\dots,N\,.$$

Resampling:

- Draw, conditionally independently given $(X^1,\dots,X^N)$, N discrete random variables $(I^1,\dots,I^N)$ taking values in the set $\{1,\dots,N\}$ with probabilities $(\omega^1,\dots,\omega^N)$, i.e.,

$$\mathbb{P}(I^1 = j) = \omega^j\,, \quad j = 1,\dots,N\,. \tag{10.39}$$

- Set, for $i = 1,\dots,N$, $\tilde{X}^i = X^{I^i}$.

10.3.1 Algorithm description

The resampling method is rooted in the *sampling importance resampling* (or SIR) method to sample a distribution μ, introduced by Rubin (1987, 1988). We discuss this procedure first and we will then explain how this procedure can be used in combination with the SIS procedure. We first consider the SIR in the simple setting of a single step importance estimator. In this setting, the SIR proceeds in two stages. In the *sampling stage*, an i.i.d. sample $\{X^i\}_{i=1}^N$ is drawn from the proposal distribution ν. The importance weights are then evaluated at particle positions,

$$\omega^i = w(X^i)\,, \tag{10.38}$$

where w is the importance weight function defined in (10.1). In the *resampling stage*, a sample of size N denoted by $\{\tilde{X}^i\}_{i=1}^N$ is drawn from the set of points $\{X^i\}_{i=1}^N$, with probability proportional to the weights (10.38). The rationale is that particles X^i associated to large importance weights ω^i are more likely under the target distribution μ and should thus be selected with higher probability during the resampling than particles with low (normalized) importance weights.

This idea is easy to extend. Assume that $\{(X^i,\omega^i)\}_{i=1}^N$ is a weighted sample consistent for ν. We may apply the resampling stage to $\{(X^i,\omega^i)\}_{i=1}^N$, i.e., draw a sample $\{\tilde{X}^i\}_{i=1}^N$ from the set of points $\{X^i\}_{i=1}^N$ with probability proportional to the weights $\{\omega^i\}_{i=1}^N$. Doing so we obtain an equally weighted sample $\{(\tilde{X}^i,1)\}_{i=1}^N$ also targeting ν. The SIR algorithm is summarized below.

10.3.2 Resampling techniques

Denoting by M^i is the number of times that the particle X^i is resampled. With these notations, we get

$$\frac{1}{N}\sum_{i=1}^{N} f(\tilde{X}^i) = \sum_{i=1}^{N} \frac{M^i}{N} f(X^i) .$$

Assume that the weighted sample $\{(X^{N,i}, \omega^{N,i})\}_{i=1}^N$ is adapted to $\mathcal{F}^N$. A resampling procedure is said to be *unbiased* if

$$\mathbb{E}\left[M^i \mid \mathcal{F}^N\right] = N\omega^i/\Omega^N , \quad i = 1, \ldots, N . \tag{10.40}$$

It is easily seen that this condition implies that the conditional expectation of $N^{-1}\sum_{i=1}^N f(\tilde{X}^i)$, with respect to the weighted sample $\{(X^i, \omega^i)\}_{i=1}^N$, is equal to the importance sampling estimator,

$$\mathbb{E}\left[N^{-1}\sum_{i=1}^{N} f(\tilde{X}^i) \mid \mathcal{F}^N\right] = \sum_{i=1}^{N} \frac{\omega^i}{\Omega^N} f(X^i) .$$

As a consequence, the mean square error of the estimator $(1/N)\sum_{i=1}^N f(\tilde{X}^i)$ after resampling is always larger than that of the importance sampling estimator (10.3):

$$\mathbb{E}\left(\frac{1}{N}\sum_{i=1}^{N} f(\tilde{X}^i) - \mu(f)\right)^2$$

$$= \mathbb{E}\left(\frac{1}{N}\sum_{i=1}^{N} f(\tilde{X}^i) - \sum_{i=1}^{N} \frac{\omega^i}{\Omega^N} f(X^i)\right)^2 + \mathbb{E}\left(\sum_{i=1}^{N} \frac{\omega^i}{\Omega^N} f(X^i) - \mu(f)\right)^2 . \tag{10.41}$$

There are several different ways to construct an unbiased sampling procedure, the most obvious approach being sampling with replacement with probability of sampling each X^i equal to the normalized importance weight ω^i/Ω^N. In this case, the $\{M^i\}_{i=1}^N$ is multinomial

$$\{M^i\}_{i=1}^N | \{X^i, \omega^i\}_{i=1}^N \sim \mathsf{Mult}\left(N, \left\{\frac{\omega^i}{\Omega^N}\right\}_{i=1}^N\right) . \tag{10.42}$$

Another possible solution is the *deterministic plus residual multinomial resampling*, introduced in Liu and Chen (1995). Denote by $\lfloor x \rfloor$ the integer part of x and by $\langle x \rangle$ the fractional part of x, $\langle x \rangle := x - \lfloor x \rfloor$. This scheme consists of retaining $\lfloor N\omega^i/\Omega^N \rfloor$, $i = 1, \ldots, N$ copies of the particles and then reallocating the remaining particles by applying the multinomial resampling procedure with the residual importance weights defined as $\langle N\omega^i/\Omega^N \rangle$. In this case, M^i may be decomposed as $M^i = \lfloor N\omega^i/\Omega^N \rfloor + H_i$ where $\{H_i\}_{i=1}^N$ are multinomial

$$\{H_i\}_{i=1}^N | \{X^i, \omega^i\}_{i=1}^N \sim \mathsf{Mult}\left(\sum_{i=1}^{N}\left\langle \frac{N\omega^i}{\Omega^N}\right\rangle, \left\{\frac{\langle N\omega^i/\Omega^N \rangle}{\sum_{i=1}^N \langle N\omega^i/\Omega^N \rangle}\right\}_{i=1}^N\right) . \tag{10.43}$$

Assuming that the weighted sample $\{(X^{N,i}, \omega^{N,i})\}_{i=1}^N$ is consistent for ν, it is a natural question to ask whether the uniformly weighted sample $\{(\tilde{X}^{N,i}, 1)\}_{i=1}^N$ is still consistent for ν.

Theorem 10.9. *Assume that the weighted sample $\{(X^{N,i}, \omega^{N,i})\}_{i=1}^N$ is adapted to $\mathcal{F}^N$ and consistent for ν. Then, the uniformly weighted sample $\{(\tilde{X}^{N,i}, 1)\}_{i=1}^N$ obtained using either* (10.42) *or* (10.43) *is consistent for ν.*

Proof. See Exercise 10.14 ■

Similarly, the resampling procedures (10.42) and (10.43) transform an asymptotically normal weighted sample for ν into an asymptotically normal sample for ν. We will discuss only the multinomial sampling case; the corresponding result for the deterministic plus multinomial residual sampling is given in Chopin (2004) and Douc and Moulines (2008).

Theorem 10.10. *Assume that $\{(X^{N,i}, \omega^{N,i})\}_{i=1}^N$ is adapted to $\mathcal{F}^N$, consistent for ν, and asymptotically normal for (ν, σ, ζ). Then the equally weighted particle system $\{(\tilde{X}^{N,i}, 1)\}_{i=1}^N$ obtained using* (10.42) *is asymptotically normal for $(\nu, \tilde{\sigma}, \tilde{\zeta})$ with $\tilde{\sigma}^2(f) = \mathrm{Var}_\nu(f) + \sigma^2(f)$ and $\tilde{\zeta} = \nu$.*

Proof. See Exercise 10.15. ■

10.4 Particle filter

10.4.1 Sequential importance sampling and resampling

The resampling step can be introduced in the sequential importance sampling framework outlined in Section 10.2. As shown in (10.41), the one-step effect of resampling seems to be negative, as it increases the variance, but, as we will show later, resampling is required to guarantee that the particle approximation does not degenerate in the long run. This remark suggests that it may be advantageous to restrict the use of resampling to cases where the importance weights are unbalanced. The criteria defined in (10.36) and (10.37), are of course helpful for that purpose. The resulting algorithm, which is generally known under the name of *sequential importance sampling with resampling* (SISR), is summarized in Algorithm 10.3.

Example 10.11 (Example 10.7, cont.). The histograms shown in Figure 10.7 are the counterparts of those shown in Figure 10.5. In this case, the resampling is applied whenever the coefficient of variation, (10.36), of the normalized weights exceeds one. The histograms of the normalized importance weights displayed in Figure 10.7 show that the weight degeneracy is avoided. ◇

The SISR algorithm combines importance sampling steps with resampling steps. By applying iteratively the one-step consistency results Theorem 10.4 and Theorem 10.9, a straightforward induction shows that, starting from a weighted sample $\{(X_0^{N,i}, \omega_0^{N,i})\}_{i=1}^N$ consistent for ϕ_0, then the SISR algorithm produces at each iteration a weighted sample $\{(X_t^{N,i}, \omega_t^{N,i})\}_{i=1}^N$ consistent for ϕ_t. Similarly, if the weighted sample $\{(X_0^{N,i}, \omega_0^{N,i})\}_{i=1}^N$ is asymptotically normal for ϕ_0, then by again

Algorithm 10.3 (SISR: Sequential Importance Sampling with Resampling)

Sampling: • Draw $\{\tilde{X}_t^i\}_{i=1}^N$ conditionally independently given $\{\{(X_s^j, \omega_s^j)\}_{j=1}^N, s \leq t-1\}$ from the proposal kernel $\tilde{X}_t^i \sim R_t(X_{t-1}^i, \cdot)$, $i = 1, \ldots, N$.

- Compute the updated importance weights

$$\tilde{\omega}_t^i = \omega_{t-1}^i w_t(X_{t-1}^i, \tilde{X}_t^i) , \qquad i = 1, \ldots, N .$$

where w_t is defined in (10.26).

Resampling (Optional):

- Draw, conditionally independently given $\{(X_s^i, \omega_s^i)\}_{i=1}^N, s \leq t-1\}$ and $\{\tilde{X}_t^i\}_{i=1}^N$, a multinomial trial $\{I_t^i\}_{i=1}^N$ with probabilities of success $\{\tilde{\omega}_t^i / \widetilde{\Omega}_t^N\}_{i=1}^N$ and set $X_t^i = \tilde{X}_t^{I_t^i}$ and $\omega_t^i = 1$ for $i = 1, \ldots, N$.

---

applying iteratively the one-step asymptotic normality results, the weighted sample $\{(X_t^{N,i}, \omega_t^{N,i})\}_{i=1}^N$ is asymptotically normal for ϕ_t, because both the importance sampling step and the resampling step preserve asymptotic normality. The limiting variance can be computed iteratively.

As an illustration, we may consider a special instance of Algorithm 10.3 for which the resampling procedure is triggered when the coefficient of variation exceeds a threshold κ, i.e., $\mathrm{CV}(\{\tilde{\omega}_t^{N,i}\}_{i=1}^N) > \kappa$, we draw $\{I_t^{N,i}\}_{i=1}^N$ conditionally independently given $\mathcal{F}_{t-1}^N \vee \sigma\left(\{\tilde{X}_t^{N,i}, \tilde{\omega}_t^{N,i}\}_{i=1}^N\right)$

$$\mathbb{P}\left(I_t^{N,i} = j \,\middle|\, \mathcal{F}_t^N\right) = \tilde{\omega}_t^{N,j} / \widetilde{\Omega}_t^N , \quad i = 1, \ldots, N , j = 1, \ldots, N \tag{10.44}$$

and we set $X_t^{N,i} = \tilde{X}_t^{N, I_t^{N,i}}$ and $\omega_t^{N,i} = 1$ for $i = 1, \ldots, N$. If $\mathrm{CV}(\tilde{\omega}_t^{N,i}) \leq \kappa$, we simply keep the updated particles and weights, i.e., we set $(X_t^{N,i}, \omega_t^{N,i}) = (\tilde{X}_t^{N,i}, \tilde{\omega}_t^{N,i})$ for $i = 1, \ldots, N$. We have only described the multinomial resampling, but the deterministic plus residual sampling, or even more sophisticated alternatives, can be considered as well; see Douc and Moulines (2008).

Theorem 10.12. *Assume that the equally weighted sample $\{(X_0^{N,i}, 1)\}_{i=1}^N$ is consistent for ϕ_0 and asymptotically normal for $(\phi_0, \sigma_0, \phi_0)$. Assume in addition that for all $(x,y) \in \mathsf{X} \times \mathsf{Y}$, $g(x,y) > 0$, $\sup_{x \in \mathsf{X}} g(x,y) < \infty$ for all $y \in \mathsf{Y}$, and $\sup_{0 \leq t \leq n} |w_t|_\infty < \infty$.*

Then, for any $1 \leq t \leq n$, $\{(X_t^{N,i}, \omega_t^{N,i})\}_{i=1}^N$ is consistent for ϕ_t and asymptotically normal for (ϕ_t, σ_t) where the function σ_t is defined by the recursion

$$\sigma_t^2(f) = \mathrm{Var}_{\phi_t}(f) + \frac{\sigma_{t-1}^2(Q_t[f - \phi_t(f)])}{\{\phi_{t-1} Q_t(\mathbf{1})\}^2} +$$
$$\iint \frac{\phi_{t-1}(\mathrm{d}x) R_t(x, \mathrm{d}x') w_t^2(x,x') \{f(x') - \phi_t(f)\}^2 - \phi_{t-1}(\{Q_t[f - \phi_t(f)]\}^2)}{\{\phi_{t-1} Q_t(\mathbf{1})\}^2} .$$

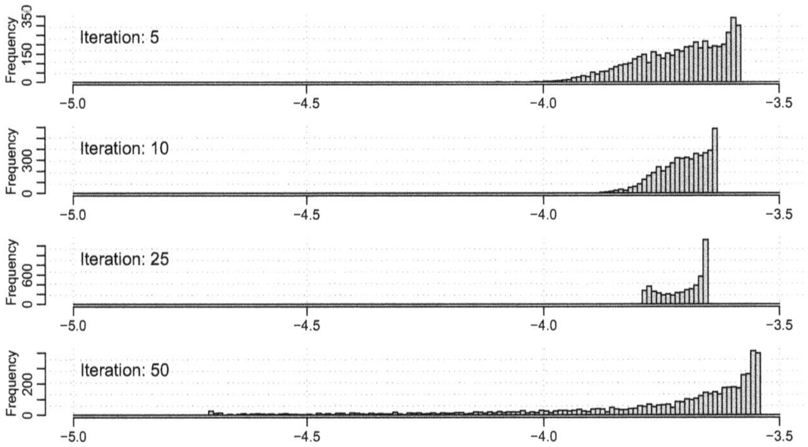

Figure 10.7 *Histograms of the base 10 logarithm of the normalized importance weights after (from top to bottom) 5, 10, 25, and 50 iterations in the noisy ARCH model of Example 10.7. Same model and data as in Figure 10.5. Resampling occurs when the coefficient of variation gets larger than 1.*

Proof. The proof is by repeated applications of Theorem 10.6 and Theorem 10.10; see Exercise 10.12. ∎

10.4.2 Auxiliary sampling

In this section, we introduce the *auxiliary particle filter* (APF) proposed by Pitt and Shephard (1999), which has proven to be one of the most useful implementations of the SMC methodology. The APF enables us to design a set of *adjustment multiplier weights* involved in the selection procedure. Assume that we at time $t-1$ have a weighted sample $\{(X_{t-1}^i, \omega_{t-1}^i)\}_{i=1}^N$ providing an approximation $\phi_{t-1}^N = \sum_{i=1}^N (\omega_{t-1}^i / \Omega_{t-1}^N) \delta_{X_{t-1}^i}$ of the filtering distribution ϕ_{t-1}. When the observation Y_t becomes available, an approximation of the filtering distribution ϕ_t may be obtained by plugging the empirical measure ϕ_{t-1}^N into the recursion (10.23), yielding, for $A \in \mathcal{X}$,

$$\phi_t^{N,\mathrm{tar}}(A) = \frac{\phi_{t-1}^N Q_t(A)}{\phi_{t-1}^N Q_t(\mathsf{X})} = \sum_{i=1}^N \frac{\omega_{t-1}^i Q_t(X_{t-1}^i, \mathsf{X})}{\sum_{j=1}^N \omega_{t-1}^j Q_t(X_{t-1}^j, \mathsf{X})} P_t^\star(X_{t-1}^i, A)\,, \tag{10.45}$$

where Q_t and $P_t^\star$ are defined in (10.22) and (10.30), respectively. Now, since we want to form a new weighted sample approximating ϕ_t, we need to find a convenient mechanism for sampling from $\phi_t^{N,\mathrm{tar}}$ given $\{(X_{t-1}^i, \omega_{t-1}^i)\}_{i=1}^N$. In most cases it is possible—but generally computationally expensive—to simulate from $\phi_t^{N,\mathrm{tar}}$ directly using *auxiliary accept-reject sampling* (see Hürzeler and Künsch, 1998, Künsch, 2005). A computationally cheaper solution consists of producing a weighted sample approximating $\phi_t^{N,\mathrm{tar}}$ by using an importance sampling procedure. Following Pitt and Shephard (1999), this may be done by considering the *auxiliary* target distribution

$$\phi_t^{N,\mathrm{aux}}(\{i\} \times A) := \frac{\omega_{t-1}^i Q_t(X_{t-1}^i, A)}{\sum_{\ell=1}^N \omega_{t-1}^\ell Q_t(X_{t-1}^\ell, \mathsf{X})}, \quad i \in \{1,\ldots,N\}, A \in \mathcal{X}, \tag{10.46}$$

on the product space $\{1,\ldots,N\} \times \mathsf{X}$. By construction, the target distribution $\phi_t^{N,\mathrm{tar}}$ is the marginal distribution with respect to the particle index of the auxiliary distribution $\phi_t^{N,\mathrm{aux}}$. Therefore, we may construct a weighted sample targeting $\phi_t^{N,\mathrm{tar}}$ on $(\mathsf{X}, \mathcal{X})$ by sampling from the auxiliary distribution, computing the associated importance weights and then discarding the indices.

To sample from $\phi_t^{N,\mathrm{aux}}$, we use an importance sampling strategy on the product space $\{1,\ldots,N\} \times \mathsf{X}$. To do this, we first draw conditionally independently pairs $\{(I_t^i, X_t^i)\}_{i=1}^N$ of indices and particles from the proposal distribution

$$\phi_t^{N,\mathrm{prop}}(\{i\} \times A) = \frac{\omega_{t-1}^i \vartheta_t(X_{t-1}^i)}{\sum_{j=1}^N \omega_{t-1}^j \vartheta_t(X_{t-1}^j)} R_t(X_{t-1}^i, A), \quad A \in \mathcal{X}, \tag{10.47}$$

on the product space $\{1,\ldots,N\} \times \mathsf{X}$, where $x \mapsto \vartheta_t(x)$ is the *adjustment multiplier weight* function and R_t is the proposal kernel. We will discuss later the choice of ϑ_t. For each draw $\{(I_t^i, X_t^i)\}_{i=1}^N$, we compute the importance weight

$$\omega_t^i = \frac{w_t(X_{t-1}^{I_t^i}, X_t^i)}{\vartheta_t(X_{t-1}^{I_t^i})}, \tag{10.48}$$

where w_t is the importance function defined in (10.26), and associate it to the corresponding particle position X_t^i. Finally, the indices $\{I_t^i\}_{i=1}^N$ are discarded. The weighted sample $\{(X_t^i, \omega_t^i)\}_{i=1}^N$ is taken as an approximation of ϕ_t. The simplest choice, yielding to the *bootstrap particle filter algorithm* proposed by Gordon et al. (1993), consists of setting, for all $x \in \mathsf{X}$, $\vartheta_t \equiv 1$ and $R_t(x, \cdot) \equiv M(x, \cdot)$. A more appealing—but often computationally costly—choice consists of using the adjustment weights $\vartheta_t(x) = \vartheta_t^\star(x) := Q_t(x, \mathsf{X})$, $x \in \mathsf{X}$, and the proposal transition kernel $P_t^\star(x, \cdot) := Q_t(x, \cdot)/Q_t(x, \mathsf{X})$. In this case, $\omega_t^i = 1$ for all $i \in \{1,\ldots,N\}$ and the auxiliary particle filter is said to be *fully adapted*. Except in some specific models, the implementation of a fully adapted sampler is computationally impractical. Heuristically, the adjustment multiplier weight function $x \mapsto \vartheta_t(x)$ should be an easy to compute proxy of $x \mapsto \vartheta_t^\star(x)$. Pitt and Shephard (1999) suggest that the adjustment multiplier weight function be set as the likelihood of the mean of the predictive distribution corresponding to each particle,

$$\vartheta_t(x) = g\left(\int x' \, M(x, \mathrm{d}x'), Y_{t+1}\right). \tag{10.49}$$

Other constructions are discussed in Douc et al. (2009b) and Cornebise et al. (2008).

Let n be an integer. Consider the following assumptions:

Assumption A10.13.

(a) For all $(x,y) \in \mathsf{X} \times \mathsf{Y}$ and $0 \leq t \leq n$, $g_t(x,y) > 0$ and $\sup_{0 \leq t \leq n} |g_t|_\infty < \infty$.

(b) $\sup_{0 \leq t \leq n} |\vartheta_t|_\infty < \infty$ and $\sup_{0 \leq t \leq n} \sup_{(x,x') \in \mathsf{X} \times \mathsf{X}} w_t(x,x')/\vartheta_t(x) < \infty$.

Theorem 10.14. *Assume A10.13. Then, for all $t \in \{0,\ldots,n\}$, the weighted sample $\{(X_t^{N,i}, \omega_t^{N,i})\}$ is consistent for ϕ_t and asymptotically normal for $(\phi_t, \sigma_t, \zeta_t)$ where*

$$\zeta_t(f) = \{\phi_{t-1} Q_t \mathbf{1}\}^{-2} \phi_{t-1}(\vartheta_t) \iint \phi_{t-1}(\mathrm{d}x) R_t(x, \mathrm{d}x') \frac{w_t^2(x,x')}{\vartheta_t(x)} f(x') , \tag{10.50a}$$

$$\sigma_t^2(f) = \{\phi_{t-1} Q_t \mathbf{1}\}^{-2} \sigma_{t-1}^2 (Q_t [f - \phi_t(f)]) + \zeta_t([f - \phi_t(f)]^2) . \tag{10.50b}$$

Proof. See Exercise 10.16 ∎

Remark 10.15. The asymptotic variance is minimized if the auxiliary weights are chosen to be equal to (see Exercise 10.19)

$$\vartheta_t^{\mathrm{opt}}(x) = \left[\int R_t(x, \mathrm{d}x') w_t^2(x,x') [f(x') - \phi_t(f)]^2 \right]^{1/2} . \tag{10.51}$$

As shown in Douc et al. (2009b), this choice of the adjustment weight can be related to the choice of the sampling weights of strata for stratified sampling estimators; see Exercise 10.21. The use of the optimal adjustment weights (10.51) provides, for a given sequence $\{R_t\}_{t \geq 0}$ of proposal kernels, the most efficient of all auxiliary particle filters. However, exact computation of the optimal weights is in general infeasible for two reasons: firstly, they depend (via $\phi_t(f)$) on the expectation $\phi_t(f)$, that is, the quantity that we aim to estimate, and, secondly, they involve the evaluation of a complicated integral; see Douc et al. (2009b) and Cornebise et al. (2008) for a discussion.

10.5 Convergence of the particle filter

10.5.1 Exponential deviation inequalities

Non-asymptotic deviation inequality provides an explicit bound on the probability that the particle estimator $\phi_t^N(h)$ deviates from its targeted value $\phi_t(h)$ by $\varepsilon > 0$: $\mathbb{P}(|\phi_t^N(h) - \phi_t(h)| > \varepsilon) \leq r(N, \varepsilon, h)$, where the rate function r is explicit. It is possible to derive such nonasymptotic bounds for the particle approximation. To make the derivations short, we concentrate in this chapter only the auxiliary particle filter introduced in Section 10.4.2 and exponential deviation inequality for bounded functions h. We preface the proof by an elementary Lemma:

Lemma 10.16. *Assume that A_N, B_N and B are random variables such that there exist positive constants β, c_1, c_2, c_3 such that*

$$|A_N/B_N| \leq c_1, \ \mathbb{P}\text{-a.s. } \textit{and } B \geq \beta, \ \mathbb{P}\text{-a.s.} \tag{10.52a}$$

$$\textit{For all } \varepsilon > 0 \textit{ and } N \geq 1, \ \mathbb{P}(|B_N - B| > \varepsilon) \leq 2\mathrm{e}^{-Nc_2\varepsilon^2} , \tag{10.52b}$$

$$\textit{For all } \varepsilon > 0 \textit{ and } N \geq 1, \ \mathbb{P}(|A_N| > \varepsilon) \leq 2\mathrm{e}^{-Nc_3(\varepsilon/c_1)^2}, \tag{10.52c}$$

Then,

$$\mathbb{P}(|A_N/B_N| > \varepsilon) \le 4\mathrm{e}^{-N(c_2 \wedge c_3)(\varepsilon\beta/2c_1)^2)} .$$

Proof. See Exercise 10.20. ■

Theorem 10.17. *Assume A10.13. For any $t \in \{0,\dots,n\}$ there exist constants $0 < c_{1,t},\ c_{2,t} < \infty$ such that, for all $N \in \mathbb{N}$, $\varepsilon > 0$, and $h \in \mathbb{F}_b(\mathsf{X},\mathcal{X})$,*

$$\mathbb{P}\left[\left|N^{-1}\sum_{i=1}^{N}\omega_t^i h(X_t^i) - \frac{\phi_{t|t-1}(g_t h)}{\phi_{t-1}(\vartheta_t)}\right| \ge \varepsilon\right] \le c_{1,t}\mathrm{e}^{-c_{2,t}N\varepsilon^2/|h|_\infty^2} , \tag{10.53}$$

$$\mathbb{P}\left[\left|\phi_t^N(h) - \phi_t(h)\right| \ge \varepsilon\right] \le c_{1,t}\mathrm{e}^{-c_{2,t}N\varepsilon^2/\mathrm{osc}^2(h)} , \tag{10.54}$$

where the weighted sample $\{(X_t^i,\omega_t^i)\}_{i=1}^N$ is defined in (10.48).

Proof. We prove (10.53) and (10.54) together by induction on $t \ge 0$. First note that, by construction, the random variables $\{(X_t^i,\omega_t^i)\}_{1\le i\le N}$ are i.i.d. conditionally to the σ-field

$$\mathcal{F}_{t-1}^N := \sigma\{(X_s^i,\omega_s^i); 0 \le s \le t-1, 1 \le i \le N\} . \tag{10.55}$$

The Hoeffding inequality implies

$$\mathbb{P}\left[\left|N^{-1}\sum_{i=1}^{N}\omega_t^i h(X_t^i) - \mathbb{E}\left[N^{-1}\sum_{i=1}^{N}\omega_t^i h(X_t^i)\,\middle|\,\mathcal{F}_{t-1}^N\right]\right| > \varepsilon\right] \le 2\mathrm{e}^{-N\varepsilon^2/(2|w_t/\vartheta_t|_\infty^2 |h|_\infty^2)} . \tag{10.56}$$

For $t = 0$, we have

$$\mathbb{E}\left[N^{-1}\sum_{i=1}^{N}\omega_0^i h(X_0^i)\,\middle|\,\mathcal{F}_{-1}^N\right] = \mathbb{E}\left[\omega_0^1 h(X_0^1)\,\middle|\,\mathcal{F}_{-1}^N\right] = \xi(g_0 h) = \phi_{0|-1}(g_0 h) .$$

Thus, (10.54) follows by Lemma 10.16 applied with $A_N = N^{-1}\sum_{i=1}^N \omega_0^i h(X_0^i)$, $B_N = N^{-1}\sum_{i=1}^N \omega_0^i$, and $B = \beta = \xi(g_0)$ ((10.52a), (10.52b) and (10.52c) are obviously satisfied). For $t \ge 1$, we prove (10.53) by deriving an exponential inequality for $\mathbb{E}\left[N^{-1}\sum_{i=1}^N \omega_t^i h(X_t^i)\,\middle|\,\mathcal{F}_{t-1}^N\right]$ thanks to the induction assumption. It follows from the definition that

$$\mathbb{E}\left[N^{-1}\sum_{i=1}^{N}\omega_t^i h(X_t^i)\,\middle|\,\mathcal{F}_{t-1}^N\right] = \frac{\sum_{i=1}^N \omega_{t-1}^i Q_t h(X_{t-1}^i)}{\sum_{\ell=1}^N \omega_{t-1}^\ell \vartheta_t(X_{t-1}^\ell)} .$$

We apply Lemma 10.16 by successively checking conditions (10.52a), (10.52b) and (10.52c) with

$$\begin{cases} A_N := \phi_{t-1}^N(Q_t h) - \frac{\phi_{t-1}(Q_t h)}{\phi_{t-1}(\vartheta_t)}\phi_{t-1}^N(\vartheta_t) \\ B_N := \phi_{t-1}^N(\vartheta_t) \\ B := \beta := \phi_{t-1}(\vartheta_t) \end{cases}$$

Note that

$$\left|\frac{\phi_{t-1}^N(Q_t h)}{\phi_{t-1}^N(\vartheta_t)}\right| = \left|\mathbb{E}\left[\omega_t^1 h(X_t^1) \,\middle|\, \mathcal{F}_{t-1}^N\right]\right| \le |w_t/\vartheta_t|_\infty \, |h|_\infty ,$$

$$\left|\frac{\phi_{t-1}(Q_t h)}{\phi_{t-1}(\vartheta_t)}\right| = \left|\phi_{t-1}\left[\vartheta_t(\cdot)\int \frac{w_t(\cdot,x)}{\vartheta_t(\cdot)} R_t(\cdot,\mathrm{d}x)h(x)\right] \Big/ \phi_{t-1}(\vartheta_t)\right| \le |w_t/\vartheta_t|_\infty \, |h|_\infty .$$

Thus, condition (10.52a) is satisfied with $c_1 = 2|w_t/\vartheta_t|_\infty \, |h|_\infty$. Now, assume that the induction assumption (10.54) holds where t is replaced by $t-1$. Then, $A_N = \phi_{t-1}^N(H_t)$ where

$$H_t(x) := Q_t h(x) - \frac{\phi_{t-1}(Q_t h)}{\phi_{t-1}(\vartheta_t)} \vartheta_t(x) .$$

By noting that $\phi_{t-1}(H_t) = 0$, exponential inequalities for A_N and $B_N - B$ are then directly derived from the induction assumption under A 10.13. Thus Lemma 10.16 applies and finally (10.53) is proved for $t \ge 1$.

Finally, we must show that (10.53) implies (10.54). Without loss of generality, we assume that $\phi_t(h) = 0$. We then apply Lemma 10.16 with $A_N := N^{-1}\sum_{i=1}^N \omega_t^i h(X_t^i)$, $B_N := N^{-1}\sum_{i=1}^N \omega_t^i$, and $B := \beta := \phi_{t-1}(Q_t \mathbf{1})/\phi_{t-1}(\vartheta_t)$. But, $\phi_t(h) = 0$ implies $\phi_{t-1}(Q_t h) = 0$, so that conditions (10.52a), (10.52b), and (10.52c) follow from (10.53). ■

10.5.2 *Time-uniform bounds*

The results above establish the convergence, as the number of particles N tends to infinity, of the particle filter for a *finite* time horizon $t \in \mathbb{N}$. For *infinite time horizons*, i.e., when t tends to infinity, the convergence is less obvious. Indeed, each recursive update of the weighted particles $\{(X_t^{N,i}, \omega_t^{N,i})\}_{i=1}^N$ is based on the implicit assumption that the empirical measure ϕ_{t-1}^N associated with the ancestor sample approximates perfectly well the filtering distribution ϕ_{t-1} at the previous time step; however, since the ancestor sample is marred by a sampling error itself, one may expect that the errors induced at the different updating steps accumulate and, consequently, that the total error propagated through the algorithm increases with t; this would for example imply that the asymptotic variance $\sigma_t^2(f)$ grows to infinity as $t \to \infty$, which would make the algorithm useless in practice.

Fortunately, as we will show below, the convergence of particle filters can be shown to be *uniform* in time under rather general conditions. To make the presentation simple, we will derive in this Section such stability results under very stringent conditions, but stability can be established for NLSS models under much milder assumptions.

We may decompose the error $\phi_t^N(h) - \phi_t(h)$ as follows

$$\phi_t^N(h) - \phi_t(h) = \underbrace{\phi_t^N(h) - \frac{\phi_{t-1}^N(Q_t h)}{\phi_{t-1}^N(Q_t \mathbf{1})}}_{\text{sampling error}} + \underbrace{\frac{\phi_{t-1}^N(Q_t h)}{\phi_{t-1}^N(Q_t \mathbf{1})} - \frac{\phi_{t-1}(Q_t h)}{\phi_{t-1}(Q_t \mathbf{1})}}_{\text{initialization error}} , \quad (10.57)$$

where we have used that $\phi_t(h) = \phi_{t-1}(Q_t h)/\phi_{t-1}(Q_t \mathbf{1})$. According to (10.57), the error $\phi_t^N(h) - \phi_t(h)$ may be decomposed into the *sampling error* introduced by replacing $\phi_{t-1}(Q_t h)/\phi_{t-1}(Q_t \mathbf{1})$ by its sampling estimate $\phi_t^N(h)$ and the *propagation error* originating from the discrepancy between empirical measure ϕ_{t-1}^N associated with the ancestor particles and the true filter ϕ_{t-1}.

By iterating the decomposition (10.57), the error $\phi_t^N(h) - \phi_t(h)$ may be written as a telescoping sum

$$\phi_t^N(h) - \phi_t(h) = \sum_{s=1}^{t} \left(\frac{\phi_s^N(Q_{s,t}h)}{\phi_s^N(Q_{s,t}\mathbf{1})} - \frac{\phi_{s-1}^N(Q_{s-1,t}h)}{\phi_{s-1}^N(Q_{s-1,t}\mathbf{1})} \right) + \frac{\phi_0^N(Q_{0,t}h)}{\phi_0^N(Q_{0,t}\mathbf{1})} - \frac{\phi_0(Q_{0,t}h)}{\phi_0(Q_{0,t}\mathbf{1})}, \tag{10.58}$$

where $Q_{t,t} = \mathrm{I}$ and for $0 \le s < t$,

$$Q_{s,t}h = Q_{s+1}Q_{s+2}\cdots Q_t h\,. \tag{10.59}$$

To prove such uniform-in-time deviation inequality, we assume that the Markov kernel M satisfies the following *strong mixing condition.*

Assumption A10.18.

(a) For all $(x,y) \in \mathsf{X} \times \mathsf{Y}$, $g(x,y) > 0$ and $\sup_{(x,y)\in \mathsf{X}\times\mathsf{Y}} g(x,y) < \infty$.

(b) $\sup_{t\ge 0}\sup_{x\in\mathsf{X}} \vartheta_t(x) < \infty$ and $\sup_{t\ge 0}\sup_{(x,x')\in\mathsf{X}\times\mathsf{X}} w_t(x,x')/\vartheta_t(x) < \infty$.

(c) There exist constants $\sigma^+ > \sigma^- > 0$ and a probability measure ν on $(\mathsf{X},\mathcal{X})$ such that for all $x \in \mathsf{X}$ and $A \in \mathcal{X}$,

$$\sigma^-\nu(A) \le M(x,A) \le \sigma^+\nu(A)\,. \tag{10.60}$$

(d) There exists a constant $c_- > 0$ such that, $\xi(g_0) \ge c_-$ and for all $t \ge 1$,

$$\inf_{x\in\mathsf{X}} Q_t\mathbf{1}(x) \ge c_- > 0\,. \tag{10.61}$$

Remark 10.19. A10.18-(b) is mild. It holds in particular under A10.18-(a) for the bootstrap filter: in this case, $\vartheta_t(x) \equiv 1$ and

$$w_t(x,x') = g_t(x') \text{ for all } x \in \mathsf{X} \text{ and } t \ge 0.$$

It automatically holds also for the fully adapted auxiliary particle filter: in this case, $\vartheta_t(x) = \int M(x,\mathrm{d}x')g_t(x') \le \sup_{t\ge 0}\sup_{x'\in\mathsf{X}} g_t(x')$ and $w_t(x,x') \equiv 1$, for all $t > 0$ and all $(x,x') \in \mathsf{X}\times\mathsf{X}$.

The key result to prove the uniform in time stability is the following uniform forgetting property.

Proposition 10.20. *Assume A10.18-(c). Then, for all distributions ξ, $\xi' \in \mathbb{M}_1(\mathcal{X})$ and for all $s \le t$ and any bounded measurable functions $h \in \mathbb{F}_b(\mathsf{X},\mathcal{X})$,*

$$\left| \frac{\xi Q_{s,t}h}{\xi Q_{s,t}\mathbf{1}} - \frac{\xi' Q_{s,t}h}{\xi' Q_{s,t}\mathbf{1}} \right| \le \rho^{t-s}\operatorname{osc}(h)\,, \tag{10.62}$$

where $\rho := 1 - \sigma_-/\sigma_+$.

Proof. Consider the Markov kernel defined for $x_s \in \mathsf{X}$ and $A \in \mathcal{X}$,

$$\bar{Q}_{s,t}(x_s, A) = \frac{Q_{s,t}(x_s, A)}{Q_{s,t}\mathbf{1}(x_s)} . \tag{10.63}$$

This kernel is obtained by normalizing the kernel $Q_{s,t}$. By construction, for any $h \in \mathbb{F}(\mathsf{X}, \mathcal{X})$, we get

$$\frac{\xi Q_{s,t} h}{\xi Q_{s,t}\mathbf{1}} = \frac{\int \xi(\mathrm{d}x_s) Q_{s,t}\mathbf{1}(x_s) \bar{Q}_{s,t} h(x_s)}{\int \xi(\mathrm{d}x_s) Q_{s,t}\mathbf{1}(x_s)} = \xi_{s,t} \bar{Q}_{s,t} h , \tag{10.64}$$

where $\xi_{s,t}$ is the probability measure defined as

$$\xi_{s,t}(A) = \frac{\int_A \xi(\mathrm{d}x_s) Q_{s,t}\mathbf{1}(x_s)}{\int_{\mathsf{X}} \xi(\mathrm{d}x_s) Q_{s,t}\mathbf{1}(x_s)} . \tag{10.65}$$

We have

$$\begin{aligned}
\bar{Q}_{s,t}(x_s, A) = \frac{Q_{s,t}(x_s, A)}{Q_{s,t}\mathbf{1}(x_s)} &= \frac{\int Q_{s+1}(x_s, \mathrm{d}x_{s+1}) Q_{s+1,t}\mathbf{1}(x_{s+1}) \bar{Q}_{s+1,t}(x_{s+1}, A)}{Q_{s,t}\mathbf{1}(x_s)} \\
&= R_{s,t} \bar{Q}_{s+1,t}(x_{s+1}, A) ,
\end{aligned}$$

where the Markov kernel $R_{s,t}$ is defined, for any $x_s \in \mathsf{X}$ and $A \in \mathcal{X}$, by

$$R_{s,t}(x_s, A) = \frac{\int_A Q_{s+1}(x_s, \mathrm{d}x_{s+1}) Q_{s+1,t}\mathbf{1}(x_{s+1})}{Q_{s,t}\mathbf{1}(x_s)} . \tag{10.66}$$

By iterating this decomposition, we may represent the Markov kernel $\bar{Q}_{s,t}$ as the product of kernels

$$\bar{Q}_{s,t} = R_{s,t} R_{s+1,t} \dots R_{t-1,t} . \tag{10.67}$$

Using A10.18-(c) the kernel $R_{s,t}$ is uniformly Doeblin: for any $x_s \in \mathsf{X}$ and $A \in \mathcal{X}$, we get

$$\begin{aligned}
R_{s,t}(x_s, A) &= \frac{\int_A Q_{s+1}(x_s, \mathrm{d}x_{s+1}) Q_{s+1,t}\mathbf{1}(x_{s+1})}{\int_{\mathsf{X}} Q_{s+1}(x_s, \mathrm{d}x_{s+1}) Q_{s+1,t}\mathbf{1}(x_{s+1})} \\
&\geq \frac{\sigma_-}{\sigma_+} \nu_{s,t}(A) ,
\end{aligned}$$

where $\nu_{s,t}$ is the probability on $(\mathsf{X}, \mathcal{X})$ given by

$$\nu_{s,t}(A) = \frac{\int_A \nu(\mathrm{d}x_{s+1}) g_{s+1}(x_{s+1}) Q_{s+1,t}\mathbf{1}(x_{s+1})}{\int_{\mathsf{X}} \nu(\mathrm{d}x_{s+1}) g_{s+1}(x_{s+1}) Q_{s+1,t}\mathbf{1}(x_{s+1})} . \tag{10.68}$$

Therefore, using Lemma 6.10, the Dobrushin coefficient $\Delta_{\mathrm{TV}}(R_{s,t}) \leq \rho$ of the Markov kernel $R_{s,t}$ is bounded by $\rho = 1 - \sigma_-/\sigma_+ < 1$ (see Definition 6.4). The submultiplicativity of the Dobrushin coefficient (6.8) and the decomposition (10.67) imply that

$$\Delta_{\mathrm{TV}}\left(\bar{Q}_{s,t}\right) \leq \Delta_{\mathrm{TV}}\left(R_{s,t}\right)\Delta_{\mathrm{TV}}\left(R_{s+1,t}\right)\dots\Delta_{\mathrm{TV}}\left(R_{t-1,t}\right) \leq \rho^{t-s} .$$

For any probability $\xi, \xi' \in \mathbb{M}_1(\mathcal{X})$, (10.64) and Lemma 6.5 imply that

$$\begin{aligned}\left\|\frac{\xi Q_{s,t}}{\xi Q_{s,t}\mathbf{1}} - \frac{\xi' Q_{s,t}}{\xi' Q_{s,t}\mathbf{1}}\right\|_{\mathrm{TV}} &= \left\|\xi_{s,t}\bar{Q}_{s,t} - \xi'_{s,t}\bar{Q}_{s,t}\right\|_{\mathrm{TV}} \\ &\leq \left\|\xi_{s,t} - \xi'_{s,t}\right\|_{\mathrm{TV}}\Delta_{\mathrm{TV}}\left(\bar{Q}_{s,t}\right) \leq \rho^{t-s}\left\|\xi_{s,t} - \xi'_{s,t}\right\|_{\mathrm{TV}} .\end{aligned}$$

The proof follows. ■

Theorem 10.21. *Assume A 10.18. Then, the filtering distribution satisfies a time-uniform exponential deviation inequality, i.e., there exist constants c_1 and c_2 such that, for all integers N and $t \geq 0$, all measurable functions h and all $\varepsilon > 0$,*

$$\mathbb{P}\left[\left|N^{-1}\sum_{i=1}^{N}\omega_t^i h(X_t^{N,i}) - \frac{\phi_{t|t-1}(g_t h)}{\phi_{t-1}(\vartheta_t)}\right| \geq \varepsilon\right] \leq c_1 \mathrm{e}^{-c_2 N\varepsilon^2/|h|_\infty^2} , \tag{10.69}$$

$$\mathbb{P}\left[\left|\phi_t^N(h) - \phi_t(h)\right| \geq \varepsilon\right] \leq c_1 \mathrm{e}^{-c_2 N\varepsilon^2/\operatorname{osc}^2(h)} . \tag{10.70}$$

Proof. We first prove (10.70). Without loss of generality, we will assume that $\phi_t(h) = 0$. Similar to (Del Moral, 2004, Eq. (7.24)), the quantity $\phi_t^N(h)$ is decomposed as,

$$\phi_t^N(h) = \sum_{s=1}^{t}\left(\frac{B_{s,t}(h)}{B_{s,t}(\mathbf{1})} - \frac{B_{s-1,t}(h)}{B_{s-1,t}(\mathbf{1})}\right) + \frac{B_{0,t}(h)}{B_{0,t}\mathbf{1}} , \tag{10.71}$$

where

$$B_{s,t}(h) = N^{-1}\sum_{i=1}^{N}\omega_s^i \frac{Q_{s,t}h(X_s^i)}{|Q_{s,t}\mathbf{1}|_\infty} . \tag{10.72}$$

We first establish an exponential inequality for $B_{0,t}(h)/B_{0,t}\mathbf{1}$ where the dependence in t will be explicitly expressed. For that purpose, we will apply Lemma 10.16 by successively checking Conditions (10.52a), (10.52b), and (10.52c), with $A_N := B_{0,t}(h)$, $B_N := B_{0,t}\mathbf{1}$, and

$$B := \int \xi(\mathrm{d}x_0)g_0(x_0)\frac{Q_{0,t}\mathbf{1}(x_0)}{|Q_{0,t}\mathbf{1}|_\infty} , \quad \beta := \frac{\sigma_-}{\sigma_+}\int \xi(\mathrm{d}x_0)g_0(x_0) .$$

Under the strong mixing condition Equation 10.60, for any $0 \leq s < t$ and $(x,x') \in \mathsf{X}\times\mathsf{X}$, we have

$$\frac{Q_{s,t}\mathbf{1}(x)}{Q_{s,t}\mathbf{1}(x')} = \frac{\int\cdots\int Q_{s+1}(x,\mathrm{d}x_{s+1})\prod_{r=s+2}^{t}Q_r(x_{r-1},\mathrm{d}x_r)}{\int\cdots\int Q_{s+1}(x',\mathrm{d}x_{s+1})\prod_{r=s+2}^{t}Q_r(x_{r-1},\mathrm{d}x_r)} \geq \frac{\sigma_-}{\sigma_+} .$$

Therefore, for any $x \in \mathsf{X}$ and $0 \leq s < t$, it holds that

$$\frac{\sigma_-}{\sigma_+} \leq \frac{Q_{s,t}\mathbf{1}(x)}{|Q_{s,t}\mathbf{1}|_\infty} \leq 1 . \tag{10.73}$$

which implies that $B \geq \beta$. Since $\phi_t(h) = 0$, the forgetting condition (10.62) implies

$$\begin{aligned}\left|\frac{A_N}{B_N}\right| &= \left|\frac{B_{0,t}(h)}{B_{0,t}\mathbf{1}} - \phi_t(h)\right| \\ &= \left|\frac{\sum_{i=1}^N \omega_0^i Q_{0,t}h(X_0^i)}{\sum_{i=1}^N \omega_0^i Q_{0,t}\mathbf{1}(X_0^i)} - \frac{\int \xi(\mathrm{d}x_0)g_0(x_0)Q_{0,t}h(x_0)}{\int \xi(\mathrm{d}x_0)g_0(x_0)Q_{0,t}\mathbf{1}(x_0)}\right| \leq \rho^t \operatorname{osc}(h) \; . \quad (10.74)\end{aligned}$$

This shows condition (10.52a) with $c_1 = \rho^t \operatorname{osc}(h)$. We now turn to condition (10.52b). We have

$$B_N - B = N^{-1}\sum_{i=1}^N \omega_0^i \frac{Q_{0,t}\mathbf{1}(X_0^i)}{|Q_{0,t}\mathbf{1}|_\infty} - \int r_0(\mathrm{d}x_0)w_0(x_0)\frac{Q_{0,t}\mathbf{1}(x_0)}{|Q_{0,t}\mathbf{1}|_\infty} \; .$$

Since $\omega_0^i Q_{0,t}\mathbf{1}(X_0^i)/|Q_{0,t}\mathbf{1}|_\infty \leq |w_0|_\infty$, we have by Hoeffding's inequality

$$\mathbb{P}\left[|B_N - B| \geq \varepsilon\right] \leq 2\exp\left(-2N\varepsilon^2/|w_0|_\infty^2\right) \; .$$

We finally check condition (10.52c). We have

$$A_N = N^{-1}\sum_{i=1}^N \omega_0^i \frac{Q_{0,t}h(X_0^i)}{|Q_{0,t}\mathbf{1}|_\infty} \; .$$

Since $\phi_t(h) = 0$ implies $\int \xi(\mathrm{d}x)g_0(x)Q_{0,t}h(x) = 0$, it holds that $\mathbb{E}\left[A_N\right] = 0$. Moreover,

$$\begin{aligned}\left|\omega_0^i \frac{Q_{0,t}h(X_0^i)}{|Q_{0,t}\mathbf{1}|_\infty}\right| &\leq |w_0|_\infty \left|\frac{Q_{0,t}\mathbf{1}(X_0^i)}{|Q_{0,t}\mathbf{1}|_\infty}\left(\frac{Q_{0,t}h(X_0^i)}{Q_{0,t}\mathbf{1}(X_0^i)} - \phi_t(h)\right)\right| \\ &\leq |w_0|_\infty \left|\frac{\delta_{X_0^i}Q_{0,t}h}{\delta_{X_0^i}Q_{0,t}\mathbf{1}} - \frac{\phi_0 Q_{0,t}h}{\phi_0 Q_{0,t}\mathbf{1}}\right| \leq |w_0|_\infty \rho^t \operatorname{osc}(h) \; ,\end{aligned}$$

using (10.62) and (10.52c) follows from Hoeffding's inequality. Then, Lemma 10.16 implies

$$\mathbb{P}\left[|B_{0,t}(h)/B_{0,t}\mathbf{1}| > \varepsilon\right] \leq b\mathrm{e}^{-cN\varepsilon^2/(\rho^t \operatorname{osc}(h))^2} \; ,$$

where the constants b and c do not depend on t. We now consider for $1 \leq s \leq t$ the difference $B_{s,t}(h)/B_{s,t}(\mathbf{1})) - B_{s-1,t}(h)/B_{s-1,t}(\mathbf{1})$, where $B_{s,t}$ is defined in (10.72). We again use Lemma 10.16 with $\mathbb{P}(\cdot) = \mathbb{P}\left(\cdot \,|\, \mathcal{F}_{s-1}^N\right)$ where $\mathcal{F}_{s-1}^N$ is defined in (10.55) and

$$\begin{cases} A_N = B_{s,t}(h) - \frac{B_{s-1,t}(h)}{B_{s-1,t}(\mathbf{1})}B_{s,t}(\mathbf{1}) \\ B_N = B_{s,t}(\mathbf{1}) \\ B = \frac{\sum_{i=1}^N \omega_{s-1}^i \int Q_s(X_{s-1}^i, \mathrm{d}x)Q_{s,t}\mathbf{1}(x)}{|Q_{s,t}\mathbf{1}|_\infty \sum_{\ell=1}^N \omega_{s-1}^\ell \vartheta_s(X_{s-1}^\ell)} \end{cases}$$

Eq. (10.73) and (10.60) show that

$$B \geq \beta := \frac{c_-\sigma_-}{\sigma_+|\vartheta_s|_\infty} \; ,$$

where σ_- and c_- are defined in (10.60) and (10.61), respectively. In addition, using the forgetting condition (10.62),

$$\left|\frac{A_N}{B_N}\right| = \left|\frac{\sum_{i=1}^N \omega_s^i Q_{s,t} h(X_s^i)}{\sum_{i=1}^N \omega_s^i Q_{s,t} \mathbf{1}(X_s^i)} - \frac{\sum_{i=1}^N \omega_{s-1}^i Q_{s-1,t} h(X_{s-1}^i)}{\sum_{i=1}^N \omega_{s-1}^i Q_{s-1,t} \mathbf{1}(X_{s-1}^i)}\right| \leq \rho^{t-s} \operatorname{osc}(h) , \quad (10.75)$$

showing condition (10.52a) with $c_1 = \rho^{t-s} \operatorname{osc}(h)$. We must now check condition (10.52b). By (10.56), we have

$$B_N - B = N^{-1} \sum_{i=1}^N \omega_s^i \frac{Q_{s,t}\mathbf{1}(X_s^i)}{|Q_{s,t}\mathbf{1}|_\infty} - \mathbb{E}\left[\omega_s^1 \frac{Q_{s,t}\mathbf{1}(X_s^1)}{|Q_{s,t}\mathbf{1}|_\infty} \,\middle|\, \mathcal{F}_{s-1}^N\right] ,$$

where $\mathcal{F}_{s-1}^N$ is defined in (10.55) Thus, since $\left|\omega_s^i Q_{s,t}\mathbf{1}(X_s^i)/|Q_{s,t}\mathbf{1}|_\infty\right| \leq \sup_t |w_t/\vartheta_t|_\infty$, we have by conditional Hoeffding's inequality

$$\mathbb{P}\left(|B_N - B| > \varepsilon \,|\, \mathcal{F}_{s-1}^N\right) \leq 2\mathrm{e}^{-Nc_2\varepsilon^2} ,$$

showing condition (10.52b) with $c_2 = (2\sup_t |w_t/\vartheta_t|_\infty)^{-2}$. Moreover, write $A_N = N^{-1} \sum_{\ell=1}^N \eta_s^\ell$ where

$$\eta_{s,t}^\ell(h) := \omega_s^\ell \frac{Q_{s,t}h(X_s^\ell)}{|Q_{s,t}\mathbf{1}|_\infty} - \frac{\phi_{s-1}^N(Q_{s-1,t}h)}{\phi_{s-1}^N(Q_{s-1,t}\mathbf{1})}\left(\omega_s^\ell \frac{Q_{s,t}\mathbf{1}(X_s^\ell)}{|Q_{s,t}\mathbf{1}|_\infty}\right) . \quad (10.76)$$

Since $\{(X_s^\ell, \omega_s^\ell)\}_{\ell=1}^N$ are i.i.d. conditionally to the σ-field $\mathcal{F}_{s-1}^N$, we have that $\{\eta^\ell\}_{\ell=1}^N$ are also i.i.d. conditionally to $\mathcal{F}_{s-1}^N$. Moreover, it can be easily checked using (10.56) that $\mathbb{E}\left[\eta_{s,t}^1(h) \,\middle|\, \mathcal{F}_{s-1}^N\right] = 0$. In order to apply the conditional Hoeffding inequality, we need to check that η_s^1 is bounded. This follows from (10.62),

$$|\eta_{s,t}^\ell(h)| = \omega_s^\ell \frac{Q_{s,t}\mathbf{1}(X_s^\ell)}{|Q_{s,t}\mathbf{1}|_\infty}\left|\frac{Q_{s,t}h(X_s^\ell)}{Q_{s,t}\mathbf{1}(X_s^\ell)} - \frac{\sum_{i=1}^N \omega_{s-1}^i Q_{s-1,t}h(X_{s-1}^i)}{\sum_{i=1}^N \omega_{s-1}^i Q_{s-1,t}\mathbf{1}(X_{s-1}^i)}\right|$$
$$\leq \sup_t |w_t/\vartheta_t|_\infty \rho^{t-s} \operatorname{osc}(h) .$$

Consequently,

$$\mathbb{P}\left(|A_N| > \varepsilon \,|\, \mathcal{F}_{s-1}^N\right) = \mathbb{P}\left(\left|N^{-1}\sum_{\ell=1}^N \eta_{s,t}^\ell(h)\right| > \varepsilon \,\middle|\, \mathcal{F}_{s-1}^N\right)$$
$$\leq 2\exp\left\{-Nc_3\left(\frac{\varepsilon}{\rho^{t-s}\operatorname{osc}(h)}\right)^2\right\} ,$$

with $c_3 = (2\sup_t |w_t/\vartheta_t|_\infty)^{-2}$. This shows condition (10.52c). Finally by Lemma 10.16,

$$\mathbb{P}\left(\left|\frac{B_{s,t}(h)}{B_{s,t}(\mathbf{1})} - \frac{B_{s-1,t}(h)}{B_{s-1,t}(\mathbf{1})}\right| > \varepsilon \,\middle|\, \mathcal{F}_{s-1}^N\right) \leq 4\exp\left\{-c_3 N\left(\frac{\varepsilon}{\rho^{t-s}\operatorname{osc}(h)}\right)^2\right\} .$$

The proof is concluded by using Lemma 10.22. ■

Lemma 10.22. *Let $\{Y_{n,i}\}_{i=1}^n$ be a triangular array of random variables such that there exist constants $b > 0$, $c > 0$ and ρ, $0 < \rho < 1$ such that, for all n, $i \in \{1,\dots,n\}$ and $\varepsilon > 0$, $\mathbb{P}(|Y_{n,i}| \geq \varepsilon) \leq b\mathrm{e}^{-c\varepsilon^2\rho^{-2i}}$. Then, there exists $\bar{b}$ and $\bar{c}$ such that, for any n and $\varepsilon > 0$,*

$$\mathbb{P}\left(\left|\sum_{i=1}^n Y_{n,i}\right| \geq \varepsilon\right) \leq \bar{b}\mathrm{e}^{-\bar{c}\varepsilon^2} .$$

Proof. Denote by $S := \sum_{i=1}^\infty \sqrt{i}\rho^i$. It is plain to see that

$$\mathbb{P}\left(\left|\sum_{i=1}^n Y_{n,i}\right| \geq \varepsilon\right) \leq \sum_{i=1}^n \mathbb{P}\left(|Y_{n,i}| \geq \varepsilon S^{-1}\sqrt{i}\rho^i\right) \leq b\sum_{i=1}^n \mathrm{e}^{-cS^{-2}\varepsilon^2 i} .$$

Set $\varepsilon_0 > 0$. The proof follows by noting that, for any $\varepsilon \geq \varepsilon_0$,

$$\sum_{i=1}^n \mathrm{e}^{-cS^{-2}i\varepsilon^2} \leq (1-\mathrm{e}^{-cS^{-2}\varepsilon_0^2})^{-1}\mathrm{e}^{-cS^{-2}\varepsilon^2} .$$

■

10.6 Endnotes

Importance sampling was introduced by Hammersley and Handscomb (1965) and has since been used in many different fields; see Glynn and Iglehart (1989), Geweke (1989), Evans and Swartz (1995), or Robert and Casella (2004), and the references therein.

Although the Sequential Importance Sampling (SIS) algorithm has been known since the early 1970s (Handschin and Mayne, 1969 and Handschin, 1970), its use in nonlinear filtering problems remained largely unnoticed until the early 1990s. Clearly, the available computational power was too limited to allow convincing applications of these methods. Another less obvious reason is that the SIS algorithm suffers from a major drawback that was not overcome and properly cured until the seminal papers of Gordon et al. (1993) and Kitagawa (1996). As the number of iterations increases, the importance weights degenerate: most of the particles have very small normalized importance weights and thus do not significantly contribute to the approximation of the target distribution. The solution proposed by Gordon et al. (1993) and Kitagawa (1996) is to rejuvenate the particles by replicating the particles with high importance weights while removing the particles with low weights.

Early applications of the particle filters are described in the book by Kitagawa and Gersch (1996), which included applications of spectral estimation and change points analysis. The collection of papers Doucet et al. (2001) provide a large number of methods and applications of the particle filters. The methodological papers Liu and Chen (1998) [see also the book by Liu (2001)] and Doucet et al. (2000) introduced variants of particle filters; these papers also showed that particle approximations can go far beyond filtering and smoothing problems for time series. The book Ristic et al. (2004) is devoted to the application of tracking. Recent methodological advances are covered in the survey papers by Cappé et al. (2007), Creal (2009) [presenting applications in economics and finance] and Doucet and Johansen (2009).

The convergence of the particle filter (and more generally of the interacting particle approximations of the Feynman-Kac semigroup) have been studied in a series of papers by P. Del Moral and co-authors. Early versions of the central limit theorems have been given in Del Moral and Guionnet (1999). Deviation inequalities are reported in Del Moral and Guionnet (1998). The book Del Moral (2004), which extends the survey paper Del Moral and Miclo (2000), provides a thorough coverage of the theoretical properties of sequential Monte Carlo algorithms. Recent theoretical results are presented in the survey papers Del Moral et al. (2011) and Del Moral et al. (2010). More elementary approaches of the convergence of the particle filter are presented in Chopin (2004), Künsch (2005), Cappé et al. (2005) and Douc and Moulines (2008).

The auxiliary particle filter was introduced in the work by Pitt and Shephard (1999). The consistency and asymptotic normality of the auxiliary particle filter is discussed in Douc et al. (2009b) and Johansen and Doucet (2008), which shows that the auxiliary particle filter can be seen as a particular instance of the Feynman-Kac formula. The concentration properties of interacting particle systems are studied in Del Moral et al. (2010). Non-asymptotic bounds for the auxiliary particle filter are given in Douc et al. (2010).

Exercises

10.1 (Some properties of the IS estimator). Let $\mu \in \mathbb{M}_1(\mathcal{X})$ be a target distribution and $\nu \in \mathbb{M}_1(\mathcal{X})$ be a proposal distribution. Assume that $\mu \ll \nu$ and denote $w = \mathrm{d}\mu/\mathrm{d}\nu$. Let $f \in \mathbb{F}(\mathsf{X}, \mathcal{X})$ be a function such that $\mu(|f|) < \infty$ and $\int f^2(x)w(x)\mu(\mathrm{d}x) < \infty$. Let $\{X^i\}_{i=1}^N$ be a sequence of i.i.d. random variables from ν. Denote by $\hat{\mu}_N(f) = N^{-1}\sum_{i=1}^N f(X^i)w(X^i)$ the importance sampling estimator.

(a) Show that $\hat{\mu}_N(f)$ is an unbiased estimator of $\mu(f)$.

(b) Show that $\hat{\mu}_N(f)$ is strongly consistent.

(c) Show that

$$\mathrm{Var}_\nu(fw) = [\mu(f)]^2\, \nu\left[\left(\frac{|f|w}{\mu(|f|)} - 1\right)^2\right].$$

(d) Under which condition does the importance sampling estimator have lower variance than the naive Monte Carlo estimator?

(e) If we choose $\nu(\mathrm{d}x) = |f(x)|\mu(\mathrm{d}x)/\mu(|f|)$, show that the variance of the IS estimator is always smaller than the variance of the naive Monte Carlo estimator.

(f) Assume that f is nonnegative. Show that the proposal distribution may be chosen in such a way that $\mathrm{Var}_\nu(fw) = 0$.

(g) Explain why this choice of the proposal distribution is only of theoretical interest.

(h) Assume that $\mathrm{Var}_\nu(fw) > 0$. Show that $\hat{\mu}_N(f)$ is asymptotically Gaussian,

$$\sqrt{N}(\hat{\mu}_N(f) - \mu(f)) \xrightarrow{\mathbb{P}} \mathrm{N}(0, \mathrm{Var}_\nu(fw)) \quad \text{as } N \to \infty\,.$$

10.2. Importance sampling is relevant to approximating a tail probability $\mathbb{P}(X \geq x)$. If the random variable X with density $g(x)$ has cumulant generating function $\kappa_X(t)$, then tilting by t gives a new density $h_t(x) = \mathrm{e}^{xt-\kappa_X(t)}g(x)$.

(a) Show that the importance weight associated with an observation X drawn from $h_t(x)$ is $w_t(x) = \mathrm{e}^{-xt+\kappa_X(t)}$.

(b) If $\mathbb{E}_t$ denotes expectation with respect to $h_t(x)$, show that the optimal tilt minimizes the second moment

$$\mathbb{E}_t[\mathbb{1}_{\{X\geq x\}}\mathrm{e}^{-2Xt+2\kappa_X(t)}] = \mathbb{E}_0[\mathbb{1}_{\{X\geq x\}}\mathrm{e}^{-Xt+\kappa_X(t)}] \leq \mathrm{e}^{-xt+\kappa_X(t)} .$$

(c) It is far simpler to minimize the displayed bound than the second moment. Show that the minimum of the upper bound is attained when $K'(t) = x$.

(d) Assume that X is normally distributed with mean μ and variance σ^2. Show that the cumulant generating function is $\kappa_X(t) = \mu t + \frac{1}{2}\sigma^2 t^2$.

(e) For a given x, show that a good tilt is therefore $t = (x-\mu)/\sigma^2$.

(f) Calculate the cumulant generating functions $\kappa_X(t)$ of the exponential and Poisson distributions. Solve the equation $\kappa_X(t) = x$ for t.

(g) Suppose X follows a standard normal distribution. Write and test a program to approximate the right-tail probability $\mathbb{P}(X \geq x)$ by tilted importance sampling. Assume that x is large and positive.

10.3. Assume that the target $\mu = \mathrm{C}(0,1)$ is a standard Cauchy distribution, and the instrumental distribution $\nu = \mathrm{N}(0,1)$ is a standard Gaussian distribution.

(a) Show that the importance weight function is given by

$$w(x) = \sqrt{2\pi}\,\frac{\exp(x^2/2)}{\pi(1+x^2)} .$$

(b) Show that

$$\frac{1}{\sqrt{2\pi}}\int_{-\infty}^{\infty} w^2(x)\exp(-x^2/2)\,\mathrm{d}x = \infty ,$$

and conclude that the IS estimator is consistent but does not converge at an asymptotic rate $\sqrt{n}$.

(c) Plot the sample quantiles of the estimator of $\mu(f)$ versus the quantiles of a standard normal distribution when $f(x) = \exp(-|x|)$.

10.4. Assume that the target distribution is a standard normal $\mu = \mathrm{N}(0,1)$ and that the proposal distribution is Cauchy $\nu = \mathrm{C}(0,1)$.

(a) Show that the importance weight is bounded by $\sqrt{2\pi/\mathrm{e}}$.

(b) Using Exercise 10.1, show that $\sqrt{N}(\hat{\mu}_N(f) - \mu(f))$ is asymptotically normal, where $f(x) = \exp(-|x|)$ and $\hat{\mu}_N(f)$ is the importance sampling estimator of $\mu(f)$. can be applied.

(c) Plot the sample quantiles of the IS estimator $\hat{\mu}_N(f)$ versus the quantile of the standard Gaussian distribution with ($N = 50, 100, 1000$).

(d) Assume now that $\nu = \mathrm{C}(0,\sigma)$ where $\sigma > 0$ is the scale parameter. Show that the importance weight function is bounded by $(\sqrt{2\pi}/\mathrm{e}\sigma)\mathrm{e}^{\sigma^2/2}$, $\sigma < \sqrt{2}$, $\sigma\sqrt{\pi/2}$, $\sigma \geq \sqrt{2}$.

(e) Show that the upper bound on the importance weight has a minimum at $\sigma = 1$.

(f) Argue that the choice $\sigma = 1$ leads to estimators that are better behaved than for $\sigma = 0.1$ and $\sigma = 10$.

10.5. Let f be a measurable function such that $\mu(|f|) < \infty$. Assume that $\mu \ll \nu$ and let $X^1, X^2, \ldots,$ be an i.i.d. sequence with distribution ν.

(a) Show that the self-normalized IS estimator $\hat{\mu}_N(f)$ given by (10.3) is a strongly consistent sequence of estimators of $\mu(f)$.

(b) Assume in addition that f satisfies $\int [1+f^2]w^2 \mathrm{d}\nu < \infty$. Show that the self-normalized IS estimator is asymptotically Gaussian: $\sqrt{N}\left(\hat{\mu}_N(f) - \mu(f)\right) \stackrel{\mathbb{P}}{\Longrightarrow} \mathrm{N}\left(0, \sigma^2(\nu, f)\right)$ where $\sigma^2(\nu, f) = \int w^2\{f - \mu(f)\}^2 \mathrm{d}\nu$.

(c) Show that the empirical variance $\hat{\sigma}_N^2(\nu, f)$ given by

$$\hat{\sigma}_N^2(\nu, f) = N\frac{\sum_{i=1}^N (f(X^i) - \hat{\mu}_N(f))^2 w^2(X^i)}{\left(\sum_{i=1}^N w(X^i)\right)^2},$$

is a consistent sequence of estimators of $\sigma^2(\nu, f)$.

(d) Construct an asymptotically valid confidence interval for the self-normalized IS.

10.6. Let μ (known up to a normalizing constant) and ν, the proposal distribution, be probability distributions on $(\mathsf{X}, \mathcal{X})$. Suppose that for some function $w \in \mathrm{L}^1(\nu)$, we have

$$\mu(f) = \nu(wf)/\nu(w). \tag{10.77}$$

Consider the weighted sample $\{(X^{N,i}, w(X^{N,i}))\}_{i=1}^N$, where for each N, $\{X^{N,i}\}_{i=1}^N$ are i.i.d. distributed according to ν.

(a) Show that, for any $\varepsilon, C > 0$,

$$N^{-1}\mathbb{E}\left[\max_{1 \leq i \leq N} \omega^{N,i}\right] \leq \varepsilon + N^{-1}\sum_{i=1}^N \mathbb{E}\left[\omega^{N,i}\mathbb{1}\{\omega^{N,i} \geq \varepsilon N\}\right] \leq \varepsilon + \nu\left(w\mathbb{1}\{w \geq C\}\right)$$

for all $N \geq C/\varepsilon$.

(b) Show that $\nu\left(w\mathbb{1}\{w \geq C\}\right)$ converges to zero as $C \to \infty$.

(c) Show that the weighted sample $\{(X^{N,i}, w(X^{N,i}))\}_{i=1}^N$ is consistent for μ.

10.7 (Importance sampling; Exercise 10.6, cont.). We use the notations and the assumptions of Exercise 10.6. Assume in addition that $\nu(w^2) < \infty$. For $f \in \mathbb{F}_b(\mathsf{X}, \mathcal{X})$, define $S_N(f) := \sum_{i=1}^N (w(X^{N,i})/\Omega^N)[f(X^{N,i}) - \mu(f)]$.

(a) Show that $\Omega^N/N \to_{\mathbb{P}} \nu(w)$ and $N^{1/2}S_N(f) \stackrel{\mathbb{P}}{\Longrightarrow} S$, where S is Gaussian random variable with zero-mean and variance

$$\sigma^2(f) := \nu\left\{w^2[f - \mu(f)]^2\right\}. \tag{10.78}$$

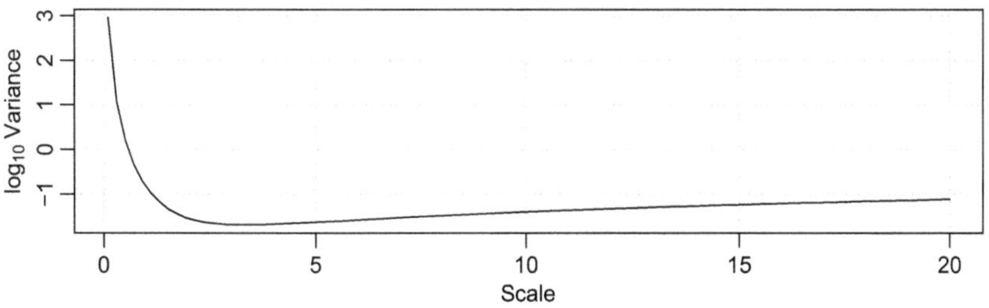

Figure 10.8 *Variance of the importance sampling estimator of the mean as a function of the scale of the t-distribution.*

(b) Show that for any $f \in \mathbb{F}_b(\mathsf{X}, \mathcal{X})$,

$$N \sum_{i=1}^{N} \left(\frac{\omega^{N,i}}{\Omega^N} \right)^2 f(X^{N,i}) = N \sum_{i=1}^{N} \left(\frac{w(X^{N,i})}{\Omega^N} \right)^2 f(X^{N,i}) \xrightarrow{\mathbb{P}} \zeta(f) = \nu \left[w^2 f \right] ,$$

(c) Show that, for any $\varepsilon, C > 0$,

$$N^{-1} \mathbb{E} \left[\max_{1 \le i \le N} \left(\omega^{N,i} \right)^2 \right] \le \varepsilon^2 + N^{-1} \sum_{i=1}^{N} \mathbb{E} \left[\left(\omega^{N,i} \right)^2 \mathbb{1}\{ \left(\omega^{N,i} \right)^2 \ge \varepsilon^2 N \} \right]$$
$$\le \varepsilon^2 + \nu \left(w^2 \mathbb{1}\{ w^2 \ge C \} \right)$$

for $N \ge C/\varepsilon$.

(d) Show that $\nu \left(w^2 \mathbb{1}\{ w \ge C \} \right)$ goes to zero as $C \to \infty$.

(e) Show that $(N^{1/2}/\Omega^N) \max_{1 \le i \le N} \omega^{N,i} \to_{\mathbb{P}} 0$.

(f) Show that $\{(X^{N,i}, w(X^{N,i}))\}_{i=1}^{N}$ is asymptotically normal for μ.

10.8 (Variance of the IS; Exercise 10.7, cont.). (a) Compute the variance of the importance sampling estimator of the mean of a Gaussian mixture using a Student-t distribution with 4 degrees of freedom for different values of the scale.

(b) Explain Figure 10.8.

10.9 (Noisy AR(1) model). Consider an AR(1) model observed in additive noise

$$X_t = \phi X_t + \sigma_W W_t , \qquad W_t \sim \mathrm{N}(0,1) ,$$
$$Y_t = X_t + \sigma_V V_t , \qquad V_t \sim \mathrm{N}(0,1) ,$$

where $|\phi| < 1$ and $\{W_t,\, t \in \mathbb{N}\}$ and $\{V_t,\, t \in \mathbb{N}\}$ are independent Gaussian white noise processes. The initial distribution ξ is the stationary distribution of the Markov chain $\{X_t,\, t \in \mathbb{N}\}$, that is, normal with zero mean and variance $\sigma_W^2/(1-\phi^2)$.

(a) Implement the particle filter with the prior kernel.

(b) Show that the optimal kernel has a density given by

$$\mathrm{N}\left(\frac{\sigma_W^2 \sigma_V^2}{\sigma_W^2 + \sigma_V^2} \left\{ \frac{\phi x}{\sigma_W^2} + \frac{Y_t}{\sigma_V^2} \right\} , \frac{\sigma_W^2 \sigma_V^2}{\sigma_W^2 + \sigma_V^2} \right) .$$

(c) Show that the weight function is

$$Q_t(x, \mathsf{X}) \propto \exp\left[-\frac{1}{2}\frac{(Y_t - \phi x)^2}{\sigma_W^2 + \sigma_V^2}\right] .$$

(d) Implement a particle filter with systematic resampling and the optimal kernel.

(e) Compare the prior and the optimal kernels when $\sigma_W \gg \sigma_V$ and $\sigma_V \ll \sigma_W$.

(f) Illustrate your conclusions by a numerical example.

10.10 (Stochastic volatility). Consider the following stochastic volatility model

$$X_t = \phi X_{t-1} + \sigma W_t , \quad W_t \sim \mathrm{N}(0,1) ,$$
$$Y_t = \beta \exp(X_t/2) V_t , \quad V_t \sim \mathrm{N}(0,1) .$$

(a) Show that

$$m(x,x') = \frac{1}{\sqrt{2\pi\sigma^2}} \exp\left[-\frac{(x' - \phi x)^2}{2\sigma^2}\right] ,$$
$$g(x', Y_t) = \frac{1}{\sqrt{2\pi\beta^2}} \exp\left[-\frac{Y_t^2}{2\beta^2}\exp(-x') - \frac{1}{2}x'\right] .$$

(b) Determine the optimal kernel $p_t^\star$ and the associated importance weight.

(c) Show that the function $x' \mapsto \ln(m(x,x')g(x',Y_t))$ is concave.

(d) Show that the mode $m_t(x)$ of $x' \mapsto p_t^\star(x,x')$ is the unique solution of the non-linear equation

$$-\frac{1}{\sigma^2}(x' - \phi x) + \frac{Y_t^2}{2\beta^2}\exp(-x') - \frac{1}{2} = 0 .$$

(e) Propose and implement a numerical method to find this maximum.

10.11 (Stochastic volatility, cont.). We consider proposal kernel t-distribution with $\eta = 5$ degrees of freedom, with location $m_{t-1}(x)$ and scale $\sigma_t(x)$ set as square-root of minus the inverse of the second-order derivative of $x' \mapsto (\ln m(x,x')g(x',Y_t))$ evaluated at the mode $m_t(x)$.

(a) Show that

$$\sigma_{t-1}^2(x) = \left\{\frac{1}{\sigma^2} + \frac{Y_t^2}{2\beta^2}\exp\left[-m_{t-1}(x)\right]\right\}^{-1} .$$

(b) Show that the incremental importance weight is given by

$$\frac{\exp\left[-\frac{(x'-\phi x)^2}{2\sigma^2} - \frac{Y_t^2}{2\beta^2}\exp(-x') - \frac{x'}{2}\right]}{\sigma_{t-1}^{-1}(x)\left\{\eta + \frac{[x'-m_{t-1}(x)]^2}{\sigma_{t-1}^2(x)}\right\}^{-(\eta+1)/2}} .$$

(c) Implement in R a particle filter with systematic resampling using this proposal kernel.

(d) Compare numerically this implementation and the particle filter with systematic resampling and the prior kernel.

10.12. We adopt the notation of Theorem 10.12. We decompose the bootstrap filter into two steps. Starting from the equally weighted sample $\{(X_{t-1}^{N,i},1)\}_{i=1}^N$ we first form an intermediate sample $\{(\tilde{X}_t^{N,i}, w(X^{N,i},\tilde{X}^{N,i}))\}_{i=1}^N$ where $\{\tilde{X}_t^{N,i}\}_{i=1}^N$ are sampled conditionally independently from the proposal kernel $\tilde{X}_t^{N,i} \sim R_t(X_{t-1}^{N,i},\cdot)$.

(a) Show that $N^{-1}\sum_{i=1}^N w(X_{t-1}^{N,i},X_t^{N,i}) \rightarrow_{\mathbb{P}} \phi_{t-1}(Q_t 1)$.

(b) Show that

$$N^{-1/2}\sum_{i=1}^{N} w(X_{t-1}^{N,i},\tilde{X}_t^{N,i}) f(\tilde{X}_t^{N,i})$$
$$= N^{-1/2}\sum_{i=1}^{N} w(X_{t-1}^{N,i},\tilde{X}_t^{N,i}) f(\tilde{X}_t^{N,i}) - Q_t f(X_{t-1}^{N,i}) + N^{-1/2}\sum_{i=1}^{N} Q_t f(X_{t-1}^{N,i}) .$$

(c) Show that $\phi_{t-1} Q_t[f-\phi_t(f)] = 0$ and that

$$N^{-1/2}\sum_{i=1}^{N}\sum_{i=1}^{N} Q_t[f(X_{t-1}^{N,i}) - \phi_t(f)] \stackrel{\mathbb{P}}{\Longrightarrow} \mathrm{N}(0, \sigma_{t-1}^2(Q_t[f-\phi_t(f)])) .$$

(d) Show that the weighted sample $\{(\tilde{X}_t^{N,i}, w(X_{t-1}^{N,i},\tilde{X}_t^{N,i}))\}_{i=1}^N$ is asymptotically normal. Compute the asymptotic variance.

(e) Prove Theorem 10.12.

10.13 (Multinomial sampling). The *ball-in-urn* method of generating multinomial variables proceeds in two stages. In an initialization phase, the cumulative probability vector is generated, where $p_*^i = \sum_{j=1}^i \omega^i/\Omega^N$, for $i \in \{1,\ldots,N\}$. In the generation phase, N uniform random variables $\{U^i\}_{i=1}^N$ on $[0,1]$ are generated and the indices $I_i = \sup\{j \in \{1,\ldots,N\}, U^i \geq p_*^j\}$ are then computed.

(a) Show that the average number of comparisons to sample N multinomial is $N\log_2 N$ comparisons.

(b) Denote by $\{U^{(i)}\}_{i=1}^N$ the ordered uniforms. Prove that starting from $\{U^{(i)}\}_{i=1}^N$ the number of comparisons required to generate $\{I_i\}_{i=1}^N$ is at most N, worst case, only N comparisons.

(c) Show that the increments $S^i = U^{(i)} - U^{(i-1)}, i \in \{1,\ldots,N\}$, (where by convention $S_1 = U_{(1)}$), referred to as the *uniform spacings*, are distributed as

$$\frac{E^1}{\sum_{i=1}^{N+1} E^i}, \ldots, \frac{E^N}{\sum_{i=1}^{N+1} E^i} ,$$

where $\{E^i\}_{i=1}^N$ is a sequence of i.i.d. exponential random variables.

(d) Propose an algorithm to sample N multinomial with a complexity growing only linearly with N.

Of course, much better algorithms are available; see Davis (1993). An efficient algorithm is implemented in R with a call to the built-in function `rmultinom()`, wrapped to repeat the indices.

10.14. (a) Show that the resampling procedures (10.42) and (10.43) are unbiased, i.e., for any $f \in \mathbb{F}_b(\mathsf{X}, \mathcal{X})$,

$$N^{-1} \sum_{i=1}^{N} \mathbb{E}\left[f(\tilde{X}^{N,i}) \,\middle|\, \mathcal{F}^N \right] = \sum_{i=1}^{N} \omega^{N,i}/\Omega^N f(X^{N,i}) \,.$$

(b) Show that

$$N^{-1} \sum_{i=1}^{N} \mathbb{E}\left[|f(\tilde{X}^{N,i})| \mathbb{1}_{\{|f(\tilde{X}^{N,i})| \geq C\}} \,\middle|\, \mathcal{F}^N \right] \\ = \sum_{i=1}^{N} \frac{\omega^{N,i}}{\Omega^N} |f(X^{N,i})| \mathbb{1}_{\{|f(X^{N,i})| \geq C\}} \xrightarrow{\mathbb{P}} \nu(|f| \mathbb{1}_{\{|f| \geq C\}}) \,. \quad (10.79)$$

(c) For any $i = 1, \ldots, N$, put $U_{N,i} := N^{-1} f(\tilde{X}^{N,i})$. Show that

$$\sum_{i=1}^{N} \mathbb{E}\left[|U_{N,i}| \,\middle|\, \mathcal{F}^N \right] = N^{-1} \sum_{i=1}^{N} \mathbb{E}\left[|f(\tilde{X}^{N,i})| \,\middle|\, \mathcal{F}^N \right] \xrightarrow{\mathbb{P}} \nu(|f|) < \infty \,,$$

(d) Show that for any $\varepsilon > 0$ and $C < \infty$, we have for all sufficiently large N,

$$\sum_{i=1}^{N} \mathbb{E}\left[|U_{N,i}| \mathbb{1}_{\{|U_{N,i}| \geq \varepsilon\}} \,\middle|\, \mathcal{F}^N \right] = \frac{1}{N} \sum_{i=1}^{N} \mathbb{E}\left[|f(\tilde{X}^{N,i})| \mathbb{1}_{\{|f|(\tilde{X}^{N,i}) \geq \varepsilon N\}} \,\middle|\, \mathcal{F}^N \right] \\ \leq N^{-1} \sum_{i=1}^{N} \mathbb{E}\left[|f(\tilde{X}^{N,i})| \mathbb{1}_{\{|f|(\tilde{X}^{N,i}) \geq C\}} \,\middle|\, \mathcal{F}^N \right] \xrightarrow{\mathbb{P}} \mu\left(|f| \mathbb{1}_{\{|f| \geq C\}} \right) \,.$$

(e) Conclude.

10.15. Assume that $\{(X^{N,i}, \omega^{N,i})\}_{i=1}^N$ is adapted to $\mathcal{F}^N$, consistent for ν, and asymptotically normal for (ν, σ, ζ). Let $\{(\tilde{X}^{N,i}, 1)\}_{i=1}^N$ denote the equally weighted sample obtained by multinomial resampling; see (10.42). Let $f \in \mathbb{F}_b(\mathsf{X}, \mathcal{X})$.

(a) Show that $N^{-1} \sum_{i=1}^N f(\tilde{X}^{N,i}) - \nu(f) = A_N + B_N$ where

$$A_N = \sum_{i=1}^{N} \omega^{N,i}/\Omega^N \{ f(X^{N,i}) - \nu(f) \} \,,$$

$$B_N = N^{-1} \sum_{i=1}^{N} \{ f(\tilde{X}^{N,i}) - \mathbb{E}\left[f(\tilde{X}^{N,i}) \,\middle|\, \mathcal{F}^N \right] \} \,.$$

Prove that $N^{1/2} A_N \overset{\mathbb{P}}{\Longrightarrow} \mathrm{N}\left(0, \sigma^2(f)\right)$.

(b) Set $U_{N,i} := N^{-1/2} f(\tilde{X}^{N,i})$. Show that

$$\sum_{j=1}^{N} \{\mathbb{E}\left[U_{N,j}^2 \,\middle|\, \mathcal{F}^N\right] - (\mathbb{E}\left[U_{N,j} \,\middle|\, \mathcal{F}^N\right])^2\} =$$

$$N^{-1} \left(\sum_{i=1}^{N} \frac{\omega^{N,i}}{\Omega^N} f^2(X^{N,i}) - \left\{ \sum_{i=1}^{N} \frac{\omega^{N,i}}{\Omega^N} f(X^{N,i}) \right\}^2 \right) \xrightarrow{\mathbb{P}} (\nu(f^2) - \{\nu(f)\}^2) .$$

(c) Pick $\varepsilon > 0$. For any $C > 0$, and N sufficiently large, show that

$$\sum_{j=1}^{N} \mathbb{E}\left[U_{N,j}^2 \mathbb{1}_{\{|U_{N,j}| \geq \varepsilon\}} \,\middle|\, \mathcal{F}^{N,j-1} \right] \leq \sum_{i=1}^{N} \frac{\omega^{N,i}}{\Omega^N} f^2(X^{N,i}) \mathbb{1}\{|f(X^{N,i})| \geq C\}$$
$$\xrightarrow{\mathbb{P}} \nu(f^2 \mathbb{1}\{|f| \geq C\}) .$$

(d) Show that $\{(\tilde{X}^{N,i}, 1)\}_{i=1}^{N}$ obtained using (10.42) is asymptotically normal for $(\nu, \tilde{\sigma}, \tilde{\zeta})$ with $\tilde{\sigma}^2(f) = \mathrm{Var}_\nu(f) + \sigma^2(f)$ and $\tilde{\zeta} = \nu$.

10.16 (Proof of Theorem 10.14). Assume A10.13. Let $f \in \mathbb{F}_+(\mathsf{X}, \mathcal{X})$. Without loss of generality, put $\phi_t(f) = 0$, and let $N^{-1} \sum_{i=1}^{N} \omega_t^i f(X_t^i) = \frac{1}{N}(A_t^N + B_t^N)$, where

$$A_t^N = \mathbb{E}\left[\omega_t^1 f(X_t^1) \,\middle|\, \mathcal{F}_{t-1}^N\right] ,$$
$$B_t^N = \frac{1}{N} \sum_{i=1}^{N} \{\omega_t^i f(X_t^i) - \mathbb{E}\left[\omega_t^i f(X_t^i) \,\middle|\, \mathcal{F}_{t-1}^N\right]\} ,$$

where $\mathcal{F}_{t-1}^N$ is defined in (10.55).

(a) Show that

$$A_t^N = \sum_{i=1}^{N} \frac{\omega_{t-1}^i}{\sum_{j=1}^{N} \omega_{t-1}^j \vartheta_t(X_{t-1}^i)} Q_t f(X_{t-1}^i) .$$

(b) By applying the induction assumption, show that

$$\sqrt{N} \sum_{i=1}^{N} \frac{\omega_{t-1}^i}{\Omega_{t-1}^N} Q_t f(X_{t-1}^i) \stackrel{\mathbb{P}}{\Longrightarrow} \mathrm{N}(0, \sigma_{t-1}^2(Q_t f)) .$$

(c) Show that

$$\sum_{i=1}^{N} \frac{\omega_{t-1}^i}{\Omega_{t-1}^N} \vartheta_t(X_{t-1}^i) \xrightarrow{\mathbb{P}} \phi_{t-1}(\vartheta_t) .$$

(d) Deduce that $\sqrt{N} A_t^N \stackrel{\mathbb{P}}{\Longrightarrow} \mathrm{N}(0, \{\phi_{t-1}(\vartheta_t)\}^{-2} \sigma_{t-1}^2(Q_t f))$.

(e) Show that

$$N\mathbb{E}\left[\{B_t^N\}^2 \,\middle|\, \mathcal{F}_{t-1}^N\right] = \mathbb{E}\left[(\omega_t^1 f(X_t^1))^2 \,\middle|\, \mathcal{F}_{t-1}^N\right] - \left(\mathbb{E}\left[\omega_t^1 f(X_t^1) \,\middle|\, \mathcal{F}_{t-1}^N\right]\right)^2 .$$

(f) Show that

$$\mathbb{E}\left[(\omega_t^1 f(X_t^1))^2 \mid \mathcal{F}_{t-1}^N\right] \xrightarrow{\mathbb{P}} \alpha_t^2 = \iint \frac{\phi_{t-1}(\mathrm{d}x)}{\phi_{t-1}(\vartheta_t)} \frac{w_t^2(x,x')}{\vartheta_t(x)} Q_t(x,\mathrm{d}x') f^2(x') \, .$$

(g) Show that $\mathbb{E}\left[\omega_t^1 f(X_t^1) \mid \mathcal{F}_{t-1}^N\right] \xrightarrow{\mathbb{P}} 0$, $\sqrt{N}B_t^N \stackrel{\mathbb{P}}{\Longrightarrow} \mathrm{N}(0,\alpha_t^2)$, and that $\sqrt{N}A_t^N$ and $\sqrt{N}B_t^N$ are asymptotically independent.

(h) Show that $\Omega_t^N/N \xrightarrow{\mathbb{P}} \phi_{t-1}(Q_t 1)/\phi_{t-1}(\vartheta_t)$ and conclude.

10.17. Assume A10.13 and that $\vartheta_s \equiv 1$ for all $t \in \{1,\ldots,n\}$ and $Q_t \equiv M$. Show that for any $t \in \{0,\ldots,n\}$ and any $f \in \mathbb{F}_b(\mathsf{X},\mathcal{X})$, $\sqrt{N}(\phi_t^N(f) - \phi_t(f))$ is asymptotically normal with zero mean and variance

$$\sigma_t^2(f) = \sum_{s=1}^{t} \frac{\mathrm{Var}_{\phi_s}\left\{Q_{s,t}[f-\phi_t(f)]\right\}}{\left[\phi_s Q_{s,t}(\mathbf{1})\right]^2} + \frac{\sigma_0^2\left\{Q_{0,t}[f-\phi_t(f)]\right\}}{[\phi_0 Q_{0,t}(\mathbf{1})]^2} \, ,$$

where $Q_{s,t}$ is defined in (10.59). Assume A10.18. Show that $\sup_{t\geq 0}\sigma_t^2(f) < \infty$.

10.18. In this exercise, we provide an alternate proof of the central limit theorem for the auxiliary particle filter, based on the decomposition (10.58). This proof allows us to obtain an explicit expression of the asymptotic variance. Assume A10.13. Set for $s \in \{0,\ldots,t\}$, $W_{s,t}^N(h) = N^{-1/2}\sum_{\ell=1}^N \eta_{s,t}^{N,\ell}(h)$ where, for $s \in \{1,\ldots,t\}$,

$$\eta_{s,t}^{N,\ell}(h) := \omega_s^\ell \left\{ Q_{s,t}h(X_s^\ell) - \frac{\phi_{s-1}^N(Q_{s-1,t}h)}{\phi_{s-1}^N(Q_{s-1,t}\mathbf{1})} Q_{s,t}\mathbf{1}(X_s^\ell) \right\} ,$$

and for $s=0$,

$$\eta_{0,t}^N := \omega_0^\ell \left\{ Q_{0,t}h(X_s^\ell) - \frac{\phi_0(Q_{0,t}h)}{\phi_0^N(Q_{0,t}\mathbf{1})} Q_{0,t}\mathbf{1}(X_0^\ell) \right\} .$$

Denote $\alpha_{s,t}^N = \{N^{-1}\sum_{\ell=1}^N Q_{s,t}\mathbf{1}(X_s^\ell)\}^{-1}$.

(a) Show that $\sqrt{N}\{\phi_t^N(h) - \phi_t(h)\} = \sum_{s=1}^t \alpha_{s,t}^N W_{s,t}^N(h)$.

(b) Show that, for all $s \in \{0,\ldots,t\}$, $\{\eta_{s,t}^\ell(h)\}_{\ell=1}^N$ are zero-mean and i.i.d. conditionally on $\mathcal{F}_{s-1}^N$, defined in (10.55).

(c) Show that, for all $s \in \{0,\ldots,t\}$, $W_{s,t}^N(h)$ converges in distribution to a zero-mean Gaussian variable, conditionally independent of $\mathcal{F}_{s-1}^N$.

(d) Show that the $[W_{0,t}^N(h), W_{1,t}^N(h),\ldots,W_{t,t}^N(h)]$ converges in distribution to a zero-mean random vector with diagonal covariance matrix.

(e) Show that, for $s \in \{1,\ldots,t\}$, $\alpha_{s,t}^N \xrightarrow{\mathbb{P}} \phi_{s-1}(\vartheta_s)/\phi_s(\mathbf{1})$.

(f) Show that, $\sqrt{N}[(\phi_0^N(h) - \phi_0(h)),\ldots,(\phi_n^N(h) - \phi_n(h))]$ converges to a multivariate Gaussian distribution with zero mean and covariance matrix Γ_n.

10.19. Show that, for any $\nu \in \mathbb{M}_1(\mathcal{X})$ and any functions $f, g \in \mathbb{F}_+(\mathsf{X}, \mathcal{X})$,

$$\left(\int \nu(\mathrm{d}x) f^{1/2}(x)\right)^2 \leq \int \nu(\mathrm{d}x) g(x) \int \nu(\mathrm{d}x) \frac{f(x)}{g(x)} .$$

Establish Eq. (10.51).

10.20. We use the notations and definitions of Lemma 10.16. Show that: $|A_N/B_N| \leq B^{-1}|A_N/B_N||B - B_N| + B^{-1}|A_N|$. Conclude.

10.21. We consider the problem of estimating, for some given measurable target function f, the expectation $\pi(f)$, where $\pi = \sum_{i=1}^d \omega_i \mu_i$. $\omega_i \geq 0$, $\sum_{i=1}^d \omega_i = 1$ and $\{\mu_i\}_{i=1}^N \in \mathbb{M}_1(\mathcal{X})$. In order to relate this to the particle filtering paradigm, we will make use of the following algorithm. Let $\{\nu_i\}_{i=1}^d \subset \mathbb{M}_1(\mathcal{X})$ be probability measures such that $\mu_i(A) = \int_A w_i(x) \nu_i(\mathrm{d}x)$ for some $w_i \in \mathbb{F}_+(\mathsf{X}, \mathcal{X})$.

For $k = 1$ to N,

(i) draw an index $I^{N,k}$ multinomially with probability proportional to the weights $\omega_i \tau_i$, $1 \leq i \leq d$;

(ii) simulate $X^k \sim \nu_{I^{N,k}}$.

Subsequently, having at hand the sample $\{X^{N,i}\}_{i=1}^N$, we use

$$\widehat{\pi}^N(f) = \frac{\sum_{k=1}^N \tau_{I^{N,k}}^{-1} w_{I^{N,k}}(X^{N,k}) f(X^{N,k})}{\sum_{k=1}^N \tau_{I^{N,k}}^{-1} w_{I^{N,k}}(X^{N,k})} .$$

(a) Show that

$$N^{1/2}\left[\widehat{\pi}^N(f) - \pi(f)\right] \stackrel{\mathbb{P}}{\Longrightarrow} \mathrm{N}\left(0, \sum_{i=1}^d \frac{\omega_i \alpha_i(f)}{\tau_i}\right) , \tag{10.80}$$

where $\alpha_i(f) := \int_{\mathsf{X}} [w_i(x)]^2 [f(x) - \pi(f)]^2 \, \nu_i(\mathrm{d}x)$.

(b) Show that the adjustment weights $\tau_i^* := \alpha_i^{1/2}(f)$, $i = 1, \ldots, d$ minimize the asymptotic variance of the limiting distribution (10.80).

Chapter 11

Particle Smoothing

In the previous chapter, we saw that particle filtering corresponds to approximating the conditional distribution of the state X_t at time t given the observations, $Y_{0:t}$, up to time t. In this chapter, we extend the idea to obtaining smoothers. *Particle smoothing* corresponds to approximating the conditional distribution of the state X_t at time t (or a subsequence of states $X_{s:t}$ for $s \leq t$) given all of the available observations, $Y_{0:n}$, up to time n, where $t \leq n$. The smoothing distribution may be thought of as a correction or an update to the filter distribution that is enhanced by the use of additional observations from time $t+1$ to n. In general, smoothing distributions are an integral part of inference for state space models. For example, we saw that in the linear Gaussian case, the smoothers were essential for MLE via the EM algorithm (see Section 2.3.2) and, of course, in fitting smoothing splines (see Section 2.4).

Smoothing distributions can rarely be computed in closed form, and only in very special cases. Principally, closed-form solutions exist when the state space model is linear and Gaussian (as in Chapter 2) or when the state space is finite (see Chapter 9). In the vast majority of cases, nonlinearity or non-Gaussianity render analytic solutions intractable. Designing algorithms for the solution to this problem is not straightforward. Designing *efficient* algorithms is even more problematic.

The basic *filtering* version of the particle filter actually provides, as a byproduct, an approximation of the joint smoothing distribution $\phi_{0:n|n}$ in the sense that the particle *ancestral paths* and their associated weights can be considered as a weighted sample approximating $\phi_{0:n|n}$; this approach is detailed in Section 11.1. From the particle paths and weights, one may readily obtain an approximation of the marginal smoothing distribution of a sequence of states (from s to t) by marginalizing over the remaining states. This technique is equivalent to extracting the required subsequence of states from the sampled particle paths and retaining the weights.

This appealingly simple scheme can be used successfully for estimating the joint smoothing distribution for small values of n or any marginal smoothing distribution $\phi_{s:t|n}$, with $s \leq t \leq n$, when s,t and n are close. Unfortunately, as is shown in Section 11.1, when $s \leq t \ll n$, this simple strategy is doomed to failure due to the degeneracy of the particle ancestor tree. For this reason, more sophisticated strategies are needed in order to obtain efficient smoothing procedures. The key to the success of these algorithms is that they rely only on the particle approximation of the marginal filtering distributions.

11.1 Poor man's smoother algorithm

The *joint smoothing distribution* $\phi_{0:n|n}$ is the posterior distribution of the sequence of states $X_{0:n}$ given the observations $Y_{0:n}$. For ease of derivations, it is assumed in this chapter that the model is fully dominated (see Section 9.1.2). The joint smoothing distribution satisfies the recursion

$$\phi_{0:n|n}(x_{0:n}) = \frac{\phi_{0:n-1|n-1}(x_{0:n-1})q_n(x_{n-1},x_n)}{\int \phi_{n-1}(x_{n-1})q_n(x_{n-1},x_n)\lambda(\mathrm{d}x_n)} \tag{11.1}$$

$$= \frac{\mathrm{p}_{n-1}(Y_{0:n-1})}{\mathrm{p}_n(Y_{0:n})}\phi_{0:n-1|n-1}(x_{0:n-1})q_n(x_{n-1},x_n) \tag{11.2}$$

where $\mathrm{p}_n(Y_{0:n})$ is the *likelihood* of the observations given by (9.16). Recall that $(x,x') \mapsto m(x,x')$ is the transition density of the underlying Markov chain and M is the associated Markov kernel, $x \mapsto g_t(x) = g(x,Y_t)$ is the conditional distribution of the observation Y_t given the current state $X_t = x$ (also referred to as the likelihood of the observation Y_t), $(x,x') \mapsto q_t(x,x')$ is the product of the state transition density and the likelihood of the observation

$$q_t(x,x') = m(x,x')g_t(x') = m(x,x')g(x',Y_t) , \tag{11.3}$$

and Q_t is the associated Markov kernel. The poor man's smoother consists of approximating the smoothing recursions by a weighted sample of *particle trajectories* (or *paths*). These paths are obtained by propagating $\{(X_t^i, I_t^i, \omega_t^i)\}_{i=1}^N$, the particles and weights targeting the filtering distribution together with $\{I_t^i\}_{i=1}^N$ the ancestor index of the particles $\{X_t^i\}_{i=1}^N$ at time $t-1$, i.e., I_t^i is the index of the particle from which X_t^i originates; see (10.47). For any $i \in \{1,\dots,N\}$ and $t \geq 0$, we can reconstruct the ancestral path of the particle X_t^i, denoted as $X_{0:t}^i$ and defined by

$$X_{0:t}^i := (X_0^{B_0^i}, \dots, X_t^{B_t^i}) , \tag{11.4}$$

where the $B_{0:t}^i$ are given recursively by the ancestor indices

$$B_t^i = i \quad \text{and} \quad B_s^i = I_{s+1}^{B_{s+1}^i}, \quad \text{for } s \in \{0,\dots,t-1\} . \tag{11.5}$$

Note that the ancestor variables B_s^i, $s \in \{0,\dots,t\}$ depend on the final time t but this dependence is omitted in the notation in order to avoid overuse of subscripts (the final time at which the ancestral tree is rooted is explicit from the context).

The weighted sample and the ancestor indices $\{(X_t^i, I_t^i, \omega_t^i)\}_{i=1}^N$ are generated using the auxiliary particle filter; see Section 10.4.2. At time 0, N particles $\{X_0^i\}_{i=1}^N$ are drawn from a common probability measure r_0. These initial particles are assigned the importance weights $\omega_0^i = w_0(X_0^i)$ producing a weighted sample $\{(X_0^i, \omega_0^i)\}_{i=1}^N$ targeting the filtering distribution ϕ_0. Assume that we have obtained at time $t-1$, $t \geq 1$, a weighted sample $\{(X_{t-1}^i, I_{t-1}^i, \omega_{t-1}^i)\}_{i=1}^N$ targeting the filtering distribution ϕ_{t-1}. This weighted sample is recursively updated according to the following procedure. We first draw conditionally independently pairs $\{(I_t^i, X_t^i)\}_{i=1}^N$ of indices and particles from the proposal distribution

$$\phi_t^{N,\text{prop}}(\{i\} \times A) = \frac{\omega_{t-1}^i \vartheta_t(X_{t-1}^i)}{\sum_{j=1}^N \omega_{t-1}^j \vartheta_t(X_{t-1}^j)} R_t(X_{t-1}^i, A) , \quad A \in \mathcal{X} , \tag{11.6}$$

on the product space $\{1,\ldots,N\} \times \mathsf{X}$, where $x \mapsto \vartheta_t(x)$ is the *adjustment multiplier weight* function and R_t is the proposal kernel. For each draw (I_t^i, X_t^i), $i \in \{1,\ldots,N\}$, we compute the importance weight

$$\omega_t^i = \frac{w_t(X_{t-1}^{I_t^i}, X_t^i)}{\vartheta_t(X_{t-1}^{I_t^i})} , \tag{11.7}$$

where w_t is the importance function defined in (10.26), and associate it to the ancestral path $X_{0:t}^i$, defined in (11.4) and (11.5) of the current particle X_t^i. At each iteration of the algorithm, the set of ancestral paths $\{X_{0:t}^i\}_{i=1}^N$ is therefore modified; some of the ancestral paths get eliminated while others are replicated.

From the joint smoothing distribution $\phi_{0:t|t}$, it is of course possible to estimate any marginal smoothing distribution $\phi_{s|t}$, for $s \in \{0,\ldots,t\}$. Indeed, the marginal smoothing distribution $\phi_{s|t}$ is obtained by marginalizing the joint smoothing distribution with respect to the state variables $X_{-s} = (X_0, \ldots, X_{s-1}, X_{s+1}, \ldots, X_t)$. This amounts to representing the marginal smoothing distribution by the weighted sample $\{(X_s^{B_s^i}, \omega_t^i)\}_{i=1}^N$, where $\{B_s^i\}_{i=1}^N$ are the ancestor variables defined in (11.5). We stress that we use for all the particles $\{X_s^{B_s^i}\}_{s=0}^t$ belonging to the i-th ancestral path the same importance weight ω_t^i.

The following theorem, which extends Theorem 10.14, shows that the weighted sample $\{(X_{0:t}^i, \omega_t^i)\}_{i=1}^N$, where the particle path $X_{0:t}^i$ is given by (11.4), targets the joint smoothing distribution $\phi_{0:t|t}$.

Theorem 11.1. *Assume A10.13. Then, for all $t \in \{0,\ldots,n\}$, the weighted sample $\{(X_{0:t}^i, \omega_t^i)\}_{i=1}^N$ is consistent for the joint smoothing distribution $\phi_{0:t|t}$ and asymptotically normal for $(\phi_{0:t|t}, \sigma_t, \zeta_t)$ where for $f \in \mathbb{F}_b(\mathsf{X}^{t+1}, \mathcal{X}^{\otimes(t+1)})$,*

$$\zeta_t(f) = \frac{\phi_{t-1}(\vartheta_t)}{\{\phi_{t-1} Q_t \mathbf{1}\}^2} \int \cdots \int \phi_{0:t|t}(\mathrm{d}x_{0:t-1}) R_t(x_{t-1}, \mathrm{d}x_t) \frac{w_t^2(x_{t-1}, x_t)}{\vartheta_t(x_{t-1})} f(x_{0:t}) , \tag{11.8a}$$

$$\sigma_t^2(f) = \frac{\sigma_{t-1}^2(Q_t[f - \phi_{0:t|t}(f)])}{\{\phi_{t-1} Q_t \mathbf{1}\}^2} + \zeta_t([f - \phi_{0:t|t}(f)]^2) , \tag{11.8b}$$

and where we extend the definition of $Q_t f$, originally given in (10.22) for $f \in \mathbb{F}_b(\mathsf{X}, \mathcal{X})$, by setting for $f \in \mathbb{F}_b(\mathsf{X}^{t+1}, \mathcal{X}^{\otimes(t+1)})$,

$$Q_t f(x_{0:t-1}) = \int Q_t(x_{t-1}, \mathrm{d}x_t) f(x_{0:t-1}, x_t) . \tag{11.9}$$

The proof of Theorem 11.1 follows along the same lines as Theorem 10.14 (see

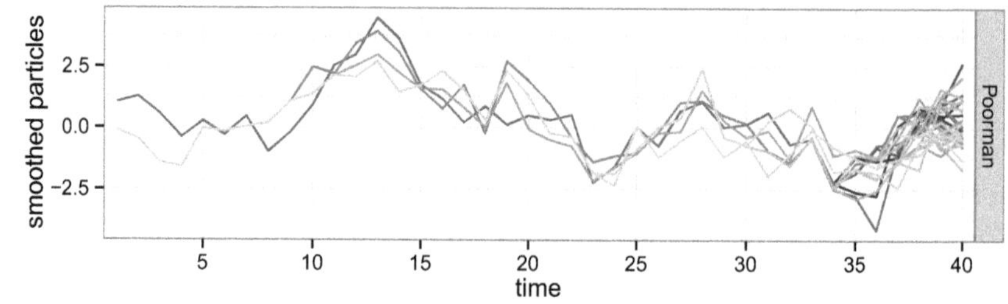

Figure 11.1 *Sampled trajectories from poor man's smoother for 50 particles in a noisy AR(1) model.*

Exercise 11.1). Despite these apparently favorable theoretical results, approximating the joint smoothing with the ancestor tree of the current generation of particles is doomed to failure. Although the convergence results are strong for an asymptotically large number of particles, we cannot expect favorable performance if the ancestor tree actually used is degenerate. As we will illustrate in Example 11.2, a large amount of degeneracy is unavoidable in all but trivial cases. Specifically, if the auxiliary particle filter is used for sufficiently many time steps, every resampling step reduces the number of unique values representing the marginal smoothing distribution at any fixed lag. For this reason, any SMC algorithm relying on the distribution of the ancestor tree will fail when the time horizon n grows for any finite number of particles, N, in spite of the asymptotic justification provided by Theorem 11.1. Consequently, it is necessary to design algorithms that do not depend upon the full ancestor tree of the current particle approximation.

Example 11.2 (Noisy AR(1) – poor man's smoother). Consider $n = 40$ simulated observations from a noisy AR(1) model (Example 2.2) with $\phi = 0.9$, and observation and state noise variances $\sigma_v^2 = 1$ and $\sigma_w^2 = 1$, respectively. We used the poor man's smoother with $N = 50$ particles. As illustrated in Figure 11.1, the path trajectories system collapses. At time 0, only 2 particles out of 50 survive; i.e., the marginal smoothing distribution $\phi_{0|n}$ is approximated with only 2 particles. At time 20, for example, only 4 particles are alive. The estimator of the joint smoothing distribution is therefore not reliable for sensible N values.

Although the resampling step is crucial in maintaining an accurate particle approximation of the filtering distribution (see Section 10.4), degeneracy of the ancestor tree is also a side-effect of the resampling step and is an unpreventable consequence of a the *curse of dimensionality*. It is not sensible to expect an accurate weighted sample representation of a distribution on a sequence of spaces of increasing dimensions while keeping the number of particles constant. It is, of course, not possible to circumvent this problem by merely increasing the number of samples at every iteration to maintain the estimation accuracy over the space as this would lead to an exponential growth in the number of samples required.

11.2 Forward filtering backward smoothing algorithm

In this section, we introduce the forward filtering backward smoothing (FFBSm) algorithm, which supplements the particle filter with a backward recursion. This algorithm shares some similarities with the Baum-Welch (or forward-backward) algorithm for discrete state-space HMM (see Algorithm 9.2) and the Kalman-filter based smoother and simulation smoother for the linear Gaussian state space models; see Section 9.2. In the forward pass, weighted sample approximations of the filtering distributions $\{\phi_t\}_{t=0}^n$ are obtained, using a classical SMC procedure. In a backward pass, the weights of the particles used to approximate the filtering distribution are updated to target the smoothing distribution.

According to Proposition 9.14, for any integer $n > 0$ and any $0 < t < n$, the posterior distribution $\phi_{t:n|n}$ may be expressed as, for any $h \in \mathbb{F}_b(\mathsf{X}^{n-t+1}, \mathcal{X}^{\otimes(n-t+1)})$,

$$\phi_{t:n|n}(h) = \int \cdots \int \phi_n(\mathrm{d}x_n) B_{\phi_{n-1}}(x_n, \mathrm{d}x_{n-1}) \cdots B_{\phi_t}(x_{t+1}, \mathrm{d}x_t) h(x_{t:n}) , \tag{11.10}$$

where B_{ϕ_s} are the backward kernels defined in (9.61). Therefore, the joint smoothing distribution may be computed recursively, backwards in time, according to

$$\phi_{t:n|n}(h) = \int \cdots \int B_{\phi_t}(x_{t+1}, \mathrm{d}x_t) \phi_{t+1:n|n}(\mathrm{d}x_{t+1:n}) h(x_{t:n}) , \tag{11.11}$$

where $h \in \mathbb{F}_b(\mathsf{X}^{(n-t+1)}, \mathcal{X}^{\otimes(n-t+1)})$.

Similarly to the Kalman smoother (see Proposition 2.7), the smoothing distribution is approximated using a two pass procedure. The basic idea underlying particle based FFBSm is to use the weighted sample $\{(X_t^i, \omega_t^i)\}_{i=1}^N$ which approximates the filtering distribution ϕ_t to approximate the backward kernel B_{ϕ_t}. Having produced, using methods described in Section 10.4, a sequence of such weighted samples $\{(X_t^i, \omega_t^i)\}_{i=1}^N$, $1 \leq t \leq n$, an approximation of the joint smoothing distribution is constructed in a backward pass by replacing in (11.10), the filtering distribution ϕ_s by their particle approximations ϕ_s^N, which yields

$$\phi_{t:n|n}^N(h) := \int \cdots \int \phi_n^N(\mathrm{d}x_n) B_{\phi_{n-1}^N}(x_n, \mathrm{d}x_{n-1}) \cdots B_{\phi_t^N}(x_{t+1}, \mathrm{d}x_t) h(x_{t:n}) \tag{11.12}$$

for any $h \in \mathbb{F}_b(\mathsf{X}^{n-t+1}, \mathcal{X}^{\otimes(n-t+1)})$. Equivalently, the particle approximation of the joint smoothing distribution $\phi_{t:n|n}^N$ above can be computed recursively backwards in time by replacing at each time step in the backward recursion (11.11) the joint smoothing distribution $\phi_{t+1:n|n}$ by its particle approximation $\phi_{t+1:n|n}^N$:

$$\phi_{t:n|n}^N(h) = \int \cdots \int B_{\phi_t^N}(x_{t+1}, \mathrm{d}x_t) \phi_{t+1:n|n}^N(\mathrm{d}x_{t+1:n}) h(x_{t:n}) . \tag{11.13}$$

In the sequel, we assume that the transition kernel M of the underlying Markov chain has a density m with respect to a reference measure λ. Using the definition (9.62) of the backward kernel, we get

$$B_{\phi_t^N}(x, A) = \sum_{i=1}^N \frac{\omega_t^i m(X_t^i, x)}{\sum_{\ell=1}^N \omega_t^\ell m(X_t^\ell, x)} \mathbb{1}_A(X_t^i) , \quad A \in \mathcal{X} . \tag{11.14}$$

For a given $x \in \mathsf{X}$, the probability $B_{\phi_t}(x, \cdot)$ is therefore approximated by a weighted sample, $\{(X_t^i, \omega_t^i m(X_t^i, x))\}_{i=1}^N$. Inserting this expression into (11.12) gives

$$\phi_{t:n|n}^N(h) = \sum_{i_t=1}^N \cdots \sum_{i_n=1}^N \left(\prod_{u=t+1}^n \frac{\omega_{u-1}^{i_{u-1}} m(X_{u-1}^{i_{u-1}}, X_u^{i_u})}{\sum_{\ell=1}^N \omega_{u-1}^\ell m(X_{u-1}^\ell, X_u^{i_u})} \right) \frac{\omega_n^{i_n}}{\Omega_n^N} h\left(X_t^{i_t}, \ldots, X_n^{i_n}\right) . \tag{11.15}$$

The FFBSm keeps the forward filter particles $\{X_t^i\}_{i=1}^N$, $t \in \{1, \ldots, n\}$ unchanged. It simply updates the importance weights of these particles, to target the joint smoothing distribution rather than the filtering distribution. The smoothing consists of computing new weights, and does not *move* the particles generated by the forward filter.

Even though (11.15) provides a closed form approximation of the joint smoothing distribution, it is impractical for any real problem of interest. The reason is, of course, that evaluating the discrete probabilities of the distribution is an $O(N^n)$ operation, both in terms of computational complexity and storage. However, even though it is not practically applicable on its own, (11.15) still provides interesting means for approximating the joint smoothing distribution. A practical approximation of this quantity is introduced in the next section.

The approximation (11.15) may also be used to estimate the sequence of marginal smoothing distributions $\phi_{t|n}$. Based on the weighted samples obtained in the forward pass of the particle filter, each marginal smoothing distribution will be approximated by a weighted particle system with no more than N particles. We thus expect that the computational complexity of approximating this sequence is lower than what is required to evaluate (11.15).

To see that this indeed is the case, we shall make use of the backward recursion for the marginal smoothing distribution. Assume that we have available a weighted particle system $\{(X_{t+1}^i, \omega_{t+1|n}^i)\}_{i=1}^N$, targeting $\phi_{t+1|n}$. At time $t = n-1$, this is provided by the particle filter by letting $\omega_{n|n}^i := \omega_n^i$ for $i = 1, \ldots, N$. Plugging the empirical distribution defined by this weighted sample and the approximation of the backward kernel (11.14) into the backward recursion for the marginal smoothing distribution yields

$$\phi_{t|n}^N(h) = \sum_{j=1}^N \omega_{t+1|n}^j \sum_{i=1}^N \frac{\omega_t^i m(X_t^i, X_{t+1}^j)}{\sum_{k=1}^N \omega_t^k m(X_t^k, X_{t+1}^j)} h(X_t^i) = \sum_{i=1}^N \omega_{t|n}^i h(X_t^i) , \tag{11.16}$$

where we have defined the marginal smoothing weights ,

$$\omega_{t|n}^i := \omega_t^i \sum_{j=1}^N \omega_{t+1|n}^j \frac{m(X_t^i, X_{t+1}^j)}{\sum_{k=1}^N \omega_t^k m(X_t^k, X_{t+1}^j)} . \tag{11.17}$$

The marginal smoothing weights are *self-normalized*, since,

$$\sum_{i=1}^N \omega_{t|n}^i = \sum_{j=1}^N \omega_{t+1|n}^j \frac{\sum_{i=1}^N \omega_t^i m(X_t^i, X_{t+1}^j)}{\sum_{k=1}^N \omega_t^k m(X_t^k, X_{t+1}^j)} = 1 .$$

Hence, we have obtained a weighted particle system $\{(X_t^i, \omega_{t|n}^i)\}_{i=1}^N$ that targets $\phi_{t|n}$.

The complexity of this algorithm is $O(nN^2)$, which is a significant reduction from the $O(N^n)$ complexity of evaluating the approximation of the joint smoothing distribution (11.15). Still, a computational cost growing quadratically with the number of particles might be prohibitive for certain problems. Similar recursion applies if instead of the marginal smoothing distribution, approximation of the joint smoothing distribution $\{\phi_{t:t+\ell-1|n}\}_{t=0}^{n}$ is considered; see Exercise 11.2.

11.3 Forward filtering backward simulation algorithm

We now introduce the forward filtering backward simulation (FFBSi) algorithm which samples, conditionally independently from the particles and the weights obtained in the forward pass, realizations of the joint smoothing distribution. We introduce an efficient algorithm, having a complexity that grows only linearly in the number of particles.

Consider, for $t \in \{0, \ldots, n-1\}$, the Markov transition matrix over the set of particle indices $\{1, \ldots, N\}$, $\{\Lambda_t^N(i,j)\}_{i,j=1}^{N}$, given by

$$\Lambda_t^N(i,j) = \frac{\omega_t^j m(X_t^j, X_{t+1}^i)}{\sum_{\ell=1}^{N} \omega_t^\ell m(X_t^\ell, X_{t+1}^i)} , \quad (i,j) \in \{1, \ldots, N\}^2 . \tag{11.18}$$

The set of transition matrices $\{\Lambda_t^N\}_{t=0}^{n}$ defined in (11.18) may be used to define a discrete state-space inhomogeneous Markov chain $\{J_u\}_{u=0}^{n}$ evolving backwards in time in the index set $\{1, \ldots, N\}$ as follows. At time n, an index J_n is sampled in the set $\{1, \ldots, N\}$ with probability proportional to filtering weights $\{\omega_n^i\}_{i=1}^{N}$ at time n. At time $t \leq n-1$, assuming that the indices $J_{t+1:n}$ have already been sampled, J_t is sampled in $\{1, \ldots, N\}$ with probability $\{\Lambda_t^N(J_{t+1}, j)\}_{j=1}^{N}$. The joint distribution of $J_{0:n}$ conditional to the particle and weights obtained in the filtering pass is therefore given by, for $j_{0:n} \in \{1, \ldots, N\}^{n+1}$,

$$\mathbb{P}\left[J_{0:n} = j_{0:n} \,\middle|\, \mathcal{F}_n^N\right] = \frac{\omega_n^{j_n}}{\Omega_n^N} \Lambda_n^N(j_n, j_{n-1}) \cdots \Lambda_0^N(j_1, j_0) . \tag{11.19}$$

Thus, the FFBSm estimator (11.15) may be written as the conditional expectation

$$\phi_{0:n|n}^N(h) = \mathbb{E}\left[h\left(X_0^{J_0}, \ldots, X_n^{J_n}\right) \,\middle|\, \mathcal{F}_n^N\right] , \quad h \in \mathbb{F}_b(\mathsf{X}^{n+1}, \mathcal{X}^{\otimes(n+1)}) \tag{11.20}$$

where $\mathcal{F}_n^N = \sigma(\{(X_t^i, \omega_t^i)\}_{i=1}^{N}, t \in \{0, \ldots, N\})$. We may therefore construct an unbiased estimator of the FFBSm estimator by drawing, conditionally independently given $\mathcal{F}_n^N$, N paths of $\{J_{0:n}^\ell\}_{\ell=1}^{N}$ of the inhomogeneous Markov chain introduced above and then forming the (practical) estimator

$$\tilde{\phi}_{0:n|n}^N(h) = N^{-1} \sum_{\ell=1}^{N} h\left(X_0^{J_0^\ell}, \ldots, X_n^{J_n^\ell}\right) , \quad h \in \mathbb{F}_b(\mathsf{X}^{n+1}, \mathcal{X}^{\otimes(n+1)}) . \tag{11.21}$$

The estimator $\phi_{0:n|n}^N$ may be seen as a Rao-Blackwellized version of $\tilde{\phi}_{0:n|n}^N$. The variance of the latter is increased, but the gain in computational complexity is large. The

---

Algorithm 11.1   FFBSi: Linear Smoothing Algorithm

1: sample $J_n^1, \ldots, J_n^N$ multinomially with probabilities proportional to $\{\omega_n^i\}_{i=1}^N$
2: **for** s from $n-1$ down to 0 **do**
3:      $L \leftarrow (1, \ldots, N)$
4:      **while** L is not empty **do**
5:          $K \leftarrow$ size(L)
6:          sample $I_1, \ldots, I_K$ with multinomial probabilities $\propto \{\omega_s^i\}_{i=1}^N$
7:          sample $U_1, \ldots, U_K$ independently and uniformly over $[0,1]$
8:          $nL \leftarrow \emptyset$
9:          **for** k from 1 to K **do**
10:             **if** $U_k \leq m\left(X_s^{I_k}, X_{s+1}^{J_{s+1}^{L_k}}\right) \big/ \sigma_+$ **then**
11:                 $J_s^{L_k} \leftarrow I_k$
12:             **else**
13:                 $nL \leftarrow nL \cup \{L_k\}$
14:             **end if**
15:         **end for**
16:         $L \leftarrow nL$
17:     **end while**
18: **end for**

---

associated algorithm is referred to in the sequel as the forward filtering backward simulation (FFBSi) algorithm.

The computational complexity for sampling a single path of $J_{0:n}$ is $O(Nn)$, i.e., it grows linearly with the number of particles and time indices; therefore, the overall computational effort required to compute the FFBSi estimator $\tilde{\phi}_{0:n|n}^N$ is $O(N^2 n)$, i.e., it grows linearly with respect to the number of time steps, but quadratically with the number of particles.

When the transition kernel m is bounded ($m(x, x') \leq \sigma_+$ for all $(x, x') \in X \times X$), the computational efficiency of the procedure above can be significantly improved. At the end of the filtering phase of the FFBSi algorithm, all weighted particle samples $\{(X_s^i, \omega_s^i)\}_{i=1}^N, 0 \leq s \leq n$, are available, and it remains to sample efficiently index paths $\{J_{0:n}^\ell\}_{\ell=1}^N$ under the distribution (11.19). These paths can be simulated recursively backwards in time, using the following acceptance-rejection procedure; see Exercise 5.23. As in the standard FFBSi algorithm, we first sample $J_n^1, \ldots, J_n^N$ multinomially with probabilities proportional to $\{\omega_n^i\}_{i=1}^N$. For $s \in \{0, \ldots, n\}$, denote by $\mathcal{G}_s^N$ the smallest σ-field containing $\mathcal{F}_n^N$ and $\sigma(J_t^\ell : 1 \leq \ell \leq N, s \leq t \leq n)$; then in order to draw J_s^ℓ conditionally on $\mathcal{G}_{s+1}^N$, we propose I_s^ℓ taking the value $\{1, \ldots, N\}$ with a probability proportional to $\{\omega_s^i\}_{i=1}^N$ and then an independent uniform random variable U_s^ℓ on $[0,1]$ is drawn. Then we set $J_s^\ell = I_s^\ell$ if $U_s^\ell \leq m(X_s^{I_s^\ell}, X_{s+1}^{J_{s+1}^\ell})/\sigma_+$; otherwise, we reject the proposed index and make another trial.

To create samples of size $K \in \{1, \ldots, N\}$ from a multinomial distribution on a set of N elements, (see Algorithm 11.1) we rely on an efficient procedure that requires

$O(N)$ elementary operations. Using this technique, the computational complexity of Algorithm 11.1 is linear in the number of particles; we call this the FFBSiLinear algorithm.

Proposition 11.3. *Assume that the transition kernel is bounded from above, i.e., $m(x,x') \le \sigma_+$ for all $(x,x') \in \mathsf{X} \times \mathsf{X}$. At each iteration $s \in \{0,\ldots,T-1\}$, let Z_s^N be the number of simulations required in the acceptance-rejection procedure of Algorithm 11.1.*

- *For the bootstrap auxiliary filter, Z_s^N/N converges in probability to*

$$\alpha(s) := \sigma_+ \phi_{s|s-1}(g_s) \frac{\int \cdots \int \lambda(\mathrm{d}x_{s+1}) \prod_{u=s+2}^{n} Q_u(x_{u-1}, \mathrm{d}x_u)}{\int \cdots \int \phi_{s|s-1}(\mathrm{d}x_s)\, g_s(x_s) \prod_{u=s+1}^{n} Q_u(x_{u-1}, \mathrm{d}x_u)}$$

  where $g_s(x_s) := g(x_s, Y_s)$.
- *In the fully adapted case, Z_s^N/N converges in probability to*

$$\beta(s) := \sigma_+ \frac{\int \cdots \int \mathrm{d}x_{s+1} \prod_{u=s+2}^{n} \int Q_u(x_{u-1}, \mathrm{d}x_u)}{\int \cdots \int \phi_s(\mathrm{d}x_s)\, g_s(x_s) \prod_{u=s+1}^{n} Q_u(x_{u-1}, \mathrm{d}x_u)} .$$

A sufficient condition for ensuring finiteness of $\alpha(s)$ and $\beta(s)$ is that $\int g(x_u, Y_u)\,\mathrm{d}x_u < \infty$ for all $u \ge 0$.

Proof. See Exercise 11.8. ■

The performance of the FFBSm and the FFBSi algorithms are studied in Section 11.7.

11.4 Smoothing functionals

Let $s_t : \mathsf{X} \times \mathsf{X} \to \mathbb{R}^d$, $t \in \mathbb{N}$, be a sequence of functions and $S_n : \mathsf{X}^n \to \mathbb{R}^d$, $n \in \mathbb{N}$, be the corresponding sequence of additive functionals given by

$$S_n(x_{0:n}) = \sum_{t=1}^{n} s_t(x_{t-1}, x_t) . \tag{11.22}$$

There are many instances (see Chapter 12) where it is required to compute recursively in time,

$$\bar{S}_n = \mathbb{E}\left[S_n(X_{0:n}) \mid Y_{0:n}\right] = \mathbb{E}\left[\sum_{t=1}^{n} s_t(X_{t-1}, X_t) \mid Y_{0:n}\right] = \phi_{0:n|n}(S_n) . \tag{11.23}$$

The quantity $\bar{S}_n$ may be computed using either the FFBSm or the FFBSi algorithms. These two algorithms require a backward pass and therefore do not yield an online algorithm.[1] Using the backward decomposition (11.10), the conditional expectation

[1] As opposed to the particle filter, which can be processed serially (*online*), the FFBSm and FFBSi algorithms require all the particles, for $t = 0, 1, \ldots, n$, to be stored in order to accomplish the backward pass.

$\bar{S}_n$ may be expressed as follows $\bar{S}_n = \int \cdots \int \phi_n(\mathrm{d}x_n)\sigma_n(x_n)$ where the function $\sigma_n : x_n \mapsto \sigma_n(x_n)$ is given by

$$\sigma_n(x_n) = \int \cdots \int B_{\phi_{n-1}}(x_n, \mathrm{d}x_{n-1}) \ldots B_{\phi_0}(x_1, \mathrm{d}x_0) S_n(x_{0:n}) \,. \tag{11.24}$$

The function may be interpreted as the conditional expectation of the statistics $S_n(X_{0:n})$ given $Y_{0:n}$ and the last state $X_n = x_n$,

$$\sigma_n(x_n) = \mathbb{E}\left[S_n(X_{0:n}) \,\middle|\, Y_{0:n}, X_n = x_n\right] \,.$$

The definition (11.22) of the additive functional S_n implies that $S_n(x_{0:n}) = S_{n-1}(x_{0:n-1}) + s_n(x_{n-1}, x_n)$, which implies that the function σ_n may de decomposed as follows

$$\begin{aligned} \sigma_n(x_n) &= \int \cdots \int B_{\phi_{n-1}}(x_n, \mathrm{d}x_{n-1}) \ldots B_{\phi_0}(x_1, \mathrm{d}x_0) \{S_{n-1}(x_{0:n-1}) + s_n(x_{n-1}, x_n)\} \\ &= \int B_{\phi_{n-1}}(x_n, \mathrm{d}x_{n-1}) \sigma_{n-1}(x_{n-1}) + \int B_{\phi_{n-1}}(x_n, \mathrm{d}x_{n-1}) s_n(x_{n-1}, x_n) \,. \end{aligned} \tag{11.25}$$

Perhaps surprisingly, the computation of smoothed additive functionals can be carried out forward in time. Denote by $\{\hat{\sigma}^N_{n-1}(X^i_{n-1})\}_{i=1}^N$ the particle approximation of the function σ_{n-1} at the particle location $\{X^i_{n-1}\}_{i=1}^N$. Plugging the particle approximation (11.14) of the backward smoothing kernel into (11.25) yields to the recursion

$$\begin{aligned} \hat{\sigma}^N_n(x_n) &= \int B_{\phi^N_{n-1}}(x_n, \mathrm{d}x_{n-1}) \left\{\hat{\sigma}^N_{n-1}(x_{n-1}) + s_n(x_{n-1}, x_n)\right\} \\ &= \sum_{j=1}^n \frac{\omega^j_{n-1} m(x_n, X^j_{n-1})}{\sum_{\ell=1}^n \omega^\ell_{n-1} m(x_n, X^\ell_{n-1})} \left\{\hat{\sigma}^N_{n-1}(X^j_{n-1}) + s_n(X^j_{n-1}, x_n)\right\} \,. \end{aligned}$$

We only need to evaluate the function $x_n \mapsto \hat{\sigma}^N_n(x_n)$ at the particle locations $\{X^i_n\}_{i=1}^N$, therefore, the computational complexity of this algorithm grows quadratically with the number N of particles. The convergence of this algorithm is studied in Section 11.7.

11.5 Particle independent Metropolis-Hastings

An alternate approach that addresses the joint smoothing problem is to target $\phi_{0:n|n}$ with a Metropolis-Hastings sampler. For large n, however, the high dimension of the space X^{n+1} creates serious problems when designing a proposal kernel for this sampler. One way to alleviate this problem is to make use of SMC in the proposal construction. Recall that the poor man's smoother generates a weighted particle system $\{(X^i_{0:n}, \omega^i_n)\}_{i=1}^N$, targeting the joint smoothing distribution $\phi_{0:n|n}$. Hence, if the random variable J is sampled from $\{1, \ldots, N\}$ with probability given by

$$\mathbb{P}\left[J = j \,\middle|\, \mathcal{F}^N_n\right] = \frac{\omega^j_n}{\Omega^N_n} \,, \quad j = 1, 2, \ldots, N \,, \tag{11.26}$$

then the ancestral path $X^J_{0:n}$ of the particle X^J_n [see (11.4)] is approximately distributed according to the joint smoothing distribution $\phi_{0:n|n}$.

Let $\mathcal{L}^N(X^J_{0:n})$ denote the law of the random vector $X^J_{0:n}$. The fact that, for large N, $\mathcal{L}^N(X^J_{0:n})$ is an approximation of the joint smoothing distribution $\phi_{0:n|n}$ suggests that ancestral particle path can be used as a proposal for an independent Metropolis-Hastings (IMH) sampler (see Example 5.38), targeting $\phi_{0:n|n}$. Unfortunately, however, a direct implementation of this IMH sampler is not possible. The problem lies in the computation of the acceptance probability, which requires a point-wise evaluation of the density of $\mathcal{L}^N(X^J_{0:n})$, which is not available. Indeed, the proposal distribution is given by

$$\mathbb{P}(X^J_{0:n} \in A) = \mathbb{E}\left[\sum_{i=1}^N \frac{\omega^i_n}{\Omega^N_n} \mathbb{1}_A(X^i_{0:n})\right] , \quad A \in \mathcal{X}^{\otimes(n+1)} , \tag{11.27}$$

where the expectation is taken with respect to the random variables generated by the SMC sampler (i.e., the particles and the weights). Computing this expectation is not feasible in practice.

A solution to this problem, as is often the case, lies in including the random variables generated by the SMC sampler as auxiliary variables, and thus avoiding having to marginalize over them. We make the derivation using the notations of the APF introduced in Section 10.4.2. Denote by

$$\begin{aligned} \boldsymbol{X}_t &= (X^1_t, \ldots, X^N_t) , \quad \boldsymbol{x}_t = (x^1_t, \ldots, x^N_t) , \\ \boldsymbol{I}_t &= (I^1_t, \ldots, I^N_t) , \quad \boldsymbol{i}_t = (i^1_t, \ldots, i^N_t) , \end{aligned}$$

all the particles and ancestor indices, respectively, generated by the APF at time t. Then, the process $\{(\boldsymbol{X}_t, \boldsymbol{I}_t)\}_{t\geq 1}$ is a second-order nonhomogeneous Markov chain. For ease of notation, assume that the model under study is fully dominated. The transition density p_t with respect to the dominating measure λ is given by, according to (10.47),

$$p_t(i_t, x_t) = \frac{\omega^{i_t}_{t-1} \vartheta_t(x^{i_t}_{t-1})}{\sum_{j=1}^N \omega^j_{t-1} \vartheta_t(x^j_{t-1})} r_t(x^{i_t}_{t-1}, x_t) , \quad t = 1, \ldots, N, , \tag{11.28}$$

where ω^j_{t-1} (see (10.48)) is the particle weight defined by

$$\omega^j_{t-1} = \frac{w_{t-1}(x^{i^j_{t-1}}_{t-2}, x^j_{t-1})}{\vartheta_{t-1}(x^{i^j_{t-1}}_{t-2})} . \tag{11.29}$$

Here, w_{t-1} is the weight function defined in (10.27). To avoid a cumbersome notation, the dependence of p_t and ω_{t-1} on the previous particles and ancestor indices $(\boldsymbol{x}_{t-2}, \boldsymbol{x}_{t-1}, \boldsymbol{i}_{t-1})$ is implicit.

We can now write down the joint density of $\boldsymbol{X}_0$ and the set of particles positions and ancestor indices $\{(\boldsymbol{X}_t, \boldsymbol{I}_t)\}_{t=1}^n$ as follows:

$$\psi_n^N(\boldsymbol{x}_{0:n}, \boldsymbol{i}_{1:n}) = \left(\prod_{\ell=1}^N r_0(x_0^\ell)\right) \prod_{t=1}^n \prod_{\ell=1}^N p_t(i_t^\ell, x_t^\ell) . \tag{11.30}$$

The sampling procedure described above, i.e., run the APF followed by the extraction of one particle trajectory according to (11.26), thus generates a family of random variables $\{\boldsymbol{X}_{0:n}, \boldsymbol{I}_{1:n}, J\}$ on the product space $\Xi \triangleq \mathsf{X}^{N(n+1)} \times \{1, \ldots, N\}^{n+1}$. Furthermore, the joint density of these variables is given by

$$(\boldsymbol{x}_{0:n}, \boldsymbol{i}_{1:n}, j) \mapsto \psi_n^N(\boldsymbol{x}_{0:n}, \boldsymbol{i}_{1:n}) \frac{\omega_n^j}{\Omega_n^N} , \quad \text{where} \quad \Omega_n^N = \sum_{\ell=1}^N \omega_n^\ell , \tag{11.31}$$

where we emphasize that the importance weights $\{\omega_n^\ell\}_{\ell=1}^N$ are deterministic functions of $(\boldsymbol{x}_{0:n}, \boldsymbol{i}_{1:n})$ according to (11.29). We are now in an unusual situation; we wish to use the proposal distribution with density (11.31) to target the joint smoothing distribution $\phi_{0:n|n}$, but these distributions are defined on different spaces. We extend the target distribution by defining an artificial target density on Ξ. Since the model is assumed to be fully dominated, the joint smoothing distribution [see (9.58)] has a density, also denoted $\phi_{0:n|n}$, with respect to the dominating measure, defined as

$$\phi_{0:n|n}(x_{0:n}) = (\mathrm{p}_n(Y_{0:n}))^{-1} \xi(x_0) g_0(x_0) \prod_{t=1}^n q_t(x_{t-1}, x_t) , \tag{11.32}$$

where $\mathrm{p}_n(Y_{0:n})$ is the likelihood of the observations $Y_{0:n}$. We may factorize the joint smoothing density as follows

$$\phi_{0:n|n}(x_{0:n}) = \phi_{0|0}(x_0) \prod_{t=1}^n \frac{\phi_{0:t|t}(x_{0:t})}{\phi_{0:t-1|t-1}(x_{0:t-1})} . \tag{11.33}$$

Using (11.32), for all $t \in \{1, \ldots, n\}$, we get, using the expression (10.27) of $w_t(x_{t-1}, x_t)$, that

$$\begin{aligned} \frac{\phi_{0:t|t}(x_{0:t})}{\phi_{0:t-1|t-1}(x_{0:t-1})} &= \frac{\mathrm{p}_{t-1}(Y_{0:t-1})}{\mathrm{p}_t(Y_{0:t})} q_t(x_{t-1}, x_t) \\ &= \frac{\mathrm{p}_{t-1}(Y_{0:t-1})}{\mathrm{p}_t(Y_{0:t})} w_t(x_{t-1}, x_t) r_t(x_{t-1}, x_t) . \end{aligned}$$

Now, using the expression of the weights (11.29) and the ancestor variables, we obtain

$$
\begin{aligned}
\frac{\phi_{0:t|t}(x_{0:t}^{b_{0:t}^j})}{\phi_{0:t-1|t-1}(x_{0:t-1}^{b_{0:t-1}^j})} &= \frac{\mathrm{p}_{t-1}(Y_{0:t-1})}{\mathrm{p}_t(Y_{0:t})} w_t(x_{t-1}^{b_{t-1}^j}, x_t^{b_t^j}) r_t(x_{t-1}^{b_{t-1}^j}, x_t^{b_t^j}) \\
&= \frac{\mathrm{p}_{t-1}(Y_{0:t-1})}{\mathrm{p}_t(Y_{0:t})} \omega_t^{b_t^j} \vartheta_t(x_{t-1}^{b_{t-1}^j}) r_t(x_{t-1}^{b_{t-1}^j}, x_t^{b_t^j}), \\
&= \frac{\mathrm{p}_{t-1}(Y_{0:t-1})}{\mathrm{p}_t(Y_{0:t})} \frac{\omega_t^{b_t^j}}{\omega_{t-1}^{b_{t-1}^j}} \left(\sum_{\ell=1}^N \omega_{t-1}^\ell \vartheta_t(x_{t-1}^\ell) \right) \frac{\omega_{t-1}^{b_{t-1}^j} \vartheta_t(x_{t-1}^{b_{t-1}^j})}{\sum_{\ell=1}^N \omega_{t-1}^\ell \vartheta_t(x_{t-1}^\ell)} r_t(x_{t-1}^{b_{t-1}^j}, x_t^{b_t^j})
\end{aligned}
$$

where the ancestor indices are defined recursively backwards in time by

$$
b_n^j = j \quad \text{and} \quad b_{t-1}^j = i_t^{b_t^j}, \quad \text{for } t = n, n-1, \ldots, 1. \tag{11.34}
$$

It follows from the definition of the joint proposal kernel (11.28), together with (11.33), that the joint distribution of the ancestral path $x_{0:n}^{b_{0:n}^j}$ is given by

$$
\phi_{0:n|n}(x_{0:n}^{b_{0:n}^j}) = \frac{\omega_n^{b_n^j}}{\mathrm{p}_n(Y_{0:n})} \left(\prod_{t=1}^n \sum_{\ell=1}^N \omega_{t-1}^\ell \vartheta_t(x_{t-1}^\ell) \right) r_0(x_0^{b_0^j}) \prod_{t=1}^n p_t(i_t^{b_t^j}, x_t^{b_t^j}). \tag{11.35}
$$

For any $\ell \in \{1, \ldots, N\}$, denote $\boldsymbol{x}_t^{-\ell} = (x_t^1, \ldots, x_t^{\ell-1}, x_t^{\ell+1}, \ldots, x_t^N)$, and for any sequence $\ell_{0:n} = (\ell_0, \ldots, \ell_n) \in \{1, \ldots, N\}^{N+1}$, define $\boldsymbol{x}_{0:n}^{-\ell_{0:n}} = (\boldsymbol{x}_0^{-\ell_0}, \ldots, \boldsymbol{x}_n^{-\ell_n})$. Now, define an extended target density on Ξ according to,

$$
\begin{aligned}
\pi_n^N(\boldsymbol{x}_{0:n}, \boldsymbol{i}_{1:n}, j) &= \pi_n^N(x_{0:n}^{b_{0:n}^j}, b_{0:n}^j) \pi_n^N(\boldsymbol{x}_{0:n}^{-b_{0:n}^j}, \boldsymbol{i}_{1:n}^{-b_{1:n}^j} \mid x_{0:n}^{b_{0:n}^j}, b_{0:n}^j) \\
&\triangleq \underbrace{\frac{\phi_{0:n|n}(x_{0:n}^{b_{0:n}^j})}{N^{n+1}}}_{\text{marginal}} \underbrace{\left(\prod_{\substack{\ell=1 \\ \ell \neq b_0^j}}^N r_0(x_0^\ell) \right) \prod_{t=1}^n \left(\prod_{\substack{\ell=1 \\ \ell \neq b_t^j}}^N p_t(i_t^\ell, x_t^\ell) \right)}_{\text{conditional}},
\end{aligned} \tag{11.36}
$$

where the backward variables are defined by (11.34). The factorization into a marginal and a conditional density (for which, by abuse of notation, we use the same symbol π_n^N) reveals the structure of the extended target density. By construction, the marginal density of the variables $\boldsymbol{X}_{0:n}^{B_{0:n}^J}$ under π_n^N is equal to the joint smoothing density. This has the important consequence that (11.36) can be used as a surrogate for the original target density (11.32) in an MCMC sampler. More precisely, if $\{(\boldsymbol{X}_{0:n}[k], \boldsymbol{I}_{1:n}[k], J[k]),\ k \in \mathbb{N}\}$ is an ergodic Markov chain on Ξ with limiting distribution π_n^N, then the sequence of ancestral path $\{X_{0:n}[k],\ k \in \mathbb{N}\}$ converges in distribution to the joint smoothing distribution $\phi_{0:n|n}$.

Consider now the IMH sampler with proposal density (11.31) and target density (11.36). The acceptance probability for a proposed move from $\{\boldsymbol{x}_{0:n}, \boldsymbol{i}_{1:n}, j\}$ to

Algorithm 11.2 (PIMH: Particle independent Metropolis-Hastings)

Initialize:

- Run an APF targeting $\phi_{0:n|n}$ and compute $\ell_n^N[0] = \widehat{\mathrm{L}}_n^N(\boldsymbol{X}_{0:n}, \boldsymbol{I}_{1:n})$.
- Sample J with $\mathbb{P}\left[J = j \mid \mathcal{F}_n^N\right] = \omega_n^j / \Omega_n^N$.

Iterate, for $k \geq 1$:

- Run an APF targeting $\phi_{0:n|n}$ and compute the likelihood estimate $\tilde{\ell}_n^N = \widehat{\mathrm{L}}_n^N(\boldsymbol{X}_{0:n}, \boldsymbol{I}_{1:n})$.
- Sample J with $\mathbb{P}\left[J = j \mid \mathcal{F}_n^N\right] = \omega_n^j / \Omega_n^N$.
- Draw U uniformly on $[0, 1]$.
- If $U \leq 1 \wedge \tilde{\ell}_n^N / \ell_n^N[k-1]$, set $X_{0:n}[k] = X_{0:n}^J$ and $\ell_n^N[k] = \tilde{\ell}_n^N$, otherwise set $X_{0:n}[k] = X_{0:n}[k-1]$ and $\ell_n^N[k] = \ell_n^N[k-1]$.

$\{\tilde{\boldsymbol{x}}_{0:n}, \tilde{\boldsymbol{i}}_{1:n}, \tilde{j}\}$ is given by a standard Metropolis-Hastings ratio,

$$\alpha(\boldsymbol{x}_{0:n}, \boldsymbol{i}_{1:n}, j; \tilde{\boldsymbol{x}}_{0:n}, \tilde{\boldsymbol{i}}_{1:n}, \tilde{j}) := 1 \wedge \frac{\pi_n^N(\tilde{\boldsymbol{x}}_{0:n}, \tilde{\boldsymbol{i}}_{1:n}, \tilde{j})}{\pi_n^N(\boldsymbol{x}_{0:n}, \boldsymbol{i}_{1:n}, j)} \frac{\psi_n^N(\boldsymbol{x}_{0:n}, \boldsymbol{i}_{1:n}) \omega_n^j \tilde{\Omega}_n^N}{\psi_n^N(\tilde{\boldsymbol{x}}_{0:n}, \tilde{\boldsymbol{i}}_{1:n}) \tilde{\omega}_n^{\tilde{j}} \Omega_n^N} . \tag{11.37}$$

Denote

$$\widehat{\mathrm{L}}_n^N(\boldsymbol{x}_{0:n}, \boldsymbol{i}_{1:n}) = \left(\frac{1}{N} \sum_{\ell=1}^N \omega_n^\ell \right) \prod_{t=1}^n \left(\frac{1}{N} \sum_{\ell=1}^N \omega_{t-1}^\ell \vartheta_t(x_{t-1}^\ell) \right) , \tag{11.38}$$

where ω_{t-1}^ℓ is given by (11.29). If follows that the extended target density (11.36) can be rewritten as,

$$\pi_n^N(\boldsymbol{x}_{0:n}, \boldsymbol{i}_{1:n}, j) = \frac{\widehat{\mathrm{L}}_n^N(\boldsymbol{x}_{0:n}, \boldsymbol{i}_{1:n})}{\mathrm{p}_n(Y_{0:n})} \frac{\omega_n^j}{\Omega_n^N} \psi_n^N(\boldsymbol{x}_{0:n}, \boldsymbol{i}_{1:n}) .$$

By plugging this expression into (11.37) we obtain the surprisingly simple expression for the Metropolis-Hastings ratio,

$$\alpha(\boldsymbol{x}_{0:n}, \boldsymbol{i}_{0:n}, j; \tilde{\boldsymbol{x}}_{0:}, \tilde{\boldsymbol{i}}_{1:n}, \tilde{j}) = 1 \wedge \frac{\widehat{\mathrm{L}}_n^N(\tilde{\boldsymbol{x}}_{0:n}, \tilde{\boldsymbol{i}}_{1:n})}{\widehat{\mathrm{L}}_n^N(\boldsymbol{x}_{0:n}, \boldsymbol{i}_{1:n})} . \tag{11.39}$$

The resulting IMH sampler, using the proposal (11.31) for the extended target (11.36), is referred to as particle independent Metropolis-Hastings (PIMH). It is worth emphasizing that, despite the rather cumbersome derivation, the PIMH sampler is very simple to implement. The PIMH sampler is summarized in Algorithm 11.2.

Standard results for IMH samplers allow us to characterize the convergence properties of PIMH.

Theorem 11.4. *Assume A 10.18. Then, for any $N \geq 1$ and for any starting point, the PIMH sampler generates a process $\{X_{0:n}[k]\}_{k \geq 0}$ whose marginal distributions converge uniformly geometrically to the target distribution $\phi_{0:n|n}$.*

Proof. The result follows from the fact that PIMH is a standard IMH sampler on the space Ξ, and that an IMH sampler is uniformly ergodic if the ratio between the target density and the proposal density is bounded (see Exercise 6.17). ■

The likelihood estimator (11.38) is consistent, which implies that the PIMH acceptance probability (11.39) converges to 1 as $N \to \infty$. We emphasize that PIMH is exact for any number of particles, in the sense that the limiting distribution of the sampler is $\phi_{0:n|n}$ for any $N \geq 1$. However, for small N, the variance of the estimator (11.38) will be large. As an effect, PIMH tends to get stuck and the convergence of the sampler will be slow.

It can seem wasteful to generate N particle trajectories at each iteration of the PIMH sampler, but keep only a single one. Indeed, since the acceptance probability is independent of J, i.e., of the specific particle trajectory $X^J_{0:n}$, it is possible to Rao-Blackwellize over J. Let $h \in \mathbb{F}_b(\mathsf{X}^{n+1}, \mathcal{X}^{\otimes(n+1)})$ and assume that we run Algorithm 11.2 for K iterations (possibly with some burn-in). PIMH then provides the natural estimator of $\phi_{0:n|n}(h)$,

$$\hat{h}_{\text{PIMH}} = \frac{1}{K} \sum_{k=1}^{K} h\left(X_{0:n}[k]\right), \tag{11.40}$$

which, by the ergodic theorem, converges almost surely to the estimand. However, Rao-Blackwellization over J allows us to reuse all the particle trajectories generated by the PIMH sampler. Let $\mathcal{F}^N_n[k] = \sigma\{\{U[j], (\boldsymbol{X}_{0:n}[j], \boldsymbol{I}_{1:n}[j])\}, j \leq k\}$ be the σ-algebra generated by the particles, the ancestor indices and the uniforms used in the accept/reject decision, sampled up to iteration k. We can then compute the Rao-Blackwellized estimator,

$$\hat{h}_{\text{PIMH-RB}} = \frac{1}{K} \sum_{k=1}^{K} \mathbb{E}\left[h\left(X_{0:n}[k]\right) \mid \mathcal{F}^N_n[k]\right] = \frac{1}{K} \sum_{k=1}^{K} \sum_{j=1}^{N} \frac{\omega^j_n[k]}{\Omega^N_n[k]} h\left(X^j_{0:n}[k]\right). \tag{11.41}$$

The possibility to make use of all the generated particles to reduce the variance of the estimator seems promising. However, a problem with the above estimator is that the particle systems $\{(X^j_{0:n}[k], \omega^j_n[k])\}^N_{j=1}$ suffer from path degeneracy. Hence, the possible benefit of Rao-Blackwellization is limited due to the low particle diversity for time points t far away from the final time n.

The problem lies in the fact that PIMH relies on the poor man's smoother as its basic building block. To obtain a larger variance reduction we can use the same idea to Metropolise a more advanced particle smoother, such as the FFBSi. Olsson and Rydén (2011) have proposed to modify the PIMH by replacing (11.26) by the run of a backward simulator, generating M backward trajectories. Interestingly, the acceptance probability for such a Metropolis-Hastings-correction of the FFBSi is the same as for the poor man's smoother. In practice, it is preferable to run the backward simulator only if the proposed sample is accepted. This is possible since, as previously pointed out, the acceptance probability (11.39) is independent of the extracted trajectories.

Let the M backward trajectories generated at iteration k of the PIMH sampler be denoted $\{\widetilde{X}^j_{0:n}[k]\}_{j=1}^M$. We obtain the FFBSi-based estimator,

$$\hat{h}_{\text{PIMH-BSi}} = \frac{1}{KM}\sum_{k=1}^{K}\sum_{j=1}^{M} h\left(\widetilde{X}^j_{0:n}[k]\right). \tag{11.42}$$

Again, by a Rao-Blackwellization type of argument, (11.42) converges almost surely to $\phi_{0:n|n}(h)$. Olsson and Rydén (2011) provide an expression for the variance of the estimator (11.42),

$$\text{Var}(\hat{h}_{\text{PIMH-BSi}}) = \frac{1}{K}\left(\frac{\sigma^2}{M} + \sigma_K^2\right) \approx \frac{1}{K}\left(\frac{\sigma^2}{M} + \sigma_\infty^2\right), \tag{11.43}$$

where

$$\sigma^2 = \mathbb{E}\left[\text{Var}\left(h\left(\widetilde{X}^1_{0:n}\right) \middle| \mathcal{F}_n^N\right)\right],$$

$$\sigma_K^2 = \frac{1}{K}\text{Var}\left(\sum_{k=1}^{K}\mathbb{E}\left[h\left(\widetilde{X}^1_{0:n}[k]\right) \middle| \mathcal{F}_n^N[k]\right]\right),$$

and $\sigma_\infty^2 = \lim_{K\to\infty}\sigma_K^2$ is the time-average variance constant. This expression can be used to find an optimal trade off between K and M, depending on the running times of the algorithm for different settings. In practice, the parameters σ^2 and σ_∞^2 are typically not known, and to be able to make use of (11.43) to tune the sampler, it is necessary to estimate these parameters from data.

11.6 Particle Gibbs

The trick behind PIMH was to recognize that it is a standard IMH sampler, targeting the extended target density (11.36). However, with (11.36) in place we can think about other MCMC samplers with the same target distribution, leading to other members of the PMCMC family. In particular, it is possible to construct a Gibbs sampler for the extended target density π_n^N, resulting in particle Gibbs (PG). In this section, we focus on the Particle Gibbs with Ancestor Sampling algorithm (PGAS). The basic building block of PGAS is a particle-filter-like procedure, referred to as a conditional particle filter with ancestor sampling (CPF-AS). The procedure is given in Algorithm 11.3. As will be seen, the CPF-AS is very similar to a standard APF. However, an important difference is that in CPF-AS, one particle at each time point is specified *a priori*. The expression *ancestor sampling* refers to the fact that the ancestor indices for the conditioned particles are also sampled using one-step backward simulations. In practice, this step mitigates the effects of path degeneracy, and improves the mixing of the sampler.

PGAS consists of iteratively running the CPF-AS, followed by the extraction of one particle trajectory as in (11.26). This sampling procedure can be interpreted as a multi-stage Gibbs sampler for the extended target density π_n^N. Denote by $X_{0:n}$ the conditioning particle path and $B_{0:n}$ the associated ancestor indices. Denote by $\tilde{\boldsymbol{X}}_t = (X_t^1, \ldots, X_t^n)$ the proposed particle set and by $\tilde{\boldsymbol{I}}_t = (I_t^1, \ldots, I_t^n)$ the ancestor index.

Algorithm 11.3 (CPF-AS: Conditional PF with ancestor sampling)

Input:

- Conditioned particles $(x'_{0:n}, b_{0:n})$.

Initialize:

- Draw $X_0^j \sim r_0(\cdot)$ for $j \neq b_0$ and set $X_0^{b_0} = x'_0$.
- Compute $\omega_0^j = w_0(X_0^j)$ for $j = 1, \ldots, N$.

For $t = 1, \ldots, n$**:**

- Draw $(I_t^j, X_t^j) \sim p_t(\cdot, \cdot)$ for $j \neq b_t$ and set $X_t^{b_t} = x'_t$.
- Draw $I_t^{b_t}$ with conditional probability of $I_t^{b_t} = j$ given by

$$\frac{\omega_{t-1}^j m(X_{t-1}^j, x'_t)}{\sum_{\ell=1}^N \omega_{t-1}^\ell m(X_{t-1}^\ell, x'_t)}.$$

- Compute $\omega_t^j = w_t(X_{t-1}^{I_t^j}, X_t^j) / \vartheta_t(X_{t-1}^{I_t^j})$ for $j = 1, \ldots, N$.

Procedure 11.5 (PGAS).

(i) Draw $\tilde{\boldsymbol{X}}_0^{-B_{0:n}^J} \sim \pi_n^N(\cdot \mid X_{0:n}, B_{0:n}^J)$;

(ii) Draw, for $t = 1, \ldots, n$,

(a) $(\tilde{\boldsymbol{X}}_t^{-B_t}, \tilde{\boldsymbol{I}}^{-B_t}) \sim \pi_n^N(\cdot \mid \tilde{\boldsymbol{X}}_{0:t-1}^{-B_{0:t-1}^J}, \tilde{\boldsymbol{I}}_{1:t-1}, X_{0:n}, B_{0:n}^J)$;

(b) $\tilde{I}_t^{B_t} \sim \pi_n^N(\cdot \mid \tilde{\boldsymbol{X}}_{0:t-1}^{-B_{0:t-1}^J}, \tilde{\boldsymbol{I}}_{1:t-1}, X_{0:n}, B_{0:n}^J)$;

(iii) Draw $\tilde{J} \sim \pi_n^N(\cdot \mid \tilde{\boldsymbol{X}}_{0:n}, \tilde{\boldsymbol{I}}_{1:n}, X_{0:n}, B_{0:n}^J)$.

Each of the steps in the sampling scheme consists of sampling one or several variables from some conditional distribution under π_n^N. It should be noted, however, that the densities involved in Procedure 11.5(i)–(ii) are not conditionals under the full joint density π_n^N, but under marginals thereof. Making use of marginalization within Gibbs sampling is commonly referred to as collapsing (see e.g., Dyk and Park, 2008). Collapsing does not violate the Gibbs sampler and the sampling scheme given by Procedure 11.5(i)–(iii) will indeed leave the extended target density π_n^N invariant.

Above, we claimed that Procedure 11.5(i)–(ii) corresponds to the sampling procedure given in Algorithm 11.3. To verify this claim, we note first that, by construction of (11.36),

$$\pi_n^N(\boldsymbol{x}_{0:n}^{-b_{0:n}^j}, \boldsymbol{i}_{1:n}^{-b_{1:n}^j} \mid x_{0:n}^{b_{0:n}^j}, b_{0:n}^j) = \left(\prod_{\substack{\ell=1 \\ \ell \neq b_0^j}}^N r_0(x_0^\ell) \right) \prod_{t=1}^n \left(\prod_{\substack{\ell=1 \\ \ell \neq b_t^j}}^N p_t(i_t^\ell, x_t^\ell) \right). \tag{11.44}$$

By marginalizing this density over $(\boldsymbol{x}_{t+1:n}^{-b^j_{t+1:n}}, \boldsymbol{i}_{t+1:n}^{-b^j_{t+1:n}})$ we get for $t \geq 1$ (see (11.28)),

$$\pi_n^N(\boldsymbol{x}_{0:t}^{-b^j_{0:t}}, \boldsymbol{i}_{1:t}^{-b^j_{1:t}} \mid x_{0:n}^{b^j_{0:n}}, b^j_{0:n}) = \left(\prod_{\substack{\ell=1 \\ \ell \neq b^j_0}}^{N} r_0(x_0^\ell) \right) \prod_{s=1}^{t} \left(\prod_{\substack{\ell=1 \\ \ell \neq b^j_s}}^{N} p_s(i_s^\ell, x_s^\ell) \right) . \tag{11.45}$$

For $t = 0$, (11.45) writes

$$\pi_n^N(\boldsymbol{x}_0^{-b^j_0} \mid x_{0:n}^{b^j_{0:n}}, b^j_{0:n}) = \prod_{\substack{\ell=1 \\ \ell \neq b^j_0}}^{N} r_0(x_0^\ell) . \tag{11.46}$$

Noting that (11.46) is the density appearing in Procedure 11.5(i), we conclude that this step corresponds to the initialization in Algorithm 11.3. Furthermore, from (11.45) it follows that, for $t \in \{1, \ldots, n\}$,

$$\begin{aligned} &\pi_n^N(\boldsymbol{x}_t^{-b^j_t}, \boldsymbol{i}_t^{-b^j_t} \mid \boldsymbol{x}_{0:t-1}^{-b^j_{0:t-1}}, \boldsymbol{i}_{1:t-1}, x_{0:n}^{b^j_{0:n}}, b^j_{t-1:n}) \\ &= \frac{\pi_n^N(\boldsymbol{x}_{0:t}^{-b^j_{0:t}}, \boldsymbol{i}_{1:t}^{-b^j_{1:t}} \mid x_{0:n}^{b^j_{0:n}}, b^j_{0:n})}{\pi_n^N(\boldsymbol{x}_{0:t-1}^{-b^j_{0:t-1}}, \boldsymbol{i}_{1:t-1}^{-b^j_{1:t-1}} \mid x_{0:n}^{b^j_{0:n}}, b^j_{0:n})} = \prod_{\substack{\ell=1 \\ \ell \neq b^j_t}}^{N} p_t(i_t^\ell, x_t^\ell) , \end{aligned} \tag{11.47}$$

which implies that Procedure 11.5(iia) equates to sampling $(\boldsymbol{X}_t^{-B^j_t}, \boldsymbol{I}_t^{-B^j_t})$.

Next, we turn to Procedure 11.5(iib) and show that this step corresponds to a one-step backward simulation. We seek the conditional distribution

$$\pi_n^N(i_t^{b^j_t} \mid \boldsymbol{x}_{0:t-1}^{-b^j_{0:t-1}}, \boldsymbol{i}_{1:t-1}, x_{0:n}^{b^j_{0:n}}, b^j_{t:n}) .$$

However, it is enough to find an expression for the distribution up to a normalizing constant. The normalization constant can then be obtained by summing over all possible values $i_t^{b^j_t} \in \{1, \ldots, N\}$, ensuring that the distribution sums to one.

$$\begin{aligned} \pi_n^N(i_t^{b^j_t} \mid \boldsymbol{x}_{0:t-1}^{-b^j_{0:t-1}}, \boldsymbol{i}_{1:t-1}, x_{0:n}^{b^j_{0:n}}, b^j_{t:n}) &\propto \pi_n^N(\boldsymbol{x}_{0:t-1}^{-b^j_{0:t-1}}, \boldsymbol{i}_{1:t-1}, x_{0:n}^{b^j_{0:n}}, b^j_{t-1:n}) \\ &= \frac{\phi_{0:n|n}(x_{0:n}^{b^j_{0:n}})}{N^{n+1}} \left(\prod_{\substack{\ell=1 \\ \ell \neq b^j_0}}^{N} r_0(x_0^\ell) \right) \prod_{s=1}^{t-1} \left(\prod_{\substack{\ell=1 \\ \ell \neq b^j_s}}^{N} p_s(i_s^\ell, x_s^\ell) \right) , \end{aligned}$$

where we have made use of the expressions in (11.36) and (11.45) and $b^j_{t-1} = i_t^{b^j_t}$. Therefore, keeping in this expression only the terms that depend on $b^j_{t-1} = i_t^{b^j_t}$, we get

$$
\begin{aligned}
&\pi_n^N(i_t^{b_t^j} \mid \boldsymbol{x}_{0:t-1}^{-b_{0:t-1}^j}, \boldsymbol{i}_{1:t-1}, x_{0:n}^{b_{0:n}^j}, b_{t:n}^j) \\
&= \frac{\phi_{0:n|n}(x_{0:n}^{b_{0:n}^j})}{N^{n+1}} \frac{\left(\prod_{\ell=1}^N r_0(x_0^\ell)\right)\prod_{s=1}^{t-1}\left(\prod_{\ell=1}^N p_s(i_s^\ell, x_s^\ell)\right)}{r_0(x_0^{b_0^j})\prod_{s=1}^{t-1} p_s(i_s^{b_s^j}, x_s^{b_s^j})} \propto \frac{\phi_{0:n|n}(x_{0:n}^{b_{0:n}^j})}{p_{t-1}(i_{t-1}^{b_{t-1}^j}, x_{t-1}^{b_{t-1}^j})} . \quad (11.48)
\end{aligned}
$$

The last step holds for $t > 1$ with an obvious modification for $t = 1$. To expand $\phi_{0:n|n}(x_{0:n})$, we note that the factorization of the joint smoothing density (11.32) implies

$$
\begin{aligned}
\phi_{0:n|n}(x_{0:n}) &\propto q_{t-1}(x_{t-2}, x_{t-1}) m(x_{t-1}, x_t) \\
&= \frac{w_{t-1}(x_{t-2}, x_{t-1})}{\vartheta_{t-1}(x_{t-2})} \vartheta_{t-1}(x_{t-2}) r_{t-1}(x_{t-2}, x_{t-1}) m(x_{t-1}, x_t) ,
\end{aligned}
$$

again discarding factors independent of x_{t-1}. By using (11.28) and (11.29), we can thus write

$$
\phi_{0:n|n}(x_{0:n}^{b_{0:n}^j}) \propto \omega_{t-1}^{b_{t-1}^j} p_{t-1}(i_{t-1}^{b_{t-1}^j}, x_{t-1}^{b_{t-1}^j}) m(x_{t-1}^{b_{t-1}^j}, x_t^{b_t^j}) . \quad (11.49)
$$

In summary, by plugging (11.49) into (11.48) and normalizing, we get (noting again that $b_{t-1}^j = i_t^{b_t^j}$) that

$$
\pi_n^N(i_t^{b_t^j} \mid \boldsymbol{x}_{0:t-1}^{-b_{0:t-1}^j}, \boldsymbol{i}_{1:t-1}, x_{0:n}^{b_{0:n}^j}, b_{t:n}) = \frac{\omega_{t-1}^{b_{t-1}^j} m(x_{t-1}^{b_{t-1}^j}, x_t^{b_t^j})}{\sum_{\ell=1}^N \omega_{t-1}^\ell m(x_{t-1}^\ell, x_t^{b_t^j})} ,
$$

which is exactly the ancestor sampling probabilities used in Algorithm 11.3.

Finally, we turn to Procedure 11.5(iii) of the sampler. Using (11.35) and (11.36) and $b_n^j = j$, we get that

$$
\pi_n^N(j \mid \boldsymbol{x}_{0:n}, \boldsymbol{i}_{1:n}) = \frac{\omega_n^j}{\sum_{\ell=1}^N \omega_n^\ell} .
$$

That is, once we have run the CPF-AS, one particle trajectory $X_{0:n}^J$ is extracted by sampling J as in (11.26).

The procedure given by Procedure 11.5(i)–(iii) is a partially collapsed Gibbs sampler, which leaves π_n^N invariant. The ordinary Gibbs sampler (see Section 5.8.2) begins with the joint target distribution of family of random variables and updates (groups) of these random variables by sampling them from their conditional distributions under the joint target distribution. Each quantity is generally updated exactly once in each iteration. The partially collapsed Gibbs sampler replaces some of these conditional distributions with conditional distributions under some marginal distributions of the joint target distribution. Such a conditional distribution that conditions on fewer unknown components is referred to as a *reduced conditional distribution*;

Algorithm 11.4 (PGAS: Particle Gibbs with ancestor sampling)

Initialize:

- Set $X_{0:n}[0]$ and $B_{0:n}[0]$ arbitrarily.

Iterate, for $k \geq 1$:

- Run CPF-AS targeting $\phi_{0:n|n}$, conditioned on $(X_{0:n}[k-1], B_{0:n}[k-1])$.
- Sample J with $\mathbb{P}\left[J = j \,\middle|\, \mathcal{F}_n^N\right] = \omega_n^j / \Omega_n^N$ and trace the ancestral path of particle X_n^J, i.e., set $X_{0:n}[k] = X_{0:n}^J$ and $B_{0:n}[k]$ to the corresponding particle indices.

see van Dyk and Park (2008). This strategy is useful because it can result in samplers with better convergence properties.

Furthermore, it holds that the PGAS sampler, summarized in Algorithm 11.4, produces an ergodic process. Since the proposal kernel used in the CPF-AS is assumed to dominate the transition kernel of the state process, sets of positive $\phi_{0:n|n}$-probability are marginally accessible by the PGAS sampler. It can then be verified that PGAS is irreducible and aperiodic, and we have the following result.

Theorem 11.6. *For any $N \geq 2$, the PGAS sampler generates a process $\{X_{0:n}[k]\}_{k\geq 0}$ whose marginal distributions converge in total variation to the posterior distribution $\phi_{0:n|n}$ for $\phi_{0:n|n}$-almost all starting points.*

Proof. See Andrieu et al. (2010, Theorem 5). ■

Example 11.7 (Noisy AR(1) – Example 11.2, cont.). In this example, we compare the various techniques described in this chapter on data simulated from a noisy AR(1) model as described in Example 11.2, but with a sample of size $n = 200$. As discussed in Example 2.2, the model is a linear state space model.

Figure 11.2 compares the marginal posterior distributions for poor man's smoothing, FFBSm, FFBSi, and FFBSiLinear, each with 100 particles, and PGAS with 100 MCMC steps and 15 particles at each MCMC step. In each plot, the solid line is the Kalman smoother obtained via Proposition 2.7. The dashed line is the median of the particles, and the shaded area represents values between the 5% and 95% empirical quantiles of the particles. Note how the only two points of the marginal estimates for poor man's smoothing effect a narrower support and lack of coverage.

Using the same model, Figure 11.3 illustrates the difference in variability of the resulting estimates, comparing multiple independent estimations of the posterior smoothing mean for each algorithm. The smoothing mean estimates distribution for poor man's smoothing, FFBSm, FFBSi, FFBSiLinear, and PGAS, are based on 20 independent runs of each algorithm, with 100 particles (100 MCMC steps with 15 particles for PGAS). The Kalman smoothing posterior mean is displayed as a solid line. The gray area represents the values between the 5% and 95% empirical quantiles of the posterior means estimate. ◇

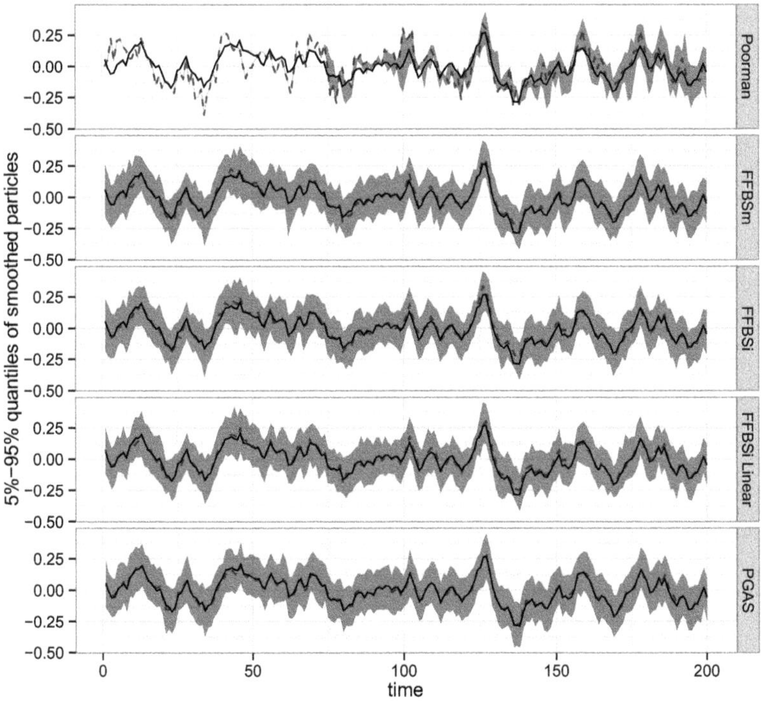

Figure 11.2 *Comparison of marginal posterior distributions for poor man's smoothing, FFBSm, FFBSi, and FFBSiLinear, each with 100 particles, and PGAS with 100 MCMC steps with 15 particles at each MCMC step for Example 11.7. In each plot, the solid line is the Kalman smoother. In the top plot, the dashed line is the median of the particles. In all plots, the shaded area represents values between the 5% and 95% empirical quantiles of the particles. Note how the only two points of the marginal estimates for poor man's smoothing effect a narrower support and consequently, lack of coverage.*

11.7 Convergence of the FFBSm and FFBSi algorithms

In this section, the convergence of the FFBSm and FFBSi algorithms is studied. For $0 \le t < n$ and $h \in \mathbb{F}_b(\mathsf{X}^{n+1}, \mathcal{X}^{\otimes(n+1)})$, define the kernel $L_{t,n} : \mathsf{X}^{t+1} \times \mathcal{X}^{\otimes n+1} \to [0,1]$ by

$$x_{0:t} \mapsto L_{t,n}h(x_{0:t}) := \int \cdots \int \left(\prod_{u=t+1}^{n} Q_u(x_{u-1}, \mathrm{d}x_u) \right) h(x_{0:n}) \tag{11.50}$$

and set $L_{n,n}h(x_{0:n}) := h(x_{0:n})$. By construction, for every $t \in \{0, \ldots, n\}$, the joint smoothing distribution may be decomposed as

$$\phi_{0:n|n}(h) = \frac{\phi_{0:t|t}(L_{t,n}h)}{\phi_{0:t|t}(L_{t,n}\mathbf{1})} \,. \tag{11.51}$$

This result suggests decomposing the error, $\phi^N_{0:n|n}(h) - \phi_{0:n|n}(h)$, of the FFBSm

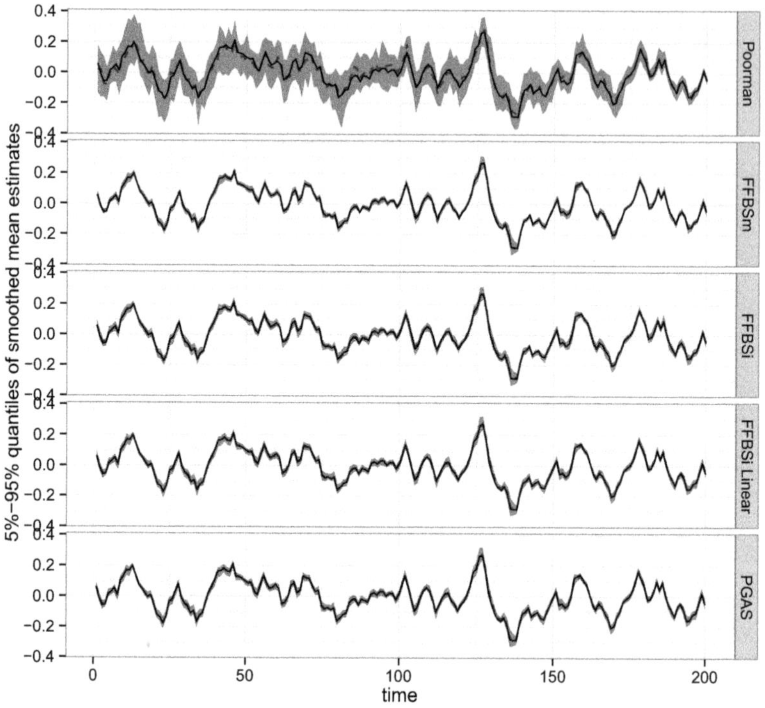

Figure 11.3 *For the noisy AR(1) model of Example 11.7, the plots show the difference in the variability of the resulting estimates, comparing multiple independent estimations of the posterior smoothing mean for each algorithm. The smoothing mean estimates distribution for poor man's smoothing, FFBSm, FFBSi, FFBSiLinear, and PGAS, are based on 20 independent runs of each algorithm, with 100 particles (100 MCMC steps with 15 particles for PGAS). The Kalman smoothing posterior mean is displayed as a solid line. The gray area represents the values between the 5% and 95% empirical quantiles of the posterior means estimate.*

estimator of the joint smoothing distribution (11.12) as the following telescoping sum:

$$\begin{aligned}\phi^N_{0:n|n}(h) - \phi_{0:n|n}(h) &= \frac{\phi^N_0(L_{0,n}h)}{\phi^N_0(L_{0,n}\mathbf{1})} - \frac{\phi_0(L_{0,n}h)}{\phi_0(L_{0,n}\mathbf{1})} \\ &\quad + \sum_{t=1}^{n} \left\{ \frac{\phi^N_{0:t|t}(L_{t,n}h)}{\phi^N_{0:t|t}(L_{t,n}\mathbf{1})} - \frac{\phi^N_{0:t-1|t-1}(L_{t-1,n}h)}{\phi^N_{0:t-1|t-1}(L_{t-1,n}\mathbf{1})} \right\} . \end{aligned} \tag{11.52}$$

The first term on RHS of the decomposition above can be easily dealt with since ϕ^N_0 is a weighted empirical distribution associated to i.i.d. random variables.

To cope with the terms in the sum of the RHS in (11.52), we introduce kernels (depending on the *past* particles) that stress the dependence with respect to the *current* particles and weights $\{(\omega^i_t, X^i_t)\}^N_{i=1}$. More precisely, $\phi^N_{0:t|t}(L_{t,n}h)$ is expressed as

$$\phi^N_{0:t|t}(L_{t,n}h) = \phi^N_t(P^N_{t,n}h) , \tag{11.53}$$

where the random kernels $P^N_{t,n} : \mathsf{X} \times \mathcal{X}^{\otimes(n+1)} \to [0,1]$ are defined by: for all $0 < t \leq n$, and $x_t \in \mathsf{X}$,

$$P^N_{t,n}h(x_t) := \int \cdots \int B_{\phi^N_{t-1}}(x_t, \mathrm{d}x_{t-1}) \cdots B_{\phi^N_0}(x_1, \mathrm{d}x_0) \, L_{t,n}h(x_{0:t}) , \tag{11.54}$$

and

$$P^N_{0,n}h(x) := L_{0,n}h(x) . \tag{11.55}$$

The kernels $P^N_{t,n}$ depend on the particles and weights $\{(X^i_s, \omega^i_s)\}^N_{i=1}$, $0 \leq s \leq t-1$, through the particle approximations $\phi^N_{t-1}, \ldots, \phi^N_0$ of the filtering distributions. To prove the CLT for the FFBSm algorithm, we need to show that for any $h \in \mathbb{F}_b(\mathsf{X}^{n+1}, \mathcal{X}^{\otimes(n+1)})$, $P^N_{t,n}h$ converges (see Lemma 11.13 below), as the number N of particles tends to infinity, to a deterministic function $P_{t,n}h$ given by

$$P_{t,n}h(x_t) := \int \cdots \int B_{\phi_{t-1}}(x_t, \mathrm{d}x_{t-1}) \cdots B_{\phi_0}(x_1, \mathrm{d}x_0) \, L_{t,n}h(x_{0:t}) . \tag{11.56}$$

In the sequel, the case where $h = \mathbf{1}$ is a constant function is of particular importance; in that case, $L_{t,n}\mathbf{1}(x_{0:t})$ does not depend on $x_{0:t-1}$, yielding

$$P^N_{t,n}\mathbf{1}(x_t) = P_{t,n}\mathbf{1}(x_t) = L_{t,n}\mathbf{1}(x_{0:t}) \tag{11.57}$$

for all $x_{0:t} \in \mathsf{X}^{t+1}$. Using these functions, the difference appearing in the sum in (11.52) may then be rewritten as

$$\frac{\phi^N_{0:t|t}(L_{t,n}h)}{\phi^N_{0:t|t}(L_{t,n}\mathbf{1})} - \frac{\phi^N_{0:t-1|t-1}(L_{t-1,n}h)}{\phi^N_{0:t-1|t-1}(L_{t-1,n}\mathbf{1})} = \frac{N^{-1}\sum_{\ell=1}^N \omega^\ell_t G^N_{t,n}h(X^\ell_t)}{N^{-1}\sum_{\ell=1}^N \omega^\ell_t P_{t,n}\mathbf{1}(X^\ell_t)} , \tag{11.58}$$

where the kernel $G^N_{t,n} : \mathsf{X} \times \mathcal{X}^{n+1} \to [0,1]$ is defined by, for $x \in \mathsf{X}$,

$$G^N_{t,n}h(x) := P^N_{t,n}h(x) - \frac{\phi^N_{t-1}(P^N_{t-1,n}h)}{\phi^N_{t-1}(P^N_{t-1,n}\mathbf{1})} P^N_{t,n}\mathbf{1}(x) . \tag{11.59}$$

Similarly to $P^N_{t,n}h$, the functions $G^N_{t,n}h$ depend on the past particles; it will however be shown (see Lemma 11.13 below) that $G^N_{t,n}h$ converges to the deterministic function given by, for $x \in \mathsf{X}$,

$$G_{t,n}h(x) := P_{t,n}h(x) - \phi_{0:n|n}(h)P_{t,n}\mathbf{1}(x) . \tag{11.60}$$

The key property of this decomposition is stated in the following lemma. Define for $t \in \{0, \ldots, N\}$ the σ-field

$$\mathcal{F}^N_t = \sigma(\{(X^i_s, \omega^i_s)\}^N_{i=1}) . \tag{11.61}$$

Lemma 11.8. *Assume A 10.13. Then, for any* $t \in \{0,\ldots,n\}$*, the variables* $\{\omega_t^\ell G_{t,n}^N h(X_t^\ell)\}_{\ell=1}^N$ *are, conditionally on the* σ*-field* $\mathcal{F}_{t-1}^N$*, i.i.d. with zero mean. Moreover, there exists a constant C (that may depend on t and n) such that, for all* $N \geq 1$*,* $\ell \in \{1,\ldots,N\}$*, and* $h \in \mathbb{F}_b(\mathsf{X}^{n+1}, \mathcal{X}^{\otimes(n+1)})$*,*

$$\left|\omega_t^\ell G_{t,n}^N h(X_t^\ell)\right| \leq |w_t/\vartheta_t|_\infty \left|G_{t,n}^N h(X_t^\ell)\right| \leq C \operatorname{osc}(h) \ .$$

Proof. The definition of the APF implies that the random variables $\{(X_t^\ell, \omega_t^\ell)\}_{\ell=1}^N$ are i.i.d. conditionally on the σ-field $\mathcal{F}_{t-1}^N$. This implies that the random variables $\{\omega_t^\ell G_{t,n}^N h(X_t^\ell)\}_{\ell=1}^N$ are also i.i.d. conditionally on the same σ-field $\mathcal{F}_{t-1}^N$. We now show that $\mathbb{E}\left[\omega_t^1 G_{t,n}^N h(X_t^1) \mid \mathcal{F}_{t-1}^N\right] = 0$. Using the definition of $G_{t,n}^N$ and the fact that $\phi_{t-1}^N(P_{t-1,n}^N h)$ and $\phi_{t-1}^N(P_{t-1,n}^N \mathbf{1})$ are $\mathcal{F}_{t-1}^N$-measurable, we have

$$\begin{aligned}
&\mathbb{E}\left[\omega_t^1 G_{t,n}^N h(X_t^1) \,\middle|\, \mathcal{F}_{t-1}^N\right] \\
&\quad = \mathbb{E}\left[\omega_t^1 P_{t,n}^N h(x) \,\middle|\, \mathcal{F}_{t-1}^N\right] - \frac{\phi_{t-1}^N(P_{t-1,n}^N h)}{\phi_{t-1}^N(P_{t-1,n}^N \mathbf{1})} \mathbb{E}\left[\omega_t^1 P_{t,n}^N \mathbf{1}(x) \,\middle|\, \mathcal{F}_{t-1}^N\right] ,
\end{aligned}$$

which is equal to zero provided that the relation

$$\mathbb{E}\left[\omega_t^1 P_{t,n}^N h(X_t^1) \,\middle|\, \mathcal{F}_{t-1}^N\right] = \frac{\phi_{t-1}^N(P_{t-1,n}^N h)}{\phi_{t-1}^N(\vartheta_t)} \tag{11.62}$$

holds for any $h \in \mathbb{F}_b(\mathsf{X}^{n+1}, \mathcal{X}^{\otimes(n+1)})$. We now turn to the proof of (11.62). Note that for any $f \in \mathbb{F}_b(\mathsf{X}, \mathcal{X})$,

$$\mathbb{E}\left[\omega_t^1 f(X_t^1) \,\middle|\, \mathcal{F}_{t-1}^N\right] = \frac{\sum_{\ell=1}^N \omega_{t-1}^\ell Q_t f(X_{t-1}^\ell)}{\sum_{\ell=1}^N \omega_{t-1}^\ell \vartheta_t(X_{t-1}^\ell)} = \frac{\phi_{t-1}^N(Q_t f)}{\phi_{t-1}^N(\vartheta_t)} . \tag{11.63}$$

It turns out that (11.62) follows from (11.63) since, as shown in Exercise 11.3,

$$\phi_{t-1}^N(P_{t-1,n}^N h) = \phi_{t-1}^N(Q_t P_{t,n}^N h) . \tag{11.64}$$

It remains to check that the random variable $\omega_t^1 G_{t,n}^N h(X_t^1)$ is bounded. But this is immediate since

$$\begin{aligned}
|\omega_t^1 G_{t,n}^N h(X_t^1)| &= |w_t/\vartheta_t|_\infty |P_{t,n}^N h - \phi_{t-1}^N(P_{t-1,n}^N h)/\phi_{t-1}^N(P_{t-1,n}^N \mathbf{1}) P_{t,n} \mathbf{1}|_\infty \\
&\leq 2|w_t/\vartheta_t|_\infty |P_{t,n}^N \mathbf{1}|_\infty \operatorname{osc}(h) \leq 2|w_t/\vartheta_t|_\infty |L_{t,n} \mathbf{1}|_\infty \operatorname{osc}(h) .
\end{aligned} \tag{11.65}$$

■

11.7.1 Exponential deviation inequality

Theorem 11.9. *Assume A 10.13. Then, there exist constants* $0 < c_{1,n}$ *and* $c_{2,n} < \infty$ *(depending on n) such that for all N,* $\varepsilon > 0$*, and all measurable functions* $h \in \mathbb{F}_b(\mathsf{X}^{n+1}, \mathcal{X}^{\otimes(n+1)})$*,*

$$\mathbb{P}\left[\left|\phi_{0:n|n}^N(h) - \phi_{0:n|n}(h)\right| \geq \varepsilon\right] \leq c_{1,n} \mathrm{e}^{-c_{2,n} N \varepsilon^2 / \operatorname{osc}^2(h)} . \tag{11.66}$$

In addition,

$$N^{-1} \sum_{\ell=1}^{N} \omega_t^\ell P_{t,n} \mathbf{1}(X_t^\ell) \xrightarrow{\mathbb{P}} \frac{\phi_{t-1}(P_{t-1,n} \mathbf{1})}{\phi_{t-1}(\vartheta_t)} . \tag{11.67}$$

Remark 11.10. As a byproduct, Theorem 11.9 provides an exponential inequality APF approximation of the filtering distribution. For any $h \in \mathbb{F}_b(\mathsf{X}, \mathcal{X})$, define the function $h_{0:n} : \mathsf{X}^{n+1} \to \mathbb{R}$ by $h_{0:n}(x_{0:n}) = h(x_n)$. By construction, $\phi_{0:n|n}(h_{0:n}) = \phi_n(h)$ and $\phi_{0:n|n}^N(h_{0:n}) = \phi_n^N(h)$. With this notation, (11.66) may be rewritten as

$$\mathbb{P}\left[\left|\phi_n^N(h) - \phi_n(h)\right| \geq \varepsilon\right] \leq c_{1,n} \mathrm{e}^{-c_{2,n} N \varepsilon^2 / \operatorname{osc}^2(h)} .$$

Theorem 10.17 can therefore be seen as a by-product of the exponential deviation inequality of the joint smoothing distribution.

Proof. We prove (11.66) by induction on n using the decomposition (11.52). Assume that (11.66) holds at time $n-1$. Let $h \in \mathbb{F}_b(\mathsf{X}^{n+1}, \mathcal{X}^{\otimes(n+1)})$ and assume without loss of generality that $\phi_{0:n|n}(h) = 0$. Then (11.51) implies that $\phi_0(L_{0,n} h) = 0$ and the first term of the decomposition (11.52) thus becomes

$$\frac{\phi_0^N(L_{0,n} h)}{\phi_0^N(L_{0,n} \mathbf{1})} = \frac{N^{-1} \sum_{\ell=0}^N w_0(X_0^\ell) L_{0,n} h(X_0^\ell)}{N^{-1} \sum_{\ell=0}^N w_0(X_0^\ell) L_{0,n} \mathbf{1}(X_0^\ell)} , \tag{11.68}$$

where $\{X_0^i\}_{i=1}^N$ is an i.i.d. sample of the proposal distribution r_0. We obtain an exponential inequality for (11.68) by applying Lemma 10.16 with

$$\begin{aligned} A_N &= N^{-1} \sum_{i=0}^N w_0(X_0^i) L_{0,n} h(X_0^i) , \\ B_N &= N^{-1} \sum_{i=0}^N w_0(X_0^i) L_{0,n} \mathbf{1}(X_0^i) , \\ B &= \beta = \xi[g_0(\cdot) L_{0,n} \mathbf{1}] , \end{aligned}$$

where ξ is the initial distribution. Condition (10.52a) is trivially satisfied and conditions (10.52b) and (10.52c) follow from the Hoeffding inequality for i.i.d. variables. By (11.52) and (11.58), it is now enough to establish an exponential inequality for

$$\frac{\phi_{0:t|t}^N(L_{t,n} h)}{\phi_{0:t|t}^N(L_{t,n} \mathbf{1})} - \frac{\phi_{0:t-1|t-1}^N(L_{t-1,n} h)}{\phi_{0:t-1|t-1}^N(L_{t-1,n} \mathbf{1})} = \frac{N^{-1} \sum_{\ell=1}^N \omega_t^\ell G_{t,n}^N h(X_t^\ell)}{N^{-1} \sum_{\ell=1}^N \omega_t^\ell P_{t,n} \mathbf{1}(X_t^\ell)} , \tag{11.69}$$

where $0 < t \leq n$. For that purpose, we use again Lemma 10.16 with

$$\begin{aligned} A_N &= N^{-1} \sum_{\ell=1}^N \omega_t^\ell G_{t,n}^N h(X_t^\ell) , \\ B_N &= N^{-1} \sum_{\ell=1}^N \omega_t^\ell P_{t,n} \mathbf{1}(X_t^\ell) , \\ B &= \beta = \phi_{t-1}(P_{t-1,n} \mathbf{1}) / \phi_{t-1}(\vartheta_t) . \end{aligned}$$

By considering the LHS of (11.69), $|A_N / B_N| \leq 2 \operatorname{osc}(h)$, verifying condition

(10.52a). By Lemma 11.8, Hoeffding's inequality implies that there exists a constant $c_{2,n}$ such that for all N, $\varepsilon > 0$, and all measurable function $h \in \mathbb{F}_b(\mathsf{X}^{n+1}, \mathcal{X}^{\otimes(n+1)})$,

$$\mathbb{P}\left[\left|N^{-1}\sum_{\ell=1}^{N}\omega_t^\ell G_{t,n}^N h(X_t^\ell)\right| \geq \varepsilon\right] \\ = \mathbb{E}\left[\mathbb{P}\left(\left|N^{-1}\sum_{\ell=1}^{N}\omega_t^\ell G_{t,n}^N h(X_t^\ell)\right| \geq \varepsilon \,\middle|\, \mathcal{F}_{t-1}^N\right)\right] \leq 2\mathrm{e}^{-c_{2,n}N\varepsilon^2/\operatorname{osc}^2(h)} ,$$

checking condition (10.52c). Condition (10.52b) is checked in Exercise 11.12.

Finally, (11.101) and (11.103) ensure that Condition (10.52b) in Lemma 10.16 is satisfied and an exponential deviation inequality for (11.69) follows. The proof of (11.66) is completed. The last statement (11.67) of the theorem is a consequence of (11.101) and (11.103). ■

The exponential inequality of Theorem 11.9 may be more or less immediately extended to the FFBSi estimator.

Corollary 11.11. *Under the assumptions of Theorem 11.9 there exist constants* $0 < c_{1,n}$ *and* $c_{2,n} < \infty$ *(depending on n) such that for all N,* $\varepsilon > 0$*, and all measurable functions h,*

$$\mathbb{P}\left[\left|\tilde{\phi}_{0:n|n}^N(h) - \phi_{0:n|n}(h)\right| \geq \varepsilon\right] \leq c_{1,n}\mathrm{e}^{-c_{2,n}N\varepsilon^2/\operatorname{osc}^2(h)} . \tag{11.70}$$

where $\tilde{\phi}_{0:n|n}^N(h)$ *is defined in* (11.21).

Proof. Using (11.20) and the definition of $\tilde{\phi}_{s:n|n}^N(h)$, we may write

$$\tilde{\phi}_{0:n|n}^N(h) - \phi_{0:n|n}^N(h) = N^{-1}\sum_{\ell=1}^{N}\left[h\left(X_0^{J_0^\ell},\ldots,X_n^{J_n^\ell}\right) - \mathbb{E}\left[h\left(X_0^{J_0},\ldots,X_n^{J_n}\right)\,\middle|\,\mathcal{F}_n^N\right]\right] ,$$

which implies (11.70) by the Hoeffding inequality and (11.66). ■

11.7.2 Asymptotic normality

We now extend the theoretical analysis of the forward-filtering backward-smoothing estimator (11.12) to a CLT. Consider the following mild assumption on the proposal distribution.

Assumption A11.12. $|m|_\infty < \infty$ and $\sup_{0\leq t\leq n}|r_t|_\infty < \infty$.

We first show that $G_{t,n}^N h$ converges to a deterministic function as $N \to \infty$.

Lemma 11.13. *Assume A10.13, A10.18, and A11.12. Then, for any* $h \in \mathbb{F}_b(\mathsf{X}, \mathcal{X})$ *and* $x \in \mathsf{X}$,

$$\lim_{N\to\infty} P_{t,n}^N h(x) = P_{t,n}h(x) , \quad \mathbb{P}\text{-a.s.},$$

$$\lim_{N\to\infty} G_{t,n}^N h(x) = G_{t,n}h(x) , \quad \mathbb{P}\text{-a.s.},$$

where $P^N_{t,n}$, $P_{t,n}$, $G^N_{t,n}$ *and* $G_{t,n}$ *are defined in* (11.54), (11.56), (11.59), *and* (11.60). *Moreover, there exists a constant C (that may depend on t and n) such that for all* $N \geq 1$, $\ell \in \{1,\ldots,N\}$, *and* $h \in \mathbb{F}_b(\mathsf{X},\mathcal{X})$,

$$\left|\omega_t^\ell G_{t,n}h(X_t^\ell)\right| \leq |w_t/\vartheta_t|_\infty \left|G_{t,n}h(X_t^\ell)\right| \leq C\operatorname{osc}(h) , \quad \mathbb{P}\text{-a.s.} \tag{11.71}$$

Proof. See Exercise 11.5. ■

Now, we may state the CLT with an asymptotic variance given by a finite sum of terms involving the limiting kernel $G_{t,n}$.

Theorem 11.14. *Assume A10.18, and A11.12. Then, for any* $h \in \mathbb{F}_b(\mathsf{X}^{n+1},\mathcal{X}^{\otimes(n+1)})$,

$$\sqrt{N}\left(\phi^N_{0:n|n}(h) - \phi_{0:n|n}(h)\right) \stackrel{\mathbb{P}}{\Longrightarrow} \mathrm{N}\left(0,\Gamma_{0:n|n}[h]\right) , \tag{11.72}$$

with

$$\Gamma_{0:n|n}[h] := \frac{r_0[w_0^2(G_{0,n}h)^2]}{\{r_0(w_0 P_{0,n}\mathbf{1})\}^2} + \sum_{t=1}^{n} \frac{\phi_{t-1}(\upsilon_{t,n}h)\phi_{t-1}(\vartheta_t)}{\phi^2_{t-1}(P_{t-1,n}\mathbf{1})} , \tag{11.73a}$$

$$\upsilon_{t,n}h(x) := \vartheta_t(x)\int R_t(x,\mathrm{d}x')\, w_t^2(x,x')\{G_{t,n}h(x')\}^2 . \tag{11.73b}$$

Proof. Without loss of generality, we assume that $\phi_{0:n|n}(h) = 0$. We show that $\sqrt{N}\phi^N_{0:n|n}(h)$ may be expressed as

$$\sqrt{N}\phi^N_{0:n|n}(h) = \sum_{t=0}^{n} \frac{V^N_{t,n}(h)}{W^N_{t,n}} , \tag{11.74}$$

where the sequence of random vectors $[V^N_{0,n}(h),\ldots,V^N_{n,n}(h)]$ is asymptotically normal and $[W^N_{0,n},\ldots,W^N_{n,n}]$ converge in probability to a deterministic vector. The proof of (11.72) then follows from Slutsky's Lemma. Actually, the decomposition (11.74) follows immediately from the backward decomposition (11.52) by setting, for $t \in \{1,\ldots,n\}$,

$$V^N_{0,n}(h) := N^{-1/2}\sum_{\ell=1}^{N} w_0(X_0^\ell)G_{0,n}h(X_0^\ell) , \quad V^N_{t,n}(h) := N^{-1/2}\sum_{\ell=1}^{N} \omega_t^\ell G^N_{t,n}h(X_t^\ell) ,$$

$$W^N_{0,n} := N^{-1}\sum_{\ell=1}^{N} w_0(X_0^\ell)P_{0,n}\mathbf{1}(X_0^\ell) , \qquad W^N_{t,n} := N^{-1}\sum_{\ell=1}^{N} \omega_t^\ell P_{t,n}\mathbf{1}(X_t^\ell) .$$

The convergence

$$W^N_{0,n} \xrightarrow{\mathbb{P}} \xi\left[g_0 P_{0,n}\mathbf{1}\right] , \quad W^N_{t,n} \xrightarrow{\mathbb{P}} \frac{\phi_{t-1}(P_{t-1,n}\mathbf{1})}{\phi_{t-1}(\vartheta_t)}$$

of $[W^N_{0,n},\ldots,W^N_{n,n}]$ to a deterministic vector follows from (11.67). We now show that the sequence of random vectors $[V^N_{0,n}(h),\ldots,V^N_{n,n}(h)]$ is asymptotically normal. We proceed recursively in time, i.e., we prove by induction over $t \in \{0,\dots,n\}$ (starting with $t=0$) that $[V_{0,n}^N(h),\dots,V_{t,n}^N(h)]$ is asymptotically normal. More precisely, using the Cramér-Wold device, it is enough to show that for all scalars $(\alpha_0,\dots,\alpha_t) \in \mathbb{R}^{t+1}$,

$$\sum_{r=0}^{t} \alpha_r V_{r,n}^N(h) \overset{\mathbb{P}}{\Longrightarrow} \mathrm{N}\left(0, \sum_{r=0}^{t} \alpha_r^2 \sigma_{r,n}^2[h]\right) , \tag{11.75}$$

where, for $r \geq 1$,

$$\sigma_{0,n}^2[h] := r_0[w_0^2 G_{0,n}^2 h] , \quad \sigma_{t,n}^2[h] := \frac{\phi_{t-1}(\upsilon_{t,n}h)}{\phi_{t-1}(\vartheta_t)} .$$

The case $t=0$ is elementary since the initial particles $\{X_0^i\}_{i=1}^N$ are i.i.d. Assume now that (11.75) holds for some $t-1 \leq n$, i.e., for all scalars $(\alpha_1,\dots,\alpha_{t-1}) \in \mathbb{R}^{t-1}$,

$$\sum_{r=s}^{t-1} \alpha_r V_{r,n}^N(h) \overset{\mathbb{P}}{\Longrightarrow} \mathrm{N}\left(0, \sum_{r=s}^{t-1} \alpha_r^2 \sigma_{r,n}^2[h]\right) . \tag{11.76}$$

The sequence of random variable $V_{t,n}^N(h)$ may be expressed as an additive function of a triangular array of random variables,

$$V_{t,n}^N(h) = \sum_{\ell=1}^{N} U_{N,\ell} , \quad U_{N,\ell} := \omega_t^\ell G_{t,n}^N h(X_t^\ell)/\sqrt{N} ,$$

where $G_{t,n}^N h(x)$ is defined in (11.59). Lemma 11.8 implies that $\mathbb{E}\left[V_{t,n}^N(h) \mid \mathcal{F}_{t-1}^N\right] = 0$, yielding

$$\mathbb{E}\left[\sum_{r=0}^{t} \alpha_r V_{r,n}^N(h) \,\middle|\, \mathcal{F}_{t-1}^N\right] = \sum_{r=0}^{t-1} \alpha_r V_{r,n}^N(h) \overset{\mathbb{P}}{\Longrightarrow} \mathrm{N}\left(0, \sum_{r=1}^{t-1} \alpha_r^2 \sigma_{r,n}^2[h]\right) ,$$

where the last limit follows by the induction assumption hypothesis (11.76). By Theorem B.20, since the random variables $\{U_{N,\ell}\}_{\ell=1}^N$ are centered and conditionally independent given $\mathcal{F}_{t-1}^N$, (11.75) holds provided that the asymptotic smallness condition

$$\sum_{\ell=1}^{N} \mathbb{E}\left[U_{N,\ell}^2 \mathbb{1}_{\{|U_{N,\ell}| \geq \varepsilon\}} \,\middle|\, \mathcal{F}_{t-1}^N\right] \overset{\mathbb{P}}{\longrightarrow} 0 \tag{11.77}$$

holds for any $\varepsilon > 0$ and that the conditional variance converges:

$$\sum_{\ell=1}^{N} \mathbb{E}\left[U_{N,\ell}^2 \,\middle|\, \mathcal{F}_{t-1}^N\right] \overset{\mathbb{P}}{\longrightarrow} \sigma_{t,n}^2[h] . \tag{11.78}$$

Lemma 11.8 implies that $|U_{N,\ell}| \leq C \operatorname{osc}(h)/\sqrt{N}$, verifying immediately the asymptotic smallness condition (11.77). To conclude the proof, we thus only need to establish the convergence (11.78) of the asymptotic variance. Using Lemma 11.8 and straightforward computations we conclude that

$$\sum_{\ell=1}^{N} \mathbb{E}\left[U_{N,\ell}^2 \,\middle|\, \mathcal{F}_{t-1}^N\right] = \mathbb{E}\left[\left(\omega_t^1 G_{t,n}^N h(X_t^1)\right)^2 \,\middle|\, \mathcal{F}_{t-1}^N\right] \tag{11.79}$$

$$= \int \sum_{\ell=1}^{N} \frac{\omega_{t-1}^\ell \vartheta_t(X_{t-1}^\ell) R_t(X_{t-1}^\ell, \mathrm{d}x)}{\sum_{j=1}^N \omega_{t-1}^j \vartheta_t(X_{t-1}^j)} \left(w_t(X_{t-1}^\ell, x) G_{t,n}^N h(x)\right)^2$$

$$= \left(\frac{\Omega_{t-1}}{\sum_{j=1}^N \omega_{t-1}^j \vartheta_t(X_{t-1}^j)}\right)\left(\frac{1}{\Omega_{t-1}} \sum_{\ell=1}^{N} \omega_{t-1}^\ell \upsilon_{t,n}^N h(X_{t-1}^\ell)\right) = \frac{\phi_{t-1}^N(\upsilon_{t,n}^N h)}{\phi_{t-1}^N(\vartheta_t)} .$$

where

$$\upsilon_{t,n}^N h := \vartheta_t \int R_t(\cdot, \mathrm{d}x)\, w_t^2(\cdot, x) \left[G_{t,n}^N h(x)\right]^2 .$$

The denominator in RHS of (11.79) converges in probability to $\phi_{t-1}(\vartheta_t)$ by Theorem 11.9. The numerator is more complex since $\upsilon_{t,n}^N$ depends on $G_{t,n}^N$ whose definition involves all the approximations $\phi_{t-1}^N, \ldots, \phi_0^N$ of the past filters. To obtain its convergence, note that, by Theorem 11.9, $\phi_{t-1}^N(\upsilon_{t,n}h) \rightarrow_{\mathbb{P}} \phi_{t-1}(\upsilon_{t,n}h)$ as N tends to infinity; hence, it only remains to prove that

$$\phi_{t-1}^N\left[\upsilon_{t,n}^N h - \upsilon_{t,n}h\right] \xrightarrow{\mathbb{P}} 0 . \tag{11.80}$$

For that purpose, consider the following notations: for all $x \in \mathsf{X}$,

$$A_N(x) := \phi_{t-1}^N\left[\vartheta_t r_t(\cdot, x) w_t^2(\cdot, x) |(G_{t,n}^N h(x))^2 - G_{t,n}^2 h(x)|\right] ,$$
$$B_N(x) := \phi_{t-1}^N\left[\vartheta_t r_t(\cdot, x)\right] .$$

Applying Fubini's theorem,

$$\lim_{N\to\infty} \mathbb{E}\left[\int A_N(x)\, \mathrm{d}x\right] = \lim_{N\to\infty} \int \mathbb{E}\left[A_N(x)\right] \mathrm{d}x = 0 , \tag{11.81}$$

where the last equality follows from the generalized Lebesgue convergence theorem (Royden, 1988, Proposition 18, p. 270) with $f_N(x) = \mathbb{E}\left[A_N(x)\right]$ and $g_N(x) = 2C \operatorname{osc}(h) \mathbb{E}\left[B_N(x)\right]$ provided that the following conditions proved in Exercise 11.4,

$$\text{for any } x \in \mathsf{X},\ \mathbb{E}\left[A_N(x)\right] \le 2C^2 \operatorname{osc}^2(h) \mathbb{E}\left[B_N(x)\right], \tag{11.82a}$$

$$\text{for any } x \in \mathsf{X},\ \lim_{N\to\infty} \mathbb{E}\left[A_N(x)\right] = 0,\ \mathbb{P}\text{-a.s.}, \tag{11.82b}$$

$$\lim_{N\to\infty} \int \mathbb{E}\left[B_N(x)\right] \mathrm{d}x = \int \lim_{N\to\infty} \mathbb{E}\left[B_N(x)\right] \mathrm{d}x. \tag{11.82c}$$

Thus, (11.81) holds, yielding that $\int A_N(x)\, \mathrm{d}x \xrightarrow{\mathbb{P}} 0$ as N tends to infinity. This in turn implies (11.80) using the inequality

$$\left|\phi_{t-1}^N\left[\upsilon_{t,n}^N h - \upsilon_{t,n}h\right]\right| \le \int A_N(x)\, \mathrm{d}x .$$

This establishes (11.75) and therefore completes the proof. ∎

Corollary 11.15. *Under the assumptions of Theorem 11.14,*

$$\sqrt{N}\left(\tilde{\phi}^N_{0:n|n}(h) - \phi_{0:n|n}(h)\right) \\ \stackrel{\mathbb{P}}{\Longrightarrow} \mathrm{N}\left(0, \phi^2_{0:n|n}\left[h - \phi_{0:n|n}(h)\right] + \Gamma_{0:n|n}\left[h - \phi_{0:n|n}(h)\right]\right) . \quad (11.83)$$

Proof. See Exercise 11.6. ■

11.7.3 Time uniform bounds

Most often, it is not required to compute the joint smoothing distribution but rather the marginal smoothing distributions $\phi_{s|n}$. In this section, we study the long-term behavior of the marginal fixed-interval smoothing distribution estimator defined in (11.16) and (11.17) under the strong mixing conditions A10.18.

The goal of this section consists in establishing, under the assumptions mentioned above, that the FFBSm approximation of the *marginal* fixed interval smoothing probability satisfies an exponential deviation inequality with constants that are uniform in time.

For any function $h \in \mathbb{F}_b(\mathsf{X}, \mathcal{X})$ and $s \leq n$, define the extension $\Pi_{s,n}h \in \mathbb{F}_b(\mathsf{X}^{n+1}, \mathcal{X}^{\otimes(n+1)})$ of h to X^{n+1} by

$$\Pi_{s,n}h(x_{0:n}) := h(x_s) , \quad x_{0:n} \in \mathsf{X}^{n+1} . \quad (11.84)$$

Lemma 11.16. *Assume A 10.18. Let $s \leq n$. Then, for all t,n, $N \geq 1$, and $h \in \mathbb{F}_b(\mathsf{X}, \mathcal{X})$,*

$$|G^N_{t,n}\Pi_{s,n}h|_\infty \leq \rho^{|t-s|} \operatorname{osc}(h) \, |P_{t,n}\mathbf{1}|_\infty , \quad (11.85)$$

where $P_{t,n}$ and $G^N_{t,n}$ are defined in (11.56) *and* (11.60), *respectively, and*

$$\rho = 1 - \sigma_-/\sigma_+ . \quad (11.86)$$

Moreover, for all t, $n \geq 1$, and $h \in \mathbb{F}_b(\mathsf{X}, \mathcal{X})$,

$$|G_{t,n}\Pi_{s,n}h|_\infty \leq \rho^{|t-s|} \operatorname{osc}(h) \, |P_{t,n}\mathbf{1}|_\infty . \quad (11.87)$$

Proof. Using (11.57) and (11.59),

$$\frac{G^N_{t,n}\Pi_{s,n}h(x)}{P_{t,n}\mathbf{1}(x)} = \frac{P^N_{t,n}\Pi_{s,n}h(x)}{P_{t,n}\mathbf{1}(x)} - \frac{\phi^N_{t-1}(P^N_{t-1,n}\Pi_{s,n}h)}{\phi^N_{t-1}(P^N_{t-1,n}\mathbf{1})} . \quad (11.88)$$

To prove (11.85), we will rewrite (11.88) and obtain an exponential bound by either using ergodicity properties of the "a posteriori" chain (when $t \leq s$), or by using ergodicity properties of the backward kernel (when $t > s$).

Assume first that $t \leq s$. The quantity $L_{t,n}\Pi_{s,n}h(x_{0:t})$ does not depend on $x_{0:t-1}$ so that by (11.54) and definition (11.50) of $L_{t,n}$,

$$P^N_{t,n}\Pi_{s,n}h(x_t) = L_{t,n}\Pi_{s,n}h(x_{0:t}) \\ = \int \cdots \int \left(\prod_{u=t+1}^{n} Q_u(x_{u-1}, \mathrm{d}x_u)\right) h(x_s) = P_{t,n}\Pi_{s,n}h(x_t) . \quad (11.89)$$

Now, by construction, for any $t \leq s$,

$$P_{t-1,n}\Pi_{s,n}h(x_{t-1}) = \int Q_t(x_{t-1}, \mathrm{d}x_t) P_{t,n}\Pi_{s,n}h(x_t) \,. \tag{11.90}$$

The relations (11.88), (11.89), and (11.90) imply that

$$\frac{G^N_{t,n}\Pi_{s,n}h(x)}{P_{t,n}\mathbf{1}(x)} = \frac{\xi(P_{t,n}\Pi_{s,n}h)}{\xi(P_{t,n}\mathbf{1})} - \frac{\xi'(P_{t,n}\Pi_{s,n}h)}{\xi'(P_{t,n}\mathbf{1})} \,, \tag{11.91}$$

where $\xi, \xi' \in \mathbb{M}_+(\mathcal{X})$, $\xi(A) := \delta_x(A)$ and $\xi'(A) := \phi^N_{t-1}Q_t(A)$, for all $A \in \mathcal{X}$. For any $\xi \in \mathbb{M}_+(\mathcal{X})$, we get

$$\begin{aligned}\frac{\xi(P_{t,n}\Pi_{s,n}h)}{\xi(P_{t,n}\mathbf{1})} &= \frac{\int\cdots\int \xi(\mathrm{d}x_t)\prod_{u=t+1}^n Q_u(x_{u-1}, \mathrm{d}x_u)h(x_s)}{\int\cdots\int \xi(\mathrm{d}x_t)\prod_{u=t+1}^n Q_u(x_{u-1}, \mathrm{d}x_u)} \\ &= \frac{\int\cdots\int \xi(\mathrm{d}x_t)\prod_{u=t+1}^s Q_u(x_{u-1}, \mathrm{d}x_u)h(x_s)P_{s,n}\mathbf{1}(x_s)}{\int\cdots\int \xi(\mathrm{d}x_t)\prod_{u=t+1}^s Q_u(x_{u-1}, \mathrm{d}x_u)P_{s,n}\mathbf{1}(x_s)} \,.\end{aligned}$$

A straightforward adaptation of Proposition 10.20 (see Exercise 11.7) implies that

$$\left| \frac{\xi(P_{t,n}\Pi_{s,n}h)}{\xi(P_{t,n}\mathbf{1})} - \frac{\xi'(P_{t,n}\Pi_{s,n}h)}{\xi'(P_{t,n}\mathbf{1})} \right| \leq \rho^{s-t}\operatorname{osc}(h) \,,$$

where ρ is defined in (11.86). This shows (11.85) when $t \leq s$. Consider now the case $s < t \leq n$. By definition,

$$\begin{aligned}P^N_{t,n}\Pi_{s,n}h(x_t) &= \int\cdots\int L_{t,n}\Pi_{s,n}h(x_{0:t}) \prod_{u=s+1}^{t} B_{\phi^N_{u-1}}(x_u, \mathrm{d}x_{u-1}) \\ &= \int\cdots\int P_{t,n}\mathbf{1}(x_t) \prod_{u=s+1}^{t} B_{\phi^N_{u-1}}(x_u, \mathrm{d}x_{u-1})\, h(x_s) \,, \end{aligned} \tag{11.92}$$

where the last expression follows from the relation: for $0 \leq s < t \leq n$,

$$L_{t,n}\Pi_{s,n}h(x_{0:t}) = h(x_s)\int\cdots\int \prod_{u=t+1}^{n} Q_u(x_{u-1}, \mathrm{d}x_u) = h(x_s)P_{t,n}\mathbf{1}(x_t) \,.$$

Moreover, combining (11.64) and (11.92),

$$\begin{aligned}\phi^N_{t-1}(P^N_{t-1,n}\Pi_{s,n}h) &= \phi^N_{t-1}(Q_t P^N_{t,n}\Pi_{s,n}h) = \iint \phi^N_{t-1}(\mathrm{d}x'_{t-1})\, Q_t(x'_{t-1}, \mathrm{d}x_t) P^N_{t,n}\Pi_{s,n}h(x_t) \\ &= \int\cdots\int \phi^N_{t-1}(\mathrm{d}x'_{t-1})\, Q_t(x'_{t-1}, \mathrm{d}x_t) P_{t,n}\mathbf{1}(x_t) \prod_{u=s+1}^{t} B_{\phi^N_{u-1}}(x_u, \mathrm{d}x_{u-1})\, h(x_s) \,.\end{aligned}$$

By plugging this expression and (11.92) into (11.88), we obtain

$$\frac{G^N_{t,n}\Pi_{s,n}h(x)}{P_{t,n}\mathbf{1}(x)} = \int\cdots\int \left\{ \frac{\xi(\mathrm{d}x_t)}{\xi(\mathsf{X})} - \frac{\xi'(\mathrm{d}x_t)}{\xi'(\mathsf{X})} \right\} \prod_{u=s+1}^{t} B_{\phi^N_{u-1}}(x_u, \mathrm{d}x_{u-1})\, h(x_s) \,,$$

with ξ and $\xi' \in \mathbb{M}_+(\mathcal{X})$ are given by

$$\xi(A) = \int_A \delta_x(\mathrm{d}x_t)\, P_{t,n}\mathbf{1}(x_t)\,,$$
$$\xi'(A) = \int \phi_{t-1}^N(m(\cdot,x_t))g_t(x_t)P_{t,n}\mathbf{1}(x_t)\mathbb{1}_A(x_t)\,\mathrm{d}x_t\,.$$

Under A10.18, for any probability measure η on $(\mathsf{X},\mathcal{X})$, and all $x \in \mathsf{X}$ and $A \in \mathcal{X}$,

$$B_\eta(x,A) = \frac{\int_A \eta(\mathrm{d}x')\, m(x',x)}{\int \eta(\mathrm{d}x')\, m(x',x)} \geq \frac{\sigma_-}{\sigma_+}\eta(A)\,.$$

Thus, the transition kernel B_η is uniformly Doeblin with minorizing constant σ_-/σ_+ and the proof of (11.85) for $s < t \leq n$ follows. (11.87) follows from (11.85) and Lemma 11.13, which shows that $\lim_{N\to\infty} G_{t,n}^N h(x) = G_{t,n}h(x)$, $\mathbb{P}$-a.s., for all $x \in \mathsf{X}$. ■

Theorem 11.17. *Assume A10.18. Then, there exist constants $0 \leq c_1,\ c_2 < \infty$ such that for all integers N, s, n, with $s \leq n$, and for all $\varepsilon > 0$,*

$$\mathbb{P}\left[\left|\phi_{s|n}^N(h) - \phi_{s|n}(h)\right| \geq \varepsilon\right] \leq c_1 \mathrm{e}^{-c_2 N\varepsilon^2/\operatorname{osc}^2(h)}\,, \tag{11.93}$$
$$\mathbb{P}\left[\left|\tilde{\phi}_{s|n}^N(h) - \phi_{s|n}(h)\right| \geq \varepsilon\right] \leq c_1 \mathrm{e}^{-c_2 N\varepsilon^2/\operatorname{osc}^2(h)}\,, \tag{11.94}$$

where $\phi_{s|n}^N(h)$ and $\tilde{\phi}_{s|n}^N(h)$ are defined in (11.12) *and* (11.21).

Proof. Combining (11.57) with the definition (11.50) yields, for all $x \in \mathsf{X}$,

$$\frac{\sigma_-}{\sigma_+} \leq \frac{P_{t,n}\mathbf{1}(x)}{|P_{t,n}\mathbf{1}|_\infty} \leq 1\,. \tag{11.95}$$

Let $h \in \mathbb{F}_b(\mathsf{X}^{n+1},\mathcal{X}^{\otimes(n+1)})$ and assume without loss of generality that $\phi_{0:n|n}(h) = 0$. Then, (11.51) implies that $\phi_0(L_{0,n}h) = 0$ and the first term of the decomposition (11.52) thus becomes

$$\frac{\phi_0^N(L_{0,n}h)}{\phi_0^N(L_{0,n}\mathbf{1})} = \frac{N^{-1}\sum_{i=0}^N w_0(X_0^i)L_{0,n}h(X_0^i)}{N^{-1}\sum_{i=0}^N w_0(X_0^i)L_{0,n}\mathbf{1}(X_0^i)}\,, \tag{11.96}$$

where the weight function w_0 is defined in (10.25) and $(X_0^i)_{i=1}^N$ are i.i.d. random variables with distribution r_0. Noting that $L_{0,n} = P_{0,n}$ we obtain an exponential deviation inequality for (11.96) by applying Lemma 10.16 with

$$\begin{cases} A_N := N^{-1}\sum_{i=0}^N w_0(X_0^i)P_{0,n}h(X_0^i)/|P_{0,n}\mathbf{1}|_\infty\,, \\ B_N := N^{-1}\sum_{i=0}^N w_0(X_0^i)P_{0,n}\mathbf{1}(X_0^i)/|P_{0,n}\mathbf{1}|_\infty\,, \\ B = \xi[g_0(\cdot)P_{0,n}\mathbf{1}]/|P_{0,n}h|_\infty\,, \\ \beta = \xi(g_0)\sigma_-/\sigma_+\,. \end{cases}$$

Here Condition (10.52a) is trivially satisfied and Conditions (10.52b) and (10.52c) follow from the Hoeffding inequality for i.i.d. variables.

According to (11.52) and (11.58), it is now required, for any $1 \leq t \leq n$, to derive an exponential inequality for

$$D_{t,n}^N := \frac{N^{-1}\sum_{\ell=1}^N \omega_t^\ell G_{t,n}^N \Pi_{s,n} h(X_t^\ell)}{N^{-1}\sum_{\ell=1}^N \omega_t^\ell P_{t,n}\mathbf{1}(X_t^\ell)} .$$

Note first that, using (11.95), we have

$$|D_{t,n}^N| \leq \left(\frac{\sigma_+}{\sigma_-}\right) \frac{N^{-1}\sum_{\ell=1}^N \omega_t^\ell G_{t,n}^N \Pi_{s,n} h(X_t^\ell)/|P_{t,n}\mathbf{1}|_\infty}{N^{-1}\sum_{\ell=1}^N \omega_t^\ell} .$$

We use again Lemma 10.16 with

$$\begin{cases} A_N = N^{-1}\sum_{\ell=1}^N \omega_t^\ell G_{t,n}^N \Pi_{s,n} h(X_t^\ell)/|P_{t,n}\mathbf{1}|_\infty , \\ B_N = N^{-1}\sum_{\ell=1}^N \omega_t^\ell , \\ B = \mathbb{E}\left[\omega_t^1 \mid \mathcal{F}_{t-1}^N\right] = \phi_{t-1}^N Q_t \mathbf{1}/\phi_{t-1}^N(\vartheta_t) , \\ \beta = c_-/|\vartheta_t|_\infty . \end{cases}$$

Assumption A 10.18 shows that $b \geq \beta$ and Lemma 11.16 shows that $|A_N/B_N| \leq \rho^{|t-s|} \operatorname{osc}(h)$, where ρ is defined in (11.86). Therefore, Condition (10.52a) is satisfied. On the other hand, the Hoeffding inequality implies

$$\begin{aligned} \mathbb{P}\left[|B_N - B| \geq \varepsilon\right] &\leq \mathbb{E}\left[\mathbb{P}\left(\left|N^{-1}\sum_{\ell=1}^N \left(\omega_t^\ell - \mathbb{E}\left[\omega_t^1 \,\middle|\, \mathcal{F}_{t-1}^N\right]\right)\right| \geq \varepsilon \,\middle|\, \mathcal{F}_{t-1}^N\right)\right] \\ &\leq 2\exp\left(-2N\varepsilon^2/|w_t/\vartheta_t|_\infty^2\right) , \end{aligned}$$

establishing Condition (10.52b). Finally, Lemma 11.8, Lemma 11.16, and the Hoeffding inequality imply that

$$\begin{aligned} \mathbb{P}\left[|A_N| \geq \varepsilon\right] &\leq \mathbb{E}\left[\mathbb{P}\left(\left|N^{-1}\sum_{\ell=1}^N \omega_t^\ell G_{t,n}^N \Pi_{s,n} h(X_t^\ell)/|P_{t,n}\mathbf{1}|_\infty\right| \geq \varepsilon \,\middle|\, \mathcal{F}_{t-1}^N\right)\right] \\ &\leq 2\exp\left(-2\frac{N\varepsilon^2}{|w_t/\vartheta_t|_\infty^2 \rho^{2|t-s|} \operatorname{osc}^2(h)}\right) . \end{aligned}$$

Lemma 10.16 therefore yields

$$\mathbb{P}\left(\left|\frac{A_N}{B_N}\right| \geq \varepsilon\right) \leq 2\exp\left(-\frac{N\varepsilon^2 c_-^2}{2\operatorname{osc}^2(h)\,\rho^{2|t-s|}|w_t|_\infty^2|\vartheta_t|_\infty}\right) ,$$

so that

$$\mathbb{P}\left(|D_{t,n}^N| \geq \varepsilon\right) \leq 2\exp\left(-\frac{N\varepsilon^2 c_-^2 \sigma_-^2}{2\operatorname{osc}^2(h)\,\rho^{2|t-s|}|w_t|_\infty^2|\vartheta_t|_\infty \sigma_+^2}\right) .$$

A time uniform exponential deviation inequality for $\sum_{t=1}^n D_{t,n}$ then follows from Lemma 10.22 and the proof is completed. ■

We conclude this section by presenting (without proof) non-asymptotic error bounds for the FFBSm and the FFBSi estimate of $\phi_{0:n|n}(S_n)$, where S_n is defined in (11.22).

Theorem 11.18. *Assume A10.18. For all $q \geq 2$, there exists a constant C (depending only on the constants appearing in A10.18) such that for any $n < \infty$, and any bounded and measurable functions $\{s_t\}_{t=1}^n$,*

$$\|\phi_{0:n|n}^N(S_n) - \phi_{0:n|n}(S_n)\|_q \leq C \frac{\left(\sum_{t=0}^n \operatorname{osc}^2(s_t)\right)^{1/2}}{\sqrt{N}} \left(1 + \sqrt{\frac{n}{N}}\right) ,$$

$$\|\tilde{\phi}_{0:n|n}^N(S_n) - \phi_{0:n|n}(S_n)\|_q \leq C \frac{\left(\sum_{t=0}^n \operatorname{osc}^2(s_t)\right)^{1/2}}{\sqrt{N}} \left(1 + \sqrt{\frac{n}{N}}\right) .$$

The proof can be found in Dubarry and Le Corff (2013, Theorem 1).

Remark 11.19. The dependence of the bound in $1/\sqrt{N}$ is hardly surprising. Under the stated strong mixing condition, it is known that the L_q-norm of the marginal smoothing estimator $\phi_{t-1:t|n}(h)$, $t \in \{1, \ldots, n\}$ is uniformly bounded in time by $\|\phi_{t-1:t|n}^N(h)\| \leq C \operatorname{osc}(h) N^{-1/2}$ (where the constant C depends only on the constants appearing in A 10.18). The dependence in $\sqrt{n}$ instead of n reflects the forgetting property of the filter and the backward smoother. As for $1 \leq s < t \leq n$, the estimators $\phi_{s-1:s|n}^N(h_s)$ and $\phi_{t-1:t|n}^{N,h_t}$ become asymptotically independent as $(t-s)$ gets large; the L_q-norm of the sum $\sum_{t=1}^T \phi_{t-1:t|n}^N(h_t)$ scales as the sum of a mixing sequence.

Remark 11.20. The scaling in $\sqrt{n/N}$ cannot in general be improved. Assume that the kernel m satisfies $m(x,x') = m(x')$ for all $(x,x') \in \mathsf{X} \times \mathsf{X}$. In this case, for any $t \in \{1, \ldots, n\}$, the filtering distribution is equal to

$$\phi_t(h_t) = \frac{\int m(x) g_t(x) h_t(x) \mathrm{d}x}{\int m(x) g_t(x) \mathrm{d}x} , \quad \text{for all } h_t \in \mathbb{F}_b(\mathsf{X}, \mathcal{X}) ,$$

and the backward kernel is the identity kernel. Hence, the fixed-interval smoothing distribution coincides with the filtering distribution. If we assume that we apply the bootstrap filter for which $r_t(x,x') = m(x')$ and $\vartheta_t(x) = 1$, the estimators $\{\phi_{t|n}^N(h_t)\}_{t \in \{0,\ldots,n\}}$ are independent random variables corresponding to importance sampling estimators. It is easily seen that

$$\|\sum_{t=0}^n \phi_{t|n}^N(h_t) - \phi_{t|n}(h_t)\| \leq C \max_{0 \leq t \leq n} \{\operatorname{osc}(h_t)\} \sqrt{\frac{n}{N}} .$$

11.8 Endnotes

Particle approximation of the fixed horizon smoothing distribution was considered in an early paper by Hürzeler and Künsch (1998). Godsill et al. (2001) describe an implementation of the forward filtering, backward smoothing algorithm to compute marginal smoothing distributions. Kitagawa and Sato (2001) proposed to approximate the fixed-interval smoothing distribution by fixed-lag smoothing distribution, which is obtained using the poor man's smoother; this algorithm was analyzed in Olsson et al. (2008). Godsill et al. (2004) described the forward filtering and backward simulation (FFBSi) algorithm. Klaas et al. (2006) argued that the N^2 complexity of the naive FFBSi can be reduced to $N\ln(N)$ by means of the fast multipole method. The gain in computational complexity, however, is obtained at the cost of introducing additional approximations causing a bias in the smoothing estimate that is not easy to quantify. The results of Godsill et al. (2004) were completed in Douc et al. (2010), who introduced an algorithm with complexity growing linearly in the number of particles N. The consistency and asymptotic normality of the FFBSi algorithm has been established in Douc et al. (2010) and Del Moral et al. (2010).

Briers et al. (2010) introduced another markedly different smoothing technique based on the two-filter approach. The 'forward' pass computes the filter while the 'backward' pass computes the *backward information filter*, which is not a probability measure on the state space. In cases where the backward information filter can be computed in closed form, this technical point is not important. However, for general state-space models where there is no closed form expression, this technicality prohibits the use of flexible numerical techniques such as Sequential Monte Carlo (SMC) to approximate the two-filter smoothing formula. Briers et al. (2010) proposed a generalized two-filter smoothing formula that requires simply approximating probability distributions and applies to any state-space model, thus removing the need to make restrictive assumptions used in previous approaches to this problem. Fearnhead et al. (2010) proposed a modification of the two-filter algorithm with computational complexity growing linearly with the number of particles N.

The recursive smoothing algorithm discussed in Section 11.4 was first described in Zeitouni and Dembo (1988) and Elliott (1993) for continuous time discrete state Markov processes observed in (Gaussian) noise. The approach is also at the core of the book by Elliott et al. (1995). The common theme of these works is the use of the EM algorithm (see Section 12.1.3), replacing forward-backward smoothing by recursive smoothing. The extension of the forward smoothing algorithm for more general functionals such as squared sums was first described in Cappé et al. (2005, Chapter 4). The particle approximation of the recursive smoothing formula was proposed and analyzed in Cappé (2001), Cappé et al. (2005), Del Moral et al. (2010) and Poyiadjis et al. (2011). Non-asymptotic error bounds for smoothed additive functionals were obtained in Del Moral et al. (2010) and later refined in Dubarry and Le Corff (2013).

The PIMH algorithm was introduced by Andrieu et al. (2010) as a member of the more general particle Markov chain Monte Carlo (PMCMC) framework. The correctness of the PIMH sampler can then be assessed by viewing the sampler as an instance of a pseudo-marginal MCMC sampler; see (Beaumont, 2003, Andrieu and Roberts, 2009). Andrieu and Vihola (2012) studied some of the theoretical properties of the pseudo-marginal algorithms in terms of the properties of the weights and those of the marginal algorithm. More precisely, the authors investigated the rate of convergence of the pseudo-marginal algorithm to equilibrium and characterized the approximation of the marginal algorithm by the pseudo-marginal algorithm in terms of the variability of their respective ergodic averages.

The asymptotics of PIMH do not rely on $N \to \infty$ to result in valid MCMC samplers. However, for the PIMH algorithm, it has been reported that the acceptance probability depends on the variance of the likelihood estimate (11.38). To get a high acceptance probability, we need to take N large so that the independent proposal distribution (11.27) is close to the target distribution $\phi_{0:n|n}$. For a fixed computational time, there is a trade-off between taking the number of particles N large to get a high acceptance probability, and to run many iterations of the MCMC sampler. This trade-off has been analyzed by Doucet et al. (2012) and Pitt et al. (2012) who, under certain assumptions, conclude that it is close to optimal to choose N so that $\mathrm{Var}(\ln(\widehat{\mathrm{L}}_n^N)) = 1$. As a rule of thumb, N should thus scale at least linearly with n to keep the variance of the likelihood estimator under control (Andrieu et al., 2010).

The particle Gibbs sampler can be designed in various different ways, leading to a variety of algorithms. The basic method, introduced in the seminal paper by Andrieu et al. (2010), is the most straightforward. However, this method is known to suffer from path degeneracy (see Section 11.1). Hence, for this method to work properly the number of particles N needs to be large enough to tackle the degeneracy. For many problems, this is unrealistic from a computational point of view.

The limitation of requiring a very large number of particles can be addressed by modifying the basic Gibbs sweep, for instance by adding a backward simulation (Whiteley, 2010, Whiteley et al., 2010, Lindsten and Schön, 2012) or ancestor sampling (Lindsten et al., 2012) step. In this chapter, we have reviewed the latter method, PG with ancestor sampling (PGAS). PGAS has been found to be robust to large n and small number N of particles. In many cases, the PGAS sampler can perform closely to the "ideal" Gibbs sampler.

PGAS offers an approach to joint smoothing that is quite different from the other SMC-based smoothers presented in this chapter. Indeed, based on the above discussion, we can say that PGAS is more MCMC-like than SMC-like. Compared to a Gibbs sampler, which updates the state variables one at a time, however, PGAS can enjoy much better mixing properties with a comparable computational complexity; see (Lindsten and Schön, 2013). PGAS is also reminiscent of the Gibbs sampler proposed by Neal et al. (2003) for inference in NLSS. In this method, a pool of candidate states is generated, followed by the simulation of a state trajectory. From this perspective, PGAS can be seen as a clever way of generating the candidate states by using SMC.

For the PGAS sampler, the dependence of the mixing rate on the number of particles N is not as obvious. Assume, for instance, that we run Algorithm 11.4 with $N = 1$. Due to the conditioning in Algorithm 11.3, we get by construction that $X_0^n[k] = X_0^n[k-1]$ for all $k \geq 1$. That is, the sampler will be stuck at the initial condition. For this reason, we require $N \geq 2$ in Theorem 11.6. On the other hand, as $N \to \infty$, the dependence on the conditioned particles in the CPF-AS will become negligible. That is, for large N, $X_0^n[k]$ will be effectively independent of $X_0^n[k-1]$. Intuitively, the larger we take N, the smaller the correlation will be between $X_0^n[k]$ and $X_0^n[k-1]$. However, it has been experienced in practice that the correlation drops off quickly as N increases, and for many models a moderate N (e.g., in the range 5–20) is enough to obtain a rapidly mixing Gibbs kernel.

The particle Gibbs sampler has been successfully applied to challenging inference problems, for instance for jump Markov linear systems by Whiteley et al.

(2010), for multiple change-point problems by Whiteley et al. (2011) and for Wiener system identification by Lindsten et al. (2013).

Exercises

11.1 (Proof of Theorem 11.1). Let $f \in \mathbb{F}_+(\mathsf{X}^{t+1}, \mathcal{X}^{\otimes(t+1)})$. Without loss of generality, we assume $\phi_{0:t|t}(f) = 0$. We assume that $\{(X^i_{0:t-1}, \omega^i_t)\}_{i=1}^N$ is consistent for $\phi_{0:t-1|t-1}$ and asymptotically normal for $(\phi_{0:t-1|t-1}, \sigma_{t-1}, \gamma_{t-1})$.

(a) Show that

$$\frac{1}{N}\sum_{i=1}^N \omega^i_t f(X^i_{0:t}) = \frac{1}{N}(A^N_t + B^N_t)\,,$$

where $A^N_t = \mathbb{E}\left[\omega^1_t f(X^1_{0:t}) \mid \mathcal{F}_{t-1}\right]$ and

$$B^N_t = \frac{1}{N}\sum_{i=1}^N \{\omega^i_t f(X^i_{0:t}) - \mathbb{E}\left[\omega^i_t f(X^i_{0:t}) \mid \mathcal{F}_{t-1}\right]\}\,.$$

(b) Show that

$$A^N_t = \sum_{i=1}^N \frac{\omega^i_{t-1}}{\sum_{j=1}^N \omega^j_{t-1}\vartheta_t(X^i_{t-1})} Q_t f(X^i_{0:t-1})\,,$$

where $\{X^i_{0:t-1}\}_{i=1}^N$ are the ancestral trajectories at time $t-1$ and $Q_t f(x_{0:t-1})$ is defined in (11.3).

(c) Show that (using the induction assumption)

$$\sqrt{N}\sum_{i=1}^N \frac{\omega^i_{t-1}}{\Omega^N_{t-1}} Q_t f(X^i_{0:t-1}) \overset{\mathbb{P}}{\Longrightarrow} \mathrm{N}(0, \sigma^2_{t-1}(Q_t f))\,.$$

(d) Show that

$$\sum_{i=1}^N \frac{\omega^i_{t-1}}{\Omega^N_{t-1}}\vartheta_t(X^i_{t-1}) \overset{\mathbb{P}}{\longrightarrow} \phi_{t-1}(\vartheta_t)\,.$$

(e) Deduce that $\sqrt{N}A^N_t \overset{\mathbb{P}}{\Longrightarrow} \mathrm{N}(0, \{\phi_{t-1}(\vartheta_t)\}^{-2}\sigma^2_{t-1}(Q_t f))$.

(f) Show that

$$N\mathbb{E}\left[\{B^N_t\}^2 \mid \mathcal{F}_{t-1}\right] = \mathbb{E}\left[(\omega^1_t f(X^1_{0:t}))^2 \mid \mathcal{F}_{t-1}\right] - \left(\mathbb{E}\left[\omega^1_t f(X^1_{0:t}) \mid \mathcal{F}_{t-1}\right]\right)^2\,.$$

(g) Show that

$$\begin{aligned}\mathbb{E}\left[(\omega^1_t f(X^1_{0:t}))^2 \mid \mathcal{F}_{t-1}\right] &\overset{\mathbb{P}}{\longrightarrow} \alpha^2_t \\ &= \iint \frac{\phi_{0:t-1|t-1}(\mathrm{d}x_{0:t-1})}{\phi_{t-1}(\vartheta_t)}\frac{w^2_t(x_{t-1},x_t)}{\vartheta_t(x_{t-1})} Q_t(x_{t-1},\mathrm{d}x_t) f^2(x_{0:t})\,.\end{aligned}$$

(h) Show that $\mathbb{E}\left[\omega^1_t f(X^1_{0:t}) \mid \mathcal{F}_{t-1}\right] \overset{\mathbb{P}}{\longrightarrow} 0$.

(i) Show that $\sqrt{N}B_t^N \overset{\mathbb{P}}{\Longrightarrow} \mathrm{N}(0, \alpha_t^2)$ and that $\sqrt{N}A_t^N$ and $\sqrt{N}B_t^N$ are asymptotically independent.

(j) Show that

$$\frac{\Omega_t^N}{N} \overset{\mathbb{P}}{\longrightarrow} \frac{\phi_{t-1}(Q_t 1)}{\phi_{t-1}(\vartheta_t)} .$$

(k) Conclude.

11.2. (a) Show that for any $\ell \geq 1$ and $0 \leq t-\ell < t \leq n$,

$$\phi^N_{t-\ell+1:t|n}(h) = \sum_{i_{t-\ell+1}=1}^{N} \cdots \sum_{i_t=1}^{N} \omega^{i_{t-\ell+1:t}}_{t-\ell+1:t|n} h\left(X^{i_{t-\ell+1}}_{t-\ell+1}, \ldots, X^{i_t}_t\right) .$$

where

$$\omega^{i_{t-\ell+1:t}}_{t-\ell+1:t|n} = \left(\prod_{u=t-\ell+2}^{t-1} \frac{\omega^{i_{u-1}}_{u-1} m(X^{i_{u-1}}_{u-1}, X^{i_u}_u)}{\sum_{\ell=1}^N \omega^\ell_{u-1} m(X^\ell_{u-1}, X^{i_u}_u)} \right) \omega^{i_t}_{t|n} .$$

(b) Propose a backward recursion for the smoothing weights.

(c) Compute the complexity of evaluating the smoothing weights.

11.3. (a) Show that

$$\begin{aligned}
&\phi^N_{t-1}(Q_t P^N_{t,n} h) \\
&= \Omega^{-1}_{t-1} \sum_{\ell=1}^N \omega^\ell_{t-1} \int \cdots \int q_t(X^\ell_{t-1}, x_t) \left(\prod_{u=1}^t B_{\phi^N_{u-1}}(x_u, \mathrm{d}x_{u-1}) \right) L_{t,n} h(x_{0:t}) \mathrm{d}x_t .
\end{aligned}$$

(b) Show that

$$\left(\sum_{\ell=1}^N \omega^\ell_{t-1} q_t(X^\ell_{t-1}, x_t) \right) B_{\phi^N_{t-1}}(x_t, \mathrm{d}x_{t-1}) = \sum_{\ell=1}^N \omega^\ell_{t-1} q_t(x_{t-1}, x_t)\, \delta_{X^\ell_{t-1}}(\mathrm{d}x_{t-1}) .$$

(c) Show that $\int Q_t(x_{t-1}, \mathrm{d}x_t) L_{t,n} h(x_{0:t}) = L_{t-1,n} h(x_{0:t-1})$.

(d) Show that

$$\begin{aligned}
&\sum_{\ell=1}^N \omega^\ell_{t-1} \int \cdots \int q_t(X^\ell_{t-1}, x_t) \prod_{u=1}^t B_{\phi^N_{u-1}}(x_u, \mathrm{d}x_{u-1}) L_{t,n} h(x_{0:t})\, \mathrm{d}x_t \\
&\qquad = \sum_{\ell=1}^N \omega^\ell_{t-1} P^N_{t-1} h(X^\ell_{t-1}) .
\end{aligned}$$

and establish (11.64).

11.4. (a) Show Eq. (11.82a) and that, for any $x \in \mathsf{X}$, $\limsup_{N\to\infty} A_N(x) = 0$.

(b) Show (11.82b). and $\lim_{N\to\infty} \int \mathbb{E}\left[B_N(x)\right] \mathrm{d}x = \int \lim_{N\to\infty} \mathbb{E}\left[B_N(x)\right] \mathrm{d}x$.

11.5. Let $h \in \mathbb{F}_b(\mathsf{X}, \mathcal{X})$ and $x_t \in \mathsf{X}$.

(a) Set $H(x_{0:t}) := m(x_{t-1}, x_t) L_{t,n} h(x_{0:t})$. Show that

$$P^N_{t,n} h(x_t) = \frac{\phi^N_{0:t-1|t-1}(H([\cdot, x_t]))}{\phi^N_{t-1}(m(\cdot, x_t))} .$$

(b) Show (11.71).

11.6. (a) Show that

$$\sqrt{N}\left(\tilde{\phi}^N_{0:n|n}(h) - \phi_{0:n|n}(h)\right) = \sqrt{N}\left(\phi^N_{0:n|n}(h) - \phi_{0:n|n}(h)\right) + N^{-1/2} \sum_{\ell=1}^{N} \left[h\left(X_0^{J_0^\ell}, \ldots, X_n^{J_n^\ell}\right) - \mathbb{E}\left[h\left(X_0^{J_0}, \ldots, X_n^{J_n}\right) \,\middle|\, \mathcal{F}^N_n \right] \right] .$$

(b) Show that

$$N^{-1} \sum_{\ell=1}^{N} \mathbb{E}\left[\left\{ h\left(X_0^{J_0^\ell}, \ldots, X_n^{J_n^\ell}\right) - \mathbb{E}\left[h\left(X_0^{J_0}, \ldots, X_n^{J_n}\right) \,\middle|\, \mathcal{F}^N_n \right] \right\}^2 \,\middle|\, \mathcal{F}^N_n \right] = \left(\phi^N_{0:n|n}\left[h - \phi^N_{0:n|n}(h)\right]\right)^2 \xrightarrow{\mathbb{P}} \left(\phi_{0:n|n}\left[h - \phi_{0:n|n}(h)\right]\right)^2 .$$

(c) Show (11.83)

11.7. Assume A 10.18-(c). Let $p \in \mathbb{F}_+(\mathsf{X}, \mathcal{X})$ be a positive function. Then, for all distributions $\xi, \xi' \in \mathbb{M}_1(\mathcal{X})$ and for all $s \le t$ and any bounded measurable functions $h \in \mathbb{F}_b(\mathsf{X}, \mathcal{X})$,

$$\left| \frac{\xi Q_{s,t}(hp)}{\xi Q_{s,t} p} - \frac{\xi' Q_{s,t}(hp)}{\xi' Q_{s,t} p} \right| \le \rho^{t-s} \operatorname{osc}(h) , \tag{11.97}$$

where $\rho := 1 - \sigma_- / \sigma_+$ and $Q_{s,t}$ is defined in (10.59).

11.8. (a) Show that the average number of simulations required to sample J_s^j conditionally on $\mathcal{G}^N_{s+1}$ is $\sigma_+ \Omega_s / \sum_{i=1}^N \omega_s^i m(X_s^i, X_{s+1}^{J_{s+1}^j})$.

(b) Show that

$$\mathbb{E}\left[Z_s^N \mid \mathcal{G}^N_{s+1}\right] = \sum_{\ell=1}^{N} \frac{\sigma_+ \Omega_s}{\sum_{i=1}^N \omega_s^i m(X_s^i, X_{s+1}^{J_{s+1}^\ell})} .$$

(c) Denote $\omega^i_{s|n} := \mathbb{P}\left[J_s^1 = i \,\middle|\, \mathcal{F}^N_n\right]$. Show that

$$\omega^{i,j}_{s:s+1|n} = \mathbb{P}\left[J_s^1 = i, J_{s+1}^1 = j \,\middle|\, \mathcal{F}^N_n\right] = \frac{\omega^j_{s+1|n} \omega_s^i m(X_s^i, X_{s+1}^j)}{\sum_{\ell=1}^N \omega_s^\ell m(X_s^\ell, X_{s+1}^j)} .$$

(d) Show that

$$\mathbb{E}\left[Z_s^N \mid \mathcal{F}^N_n\right] = \sigma_+ \Omega_s \sum_{i=1}^{N} \sum_{j=1}^{N} \omega^{i,j}_{s:s+1|n} \frac{1}{\omega_s^i m(X_s^i, X_{s+1}^j)} .$$

(e) Show that, for the bootstrap particle filter, $\omega_s^i \equiv g(X_s^i, Y_s)$ and $\Omega_s/N \xrightarrow{\mathbb{P}} \phi_{s|s-1}(g_s)$.

(f) Show that, for the bootstrap filter,

$$\sum_{i=1}^N \sum_{j=1}^N \omega_{s:s+1|n}^{i,j} \frac{1}{\omega_s^i m(X_s^i, X_{s+1}^j)} \xrightarrow{\mathbb{P}} \iint \phi_{s:s+1|n}(\mathrm{d}x_{s:s+1}) \frac{1}{q_{s+1}(x_s, x_{s+1})} .$$

(g) Show that

$$\iint \phi_{s:s+1|n}(\mathrm{d}x_{s:s+1}) \frac{1}{q_{s+1}(x_s, x_{s+1})} = \frac{\int \cdots \int \lambda(\mathrm{d}x_{s+1}) \prod_{u=s+2}^n \int Q_u(x_{u-1}, \mathrm{d}x_u)}{\int \cdots \int \phi_{s|s-1}(\mathrm{d}x_s)\, g_s(x_s) \prod_{u=s+1}^n Q_u(x_{u-1}, \mathrm{d}x_u)} .$$

(h) In the fully adapted case we have $\omega_s^i \equiv 1$ for all $i \in \{1, \ldots, N\}$; show that $\Omega_s = N$ and

$$\sum_{i=1}^N \sum_{j=1}^N \omega_{s:s+1|n}^{i,j} \frac{1}{\omega_s^i m(X_s^i, X_{s+1}^j)} \xrightarrow{\mathbb{P}} \frac{\int \cdots \int g_{s+1}(x_{s+1})\, \lambda(\mathrm{d}x_{s+1}) \prod_{u=s+2}^n Q_u(x_{u-1}, \mathrm{d}x_u)}{\int \cdots \int \phi_s(\mathrm{d}x_s) \prod_{u=s+1}^n Q_u(x_{u-1}, \mathrm{d}x_u)} .$$

(i) Conclude.

11.9 (Gibbs sampling for a discrete valued HMM). Consider an HMM on a finite state space X. Without loss of generality, we assume that the model is fully dominated.

(a) Show that the conditional probability density function of a *single* variable in the hidden chain, X_k, $k \in \{1, \ldots, n-1\}$, given $Y_{0:n}$ *and* its two neighbors X_{k-1} and X_{k+1} is such that

$$\phi_{k-1:k+1|n}(x_k | x_{k-1}, x_{k+1}) \propto \phi_{k-1:k+1|n}(x_{k-1}, x_k, x_{k+1}) \propto m(x_{k-1}, x_k) m(x_k, x_{k+1}) g_k(x_k) .$$

(b) Show that, at the two endpoints $k = 0$ and $k = n$, we have $\phi_{0:1|n}(x_0|x_1) \propto \xi(x_0) m(x_0, x_1)$ $\phi_{n-1:n|n}(x_n|x_{n-1}) \propto m(x_{n-1}, x_n) g_n(x_n)$.

(c) Write the Gibbs sampler; analyze the complexity of a single-sweep.

11.10. Assume that the state space X is finite. Consider the following algorithm to sample from the joint smoothing algorithm. Given the stored values of $\phi_{\xi,0}, \ldots, \phi_{\xi,n}$ computed by forward recursion according to Algorithm 9.1, do the following.

Final State: Simulate X_n from $\phi_{\xi,n}$.

Backward Simulation: For $t = n-1$ down to 0, sample X_k from $B_{\phi_{\xi,t}}(X_{t+1}, \cdot)$.

(a) Show that $X_{0:n}$ is distributed under the joint smoothing distribution.

(b) Show that the complexity of this algorithm is linear in n and quadratic in the cardinal of the state space.

(c) Consider a two-states hidden Markov chain $Y_t = \mu_{X_t} + \sigma_{X_t} V_t$, where $\{V_t,\ t \in \mathbb{N}\} \sim \text{iid}\ \mathrm{N}(0,1)$ and $\{X_t,\ t \in \mathbb{N}\}$ is a Markov chain on $\mathsf{X} = \{0,1\}$. Implement this algorithm and discuss its relative merits with respect to the Gibbs sampler Exercise 11.9.

11.11. The purpose of this exercise is to show that we may compute forward in time $\phi_{0:n|n}(S_n(x_{0:n})S_n'(x_{0:n}))$, where $S_n(x_{0:n})$ is defined in (11.22).

(a) Using n $S_n(x_{0:n}) = S_{n-1}(x_{0:n-1}) + s_n(x_{n-1:n})$, show that

$$\begin{aligned}\mathrm{Vec}(S_nS_n'(x_{0:n})) &= \mathrm{Vec}(S_{n-1}S_{n-1}'(x_{0:n-1})) + \mathrm{Vec}(s_ns_n'(x_{n-1:n})) \\ &\quad + \{s_n(x_{n-1:n}) \otimes \mathrm{I}_{d\times d} + \mathrm{I}_{d\times d} \otimes s_n(x_{n-1:n})\}S_{n-1}(x_{0:n-1})\,, \end{aligned} \tag{11.98}$$

where we used $\mathrm{Vec}(S_nS_n'(x_{0:n}))$ and $\mathrm{Vec}(s_ns_n'(x_{n-1},x_n))$ as shorthand notation for $\mathrm{Vec}(S_n(x_{0:n})S_n'(x_{0:n}))$ and $\mathrm{Vec}(s_n(x_{n-1},x_n)s_n'(x_{n-1},x_n))$

(b) Define the function $T_n : \mathsf{X}^{n+1} \to \mathbb{R}^{d+d^2}$ by

$$T_n(x_{0:n}) = \begin{pmatrix} S_n(x_{0:n}) \\ \mathrm{Vec}(S_nS_n'(x_{0:n})) \end{pmatrix}. \tag{11.99}$$

Define

$$\tau_n(x_n) = \int \cdots \int B_{\phi_{n-1}}(x_n, \mathrm{d}x_{n-1}) \ldots B_{\phi_1}(x_1, \mathrm{d}x_0) T_n(x_{0:n})\,. \tag{11.100}$$

Show that

$$\phi_{0:n|n}(T_n) = \int \phi_n(\mathrm{d}x_n)\tau_n(x_n)\,.$$

(c) Denote

$$A_n(x_{n-1:n}) = \begin{pmatrix} \mathrm{I}_{d\times d} & 0 \\ s_n(x_{n-1},x_n) \otimes \mathrm{I}_{d\times d} + \mathrm{I}_{d\times d} \otimes s_n(x_{n-1},x_n) & \mathrm{I}_{d^2\times d^2} \end{pmatrix},$$

$$b_n(x_{n-1:n}) = \begin{pmatrix} s_n(x_{n-1},x_n) \\ \mathrm{Vec}(s_ns_n'(x_{n-1},x_n)) \end{pmatrix}.$$

Show that

$$\tau_n(x_n) = \int B_{\phi_{n-1}}(x_n, \mathrm{d}x_{n-1}) A_n(x_{n-1:n}) \tau_{n-1}(x_{n-1}) + \int B_{\phi_{n-1}}(x_n, \mathrm{d}x_{n-1}) b_n(x_{n-1:n})\,.$$

(d) Denote

$$\tau_{n,1}(x_n) = \int \cdots \int B_{\phi_{n-1}}(x_n, \mathrm{d}x_{n-1}) \ldots B_{\phi_1}(x_1, \mathrm{d}x_0) S_n(x_{0:n})\,,$$

$$\tau_{n,2}(x_n) = \int \cdots \int B_{\phi_{n-1}}(x_n, \mathrm{d}x_{n-1}) \ldots B_{\phi_1}(x_1, \mathrm{d}x_0) S_nS_n'(x_{0:n})\,.$$

Show that

$$\tau_{n,1}(x_n) = \int B_{\phi_{n-1}}(x_n, \mathrm{d}x_{n-1})\{\tau_{n-1,1}(x_{n-1}) + s_n(x_{n-1}, x_n)\}$$

$$\tau_{n,2}(x_n) = \int B_{\phi_{n-1}}(x_n, \mathrm{d}x_{n-1})\{\tau_{n-1,2}(x_{n-1}) + s_n s_n'(x_{n-1}, x_n)\}$$

$$+ \int B_{\phi_{n-1}}(x_n, \mathrm{d}x_{n-1})\left\{s_n(x_{n-1}, x_n)\tau_{n-1,1}'(x_{n-1}) + \tau_{n-1,1}(x_{n-1})s_n'(x_{n-1}, x_n)\right\} .$$

(e) Denote by $\tau_{n-1,1}^N$ and $\tau_{n-1,2}^N$ particles approximations of the smoothing functionals at time $n-1$. Show that

$$\hat{\tau}_{n,1}^N(x_n) = \sum_{i=1}^N \frac{\omega_{n-1}^i m(X_{n-1}^i, x_n)}{\sum_{j=1}^N \omega_{n-1}^j m(X_{n-1}^j, x_n)} \{\hat{\tau}_{n-1,1}^N(X_{n-1}^i) + s_n(X_{n-1}^i, x_n)\}$$

$$\hat{\tau}_{n,2}(x_n) = \sum_{i=1}^N \frac{\omega_{n-1}^i m(X_{n-1}^i, x_n)}{\sum_{j=1}^N \omega_{n-1}^j m(X_{n-1}^j, x_n)} \{\hat{\tau}_{n-1,2}^N(X_{n-1}^i) + s_n s_n'(X_{n-1}^i, x_n)\}$$

$$+ \sum_{i=1}^N \frac{\omega_{n-1}^i m(X_{n-1}^i, x_n)}{\sum_{j=1}^N \omega_{n-1}^j m(X_{n-1}^j, x_n)} \{s_n(X_{n-1}^i, x_n)\hat{\tau}_{n-1,1}^{N\prime}(X_{n-1}^i)$$

$$+ \hat{\tau}_{n-1,1}^N(X_{n-1}^i)s_n'(X_{n-1}^i, x_n)\} .$$

(f) Show that

$$\hat{\tau}_{n,2}(x_n)$$

$$= \sum_{i=1}^N \frac{\omega_{n-1}^i m(X_{n-1}^i, x_n)}{\sum_{j=1}^N \omega_{n-1}^j m(X_{n-1}^j, x_n)} \{\hat{\tau}_{n-1,2}^N(X_{n-1}^i) - \hat{\tau}_{n-1,1}^N(X_{n-1}^i)\{\hat{\tau}_{n-1}^N(X_{n-1}^i)\}'\}$$

$$+ \sum_{i=1}^N \frac{\omega_{n-1}^i m(X_{n-1}^i, x_n)}{\sum_{j=1}^N \omega_{n-1}^j m(X_{n-1}^j, x_n)} \{\hat{\tau}_{n-1,1}^N(X_{n-1}^i) + s_n(X_{n-1}^i, x_n)\}$$

$$\times \{\hat{\tau}_{n-1,1}^N(X_{n-1}^i) + s_n(X_{n-1}^i, x_n)\}' .$$

(g) Write the algorithm and determine its numerical complexity.

11.12. The notations and definitions of Theorem 11.9 are used.

(a) Show that

$$\mathbb{P}\left[\left|B_N - \mathbb{E}\left[N^{-1}\sum_{\ell=1}^N \omega_t^\ell P_{t,n}\mathbf{1}(X_t^\ell) \,\middle|\, \mathcal{F}_{t-1}^N\right]\right| \geq \varepsilon\right] \leq 2\mathrm{e}^{-CN\varepsilon^2} . \tag{11.101}$$

(b) Show that

$$\mathbb{E}\left[N^{-1}\sum_{\ell=1}^N \omega_t^\ell P_{t,n}\mathbf{1}(X_t^\ell) \,\middle|\, \mathcal{F}_{t-1}^N\right] - B$$

$$= \frac{\phi_{t-1}^N(P_{t-1,n}\mathbf{1})}{\phi_{t-1}^N(\vartheta_t)} - \frac{\phi_{t-1}(P_{t-1,n}\mathbf{1})}{\phi_{t-1}(\vartheta_t)} = \frac{\phi_{t-1}^N(H)}{\phi_{t-1}^N(\vartheta_t)} , \tag{11.102}$$

with $H := P_{t-1,n}\mathbf{1} - \phi_{t-1}(P_{t-1,n}\mathbf{1})/\phi_{t-1}(\vartheta_t)\vartheta_t$.

(c) Show that

$$\mathbb{P}\left[\left|\mathbb{E}\left[N^{-1}\sum_{\ell=1}^{N}\omega_t^\ell P_{t,n}\mathbf{1}(X_t^\ell)\,\middle|\,\mathcal{F}_{t-1}^N\right]-B\right|>\varepsilon\right]\leq 2\mathrm{e}^{-CN\varepsilon^2}\,. \tag{11.103}$$

11.13. The PMCMC samplers of Algorithm 11.2 and Algorithm 11.4 can be specialized to the case where the observations are independent. The SMC components of the algorithms will then be reduced to importance sampling. Let the target density on X be given by π and assume that we use a proposal density r in an importance sampler for π. The importance weight function is thus given by

$$w(x)=\frac{\pi(x)}{r(x)}\,.$$

This importance sampler can be used to construct Markov chains with invariant density π.

(a) Repeat the derivation of the PIMH sampler for the static problem by defining an extended target density on $\mathsf{X}^N\times\{1,\ldots,N\}$ analogously to (11.36). Show that

$$\pi^N(\boldsymbol{x},j)=\frac{\pi(x^j)}{N}\prod_{\substack{\ell=1\\ \ell\neq j}}^{N}r(x^\ell)$$

$$\psi^N(\boldsymbol{x},j)=\prod_{\ell=1}^{N}r(x^\ell)\frac{\omega^j}{\Omega^N}\,,\quad\text{where}\quad \omega^j=w(x^j)\,.$$

(b) Show that the acceptance ratio in the IMH sampler is given by

$$\alpha(\boldsymbol{x},j;\tilde{\boldsymbol{x}},\tilde{j})=1\wedge\frac{\sum_{\ell=1}^N w(\tilde{x}^\ell)}{\sum_{\ell=1}^N w(x^\ell)}\,.$$

(c) Show that if $\inf_{x\in\mathsf{X}}\pi(x)/r(x)>0$, then this sampler is geometrically ergodic.

(d) Let $\pi(x)=\mathfrak{g}(x;0.5,0.5)$ and $r(x)=\mathfrak{g}(x;0,1)$. Implement and compare this sampler for different values of N.

(e) Consider the further specialization $N=1$. Verify that the PIMH sampler reduces to a standard IMH sampler for $\pi(x)$ with proposal density $r(x)$.

11.14. We use the same notation as in Exercise 11.13.

(a) Show that the conditional distribution of the particle index given the all the particles is given by

$$\pi^N(j|\boldsymbol{x})=\frac{w(x^j)}{\sum_{\ell=1}^N w(x^\ell)}\,.$$

(b) Show that

$$\pi^N(\boldsymbol{x}^{-j}|x^j,j)=\prod_{\ell\neq j}r(x^\ell)\,,$$

where $\boldsymbol{x}^{-j}=[x^1,\ldots,x^{j-1},x^{j+1},\ldots,x^n]$.

(c) Write a particle Gibbs algorithm targeting π.

(d) Show that if $\inf_{x \in \mathsf{X}} \pi(x)/r(x) > 0$, then this sampler is geometrically ergodic.

(e) Let $\pi(x) = \mathfrak{g}(x; 0.5, 0.5)$ and $r(x) = \mathfrak{g}(x; 0, 1)$. Implement the PIMH and the PG samplers for different values of N.

Chapter 12

Inference for Nonlinear State Space Models

In this chapter, we consider maximum likelihood estimation (MLE) and Bayesian inference for general nonlinear state space models (NLSS). MLE is, of course, a fundamental inference tool in classical statistics; we have already discussed the procedure in Section 2.3 for linear Gaussian state space models and in Section 8.1 for Markovian models. In NLSS, the maximization of the likelihood function is computationally involved. The likelihood and its gradient are seldom available in closed form, and Monte Carlo techniques are required to approximate these quantities. A first solution consists of using a derivative-free optimization method for noisy function; see Section 12.1.1. This approach is generally slow, and is limited to the case where the likelihood depends on only a few parameters. Another solution is to use a gradient-based search technique (for example, the steepest descent algorithm or a damped Gauss-Newton method) to compute estimates. This requires the computation of a gradient of the likelihood and the *score* function; for linear Gaussian state space models, the likelihood can be computed by deriving the recursions defining the Kalman filter (see Section 2.3.1). For general state space, the score function should also be approximated by Monte Carlo integration using Fisher's identity (see Section 12.1.2).

Unlike gradient-based searches, which are applicable to maximization of any differentiable cost function, the EM approach (see Section 2.3.2) is tailored to the maximization of likelihood functions in incomplete data models (see Appendix D). This method is clearly of interest for MLE in NLSS; in this case, the latent state sequence is considered as missing data. EM methods are known to be numerically stable and typically require little expertise to make fine tunings of the algorithm. Since its introduction in the early works in HMM by Baum and Petrie (1966), the EM algorithm has been the main vehicle for ML inference in HMMs. The EM is an iterative algorithm: at each step, a surrogate function is maximized. This surrogate function is defined as a conditional expectation (with respect to the observations) of the joint smoothing distribution of the complete data log-likelihood (i.e., the likelihood of the observations augmented by the latent state sequence). For Gaussian linear state-space models, the surrogate is obtained by standard linear smoothing (see Section 2.2). For general state-space models, the required smoothed state estimates are approximated by Monte Carlo based empirical averages. We primarily consider particle-based approximations of the smoothing distribution that were introduced in Chapter 11; however, other Monte Carlo techniques such as MCMC based algorithms can also be used as alternatives.

Recent years have witnessed an increased interest in Bayesian inference for NLSS. The most popular method for sampling from the posterior distribution is to include the latent state variables $X_{0:n}$ in the MCMC state space and to run the full Gibbs sampler, alternating between sampling model parameters and latent data from their respective full conditional distributions (see Section 5.8.2). The reasoning behind this is that, given the latent Markov chain and the data, the parameters are conditionally independent of distributions from standard parametric families (at least as long as the prior distribution for the model parameters is conjugate relative to the model specification). In addition, given the parameters and the data, the latent process is a non-homogeneous Markov chain. For linear Gaussian state-space models, this non-homogeneous Markov chain can be sampled directly; see Section 12.2.2. For general state space models, several options are available. A classical approach is to use a *one-at-a-time* approach, in which each individual latent state is updated conditionally to the others. This algorithm is straightforward to implement, but is known to suffer from poor mixing, because the distribution of the latent state sequence given the observations and the current fit of the parameter is often highly correlated. A more appealing approach consists of using the particle Gibbs algorithm; see Section 12.2.4.

Finally, in some cases, only the marginal distribution of the parameters given the observations is of interest. For linear Gaussian linear state space, sampling from the marginal distribution may be achieved by using a Metropolis-Hastings algorithm, because the likelihood of the observations can be computed in closed-form. In the general state space case, the likelihood is not explicitly available. However, an unbiased estimator of this quantity can be obtained through an adaptation of the Metropolis-Hastings algorithm, which yields the particle MCMC algorithm; see Section 12.2.3.

We denote by $\theta \in \Theta \subset \mathbb{R}^d$ the vector containing some components that parameterize the transition kernel of the NLSS and other components that parameterize the conditional distributions of the observations given the states. Throughout the chapter, it is assumed that the HMM model is, for all θ, fully dominated.

Assumption A 12.1.

(a) There exists a probability measure λ on $(\mathsf{X},\mathcal{X})$ such that for any $x \in \mathsf{X}$ and any $\theta \in \Theta$, $M^\theta(x,\cdot) \ll \lambda$ with transition density m^θ. That is, $M^\theta(x,A) = \int m^\theta(x,x')\,\lambda(\mathrm{d}x')$ for $A \in \mathcal{X}$.

(b) There exists a probability measure μ on $(\mathsf{Y},\mathcal{Y})$ such that for any $x \in \mathsf{X}$ and any $\theta \in \Theta$, $G^\theta(x,\cdot) \ll \mu$ with transition density function g^θ. That is, $G^\theta(x,A) = \int g^\theta(x,y)\,\mu(\mathrm{d}y)$ for $A \in \mathcal{Y}$.

For each parameter $\theta \in \Theta$, the distribution of the HMM is specified by the transition kernel M^θ, or equivalently the density m^θ of the Markov chain $\{X_t,\ t \in \mathbb{N}\}$, and by the conditional distribution g^θ of the observation Y_t given the hidden state X_t, referred to as the likelihood of the observation. As is common practice, we will also use the symbol p^θ generically for probability density functions that depend on the parameters θ. For example, we will write $m^\theta(X_{t-1}, X_t) = \mathrm{p}^\theta(X_t \mid X_{t-1})$, $g^\theta(X_t, Y_t) = \mathrm{p}^\theta(Y_t \mid X_t)$ or $\phi_t^\theta(X_t) = \mathrm{p}^\theta(X_t \mid Y_{0:t})$ when the notation is more convenient.

12.1 Monte Carlo maximum likelihood estimation

For any initial probability measure ξ on $(\mathsf{X}, \mathcal{X})$ and any $n \geq 0$, the likelihood of the observations $Y_{0:n}$ is given by

$$\mathrm{L}(\theta; Y_{0:n}) := \int \cdots \int \xi(\mathrm{d}x_0) g^\theta(x_0, Y_0) \prod_{s=1}^{n} M^\theta(x_{s-1}, \mathrm{d}x_s) g^\theta(x_s, Y_s) \ . \tag{12.1}$$

For notational convenience, the dependence on the initial distribution ξ is dropped. By conditioning, we may write the likelihood as the product of the predictive densities

$$\mathrm{L}(\theta; Y_{0:n}) = \mathrm{p}^\theta(Y_0) \prod_{t=1}^{n} \mathrm{p}^\theta(Y_t \mid Y_{0:t-1}) \ . \tag{12.2}$$

For general state space models, the likelihood $\theta \mapsto \mathrm{L}(\theta; Y_{0:n})$ cannot be expressed in closed-form and alternate solution must be found.

By conditioning with respect to the latent state X_t and then using the conditional independence of the observations and the state sequence, the predictive density $\mathrm{p}^\theta(X_t \mid Y_{0:t-1})$ may be expressed as a function of the filtering distribution of the latent state X_{t-1} as follows

$$\begin{aligned} \mathrm{p}^\theta(Y_t \mid Y_{0:t-1}) &= \iint \phi_{t-1}^\theta(\mathrm{d}x_{t-1}) M^\theta(x_{t-1}, \mathrm{d}x_t) g^\theta(x_t, Y_t) \\ &= \phi_{t-1}^\theta(Q_t^\theta \mathbf{1}) = \frac{\gamma_{t-1}^\theta(Q_t^\theta \mathbf{1})}{\gamma_{t-1}^\theta(\mathbf{1})} = \frac{\gamma_t^\theta(\mathbf{1})}{\gamma_{t-1}^\theta(\mathbf{1})} \ , \end{aligned} \tag{12.3}$$

where γ_t^θ, ϕ_t^θ, and Q_t^θ are defined in Section 9.2.2. For Gaussian linear state space models, the associated integrals have closed form solutions that lead to the Kalman filter (see Chapter 2). For finite-valued state spaces, the integrals can be transformed into finite sums, which also leads to explicit computations (see Chapter 9). In general, such recursions are numerically intractable; numerical integration is an option only when the dimension of the state space is small. Therefore, whereas in principle (12.2) and (12.3) provide a solution to the computation of (12.1), in practice there is a remaining obstacle. In the sequel, we will see how particle filtering techniques can be used to address this issue.

12.1.1 *Particle approximation of the likelihood function*

For $t \in \mathbb{N}$ and $f \in \mathbb{F}_b(\mathsf{X}, \mathcal{X})$, define

$$\kappa_t^N(f) = \frac{1}{N} \sum_{i=1}^{N} \omega_t^i f(X_t^i) \ , \tag{12.4}$$

where $\{(X_t^i, \omega_t^i)\}_{i=1}^N$ is the weighted particle approximation of the filtering distribution obtained using the auxiliary particle filter; see Section 10.4.2. To avoid overloading the notations, the dependence of the particles and the weights in the parameter θ is implicit. For example, the importance weight is

$$\omega_t^i = \frac{w_t^\theta(X_{t-1}^{I_t^i}, X_t^i)}{\vartheta_t^\theta(X_{t-1}^{I_t^i})} ,$$

where w_t^θ is the weight function associated to $Q_t^\theta(x_{t-1}, \mathrm{d}x_t) = M^\theta(x_{t-1}, \mathrm{d}x_t) g^\theta(x_t, Y_t)$ and the proposal kernel R_t^θ and ϑ_t^θ is the adjustment multiplier weight. We consider the following estimator of the likelihood function:

$$\widehat{\mathrm{L}}(\theta; Y_{0:n}) = \kappa_n^N(\mathbf{1}) \prod_{t=0}^{n-1} \kappa_t^N(\vartheta_{t+1}^\theta) . \tag{12.5}$$

Consider the following assumption.

Assumption A12.2.

(a) For all $(x,y) \in \mathsf{X} \times \mathsf{Y}$, $\theta \in \Theta$ and $t \in \{0, \ldots, n\}$, $g_t^\theta(x,y) > 0$ and $|g_t^\theta|_\infty < \infty$.

(b) For all $\theta \in \Theta$, $\sup_{0 \le t \le n} |\vartheta_t^\theta|_\infty < \infty$ and $\sup_{0 \le t \le n} \sup_{(x,x') \in \mathsf{X} \times \mathsf{X}} w_t^\theta(x,x') / \vartheta_t^\theta(x) < \infty$.

Theorem 12.3. *Assume A12.2. For all $n \in \mathbb{N}$, and $\theta \in \Theta$, $\widehat{\mathrm{L}}(\theta; Y_{0:n})$ given by* (12.5) *is an unbiased estimator of the likelihood of the observations. In addition, for any given sample size n, as the number of particles $N \to \infty$, the estimator is consistent and asymptotically normal.*

Proof. For all $f \in \mathbb{F}_+(\mathsf{X}, \mathcal{X})$ and $n \in \mathbb{N}$, we now show that

$$H_n^N(f) := \kappa_n^N(f) \prod_{t=0}^{n-1} \kappa_t^N(\vartheta_{t+1}^\theta) \tag{12.6}$$

is an unbiased estimator of $\gamma_0^\theta(Q_1^\theta \ldots Q_n^\theta f)$, i.e.,

$$\mathbb{E}\left[H_n^N(f)\right] = \gamma_0^\theta(Q_1^\theta \ldots Q_n^\theta f) . \tag{12.7}$$

We proceed by induction. For $n = 0$, the result is obvious. Assume that the result is true for $n-1$, where $n \ge 1$. Denote by $\mathcal{F}_n^N := \sigma\{Y_{0:n}, (X_s^i, \omega_s^i); 0 \le s \le n, 1 \le i \le N\}$, for $0 \le t \le n$, the σ-algebra generated by the observations as well as the particles and importance weights produced up to time n. By construction, for all $f \in \mathbb{F}_+(\mathsf{X}, \mathcal{X})$, we get

$$\mathbb{E}\left[\kappa_n^N(f) \mid \mathcal{F}_{n-1}^N\right] = \sum_{j=1}^N \frac{\omega_{n-1}^j \vartheta_n^\theta(X_{n-1}^j)}{\sum_{\ell=1}^N \omega_{n-1}^\ell \vartheta_n^\theta(X_{n-1}^\ell)} \int \frac{w_n^\theta(X_{n-1}^j, x)}{\vartheta_n^\theta(X_{n-1}^j)} R_n^\theta(X_{n-1}^j, \mathrm{d}x) f(x)$$
$$= \sum_{j=1}^N \frac{\omega_{n-1}^j}{\sum_{\ell=1}^N \omega_{n-1}^\ell \vartheta_n^\theta(X_{n-1}^\ell)} Q_n^\theta f(X_{n-1}^j) = \frac{\kappa_{n-1}^N(Q_n^\theta f)}{\kappa_{n-1}^N(\vartheta_n^\theta)} = \frac{\phi_{n-1}^N(Q_n^\theta f)}{\phi_{n-1}^N(\vartheta_n^\theta)} . \tag{12.8}$$

By applying (12.8), we get

$$\begin{aligned}\mathbb{E}\left[H_n^N(f) \mid \mathcal{F}_{n-1}^N\right] &= \prod_{t=0}^{n-1} \kappa_t^N(\vartheta_{t+1}^\theta)\mathbb{E}\left[\kappa_n^N(f) \mid \mathcal{F}_{n-1}^N\right] \\ &= \prod_{t=0}^{n-1} \kappa_t^N(\vartheta_{t+1}^\theta)\frac{\kappa_{n-1}^N(Q_n^\theta f)}{\kappa_{n-1}^N(\vartheta_n^\theta)} = \prod_{t=0}^{n-2} \kappa_t^N(\vartheta_{t+1}^\theta)\kappa_{n-1}^N(Q_n^\theta f) = H_{n-1}^N(Q_n^\theta f)\,.\end{aligned}$$

Since $\mathbb{E}\left[H_n^N(f)\right] = \mathbb{E}\left[\mathbb{E}\left[H_n^N(f) \mid \mathcal{F}_{n-1}^N\right]\right]$, the proof of (12.7) follows from the induction assumption. The proofs of consistency and asymptotic normality are given in Exercise 12.1 and Exercise 12.2. ■

To obtain $\widehat{\mathrm{L}}(\theta; Y_{0:n})$ for different $\theta \in \Theta$, a naïve use of (12.5) consists in generating new particle approximations using the model dynamics determined by these parameters. The errors for different θ are independent and therefore the estimated likelihood surface $\theta \mapsto \widehat{\mathrm{L}}(\theta; Y_{0:n})$ is noisy. This problem precludes the use of derivative-free deterministic optimization routines such as the Nelder-Mead method implemented in `optim` in R . Instead, one should use one of the derivative-free *noisy* function optimization algorithms; see Spall (2005), Devroye and Krzyzak (2005), Conn et al. (2009) and the references therein. These methods require the evaluation of the objective function at a finite number of points at each iteration and deciding which actions to take next solely based on those noisy function values.

A classical approach for derivative-free optimization of noisy function approximates the gradient of the function to be optimized by function difference; see Spall (2005, Chapter 5) and Section C.4.[1] The oldest method is the finite-difference (FD) approximation. The FD approximation relies on small one-at-a-time changes to each of the individual elements. After each change, the (possibly noisy) value of $\widehat{\mathrm{L}}(\theta; Y_{0:n})$ is measured. When measurements of $\widehat{\mathrm{L}}(\theta; Y_{0:n})$ have been collected for perturbations in each of the elements of θ, the gradient approximation may be formed. Note that, using noisy gradient-free methods, optimizing likelihood with more than a few parameters is usually very computationally expensive. In addition, even on relatively simple and well-conditioned problems, it is not realistic to expect accurate solutions. These types of methods are seldom used for MLE of NLSS models; see Olsson and Rydén (2008).

As mentioned above, in the estimator (12.5) of the likelihood function suggested by Theorem 12.3, a new set of particles and weights is generated for every value of the parameter θ. This can be at least partially avoided using a method introduced by Hürzeler and Künsch (2001), which extends the simulated likelihood method summarized in Geyer and Thompson (1992) and expanded upon in Geyer (1994). The technique makes it possible to use the same set of particles for different parameter values. This is done by multiplying the normalized weights in (12.4) by the ratio of the true predictive distribution at θ to the one at θ_0 (which may be updated during the

[1]The evaluation of the gradient can be avoided, as in the Nelder-Mead method for deterministic optimization of functions. Noisy direct-search methods (see Conn et al., 2009, Chapter 7) decide which parameters to choose without any explicit or implicit derivative approximation. We do not discuss these methods.

optimization procedure). The ratios are evaluated at the particles $\{(X_s^{N,i}, \omega_s^{N,i})\}_{i=1}^N$, which in this case are produced at some parameter θ_0. The disadvantages of this method are, firstly, that the weights degenerate if the current parameter θ value and the reference parameter value θ_0 for which the particle set has been drawn are far from each other and, secondly, that the weights themselves have to be recursively computed.

12.1.2 Particle stochastic gradient

Using the conditional independence structure of an NLSS, the complete data log-likelihood may be expressed as

$$\begin{aligned}\ln \mathrm{p}^\theta(X_{0:n}, Y_{0:n}) &= \ln \mathrm{p}^\theta(Y_{0:n}|X_{0:n}) + \ln \mathrm{p}^\theta(X_{0:n}) \\ &= \ln \mathrm{p}^\theta(X_0) + \sum_{t=1}^n \ln m^\theta(X_{t-1}, X_t) + \sum_{t=0}^n \ln g^\theta(X_t, Y_t) . \end{aligned} \tag{12.9}$$

Using the Fisher identity (see Section D.2), the score may be computed by smoothing the complete data score, i.e.,

$$\begin{aligned}\nabla \ln \mathrm{L}(\theta; Y_{0:n}) = \sum_{t=1}^n \iint \nabla \ln m^\theta(x_{t-1}, x_t) \phi^\theta_{t-1:t|n}(\mathrm{d}x_{t-1:t}) \\ + \sum_{t=0}^n \int \nabla \ln g^\theta(x_t, Y_t) \phi^\theta_{t|n}(\mathrm{d}x_t) ,\end{aligned} \tag{12.10}$$

assuming implicitly that the technical conditions required for Fisher's identity to be valid are satisfied. The evaluation of the incomplete data score only requires the computation of expectations under the marginal $\phi^\theta_{t|n}$ and the pairwise $\phi^\theta_{t-1:t|n}$ smoothing distributions. For Gaussian linear state-space models, the gradients $\nabla \ln m^\theta(X_{t-1}, X_t)$ and $\nabla \ln g^\theta(X_t, Y_t)$ are either linear or quadratic functions of the state X_{t-1} and X_t.

Computing the conditional expectation of the complete data score can therefore be achieved by using the Kalman smoother (see Section 2.3.2). The score may also be computed explicitly in cases where the state-space is finite, in which case the expectations can be replaced by finite sums. For general nonlinear state space models, the computation of (12.10) is intractable. The score function is a specific instance of a smoothed additive functional; see Section 11.4. To make the connection with (11.22), we simply rewrite $\nabla \ln \mathrm{L}(\theta; Y_{0:n})$ as $\sum_{t=0}^n \mathbb{E}^\theta \left[s_t^\theta(X_{t-1}, X_t) \,\middle|\, Y_{0:n} \right]$ where

$$s_0^\theta(X_0) := s_0^\theta(X_{-1}, X_0) := \nabla \ln g^\theta(X_0, Y_0) , \tag{12.11}$$
$$s_t^\theta(X_{t-1}, X_t) := \nabla \ln m^\theta(X_{t-1}, X_t) + \nabla \ln g^\theta(X_t, Y_t) . \tag{12.12}$$

Using the FFBSm algorithm, $\sum_{t=0}^n \mathbb{E}^\theta \left[s_t^\theta(X_{t-1}, X_t) \,\middle|\, Y_{0:n} \right]$ may be approximated by

$$\sum_{t=0}^n \sum_{i,j=1}^N \omega^{ij}_{t|N} s_t^\theta(X^i_{t-1}, X^j_t) ,$$

where the pairwise smoothing weight is given by (see Exercise 12.7)

$$\omega_{t|n}^{ij} = \frac{\omega_{t-1}^{i} \omega_{t|n}^{j} m^{\theta}(X_{t-1}^{i}, X_{t}^{j})}{\sum_{\ell=1}^{N} \omega_{t-1}^{\ell} m^{\theta}(X_{t-1}^{\ell}, X_{t}^{j})} . \tag{12.13}$$

In addition to the gradient, it is also possible to approximate the Hessian of the likelihood; we do not pursue this here, but encourage the reader to work out the details in Exercise 12.8. Using this estimate of the score function, maximum likelihood estimation can be tackled using the stochastic gradient algorithm; see Section C.2. Denote by $\theta[k]$ the k-th iterate of the parameter estimate. The k-th iteration of the stochastic gradient algorithm may be written as

$$\theta[k] = \theta[k-1] + a_k \left\{ \sum_{i=1}^{m_k} \omega_{0|n}^{k,i} s_0^{\theta[k-1]}(X_0^{k,i}) + \sum_{t=1}^{n} \sum_{i,j=1}^{m_k} \omega_{t|n}^{k,ij} s_t^{\theta[k-1]}(X_{t-1}^{k,i}, X_t^{k,j}) \right\} , \tag{12.14}$$

where $\{m_k,\ k \in \mathbb{N}\}$ is the number of particles used at each iteration of the algorithm, $\{a_k,\ k \in \mathbb{N}\}$ is a sequence of positive stepsizes and $\{(X_t^{k,i}, \omega_{t|n}^{k,i}),\ t = 0, \ldots, n\}_{i=1}^{m_k}$ is a set of particles and weights targeting the sequence of marginal smoothing distributions $\{\phi_{t|n}^{\theta[k-1]},\ t = 0, \ldots, n\}$. There are several possible settings. We may use an increasing number of particles m_k i.e., $\lim_{k \to \infty} m_k = \infty$. Intuitively, for the first few iterations, only a rather crude estimate of the gradient is required. As the algorithm approaches convergence, however, the number of particles used to approximate the gradient must be increased to mitigate the errors. The selection of an appropriate stepsize is, in this setting, a difficult issue. The classical techniques such as the limited line search (with golden section rules or cubic interpolation) or the Armijo rule see (see Lange, 2010, Chapters 5 and 15) require the evaluation of the likelihood, which is numerically costly. Also, in practice, these methods are tricky to use because both the score and the likelihood are noisy approximations of the true quantities (the convergence of such schemes has never been carefully established).

Another possibility consists of using a fixed number of particles $m_k \equiv m$, in which case a sequence of decreasing stepsize must be used to mitigate the noise in the gradient estimator. A typical condition on the stepsize to ensure $\mathbb{P}$-a.s. convergence of the sequence of estimators (see Section C.1) is $\sum_{k=1}^{\infty} a_k = +\infty$ and $\sum_{k=1}^{\infty} a_k^2 < \infty$; this condition is often fulfilled by taking $a_k = Ck^{-\alpha}$ where C is a constant and $\alpha \in (1/2, 1)$. The number of particles, m, cannot be chosen too small, otherwise the score approximation will be both very noisy and biased. The presence of additive noise is (at least theoretically) not a serious impediment when using stochastic approximation because the use of decreasing stepsizes is designed to mitigate these fluctuations; see Section C.1. On the contrary, the bias is a problem because its effect will not be corrected by the stochastic approximation procedure. The bias reduction would require an increase in the number of simulations m_k as $k \to \infty$ and to adjust accordingly the rate at which a_k approaches zero: we do not investigate this possibility here.

12.1.3 Particle Monte Carlo EM algorithms

The main drawback of the stochastic gradient algorithm is that it is difficult to properly scale the components of the computed gradient vector. For this reason, the EM algorithm is usually favored by practitioners whenever it is applicable; see Section D.1.

The first challenge in implementing the EM algorithm for NLSS is the computation of the EM surrogate function $\mathcal{Q}(\theta\,;\theta')$ defined as the conditional expectation of the complete data log-likelihood, see (D.4). As above, for Gaussian linear state-space models, the $\ln m^{\theta}(X_{t-1}, X_t)$ and $\ln g^{\theta}(X_t, Y_t)$ terms are either linear or quadratic functions of the state X_{t-1} and X_t. Taking the conditional expectation of the complete data likelihood is simply achieved by invoking the Kalman smoother (see Section 2.3.2). The surrogate function may also be computed explicitly in cases where the state-space is finite, in which case the expectations can be replaced by sums.

Example 12.4 (Normal hidden Markov models). In the frequently used normal HMM, X is a finite set, identified with $\{1, \ldots, r\}$, $\mathsf{Y} = \mathbb{R}$, and g is a Gaussian probability density function (with respect to Lebesgue measure) given by

$$g^{\theta}(x,y) = \frac{1}{\sqrt{2\pi \upsilon_x}} \exp\left\{ -\frac{(y-\mu_x)^2}{2\upsilon_x} \right\} .$$

We first assume that the initial distribution ξ is known and fixed. The parameter vector θ thus encompasses the transition probabilities m_{ij} for $i, j = 1, \ldots, r$ and the means μ_i and the variances υ_i for $i = 1, \ldots, r$. Because we will often need to differentiate with respect to υ_i, it is simpler to use the variances $\upsilon_i = \sigma_i^2$ rather than the standard deviations σ_i as parameters. The means and variances are unconstrained, except for the positivity of the latter, but the transition probabilities are subject to the constraints $m_{ij} \geq 0$ and $\sum_{j=1}^{r} m_{ij} = 1$ for $i = 1, \ldots, r$.

For the model under consideration, $\mathcal{Q}(\theta\,;\theta')$ may be rewritten as, up to a constant,

$$\mathcal{Q}(\theta\,;\theta') = -\frac{1}{2}\sum_{t=0}^{n} \mathbb{E}_{\theta'}\left[\sum_{i=1}^{r} \mathbb{1}\{X_t = i\} \left(\ln \upsilon_i + \frac{(Y_t-\mu_i)^2}{\upsilon_i} \right) \middle| \, Y_{0:n} \right] \\ + \sum_{t=1}^{n} \mathbb{E}_{\theta'}\left[\sum_{i=1}^{r}\sum_{j=1}^{r} \mathbb{1}\{(X_{t-1}, X_t) = (i,j)\} \ln m_{ij} \middle| \, Y_{0:n} \right] .$$

Using the notations introduced in Chapter 11 for the smoothing distributions, we may write

$$\mathcal{Q}(\theta\,;\theta') = -\frac{1}{2}\sum_{t=0}^{n}\sum_{i=1}^{r} \phi_{t|n}^{\theta'}(i) \left[\ln \upsilon_i + \frac{(Y_t-\mu_i)^2}{\upsilon_i} \right] \\ + \sum_{t=1}^{n}\sum_{i=1}^{r}\sum_{j=1}^{r} \phi_{t-1:t|n}^{\theta'}(i,j) \ln m_{ij} . \quad (12.15)$$

Given the initial distribution ξ and parameter θ', the smoothing distributions appearing in (12.15) can be evaluated by any of the variants of forward-backward smoothing discussed in Chapter 11. The E-step of EM thus reduces to solving the smoothing problem. The M-step is specific and depends on the model parameterization: the task consists of finding a global optimum of $\mathcal{Q}(\theta\,;\theta')$ that satisfies the constraints mentioned above; see Appendix D. For this, simply introduce the Lagrange multipliers $\lambda = (\lambda_1, \ldots, \lambda_r)$ that correspond to the equality constraints $\sum_{j=1}^{r} m_{ij} = 1$ for $i = 1, \ldots, r$ (Luenberger, 1984, Chapter 10). The first-order partial derivatives of the Lagrangian

$$\mathcal{L}(\theta, \lambda\,;\theta') = \mathcal{Q}(\theta\,;\theta') + \sum_{i=1}^{r} \lambda_i \left(1 - \sum_{j=1}^{r} m_{ij}\right)$$

are given by

$$\begin{aligned}
\frac{\partial}{\partial \mu_i}\mathcal{L}(\theta, \lambda\,;\theta') &= \frac{1}{\upsilon_i}\sum_{t=0}^{n} \phi_{t|n}^{\theta'}(i)(Y_t - \mu_i)\,, \\
\frac{\partial}{\partial \upsilon_i}\mathcal{L}(\theta, \lambda\,;\theta') &= -\frac{1}{2}\sum_{t=0}^{n} \phi_{t|n}^{\theta'}(i)\left[\frac{1}{\upsilon_i} - \frac{(Y_t - \mu_i)^2}{\upsilon_i^2}\right]\,, \\
\frac{\partial}{\partial m_{ij}}\mathcal{L}(\theta, \lambda\,;\theta') &= \sum_{t=1}^{n} \frac{\phi_{t-1:t|n}(i, j\,;\theta')}{m_{ij}} - \lambda_i\,, \\
\frac{\partial}{\partial \lambda_i}\mathcal{L}(\theta, \lambda\,;\theta') &= 1 - \sum_{j=1}^{r} m_{ij}\,.
\end{aligned} \tag{12.16}$$

Equating all expressions in (12.16) to zero yields the parameter vector

$$\theta^* = \left[(\mu_i^*)_{i=1,\ldots,r}, (\upsilon_i^*)_{i=1,\ldots,r}, (m_{ij}^*)_{i,j=1,\ldots,r}\right]\,,$$

which achieves the maximum of $\mathcal{Q}(\theta\,;\theta')$ under the applicable parameter constraints:

$$\mu_i^* = \frac{\sum_{t=0}^{n} \phi_{t|n}^{\theta'}(i) Y_t}{\sum_{t=0}^{n} \phi_{t|n}^{\theta'}(i)}\,, \tag{12.17}$$

$$\upsilon_i^* = \frac{\sum_{t=0}^{n} \phi_{t|n}^{\theta'}(i)(Y_t - \mu_i^*)^2}{\sum_{t=0}^{n} \phi_{t|n}^{\theta'}(i)}\,, \tag{12.18}$$

$$m_{ij}^* = \frac{\sum_{t=1}^{n} \phi_{t-1:t|n}^{\theta'}(i, j)}{\sum_{t=1}^{n}\sum_{\ell=1}^{r} \phi_{t-1:t|n}^{\theta'}(i, \ell)}\,, \tag{12.19}$$

for $i, j = 1, \ldots, r$, where the last equation may be rewritten more concisely as

$$m_{ij}^* = \frac{\sum_{t=1}^{n} \phi_{t-1:t|n}^{\theta'}(i, j)}{\sum_{t=1}^{n} \phi_{t-1|n}^{\theta'}(i)}\,. \tag{12.20}$$

◇

For nonlinear state space models, the situation is more delicate. The surrogate function $\mathcal{Q}(\theta;\theta')$ may be decomposed as $\mathcal{Q}(\theta;\theta') = \sum_{i=1}^{3} \mathcal{Q}_i(\theta;\theta')$, where

$$\mathcal{Q}_1(\theta;\theta') = \int \ln \mathrm{p}^{\theta}(x_0)\phi_{0|n}^{\theta'}(\mathrm{d}x_0) , \tag{12.21a}$$

$$\mathcal{Q}_2(\theta;\theta') = \sum_{t=1}^{n} \iint \ln m^{\theta}(x_{t-1},x_t)\phi_{t-1:t|n}^{\theta'}(\mathrm{d}x_{t-1:t}) , \tag{12.21b}$$

$$\mathcal{Q}_3(\theta;\theta') = \sum_{t=0}^{n} \int \ln g^{\theta}(x_t,Y_t)\phi_{t|n}^{\theta'}(\mathrm{d}x_t) . \tag{12.21c}$$

One way to address this issue is to compute the E-step using particle smoothing (see Chapter 11), leading to the Particle EM algorithm, which is a special instance of the Monte Carlo EM (MCEM) algorithm; see Section D.3.

The FFBSm algorithm, (11.16)–(11.17), directly provides the importance sampling approximations of (12.21), namely,

$$\hat{\mathcal{Q}}_1(\theta;\theta') := \sum_{i=1}^{N} \omega_{0|n}^{i} \ln \mathrm{p}^{\theta}(X_0^i) , \tag{12.22a}$$

$$\hat{\mathcal{Q}}_2(\theta;\theta') := \sum_{t=1}^{n}\sum_{i=1}^{N}\sum_{j=1}^{N} \omega_{t|N}^{ij} \ln m^{\theta}(X_{t-1}^i,X_t^j) , \tag{12.22b}$$

$$\hat{\mathcal{Q}}_3(\theta;\theta') := \sum_{t=0}^{n}\sum_{i=1}^{N} \omega_{t|N}^{i} \ln g^{\theta}(X_t^i,Y_t) . \tag{12.22c}$$

where the smoothing weights $\{\omega_{t|n}^i\}_{i=1}^N$ and $\{\omega_{t|N}^{ij}\}_{i,j=1}^N$ are given by the recursions (11.17) and (12.13), respectively under the parameter θ'.

Using these estimates, the surrogate function $\mathcal{Q}(\theta;\theta')$ may be approximately computed as $\hat{\mathcal{Q}}(\theta;\theta') := \sum_{i=1}^{3} \hat{\mathcal{Q}}_i(\theta;\theta')$. The M-step of the algorithm requires that the surrogate $\hat{\mathcal{Q}}(\theta;\theta')$ is maximized with respect to θ. In certain cases, it is possible to maximize $\hat{\mathcal{Q}}(\theta;\theta')$ using closed-form expressions. When a closed form maximizer is not available, a gradient-based search technique may be used to maximize $\theta \mapsto \hat{\mathcal{Q}}(\theta;\theta')$. With this gradient available, a wide variety of algorithms can be employed to develop a sequence of iterates that terminate at a value that approximates the maximizer $\theta \mapsto \hat{\mathcal{Q}}(\theta;\theta')$. We note that it is not necessary to find a global optimizer of $\theta \mapsto \hat{\mathcal{Q}}(\theta;\theta')$. All that is necessary is to find a value θ for which $\mathcal{Q}(\theta;\theta') > \mathcal{Q}(\theta';\theta')$ at every point θ' that is not a stationary point of the incomplete data likelihood, i.e., $\nabla\mathcal{Q}(\theta';\theta') \neq 0$. The method is summarized in Algorithm 12.1.

Heuristically, there is no need to use a large number of simulations in the early stage of the iterations. Even a crude estimate of $\theta \mapsto \mathcal{Q}(\theta;\hat{\theta}^{k-1})$ might suffice to drive the parameters toward the region of interest. Thus, in making the trade-off between improving accuracy and reducing the computational cost associated with a large sample size, one should favor increasing the sample size m_i as $\hat{\theta}^i$ approaches its limit. Determining exactly how this increase should be accomplished to produce the "best" possible result is a topic that still attracts much research interest (Booth and Hobert, 1999, Levine and Casella, 2001, Levine and Fan, 2004).

Algorithm 12.1 (Particle MCEM Algorithm)

Iterate, for $k \geq 1$:

- Generate the particle approximation of the marginal smoothing distribution $\{(X_t^j, \omega_{t|n}^j)\}_{j=1}^N$, for $t \in \{0, \ldots, n\}$ by running the FFBSm algorithm.
- Compute $\theta \mapsto \hat{\mathcal{Q}}(\theta\,; \theta_{k-1})$ using (12.22).
- Compute $\theta_k = \arg\max_{\theta \in \Theta} \hat{\mathcal{Q}}(\theta\,; \theta_{k-1})$.

Example 12.5 (Particle MCEM for the NGM model). Consider the analysis of data generated from the nonlinear NGM model presented in Example 9.6,

$$X_t = F_t^\theta(X_{t-1}) + W_t \quad \text{and} \quad Y_t = H_t(X_t) + V_t\,; \tag{12.23a}$$

$$F_t^\theta(X_{t-1}) = \alpha X_{t-1} + \beta X_{t-1}/(1 + X_{t-1}^2) + \gamma \cos[1.2(t-1)], \tag{12.23b}$$

$$H_t(X_t) = X_t^2/20, \tag{12.23c}$$

where $X_0 \sim \mathrm{N}(\mu_0, \sigma_0^2)$. In this example, we take $n = 100$, $\{W_t,\, t \in \mathbb{N}\} \sim \text{iid}\, \mathrm{N}(0, \sigma_w^2 = 10)$, independent of $\{V_t,\, t \in \mathbb{N}\} \sim \text{iid}\, \mathrm{N}(0, \sigma_v^2 = 1)$ and each sequence independent of X_0. As in Kitagawa (1987) and Carlin et al. (1992), we take $\theta = (\alpha = .5, \beta = 25, \gamma = 8)'$ and write $\Phi_t(x) = \big(x, x/(1+x^2), \cos[1.2(t-1)]\big)'$. The complete data likelihood is, up to an additive constant,

$$\ln m_t^{\theta, \sigma_W^2}(x_{t-1}, x_t) = -\frac{1}{2}\ln(\sigma_W^2) - \frac{1}{2\sigma_W^2}\left[x_t - \theta' \Phi_t(x_{t-1})\right]^2,$$

$$\ln g^{\sigma_V^2}(x_t, Y_t) = -\frac{1}{2}\ln(\sigma_V^2) - \frac{1}{2\sigma_V^2}(Y_t - x_t^2/20)^2.$$

The particle E-step amounts to approximate the vector of complete data sufficient statistics (see Section D.3)

$$S(x_{0:n}, Y_{0:n}) = \left[\sum_{t=1}^n x_t^2, \sum_{t=1}^n x_t \Phi_t(x_{t-1}), \sum_{t=0}^{n-1} \Phi_{t+1}(x_t)\Phi_{t+1}(x_t)', \sum_{t=0}^n (Y_t - x_t^2/20)^2\right].$$

The M-step is explicit. Figure 12.1 shows 30 independent realizations of 50 iterations of the particle EM algorithm, with random initialization. The number of particles is increased quadratically with the iteration index (from 50 particles initially) to a maximum of 500 particles. The convergence of the algorithm may be studied using the results in Section D.5. ◇

12.1.4 Particle stochastic approximation EM (SAEM)

One problem with the MCEM algorithm discussed in the previous section is that it relies on the number of simulations $\{m_k,\, k \in \mathbb{N}\}$ to increase with k to be convergent (see Section D.5). Another problem is that a complete weighted sample $\{(X_t^{k,i}, \omega_{t|n}^i)\}_{i=1}^{m_k}$,

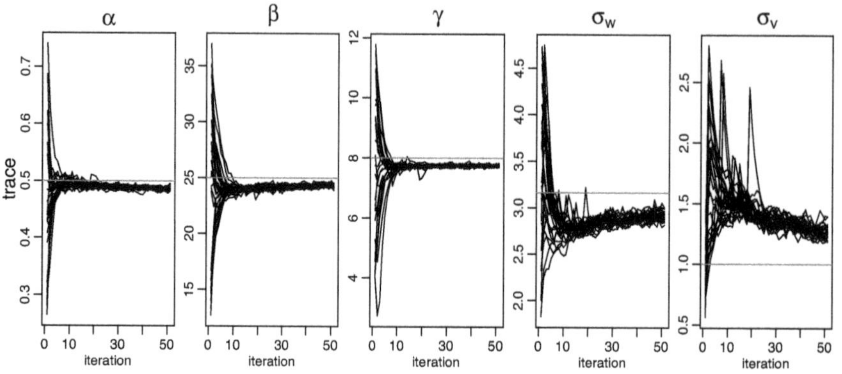

Figure 12.1 *Display for Example 12.5. Each parameter in the NGM model,* (12.23)*, is estimated via particle MCEM from a random starting point. The EM procedure is based on 50 iterations with the number of particles increasing quadratically in iteration index to a maximum of 500 particles. The display shows the results of 30 independent runs.*

for $t \in \{0,\ldots,n\}$ approximating $\mathrm{L}(\theta_{k-1}; Y_{0:n})$ has to be generated at each iteration of the algorithm. After making an update of the parameter, these values are discarded and a new set has to be simulated at the next iteration. Consequently, the particle MCEM algorithm can be computationally expensive.

An alternative is to make use of the SAEM algorithm (see Section D.3.1). When the simulation step is computationally involved, which is the case when using SMC, there is a considerable computational advantage of SAEM over MCEM (Delyon et al., 1999). As for MCEM, the SAEM algorithm requires a weighted sample from the joint smoothing distribution; see (D.15). One option is, again, to apply the FFBSi to generate a collection of backward trajectories $\{X^i_{0:n}\}^m_{i=1}$, which are used to approximate the auxiliary quantity in (D.15). Due to the fact that we now use a stochastic approximation update, a small number of backward trajectories is sufficient. However, we still require these trajectories to be generated from a distribution with a small (and diminishing) bias from the joint smoothing distribution. In other words, the SAEM algorithm relies on asymptotics in the number of forward filter particles and it is not clear how the approximation error for a finite number of particles affects the parameter estimates.

An elegant way to circumvent this issue is to use a Markovian version of stochastic approximation (Benveniste et al., 1990, Andrieu et al., 2005b). Instead of sampling exactly the posterior distribution of the latent variables, it is enough to sample from a family of ergodic Markov kernels $\{K^\theta : \theta \in \Theta\}$, leaving the family of posteriors invariant (Kuhn and Lavielle, 2004). At iteration k of the SAEM algorithm, let $\theta[k-1]$ be the previous value of the parameter estimate and let $X_{0:n}[k-1]$ be the previous draw from the Markov kernel. We then proceed by sampling

$$X_{0:n}[k] \sim K^{\theta[k-1]}(X_{0:n}[k-1], \cdot) ,$$

Algorithm 12.2 (PGAS-SAEM algorithm)

Initialize:

- Set θ_0 and $X_{0:n}^{0,\star}$ arbitrarily. Set $\hat{Q}_0(\theta) \equiv 0$.

Iterate, for $k \geq 1$:

- Generate $\{(X_{0:n}^{k,j}, \omega_n^{k,j})\}_{j=1}^N$ by running the PGAS (Algorithm 11.3), targeting $\phi_{0:n|n}^{\theta[k-1]}$ and conditioned on $(X_{0:n}^{k-1,\star}, (1, \ldots, 1))$.
- Compute $\hat{Q}_k(\theta)$ according to (12.25).
- Compute $\theta[k] = \arg\max_{\theta \in \Theta} \hat{Q}_k(\theta)$.
- Sample J_k with probability proportional to $\{\omega_n^{k,j}/\Omega_n^{k,N}\}_{j=1}^N$ and set $X_{0:n}^{k,\star} = X_{0:n}^{k,J_k}$.

---

and update the auxiliary quantity according to

$$\hat{Q}_k(\theta) = \hat{Q}_{k-1}(\theta) + \gamma_k \left\{ \ln \mathrm{L}(\theta; X_{0:n}[k], Y_{0:n}) - \hat{Q}_{k-1}(\theta) \right\} . \tag{12.24}$$

This quantity is then maximized with respect to θ in the M-step.

To construct the required Markov kernels, we make use of PMCMC theory; see Section 11.5 and Section 11.6. The PMCMC framework allows us to construct a family of Markov kernels on X^{n+1}, such that, for each $\theta \in \Theta$, K^θ leaves the joint smoothing distribution $\phi_{0:n|n}^\theta$ invariant. Different PMCMC strategies give rise to different algorithms.

To be more specific we review the latter method in more detail below. We define the Markov kernel K^θ as follows:

(a) Run the PGAS algorithm (Algorithm 11.3) targeting $\phi_{0:n|n}^{\theta[k-1]}$, conditioned on $(X_{0:n}[k-1], (1, \ldots, 1))$;

(b) Sample $J[k]$ with probability proportional to $\{\omega_n^{k,j}/\Omega_n^k\}_{j=1}^N$ and trace the ancestral path of particle $X_n^{k,J[k]}$, i.e., set $X_{0:n}[k] = X_{0:n}^{k,J[k]}$

Note that we may set the ancestors' indices of the conditioned particles deterministically: we always place the conditioned particle on the 1st position. This is possible since the law of the extracted trajectory $X_{0:n}^{k,\star}$ is independent of permutations of the particle indices. It is clear that the above procedure defines a Markov kernel on X^{n+1}. Furthermore, from the construction of the PGAS in Section 11.6, this kernel leaves $\phi_{0:n|n}^{\theta[k-1]}$ invariant, as required. It has been observed that the PGAS kernel for many models of interest mixes rapidly, even with a small number of particles. The SAEM algorithm using PGAS kernels thus offers a significant reduction in computational complexity over MCEM algorithm.

Lindsten (2013) suggests using all the particles generated by Algorithm 11.3 to approximate the auxiliary quantity of the EM algorithm. That is, $\hat{Q}_k(\theta)$ is updated

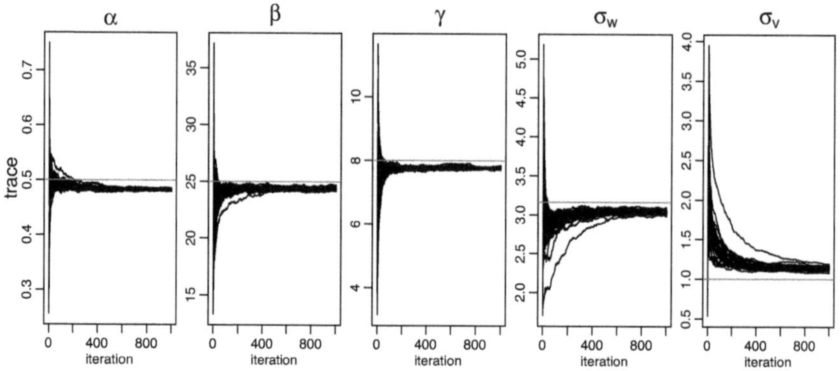

Figure 12.2 *Display for Example 12.6. Each parameter in the NGM model,* (12.23), *is estimated via Algorithm 12.2 from a random starting point. The EM procedure is based on 1000 iterations with 15 particles. The display shows the results of 30 independent runs.*

according to,

$$\hat{Q}_k(\theta) = \hat{Q}_{k-1}(\theta) + \gamma_k \left\{ \sum_{j=1}^{N} \frac{\omega_n^{k,j}}{\sum_{\ell=1}^{N} \omega_n^{k,\ell}} \ln \mathrm{L}(\theta; X_{0:n}^{k,j}, Y_{0:n}) - \hat{Q}_{k-1}(\theta) \right\}, \qquad (12.25)$$

where $\{(X_{0:n}^{k,j}, \omega_n^{k,j})\}_{j=1}^{N}$ is the weighted particle system generated at iteration k. Using (12.25) instead of (12.24) simply amounts to a Rao-Blackwellization over the random index J_k. The algorithm is summarized in Algorithm 12.2, and we illustrate the algorithm using the NGM model used in Example 12.5 and given by (12.23).

Example 12.6 (PGAS-SAEM; Example 12.5 cont.). In this example, we use Algorithm 12.2 for inference for the NGM model, (12.23). Whereas the performance of the two implementations in this example and in Example 12.5 are roughly equivalent, the PGAS-SAEM algorithm is computationally much more efficient, because only a small number of particles are required at each iteration.

Figure 12.2 displays the results of running the algorithm for the model. The EM procedure is based on 1000 iterations with 15 particles each. The display shows the results of 30 independent runs. The time to run this example was about half of the time it took to run the algorithm in Example 12.5; moreover, in Example 12.5, we used only 50 iterations as compared to 1000 iterations in this example. ◇

12.2 Bayesian analysis

We now consider some Bayesian approaches to fitting NLSS models via MCMC methods. Throughout this section, we assume we have n observations, $Y_{1:n}$, whereas the states are $X_{0:n}$, with X_0 being the initial state value. A main object of interest in

Bayesian analysis is the posterior density of the parameters[2] $p(\theta \mid Y_{1:n})$; we refer to this as *marginal sampling*.

The by-far most popular method for sampling from the posterior distribution is to include the latent state sequence $X_{0:n}$ in the MCMC state space and to run the full Gibbs sampler, i.e., alternating between sampling model parameters and latent state sequences from their respective full conditional distributions:

Procedure 12.7 (Idealized Gibbs algorithm).

(i) Draw $\theta' \sim p(\theta \mid X_{0:n}, Y_{1:n})$

(ii) Draw $X'_{0:n} \sim p(X_{0:n} \mid \theta', Y_{1:n})$

Sampling θ in Procedure 12.7-(i) is generally much easier than sampling θ conditionally only on the observed data. The parameters are conditionally independent with distributions from standard parametric families (at least as long as the prior distribution is conjugate relative to the Bayesian model specification). For non-conjugate models, one option is to replace Procedure 12.7-(i) with a Metropolis-Hastings step, which is feasible since the complete data density $p(\theta, X_{0:n}, Y_{1:n})$ can be evaluated pointwise.

Procedure 12.7-(ii) amounts to sampling from the joint smoothing distribution of the latent state sequence. Given the parameters and the observations, the latent state-sequence is a non-homogeneous Markov chain (see Chapter 11). Thus, in Procedure 12.7-(ii), we must sample from a nonhomogeneous Markov chain. For discrete-valued state processes, HMM, or Gaussian linear state space models, the transition kernel of this nonhomogeneous Markov chain can be computed explicitly and the state sequence can be sampled exactly. For a general NLSS, Procedure 12.7-(ii) is less straightforward. A common approach is to sample the state variables one-at-a-time in separate Metropolis within Gibbs steps. However, since there is often strong dependencies between consecutive state variables, this approach can lead to a sampler with poor mixing properties. To improve mixing, particle smoothing techniques, such as the Particle Independent Metropolis-Hastings (Section 11.5) or the Particle Gibbs (Section 11.6) can be used instead.

12.2.1 Gaussian linear state space models

Before discussing the general case of nonlinear state-space models, it is instructive to briefly introduce MCMC methods for the linear Gaussian model, as presented in Frühwirth-Schnatter (1994) and Carter and Kohn (1994); see Petris et al. (2009) and the references therein for a thorough discussion. Consider a Gaussian DLM given by

$$X_t = \Phi^\theta X_{t-1} + W_t \quad \text{and} \quad Y_t = A^\theta X_t + V_t \,. \tag{12.26}$$

This DLM includes the structural models presented in Section 2.7. The prior ξ on the initial state is $X_0 \sim \mathrm{N}(\mu_0, \Sigma_0)$, and we assume that $\{W_t,\ t \in \mathbb{N}\} \sim \text{iid}\, \mathrm{N}(0, Q^\theta)$,

[2]The prior distribution of θ often depends on "hyperparameters" that add another level in the hierarchy. For simplicity, these hyperparameters are assumed to be known. Some authors consider the state sequence $X_{0:n}$ as the first level of parameters because they are unobserved. In this case, the values in θ are regarded as the hyperparameters, and the parameters of their distributions are regarded as hyper-hyperparameters.

independent of $\{V_t, t \in \mathbb{N}\} \sim \text{iid } \mathrm{N}(0, R^\theta)$ and of the initial state X_0. We denote by π the prior distribution of the parameters θ.

Marginal sampling is conceptually straightforward because the posterior distribution of θ given $Y_{1:n}$ is proportional to the product of the prior π and the likelihood of the observations $\mathrm{L}(\theta; Y_{1:n})$, which is given in (2.34). When θ is multidimensional, it is often necessary to perform a Gibbs step for each component (or block of components) of θ instead of drawing all the components at once. When sampling the (block of) parameter(s) from the full conditional distributions is difficult, a Metropolis-Hastings step can be applied; another possibility is to use an acceptance-rejection algorithm, like the Adaptive Rejection Metropolis Sampling (ARMS) algorithm (see Gilks et al., 1995).

Consider now the problem of jointly sampling the unknown parameters and the states, i.e., $\mathrm{p}(\theta, X_{0:n} \mid Y_{1:n})$. A first possibility amounts to using the Gibbs sampler, which requires first sampling the parameters from $\mathrm{p}(\theta \mid X_{0:n}, Y_{1:n})$ and then from $p(X_{0:n} \mid \theta, Y_{1:n})$. From Proposition 9.14 and Proposition 9.14, the joint smoothing posterior can be decomposed as

$$\mathrm{p}(X_{0:n} \mid \theta, Y_{1:n}) = \phi_n^\theta(X_n) b^\theta_{\phi^\theta_{n-1}}(X_n, X_{n-1}) \cdots b^\theta_{\phi^\theta_0}(X_1, X_0) , \tag{12.27}$$

where $b^\theta_{\phi^\theta_t}(X_{t+1}, X_t)$ are the backward transition densities defined in (9.61)-(9.62). Using (9.65) and (9.66), $b_{\phi^\theta_t}(X_{t+1}, \cdot)$ is a Gaussian distribution with mean and variance given by

$$\begin{aligned} m_t^\theta(X_{t+1}) &= X^\theta_{t|t} + J_t^\theta(X_{t+1} - X^\theta_{t|t}) , \\ V_t^\theta &= P^\theta_{t|t} - J_t^\theta P^\theta_{t+1|t}[J_t^\theta]', \end{aligned} \tag{12.28}$$

where the filtering mean $X^\theta_{t|t}$, the one-step ahead prediction covariance $P^\theta_{t+1|t}$ and the Kalman gain J_t^θ are defined in (2.11), (2.12), and (2.25), respectively. In view of the backward decomposition of the smoothing density (12.27), it is possible to sample the entire set of state vectors, $X_{0:n}$ conditionally to the observation, by sampling from the filtering distribution ϕ_n^θ at time n and then sequentially simulating the individual states backward from the backward transition densities. This process yields a simulation method that Frühwirth-Schnatter (1994) called the *forward-filtering, backward-sampling algorithm*, which is similar to the FFBSi algorithm introduced and analyzed in Section 11.3.

A second option to obtain the joint posterior is to use a *partially collapsed Gibbs sampler*, as discussed in van Dyk and Park (2008). In this case, we can write

$$\begin{aligned} &\mathrm{p}(\theta, X_{0:n} \mid Y_{0:n}) \propto \mathrm{p}(X_{0:n} \mid \theta, Y_{1:n}) \mathrm{p}(\theta \mid Y_{1:n}) \\ &\qquad \propto \mathrm{p}(\theta \mid Y_{1:n}) \xi(X_0) g^\theta(X_0, Y_0) \prod_{t=1}^n m^\theta(X_{t-1}, X_t)\, g^\theta(X_t, Y_t) , \end{aligned} \tag{12.29}$$

where $m_t^\theta(X_{t-1}, X_t) = \mathfrak{g}(X_t; \Phi_t^\theta X_{t-1}, Q_t^\theta)$ and $g_t^\theta(X_t, Y_t) = \mathfrak{g}(Y_t; A_t^\theta X_t, R_t^\theta)$. A sample from the joint posterior of the parameters and the state variables given the observations may be obtained by sampling first the unknown parameters using the marginal sampler, and then the joint distribution of the states given the current fit of the parameters and the observations. Because we are dealing with a Gaussian linear model, we can rely on the existing theory of the Kalman filter/smoother to accomplish these two steps. The first step amounts to computing the likelihood of the observations, which can be done using (2.34). The forward filtering backward sampling algorithm amounts to first running the Kalman filter in the forward direction (see Proposition 2.3) and then to store $X^{\theta}_{t|t}$ and $P^{\theta}_{t+1|t}$, for $t = 0,\ldots,n$. Then, sample X_n from $\mathrm{N}(X^{\theta}_{n|n}, P^{\theta}_{n|n})$, and then sample X_t from $\mathrm{N}(m^{\theta}_t(X_{t+1}), V^{\theta}_t)$, for $t = n-1, n-2, \ldots, 0$, where the conditioning value of X_{t+1} is the value previously sampled.

Example 12.8 (Gaussian DLM). Consider the Johnson & Johnson quarterly earnings per share series that was discussed in Example 2.16. Recall that the model is

$$Y_t = \begin{pmatrix} 1 & 1 & 0 & 0 \end{pmatrix} X_t + V_t,$$

$$X_t = \begin{pmatrix} T_t \\ S_t \\ S_{t-1} \\ S_{t-2} \end{pmatrix} = \begin{pmatrix} \phi & 0 & 0 & 0 \\ 0 & -1 & -1 & -1 \\ 0 & 1 & 0 & 0 \\ 0 & 0 & 1 & 0 \end{pmatrix} \begin{pmatrix} T_{t-1} \\ S_{t-1} \\ S_{t-2} \\ S_{t-3} \end{pmatrix} + \begin{pmatrix} W_{t1} \\ W_{t2} \\ 0 \\ 0 \end{pmatrix}$$

where $R^{\theta} = \sigma_v^2$ and

$$Q^{\theta} = \begin{pmatrix} \sigma^2_{w,11} & 0 & 0 & 0 \\ 0 & \sigma^2_{w,22} & 0 & 0 \\ 0 & 0 & 0 & 0 \\ 0 & 0 & 0 & 0 \end{pmatrix}.$$

The parameters to be estimated are the transition parameter associated with the growth rate, $\phi > 1$, the observation noise variance, σ_v^2, and the state noise variances associated with the trend and the seasonal components, $\sigma^2_{w,11}$ and $\sigma^2_{w,22}$, respectively.

In this case, sampling from $p(X_{0:n} \mid \theta, Y_{1:n})$ follows directly from (12.27)–(12.28). Next, we discuss how to sample from $p(\theta \mid X_{0:n}, Y_{1:n})$. For the transition parameter, write $\phi = 1+\beta$, where $0 < \beta \ll 1$; recall that in Example 2.16, ϕ was estimated to be 1.035, which indicated a growth rate, β, of 3.5%. Note that the trend component may be rewritten as

$$\nabla T_t = T_t - T_{t-1} = \beta T_{t-1} + W_{t1} \,.$$

Consequently, conditional on the states, the parameter β is the slope in the linear regression (through the origin) of ∇T_t on T_{t-1}, for $t = 1,\ldots,n$, and W_{t1} is the error. As is typical, we put a Normal/Inverse Gamma (IG) prior on $\beta/\sigma^2_{w,11}$, i.e., $\beta \mid \sigma^2_{w,11} \sim \mathrm{N}(b_0, \sigma^2_{w,11} B_0)$ and $\sigma^2_{w,11} \sim \mathrm{IG}(n_0/2, n_0 s_0^2/2)$, with known hyperparameters b_0, B_0, n_0, s_0^2. Using standard normal-gamma model theory, we can show

$$\beta \mid X_{0:n}, Y_{1:n} \sim t_{n_1}(b_1, s_1^2 B_1)$$
$$\sigma^2_{w,11} \mid \beta, X_{0:n}, Y_{1:n} \sim \mathrm{IG}(n_1/2, n_1 s_1^2/2)$$

where $n_1 = n_0 + (n-1)$, $B_1^{-1} = B_0^{-1} + \sum_{t=1}^n T_{t-1}^2$, $B_1^{-1} b_1 = B_0^{-1} b_0 + \sum_{t=1}^n (\nabla T_t) T_{t-1}$, and $n_1 s_1^2 = n_0 s_0^2 + \sum_{t=1}^n (\nabla T_t - \beta T_{t-1})^2$.

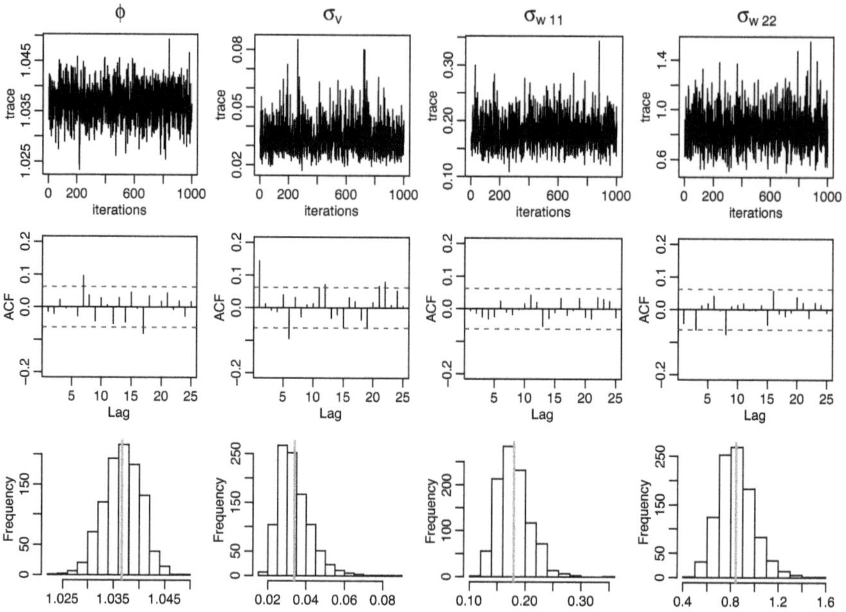

Figure 12.3 *Parameter estimation results for Example 12.8. The top row displays the traces of 1000 draws, after a burn-in of 1000, with a step size of 20. The middle row displays the ACF of the traces. The sampled posteriors are displayed in the last row (the mean is marked by a solid vertical line).*

We also used IG priors for the other two variance components, σ_v^2 and $\sigma_{w,22}^2$; recall the relationships (2.57) and (2.59). Thus, if the prior $\sigma_v^2 \sim \mathrm{IG}(n_0/2, n_0 s_0^2/2)$, then the posterior $\sigma_v^2 \mid X_{0:n}, Y_{1:n} \sim \mathrm{IG}(n_v/2, n_v s_v^2/2)$, where $n_v = n_0 + n$, and $n_v s_v^2 = n_0 s_0^2 + \sum_{t=1}^{n}(Y_t - T_t - S_t)^2$. Similarly, if the prior $\sigma_{w,22}^2 \sim \mathrm{IG}(n_0/2, n_0 s_0^2/2)$, then the posterior $\sigma_{w,22}^2 \mid X_{0:n}, Y_{1:n} \sim \mathrm{IG}(n_w/2, n_w s_w^2/2)$, where $n_w = n_0 + (n-3)$, and $n_w s_w^2 = n_0 s_0^2 + \sum_{t=1}^{n-3}(S_t - S_{t-1} - S_{t-2} - S_{t-3})^2$.

Figure 12.3 displays the results of the posterior estimates of the parameters. The top row of the figure displays the traces of 1000 draws, after a burn-in of 1000, with a step size of 20 (i.e., every 20th sampled value is retained). The middle row of the figure displays the ACF of the traces, and the sampled posteriors are displayed in the last row of the figure. The results of this analysis are comparable to the results obtained in Example 2.16; the posterior mean and median for ϕ indicates a 3.7% growth rate in the Johnson & Johnson quarterly earnings over this time period. It should be apparent that $W_{t,2}$ is not stationary with constant variance $\sigma_{w,2}$. In Exercise 12.10, we explore an alternate model.

Figure 12.4 displays the smoothers of trend (T_t) and season (S_t) along with 99% credible intervals. Again, these results are comparable to the results obtained in Example 2.16. ◇

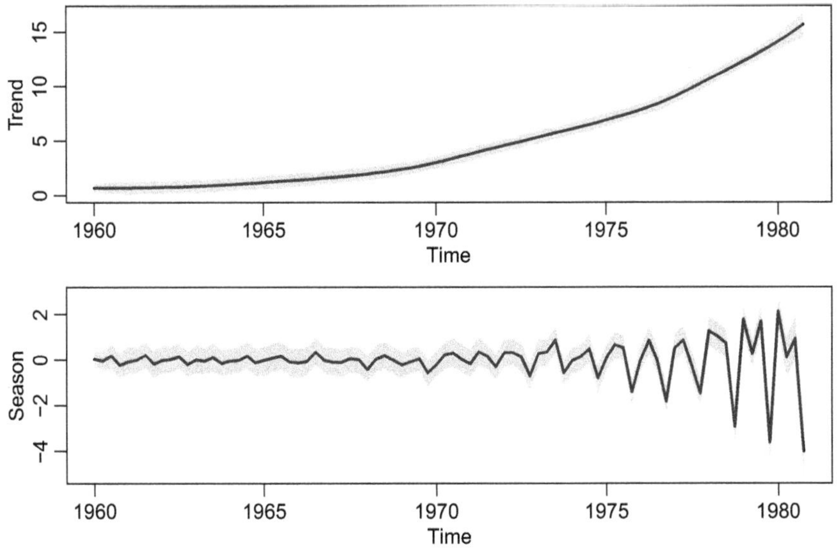

Figure 12.4 *Example 12.8 smoother estimates of trend (T_t) and season (S_t) along with corresponding 99% credible intervals.*

12.2.2 Gibbs sampling for NLSS model

For general NLSS models, the main difficulty lies in the simulation of the latent state sequence given the observations and the current fit of the parameters. The particular structure of the NLSS implies, however, a simple decomposition of the full marginal distribution of the state, which can be exploited to develop a Gibbs sampler (or, more generally, a Metropolis-within-Gibbs sampler). Starting from the joint conditional distribution of $p(X_{0:n} \mid Y_{0:n}, \theta)$ defined (up to a proportionality constant) by

$$p(X_{0:n}, Y_{0:n} \mid \theta) = \xi(X_0)g(X_0, Y_0)\prod_{t=1}^{n} m_t^{\theta}(X_{t-1}, X_t)g^{\theta}(X_t, Y_t)\,, \tag{12.30}$$

the conditional probability density function of a *single* latent state X_t given $Y_{0:n}$ and the remaining latent states $X_{-t} := \{X_s\}_{s \neq t}$ is given by

$$\begin{aligned} p(X_t \mid X_{-t}, Y_{0:n}, \theta) &\propto p(X_t \mid X_{t-1}, X_{t+1}, Y_t, \theta) \\ &\propto m_t^{\theta}(X_{t-1}, X_t)m_t^{\theta}(X_t, X_{t+1})g^{\theta}(X_t, Y_t)\,. \end{aligned} \tag{12.31}$$

Note that this conditional distribution depends only on the two neighboring latent states, which makes the computations straightforward. At the two endpoints $t = 0$ and $t = n$, we have the obvious corrections $p(X_0 \mid X_{-0}, \theta) \propto \xi^{\theta}(x_0)m_0^{\theta}(X_0, X_1)$ and $p(X_n \mid X_{-n}) \propto m_n^{\theta}(X_{n-1}, X_n)g^{\theta}(X_n)$. The most basic Gibbs sampler (see Section 5.8.2) for the whole latent state sequence $X_{0:n}$ simulates one component of the latent sequence at a time. The full conditional probability for each latent state given by (12.31) is in a simple closed-form. Note that this expression is simple only because the knowledge of the normalization factor is not required for performing MCMC simulations.

An important point to stress here is that replacing an exact simulation by a Metropolis-Hastings step in a general MCMC algorithm does not jeopardize its validity as long as the Metropolis-Hastings step is associated with the correct stationary distribution. Hence, the most natural alternative to the Gibbs sampler in cases where sampling from the full conditional distribution is not directly feasible is the *one-at-a-time Metropolis-Hastings* algorithm that combines successive Metropolis-Hastings steps that update only one of the variables. For $k = 0,\ldots,n$, we thus update the t-th component X_t^i of the current simulated sequence of states $X_{0:n}$ by proposing a new candidate for X_t^{i+1} and accepting it according to (5.37), using (12.31) as the target.

Example 12.9 (An approach to NLSS modeling). Here, we address an MCMC approach to nonlinear and non-Gaussian state-space modeling that was first presented in Carlin et al. (1992). We consider the general model

$$X_t = F^\theta(X_{t-1}) + W_t \quad \text{and} \quad Y_t = H^\theta(X_t) + V_t\,, \tag{12.32}$$

where F^θ and H^θ are given, but may also depend on unknown regression parameters, θ. The state and measurement errors $\{W_t,\, t\in\mathbb{N}\}$ and $\{V_t,\, t\in\mathbb{N}\}$ are assumed to be independent noise sequences, with zero-means and respective covariances $\mathrm{Var}(W) = Q_t$ and $\mathrm{Var}(V_t) = R_t$. Although time-varying variance–covariance matrices are easily incorporated in this framework, to ease the discussion we focus on the case $Q_t \equiv Q$ and $R_t \equiv R$. It is possible that some elements of Q or R may be fixed (e.g., R may be diagonal), but we just write Q or R to denote any of the parameters contained in these matrices. Also, while it is not necessary, we assume the initial state condition X_0 is fixed and known; this is merely for notational convenience, so we do not have to carry along the additional terms involving X_0 throughout the discussion.

Tractability in the non-normal case is introduced through augmentation as follows. In general, the likelihood specification for the model is given by

$$\mathrm{L}(\theta,Q,R;X_{1:n},Y_{1:n}) = \prod_{t=1}^{n} \mathrm{p}_1(X_t \mid X_{t-1},\theta,Q)\,\mathrm{p}_2(Y_t \mid X_t,\theta,R)\,, \tag{12.33}$$

where the densities $\mathrm{p}_1(\cdot)$ and $\mathrm{p}_2(\cdot)$ are scale mixtures of normals. Specifically, for $t = 1,\ldots,n$,

$$\mathrm{p}_1(X_t \mid X_{t-1},\theta,Q) = \int h_1(X_t \mid X_{t-1},\theta,Q,\lambda_t)\, g_1(\lambda_t)\, d\lambda_t\,, \tag{12.34}$$

$$\mathrm{p}_2(Y_t \mid X_t,\theta,R) = \int h_2(Y_t \mid X_t,\theta,R,\omega_t)\, g_2(\omega_t)\, d\omega_t\,, \tag{12.35}$$

where conditional on the independent sequences of nuisance parameters $\lambda = (\lambda_t; t = 1,\ldots,n)$ and $\omega = (\omega_t; t = 1,\ldots,n)$,

$$X_t \mid X_{t-1},\theta,Q,\lambda_t \sim \mathrm{N}\Big(F^\theta(X_{t-1}),\ \lambda_t Q\Big)\,, \tag{12.36}$$

$$Y_t \mid X_t,\theta,R,\omega_t \sim \mathrm{N}\Big(H^\theta(X_t),\ \omega_t R\Big)\,. \tag{12.37}$$

By varying $g_1(\lambda_t)$ and $g_2(\omega_t)$, we can have a wide variety of non-Gaussian error densities. These densities include, for example, double exponential, logistic, and t distributions in the univariate case and a rich class of multivariate distributions; this is discussed further in Carlin et al. (1992). The key to the approach is augmentation via the nuisance parameters λ and ω and the structure of (12.36) and(12.37) that lends itself naturally to MCMC and allows for the analysis of this general nonlinear and non-Gaussian problem. Note that λ or ω may be suppressed if either $\mathrm{p}_1(\cdot)$ in (12.34) or $\mathrm{p}_2(\cdot)$ in (12.35), respectively, are Gaussian. In either case, we would simply ignore λ in (12.36) or ω in (12.37).

To implement a sampler, the following complete conditional distributions need to be specified:

- $X_t \mid X_{s\neq t}, \lambda, \omega, \theta, Q, R, Y_{1:n} \sim X_t \mid X_{s\neq t}, \lambda, \omega, \theta, Q, R, Y_t \quad t = 0, 1, \ldots, n,$
- $\lambda_t \mid \lambda_{s\neq t}, \omega, \theta, Q, R, Y_{1:n}, X_{0:n} \sim \lambda_t \mid \theta, Q, X_t, X_{t-1} \quad t = 1, \ldots, n,$
- $\omega_t \mid \omega_{s\neq t}, \lambda, \theta, Q, R, Y_{1:n}, X_{0:n} \sim \omega_t \mid \theta, R, Y_t, X_t \quad t = 1, \ldots, n,$
- $Q \mid \lambda, \omega, \theta, R, Y_{1:n}, X_{0:n} \sim Q \mid \lambda, Y_{1:n}, X_{0:n},$
- $R \mid \lambda, \omega, \theta, Q, Y_{1:n}, X_{0:n} \sim R \mid \omega, Y_{1:n}, X_{0:n},$
- $\theta \mid \lambda, \omega, Q, R, Y_{1:n}, X_{0:n} \sim \theta \mid Y_{1:n}, X_{0:n},$

where $X_{0:n} = \{X_0, \ldots, X_n\}$, $Y_{1:n} = \{Y_1, \ldots, Y_n\}$, $Y_0 = 0$ and $X_{s\neq t} = \{X_s; 0 \le s \neq t \le n\}$. As previously indicated, the main difference between this method and the linear Gaussian case is that, because of the generality, we sample the states one-at-a-time rather than simultaneously generating all of them. ◇

Next, we provide a numerical example to examine the use of the results in Example 12.9. In this particular example, the noise processes are Gaussian, so that we take $\lambda \equiv 1$ and $\omega \equiv 1$, which suppresses these values throughout the example. Again, we refer the reader to Carlin et al. (1992) for examples of models with non-normal noise.

Example 12.10 (NGM model). Consider the analysis of data generated from the nonlinear NGM model presented in Example 9.6 and used throughout Section 12.1. Recall that the model is

$$X_t = F_t^\theta(X_{t-1}) + W_t \quad \text{and} \quad Y_t = H_t(X_t) + V_t\,; \tag{12.38a}$$

$$F_t^\theta(X_{t-1}) = \alpha X_{t-1} + \beta X_{t-1}/(1 + X_{t-1}^2) + \gamma\cos[1.2(t-1)], \tag{12.38b}$$

$$H_t(X_t) = X_t^2/20, \tag{12.38c}$$

where $X_0 \sim \mathrm{N}(\mu_0, \sigma_0^2)$, with $W_t \sim \text{iid N}(0, \sigma_w^2)$ independent of $V_t \sim \text{iid N}(0, \sigma_v^2)$ and each sequence independent of X_0.

The priors on the variance components are chosen from a conjugate family, that is, $\sigma_w^2 \sim \mathrm{IG}(a_0/2, b_0/2)$ independent of $\sigma_v^2 \sim \mathrm{IG}(c_0/2, d_0/2)$, where IG denotes the inverse (reciprocal) gamma distribution. Then,

$$\sigma_w^2 \mid Y_{1:n}, X_{0:n} \sim \mathrm{IG}\Big(\tfrac{1}{2}(a_0 + n),\ \tfrac{1}{2}\{b_0 + \sum_{t=1}^{n}[X_t - F^\theta(X_{t-1})]^2\}\Big)\,, \tag{12.39}$$

$$\sigma_v^2 \mid Y_{1:n}, X_{0:n} \sim \text{IG}\Big(\tfrac{1}{2}(c_0+n),\ \tfrac{1}{2}\{d_0 + \sum_{t=1}^{n}[X_t - H(X_t)]^2\}\Big)\,. \tag{12.40}$$

The prior on $\theta = (\alpha,\beta,\gamma)'$ is taken to be trivariate normal with mean $(\mu_\alpha,\mu_\beta,\mu_\gamma)'$ and diagonal variance–covariance matrix $\text{diag}\{\sigma_\alpha^2,\sigma_\beta^2,\sigma_\gamma^2\}$. The necessary conditionals can be found using standard normal theory. For example, the complete conditional distribution of α is of the form $\text{N}(Bb,B)$, where

$$B^{-1} = \frac{1}{\sigma_\alpha^2} + \frac{1}{\sigma_w^2}\sum_{t=1}^{n} X_{t-1}^2\,,$$

$$b = \frac{\mu_\alpha}{\sigma_\alpha^2} + \frac{1}{\sigma_w^2}\sum_{t=1}^{n} X_{t-1}\left(X_t - \beta\frac{X_{t-1}}{1+X_{t-1}^2} - \gamma\cos[1.2(t-1)]\right).$$

The complete conditional for β has the same form, with

$$B^{-1} = \frac{1}{\sigma_\beta^2} + \frac{1}{\sigma_w^2}\sum_{t=1}^{n} \frac{X_{t-1}^2}{(1+X_{t-1}^2)^2}\,,$$

$$b = \frac{\mu_\beta}{\sigma_\beta^2} + \frac{1}{\sigma_w^2}\sum_{t=1}^{n} \frac{X_{t-1}}{(1+X_{t-1}^2)}\left(X_t - \alpha X_{t-1} - \gamma\cos[1.2(t-1)]\right),$$

and for γ the values are

$$B^{-1} = \frac{1}{\sigma_\gamma^2} + \frac{1}{\sigma_w^2}\sum_{t=1}^{n} \cos^2[1.2(t-1)]\,,$$

$$b = \frac{\mu_\gamma}{\sigma_\gamma^2} + \frac{1}{\sigma_w^2}\sum_{t=1}^{n} \cos[1.2(t-1)]\left(X_t - \alpha X_{t-1} - \beta\frac{X_{t-1}}{1+X_{t-1}^2}\right).$$

Finally, to sample the states, a simple random walk Metropolis algorithm with tuning variance v_x^2 works as follows. For $t = 0,1,\ldots,n$:

(a) At iteration j, let the current state value be $X_t^{(j)}$.

(b) Sample $X_t^\star$ from $\text{N}(X_t^{(j)}, v_x^2)$.

(c) Determine the acceptance probability

$$a = 1 \wedge \frac{\text{p}(X_t^\star \mid X_{s\neq t}, Y_t, \theta, \sigma_w^2, \sigma_v^2)}{\text{p}(X_t^{(j)} \mid X_{s\neq t}, Y_t, \theta, \sigma_w^2, \sigma_v^2)}\,,$$

with

$$\text{p}(X_t \mid X_{s\neq t}, Y_t, \theta, \sigma_w^2, \sigma_v^2) \propto \mathfrak{g}\big(Y_t; H(X_t), \sigma_v^2\big)\,\mathfrak{g}\big(X_t; F^\theta(X_{t-1}), \sigma_w^2\big)\,\mathfrak{g}\big(X_{t+1}; F^\theta(X_t), \sigma_w^2\big)\,, \tag{12.41}$$

for $t = 1,\ldots,n-1$, where $\mathfrak{g}(x;\mu,\sigma^2)$ denotes the Gaussian density in variable x with mean μ and variance σ^2. When $t = n$, drop the third term in (12.41). When $t = 0$, replace the first two terms in (12.41) with $\mathfrak{g}(X_0; \mu_0, \sigma_0^2)$.

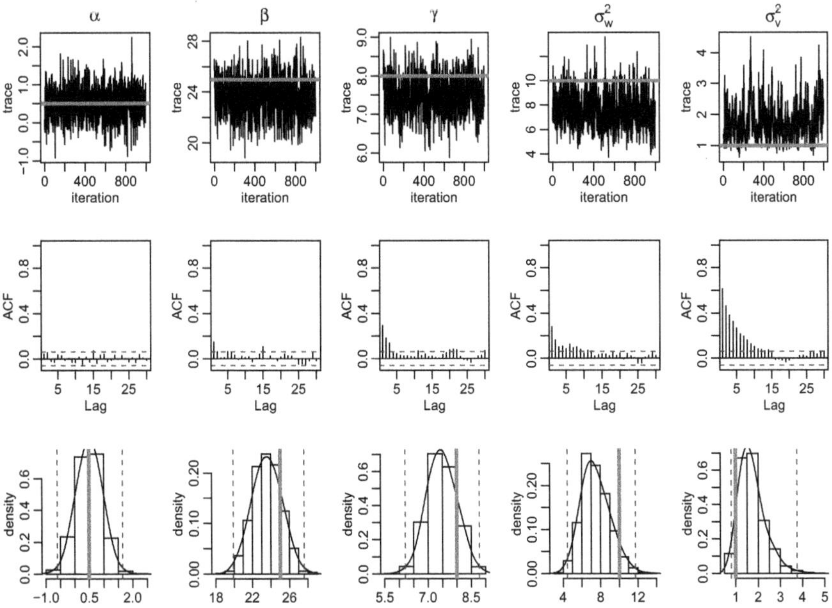

Figure 12.5 *Parameter Inference for Example 12.10: The top row shows the output of the sampling traces for each parameter; the actual value is shown as a horizontal line. The middle row displays the ACF of the traces, and the bottom row shows the sampled posteriors of each parameter. The actual values are shown as thick vertical lines and 99% credible intervals are represented as vertical dashed lines.*

(d) Select the new value as

$$X_t^{(j+1)} = \begin{cases} X_t^\star & \text{w.p. } a, \\ X_t^{(j)} & \text{w.p. } 1-a. \end{cases}$$

In this example, we set $v_x = .05$, $\mu_0 = 0$, $\sigma_0^2 = 10$, $a_0/2 = 6$, $b_0/2 = 50$, and $c_0/2 = 6$, $d_0/2 = 5$. The normal prior on $\theta = (\alpha, \beta, \gamma)'$ has corresponding mean vector equal to $(\mu_\alpha = .5, \mu_\beta = 25, \mu_\gamma = 8)'$ and diagonal variance matrix equal to $\text{diag}\{\sigma_\alpha^2 = .25, \sigma_\beta^2 = 10, \sigma_\gamma^2 = 4\}$. The burn-in size was 500, the step size was 25 (i.e., every 25th sampled value is retained), and we obtained 1000 samples, so that the entire procedure consisted of 25,500 ($1000 \times 25 + 500$) draws. The time to complete the run was 10.2 minutes using R3.0.0 on a Falcon Northwest Mach V™ PC running Windows™ 7 x64, with an Intel™ Core i7 CPU 960 3.20 GHz and 6 GB of RAM; we will use this information to compare computational burden of the examples that will follow.

Parameter estimation results are displayed in Figure 12.5. The top row shows the output of the sampling traces for each parameter; the actual value is shown as a horizontal line. The middle row displays the ACF of the traces, and the bottom row shows the sampled posteriors of each parameter; the actual values are shown as vertical lines. State estimation is displayed in Figure 12.6. In that figure, the actual values

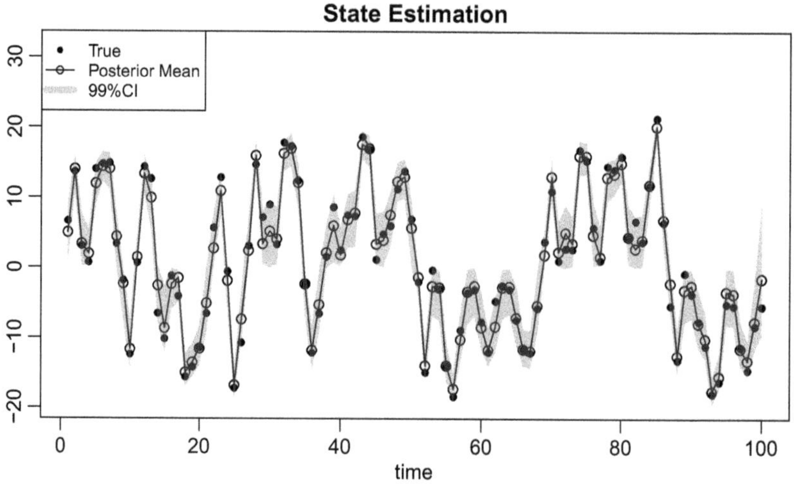

Figure 12.6 *State Estimation for Example 12.10: The actual values of the states are displayed as solid points, whereas the posterior means are plotted as circles connected by a solid line. 99% credible intervals are displayed as dashed lines.*

of the states are displayed as solid points, whereas the posterior means are plotted as circles connected by a solid line. Finally, 99% credible intervals are displayed as a gray area. ◇

12.2.3 Particle marginal Markov chain Monte Carlo

We have already seen the usefulness of PMCMC for joint smoothing (Section 11.5 and Section 11.6) and for maximum likelihood inference (Section 12.1.4). However, the application of PMCMC that has received the most attention is Bayesian parameter inference. Because PMCMC provides a systematic way of combining SMC and MCMC, it lends itself naturally to Bayesian inference in NLSS.

Suppose we are interested in sampling from a probability distribution π defined on some measurable space Θ. As outlined in Section 5.8.1, a practical algorithm to achieve this when the distribution π is known up to a normalizing constant consists of using Markov chain Monte Carlo (MCMC) methods, of which the Metropolis-Hastings algorithm is the main workhorse. The Metropolis-Hastings kernel may be written as

$$P(\theta, \mathrm{d}\theta') := \min\{1, r(\theta, \theta')\} q(\theta, \mathrm{d}\theta') + \delta_\theta(\mathrm{d}\theta') \rho(\theta) , \tag{12.42}$$

where $r(\theta, \theta')$ is the Radon-Nikodym derivative

$$r(\theta, \theta') := \frac{\pi(\mathrm{d}\theta') q(\theta', \mathrm{d}\theta)}{\pi(\mathrm{d}\theta) q(\theta, \mathrm{d}\theta')} \quad \text{and} \quad \rho(\theta) := 1 - \int \min\{1, r(\theta, \theta')\} q(\theta, \mathrm{d}\theta') , \tag{12.43}$$

where q is the proposal kernel. In this section, we call this algorithm the *marginal algorithm*.

In some situations, the marginal algorithm cannot be implemented because the target distribution π is intractable. For example, assuming that π and Q have densities (denoted π and q) with respect to some σ-finite measure, it may happen that the target density π cannot be evaluated pointwise, and although the acceptance ratio $r(\theta,\theta')$ may be well defined theoretically, it cannot be computed. It is sometimes the case that unbiased non-negative estimates $\hat{\pi}(\theta)$ are available. Denote by $W_\theta = \hat{\pi}(\theta)/\pi(\theta)$ the *weight* (assuming for simplicity that $\pi(\theta) > 0$ for all $\theta \in \Theta$). The distribution of the weight is denoted Q_θ, i.e., $W_\theta \sim Q_\theta(\cdot) \geq 0$. The unbiasedness property imposes that $\mathbb{E}[W_\theta] = 1$ for any $\theta \in \Theta$. An apparently ad-hoc idea may be to use such estimates in place of the true values in order to compute the acceptance probability. A remarkable property is that such an algorithm is in fact correct. This can be seen by considering the following probability distribution

$$\tilde{\pi}(\mathrm{d}\theta,\mathrm{d}w) := \pi(\mathrm{d}\theta)\pi_\theta(\mathrm{d}w) \quad \text{with} \quad \pi_\theta(\mathrm{d}w) := Q_\theta(\mathrm{d}w)w\,, \tag{12.44}$$

on the product space $(\Theta \times \mathbb{R}_+, \mathcal{B}(\Theta) \times \mathcal{B}(\mathbb{R}_+))$. For all $\theta \in \Theta$, $\pi_\theta(\mathrm{d}w)$ is a probability measure, therefore π is a marginal distribution of the product measure $\tilde{\pi}$. Consider the Metropolis-Hastings (MH) algorithm targeting $\tilde{\pi}(\mathrm{d}\theta,\mathrm{d}w)$ using a proposal kernel $\tilde{q}(\theta,w;\mathrm{d}\theta',\mathrm{d}w') := q(\theta,\mathrm{d}\theta')Q_{\theta'}(\mathrm{d}w')$. The acceptance ratio of this MH algorithm may be expressed as

$$\frac{\tilde{\pi}(\mathrm{d}\theta',\mathrm{d}w')\tilde{q}(\theta',w';\mathrm{d}\theta,\mathrm{d}w)}{\tilde{\pi}(\mathrm{d}\theta,\mathrm{d}w)\tilde{q}(\theta,w;\mathrm{d}\theta',\mathrm{d}w')} = \frac{\pi(\theta')Q_{\theta'}(\mathrm{d}w')w'q(\theta',\mathrm{d}\theta)Q_\theta(\mathrm{d}w)}{\pi(\theta)Q_\theta(\mathrm{d}w)wq(\theta,\mathrm{d}\theta')Q_{\theta'}(\mathrm{d}w')} = r(\theta,\theta')\frac{w'}{w}\,, \tag{12.45}$$

showing that the distribution of the weight W cancels. On the other hand, by construction, $\hat{\pi}(\theta) = w\pi(\theta)$, which implies that the MH kernel can be explicitly computed

$$\tilde{P}(\theta,w;\mathrm{d}\theta',\mathrm{d}w') := \min\left\{1, r(\theta,\theta')\frac{w'}{w}\right\}q(\theta,\mathrm{d}\theta')Q_{\theta'}(\mathrm{d}w') + \delta_{\theta,w}(\mathrm{d}\theta',\mathrm{d}w')\tilde{\rho}(\theta,w)\,, \tag{12.46}$$

where the rejection probability is given by

$$\tilde{\rho}(\theta,w) := 1 - \iint \min\left\{1, r(\theta,\theta')\frac{w'}{w}\right\}q(\theta,\mathrm{d}\theta')Q_{\theta'}(\mathrm{d}w')\,. \tag{12.47}$$

This is the *pseudo-marginal algorithm* introduced in Andrieu and Roberts (2009), which targets $\tilde{\pi}$ whose marginal distribution is π. This algorithm may be implemented as soon as we can construct an unbiased estimator of $\pi(\theta)$. As a particular instance of the Metropolis-Hastings algorithm, the pseudo-marginal algorithm converges to $\tilde{\pi}$ under mild assumptions, and although it may be seen as an approximate (noisy) version of the marginal algorithm, it targets the (intractable) target distribution π.

The structure of the pseudo-marginal algorithm is in fact shared by several practical algorithms that have been proposed in order to sample from intractable distributions. In most cases, the distribution of the weight W is not explicitly known. Assume for simplicity that the space Θ is (a Borel subset of) $\mathbb{R}^d$ and that π and $q(\theta, \mathrm{d}\theta')$, for any $\theta \in \Theta$, have densities with respect to the Lebesgue measure. Consider a situation in which $\pi(\theta)$ is analytically intractable or too complex to evaluate, whereas the introduction of an auxiliary variable x might lead to an analytical expression, or ease the implementation of numerical methods. In the sequel, assume that the target density is of the form $\pi(\theta) = \int \pi(\theta, x) \mathrm{Leb}(\mathrm{d}x)$ where the integral cannot be computed analytically. A straightforward approach to obtain an unbiased estimator of this quantity consists of using an importance sampling estimator. For $\theta \in \Theta$, let h_θ be the proposal density satisfying the usual support assumption. In this case we get

$$W_\theta \pi(\theta) = \hat{\pi}(\theta) = \frac{1}{N} \sum_{n=1}^{N} \frac{\pi(\theta, X^k)}{h_\theta(X^k)} , \tag{12.48}$$

where $\{X^k\}_{k=1}^N$ are i.i.d. sampled from h_θ.

There are more involved applications of this idea. In the context of state-space models, Andrieu et al. (2010) proposed sampling from W_θ with a particle filter, resulting in particle MCMC algorithms. We can, for example, use the PIMH method (Algorithm 11.2) for Bayesian inference. The key property that we will make use of is the unbiasedness of the likelihood estimator (11.38). A major objective in the Bayesian setting is obtaining the posterior parameter density,

$$p(\theta \mid Y_{1:n}) = \frac{\mathrm{L}(\theta; Y_{0:n}) \pi(\theta)}{\int_\Theta \mathrm{L}(\theta; Y_{1:n}) \pi(\theta) \mathrm{Leb}(\mathrm{d}\theta)} , \tag{12.49}$$

where π stands here for the prior distribution and $\mathrm{L}(\theta; Y_{1:n})$ is the likelihood of the data under the model parameterized by θ. For a general NLSS, it is not possible to target this density directly using Metropolis-Hastings sampling since the likelihood function is not available in closed form. However, by running an SMC sampler, we can construct an unbiased estimate of the likelihood according to (11.38). Let $\widehat{\mathrm{L}}^N(\theta; Y_{1:n})$ be an unbiased estimator of the likelihood for a given value of θ. We denote by $W_{n,\theta}^N = \widehat{\mathrm{L}}(\theta; Y_{1:n}) / \mathrm{L}(\theta; Y_{1:n})$, the weight, and by $Q_\theta^N(w)$, the distribution, of this random variable for a given value of θ (we do not require this distribution to be explicitly available). Now, construct an extended target density on $\Theta \times \mathbb{R}_+$ according to,

$$\tilde{p}(\theta, w \mid Y_{1:n}) = p(\theta \mid Y_{1:n}) w Q_\theta^N(w) . \tag{12.50}$$

By construction, the unbiasedness property implies that $\int w Q_\theta^N(w) \mathrm{Leb}(\mathrm{d}w) = 1$. It follows that $\tilde{p}(\theta, w \mid Y_{1:n})$ is a proper probability density function and, furthermore, that the marginal of θ under $\tilde{p}(\theta, w \mid Y_{1:n})$ is the posterior distribution $p(\theta \mid Y_{1:n})$.

Thus, we can use the extended target as a surrogate for $p(\theta \mid Y_{1:n})$ in an MCMC algorithm. To target (12.50) with a Metropolis-Hastings sampler, we first propose θ' from a proposal kernel on the marginal space, $\theta' \sim q(\theta, \cdot)$. We then propose $W_{n,\theta}^N$ by sampling from the conditional distribution $Q_\theta^N(\cdot)$. Note that we in general do not

Algorithm 12.3 (PGAS: Particle Gibbs with Ancestor Sampling)

Initialize:

- Set $X_{0:n}[0]$ arbitrarily.

Iterate, for $k \geq 1$:

- Sample $\theta[k] \sim p(\theta \mid X_{0:n}[k-1], Y_{0:n})$.
- Run the PGAS sampler (Algorithm 11.3) targeting $\phi^{\theta[k]}_{0:n|n}$, conditioned on $(X_{0:n}[k-1], (1, \ldots, 1))$.
- Sample $J[k]$ with probability proportional to $\{\omega_n^{k,j} / \Omega_n^k\}_{j=1}^N$ and trace the ancestral path of particle $X_n^{k,J[k]}$, i.e., set $X_{0:n}[k] = X_{0:n}^{k,J[k]}$.

have a closed form expression for this distribution. However, we can still generate such a sample implicitly by running an SMC sampler targeting $\phi^{\theta'}_{0:n|n}$ and computing (11.38). An aesthetically appealing property of the method is that it looks very much like an "ideal" marginal Metropolis-Hastings sampler for $p(\theta \mid Y_{1:n})$, with the only difference being that the exact likelihood is replaced by its unbiased estimator. Consequently, the algorithm is referred to as particle marginal Metropolis-Hastings (PMMH).

Example 12.11 (PMMH for the NGM model; Example 12.10, cont.). In this example, we used the PMMH algorithm on the model specified in Example 12.10 using the same priors, and the means of the priors for starting values for the parameters. After a burn-in of 500 iterations, we kept every 25th sample for a total of 1000 values and 25,500 ($1000 \times 25 + 500$) iterations. The number of particles generated was 1000.

Parameter estimation results are displayed in Figure 12.7. The top row shows the output of the sampling traces for each parameter; the actual value is shown as a horizontal line. The middle row displays the ACF of the traces, and the bottom row shows the sampled posteriors of each parameter; the actual values are shown as vertical lines. State estimation is displayed in Figure 12.8. In that figure, the actual values of the states are displayed as solid points, whereas the posterior means are plotted as circles connected by a solid line. Finally, 99% credible intervals are displayed as a gray area.

We note that there is considerable autocorrelation among the sampled values. Also, the time to run this example was about 3 times longer than Example 12.10 (30.4 minutes in this case). In addition, as seen in Figure 12.8, the particles can be quite off the mark from the true state values. ◇

12.2.4 *Particle Gibbs algorithm*

Recall that the PGAS algorithm (Algorithm 11.4) defines a family of Markov kernels on X^{n+1}, leaving the family of joint smoothing distributions invariant. Consequently, we can simply replace Procedure 12.7-(ii) of the "ideal" Gibbs sampler with a run

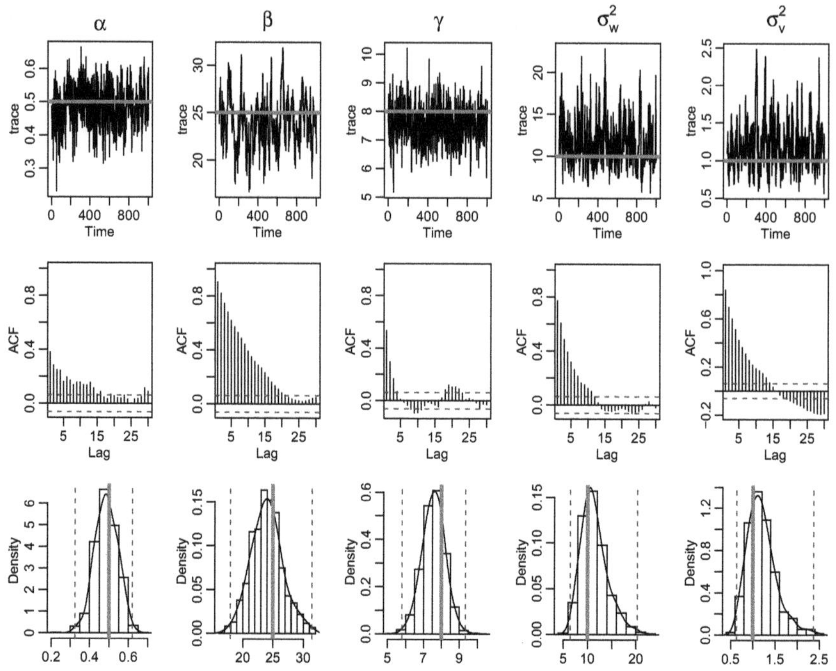

Figure 12.7 *Parameter Inference for Example 12.11: The top row shows the output of the sampling traces for each parameter; the actual value is shown as a horizontal line. The middle row displays the ACF of the traces, and the bottom row shows the sampled posteriors of each parameter. The actual values are shown as thick vertical lines and 99% credible intervals are represented as vertical dashed lines.*

of the PGAS algorithm. Due to the invariance properties of PGAS, this will indeed result in a valid MCMC method for the joint density $p(\theta, X_{0:n} \mid Y_{1:n})$. By this simple modification, we thus obtain a method capable of joint Bayesian state and parameter inference in NLSS. For clarity, the resulting method is summarized in Algorithm 12.3.

Example 12.12 (PGAS for the NGM model; Example 12.10 cont.). In this example, we used the PGAS algorithm on the model specified in Example 12.10 using the same priors, and the means of the priors for starting values for the parameters. As in the other examples, after a burn-in of 500 iterations, we kept every 25th sample for a total of 1000 values and 25,500 ($1000 \times 25 + 500$) iterations. The number of particles generated was 20.

Parameter estimation results are displayed in Figure 12.9. The top row shows the output of the sampling traces for each parameter; the actual value is shown as a horizontal line. The middle row displays the ACF of the traces, and the bottom row shows the sampled posteriors of each parameter; the actual values are shown as vertical lines. State estimation is displayed in Figure 12.10. In that figure, the actual values of the states are displayed as solid points, whereas the posterior means

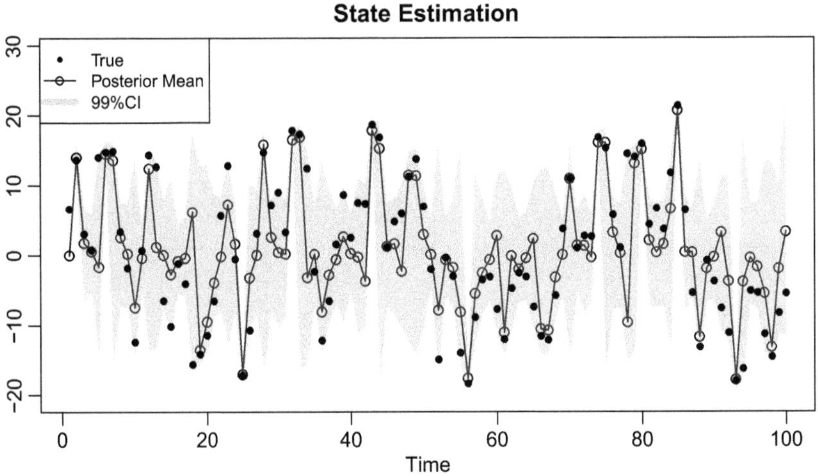

Figure 12.8 *State Estimation for Example 12.11: The actual values of the states are displayed as solid points, whereas the posterior means are plotted as circles connected by a solid line. 99% credible intervals are displayed as dashed lines.*

are plotted as circles connected by a solid line. Finally, 99% credible intervals are displayed as a gray area.

As opposed to previous examples, we note that there is little autocorrelation among the sampled values. Also, the time to run this example was shorter (8.1 minutes) than in Example 12.10. In addition, as seen in Figure 12.10, the particles are close to the true state values. Finally, we note that it was not necessary to have a step size of 25; this was done to keep the examples comparable. The methodology works fine without skipping sampled values (although there is more autocorrelation among the samples). In that case, the run time for this example is less than one minute. ◇

12.3 Endnotes

The approximation of the likelihood is studied in Olsson and Rydén (2008). These authors have suggested running the particle filter for all parameter values on a fixed grid in the parameter space, whereupon the approximation (12.5) is computed at each grid point. This pointwise approximation is extended from the grid to the whole parameter space by means of piecewise constant functions or smoothing B-splines. Olsson and Rydén (2008) discuss how to, as the number of observations increases, optimally vary the grid size, the size N of the particle population, and how to balance these quantities to guarantee that the estimates obtained when maximizing the log-likelihood approximations are consistent and asymptotically normal.

Particle approximation of the complete data score and the stochastic gradient algorithm has been considered in in the survey paper of Doucet et al. (2009); see also Andrieu et al. (2005a) and Cappé et al. (2005) for earlier references. Poyiadjis et al. (2011) discuss a method based on evaluating the incomplete data score based on the

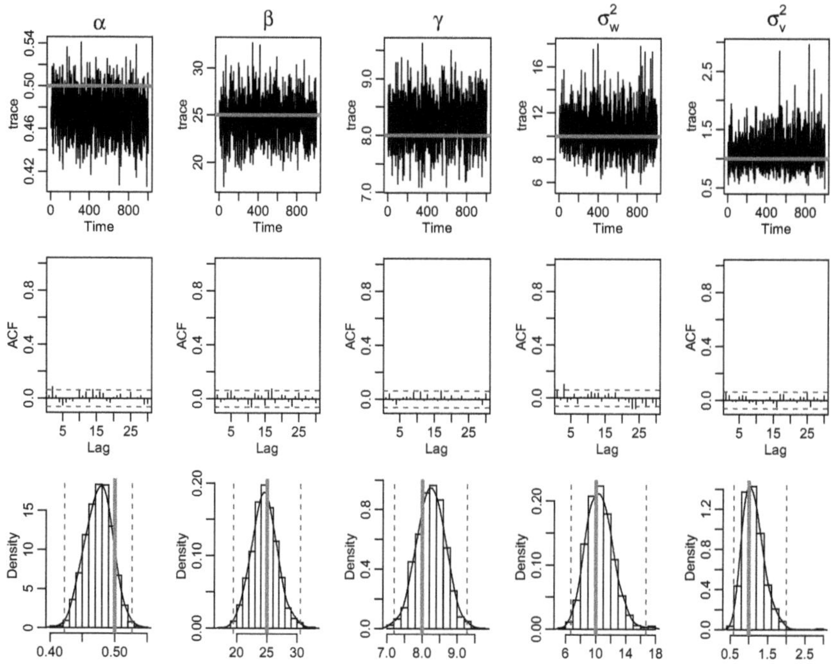

Figure 12.9 *Parameter Inference for Example 12.12: The top row shows the output of the sampling traces for each parameter; the actual value is shown as a horizontal line. The middle row displays the ACF of the traces, and the bottom row shows the sampled posteriors of each parameter. The actual values are shown as thick vertical lines and 99% credible intervals are represented as vertical dashed lines.*

Fisher identity and the forward-only evaluation of smoothing functionals discussed in Section 11.4. The use of the MCEM algorithm for NLSS has been discussed in Fort and Moulines (2003) (using MCMC techniques). Particle approximation for the MCEM algorithm has been studied in Olsson et al. (2008), using fixed-lag smoothing (this work quantifies the bias introduced by the fixed-lag smoothing approximation of Kitagawa and Sato, 2001).

The SAEM algorithm has been introduced by Delyon et al. (1999) for general hierarchical models and generalized in Kuhn and Lavielle (2004). The use of the Particle MCMC algorithm in conjunction with the SAEM algorithm was first investigated by Donnet and Samson (2011) using the PIMH kernel (Algorithm 11.2); see also Andrieu and Vihola (2011). Lindsten (2013) suggested using the PGAS kernel (Algorithm 11.4).

Carlin et al. (1992) were the first to develop an approach based on the Gibbs sampler to fit general nonlinear and non-Gaussian state-space models. Many examples of the technique presented in Example 12.9 are given in Carlin et al. (1992), including the problem of model choice. In that paper, the authors used a rejection method that could be costly. Frühwirth-Schnatter (1994) and Carter and Kohn (1994) built on

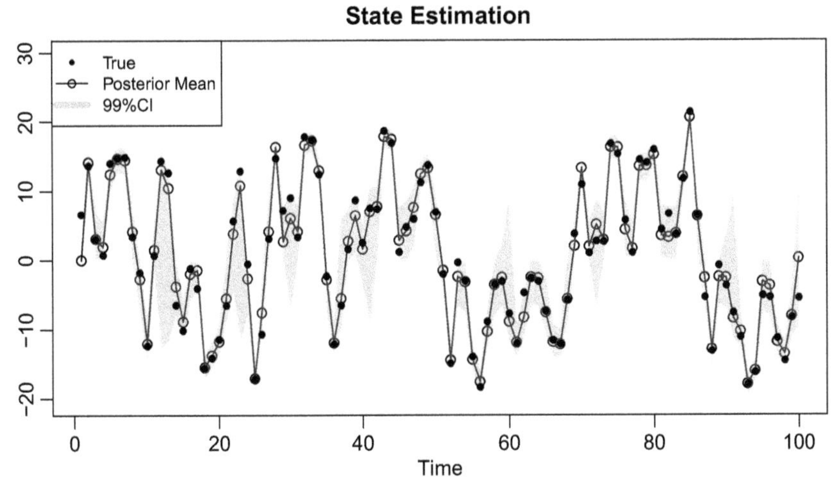

Figure 12.10 *State Estimation for Example 12.12: The actual values of the states are displayed as solid points, whereas the posterior means are plotted as circles connected by a solid line. 99% credible intervals are displayed as dashed lines.*

these ideas to develop efficient Gibbs sampling schemes for Gaussian Linear State Space; see Petris et al. (2009) and the references therein. For a further discussion, the reader may see Geweke and Tanizaki (2001).

Particle MCMC was introduced in Andrieu et al. (2010). This paper shows that it is possible to build high-dimensional proposal distributions for MCMC using SMC; as shown in this chapter, this algorithm can be used to develop algorithms to sample from the joint posterior distribution of states and parameters. This paper establishes the ergodicity of the PIMH algorithm from the unbiasedness of the marginal likelihood estimator under restrictive assumptions; improved conditions have been given in Andrieu and Vihola (2012). Andrieu et al. (2010) also introduced the conditional SMC/particle Gibbs sampler. Lindsten et al. (2012) and Lindsten and Schön (2013) proposed an interesting variant of the particle Gibbs sampler where ancestors are resampled in a forward pass. Chopin and Singh (2013) established the uniform ergodicity of the particle Gibbs sampler and discussed some algorithmic variants.

Chopin et al. (2011) substitutes an SMC algorithm for the MCMC used in the particle MCMC paper, which yields to a two-stage hierarchical SMC algorithm. A similar algorithm was described in Fulop and Li (2012).

Exercises

12.1. Assume A12.2.

(a) Show that, for any $f \in \mathbb{F}_b(\mathsf{X}, \mathcal{X})$ and $t \in \{1, \dots, n\}$,

$$\mathbb{E}\left[\omega_t^i f(X_t^i) \mid \mathcal{F}_{t-1}\right] = \frac{\kappa_{t-1}^N(Q_t^\theta f)}{\kappa_{t-1}^N(\vartheta_t^\theta)} ,$$

where κ_t^N is defined in (12.4).

(b) Show that

$$\frac{1}{N}\sum_{i=1}^{N} \omega_t^i f(X_t^i) \xrightarrow{\mathbb{P}} \frac{\gamma_t^\theta(f)}{\gamma_{t-1}^\theta(\vartheta_t^\theta)} = \frac{\phi_t^\theta(f)}{\phi_{t-1}^\theta(\vartheta_t^\theta)} := \kappa_t^\theta(f) . \tag{12.51}$$

(c) Show that $\kappa_0^N(\vartheta_1^\theta) \xrightarrow{\mathbb{P}} \gamma_0^\theta(\vartheta_1^\theta)$ and deduce that

$$\widehat{\mathrm{L}}(\theta; Y_{0:n}) \xrightarrow{\mathbb{P}} \gamma_0^\theta(\vartheta_1^\theta) \prod_{t=1}^{n-1} \frac{\gamma_t^\theta(\vartheta_{t+1}^\theta)}{\gamma_{t-1}^\theta(\vartheta_t^\theta)} \frac{\gamma_n^\theta(\mathbf{1})}{\gamma_{n-1}^\theta(\vartheta_n^\theta)} .$$

(d) Conclude that $\widehat{\mathrm{L}}(\theta; Y_{0:n})$ is a consistent sequence of estimators of the likelihood of the observations.

12.2. Assume A12.2.

(a) By adapting the arguments of Theorem 10.14, show that for any bounded $t \in \{0,\ldots,n\}$ and any functions $f_0,\ldots,f_t \in \mathbb{F}_b(\mathsf{X},\mathcal{X})$, the random vector

$$\sqrt{N}\left(\kappa_0^N(f_0) - \kappa_0^\theta(f_0),\ldots,\kappa_t^N(f_t) - \kappa_t^\theta(f_t)\right)$$

where κ_s^N and κ_s^θ are defined in (12.4) and (12.51), respectively, is asymptotically Gaussian.

(b) Show that $\sqrt{N}(\widehat{\mathrm{L}}(\theta; Y_{0:n}) - \mathrm{L}(\theta; Y_{0:n}))$ is asymptotically normal

12.3. Let f be a twice continuously differentiable function. Explain why the gradient is the line of steepest ascent.

12.4. Assume that f is a twice continuously differentiable function.

(a) Show that

$$f(x_k + p) \approx f_k + p'\nabla f_k + \frac{1}{2}p'\nabla^2 f_k p + o(|p|^2) := m_k(p) + o(|p|^2)$$

(b) Assume that $\nabla^2 f_k$ is positive definite. Show that the vector p that minimizes $m_k(p)$ is given by

$$p_k^{\mathrm{N}} = -(\nabla^2 f_k)^{-1}\nabla f_k .$$

12.5 (Another derivation of Beaumont's algorithm). We use the notations of Section 12.2.3. Let $\{X^k\}_{k=1}^N$ be N i.i.d. random variables from the proposal distribution h_θ. Denote by

$$\tilde{\pi}(\theta, X^{1:N}) = N^{-1}\sum_{k=1}^{n} \pi(\theta, X^k) \prod_{\ell \neq k} h_\theta(X^\ell) \tag{12.52}$$

$$= \hat{\pi}(\theta) \prod_{\ell=1}^{N} h_\theta(X^\ell) , \tag{12.53}$$

where $\hat{\pi}(\theta)$ is the importance sampling estimator of $\pi(\theta)$.

(a) Show that $\pi(\theta)$ is the marginal distribution of $\tilde{\pi}(\theta, X^{1:N})$.

(b) Denote by

$$\tilde{q}(\theta, X^{1:N}, \theta', \tilde{X}^{1:N}) = q(\theta, \theta') \prod_{\ell=1}^{N} h_\theta(X^\ell) .$$

Show that the acceptance ratio of the Metropolis Hastings algorithm targeting $\tilde{\pi}(\theta, X^{1:N})$ is given by $1 \wedge (\hat{\pi}(\theta')q(\theta', \theta)/(\hat{\pi}(\theta)q(\theta, \theta'))$ and conclude

12.6 (Stochastic volatility model). We consider maximum likelihood estimation in the stochastic volatility model of Example 9.5,

$$X_t = \phi X_{t-1} + \sigma U_t , \qquad U_t \sim \mathrm{N}(0,1) ,$$
$$Y_t = \beta \exp(X_t/2) V_t , \qquad V_t \sim \mathrm{N}(0,1) ,$$

where $\{U_t,\, t \in \mathbb{N}\}$ and $\{V_t,\, t \in \mathbb{N}\}$ are independent. In our analysis, we will assume that the log-volatility process $\{X_t,\, t \in \mathbb{N}\}$ is stationary ($|\phi| < 1$) so that the initial distribution ξ is given by $X_0 \sim \mathrm{N}(0, \sigma^2/(1-\phi^2))$.

Define $S(X_{0:n}, Y_{0:n}) = \{S_i(X_{0:n}, Y_{0:n})\}_{0 \le i \le 4}$ by $S_0(x_{0:n}) = x_0^2$,

$$S_1(x_{0:n}) = \sum_{t=0}^{n-1} x_t^2 , \qquad S_2(x_{0:n}) = \sum_{t=1}^{n} x_t^2 ,$$
$$S_3(x_{0:n}) = \sum_{t=1}^{n} x_t x_{t-1} , \qquad S_4(x_{0:n}, y_{0:n}) = \sum_{t=0}^{n} y_t^2 \exp(-x_t) . \tag{12.54}$$

(a) Show that the complete data log-likelihood may be written as

$$\ln \mathrm{L}(\beta, \phi, \sigma; X_{0:n}, Y_{0:n}) = \ln \mathrm{L}(\beta, \phi, \sigma; S(X_{0:n}, Y_{0:n})) , \tag{12.55}$$

where the function $s = (s_i)_{0 \le i \le 4} \mapsto \ln \mathrm{L}(\beta, \phi, \sigma; s)$ is given by

$$\ln \mathrm{L}(\beta, \phi, \sigma; s) = -\frac{n+1}{2} \ln \beta^2 - \frac{1}{2\beta^2} s_4 - \frac{n+1}{2} \ln \sigma^2 + \frac{1}{2} \ln(1-\phi^2)$$
$$- \frac{(1-\phi^2) s_0}{2\sigma^2} - \frac{1}{2\sigma^2} \left(s_2 - 2\phi s_3 + \phi^2 s_1 \right) . \tag{12.56}$$

(b) Use these expressions to obtain expressions of the incomplete data score.

12.7. For any function $f \in \mathbb{F}_b(\mathsf{X}, \mathcal{X})$, show that

$$\phi_{t:t+1|n}(f) = \iint f(x_t, x_{t+1}) B_{\phi_t}(x_t, \mathrm{d}x_{t+1}) \phi_{t+1|n}(\mathrm{d}x_{t+1}) .$$

and justify why

$$\sum_{i=1}^{N} \frac{\omega_t^i m(X_t^i, X_{t+1}^j)}{\sum_{\ell=1}^{N} \omega_t^\ell m(X_t^\ell, X_{t+1}^j)} \omega_{t|n}^j f(X_t^i, X_{t+1}^j)$$

is a consistent estimate of $\phi_{t:t+1|n}(f)$.

12.8 (Hessian computation). Show that

$$\nabla^2 \ln \mathrm{L}(\theta; Y_{0:n}) + \{\nabla \ln \mathrm{L}(\theta; Y_{0:n})\}\{\nabla \ln \mathrm{L}(\theta; Y_{0:n})\}' \\ = \phi^{\theta}_{0:n|n}\left(\sum_{t=0}^{n} \nabla s^{\theta}_t\right) + \phi^{\theta}_{0:n|n}\left[\left(\sum_{t=0}^{n} s^{\theta}_t\right)\left(\sum_{t=0}^{n} s^{\theta}_t\right)'\right], \quad (12.57)$$

where s^{θ}_t is defined in (12.12); give the recursions for $\alpha^{\theta,N}_n$ and $\beta^{\theta,N}_n$.

12.9. We use the notations of Exercise 12.6.

(a) Show that the auxiliary function $\theta \mapsto \mathcal{Q}(\theta\,;\theta_{k-1})$ of the EM algorithm may be expressed as

$$\mathcal{Q}(\theta\,;\theta_{k-1}) = -\frac{n+1}{2}\ln\beta^2 - \frac{1}{2\beta^2}\bar{s}_4(\theta_{k-1}) - \frac{n+1}{2}\ln\sigma^2 + \frac{1}{2}\ln(1-\phi^2) \\ -\frac{(1-\phi^2)}{2\sigma^2}\bar{s}_0(\theta_{k-1}) - \frac{1}{2\sigma^2}\left(\bar{s}_2(\theta_{k-1}) - 2\phi\bar{s}_3(\theta_{k-1}) + \phi^2\bar{s}_1(\theta_{k-1})\right), \quad (12.58)$$

where $\theta = (\beta, \phi, \sigma)$ and $\bar{s}_i(\theta) = \mathbb{E}_{\theta}\left[S_i(X_{0:n}) \mid Y_{0:n}\right]$, $i = 1, \ldots, 4$, are the conditional expectations of the complete data sufficient statistics S_i, $i = 1, 4$, defined in (12.54).

(b) Show that the maximization with respect to β yields the update

$$\beta_k = \sqrt{\frac{\bar{s}_4(\theta_{k-1})}{n+1}}. \quad (12.59)$$

(c) Show that the update for the variance term is given by

$$\sigma^2(s;\phi) = \frac{1}{n+1}\left\{(s_0 + s_2) - 2\phi s_3 + \phi^2(s_1 - s_0)\right\}. \quad (12.60)$$

(d) Show that the maximization with respect to the autoregression coefficient ϕ is equivalent to the cubic equation

$$\phi^3[n(s_1 - s_0)] + \phi^2[-(n-1)s_3] \\ + \phi[-s_2 + n s_0 - (n+1)s_1] + (n+1)s_3 = 0. \quad (12.61)$$

(e) Justify that the M-step implies the following computations: find ϕ_k as the root of (12.61), selecting the one that is, in absolute value, smaller than one; determine σ^2_k using (12.60); β_k is given by (12.59).

12.10. Using Example 12.8 as a guide, now let Y_t be the *logarithm* of the Johnson & Johnson quarterly earnings per share series, and consider a similar model, but where the trend is a random walk, $T_t = T_{t-1} + W_{t1}$. Plot Y_t and then repeat the analysis of Example 12.8 with the appropriate model changes.

12.11. The Ricker (1954) model is a classical population model that gives the expected number of individuals N_t in generation $t \in \mathbb{N}$ as a function of the number of individuals in the previous generation. This model is described by a difference equation, $N_t = N_{t-1} \exp\{r(1 - N_{t-1}/\kappa)\}$, where r is the maximum per capita growth rate and κ is the environmental carrying capacity (the maximum population size of a species that the given environment can sustain indefinitely). If we let X_t be the logarithm of the number of individuals in generation t, then the Ricker model suggests the state space model

$$X_t = X_{t-1} + \alpha + \beta \exp(X_{t-1}) + W_t \quad \text{and} \quad Y_t = X_t + V_t$$

where $\alpha = r$ and $\beta = -r/\kappa$ are transition parameters. Assume that W_t and V_t are independent Gaussian noise sequences with respective variances σ_w^2 and σ_v^2. The model for the observations, Y_t, indicates that we observe the population size with error.

The data `salmon` in `nltsa` are the annual number of salmon (any species) counted over the John Day Dam for 30 years, beginning in 1968.

(a) Examine the data, and with justification, decide if a Ricker model is appropriate.

(b) Use one or two of the methods discussed in this chapter to fit the model. State your conclusions.

Chapter 13

Asymptotics of Maximum Likelihood Estimation for Nonlinear State Space Models

We consider a parametrized family of NLSS with parameter space Θ. For each parameter $\theta \in \Theta$, the distribution of the NLSS is specified by the transition kernel M^θ of the underlying Markov chain $\{X_t,\ t \in \mathbb{N}\}$, and by the conditional density $g^\theta(X_t, Y_t)$ of the observation Y_t given the state X_t with respect to some dominating measure μ.

We observe a single path of the process $\{Y_t,\ t \in \mathbb{N}\}$. We assume that this process is strict-sense stationary and ergodic so that it may be extended on $\mathbb{Z}$ and we denote by $\mathbb{P}$ and $\mathbb{E}$ the probability and expectation induced by the sequence $\{Y_t,\ t \in \mathbb{Z}\}$ on $(\mathsf{Y}^{\mathbb{Z}}, \mathcal{Y}^{\otimes \mathbb{Z}})$. If the model is well-specified, the law of the observations is assumed to belong to the parametric family, that is, there exists some $\theta_\star \in \Theta$, such that $\mathbb{P}$ is the law of the observation process of an NLSS associated to $(M^{\theta_\star}, g^{\theta_\star})$ and $\theta_\star$ is interpreted as the *true* parameter value, which is not known a priori. Our basic problem is to form a consistent estimate of $\theta_\star$ on the basis of the observations $\{Y_t,\ t \in \mathbb{N}\}$ only, i.e., without access to the hidden process $\{X_t,\ t \in \mathbb{N}\}$. This will be accomplished by means of the maximum likelihood method. This chapter is devoted to the asymptotic properties, as the number of observations n goes to infinity, of the maximum likelihood estimator (MLE), $\hat{\theta}_{\xi,n}$, defined by

$$\hat{\theta}_{\xi,n} \in \underset{\theta \in \Theta}{\arg\max} \left\{ \mathrm{p}_\xi^\theta(Y_{0:n-1}) \right\} ,$$

where $\mathrm{p}_\xi^\theta(\cdot)$ is the likelihood function of an NLSS associated to (M^θ, g^θ) with initial distribution $X_0 \sim \xi$. More precisely, for well-specified models, we aim at finding conditions under which $\hat{\theta}_{\xi,n}$ turns out to be *strongly consistent*, that is:

$$\lim_{n \to \infty} \hat{\theta}_{\xi,n} = \theta_\star , \quad \mathbb{P}\text{-a.s.}$$

and *asymptotically normal*, that is:

$$\sqrt{n}(\hat{\theta}_{\xi,n} - \theta_\star) \overset{\mathbb{P}}{\Longrightarrow} \mathrm{N}(0, \mathcal{J}^{-1}(\theta_\star)) ,$$

for some nonsingular $d \times d$-matrix $\mathcal{J}(\theta_\star)$.

13.1 Strong consistency of the MLE

Let X and Y be Polish spaces endowed with their Borel field $\mathcal{X}$, $\mathcal{Y}$. Let (Θ, d) be a compact metric space. The proof is in two steps. In a first step, we show that for any initial distribution ξ and parameter $\theta \in \Theta$, there is a constant $\ell(\theta)$ such that

$$\lim_{n\to\infty} n^{-1} \ln \mathrm{p}_\xi^\theta(Y_{0:n-1}) = \ell(\theta) \,, \quad \mathbb{P}\text{-a.s.}$$

Up to an additive constant, $\ell(\theta)$ is the negated relative entropy rate between the distribution of the observations and $\mathrm{p}_\xi^\theta(\cdot)$ respectively. When the model is well-specified and $\theta = \theta_\star$ is the true value of the parameter, this convergence follows from the generalized Shannon-Breiman-McMillan theorem, see Barron (1985); for misspecified models or for well-specified models with $\theta \neq \theta_\star$ the existence of this limit is far from obvious.

In the second step, we will prove that the maximizer of the likelihood $\theta \mapsto n^{-1} \ln \mathrm{p}_\xi^\theta(Y_{0:n})$ converges $\mathbb{P}$-a.s. to the maximizer of $\theta \mapsto \ell(\theta)$.

13.1.1 Forgetting the initial distribution for the filter distribution

For any probability measure ξ on $(\mathsf{X}, \mathcal{X})$, define the likelihood and the conditional likelihood of the observations by

$$\mathrm{p}_\xi^\theta(y_{r:s}) := \int \cdots \int \xi(\mathrm{d}x_r) g^\theta(x_r, y_r) \prod_{p=r+1}^{s} M^\theta(x_{p-1}, \mathrm{d}x_p) g^\theta(x_p, y_p) \,, \tag{13.1}$$

$$\mathrm{p}_\xi^\theta(y_{t:s} | y_{r:t-1}) := \mathrm{p}_\xi^\theta(y_{r:s}) / \mathrm{p}_\xi^\theta(y_{r:t-1}) \,, \quad r < t \leq s \,. \tag{13.2}$$

For any probability measure ξ, the log-likelihood of the observations $\ell_{\xi,n}(\theta) = \ln \mathrm{p}_\xi^\theta(Y_{0:n-1})$ may be decomposed as follows:

$$\ell_{\xi,n}(\theta) = \sum_{t=0}^{n-1} \ln \mathrm{p}_\xi^\theta(Y_t | Y_{0:t-1}) \,. \tag{13.3}$$

The summands do not form a stationary ergodic sequence so that the Birkhoff theorem (Theorem 7.10 and Corollary 7.12) does not directly apply. Instead, we first approximate $\ell_{\xi,n}(\theta)$ by the partial sum of a stationary ergodic sequence. To this aim, we first show that $\lim_{m\to\infty} \mathrm{p}_\xi^\theta(Y_t \mid Y_{-m:t-1})$ exists $\mathbb{P}$-a.s. and does not depend on the initial distribution ξ. This limit will be denoted by $\mathrm{p}^\theta(Y_t \mid Y_{-\infty:t-1})$.

A key feature of these variables is that they now form an ergodic stationary sequence, whence the Birkhoff theorem applies as soon as this sequence is integrable. Furthermore we will approximate $\ell_{\xi,n}(\theta)$ by

$$\ell_n^s(\theta) = \sum_{t=0}^{n-1} \ln \mathrm{p}^\theta(Y_t \mid Y_{-\infty:t-1}) \,, \tag{13.4}$$

where superscript s stands for "stationary." The cornerstone in this analysis is the strong mixing condition.

Assumption A 13.1.

(a) There exists a probability measure λ on $(\mathsf{X},\mathcal{X})$ that dominates $M^\theta(x,\cdot)$ for all $x \in \mathsf{X}$ and $\theta \in \Theta$. We denote $m^\theta(x,x') := (\mathrm{d}M^\theta(x,\cdot)/\mathrm{d}\lambda)(x')$.

(b) The transition density $m^\theta(x,x')$ satisfies $0 < \sigma^- \le m^\theta(x,x') \le \sigma^+ < \infty$ for all $x,x' \in \mathsf{X}$ and all $\theta \in \Theta$.

(c) For all $y \in \mathsf{Y}$, the integral $\int_{\mathsf{X}} g^\theta(x,y)\,\lambda(\mathrm{d}x)$ is bounded away from 0 and ∞ on Θ.

A13.1-(b) typically requires that the state space X is finite or compact. It implies that the Markov kernel M^θ is uniformly geometrically ergodic (see Definition 6.7 and Proposition 6.8): therefore, there exists a unique stationary distribution π^θ for M^θ and for all probability distributions ξ and ξ' and all $n \ge 0$, $d_{\mathrm{TV}}(\xi M^{\theta^n}, \xi' M^{\theta^n}) \le (1-\sigma^-)^n$.

Of particular interest when dealing with $\mathrm{p}^\theta_\xi(y_t \mid y_{-m:t})$ is the filter distribution $A \mapsto \mathbb{P}^\theta_{\xi,-m}(X_t \in A \mid Y_{-m:t} = y_{-m:t})$ defined as the distribution under the parameter θ of the state X_t given the observations $Y_{-m:t} = y_{-m:t}$ and starting at time $-m$ with $X_{-m} \sim \xi$:

$$\mathbb{P}^\theta_{\xi,-m}(X_t \in A \mid Y_{-m:t} = y_{-m:t}) \\ = \frac{\int \xi(\mathrm{d}x_{-m}) g^\theta(x_{-m},y_{-m}) \prod_{p=-m+1}^t M^\theta(x_{p-1},\mathrm{d}x_p) g^\theta(x_p,y_p) \mathbb{1}_A(x_t)}{\int \xi(\mathrm{d}x_{-m}) g^\theta(x_{-m},y_{-m}) \prod_{p=-m+1}^t M^\theta(x_{p-1},\mathrm{d}x_p) g^\theta(x_p,y_p)} . \tag{13.5}$$

More generally, we denote by $\mathbb{P}^\theta_{\xi,-m}$ (resp. $\mathbb{E}^\theta_{\xi,-m}$) the probability (resp. expectation) induced by a Markov chain associated to (M^θ, g^θ) starting at time $-m$ with the initial distribution $X_{-m} \sim \xi$. When $m = 0$, we simply write $\mathbb{P}^\theta_\xi = \mathbb{P}^\theta_{\xi,0}$ and $\mathbb{E}^\theta_\xi = \mathbb{E}^\theta_{\xi,0}$. In the particular case where the kernel of the hidden chain is uniformly geometrically ergodic (see A 13.1-(b)), it is easy to establish the forgetting of the initial distribution (perhaps surprisingly, this can be done *without* any assumption of the likelihood function g^θ beyond positivity).

Lemma 13.2. *Under A13.1, for all probability measures ξ, ξ' and $-\infty < r \le s < \infty$,*

$$d_{\mathrm{TV}}(\mathbb{P}^\theta_{\xi,r}(X_s \in \cdot \mid Y_{r:s}), \mathbb{P}^\theta_{\xi',r}(X_t \in \cdot \mid Y_{r:s})) \le \rho^{s-r} , \tag{13.6}$$

$$\sup_{\theta \in \Theta} d_{\mathrm{TV}}(\mathbb{P}^\theta_{\xi,r}(X_s \in \cdot \mid Y_{r:s}), \mathbb{P}^\theta_{\xi,r-1}(X_s \in \cdot \mid Y_{r-1:s})) \le \rho^{s-r} , \quad \mathbb{P}\text{-a.s.} \tag{13.7}$$

where the rate ρ is equal to $1 - \sigma^-/\sigma^+$ (defined in A13.1).

Proof. The proof of (13.6) follows from Proposition 10.20. According to (13.6), for all probability measures ξ and ξ' and all $-\infty < r \le s < \infty$,

$$\sup_{\theta \in \Theta} d_{\mathrm{TV}}(\mathbb{P}^\theta_{\xi,r}(X_s \in \cdot \mid Y_{r:s}), \mathbb{P}^\theta_{\xi',r}(X_s \in \cdot \mid Y_{r:s})) \le \rho^{s-r} , \quad \mathbb{P}\text{-a.s.}$$

Now, choose $\xi'(\cdot) \propto \int \xi(\mathrm{d}x) g^\theta(x, Y_{r-1}) M^\theta(x,\cdot)$ so that by using (13.5),

$$\mathbb{P}^\theta_{\xi',r}(X_s \in \cdot \mid Y_{r:s}) = \mathbb{P}^\theta_{\xi,r-1}(X_s \in \cdot \mid Y_{r-1:s}) .$$

The proof follows. ∎

13.1.2 Approximation by a stationary conditional log-likelihood

The predictive distribution $y_t \mapsto \mathrm{p}^\theta_\xi(y_t \mid Y_{-m:t-1})$ and the filter distribution $A \mapsto \mathbb{P}^\theta_\xi[X_t \in A \mid Y_{-m:t-1}]$ satisfy the following relation

$$\mathrm{p}^\theta_\xi(Y_t \mid Y_{-m:t-1}) = \iint \mathbb{P}^\theta_\xi[X_{t-1} \in \mathrm{d}x \mid Y_{-m:t-1}] M^\theta(x, \mathrm{d}x') g^\theta(x', Y_t) \,. \tag{13.8}$$

Thanks to (13.8), the predictive distribution $\mathrm{p}^\theta_\xi(Y_t \mid Y_{-m:t-1})$ inherits the forgetting of the initial distribution of the filter. Some additional assumptions on g^θ are required:

Assumption A13.3. $b^+ := \sup_\theta \sup_{x,y} g^\theta(x,y) < \infty$ and $\mathbb{E}^{\theta_\star}|\ln b^-(Y_0)| < \infty$, where $b^-(y) := \inf_\theta \int_\mathsf{X} g^\theta(x,y)\,\lambda(\mathrm{d}x)$.

Lemma 13.4. *Assume A13.1 and A13.3. Then, for all probability measures ξ and ξ', all indices $t \geq 1$ and $m \geq 0$ and all sequences $y_{-m:t} \in \mathsf{Y}^{t+m+1}$:*

$$\sup_{\theta \in \Theta} |\ln \mathrm{p}^\theta_\xi(y_t \mid y_{-m:t-1}) - \ln \mathrm{p}^\theta_{\xi'}(y_t \mid y_{-m:t-1})| \leq \frac{\rho^{t+m-1}}{1-\rho} \,, \tag{13.9}$$

$$\sup_{\theta \in \Theta} |\ln \mathrm{p}^\theta_\xi(y_t \mid y_{-m:t-1}) - \ln \mathrm{p}^\theta_\xi(y_t \mid y_{-m-1:t-1})| \leq \frac{\rho^{t+m-1}}{1-\rho} \,, \tag{13.10}$$

$$\sup_{\theta \in \Theta} \sup_{m \geq 0} \mathrm{p}^\theta_\xi(y_t \mid y_{-m:t-1}) \leq b^+ \,. \tag{13.11}$$

Proof. (13.11) follows directly from (13.8), A13.3. (13.10) follows from (13.9) by choosing $\xi'(\cdot) \propto \int \xi(\mathrm{d}u) g^\theta(x, y_{-m-1}) M^\theta(x, \cdot)$ so that by (13.1),

$$\mathrm{p}^\theta_{\xi'}(y_t \mid y_{-m:t-1}) = \mathrm{p}^\theta_\xi(y_t \mid y_{-m-1:t-1}) \,.$$

It thus remains to check (13.9). By (13.8) and Lemma 13.2, we have under A13.1,

$$|\mathrm{p}^\theta_\xi(y_t \mid y_{-m:t-1}) - \mathrm{p}^\theta_{\xi'}(y_t \mid y_{-m:t-1})| \leq \rho^{m+t-1} \sigma^+ \int g^\theta(x, y_t) \lambda(\mathrm{d}x) \,.$$

Moreover, combining (13.8) with A13.1, we have

$$\mathrm{p}^\theta_\xi(y_t \mid y_{-m:t-1}) \vee \mathrm{p}^\theta_{\xi'}(y_t \mid y_{-m:t-1}) \geq \sigma^- \int g^\theta(x, y_t) \lambda(\mathrm{d}x) \,.$$

Finally, (13.9) follows from these two bounds, using the inequality $|\ln u - \ln v| \leq |u - v|/(u \vee v)$ and noting that $\sigma^+/\sigma^- = (1-\rho)^{-1}$. ■

Now, consider (13.10) for a fixed integer t. The sequence $\{\ln \mathrm{p}^\theta_\xi(Y_t \mid Y_{-m:t-1})\}_{m \geq 0}$ is a uniform (in θ) Cauchy sequence as $m \to \infty$, $\mathbb{P}$-a.s.; therefore, this sequence has an $\mathbb{P}$-a.s. finite limit. Moreover, using again (13.9), this limit does not depend on ξ; it is denoted $\ln \mathrm{p}^\theta(Y_t \mid Y_{-\infty:t-1})$. Since the observation process is strict-sense stationary and ergodic, the Birkhoff ergodic theorem (Theorem 7.44) shows that $n^{-1} \sum_{t=0}^{n-1} \ln \mathrm{p}^\theta(Y_t \mid Y_{-\infty:t-1})$ exists $\mathbb{P}$-a.s., provided that $\mathbb{E}[|\ln \mathrm{p}^\theta(Y_t \mid Y_{-\infty:t-1})|] < \infty$, which follows from (13.8):

$$\sigma^- b^-(Y_t) \leq \mathrm{p}^\theta_\xi(Y_t \mid Y_{-m:t-1}) \leq b^+ \,, \text{ for all } \theta \in \Theta,$$

where σ^- is defined in A 13.1 and (b^-, b^+) in A 13.3. Then, letting $m \to \infty$, A 13.3 shows that $\mathbb{E}|\ln \mathrm{p}^\theta(Y_t \mid Y_{-\infty:t-1})| < \infty$. We summarize our findings in the following proposition.

Proposition 13.5. *Under A 13.1 and A 13.3, for all $\theta \in \Theta$ and all probability measures ξ, the sequence $\{\ln \mathrm{p}_\xi^\theta(Y_t \mid Y_{-m:t-1})\}_{m\geq 0}$ has, $\mathbb{P}$-a.s., a limit $\ln \mathrm{p}^\theta(Y_t \mid Y_{-\infty:t-1})$ as $m \to \infty$. This limit does not depend on ξ. In addition, for all $\theta \in \Theta$, $\ln \mathrm{p}^\theta(Y_t \mid Y_{-\infty:t-1})$ belongs to $\mathrm{L}^1(\mathbb{P})$. For all probability distribution ξ,*

$$\lim_{n\to\infty} \sup_{\theta\in\Theta} n^{-1} |\ell_{\xi,n}(\theta) - \ell_n^s(\theta)| = 0, \qquad \mathbb{P}\text{-a.s.}$$

Proof. By (13.10), we have by summing for $m = 0$ to ∞,

$$\sup_{\theta\in\Theta} |\ln \mathrm{p}_\xi^\theta(Y_t \mid Y_{0:t-1}) - \ln \mathrm{p}^\theta(Y_t \mid Y_{-\infty:t-1})| \leq \frac{\rho^{t-1}}{(1-\rho)^2} .$$

The proof follows from the definitions of $\ell_{\xi,n}(\theta)$ and $\ell_n^s(\theta)$ in (13.3) and (13.4), respectively. ■

13.1.3 Strong consistency in misspecified and well-specified models

Misspecified models

In misspecified models, we *do not* assume that the distribution of the observations belongs to the set of distributions where the maximization takes place. In particular, $\{Y_t,\ t \in \mathbb{Z}\}$ is a strict sense stationary and ergodic but is not necessarily the observation process of an NLSS. In Section 13.1.2, we obtain stationary approximations of the loglikelihood and we are now ready to get the strong consistency of the MLE in misspecified models by applying the general result of Pfanzagl (1969), recalled in Theorem 8.42. In order to check the assumptions of Theorem 8.42, note that

$$\hat{\theta}_{\xi,n} \in \arg\max_{\theta\in\Theta} \left\{ n^{-1} \sum_{t=0}^{n-1} \ln \mathrm{p}_\xi^\theta(Y_t \mid Y_{0:t-1}) \right\} .$$

Consider the following assumption:

Assumption A 13.6. For all $(x, x') \in \mathsf{X} \times \mathsf{X}$ and $y \in \mathsf{Y}$, the functions $\theta \mapsto m^\theta(x, x')$ and $\theta \mapsto g^\theta(x, y)$ are continuous.

Theorem 13.7. *Under A 13.1–A 13.6, the MLE, $\hat{\theta}_{\xi,n}$, is strongly consistent for all probability measures ξ, i.e.*

$$\lim_{n\to\infty} \mathsf{d}(\hat{\theta}_{\xi,n}, \Theta_\star) = 0, \quad \mathbb{P}\text{-a.s.}$$

where $\Theta_\star = \arg\max_{\theta\in\Theta} \ell(\theta)$ and $\ell(\theta) = \mathbb{E}[\ln \mathrm{p}^\theta(Y_0 \mid Y_{-\infty:-1})]$. Moreover, for all $\theta \in \Theta$ and all probability measures ξ,

$$\lim_{n\to\infty} n^{-1} \ell_{\xi,n}(\theta) = \ell(\theta), \quad \mathbb{P}\text{-a.s.}$$

Proof. The stationary approximation of $\ln \mathrm{p}^{\theta}_{\xi}(Y_t \mid Y_{0:t-1})$ is $\ln \mathrm{p}^{\theta}(Y_t \mid Y_{-\infty:t-1})$ which is defined, according to Proposition 13.5, as the $\mathbb{P}$-a.s. limit of $\ln \mathrm{p}^{\theta}_{\xi}(Y_t \mid Y_{-m:t-1})$ as m goes to infinity. Moreover, $\mathbb{E}[\sup_{\theta\in\Theta} \left(\ln \mathrm{p}^{\theta}(Y_t \mid Y_{-\infty:t-1})\right)_+] < \infty$ by using (13.11) so that A8.39 holds.

By using the continuity of the functions involved in A 13.6 and by noting that $M^{\theta}(x, \mathrm{d}x') \leq \sigma^+ \lambda(\mathrm{d}x')$ and $g^{\theta}(x,y) \leq b^+$, we see that the dominated convergence theorem implies that for all $m \leq t$ and all probability measures ξ, the function $\theta \mapsto \ln \mathrm{p}^{\theta}_{\xi}(Y_{-m:t})$ is continuous, $\mathbb{P}$-a.s. and consequently, the uniform (in θ) bound in (13.10) implies that the function $\theta \mapsto \ln \mathrm{p}^{\theta}(Y_t \mid Y_{-\infty:t-1})$ is $\mathbb{P}$-a.s. continuous. Thus, A8.40 holds. And finally, since Proposition 13.5 can be rewritten as

$$\lim_{n\to\infty} \sup_{\theta\in\Theta} \left| n^{-1} \sum_{t=0}^{n-1} \ln \mathrm{p}^{\theta}_{\xi}(Y_t \mid Y_{0:t-1}) - \ln \mathrm{p}^{\theta}(Y_t \mid Y_{-\infty:t-1}) \right| = 0, \quad \mathbb{P}\text{-a.s.}$$

A8.41 follows. Theorem 8.42 therefore applies. ∎

Actually, under A 13.1–A 13.6, we can get some additional results as, for example, the L^1-convergence of $n^{-1}\ell_{\xi,n}(\theta)$ to $\ell(\theta)$. Nevertheless, this L^1-convergence is not needed to establish the consistency result and we choose here to only highlight the "almost necessary" steps towards the consistency of the MLE that should be successively shown when trying to weaken A13.1.

Identifiability and well-specified models

As previously observed, A13.1 implies that for all $\theta \in \Theta$, the Markov kernel M^{θ} admits a unique invariant probability measure that we denote by π^{θ}. For all probability measures ξ, denote by $\mathbb{P}^{\theta}_{\xi}$ and $\mathbb{E}^{\theta}_{\xi}$ the probability and expectation on $(\mathsf{Y}^{\mathbb{N}}, \mathcal{Y}^{\otimes\mathbb{N}})$ induced by the observation process of an NLSS associated to (M^{θ}, g^{θ}) starting from $X_0 \sim \xi$. If the chain starts from its stationary distribution, i.e. when $\xi = \pi^{\theta}$, we simply set $\mathbb{P}^{\theta} = \mathbb{P}^{\theta}_{\pi^{\theta}}$ and $\mathbb{E}^{\theta} = \mathbb{E}^{\theta}_{\pi^{\theta}}$. In such a case, the process may be extended on $\mathbb{Z}$ and by abuse of notation, we also write as $\mathbb{P}^{\theta}$ and $\mathbb{E}^{\theta}$ the associated probability and expectation on $(\mathsf{Y}^{\mathbb{Z}}, \mathcal{Y}^{\otimes\mathbb{Z}})$.

In this section, we consider well-specified models: there exists $\theta_\star \in \Theta$ such that $\{Y_t,\ t \in \mathbb{Z}\}$ is the observation process associated to a strict-sense stationary and ergodic version of the NLSS with parameter $\theta_\star$. In such a case, we have $\mathbb{P} = \mathbb{P}^{\theta_\star}$ and $\mathbb{E} = \mathbb{E}^{\theta_\star}$. To avoid complicated notations, we use as much as possible $\mathbb{P}$ and $\mathbb{E}$ when there is no ambiguity and use $\mathbb{P}^{\theta_\star}$ and $\mathbb{E}^{\theta_\star}$ otherwise.

We first introduce the notion of *equivalence of parameters*.

Definition 13.8. *The parameters $\theta, \theta' \in \Theta$ are* equivalent *if $\mathbb{P}^{\theta} = \mathbb{P}^{\theta'}$. In such a case, we write $\theta \sim \theta'$.*

We note from Kolmogorov's extension theorem (see Theorem 1.5), that θ and θ' are equivalent if and only if the finite-dimensional distributions $\mathbb{P}^{\theta}(Y_0 \in \cdot, Y_1 \in \cdot, \ldots, Y_{n-1} \in \cdot)$ and $\mathbb{P}^{\theta'}(Y_0 \in \cdot, Y_1 \in \cdot, \ldots, Y_{n-1} \in \cdot)$ agree for all $n \geq 1$.

We say that the MLE is strongly consistent *in the quotient topology* for the probability measure ξ if

$$\lim_{n\to\infty} \mathsf{d}(\hat{\theta}_{\xi,n}, \{\theta \in \Theta : \theta \sim \theta_\star\}) = 0, \quad \mathbb{P}^{\theta_\star}\text{-a.s.}$$

In order to obtain the strong consistency of the MLE, we apply Theorem 13.7. For that purpose, we first establish that the set $\Theta_\star = \arg\max_\theta \ell(\theta)$ defined in Theorem 13.7 is equal to $\{\theta \in \Theta : \theta \sim \theta_\star\}$. To obtain such a result, we show $\ell(\theta) < \ell(\theta_\star)$ for all parameters $\theta \not\sim \theta_\star$. According to Theorem 13.7, the relative entropy rate $\ell(\theta)$ may be expressed as:

$$\ell(\theta) = \lim_{n\to\infty} n^{-1} \ln \mathrm{p}^{\theta}_{\xi}(Y_{0:n}), \quad \mathbb{P}^{\theta_\star}\text{-a.s.} \tag{13.12}$$

$$\ell(\theta) = \mathbb{E}[\ln \mathrm{p}^{\theta}(Y_0 \mid Y_{-\infty:-1})] . \tag{13.13}$$

We present two different approaches, each one being based on one of these two expressions. Consider first (13.12): for all probability measures ξ,

$$\ell(\theta) - \ell(\theta_\star) = \lim_{n\to\infty} n^{-1} \ln M_n , \ \mathbb{P}\text{-a.s.} \quad \text{where} \quad M_n = \frac{\mathrm{p}^{\theta}_{\xi}(Y_{0:n-1})}{\mathrm{p}^{\theta_\star}_{\pi_{\theta_\star}}(Y_{0:n-1})} , \tag{13.14}$$

Without any additional assumption, $\theta_\star$ is a maximizer of $\theta \mapsto \ell(\theta)$; see Exercise 13.1. To show that this maximizer is unique up to the equivalence relation, consider the following exponential separation assumption:

Assumption A13.9. For all $\theta \in \Theta$ such that $\theta \not\sim \theta_\star$, there exist a sequence of sets $\{A_n,\ n \in \mathbb{N}\}$ and a probability distribution ξ such that

$$\liminf_{n\to\infty} \mathbb{1}_{A_n}(Y_{0:n-1}) > 0, \quad \mathbb{P}^{\theta_\star}\text{-a.s.} \tag{13.15}$$

$$\limsup_{n\to\infty} n^{-1} \ln \mathbb{P}^{\theta}_{\xi}(Y_{0:n-1} \in A_n) < 0 . \tag{13.16}$$

Lemma 13.10. *Under A13.9, for all $\theta \not\sim \theta_\star$,*

$$\ell(\theta) - \ell(\theta_\star) = \limsup_{n\to\infty} n^{-1} \ln M_n < 0, \quad \mathbb{P}^{\theta_\star}\text{-a.s.} .$$

Proof. Under A13.9, for all $\theta \neq \theta_\star$, there exists some constant $\delta > 0$ such that

$$\sum_{t=1}^{\infty} \mathbb{E}^{\theta_\star}\left[\frac{\mathrm{p}^{\theta}_{\xi}(Y_{0:t-1})}{\mathrm{p}^{\theta_\star}_{\pi_{\theta_\star}}(Y_{0:t-1})} \mathrm{e}^{\delta t} \mathbb{1}_{A_t}(Y_{0:t-1})\right] = \sum_{t=0}^{\infty} \mathbb{P}^{\theta}_{\xi}(Y_{0:t-1} \in A_t) \mathrm{e}^{\delta t} < \infty ,$$

which implies that

$$\sum_{t=0}^{\infty} \frac{\mathrm{p}^{\theta}_{\xi}(Y_{0:t-1})}{\mathrm{p}^{\theta_\star}_{\pi_{\theta_\star}}(Y_{0:t-1})} \mathrm{e}^{\delta t} \mathbb{1}_{A_t}(Y_{0:t-1}) < \infty, \quad \mathbb{P}^{\theta_\star}\text{-a.s.}$$

Using again A13.9, $\mathbb{1}_{A_t}(Y_{0:t-1}) = 1$ for all but an $\mathbb{P}$-a.s.-finite number of t and we then obtain that $\limsup_{t\to\infty} t^{-1} \ln\left(\mathrm{p}^{\theta}_{\xi}(Y_{0:t-1}) / \mathrm{p}^{\theta_\star}_{\pi_{\theta_\star}}(Y_{0:t-1})\right) \leq -\delta < 0$, which implies that $\ell(\theta) < \ell(\theta_\star)$. ■

The set A_n in A13.9 may be chosen typically in the following way. For $\theta \not\sim \theta_\star$, there exist an integer ℓ and a set $A \in \mathcal{Y}^{\otimes \ell}$ satisfying $\Delta_A = |\mathbb{P}^\theta(Y_{0:\ell-1} \in A) - \mathbb{P}^{\theta_\star}(Y_{0:\ell-1} \in A)| \neq 0$. Now, pick $0 < \varepsilon < \Delta_A$ and define

$$A_n = \left\{ y_{0:n-1} \in \mathsf{Y}^n : \left| (n-\ell+1)^{-1} \sum_{k=0}^{n-\ell} (\mathbb{1}_A(y_{t:t+\ell-1}) - \mathbb{P}^\theta(Y_{0:\ell-1} \in A)) \right| > \varepsilon \right\} .$$

The first condition (13.15) of A13.9 follows directly from the Birkhoff ergodic theorem (see Theorem 7.44). The second condition (13.16) of A13.9 may be obtained by an exponential deviation inequality. Note that the result of Lemma 13.10 is valid for any process $\{Y_t,\, t \in \mathbb{N}\}$ satisfying A13.9.

Lemma 13.11. *A13.1 implies A13.9.*

Proof. Note that $\{Y_t,\, t \in \mathbb{N}\}$ is not itself a Markov chain, but (13.16) follows by considering Y_t as component of $Z_t = (X_t, Y_t)$ which is a uniformly ergodic Markov chain under A13.1 and thus satisfies a Hoeffding type inequality (see Corollary 7.37). ∎

We now turn to another approach based on the expression of $\ell(\theta)$ as an expectation, given by (13.13). In comparison with the previous approach, no exponential inequality is needed. On the other hand, it relies strongly on the NLSS structure of the model. Write, using (13.13),

$$\begin{aligned} \ell(\theta_\star) - \ell(\theta) &= \mathbb{E}^{\theta_\star}\left[\ln \frac{\mathrm{p}^{\theta_\star}(Y_0 \mid Y_{-\infty:-1})}{\mathrm{p}^\theta(Y_0 \mid Y_{-\infty:-1})} \right] \\ &= \mathbb{E}^{\theta_\star}\left[\mathbb{E}^{\theta_\star}\left[\ln \frac{\mathrm{p}^{\theta_\star}(Y_0 \mid Y_{-\infty:-1})}{\mathrm{p}^\theta(Y_0 \mid Y_{-\infty:-1})} \,\middle|\, Y_{-\infty:-1} \right]\right] . \end{aligned} \tag{13.17}$$

The inner term can thus be interpreted as a conditional Kullback-Leibler divergence provided that the ratio is shown to be a ratio of densities, the numerator being the "true" density of Y_0 conditionally on $Y_{-\infty:-1}$. This is not straightforward since Proposition 13.5 only allows us to define $\mathrm{p}^\theta(Y_0|Y_{-\infty:-1})$ as the $\mathbb{P}^{\theta_\star}$-a.s. limit of the random variables:

$$\mathrm{p}^\theta(Y_0 \mid Y_{-\infty:-1}) = \lim_{m\to\infty} \mathrm{p}^\theta(Y_0 \mid Y_{-m:-1}), \quad \mathbb{P}^{\theta_\star}\text{-a.s.}, \tag{13.18}$$

and thus, does not provide the existence of the limit of $\mathrm{p}^\theta(y_0 \mid Y_{-m:-1})$ for all $y_0 \in \mathsf{Y}$. The following Lemma gathers the essential properties about the existence of these densities.

Lemma 13.12. *Under A 13.1–A 13.6, for all $\theta \in \Theta$, there exists $\mathbb{P}^{\theta_\star}$-a.s. a continuous density function $x_0 \mapsto \mathrm{p}^\theta(x_0|Y_{-\infty:-1})$ with respect to λ, for all $x_0 \in \mathsf{X}$, $\mathrm{p}^\theta(x_0|Y_{-\infty:-1}) = \lim_{m\to\infty} \mathrm{p}^\theta(x_0|Y_{-m:-1})$, $\mathbb{P}^{\theta_\star}$-a.s.. Define the function*

$$y_0 \mapsto \mathrm{p}^\theta(y_0|Y_{-\infty:-1}) = \int g^\theta(x_0, y_0)\, \mathrm{p}^\theta(x_0|Y_{-\infty:-1}) \lambda(\mathrm{d}x_0) .$$

Then, $\mathbb{P}^{\theta_\star}$-a.s., $\mathrm{p}^\theta(Y_0 \mid Y_{-\infty:-1}) = \mathrm{p}^\theta(y_0 \mid Y_{-\infty:-1})\big|_{y_0=Y_0}$. In addition, $y_0 \mapsto \mathrm{p}^{\theta_\star}(y_0|Y_{-\infty:-1})$ is the conditional density with respect to μ of Y_0 conditionally to $\sigma(Y_t, t < 0)$ under $\mathbb{P}^{\theta_\star}$.

Proof. Write

$$\mathrm{p}^{\theta}(x_0|Y_{-m:-1}) = \int m^{\theta}(x_{-1},x_0)\mathbb{P}^{\theta}(\mathrm{d}x_{-1}|Y_{-m:-1}) , \tag{13.19}$$

and note that for all $(x,x') \in \mathsf{X}^2$, $m^{\theta}(x,x') \leq \sigma^+$. We can then apply Lemma 13.2 so that

$$\mathbb{P}^{\theta_\star}\left[\sum_{m=0}^{\infty} \sup_{x_0 \in \mathsf{X}} |\mathrm{p}^{\theta}(x_0|Y_{-m:-1}) - \mathrm{p}^{\theta}(x_0|Y_{-m-1:-1})| < \infty\right] = 1 . \tag{13.20}$$

This implies that $\mathbb{P}^{\theta_\star}$-a.s., there exists a nonnegative real-valued function $x_0 \mapsto \mathrm{p}^{\theta}(x_0|Y_{-\infty:-1})$ such that for all $x_0 \in \mathsf{X}$, $\mathrm{p}^{\theta}(x_0|Y_{-\infty:-1}) = \lim_{m\to\infty} \mathrm{p}^{\theta}(x_0|Y_{-m:-1})$. The function $x_0 \mapsto \mathrm{p}^{\theta}(x_0|Y_{-\infty:-1})$ is continuous as a uniform limit of continuous functions. We now check that this function is indeed a density function. Applying the Fatou Lemma to the nonnegative functions $\mathrm{p}^{\theta}(x_0|Y_{-m:-1})$, we obtain

$$\begin{aligned}\int \mathrm{p}^{\theta}(x_0|Y_{-\infty:-1})\lambda(\mathrm{d}x_0) &= \int \liminf_{m\to\infty} \mathrm{p}^{\theta}(x_0|Y_{-m:-1})\lambda(\mathrm{d}x_0) \\ &\leq \liminf_{m\to\infty} \int \mathrm{p}^{\theta}(x_0|Y_{-m:-1})\lambda(\mathrm{d}x_0) = 1 ,\end{aligned}$$

According to (13.19), $0 \leq \mathrm{p}^{\theta}(x_0|Y_{-m:-1}) \leq \sigma^+$. Applying again the Fatou Lemma to $\sigma^+ - \mathrm{p}^{\theta}(x_0|Y_{-m:-1})$, we get

$$\begin{aligned}1 &= \limsup_{m\to\infty} \int \mathrm{p}^{\theta}(x_0|Y_{-m:-1})\lambda(\mathrm{d}x_0) \\ &\leq \int \limsup_{m\to\infty} \mathrm{p}^{\theta}(x_0|Y_{-m:-1})\lambda(\mathrm{d}x_0) = \int \mathrm{p}^{\theta}(x_0|Y_{-\infty:-1})\lambda(\mathrm{d}x_0) ,\end{aligned}$$

Thus, $\int \mathrm{p}^{\theta}(x_0|Y_{-\infty:-1})\lambda(\mathrm{d}x_0) = 1$ and $x_0 \mapsto \mathrm{p}^{\theta}(x_0|Y_{-\infty:-1})$ is therefore a density function with respect to λ. Moreover, (13.20) combined with A13.3 yields

$$\mathbb{P}^{\theta_\star}\left[\sum_{m=0}^{\infty} \sup_{y_0 \in \mathsf{Y}} |\mathrm{p}^{\theta}(y_0|Y_{-m:-1}) - \mathrm{p}^{\theta}(y_0|Y_{-m-1:-1})| < \infty\right] = 1 ,$$

so that $\mathrm{p}^{\theta}(Y_0 \mid Y_{-\infty:-1}) = \mathrm{p}^{\theta}(y_0 \mid Y_{-\infty:-1})\big|_{y_0=Y_0}$. Now, let $K \in \mathcal{Y}$ be a set such that $\mu(K) < \infty$. By (B.13) we get $\mathbb{E}^{\theta_\star}[\mathbb{1}_K(Y_0)|Y_{-\infty:-1}] = \lim_{m\to\infty} \mathbb{E}^{\theta_\star}[\mathbb{1}_K(Y_0)|Y_{-m:-1}]$. The Lebesgue theorem then implies that, $\mathbb{P}^{\theta_\star}$-a.s.,

$$\begin{aligned}\lim_{m\to\infty} \mathbb{E}^{\theta_\star}[\mathbb{1}_K(Y_0)|Y_{-m:-1}] &= \lim_{m\to\infty} \int \mathbb{1}_K(y_0)\,\mathrm{p}^{\theta_\star}(y_0|Y_{-m:-1})\mu(\mathrm{d}y_0) \\ &= \int \mathbb{1}_K(y_0) \lim_{m\to\infty} \mathrm{p}^{\theta_\star}(y_0|Y_{-m:-1})\mu(\mathrm{d}y_0) = \int \mathbb{1}_K(y_0)\,\mathrm{p}^{\theta_\star}(y_0|Y_{-\infty:-1})\mu(\mathrm{d}y_0) ,\end{aligned}$$

showing that $y_0 \mapsto \mathrm{p}^{\theta_\star}(y_0|Y_{-\infty:-1})$ is indeed the density of Y_0 conditionally on $Y_{-\infty:-1}$. ■

According to Lemma 13.12, the inner term of (13.17) is a conditional Kullback-Leibler divergence and thus, $\ell(\theta) \le \ell(\theta_\star)$. Therefore, to obtain a strict inequality for $\theta \not\sim \theta_\star$, we only need to show that $\ell(\theta) \neq \ell(\theta_\star)$. This will be done by contradiction. Assume indeed that there exists $\theta \not\sim \theta_\star$ such that $\ell(\theta) = \ell(\theta_\star)$. Then, the stationarity of the observation process under $\mathbb{P}^{\theta_\star}$ implies

$$0 = n(\ell(\theta) - \ell(\theta_\star)) = \sum_{t=0}^{n-1} \mathbb{E}^{\theta_\star}\left[\ln \frac{\mathrm{p}^\theta(Y_t|Y_{-\infty:t-1})}{\mathrm{p}^{\theta_\star}(Y_t|Y_{-\infty:t-1})}\right] = \mathbb{E}^{\theta_\star}\left[\ln \frac{\mathrm{p}^\theta(Y_{0:n-1}|Y_{-\infty:-1})}{\mathrm{p}^{\theta_\star}(Y_{0:n-1}|Y_{-\infty:-1})}\right] .$$

Moreover, by the tower property of the conditional expectation,

$$\begin{aligned}\mathbb{E}^{\theta_\star}\left[\frac{\mathrm{p}^\theta(Y_{0:n-1}|Y_{-\infty:-1})}{\mathrm{p}^{\theta_\star}(Y_{0:n-1}|Y_{-\infty:-1})}\right] &= \mathbb{E}^{\theta_\star}\left[\mathbb{E}^{\theta_\star}\left[\frac{\mathrm{p}^\theta(Y_{0:n-1} \mid Y_{-\infty:-1})}{\mathrm{p}^{\theta_\star}(Y_{0:n-1} \mid Y_{-\infty:-1})}\,\middle|\, Y_{-\infty:-1}\right]\right] \\ &= \mathbb{E}^{\theta_\star}\left[\int \mathrm{p}^\theta(y_{0:n-1}|Y_{-\infty:-1})\mathrm{d}\mu^{\otimes n}(y_{0:n-1})\right] = 1 ,\end{aligned}$$

so that

$$0 = \mathbb{E}^{\theta_\star}\left[\varphi\left(\frac{\mathrm{p}^\theta(Y_{0:n-1}|Y_{-\infty:-1})}{\mathrm{p}^{\theta_\star}(Y_{0:n-1}|Y_{-\infty:-1})}\right)\right], \quad \text{with} \quad \varphi(u) = \ln u - u + 1 .$$

Since $\varphi(u) < 0$ for $u \neq 1$ and $\varphi(1) = 0$, the previous equality yields

$$\mathrm{p}^\theta(Y_{0:n-1}|Y_{-\infty:-1}) = \mathrm{p}^{\theta_\star}(Y_{0:n-1}|Y_{-\infty:-1}), \quad \mathbb{P}^{\theta_\star}\text{-a.s.} \tag{13.21}$$

We are not far from a contradiction since we know that $\mathbb{P}^{\theta_\star} \neq \mathbb{P}^\theta$ for $\theta \not\sim \theta_\star$ but we have first to get rid of $Y_{-\infty:-1}$. This will be done by integrating with respect to $\mathbb{P}^{\theta_\star}[\mathrm{d}Y_{-\infty:-1}]$. More precisely, according to (13.21), we have for all nonnegative functions $h : \mathsf{Y}^n \to \mathbb{R}^+$,

$$\begin{aligned}\mathbb{E}^{\theta_\star}[h(Y_{0:n-1})] &= \mathbb{E}^{\theta_\star}\left[h(Y_{0:n-1})\frac{\mathrm{p}^\theta(Y_{0:n-1}|Y_{-\infty:-1})}{\mathrm{p}^{\theta_\star}(Y_{0:n-1}|Y_{-\infty:-1})}\right] \\ &= \mathbb{E}^{\theta_\star}\left[\int h(y_{0:n-1})\mathrm{p}^\theta(y_{0:n-1}|Y_{-\infty:-1})\mathrm{d}\mu^{\otimes n}(y_{0:n-1})\right] \\ &= \int h(y_{0:n-1})\mathbb{E}^{\theta_\star}\left[\mathrm{p}^\theta(y_{0:n-1}|Y_{-\infty:-1})\right]\mathrm{d}\mu^{\otimes n}(y_{0:n-1}) ,\end{aligned}$$

where the last equality follows from Fubini's theorem. Combining it with

$$\mathrm{p}^\theta(y_{0:n-1}|Y_{-\infty:-1}) = \int \mathrm{p}^\theta_{x_0}(y_{0:n-1})\mathrm{p}^\theta(x_0|Y_{-\infty:-1})\lambda(\mathrm{d}x_0), \quad \mathbb{P}^{\theta_\star}\text{-a.s.}$$

and using again Fubini's theorem, we obtain

$$\begin{aligned}\mathbb{E}^{\theta_\star}[h(Y_{0:n-1})] &= \iint h(y_{0:n-1})\mathrm{p}^\theta_{x_0}(y_{0:n-1})\xi^\theta(\mathrm{d}x_0)\mathrm{d}\mu^{\otimes n}(y_{0:n-1}) \\ &= \mathbb{E}^\theta_{\xi^\theta}[h(Y_{0:n-1})] ,\end{aligned} \tag{13.22}$$

where

$$\xi^{\theta}(\mathrm{d}x_0) := \lambda(\mathrm{d}x_0)\mathbb{E}^{\theta_\star}[\mathrm{p}^{\theta}(x_0|Y_{-\infty:-1})] . \tag{13.23}$$

The nonnegative function h being arbitrary, (13.22) implies that $\mathbb{P}^{\theta}_{\xi^\theta} = \mathbb{P}^{\theta_\star} = \mathbb{P}^{\theta_\star}_{\pi^{\theta_\star}}$ so that $\theta \sim \theta_\star$ provided that $\xi^\theta = \pi^\theta$. To obtain this, we only need to show that ξ^θ is invariant for the kernel M^θ since under A13.1, the Markov kernel M^θ admits π^θ as its unique invariant probability. Now, for all nonnegative function f on X, we have, by (13.23) and by stationarity of $\{Y_t,\, t \in \mathbb{Z}\}$ under $\mathbb{P}^{\theta_\star}$,

$$\begin{aligned}
\xi^{\theta}(f) &= \mathbb{E}^{\theta_\star}\left[\int f(x_1)\mathbb{P}^{\theta}[X_1 \in \mathrm{d}x_1|Y_{-\infty:0}]\right] \\
&= \mathbb{E}^{\theta_\star}\left[\frac{\iint \mathbb{P}^{\theta}[X_0 \in \mathrm{d}x_0|Y_{-\infty:-1}]g^{\theta}(x_0,Y_0)M^{\theta}(x_0,\mathrm{d}x_1)f(x_1)}{\mathrm{p}^{\theta}(Y_0|Y_{-\infty:-1})}\right] \\
&= \mathbb{E}^{\theta_\star}\left[\frac{\iint \mathbb{P}^{\theta}[X_0 \in \mathrm{d}x_0|Y_{-\infty:-1}]g^{\theta}(x_0,Y_0)M^{\theta}(x_0,\mathrm{d}x_1)f(x_1)}{\mathrm{p}^{\theta_\star}(Y_0|Y_{-\infty:-1})}\right] ,
\end{aligned}$$

where the last equality follows from (13.21) with $n = 1$. By taking the expectation of the inner term conditionally on $\sigma(Y_{-\infty:-1})$, the tower property shows that

$$\xi^{\theta}(f) = \mathbb{E}^{\theta_\star}\left[\int \mathbb{P}^{\theta}[X_0 \in \mathrm{d}x_0|Y_{-\infty:-1}]M^{\theta}(f)(x_0)\int g^{\theta}(x_0,y_0)\mu(\mathrm{d}y_0)\right] = \xi^{\theta}M^{\theta}(f) ,$$

so that ξ^θ is invariant for M^θ and thus, $\xi^\theta = \pi^\theta$. The proof is completed.

Finally, we obtain:

Proposition 13.13. *Under A13.1–A13.6, for all $\theta \not\sim \theta_\star$,*

$$\ell(\theta) - \ell(\theta_\star) = \mathbb{E}^{\theta_\star}\left[\ln \frac{\mathrm{p}^{\theta}(Y_0|Y_{-\infty:-1})}{\mathrm{p}^{\theta_\star}(Y_0|Y_{-\infty:-1})}\right] < 0, \quad \mathbb{P}^{\theta_\star}\text{-a.s.}$$

Taking any of these two approaches for proving $\ell(\theta) < \ell(\theta_\star)$ for $\theta \not\sim \theta_\star$ yields the following theorem:

Theorem 13.14. *Under A13.1–A13.6, the MLE, $\hat{\theta}_{\xi,n}$, in a well-specified model, is strongly consistent in the quotient topology for all probability measures ξ, i.e.,*

$$\lim_{n\to\infty} \mathsf{d}(\hat{\theta}_{\xi,n}, \{\theta \in \Theta : \theta \sim \theta_\star\}) = 0, \quad \mathbb{P}^{\theta_\star}\text{-a.s.}$$

The above theorem shows that the points of global maxima of ℓ—forming the set of possible limit points of the MLE—are those that are statistically equivalent to $\theta_\star$. This result, although natural and important (but not trivial!), is however yet of a somewhat "high level" character, that is, not verifiable in terms of "low level" conditions. We would like to provide some conditions, expressed directly in terms of the Markov chain and the conditional distributions $g^\theta(x,y)$, that give information about parameters that are equivalent to $\theta_\star$ and, in particular, when there is no other such parameter than $\theta_\star$. We will do this using the framework of mixtures of distributions.

13.1.4 Identifiability of mixture densities

Definition 13.15. *Let $\{f_\phi : \phi \in \Phi\}$ be a parametric family of densities on* Y *with respect to a common dominating measure μ and parameter ϕ in some measurable space $(\Phi, \mathcal{F})$. Assume that $(y, \phi) \mapsto f_\phi(y)$ is measurable on the product space $(\mathsf{Y} \times \Phi, \mathcal{Y} \otimes \mathcal{F})$. If π is a probability measure on Φ, then the density*

$$h_\pi(y) = \int_\Phi f_\phi(y)\, \pi(\mathrm{d}\phi)$$

is called a mixture density*; the distribution π is called a* mixing distribution.

Let $\mathcal{P}$ be a subset of $\mathbb{M}_1(\Phi, \mathcal{F})$ and $\mathcal{H}$ the induced class of mixtures. The class $\mathcal{H}$ is called identifiable in $\mathcal{P}$ if $h_\pi = h_{\pi'}$ μ-a.e. if and only if $\pi = \pi'$, $(\pi, \pi') \in \mathcal{P}$. If $\mathcal{H}$ is identifiable in the class of all non-degenerated probabilities, then $\mathcal{H}$ is said to be identifiable.

Furthermore we say that the class of finite mixtures of $\{f_\phi : \phi \in \Phi\}$ is identifiable *if for all probability measures π and π' with finite support, $f_\pi = f_{\pi'}$ μ-a.e. if and only if $\pi = \pi'$.*

Many important and commonly used parametric classes of densities are identifiable. We mention the following examples.

1. The Poisson family (Feller, 1943). In this case, $\mathsf{Y} = \mathbb{Z}_+$, $\Phi = \mathbb{R}_+$, ϕ is the mean of the Poisson distribution, μ is counting measure, and $f_\phi(y) = \phi^y \mathrm{e}^{-\phi}/y!$.
2. The Normal family (Teicher, 1960), with the mixture being either on the mean (with fixed variance) or on the variance (with fixed mean). The class of joint mixtures over both mean and variance is not identifiable, but the class of joint *finite* mixtures is identifiable.
3. The Binomial family $\mathrm{Bin}(N, p)$ (Teicher, 1963), with the mixture being on the probability p. The class of finite mixtures is identifiable, provided the number of components k of the mixture satisfies $2k - 1 \le N$.

Further reading on identifiability of mixtures is found, for instance, in Titterington et al. (1985, Section 3.1).

A very useful result on mixtures, taking identifiability in one dimension into several dimensions, is the following.

Theorem 13.16 (Teicher, 1967). *Assume that the class of all mixtures of the family $\{f_\phi : \phi \in \Phi\}$ of densities on* Y *with parameter $\phi \in \Phi$ is identifiable. Then the class of all mixtures of the n-fold product densities $f_\phi^{(n)}(y) = f_{\phi_1}(y_1) \cdots f_{\phi_n}(y_n)$ on $y = (y_1, \ldots, y_n) \in \mathsf{Y}^n$ with parameter $\phi \in \Phi^n$ is identifiable. The same conclusion holds true when "all mixtures" is replaced by "finite mixtures."*

Let us now explain how identifiability of mixture densities applies to NLSS. Assume that $\{(X_t, Y_t),\ t \in \mathbb{N}\}$ is an NLSS such that the conditional densities $g^\theta(x, y)$ all belong to a single parametric family. Then given $X_t = x$, Y_t has conditional density $g_{\phi(x)}$ say, where $\phi(x)$ is a function mapping the current state x into the parameter space Φ of the parametric family of densities. Now assume that the class of all mixtures of this family of densities is identifiable, and that we are given a true parameter

$\theta_\star$ of the model as well as an equivalent other parameter θ. Associated with these two parameters are two mappings, $\phi_\star(x)$ and $\phi(x)$, respectively, as above. As $\theta_\star$ and θ are equivalent, the n-dimensional restrictions of $\mathbb{P}$ and $\mathbb{P}^\theta$ coincide; that is, $\mathbb{P}(Y_{1:n} \in \cdot)$ and $\mathbb{P}^\theta(Y_{1:n} \in \cdot)$ agree. Because the class of all mixtures of (g_ϕ) is identifiable, Theorem 13.16 shows that the n-dimensional distributions of the processes $\{\phi_\star(X_t)\}$ and $\{\phi(X_t)\}$ agree. That is, for all subsets $A \subseteq \Phi^T$,

$$\begin{aligned} &\mathbb{P}\{(\phi_\star(X_1), \phi_\star(X_2), \ldots, \phi_\star(X_n)) \in A\} \\ &\quad = \quad \mathbb{P}^\theta\{(\phi(X_1), \phi(X_2), \ldots, \phi(X_n)) \in A\} . \end{aligned}$$

Example 13.17 (Normal NLSS). Assume that X is finite, say $\mathsf{X} = \{1, 2, \ldots, r\}$, and that $Y_t | X_t = i \sim \mathrm{N}(\mu_i, \sigma^2)$. The parameters of the model are the transition probabilities m_{ij} , the μ_i and σ^2. We thus identify $\phi(x) = \mu_x$. If $\theta_\star$ and θ are two equivalent parameters, the laws of the processes $\{\mu_{\star X_t}\}$ and $\{\mu_{X_t}\}$ are thus the same, and in addition $\sigma_\star^2 = \sigma^2$. Here $\mu_{\star i}$ denotes the μ_i-component of $\theta_\star$, etc. Assuming the $\mu_{\star i}$ to be distinct, this can only happen if the *sets* $\{\mu_{\star 1}, \ldots, \mu_{\star r}\}$ and $\{\mu_1, \ldots, \mu_r\}$ are identical. We may thus conclude that the sets of means must be the same for both parameters, but they need not be enumerated in the same order. Thus there is a permutation $\{c(1), c(2), \ldots, c(r)\}$ of $\{1, 2, \ldots, r\}$ such that $\mu_{c(i)} = \mu_{\star i}$ for all $i \in \mathsf{X}$. Now because the laws of $\{\mu_{\star X_t}\}$ under $\mathbb{P}$ and $\{\mu_{c(X_t)}\}$ under $\mathbb{P}^\theta$ coincide with the μ_is being distinct, we conclude that the laws of $\{X_t,\ t \in \mathbb{Z}\}$ under $\mathbb{P}$ and of $\{c(X_t), t \in \mathbb{N}\}$ under $\mathbb{P}^\theta$ also agree, which in turn implies $m_{\star ij} = m_{c(i),c(j)}$ for all $i, j \in \mathsf{X}$.

Hence any parameter θ that is equivalent to $\theta_\star$ is in fact identical, up to a permutation of state indices. ◇

Example 13.18 (General stochastic volatility). In this example, we will consider a stochastic volatility model of the form $Y_t | X_t = x \in \mathrm{N}(0, \sigma^2(x))$, where $\sigma^2(x)$ is a mapping from X to $\mathbb{R}_+$. Thus, we identify $\phi(x) = \sigma^2(x)$. Again assume that we are given a true parameter $\theta_\star$ as well as another parameter θ, which is equivalent to $\theta_\star$. Because all variance mixtures of normal distributions are identifiable, the laws of $\{\sigma_\star^2(X_t)\}$ under $\mathbb{P}$ and of $\{\sigma^2(X_t)\}$ under $\mathbb{P}^\theta$ agree. Assuming for instance that $\sigma_\star^2(x) = \sigma^2(x) = x$ (and hence also $\mathsf{X} \subseteq \mathbb{R}_+$), we conclude that the laws of $\{X_t\}$ under $\mathbb{P}$ and $\mathbb{P}^\theta$, respectively, agree. For particular models of the transition kernel M, such as the finite case of the previous example, we may then be able to show that $\theta = \theta_\star$, possibly up to a permutation of state indices. ◇

13.2 Asymptotic normality of the score and convergence of the observed information

In this section, we consider a well-specified model and denote by $\theta_\star$ the true value of the parameter. We assume that Θ is a subset of $\mathbb{R}^d$ whose interior Θ^o is non empty and contains $\theta_\star$. As in the consistency section for well-specified models, we denote $\mathbb{P} = \mathbb{P}^{\theta_\star}$ and $\mathbb{E} = \mathbb{E}^{\theta_\star}$; to keep the notations concise, we use $\mathbb{P}$ and $\mathbb{E}$ when there is no ambiguity. The asymptotic normality of the MLE is a consequence of the two following convergence results of the score function and the observed information (see Section 8.2.2):

$$n^{-1/2}\nabla\ell_{\xi,n}(\theta_\star) \xrightarrow{\mathbb{P}} \mathrm{N}(0,\mathcal{J}(\theta_\star)) ,$$
$$-n^{-1}\nabla^2\ell_{\xi,n}(\tilde{\theta}_n) \xrightarrow{\mathbb{P}} \mathcal{J}(\theta_\star) ,$$

where $\mathcal{J}(\theta_\star)$ is the asymptotic Fisher information matrix whose exact expression will be given afterwards and $\{\tilde{\theta}_n,\, n \in \mathbb{N}\}$ is a sequence of random variables that converges in probability to $\theta_\star$. We will successively study the asymptotic properties of the score function and the observed information.

For any initial distribution $\xi \in \mathbb{M}_1(\mathcal{X})$, define the score function by

$$\nabla\ell_{\xi,n}(\theta) = \sum_{t=0}^{n-1} \nabla \ln\left[\int g^\theta(x_t,Y_t)\mathbb{P}_\xi^\theta(\mathrm{d}x_t \mid Y_{0:t-1})\right] , \tag{13.24}$$

where $\mathbb{P}_\xi^\theta(\mathrm{d}x_t \mid Y_{0:t-1}) = \int M^\theta(x_{t-1},\mathrm{d}x_t)\mathbb{P}_\xi^\theta(X_{t-1} \in \mathrm{d}x_{t-1} \mid Y_{0:t-1})$ and $\mathbb{P}_\xi^\theta$ is defined in (13.5) for $t \geq 1$ and where we use the convention when $t = 0$, $\mathbb{P}_\xi^\theta(A \mid Y_{0:t-1}) = \xi(A)$. Consider the following assumptions.

Assumption A13.19. There is an open neighborhood, $\mathcal{U} = \{\theta \in \Theta : |\theta - \theta_\star| < \delta\}$, of $\theta_\star$ such that the following hold.

(a) For all $(x,x') \in \mathsf{X} \times \mathsf{X}$ and all $y \in \mathsf{Y}$, the functions $\theta \mapsto m^\theta(x,x')$ and $\theta \mapsto g^\theta(x,y)$ are twice continuously differentiable on $\mathcal{U}$.

(b) $\sup_{\theta\in\mathcal{U}} \sup_{x,x'} \|\nabla \ln m^\theta(x,x')\| < \infty$ and $\sup_{\theta\in\mathcal{U}} \sup_{x,x'} \|\nabla^2 \ln m^\theta(x,x')\| < \infty$.

(c) $\mathbb{E}^{\theta_\star}\left[\sup_{\theta\in\mathcal{U}} \sup_x \|\nabla \ln g^\theta(x,Y_1)\|^2\right] < \infty$ and $\mathbb{E}^{\theta_\star}\left[\sup_{\theta\in\mathcal{U}} \sup_x \|\nabla^2 \ln g^\theta(x,Y_1)\|\right] < \infty$.

(d) For μ-almost all $y \in \mathsf{Y}$, there exists a function $f_y : \mathsf{X} \to \mathbb{R}_+$ in $\mathrm{L}^1(\lambda)$ such that $\sup_{\theta\in\mathcal{U}} g^\theta(x,y) \leq f_y(x)$.

(e) For λ-almost all $x \in \mathsf{X}$, there exist functions $f_x^1 : \mathsf{Y} \to \mathbb{R}_+$ and $f_x^2 : \mathsf{Y} \to \mathbb{R}_+$ in $\mathrm{L}^1(\mu)$ such that $\|\nabla g^\theta(x,y)\| \leq f_x^1(y)$ and $\|\nabla_\theta^2 g^\theta(x,y)\| \leq f_x^2(y)$ for all $\theta \in \mathcal{U}$.

In A13.19, we have used the notation $\|\cdot\|$ to denote any norm on the space of vectors $\mathbb{R}^d$ and $d \times d$ real-valued entries. These assumptions imply that the loglikelihood is twice continuously differentiable, and also that the score function and observed information have finite moments of order two and one, respectively, under $\mathbb{P}$. They are natural extensions of standard assumptions that are used to prove asymptotic normality of the MLE for i.i.d. observations.

A tool that is often used in the context of models with incompletely observed data is the *Fisher identity* (see (D.10)). We recall that this identity states that if X and Y are some random variables (or vectors) such that Y is observed but X is not, then, under some regularity conditions, $S_Y(y;\theta) = \mathbb{E}^\theta[S_{X,Y}(X,y;\theta) \mid Y = y]$. Here S_Y is the score function of Y and $S_{X,Y}$ is the joint score function of X and Y, often called the *complete score*. This expression is useful when the complete score has a functional form that is much simpler than that of the marginal score of Y. For NLSS this is indeed the case: the score function is involved while the complete score is a simple sum. Invoking the Fisher identity, which holds in a neighborhood of $\theta_\star$ under A13.19, we find that for all $n \geq 1$,

$$\nabla \ell_{\xi,n}(\theta) = \nabla \ln \int \xi(\mathrm{d}x_0) g^\theta(x_0, Y_0) + \mathbb{E}^\theta_\xi \left[\sum_{t=1}^{n-1} d^\theta(X_{t-1}, X_t, Y_t) \,\middle|\, Y_{0:n} \right] , \tag{13.25}$$

where $d^\theta(x, x', y') = \nabla \ln[m^\theta(x, x') g^\theta(x', y')]$. The score function may be expressed as a sum of conditional scores. For that purpose we write for all $n \geq 1$,

$$\nabla \ell_{\xi,n}(\theta) = \nabla \ell_{\xi,0}(\theta) + \sum_{t=1}^{n-1} \{\nabla \ell_{\xi,t}(\theta) - \nabla \ell_{\xi,t-1}(\theta)\} = \sum_{t=0}^{n-1} \dot{h}_{\xi,0,t}(\theta) , \tag{13.26}$$

where $\dot{h}_{\xi,0,0} = \nabla \ln \int \xi(\mathrm{d}x_0) g^\theta(x_0, Y_0)$ and, for $t \geq 1$,

$$\dot{h}_{\xi,0,t}(\theta) = \mathbb{E}^\theta_\xi \left[\sum_{s=1}^{t} d^\theta(X_{s-1}, X_s, Y_s) \,\middle|\, Y_{0:t} \right] - \mathbb{E}^\theta_\xi \left[\sum_{s=1}^{t-1} d^\theta(X_{s-1}, X_s, Y_s) \,\middle|\, Y_{0:t-1} \right] .$$

Recall that $\mathbb{P}^\theta_{\xi,-m}$ (resp. $\mathbb{E}^\theta_{\xi,-m}$) denotes the probability (resp. expectation) induced by a Markov chain associated to (M^θ, g^θ) starting at time $-m$ with the initial distribution $X_{-m} \sim \xi$. We can extend, for all integers $t \geq 1$ and $m \geq 0$, the definition of $\dot{h}_{\xi,0,t}(\theta)$ to

$$\begin{aligned} \dot{h}_{\xi,-m,t}(\theta) = \mathbb{E}^\theta_{\xi,-m} &\left[\sum_{s=-m+1}^{t} d^\theta(X_{s-1}, X_s, Y_s) \,\middle|\, Y_{-m:t} \right] \\ &- \mathbb{E}^\theta_{\xi,-m} \left[\sum_{s=-m+1}^{t-1} d^\theta(X_{s-1}, X_s, Y_s) \,\middle|\, Y_{-m:t-1} \right] , \end{aligned}$$

with the aim, just as before, to let $m \to \infty$. We will prove that $\dot{h}_{\xi,-m,t}(\theta) \xrightarrow{\mathbb{P}\text{-a.s.}} \dot{h}_{-\infty,t}(\theta)$, where the dependence on ξ will vanish in the limit.

We want to prove a central limit theorem (CLT) for the score function evaluated at the true parameter. The score increments form, under reasonable assumptions, a martingale increment sequence. This sequence is not stationary though, so one must either use a general martingale CLT or first approximate the sequence by a stationary martingale increment sequence. We will take the latter approach, and our approximating sequence is nothing but $\{\dot{h}_{-\infty,t}(\theta_\star), t \in \mathbb{N}\}$.

We now proceed to the construction of $\dot{h}_{-\infty,t}(\theta)$. First write $\dot{h}_{\xi,-m,t}(\theta)$ as

$$\begin{aligned} \dot{h}_{\xi,-m,t}(\theta) &= \mathbb{E}^\theta_{\xi,-m}[d^\theta(X_{t-1}, X_t, Y_t) \mid Y_{-m:t}] \\ &+ \sum_{s=-m+1}^{t-1} \left(\mathbb{E}^\theta_{\xi,-m}[d^\theta(X_{s-1:s}, Y_s) \mid Y_{-m:t}] - \mathbb{E}^\theta_{\xi,-m}[d^\theta(X_{s-1:s}, Y_s) \mid Y_{-m:t-1}] \right) . \end{aligned} \tag{13.27}$$

The following result shows that it makes sense to take the limit as $m \to \infty$ in the previous display.

Proposition 13.20. *Assume A 13.1 and A 13.19 hold. Then for all integers $s \le t$, the sequence $\{\mathbb{E}^\theta_{\xi,-m}[d^\theta(X_{s-1}, X_s, Y_s) \mid Y_{-m:t}]\}_{m\ge 0}$ converges $\mathbb{P}$-a.s. and in $\mathrm{L}^2(\mathbb{P})$, uniformly with respect to $\theta \in \mathcal{U}$ as $m \to \infty$. The limit does not depend on ξ and will be written $\mathbb{E}^\theta[d^\theta(X_{s-1}, X_s, Y_s) \mid Y_{-\infty:t}]$.*

Proof. See Exercise 13.2 ∎

With the sums arranged as in (13.27), we can let $m \to \infty$ and define, for $t \ge 1$,

$$\dot{h}_{-\infty,t}(\theta) = \mathbb{E}^\theta[d^\theta(X_{t-1}, X_t, Y_t) \mid Y_{-\infty:t}] + \sum_{s=-\infty}^{t-1} \left(\mathbb{E}^\theta[d^\theta(X_{s-1}, X_s, Y_s) \mid Y_{-\infty:t}] - \mathbb{E}^\theta[d^\theta(X_{s-1}, X_s, Y_s) \mid Y_{-\infty:t-1}]\right).$$

The following result gives an L^2-bound on the difference between $\dot{h}_{\xi,-m,t}(\theta)$ and $\dot{h}_{-\infty,t}(\theta)$.

Lemma 13.21. *Assume A13.1, A13.3, and A13.19 hold. Then for $t \ge 1$,*

$$\left(\mathbb{E}^\theta \|\dot{h}_{\xi,-m,t}(\theta) - \dot{h}_{-\infty,t}(\theta)\|^2\right)^{1/2} \le 12 \left(\mathbb{E}^\theta\left[\sup_{x,x' \in \mathsf{X}} \|d^\theta(x, x', Y_1)\|^2\right]\right)^{1/2} \frac{\rho^{(t+m)/2-1}}{1-\rho}.$$

Proof. The idea of the proof is to match, for each index s of the sums expressing $\dot{h}_{\xi,-m,t}(\theta)$ and $\dot{h}_{-\infty,t}(\theta)$, pairs of terms that are close. To be more precise, we match

1. The first terms of $\dot{h}_{\xi,-m,t}(\theta)$ and $\dot{h}_{-\infty,t}(\theta)$;
2. For i close to t,
$$\mathbb{E}^\theta_{\xi,-m}[d^\theta(X_{i-1}, X_i, Y_i) \mid Y_{-m:t}]$$
and
$$\mathbb{E}^\theta[d^\theta(X_{i-1}, X_i, Y_i) \mid Y_{-\infty:t}],$$
and similarly for the corresponding terms conditioned on $Y_{-m:t-1}$ and $Y_{-\infty:t-1}$, respectively;
3. For s far from t,
$$\mathbb{E}^\theta_{\xi,-m}[d^\theta(X_{s-1}, X_s, Y_s) \mid Y_{-m:t}]$$
and
$$\mathbb{E}^\theta_{\xi,-m}[d^\theta(X_{s-1}, X_s, Y_s) \mid Y_{-m:t-1}],$$
and similarly for the corresponding terms conditioned on $Y_{-\infty:t}$ and $Y_{-\infty:t-1}$, respectively.

We start with the second kind of matches (of which the first terms are a special case). Taking the limit in $m' \to \infty$ in (13.30) (Exercise 13.2), we see that

$$\|\mathbb{E}^\theta_{\xi,-m}[d^\theta(X_{s-1}, X_s, Y_s) \mid Y_{-m:t}] - \mathbb{E}^\theta[d^\theta(X_{s-1}, X_s, Y_s) \mid Y_{-\infty:t}]\| \le 2 \sup_{x,x' \in \mathsf{X}} \|d^\theta(x, x', Y_s)\| \rho^{(s-1)+m}.$$

This bound remains the same if t is replaced by $t-1$. Obviously, it is small if s is far away from $-m$, that is, close to t.

For the third kind of matches, we need a total variation bound that works "backwards in time." Such a bound reads

$$d_{\mathrm{TV}}(\mathbb{P}^{\theta}_{\xi,-m}(X_s \in \cdot \mid Y_{-m:t}), \mathbb{P}^{\theta}_{\xi,-m}(X_s \in \cdot \mid Y_{-m:t-1})) \leq \rho^{t-1-s} .$$

The proof of this bound is similar to that of (13.6) and uses the time-reversed process. We postpone the proof to the end of this section. We may also let $m \to \infty$ and omit the condition on $X_{-m} \sim \xi$ without affecting the bound. As a result of these bounds, we have

$$\begin{aligned}\|\mathbb{E}^{\theta}_{\xi,-m}[d^{\theta}(X_{s-1},X_s,Y_s) \mid Y_{-m:t}] - \mathbb{E}^{\theta}_{\xi,-m}[d^{\theta}(X_{s-1},X_s,Y_s) \mid Y_{-m:t-1}]\| \\ \leq 2 \sup_{x,x' \in \mathsf{X}} \|d^{\theta}(x,x',Y_s)\| \rho^{t-1-s} ,\end{aligned}$$

with the same bound being valid if the conditioning is on $Y_{-\infty:t}$ and $Y_{-\infty:t-1}$, respectively. This bound is small if s is far away from t.

Combining these two kinds of bounds and using Minkowski's inequality for the L^2-norm, we find that $(\mathbb{E}^{\theta}\|\dot{h}_{\xi,-m,t}(\theta) - \dot{h}_{-\infty,t}(\theta)\|^2)^{1/2}$ is bounded by

$$\begin{aligned}&2\rho^{t+m-1} + 2 \times 2 \sum_{s=-m}^{t-1} (\rho^{t-s-1} \wedge \rho^{s+m-1}) + 2 \sum_{s=-\infty}^{-m} \rho^{t-s-1} \\ &\leq \ 4\frac{\rho^{t+m-1}}{1-\rho} + 4 \sum_{-\infty < s \leq (t-m)/2} \rho^{t-s-1} + 4 \sum_{(t-m)/2 \leq s < \infty} \rho^{s+m-1} \\ &\leq \ 12\frac{\rho^{(t+m)/2-1}}{1-\rho}\end{aligned}$$

up to the factor $(\mathbb{E}^{\theta} \sup_{x,x' \in \mathsf{X}} \|d^{\theta}(x,x',Y_s)\|^2)^{1/2}$. ■

We now state the "backwards in time" uniform forgetting property, which played a key role in the above proof.

Proposition 13.22. *Assume A13.1, and A13.3 hold. Then for any integers s, t, and m such that $m \geq 0$ and $-m \leq s < t$, any $y_{-m:t} \in \mathsf{Y}^{t+m+1}$, and $\theta \in \mathcal{U}$,*

$$d_{\mathrm{TV}}(\mathbb{P}^{\theta}_{\xi,-m}(X_s \in \cdot \mid Y_{-m:t}), \mathbb{P}^{\theta}_{\xi,-m}(X_s \in \cdot \mid Y_{-m:t-1})) \leq \rho^{t-1-s} .$$

Proof. See Exercise 13.3 ■

We now return to the question of a weak limit of the normalized score $n^{-1/2}\sum_{t=0}^{n-1} \dot{h}_{\xi,0,t}(\theta_\star)$. Using Lemma 13.21 and Minkowski's inequality, we see that

$$\begin{aligned}&\left[\mathbb{E}^{\theta_\star}\left\|n^{-1/2}\sum_{t=0}^{n-1}(\dot{h}_{\xi,0,t}(\theta_\star) - \dot{h}_{-\infty,t}(\theta_\star))\right\|^2\right]^{1/2} \\ &\qquad \leq n^{-1/2}\sum_{t=0}^{n-1}\left[\mathbb{E}^{\theta_\star}\|\dot{h}_{\xi,0,t}(\theta_\star) - \dot{h}_{-\infty,t}(\theta_\star)\|^2\right]^{1/2} \to 0 \quad \text{as } n \to \infty ,\end{aligned}$$

and consequently, the limiting behavior of the normalized score agrees with that of $n^{-1/2}\sum_{t=0}^{n-1}\dot{h}_{-\infty,t}(\theta_\star)$. Now define the filtration $\mathcal{F}=\{\mathcal{F}_t,\, t\in\mathbb{N}\}$ by $\mathcal{F}_t=\sigma(Y_s,\, -\infty<s\leq t)$ for all integers k. By conditional dominated convergence,

$$\mathbb{E}^{\theta_\star}\left[\sum_{i=-\infty}^{t-1}(\mathbb{E}^{\theta_\star}[d^{\theta_\star}(X_{i-1},X_i,Y_i)\mid Y_{-\infty:t}] - \mathbb{E}^{\theta_\star}[d^{\theta_\star}(X_{i-1},X_i,Y_i)\mid Y_{-\infty:t-1}])\,\middle|\,\mathcal{F}_{t-1}\right]=0\,,$$

and Assumption A13.19 implies that

$$\mathbb{E}^{\theta_\star}[d^{\theta_\star}(X_{t-1},X_t,Y_t)\mid Y_{-\infty:t-1}] = \mathbb{E}^{\theta_\star}[\mathbb{E}^{\theta_\star}[d^{\theta_\star}(X_{t-1},X_t,Y_t)\mid Y_{-\infty:t-1},X_{t-1}]\mid \mathcal{F}_{t-1}]=0\,.$$

It follows immediately that $\dot{h}_{-\infty,t}(\theta_\star)$ is $\mathcal{F}_t$-measurable. Hence, the sequence $\{\dot{h}_{-\infty,t}(\theta_\star)\}_{t\geq 0}$ is a $\mathbb{P}$-martingale increment sequence with respect to the filtration $\{\mathcal{F}_t,\, t\in\mathbb{N}\}$ in $\mathrm{L}^2(\mathbb{P})$. Moreover, this sequence is stationary because $\{Y_t,\, t\in\mathbb{N}\}$ is. Any stationary martingale increment sequence in $\mathrm{L}^2(\mathbb{P})$ satisfies a CLT (Durrett, 1999, p. 418), that is, $n^{-1/2}\sum_{t=0}^{n-1}\dot{h}_{-\infty,t}(\theta_\star)\Rightarrow_{\mathbb{P}}\mathrm{N}(0,\mathcal{J}(\theta_\star))$, where

$$\mathcal{J}(\theta_\star):=\mathbb{E}^{\theta_\star}[\dot{h}_{-\infty,1}(\theta_\star)\dot{h}'_{-\infty,1}(\theta_\star)] \tag{13.28}$$

is the limiting Fisher information.

Because the normalized score function has the same limiting behavior, the following result is immediate.

Theorem 13.23. *Under Assumptions A13.1, A13.3, and A13.19,*

$$n^{-1/2}\nabla\ell_{\xi,n}(\theta_\star)\overset{\mathbb{P}}{\Longrightarrow}\mathrm{N}(0,\mathcal{J}(\theta_\star))\,,$$

for all initial distribution ξ, where $\mathcal{J}(\theta_\star)$ is the limiting Fisher information as defined above.

13.2.1 Convergence of the normalized observed information

We shall now very briefly discuss the asymptotics of the observed information matrix, $-\nabla^2\ell_{\xi,n}(\theta)$. To handle this matrix, one can employ the *missing information principle* (see (D.11)). This principle states that, if X and Y again are two random variables (or vectors) such that Y is observed but X is not, then $I_Y(\theta)=\mathbb{E}^\theta[I_{X,Y}(\theta)\mid Y]-\mathrm{var}^\theta[S_{X,Y}(\theta)\mid Y]$. Here $I_Y(\theta)$ and $I_{X,Y}(\theta)$ are the observed information matrices for Y and (X,Y), respectively, and $S_{X,Y}$ is, as above, the complete score function.

Because the complete information matrix has a simple form, this principle allows us to establish the convergence of the observed information in a way similar to what was done above for the score function. The analysis becomes more difficult, as covariance terms, arising from the conditional variance of the complete score, also need to be accounted for. In addition, we need the convergence to be uniform in a certain sense. We state the following theorem, whose proof can be found in Douc et al. (2004b).

Theorem 13.24. *Under Assumptions A13.1, A13.3, and A13.19,*

$$\lim_{\delta\to 0}\lim_{n\to\infty}\sup_{|\theta-\theta_\star|\le\delta}\|(-n^{-1}\nabla^2\ell_{\xi,n}(\theta))-\mathcal{J}(\theta_\star)\|=0 \quad \mathbb{P}\text{-a.s.}$$

for all initial distribution ξ.

We almost have all the tools for obtaining the asymptotic normality of the MLE. The last step consists of providing conditions under which the asymptotic Fisher information matrix $\mathcal{J}(\theta_\star)$ is nonsingular. This is far from obvious since this matrix using (13.28) and Lemma 13.21 can be expressed as:

$$\mathcal{J}(\theta_\star)=\lim_{n\to\infty}\mathbb{E}^{\theta_\star}[n^{-1/2}\nabla\ell_{\xi,n}(\theta_\star)(n^{-1/2}\nabla\ell_{\xi,n}(\theta_\star))']$$
$$=\lim_{n\to\infty}\mathbb{E}^{\theta_\star}[-n^{-1}\nabla^2\ell_{\xi,n}(\theta_\star)]\,, \quad (13.29)$$

for all initial distribution ξ. Define $\mathcal{J}_{Y_{0:n-1}}(\theta_\star)=\mathbb{E}^{\theta_\star}[-n^{-1}\nabla^2\ell_{\xi,n}(\theta)]\big|_{\xi=\pi^{\theta_\star}}$.

Theorem 13.25. *Assume A13.1, A13.3 and A13.19. Then, $\mathcal{J}(\theta_\star)$ is non singular if and only if there exists $n\ge 1$, such that $\mathcal{J}_{Y_{0:n-1}}(\theta_\star)$ is non singular.*

Before proving the necessary and sufficient condition of Theorem 13.25, we need a technical proposition about the convergence of the Fisher information matrix. Let

$$\mathcal{J}_{Y_{0:n-1}|Y_{-k-m:-k}}(\theta_\star):=-\,\mathbb{E}^{\theta_\star}[\nabla^2\ln \mathrm{p}_\xi^{\theta_\star}(Y_{0:n-1}|Y_{-k-m:-k})]\Big|_{\xi=\pi^{\theta_\star}}\,.$$

Proposition 13.26. *Assume A13.1, A13.3 and A13.19. Then, for all $n\ge 1$,*

$$\lim_{k\to\infty}\sup_m\|\mathcal{J}_{Y_{0:n-1}|Y_{-k-m:-k}}(\theta_\star)-\mathcal{J}_{Y_{0:n-1}}(\theta_\star)\|=0.$$

Although this result may seem intuitive because the ergodicity of $\{X_t,\,t\in\mathbb{Z}\}$ implies the "asymptotic independence" of $Y_{0:n-1}$ (where n is fixed) with respect to $\sigma\{Y_{-l:-k}:l\ge k\ge 1\}$, the rigorous proof of this proposition is rather technical and can be found in Douc et al. (2004b). Using this proposition, Theorem 13.25 may now be proved with elementary arguments.

Proof (of Theorem 13.25). By (13.29),

$$\det\mathcal{J}(\theta_\star)=\det\left(\lim_n\frac{\mathcal{J}_{Y_{0:n-1}}(\theta_\star)}{n}\right)=\lim_n\det\left(\frac{\mathcal{J}_{Y_{0:n-1}}(\theta_\star)}{n}\right).$$

And thus if for all n, $\mathcal{J}_{Y_{0:n-1}}(\theta_\star)$ is singular then $\mathcal{J}(\theta_\star)$ is singular. Now, assume that $\mathcal{J}(\theta_\star)$ is singular. Fix some $n\ge 1$ and let $k\ge n$. By stationarity of the sequence $\{Y_t,\,t\in\mathbb{Z}\}$ under $\mathbb{P}^{\theta_\star}$ and elementary properties of the Fisher information matrix,

$$
\begin{aligned}
k\mathcal{J}(\theta_\star) &= k \lim_{m\to\infty} \mathbb{E}^{\theta_\star}[-\nabla^2 \ln \mathrm{p}^{\theta_\star}(Y_1|Y_{-m:0})] \\
&= \lim_{m\to\infty} \sum_{i=1}^{k} \mathbb{E}^{\theta_\star}[-\nabla^2 \ln \mathrm{p}^{\theta_\star}(Y_i|Y_{-m+i-1:i-1})] \\
&= \lim_{m\to\infty} \mathbb{E}^{\theta_\star}[-\nabla^2 \ln \mathrm{p}^{\theta_\star}(Y_{1:k}|Y_{-m:0})] \\
&= \lim_{m\to\infty} \{\mathbb{E}^{\theta_\star}[-\nabla^2 \ln \mathrm{p}^{\theta_\star}(Y_{k-n+1:k}|Y_{-m:0})] \\
&\qquad + \mathbb{E}^{\theta_\star}[-\nabla^2 \ln \mathrm{p}^{\theta_\star}(Y_{1:k-n}|Y_{k-n+1:k},\, Y_{-m:0})]\} \\
&\geq \limsup_{m\to\infty} \mathbb{E}^{\theta_\star}[-\nabla^2 \ln \mathrm{p}^{\theta_\star}(Y_{k-n+1:k}|Y_{-m:0})] \\
&= \limsup_{m\to\infty} \mathbb{E}^{\theta_\star}[-\nabla^2 \ln \mathrm{p}^{\theta_\star}(Y_{1:n}|Y_{-k+n-m:-k+n})] .
\end{aligned}
$$

Let $\varphi \in \mathbb{R}^d$ such that $\mathcal{J}(\theta_\star)\varphi = 0$. Then, by the above inequality, for all $k \geq n$,

$$
\varphi^T \limsup_{m\to\infty} \mathbb{E}^{\theta_\star}[-\nabla^2 \ln \mathrm{p}^{\theta_\star}(Y_{0:n-1}|Y_{-k+n-m:-k+n})]\varphi = 0.
$$

Now, letting $k \to \infty$, and using Proposition 13.26, we get $\mathcal{J}_{Y_{0:n-1}}(\theta_\star)\varphi = 0$. Thus, $\mathcal{J}_{Y_{0:n-1}}$ is singular for any $n \geq 1$. The proof is completed. ∎

13.2.2 *Limit distribution of the MLE*

The theorems above prove the following result.

Theorem 13.27. *Assume A13.1, A13.3, A13.6, and A13.19, and that $\theta_\star$ is identifiable, that is, θ is equivalent to $\theta_\star$ only if $\theta = \theta_\star$ (possibly up to a permutation of states if X is finite). Then, for all initial distribution ξ, the following hold true.*

(i) The MLE $\hat{\theta}_{\xi,n}$ is strongly consistent: $\hat{\theta}_{\xi,n} \overset{\mathbb{P}\text{-a.s.}}{\longrightarrow} \theta_\star$ as $n \to \infty$.

(ii) If the Fisher information matrix $\mathcal{J}(\theta_\star)$ defined above is nonsingular and $\theta_\star$ is an interior point of Θ, then the MLE is asymptotically normal:

$$
n^{1/2}(\hat{\theta}_{\xi,n} - \theta_\star) \overset{\mathbb{P}}{\Longrightarrow} \mathrm{N}(0, \mathcal{J}(\theta_\star)^{-1}) .
$$

(iii) The normalized observed information at the MLE is a strongly consistent estimator of $\mathcal{J}(\theta_\star)$:

$$
-n^{-1}\nabla^2 \ell_{\xi,n}(\hat{\theta}_{\xi,n}) \overset{\mathbb{P}\text{-a.s.}}{\longrightarrow} \mathcal{J}(\theta_\star) .
$$

The last part of the result is important, as is says that confidence intervals or regions and hypothesis tests based on the estimate $-n^{-1}\nabla^2\ell_{\xi,n}(\widehat{\theta}_n)$ of $\mathcal{J}(\theta_\star)$ will asymptotically be of correct size. In general, there is no closed-form expression for $\mathcal{J}(\theta_\star)$, so that it needs to be estimated in one way or another. The observed information is obviously one way to do that, while another one is to use a parametric bootstrap approach. Yet another approach, motivated by (13.28), is to estimate the Fisher information by the empirical covariance matrix of the conditional scores of (13.24) at the MLE, that is, by $(n+1)^{-1}\sum_0^n [S_{k|k-1}(\widehat{\theta}_n) - \bar{S}(\widehat{\theta}_n)][S_{k|k-1}(\widehat{\theta}_n) - \bar{S}(\widehat{\theta}_n)]^t$ with $S_{k|k-1}(\theta) = \nabla \ln \int g^\theta(x, Y_k)\, \phi_{\xi,k|k-1}^{Y_{0:k-1}}(\mathrm{d}x; \theta)$ and $\bar{S}(\theta) = (n+1)^{-1}\sum_0^n S_{k|k-1}(\theta)$.

13.3 Endnotes

The study of asymptotic properties of the MLE in NLSS was initiated in the seminal work of Baum and Petrie (1966) and Petrie (1969). In these papers, the model is assumed to be well-specified and the state space X and the observation space Y were both finite. More than two decades later, Leroux (1992) proved consistency for well-specified models in the case that X is a finite set and Y is a general state space. The consistency of the MLE in more general NLSS was subsequently investigated for well-specified models in a series of contributions by Le Gland and Mevel (2000a,b), Douc and Matias (2001), Douc et al. (2004b), and Genon-Catalot and Laredo (2006) using different methods. A general consistency result for NLSS has been developed in Douc et al. (2011); these results were later improved and completed in Douc and Moulines (2012).

The predictive distribution $\mathrm{p}_{\xi}^{\theta}(Y_k|Y_{0:k-1})$ can be expressed as a component of the state of a measure-valued Markov chain; in this approach, the existence of the limiting relative entropy rate, $\ell(\theta)$, follows from the ergodic theorem for Markov chains. This approach was used in Le Gland and Mevel (2000a,b), Douc and Matias (2001) and later extended to misspecified models by Mevel and Finesso (2004). Although adequate for finite state-space Markov chains, it does not extend easily to general state-spaces; see Douc and Matias (2001).

In Leroux (1992), the existence of the relative entropy rate is established by means of Kingman's subadditive ergodic theorem; the same approach is used indirectly in Petrie (1969), which invokes the Furstenberg-Kesten theory for products of random matrices. After some additional work, an explicit representation of the relative enropy rate is again obtained. However, as is noted in Leroux (1992, p. 136), the latter is surprisingly difficult, as Kingman's ergodic theorem does not directly yield a representation of the limit as an expectation.

When the state space is not finite or compact, the geometric forgetting of the filter may also be established (see Kleptsyna and Veretennikov, 2008 or Douc et al., 2009a) but the upper-bound explicitly depends on the initial distributions and on the past observations. Fortunately, this dependence takes a simple form as the number of observations tends to infinity and consistency of the MLE can then be established in general state space (see Douc et al., 2011 and Douc and Moulines, 2012).

All of the results above can be extended to Markov-switching autoregressions; see Douc et al. (2004b). Under A13.1, the conditional chain still satisfies the same favorable mixing properties. The loglikelihood, the score function, and observed observation can be analyzed along the same lines as above. Other papers that examine consistency and asymptotic normality of estimators in Markov-switching autoregressions include Francq and Roussignol (1997), Krishnamurthy and Rydén (1998), and Francq and Roussignol (1998). Markov-switching GARCH models were studied by Francq et al. (2001), Liu (2012), and Elliott et al. (2012).

Exercises

13.1. (a) Show that for all $t \in \mathbb{N}^*$, $\mathbb{E}[M_t] = 1 < \infty$ and that $\{M_t,\ t \in \mathbb{N}^*\}$ is a nonnegative martingale with respect to the filtration $\{\mathcal{F}_t,\ t \in \mathbb{N}^*\} = \sigma\{Y_s : 0 \le s \le t-1\}$.

(b) Show that $\{M_t,\ t \in \mathbb{N}\}$ converges $\mathbb{P}$-a.s. to an integrable random variable

(c) Show that $\limsup_{n\to\infty} n^{-1} \ln M_n \le 0$.

(d) Show that $\ell(\theta) \le \ell(\theta_\star)$ so that $\theta_\star$ is a maximizer of $\theta \mapsto \ell(\theta)$.

13.2 (Proof of Proposition 13.20). (a) Show that, for any $(x,x') \in \mathsf{X} \times \mathsf{X}$ and nonnegative integers $m' \ge m$,

$$\left|\mathbb{E}^{\theta}_{\xi,-m}[d^{\theta}(X_{i-1},X_i,Y_i) \mid Y_{-m:t}] - \mathbb{E}^{\theta}_{\xi',-m'}[d^{\theta}(X_{i-1},X_i,Y_i) \mid Y_{-m':t}]\right| \\ \le 2\sup_{x,x'} \|d^{\theta}(x,x',Y_i)\| \rho^{(i-1)+m} \,. \tag{13.30}$$

(b) Show that $\{\mathbb{E}^{\theta}_{\xi,-m}[d^{\theta}(X_{i-1},X_i,Y_i) \mid Y_{-m:t}]\}_{m\ge 0}$ converges $\mathbb{P}_{\theta_\star}$-a.s.. to a limit that does not depend on ξ.

(c) Show that for all nonnegative integers m and $i \le t$, all initial distribution ξ and all $\theta \in \mathcal{U}$,

$$\left\|\mathbb{E}^{\theta}_{\xi,-m}[d^{\theta}(X_{i-1},X_i,Y_i) \mid Y_{-m:t}]\right\| \le \sup_{x,x'} \|d^{\theta}(x,x',Y_i)\|$$

with the right-hand side belonging to $\mathrm{L}^2(\mathbb{P}_{\theta_\star})$,

(d) Show that $\{\mathbb{E}^{\theta}_{\xi,-m}[d^{\theta}(X_{i-1},X_i,Y_i) \mid Y_{-m:t}\}_{m\ge 0}$ converges in $\mathrm{L}^2(\mathbb{P}^{\theta_\star})$. Conclude.

13.3 (Proof of Proposition 13.22). Denote by $\phi^{\theta}_{\xi,-m,i}(A) = \mathbb{P}^{\theta}_{\xi,-m}(X_i \in A \mid Y_{-m:i})$ the filter distribution at time i starting with the initial distribution ξ at time $-m$ as defined in (13.5).

(a) Show that the backward kernels (see Proposition 9.14) satisfy a uniform Doeblin condition,

$$B^{\theta}_{\phi^{\theta}_{\xi,-m,i}}(x_{i+1}, \mathrm{d}x_i) \ge \frac{\sigma^-}{\sigma^+} \phi^{\theta}_{\xi,-m,i}(\mathrm{d}x_i) \,.$$

(b) Show that the Dobrushin coefficient of each backward kernel is bounded by $\rho = 1 - \sigma^-/\sigma^+$.

(c) Show that

$$\mathbb{P}^{\theta}_{\xi,-m}(X_i \in \cdot \mid Y_{-m:t-1}) = \int \mathbb{P}^{\theta}_{\xi,-m}(X_i \in \cdot \mid Y_{-m:t-1}, X_{t-1} = x_{t-1}) \\ \times \mathbb{P}^{\theta}_{\xi,-m}(X_{t-1} \in \mathrm{d}x_{t-1} \mid Y_{-m:t-1})$$

and

$$\mathbb{P}^{\theta}_{\xi,-m}(X_i \in \cdot \mid Y_{-m:t-1}) = \int \mathbb{P}^{\theta}_{\xi,-m}(X_i \in \cdot \mid Y_{-m:t-1}, X_{t-1} = x_{t-1}) \\ \times \mathbb{P}^{\theta}_{\xi,-m}(X_{t-1} \in \mathrm{d}x_{t-1} \mid Y_{-m:t}) \,.$$

(d) Prove Proposition 13.22.

13.4. Consider the following observation-driven time series model: (see Definition 5.21)

$$\mathbb{P}^\theta\left[Y_t \in A \mid \mathcal{F}_{t-1}\right] = Q^\theta(X_{t-1}, A) = \int_A q^\theta(X_{t-1}, y)\mu(\mathrm{d}y)\,, \quad \text{for any } A \in \mathcal{Y}\,,$$
$$X_t = f^\theta_{Y_t}(X_{t-1})\,.$$

We assume that the model is well-defined that is, $\{Y_t,\, t \in \mathbb{Z}\}$ is the observation process of an observation-driven time series model associated to the parameter $\theta = \theta_\star$. The aim of this exercise is to show that the identifiablity result in Theorem 13.14, Section 13.1.3, is also valid in this model. Define $\bar{\ell}^\theta(Y_{-\infty:t})$ as in (8.39):

$$\bar{\ell}^\theta(Y_{-\infty:t}) := \ln q^\theta(f^\theta_{Y_{-\infty:t-1}}, Y_t)\,,$$

and assume that for some θ, $\mathbb{E}[\bar{\ell}^\theta(Y_{-\infty:1})] = \mathbb{E}[\bar{\ell}^{\theta_\star}(Y_{-\infty:1})]$. Define $p^\theta_x(y_{1:n}) = \prod_{i=1}^n q^\theta(f^\theta_{y_{1:i-1}}(x), y_i)$ with the convention $f^\theta_{y_{1:0}}(x) = x$ and denote $\mathbb{P}^\theta_x(\mathrm{d}y_{1:n}) = p^\theta_x(\mathrm{d}y_{1:n})\mu(\mathrm{d}y_1)\ldots\mu(\mathrm{d}y_n)$.

(a) Show that

$$n\mathbb{E}[\bar{\ell}^\theta(Y_{-\infty:1})] = \mathbb{E}\left[\ln p^\theta_{f^\theta_{Y_{-\infty:0}}}(y_{1:n})\right]\,.$$

(b) Deduce that for all nonnegative function f, we have $\mathbb{P}$-a.s.,

$$\int \mathbb{P}^\theta_{f^\theta_{Y_{-\infty:0}}}(\mathrm{d}y_{1:n})f(y_{1:n}) = \int \mathbb{P}^{\theta_\star}_{f^{\theta_\star}_{Y_{-\infty:0}}}(\mathrm{d}y_{1:n})f(y_{1:n})\,.$$

(c) In particular, show that $\mathbb{P}$-a.s., we have $Q^\theta(f^\theta_{Y_{-\infty:0}}, \cdot) = Q^{\theta_\star}(f^{\theta_\star}_{Y_{-\infty:0}}, \cdot)$.

(d) Denote by M^θ the Markov kernel of the Markov chain $\{X_t,\, t \in \mathbb{N}\}$ and denote by $\pi^{\theta,\theta_\star}$ the law of $f^\theta_{Y_{-\infty:0}}$ under $\mathbb{P}$. Using the previous question, show that

$$\pi^{\theta,\theta_\star}M^\theta(A) = \mathbb{E}\left[\int Q^{\theta_\star}(f^{\theta_\star}_{Y_{-\infty:0}}, \mathrm{d}y_1)\mathbb{1}_A(f^\theta_{y_1} \circ f^\theta_{Y_{-\infty:0}})\right] = \mathbb{E}[\mathbb{1}_A(f^\theta_{Y_{-\infty:1}})]\,.$$

(e) Deduce that $\pi^{\theta,\theta_\star}$ is the invariant probability measure π^θ of the Markov kernel M^θ.

(f) Deduce that $\int \pi^\theta(\mathrm{d}x)\mathbb{P}^\theta_x(\mathrm{d}y_{1:n})1_A(y_{1:n}) = \int \pi^{\theta_\star}(\mathrm{d}x)\mathbb{P}^{\theta_\star}_x(\mathrm{d}y_{1:n})1_A(y_{1:n})$.

(g) Conclude.

13.5. This exercise provides an example of Hidden Markov model for which the Maximum Likelihood Estimator is not consistent. Consider the state spaces $\mathsf{X} = \mathsf{Y} = \{1,2\}$. Let Q_θ be the Markov kernel associated to the transition probability matrix, Q, with invariant probability measure π:

$$Q = \begin{pmatrix} 0 & 1 \\ 1 & 0 \end{pmatrix}\,, \quad \pi = \begin{pmatrix} 1/2 \\ 1/2 \end{pmatrix}\,.$$

Set $\Theta := [0.5, 0.9]$ and define the observation density $g^\theta(x,y)$ with respect to the counting measure:

$$g^\theta(x,y) = \theta \mathbb{1}_{y=x} + (1-\theta)\mathbb{1}_{y \neq x} .$$

(a) Let $\xi = \delta_1$ and show that the likelihood associated to the first $2n+1$ observations can be written as

$$\ln \mathrm{p}_\xi^\theta(Y_{0:2n}) = \sum_{k=0}^{n} \{\mathbb{1}_{Y_{2k}=1} \ln\theta + \mathbb{1}_{Y_{2k}=2}\ln(1-\theta)\} \\ + \sum_{k=1}^{n} \{\mathbb{1}_{Y_{2k-1}=2}\ln\theta + \mathbb{1}_{Y_{2k-1}=1}\ln(1-\theta)\} .$$

(b) For all initial distribution ξ, show that $\mathbb{P}$-a.s., $\lim_{n\to\infty}(2n)^{-1}\ln \mathrm{p}_\xi^\theta(Y_{0:2n})$ exists and is equal to

$$\{\theta_\star \ln\theta + (1-\theta_\star)\ln(1-\theta)\}\mathbb{1}_{X_0=1} + \{(1-\theta_\star)\ln\theta + \theta_\star\ln(1-\theta)\}\mathbb{1}_{X_0=2} .$$

Deduce that $\lim_{n\to\infty}(2n)^{-1}\ln \mathrm{p}_\xi^\theta(Y_{0:2n})$ is not a constant.

(c) Show directly that $\lim_{n\to\infty}\hat{\theta}_{\xi,n} = \theta_\star \mathbb{1}_{X_0=1} + 0.5\mathbb{1}_{X_0=2}$, $\mathbb{P}$-a.s.

(d) Conclude.

13.6. In this exercise, we replace the assumption A 13.9 by the following assumption:

Assumption A 13.28. For all $\theta \in \theta$ such that $\theta \not\sim \theta_\star$, there exists a sequence of sets $\{A_n,\ n \in \mathbb{N}\}$ and a probability distribution ξ such that

$$\liminf_{n\to\infty} \mathbb{P}^{\theta_\star}(Y_{0:n-1} \in A_n) > 0 , \tag{13.31}$$

$$\limsup_{n\to\infty} n^{-1}\ln \mathbb{P}_\xi^\theta(Y_{0:n-1} \in A_n) < 0 , \tag{13.32}$$

and we wish to show that

$$\liminf_{n\to\infty} n^{-1}\mathbb{E}\left[\ln \frac{\mathrm{p}_\xi^\theta(Y_{0:n-1})}{\mathrm{p}_{\pi^{\theta_\star}}^{\theta_\star}(Y_{0:n-1})}\right] > 0 , \tag{13.33}$$

which is a crucial property for showing the identifiability result in Section 13.1.3 (see Douc et al., 2011 for more details).

(a) Let $\mathbb{P}$ and $\mathbb{Q}$ be two probability measures such that $\mathbb{P}$ is dominated by $\mathbb{Q}$, define the Kullback divergence between $\mathbb{P}$ and $\mathbb{Q}$ by:

$$K(\mathbb{P}\|\mathbb{Q}) = \int \ln \frac{d\mathbb{P}}{d\mathbb{Q}}(x)\mathbb{P}(dx) .$$

Show that for all measurable set A,

$$K(\mathbb{P}\|\mathbb{Q}) = \mathbb{P}(A)K(\mathbb{P}_A\|\mathbb{Q}_A) + \mathbb{P}(A^c)K(\mathbb{P}_{A^c}\|\mathbb{Q}_{A^c}) + K(\mathrm{Ber}(\mathbb{P}(A))\|\mathrm{Ber}(\mathbb{Q}(A))) ,$$

where $\mathbb{P}_A(B) = \mathbb{P}(A \cap B)/\mathbb{P}(A)$ and $\mathrm{Ber}(\alpha)$ is the Bernoulli distribution with probability of success α.

(b) Deduce that $K(\mathbb{P}||\mathbb{Q}) \geq -\mathbb{P}(A)\ln\mathbb{Q}(A) - 2e^{-1}$.

(c) Show that (13.31) and (13.32) imply (13.33).

13.7. In this exercise, we consider a Hidden Markov Model associated to (M,G) as defined in (9.3). We note $(x,y) \mapsto g(x,y)$ the transition kernel density associated to G with respect to some common dominating measure μ. For all nonnegative measure ξ on $(\mathsf{X},\mathcal{X})$ and all $y_{0:n-1} \in \mathsf{Y}^n$, denote by $p_\xi(y_{0:n-1})$ the likelihood associated to the observations $y_{0:n-1}$:

$$p_\xi(y_{0:n-1}) = \int \cdots \int \xi(\mathrm{d}x_0) g(x_0,y_0) \prod_{i=1}^{n-1} M(x_{i-1},\mathrm{d}x_i) g(x_i,y_i) ,$$

and by abuse of notation, we simply write $p_x(y_{0:n-1}) = p_{\delta_x}(y_{0:n-1})$. Moreover, we assume that M is dominated by some nonnegative measure ν and we note $(x,x') \mapsto m(x,x')$ its transition density. The dominating measure ν may be improper, that is $\int \nu(\mathrm{d}x) = \infty$. Nevertheless it is assumed that: for all $y \in \mathsf{Y}$,

$$\int g(x,y)\nu(\mathrm{d}x) < \infty, \quad \sup_{x\in\mathsf{X}} g(x,y) < \infty, \qquad \sup_{(x,y)\in\mathsf{X}\times\mathsf{Y}} m(x,y) < \infty . \tag{13.34}$$

We will show in this exercise that under (13.34) the normalized log-likelihood $n^{-1}\ln p_\nu(Y_{0:n})$ starting with the (possibly improper) measure ν converges to a constant as n goes to infinity provided that the sequence $\{Y_t,\ t \in \mathbb{N}\}$ is strictly stationary and ergodic.

(a) For all $r < s$, write $W_{r,s} = \sup_{x\in\mathsf{X}} \ln p_x(Y_{r:s})$. Show that for all $0 \leq k < n$,

$$W_{0,n} \leq W_{0,k} + W_{k+1,n} .$$

Using Kingman's subadditive theorem, show that $\lim_{n\to\infty} n^{-1} \ln \sup_{x\in\mathsf{X}} p_x(Y_{0:n})$ exists and is equal to

$$\lim_{n\to\infty} n^{-1} \ln \sup_{x\in\mathsf{X}} p_x(Y_{0:n}) = \inf_{n\in\mathbb{N}} \mathbb{E}[\ln \sup_{x\in\mathsf{X}} p_x(Y_{0:n})]/n .$$

(b) Show that

$$\sup_x p_x(Y_{0:n-1}) \leq \sup_x g(Y_0|x) \times \sup_{x,x'\in\mathsf{X}} m(x,x') \times p_\nu(Y_{1:n-1}) , \tag{13.35}$$

$$p_\nu(Y_{0:n-1}) \leq \int g(u,Y_0)\nu(\mathrm{d}u) \times \sup_x p_x(Y_{1:n-1}) . \tag{13.36}$$

(c) Conclude.

Part IV

Appendices

Appendix A

Some Mathematical Background

A.1 Some measure theory

A *measurable space* is a pair $(\mathsf{X}, \mathcal{X})$ with X a set and $\mathcal{X}$ a σ-field of subsets of X, i.e.,

(i) $\mathsf{X} \in \mathcal{X}$;

(ii) if $A \in \mathcal{X}$, then $A^c \in \mathcal{X}$;

(iii) if $A_k \in \mathcal{X}$, $k = 1, 2, 3, \ldots$, then $\bigcup_{k=1}^{\infty} A_k \in \mathcal{X}$.

A σ-field $\mathcal{B}$ is *generated* by a collection of sets $\mathcal{A} \subset \mathcal{B}$ if $\mathcal{B}$ is the smallest σ-field containing the sets $\mathcal{A}$, and then we write $\mathcal{B} = \sigma(\mathcal{A})$; a σ-field $\mathcal{B}$ is *countably generated* if it is generated by a countable collection $\mathcal{A}$ of sets in $\mathcal{B}$. The σ-fields $\mathcal{X}$ we use are always assumed to be countably generated.

On the real line $\mathbb{R} := (-\infty, \infty)$ the Borel σ-field $\mathcal{B}(\mathbb{R})$ is generated by the countable collection of sets $\mathcal{A} = (a, b]$ where a, b range over the rationals $\mathbb{Q}$. The spaces $\mathbb{R}$, $\bar{\mathbb{R}}$, $\mathbb{R}^d$, $\bar{\mathbb{R}}^d$, ... are always equipped with their Borel σ-fields.

If $(\mathsf{X}_1, \mathcal{X}_1)$ is a measurable space and $(\mathsf{X}_2, \mathcal{X}_2)$ is another measurable space, then a mapping $h : \mathsf{X}_1 \to \mathsf{X}_2$ is called a *measurable function* if $h^{-1}\{B\} := \{x : h(x) \in B\} \in \mathcal{X}_1$ for all $B \in \mathsf{X}_2$.

As a convention, functions on $(\mathsf{X}, \mathcal{X})$ are always assumed to be measurable, and this is often omitted in the statements.

Definition A.1 (Signed measures). *A signed measure on $(\mathsf{X}, \mathcal{X})$ is a function $\nu : \mathcal{X} \to [-\infty, \infty]$ such that*

(i) $\nu(\emptyset) = 0$.

(ii) ν *assumes at most one of the values* $\pm\infty$.

(iii) If $\{E_n,\ n \in \mathbb{N}\} \subset \mathcal{X}$ is a sequence of pairwise disjoints sets, then $\nu(\bigcup_{n=1}^{\infty} E_n) = \sum_{n=1}^{\infty} \nu(E_n)$ *with the sum converging absolutely if* $\nu(\bigcup_{n=1}^{\infty} E_n)$ *is finite.*

A signed measure ν is finite *if* $|\nu(A)| < \infty$ *for all* $A \in \mathcal{X}$.

Definition A.2 (Positive sets). *A set $E \in \mathcal{X}$ is said to be positive (negative, null) for the signed measure ν if*

$$F \in \mathcal{X},\ F \subset E \ \Rightarrow\ \nu(F) \geq 0 (\leq 0, = 0)\,.$$

If μ is a signed measure and X is a positive set for μ, then μ is said to be a measure.

Theorem A.3 (The Jordan decomposition theorem). *If ν is a signed measure on $(\mathsf{X}, \mathcal{X})$, then there are unique positive measures ν_+, ν_- on $(\mathsf{X}, \mathcal{X})$ such that ν_+ and ν_- are singular and $\nu = \nu_+ - \nu_-$. If ν is finite, then ν_+ and ν_- are both finite.*

Definition A.4. *Let Ω be a set. A collection $\mathcal{M}$ of subsets of Ω is called a* monotone class *if*

(i) $\Omega \in \mathcal{M}$,

(ii) $A, B \in \mathcal{M}$, $A \subset B \Longrightarrow B \setminus A \in \mathcal{M}$,

(iii) $\{A_n,\ n \in \mathbb{N}\} \subset \mathcal{M}$, $A_n \subset A_{n+1} \Longrightarrow \bigcup_{n=1}^{\infty} A_n \in \mathcal{M}$.

A σ-algebra is a monotone class. The intersection of an arbitrary family of monotone classes is also a monotone class. Hence for any family of subsets $\mathcal{C}$ of Ω, there is a smallest monotone class containing $\mathcal{C}$, which is the intersection of all monotone classes containing $\mathcal{C}$.

If $\mathcal{N}$ is a monotone class that is stable by finite intersection, then $\mathcal{N}$ is a σ-algebra. Since $\Omega \subset \mathcal{N}$, $\mathcal{N}$ is stable by proper difference, if $A \in \mathcal{N}$, then $A^c = \Omega \setminus A \in \mathcal{N}$. The stability by finite intersection implies the stability by finite union.

Theorem A.5. *Let $\mathcal{M}$ be a monotone class and $\mathcal{C} \subset \mathcal{M}$ be a class of sets stable by finite intersection. Then $\sigma(\mathcal{C}) \subset \mathcal{M}$.*

Theorem A.6. *Let $\mathcal{H}$ be a vector space of bounded functions on Ω and $\mathcal{C}$ a class of subsets of Ω stable by finite intersection. We assume that $\mathcal{H}$ satisfies*

(i) $\mathbb{1}_\Omega \in \mathcal{H}$ *and for all* $A \in \mathcal{C}$, $\mathbb{1}_A \in \mathcal{H}$.

(ii) If $\{f_n,\ n \in \mathbb{N}\}$ *is a nondecreasing sequence of nonnegative functions of* $\mathcal{H}$, *such that* $\sup_{n \in \mathbb{N}} f_n = \lim_{n \to \infty} f_n = f$ *is bounded, then* $f \in \mathcal{H}$.

Then $\mathcal{H}$ contains all the bounded $\sigma(\mathcal{C})$-measurable functions.

Let Ω be a set and $\{(\mathsf{X}_i, \mathcal{X}_i),\ i \in I\}$ be a family of measurable spaces indexed by a set I. For each $i \in I$, let $\mathcal{F}_i$ be a class of subsets of X_i that is closed under finite intersections, and is such that $\mathcal{X}_i = \sigma(\mathcal{F}_i)$. Finally, let $\{X_i,\ i \in I\}$ be a collection of functions from Ω to X_i. The collection $\mathcal{G}$ of all sets of the form $\bigcap_{i \in J} X_i^{-1}(A_i)$, where $A_i \in \mathcal{F}_i$ and J ranges over all finite subsets of I, is closed under finite intersection. The smallest σ-algebra containing $\mathcal{G}$ is denoted by $\sigma(X_i, i \in I)$.

Theorem A.7. *Let $\mathcal{H}$ be a vector space of bounded numerical functions on Ω and $\{X_i,\ i \in I\}$ be a family of applications, $X_i : \Omega \to \mathsf{X}_i$. Assume that $\mathcal{H}$ satisfies the assumptions of Theorem A.6 and that, for all J a finite subset of I and $A_i \in \mathcal{F}_i$, $i \in J$,*

$$\prod_{i \in J} \mathbb{1}_{A_i} \circ X_i \in \mathcal{H}.$$

Then $\mathcal{H}$ contains all the bounded $\sigma(X_i, i \in I)$-measurable functions.

Corollary A.8. *Let $I \subset \mathbb{N}$, $(X_i)_{i \in I}$ be a family of random variables $(\mathsf{X}_i, \mathcal{X}_i)$-valued and $\mathcal{F}_I^X = \sigma(X_i, i \in I)$. Then a bounded random variable Y is $\mathcal{F}_I^X$-measurable if and only if $Y = f((X_i)_{i \in I})$ with $f \in \mathbb{F}_b(\times_{i \in I} \mathsf{X}_i, \otimes_{i \in I} \mathcal{X}_i)$.*

A.2 Some probability theory

Let $(\Omega,\mathcal{F})$ and $(\mathsf{X},\mathcal{X})$ be two measurable spaces. A function $X : \Omega \to \mathsf{X}$ is said to be *measurable* if the set $X^{-1}(A) \in \mathcal{F}$ for all $A \in \mathcal{X}$. If $(\mathsf{X},\mathcal{X}) = (\mathbb{R},\mathcal{B}(\mathbb{R}))$ where $\mathcal{B}(\mathbb{R})$ is the Borel σ-field, X is said to be a real-valued random variable. In accordance with well-established traditions, the phrase "random variable" usually refers to a real-valued random variable. If X is not the real numbers $\mathbb{R}$, we often write "X-valued random variable."

A σ-field $\mathcal{G}$ on Ω such that $\mathcal{G} \subseteq \mathcal{F}$ is called a *sub-σ-field of* $\mathcal{F}$. If X is a random variable (real-valued or not) such that $X^{-1}(A) \in \mathcal{G}$ for all $A \in \mathcal{X}$ for such a sub-σ-field $\mathcal{G}$, then X is said to be $\mathcal{G}$*-measurable.* If X denotes an X-valued mapping on Ω, then the *σ-field generated by* X, denoted by $\sigma(X)$, is the smallest σ-field on Ω that makes X measurable.

If $(\Omega,\mathcal{F})$ is a measurable space and $\mathbb{P}$ is a probability measure on $\mathcal{F}$, the triplet $(\Omega,\mathcal{F},\mathbb{P})$ is called a *probability space*. We then write $\mathbb{E}[X]$ for the expectation of a random variable X on $(\Omega,\mathcal{F})$, meaning the (Lebesgue) integral $\int_\Omega X\,d\mathbb{P}$. The *image of* $\mathbb{P}$ *by* X, denoted by $\mathbb{P}^X$, is the probability measure defined by $\mathbb{P}^X(B) = \mathbb{P}(X^{-1}(B))$.

Let $(\Omega,\mathcal{F},\mathbb{P})$ be a probability space. For $p > 0$ we denote by $\mathcal{L}^p(\Omega,\mathcal{F},\mathbb{P})$ the space of random variables X such that $\mathbb{E}|X|^p < \infty$, and by $\mathcal{L}^+(\Omega,\mathcal{F},\mathbb{P})$ the space of random variables X such that $X \geq 0$ $\mathbb{P}$-a.s. If we identify random variables that are equal $\mathbb{P}$-a.s., we get respectively the spaces $L^p(\Omega,\mathcal{F},\mathbb{P})$ and $L^+(\Omega,\mathcal{F},\mathbb{P})$. We allow random variables to assume the values $\pm\infty$.

Lemma A.9. *Let $(\Omega,\mathcal{F},\mathbb{P})$ be a probability space, let $X \in \mathcal{L}^+(\Omega,\mathcal{F},\mathbb{P})$, and let $\mathcal{G}$ be a sub-σ-field of $\mathcal{F}$. Then there exists $Y \in \mathcal{L}^+(\Omega,\mathcal{G},\mathbb{P})$ such that*

$$\mathbb{E}[XZ] = \mathbb{E}[YZ] \tag{A.1}$$

for all $Z \in \mathcal{L}^+(\Omega,\mathcal{G},\mathbb{P})$. If $Y' \in \mathcal{L}^+(\Omega,\mathcal{G},\mathbb{P})$ also satisfies (A.1)*, then $Y = Y'$* $\mathbb{P}$-a.s.

A random variable with the above properties is called a *version of the conditional expectation* of X given $\mathcal{G}$, and we write $Y = \mathbb{E}\left[X \mid \mathcal{G}\right]$. Conditional expectations are thus defined up to $\mathbb{P}$-almost sure equality.

One can indeed extend the definition of the conditional expectation to random variables that do not belong to $\mathcal{L}^+(\Omega,\mathcal{F},\mathbb{P})$. We follow here the approach outlined in Shiryaev (1996, Section II.7).

Definition A.10 (Conditional expectation). *Let $(\Omega,\mathcal{F},\mathbb{P})$ be a probability space, let X be a random variable and let $\mathcal{G}$ be a a sub-σ field of $\mathcal{F}$. Define $X^+ := \max(X,0)$ and $X^- := -\min(X,0)$. If*

$$\min\{\mathbb{E}\left[X^+ \mid \mathcal{G}\right], \mathbb{E}\left[X^- \mid \mathcal{G}\right]\} < \infty \quad \mathbb{P}\text{-a.s.},$$

then (a version of) the conditional expectation of X given $\mathcal{G}$ is defined by

$$\mathbb{E}\left[X \mid \mathcal{G}\right] = \mathbb{E}\left[X^+ \mid \mathcal{G}\right] - \mathbb{E}\left[X^- \mid \mathcal{G}\right];$$

on the set of probability 0 of sample points where $\mathbb{E}\left[X^+ \mid \mathcal{G}\right]$ and $\mathbb{E}\left[X^- \mid \mathcal{G}\right]$ are both infinite, the above difference is assigned an arbitrary value, for instance, zero.

In particular, if $\mathbb{E}\left[|X| \mid \mathcal{G}\right] < \infty$ $\mathbb{P}$-a.s., then $\mathbb{E}\left[X^+ \mid \mathcal{G}\right] < \infty$ and $\mathbb{E}\left[X^- \mid \mathcal{G}\right] < \infty$ $\mathbb{P}$-a.s., and we may always define the conditional expectation in this context. Note that for $X \in \mathcal{L}^1(\Omega, \mathcal{F}, \mathbb{P})$, $\mathbb{E}[X^+] < \infty$ and $\mathbb{E}[X^-] < \infty$. By applying (A.1) with $Z \equiv 1$, $\mathbb{E}\left[\mathbb{E}\left[X^+ \mid \mathcal{G}\right]\right] = \mathbb{E}[X^+] < \infty$ and $\mathbb{E}\left[\mathbb{E}\left[X^- \mid \mathcal{G}\right]\right] = \mathbb{E}[X^-] < \infty$. Therefore, $\mathbb{E}\left[X^+ \mid \mathcal{G}\right] < \infty$ and $\mathbb{E}\left[X^- \mid \mathcal{G}\right] < \infty$, and thus the conditional expectation is always defined for $X \in \mathcal{L}^1(\Omega, \mathcal{F}, \mathbb{P})$.

Many of the useful properties of expectations extend to conditional expectations. We state below some these useful properties. In the following statements, all equalities and inequalities between random variables, and convergence of such, should be understood to hold $\mathbb{P}$-a.s.

Proposition A.11 (Elementary properties of conditional expectation).

(i) *If $X \leq Y$ and, either, $X \geq 0$ and $Y \geq 0$, or $\mathbb{E}\left[|X| \mid \mathcal{G}\right] < \infty$ and $\mathbb{E}\left[|Y| \mid \mathcal{G}\right] < \infty$, then $\mathbb{E}\left[X \mid \mathcal{G}\right] \leq \mathbb{E}\left[Y \mid \mathcal{G}\right]$.*

(ii) *If $\mathbb{E}\left[|X| \mid \mathcal{G}\right] < \infty$, then $\left|\mathbb{E}\left[X \mid \mathcal{G}\right]\right| \leq \mathbb{E}\left[|X| \mid \mathcal{G}\right]$.*

(iii) *If $X \geq 0$ and $Y \geq 0$, then for any non-negative real numbers a and b,*

$$\mathbb{E}\left[aX + bY \mid \mathcal{G}\right] = a\mathbb{E}\left[X \mid \mathcal{G}\right] + b\mathbb{E}\left[Y \mid \mathcal{G}\right] .$$

If $\mathbb{E}\left[|X| \mid \mathcal{G}\right] < \infty$ and $\mathbb{E}\left[|Y| \mid \mathcal{G}\right] < \infty$, the same equality holds for arbitrary real numbers a and b.

(iv) *If $\mathcal{G} = \{\emptyset, \Omega\}$ is the trivial σ-field and $X \geq 0$ or $\mathbb{E}[|X|] < \infty$, then $\mathbb{E}\left[X \mid \mathcal{G}\right] = \mathbb{E}[X]$.*

(v) *If $\mathcal{H}$ is a sub-σ-field of $\mathcal{F}$ such that $\mathcal{G} \subseteq \mathcal{H}$ and $X \geq 0$, then*

$$\mathbb{E}\left[\mathbb{E}\left[X \mid \mathcal{H}\right] \mid \mathcal{G}\right] = \mathbb{E}\left[X \mid \mathcal{G}\right] . \tag{A.2}$$

If $\mathbb{E}\left[X| \mid \mathcal{G}\right] < \infty$, then $\mathbb{E}\left[|X| \mid \mathcal{H}\right] < \infty$ and (A.2) *holds.*

(vi) *Assume that X is independent of $\mathcal{G}$, in the sense that $\mathbb{E}[XY] = \mathbb{E}[X]\mathbb{E}[Y]$ for all $\mathcal{G}$-measurable random variables Y. If, in addition, either $X \geq 0$ or $\mathbb{E}[|X|] < \infty$, then*

$$\mathbb{E}\left[X \mid \mathcal{G}\right] = \mathbb{E}[X] . \tag{A.3}$$

(vii) *If X is $\mathcal{G}$-measurable, $X \geq 0$, and $Y \geq 0$, then*

$$\mathbb{E}\left[XY \mid \mathcal{G}\right] = X\mathbb{E}\left[Y \mid \mathcal{G}\right] . \tag{A.4}$$

The same conclusion holds if $\mathbb{E}\left[|XY| \mid \mathcal{G}\right]$, $|X|$, and $\mathbb{E}\left[|Y| \mid \mathcal{G}\right]$ are all finite.

Proposition A.12. *Let $\{X_n\}_{n \geq 0}$ be a sequence of random variables.*

(i) *If $X_n \geq 0$ and $X_n \uparrow X$, then $\mathbb{E}\left[X_n \mid \mathcal{G}\right] \uparrow \mathbb{E}\left[X \mid \mathcal{G}\right]$.*

(ii) *If $X_n \leq Y$, $\mathbb{E}\left[|Y| \mid \mathcal{G}\right] < \infty$, and $X_n \downarrow X$ with $\mathbb{E}\left[|X| \mid \mathcal{G}\right] < \infty$, then $\mathbb{E}\left[X_n \mid \mathcal{G}\right] \downarrow \mathbb{E}\left[X \mid \mathcal{G}\right]$.*

(iii) *If $|X_n| \leq Z$, $\mathbb{E}\left[Z \mid \mathcal{G}\right] < \infty$, and $X_n \to X$, then $\mathbb{E}\left[X_n \mid \mathcal{G}\right] \to \mathbb{E}\left[X \mid \mathcal{G}\right]$ and $\mathbb{E}\left[|X_n - X| \mid \mathcal{G}\right] \to 0$.*

Let $(\Omega, \mathcal{F}, \mathbb{P})$ be a probability space and $(\mathsf{X}, \mathcal{X})$ be a measurable space. A sequence $\{X_t,\ t \in \mathbb{N}\}$ of random variables with values in $(\mathsf{X}, \mathcal{X})$ is called a *stochastic process* on $(\mathsf{X}, \mathcal{X})$. It is often of interest to use the information contained in a stochastic process up to a certain time instant. To formalize this notion, it is convenient to introduce the notion of filtration and filtered space.

Definition A.13 (Filtration). *A probability space* $(\Omega, \mathcal{F}, \mathbb{P})$ *is a* filtered probability space *if there exists an increasing sequence* $\{\mathcal{F}_t,\ t \in \mathbb{N}\}$ *of* σ*-algebras included in* $\mathcal{F}$ *(*$\mathcal{F}_s \subset \mathcal{F}_t$ *for all* $s \leq t$*). This is denoted:* $(\Omega,\ \mathcal{F}, \{\mathcal{F}_t,\ t \in \mathbb{N}\}, \mathbb{P})$.

Definition A.14 (Adapted stochastic process). *A stochastic process* $\{X_t,\ t \in \mathbb{N}\}$ *defined on a filtered probability space* $(\Omega,\ \mathcal{F}, \{\mathcal{F}_t,\ t \in \mathbb{N}\}, \mathbb{P})$ *is called* adapted *if and only if* X_t *is* $\mathcal{F}_t$*-measurable for any* $t \in \mathbb{N}$*. This is denoted* $\{(X_t, \mathcal{F}_t),\ t \in \mathbb{N}\}$.

By requiring the process to be adapted, we ensure that we can calculate probabilities related to X_t based solely on the information available at time t.

In some cases we are only given a probability space $(\Omega, \mathcal{F}, \mathbb{P})$ (and not a filtered space). This usually corresponds to the case where we assume that all the information available at time t on X_t comes from the stochastic process X_t itself. Loosely speaking, there is no additional source of information available. In this case we introduce the natural filtration generated by the process $\{X_t,\ t \in \mathbb{N}\}$.

Definition A.15 (Natural filtration). *Let* $(\Omega, \mathcal{F}, \mathbb{P})$ *be a probability space and* $\{X_t,\ t \in \mathbb{N}\}$ *be a stochastic process. The* natural filtration *of the process* $\{X_t,\ t \in \mathbb{N}\}$ *is the family of* σ*-algebra*

$$\mathcal{F}_t^X = \sigma(\{X_s : 0 \leq s \leq t\}), \quad t \in \mathbb{N}.$$

With this definition the collection of sigma algebras $\{\mathcal{F}_t^X,\ t \in \mathbb{N}\}$ is increasing and obviously the process $\{X_t,\ t \in \mathbb{N}\}$ is adapted with respect to it.

Appendix B

Martingales

B.1 Definitions and elementary properties

Definition B.1 (Martingale). *Let $(\Omega, \mathcal{F}, \{\mathcal{F}_n,\ n \in \mathbb{N}\}, \mathbb{P})$ be a filtered probability space and $\{(X_n, \mathcal{F}_n),\ n \in \mathbb{N}\}$ a real integrable adapted process. We say that $\{(X_n, \mathcal{F}_n),\ n \in \mathbb{N}\}$ is*

(i) *a* martingale *if for all $0 \le m < n$, $\mathbb{E}[X_n \,|\, \mathcal{F}_m] = X_m$, $\mathbb{P}$-a.s.*

(ii) *a* submartingale *if for all $0 \le m < n$, $\mathbb{E}[X_n \,|\, \mathcal{F}_m] \ge X_m$, $\mathbb{P}$-a.s.*

(iii) *a* supermartingale *if for all $0 \le m < n$, $\mathbb{E}[X_n \,|\, \mathcal{F}_m] \le X_m$, $\mathbb{P}$-a.s.*

Definition B.2 (Martingale difference). *A process $\{(Z_n, \mathcal{F}_n),\ n \in \mathbb{N}^*\}$ is a* martingale difference *if for all $n \in \mathbb{N}^*$, $\mathbb{E}(|Z_n|) < \infty$ and $\mathbb{E}[Z_n \,|\, \mathcal{F}_{n-1}] = 0$, $\mathbb{P}$-a.s.*

If $\{(Z_n, \mathcal{F}_n),\ n \in \mathbb{N}^*\}$ is a martingale difference, then $X_n = X_0 + \sum_{k=1}^n Z_k$ is a martingale; conversely, if $\{(X_n, \mathcal{F}_n),\ n \in \mathbb{N}\}$ is a martingale, then $Z_n = X_n - X_{n-1}$ is a martingale difference.

Proposition B.3. *Let $\{(X_n, \mathcal{F}_n),\ n \in \mathbb{N}\}$ be a martingale (resp. a submartingale) and f be a convex function on $\mathbb{R}$ (resp. a convex increasing) such that $\mathbb{E}|f(X_n)| < \infty$. Then $\{f(X_n), \mathcal{F}_n,\ n \in \mathbb{N}\}$ is a submartingale.*

If $\{(X_n, \mathcal{F}_n),\ n \in \mathbb{N}\}$ is a martingale, $\{(|X_n|^p, \mathcal{F}_n),\ n \in \mathbb{N}\}$, $p \ge 1$, is a submartingale. If $\{(X_n, \mathcal{F}_n),\ n \in \mathbb{N}\}$ is a submartingale, $\{((X_n - a)^+, \mathcal{F}_n),\ n \in \mathbb{N}\}$ is a submartingale.

Definition B.4 (Previsible process). *The adapted process $\{(H_n, \mathcal{F}_n),\ n \in \mathbb{N}^*\}$ is* previsible *if for all $n \ge 1$, H_n est $\mathcal{F}_{n-1}$ -measurable.*

Proposition B.5. *Let $\{(X_n, \mathcal{F}_n),\ n \in \mathbb{N}\}$ be a submartingale. Then there exists a unique martingale $\{(M_n, \mathcal{F}_n),\ n \in \mathbb{N}\}$ and a unique nondecreasing integrable previsible process $\{(A_n, \mathcal{F}_n),\ n \in \mathbb{N}\}$ with $A_0 = 0$ such that $X_n = M_n + A_n$. The process A_n is given by $A_{n+1} - A_n = \mathbb{E}[(X_{n+1} - X_n) \,|\, \mathcal{F}_n]$.*

This decomposition is referred to as the *Doob decomposition*. This decomposition plays an important role in the study of square integrable martingales, i.e., martingales $\{(M_n, \mathcal{F}_n),\ n \in \mathbb{N}\}$ that satisfy $\mathbb{E}(M_n^2) < \infty$ for all $n \in \mathbb{N}$. For such a martingale, according to Proposition B.3, $\{(M_n^2, \mathcal{F}_n),\ n \in \mathbb{N}\}$ is a submartingale and the Doob decomposition then shows that there exists a unique martingale $\{(N_n, \mathcal{F}_n),\ n \in \mathbb{N}\}$

and a nondecreasing integrable previsible process $\{(\langle M\rangle_n, \mathcal{F}_n),\ n\in\mathbb{N}\}$ such that

$$M_n^2 = N_n + \langle M\rangle_n\,. \tag{B.1}$$

The process $\{(\langle M\rangle_n, \mathcal{F}_n),\ n\in\mathbb{N}\}$ is referred to as the previsible quadratic variation of the square integrable martingale $\{(M_n, \mathcal{F}_n),\ n\in\mathbb{N}\}$. By construction, we get

$$\langle M\rangle_n = \sum_{j=1}^{n} \mathbb{E}\left[(\Delta M_j)^2 \,\middle|\, \mathcal{F}_{j-1}\right], \tag{B.2}$$

and for all $\ell \le k$, $\mathbb{E}\left[(M_k - M_\ell)^2 \,\middle|\, \mathcal{F}_\ell\right] = \mathbb{E}\left[\langle M\rangle_k - \langle M\rangle_\ell \,|\, \mathcal{F}_\ell\right]$. If $M_0 = 0$ $\mathbb{P}$-a.s. , then $\mathbb{E}\left[M_k^2\right] = \mathbb{E}\left[\langle M\rangle_k\right]$.

Theorem B.6 (Doob inequalities). *Let $\{(X_n, \mathcal{F}_n),\ n\in\mathbb{N}\}$ be a martingale (resp. a positive submartingale). Assume that for for some $p>1$, $\sup_{n\ge 0}\|X_n\|_p < \infty$. Then, for all $m>0$,*

$$\|\max_{k\le m}|X_k|\|_p \le \frac{p}{p-1}\|X_m\|_p \quad \textit{and} \quad \|\sup_n|X_n|\|_p \le \frac{p}{p-1}\sup_n\|X_n\|_p\,.$$

The following result extends to martingales the Rosenthal inequality for independent random variables.

Theorem B.7 (Rosenthal's inequality). *Assume that $\{(X_k, \mathcal{F}_k),\ k\in\mathbb{N}\}$ is a martingale. Suppose that for some $p\ge 2$, $\mathbb{E}(|X_k|^p) < \infty$, for all $k\ge 1$. Then there exist constants $c_1(p)$ and $c_2(p)$ such that*

$$c_1(p)\left\{\|\langle X\rangle_n^{1/2}\|_p + \|\max_{k\le n}\Delta X_k\|_p\right\} \\ \le \|\max_{k\le n}X_k\|_p \le c_2(p)\left\{\|\langle X\rangle_n^{1/2}\|_p + \|\max_{k\le n}\Delta X_k\|_p\right\},$$

where $\Delta X_k = X_k - X_{k-1}$ and $\langle X\rangle_n$ is the nondecreasing previsible quadratic variation (B.2).

This inequality is established in Burkholder (1973). It is known that c_1^{-1} grows like $\sqrt{p}$ and c_2 grows like $p/\ln(p)$ as $p\to\infty$; see Hitczenko (1990).

Theorem B.8 (Martingale convergence theorem). *If $\{(X_n, \mathcal{F}_n),\ n\in\mathbb{N}\}$ is a submartingale that, in addition, satisfies $\sup\mathbb{E}\left[X_n^+\right] < \infty$, then $X_n \stackrel{\mathbb{P}\text{-a.s.}}{\longrightarrow} X$ and $\mathbb{E}\left[|X|\right] < \infty$. If $\{(X_n, \mathcal{F}_n),\ n\in\mathbb{N}\}$ is a positive supermartingale, then $X_n \stackrel{\mathbb{P}\text{-a.s.}}{\longrightarrow} X$ as $n\to\infty$ and $\mathbb{E}\left[X\right] \le \mathbb{E}\left[X_0\right]$.*

A family $\{X_n,\ n\in\mathbb{N}\}$ of random variables is said to be *uniformly integrable* (U.I.) if

$$\lim_{c\to\infty}\sup_{n\in\mathbb{N}}\mathbb{E}\left[|X_n|\mathbb{1}\{|X_n|\ge c\}\right] = 0\,.$$

Proposition B.9. *Let $\{X_n,\ n\in\mathbb{N}\}$ be a uniformly integrable sequence of random variables. Then,*

(i) $\mathbb{E}[\liminf_{n\to\infty} X_n] \le \liminf_{n\to\infty} \mathbb{E}[X_n] \le \limsup_{n\to\infty} \mathbb{E}[X_n] \le \mathbb{E}(\limsup_{n\to\infty} X_n)$

(ii) if $X_n \xrightarrow{\mathbb{P}\text{-a.s.}} X_\infty$, then $\mathbb{E}[|X_\infty|] < \infty$ and $\lim_{n\to\infty} \mathbb{E}[|X_n - X_\infty|] = 0$.

Theorem B.10. *Let $\{(X_n, \mathcal{F}_n),\ n \in \mathbb{N}\}$ be an uniformly integrable submartingale (resp. martingale). There exists a random variable X_∞ such that $\mathbb{E}[|X_\infty|] < \infty$, $\lim_{n\to\infty} X_n = X_\infty$ $\mathbb{P}$-a.s. and $\lim_{n\to\infty} \mathbb{E}[|X_n - X_\infty|] = 0$. In addition, for any $n \in \mathbb{N}$, $X_n \le \mathbb{E}[X_\infty \mid \mathcal{F}_n]$, $\mathbb{P}$-a.s. (resp. $X_n = \mathbb{E}[X_\infty \mid \mathcal{F}_n]$, $\mathbb{P}$-a.s.).*

Corollary B.11. *Let $\{(X_n, \mathcal{F}_n),\ n \in \mathbb{N}\}$ be a submartingale. Assume that for some $p > 1$,*

$$\sup_{n\ge 0} \mathbb{E}[|X_n|^p] < \infty . \tag{B.3}$$

Then there exists a random variable X_∞ such that $\|X_\infty\|_p < \infty$, $\lim_{n\to\infty} X_n = X_\infty$ $\mathbb{P}$-a.s. and $\lim_{n\to\infty} \mathbb{E}[|X_n - X_\infty|^p] = 0$.

Theorem B.12. *Let $\{(X_n, \mathcal{F}_n),\ n \in \mathbb{N}\}$ be a martingale. The three following properties are equivalent*

(i) *The sequence $\{X_n,\ n \in \mathbb{N}\}$ is uniformly integrable.*

(ii) *The sequence $\{X_n,\ n \in \mathbb{N}\}$ converges in L^1.*

(iii) *There exists $X \in \mathrm{L}^1$ such that for all $n \in \mathbb{N}$, $X_n = \mathbb{E}[X \mid \mathcal{F}_n]$ $\mathbb{P}$-a.s.*

Corollary B.13. *Let $X \in \mathrm{L}^1$ and $\{\mathcal{F}_n,\ n \in \mathbb{N}\}$ be a filtration. Then the sequence $\{\mathbb{E}[X \mid \mathcal{F}_n], n \in \mathbb{N}\}$ converges $\mathbb{P}$-a.s. and in L^1 to $\mathbb{E}[X \mid \mathcal{F}_\infty]$ where $\mathcal{F}_\infty = \sigma(\bigcup_{n=0}^\infty \mathcal{F}_n)$.*

Corollary B.14 (Conditional Borel-Cantelli lemma). *Let $\{\mathcal{G}_k\}$ be a filtration and let $\{\zeta_k\}$ be a $\{\mathcal{G}_k\}$-adapted sequence of random variables. Assume that there exists a constant C such that for any k, $0 \le \zeta_k \le C$. Then if $\sum_{k=1}^\infty \mathbb{E}[\zeta_k \mid \mathcal{G}_{k-1}] < \infty$ $\mathbb{P}$-a.s., it holds that $\sum_{k=1}^\infty \zeta_k < \infty$ $\mathbb{P}$-a.s.*

B.2 Limits theorems

In this section, we derive limit theorems for triangular arrays of dependent random variables. Let $(\Omega, \mathcal{F}, \mathbb{P})$ be a probability space, let X be a random variable and let $\mathcal{G}$ be a a sub-σ field of $\mathcal{F}$. Let $\{M_N\}_{N\ge 0}$ be a sequence of positive integers, $\{U_{N,i}\}_{i=1}^\infty$ be a triangular array of random variables, and $\{\mathcal{F}_{N,i}\}_{0\le i\le\infty}$ be a triangular array of sub-sigma-fields of $\mathcal{F}$. Throughout this section, it is assumed that $\mathcal{F}_{N,i-1} \subseteq \mathcal{F}_{N,i}$ and for each N and $i = 1, \ldots, M_N$, $U_{N,i}$ is $\mathcal{F}_{N,i}$-measurable. We preface the proof with some technical lemmas.

Lemma B.15. *Let $\{(Z_n, \mathcal{F}_n),\ n \in \mathbb{N}\}$ be an adapted sequence of nonnegative random variables. Then, for all $\varepsilon > 0$,*

$$\xi_n = \mathbb{E}\left[\sum_{i=1}^n Z_i \mathbb{1}(\sum_{j=1}^i \mathbb{E}[Z_j \mid \mathcal{F}_{j-1}]) \le \varepsilon\right] \le \varepsilon .$$

Proof. We set $\mu_i = \sum_{j=1}^i \mathbb{E}[Z_j \mid \mathcal{F}_{j-1}]$. Then,

$$\xi_n = \sum_{i=1}^n \mathbb{E}[Z_i \mathbb{1}(\{\mu_i \le \varepsilon\})] = \mathbb{E}\left[\sum_{i=1}^n \mathbb{E}[Z_i \mid \mathcal{F}_{i-1}] \mathbb{1}(\{\mu_i \le \varepsilon\})\right] .$$

Set $\tau := \max\{1 \le i \le n,\ \mu_i \le \varepsilon\}$ and $\tau = 0$ on $\{\min 1 \le i \le n,\ \mu_i > \varepsilon\}$. On $\{\tau = 0\}$ we get $\sum_{i=1}^n \mathbb{E}\left[Z_i \mid \mathcal{F}_{i-1}\right] \mathbb{1}(\{\mu_i \le \varepsilon\}) = 0$ and, on$\{\tau > 0\}$, since $\mu_1 \le \mu_2 \le \cdots \le \mu_n$,

$$\sum_{i=1}^n \mathbb{E}\left[Z_i \mid \mathcal{F}_{i-1}\right] \mathbb{1}(\{\mu_i \le \varepsilon\}) = \sum_{i=1}^\tau \mathbb{E}\left[Z_i \mid \mathcal{F}_{i-1}\right] \le \varepsilon\,.$$

The following Lemma has been established in Dvoretzky (1972, Lemma 3.5).

Lemma B.16. *Let $\{(Z_n, \mathcal{F}_n),\ n \in \mathbb{N}\}$ be an adapted sequence of nonnegative random variables. Then, for all $\varepsilon > 0$, and $\alpha > 0$,*

$$\mathbb{P}(\max_{1\le i\le n} Z_i > \varepsilon) \le \alpha + \mathbb{P}\left(\sum_{i=1}^n \mathbb{P}\left[Z_i > \varepsilon \mid \mathcal{F}_{i-1}\right] > \alpha\right)\,.$$

Proof. We set $\nu_i := \sum_{j=1}^i \mathbb{P}\left[Z_j > \varepsilon \mid \mathcal{F}_{j-1}\right]$. Then,

$$\mathbb{P}(\max_{1\le i\le n} Z_i > \varepsilon) \le \mathbb{P}(\max_{1\le i\le n} Z_i > \varepsilon,\ \nu_n \le \alpha) + \mathbb{P}(\nu_n > \alpha)\,,$$

and

$$\begin{aligned}\mathbb{P}(\max_{1\le i\le n} Z_i > \varepsilon,\ \nu_n \le \alpha) &\le \sum_{i=1}^n \mathbb{P}(Z_i > \varepsilon,\ \nu_n \le \alpha)\\ &\le \sum_{i=1}^n \mathbb{P}(Z_i > \varepsilon,\ \nu_i \le \alpha) = \mathbb{E}\left[\sum_{i=1}^n \mathbb{1}_{\{Z_i>\varepsilon\}} \mathbb{1}_{\{\nu_i\le\alpha\}}\right] \le \alpha\,,\end{aligned}$$

using Lemma B.15. ■

Lemma B.17. *Let $\mathcal{G}$ be a σ-field and X a random variable such that $\mathbb{E}\left[X^2 \mid \mathcal{G}\right] < \infty$. Then, for any $\varepsilon > 0$,*

$$4\mathbb{E}\left[|X|^2 \mathbb{1}\{|X| \ge \varepsilon\} \mid \mathcal{G}\right] \ge \mathbb{E}\left[|X - \mathbb{E}[X \mid \mathcal{G}]|^2 \mathbb{1}\{|X - \mathbb{E}[X \mid \mathcal{G}]| \ge 2\varepsilon\} \mid \mathcal{G}\right]\,.$$

Proof. Let $Y = X - \mathbb{E}[X \mid \mathcal{G}]$. We have $\mathbb{E}[Y \mid \mathcal{G}] = 0$. It is equivalent to show that for any $\mathcal{G}$-measurable random variable Z,

$$\mathbb{E}\left[Y^2 \mathbb{1}\{|Y| \ge 2\varepsilon\} \mid \mathcal{G}\right] \le 4\mathbb{E}\left[|Y+Z|^2 \mathbb{1}\{|Y+Z| \ge \varepsilon\} \mid \mathcal{G}\right]\,.$$

On the set $\{|Z| < \varepsilon\}$,

$$\begin{aligned}\mathbb{E}\left[Y^2 \mathbb{1}\{|Y| \ge 2\varepsilon\} \mid \mathcal{G}\right] &\le 2\mathbb{E}\left[((Y+Z)^2 + Z^2)\mathbb{1}\{|Y+Z| \ge \varepsilon\} \mid \mathcal{G}\right]\\ &\le 2(1 + Z^2/\varepsilon^2)\mathbb{E}\left[(Y+Z)^2 \mathbb{1}\{|Y+Z| \ge \varepsilon\} \mid \mathcal{G}\right]\\ &\le 4\mathbb{E}\left[(Y+Z)^2 \mathbb{1}\{|Y+Z| \ge \varepsilon\} \mid \mathcal{G}\right]\,.\end{aligned}$$

Moreover, on the set $\{|Z| \ge \varepsilon\}$, using that $\mathbb{E}[ZY \mid \mathcal{G}] = Z\mathbb{E}[Y \mid \mathcal{G}] = 0$.

$$\begin{aligned}\mathbb{E}\left[Y^2 \mathbb{1}\{|Y| \ge 2\varepsilon\} \mid \mathcal{G}\right] &\le \mathbb{E}\left[Y^2 + Z^2 - \varepsilon^2 \mid \mathcal{G}\right]\\ &\le \mathbb{E}\left[(Y+Z)^2 - \varepsilon^2 \mid \mathcal{G}\right] \le \mathbb{E}\left[(Y+Z)^2 \mathbb{1}\{|Y+Z| \ge \varepsilon\} \mid \mathcal{G}\right]\,.\end{aligned}$$

The proof is completed. ■

Theorem B.18. *Assume that* $\mathbb{E}\left[|U_{N,j}|\,\middle|\,\mathcal{F}_{N,j-1}\right] < \infty$ $\mathbb{P}$-a.s. *for any N and any $j = 1,\ldots,M_N$, and*

$$\sup_N \mathbb{P}\left(\sum_{j=1}^{M_N} \mathbb{E}\left[|U_{N,j}|\,\middle|\,\mathcal{F}_{N,j-1}\right] \geq \lambda\right) \to 0 \qquad \text{as } \lambda \to \infty \tag{B.4}$$

$$\sum_{j=1}^{M_N} \mathbb{E}\left[|U_{N,j}|\mathbb{1}\{|U_{N,j}| \geq \varepsilon\}\,\middle|\,\mathcal{F}_{N,j-1}\right] \xrightarrow{\mathbb{P}} 0 \qquad \text{for any } \varepsilon > 0\,. \tag{B.5}$$

Then, $\max_{1\leq i\leq M_N}\left|\sum_{j=1}^{i} U_{N,j} - \sum_{j=1}^{i}\mathbb{E}\left[U_{N,j}\,\middle|\,\mathcal{F}_{N,j-1}\right]\right| \xrightarrow{\mathbb{P}} 0\,.$

Proof. Assume first that for each N and each $i = 1,\ldots,M_N$, $U_{N,i} \geq 0$, $\mathbb{P}$-a.s. By Lemma B.16, we have that for any constants ε and $\eta > 0$,

$$\mathbb{P}\left[\max_{1\leq i\leq M_N} U_{N,i} \geq \varepsilon\right] \leq \eta + \mathbb{P}\left[\sum_{i=1}^{M_N} \mathbb{P}(U_{N,i} \geq \varepsilon \,|\, \mathcal{F}_{N,i-1}) \geq \eta\right]\,.$$

From the conditional version of the Chebyshev identity,

$$\mathbb{P}\left[\max_{1\leq i\leq M_N} U_{N,i} \geq \varepsilon\right] \leq \eta + \mathbb{P}\left[\sum_{i=1}^{M_N} \mathbb{E}[U_{N,i}\mathbb{1}\{U_{N,i} \geq \varepsilon\} \,|\, \mathcal{F}_{N,i-1}] \geq \eta\varepsilon\right]\,. \tag{B.6}$$

Let ε and $\lambda > 0$ and define

$$\bar{U}_{N,i} := U_{N,i}\mathbb{1}\{U_{N,i} < \varepsilon\}\,\mathbb{1}\left\{\sum_{j=1}^{i}\mathbb{E}\left[U_{N,j}\,\middle|\,\mathcal{F}_{N,j-1}\right] < \lambda\right\}\,.$$

For any $\delta > 0$,

$$\begin{aligned}
&\mathbb{P}\left(\max_{1\leq i\leq M_N}\left|\sum_{j=1}^{i} U_{N,j} - \sum_{j=1}^{i}\mathbb{E}\left[U_{N,j}\,\middle|\,\mathcal{F}_{N,j-1}\right]\right| \geq 2\delta\right) \\
&\quad \leq \mathbb{P}\left(\max_{1\leq i\leq M_N}\left|\sum_{j=1}^{i} \bar{U}_{N,j} - \sum_{j=1}^{i}\mathbb{E}\left[\bar{U}_{N,j}\,\middle|\,\mathcal{F}_{N,j-1}\right]\right| \geq \delta\right) \\
&\quad + \mathbb{P}\left(\max_{1\leq i\leq M_N}\left|\sum_{j=1}^{i} U_{N,j} - \bar{U}_{N,j} - \sum_{j=1}^{i}\mathbb{E}\left[U_{N,j} - \bar{U}_{N,j}\,\middle|\,\mathcal{F}_{N,j-1}\right]\right| \geq \delta\right)\,.
\end{aligned}$$

The second term in the right-hand side is bounded by

$$\mathbb{P}\left(\max_{1\leq i\leq M_N} U_{N,i} \geq \varepsilon\right) + \mathbb{P}\left(\sum_{j=1}^{M_N}\mathbb{E}\left[U_{N,j}\,\middle|\,\mathcal{F}_{N,j-1}\right] \geq \lambda\right) + \\
\mathbb{P}\left(\sum_{j=1}^{M_N}\mathbb{E}\left[U_{N,j}\mathbb{1}\{U_{N,j} \geq \varepsilon\}\,\middle|\,\mathcal{F}_{N,j-1}\right] \geq \delta\right)\,.$$

Eqs. (B.5) and (B.6) imply that the first and last terms in the last expression converge to zero for any $\varepsilon > 0$ and (B.4) implies that the second term may be arbitrarily small by choosing for λ sufficiently large. Now, by the Doob maximal inequality,

$$\mathbb{P}\left(\max_{1\le i\le M_N}\left|\sum_{j=1}^{i}\bar{U}_{N,j}-\mathbb{E}\left[\bar{U}_{N,j}\,|\,\mathcal{F}_{N,j-1}\right]\right|\ge\delta\right)\\ \le\delta^{-2}\mathbb{E}\left[\sum_{j=1}^{M_N}\mathbb{E}\left[\left(\bar{U}_{N,j}-\mathbb{E}\left[\bar{U}_{N,j}\,|\,\mathcal{F}_{N,j-1}\right]\right)^2\,|\,\mathcal{F}_{N,0}\right]\right].$$

This last term does not exceed

$$\delta^{-2}\mathbb{E}\left[\sum_{i=1}^{M_N}\mathbb{E}\left[\bar{U}_{N,j}^2\mid\mathcal{F}_{N,0}\right]\right]\le\delta^{-2}\varepsilon\mathbb{E}\left[\sum_{j=1}^{M_N}\mathbb{E}\left[\bar{U}_{N,j}\mid\mathcal{F}_{N,0}\right]\right]\\ \le\delta^{-2}\varepsilon\mathbb{E}\left[\sum_{j=1}^{M_N}\mathbb{E}\left[\bar{U}_{N,j}\,|\,\mathcal{F}_{N,j-1}\right]\right]\le\delta^{-2}\varepsilon\lambda\,.$$

Since ε is arbitrary, the proof follows for $U_{N,j}\ge 0$, $\mathbb{P}$-a.s., for each N and $j = 1,\ldots,M_N$. The proof extends to an arbitrary triangular array $\{U_{N,j}\}_{i=1}^{M_N}$ by applying the preceding result to $\{U_{N,j}^{+}\}_{1\le j\le M_N}$ and $\{U_{N,j}^{-}\}_{1\le j\le M_N}$. ■

Lemma B.19. *Assume that for all* N, $\sum_{i=1}^{M_N}\mathbb{E}\left[U_{N,i}^2\,\middle|\,\mathcal{F}_{N,i-1}\right]=1$, $\mathbb{E}\left[U_{N,i}\,|\,\mathcal{F}_{N,i-1}\right]=0$ *for* $i=1,\ldots,M_N$, *and for all* $\varepsilon>0$,

$$\sum_{i=1}^{M_N}\mathbb{E}\left[U_{N,i}^2\mathbb{1}\{|U_{N,i}|\ge\varepsilon\}\,|\,\mathcal{F}_{N,0}\right]\xrightarrow{\mathbb{P}}0\,. \tag{B.7}$$

Then, for any real u, $\mathbb{E}\left[\exp\left(\mathrm{i}u\sum_{j=1}^{M_N}U_{N,j}\right)\,\middle|\,\mathcal{F}_{N,0}\right]-\exp\left(-u^2/2\right)\xrightarrow{\mathbb{P}}0$.

Proof. Denote $\sigma_{N,i}^2:=\mathbb{E}\left[U_{N,i}^2\,\middle|\,\mathcal{F}_{N,i-1}\right]$. Write the following decomposition (with the convention $\sum_{j=a}^{b}=0$ if $a>b$):

$$\mathrm{e}^{\mathrm{i}u\sum_{j=1}^{M_N}U_{N,j}}-\mathrm{e}^{-\frac{u^2}{2}\sum_{j=1}^{M_N}\sigma_{N,j}^2}=\sum_{l=1}^{M_N}\mathrm{e}^{\mathrm{i}u\sum_{j=1}^{l-1}U_{N,j}}\left(\mathrm{e}^{\mathrm{i}uU_{N,l}}-\mathrm{e}^{-\frac{u^2}{2}\sigma_{N,l}^2}\right)\mathrm{e}^{-\frac{u^2}{2}\sum_{j=l+1}^{M_N}\sigma_{N,j}^2}\,.$$

Since $\sum_{j=1}^{l-1}U_{N,j}$ and $\sum_{j=l+1}^{M_N}\sigma_{N,j}^2=1-\sum_{j=1}^{l}\sigma_{N,j}^2$ are $\mathcal{F}_{N,l-1}$-measurable,

$$\left|\mathbb{E}\left[\exp\left(\mathrm{i}u\sum_{j=1}^{M_N}U_{N,j}\right)-\exp\left(-(u^2/2)\sum_{j=1}^{M_N}\sigma_{N,j}^2\right)\,\middle|\,\mathcal{F}_{N,0}\right]\right|\\ \le\sum_{l=1}^{M_N}\mathbb{E}\left[\left|\mathbb{E}\left[\exp(\mathrm{i}uU_{N,l})\,|\,\mathcal{F}_{N,l-1}\right]-\exp(-u^2\sigma_{N,l}^2/2)\right|\,|\,\mathcal{F}_{N,0}\right]\,. \tag{B.8}$$

For any $\varepsilon > 0$, it is easily shown that

$$\mathbb{E}\left[\sum_{l=1}^{M_N}\left|\mathbb{E}\left[\exp(\mathrm{i}uU_{N,l}) - 1 + \frac{1}{2}u^2\sigma_{N,l}^2 \,\middle|\, \mathcal{F}_{N,l-1}\right]\right| \,\middle|\, \mathcal{F}_{N,0}\right] \le \frac{1}{6}\varepsilon|u|^3 + u^2\sum_{l=1}^{M_N}\mathbb{E}\left[U_{N,l}^2\mathbb{1}\{|U_{N,l}| \ge \varepsilon\} \,\middle|\, \mathcal{F}_{N,0}\right]. \quad \text{(B.9)}$$

Since $\varepsilon > 0$ is arbitrary, it follows from (B.7) that the right-hand side tends in probability to 0 as $N \to \infty$. Finally, for all $\varepsilon > 0$,

$$\mathbb{E}\left[\sum_{l=1}^{M_N}\left|\mathbb{E}\left[\exp(-u^2\sigma_{N,l}^2/2) - 1 + \frac{1}{2}u^2\sigma_{N,l}^2 \,\middle|\, \mathcal{F}_{N,l-1}\right]\right| \,\middle|\, \mathcal{F}_{N,0}\right] \le \frac{u^4}{8}\sum_{l=1}^{M_N}\mathbb{E}\left[\sigma_{N,l}^4 \,\middle|\, \mathcal{F}_{N,0}\right] \le \frac{u^4}{8}\left(\varepsilon^2 + \sum_{j=1}^{M_N}\mathbb{E}\left[U_{N,j}^2\mathbb{1}\{|U_{N,j}| \ge \varepsilon\} \,\middle|\, \mathcal{F}_{N,0}\right]\right).$$

(B.7) shows that the right-hand side of previous equation tends in probability to 0 as $N \to \infty$. The proof follows. ■

Theorem B.20. *Assume that for each N and $i = 1,\ldots,M_N$, $\mathbb{E}\left[U_{N,i}^2 \,\middle|\, \mathcal{F}_{N,i-1}\right] < \infty$ and*

$$\sum_{i=1}^{M_N}\{\mathbb{E}\left[U_{N,i}^2 \,|\, \mathcal{F}_{N,i-1}\right] - (\mathbb{E}\left[U_{N,i} \,|\, \mathcal{F}_{N,i-1}\right])^2\} \xrightarrow{\mathbb{P}} \sigma^2 \quad \textit{for some } \sigma^2 > 0\,, \quad \text{(B.10)}$$

$$\sum_{i=1}^{M_N}\mathbb{E}\left[U_{N,i}^2\mathbb{1}_{\{|U_{N,i}|\ge\varepsilon\}} \,\middle|\, \mathcal{F}_{N,i-1}\right] \xrightarrow{\mathbb{P}} 0 \quad \textit{for any } \varepsilon > 0\,. \quad \text{(B.11)}$$

Then, for any real u,

$$\mathbb{E}\left[\exp\left(\mathrm{i}u\sum_{i=1}^{M_N}\{U_{N,i} - \mathbb{E}\left[U_{N,i} \,|\, \mathcal{F}_{N,i-1}\right]\}\right) \,\middle|\, \mathcal{F}_{N,0}\right] \xrightarrow{\mathbb{P}} \exp(-(u^2/2)\sigma^2)\,. \quad \text{(B.12)}$$

Proof. We first assume that $\mathbb{E}[U_{N,i} \,|\, \mathcal{F}_{N,i-1}] = 0$ for all $i = 1,\ldots,M_N$, and $\sigma^2 = 1$. Define $\tau_N := \max\left\{1 \le k \le M_N : \sum_{j=1}^{k}\sigma_{N,j}^2 \le 1\right\}$, with the convention $\max\emptyset = 0$. Put $\bar{U}_{N,k} = U_{N,k}$ for $k \le \tau_N$, $\bar{U}_{N,k} = 0$ for $\tau_N < k \le M_N$ and $\bar{U}_{N,M_N+1} = \left(1 - \sum_{j=1}^{\tau_N}\sigma_{N,j}^2\right)^{1/2} Y_N$, where $\{Y_N\}$ are $\mathcal{N}(0,1)$ independent and independent of $\mathcal{F}_{N,M_N}$. Put

$$\sum_{j=1}^{M_N} U_{N,j} = \sum_{j=1}^{M_N+1}\bar{U}_{N,j} - \bar{U}_{N,M_N+1} + \sum_{j=\tau_N+1}^{M_N} U_{N,j}\,. \quad \text{(B.13)}$$

We will prove that (a) $\{\bar{U}_{N,j}\}_{1\le j\le M_N+1}$ satisfies the assumptions of Lemma B.19, (b) $\bar{U}_{N,M_N+1} \xrightarrow{\mathbb{P}} 0$, (c) $\sum_{j=\tau_N+1}^{M_N} U_{N,j} \xrightarrow{\mathbb{P}} 0$. If $\tau_N < M_N$, then for any $\varepsilon > 0$,

$$0 \le 1 - \sum_{j=1}^{\tau_N}\sigma_{N,j}^2 \le \sigma_{N,\tau_N+1}^2 \le \max_{1\le j\le M_N}\sigma_{N,j}^2 \le \varepsilon^2 + \sum_{j=1}^{M_N}\mathbb{E}\left[U_{N,j}^2\mathbb{1}\{|U_{N,j}| \ge \varepsilon\} \,\middle|\, \mathcal{F}_{N,j-1}\right]$$

Since $\varepsilon > 0$ is arbitrary, it follows from (B.11) that $1 - \sum_{j=1}^{\tau_N} \sigma_{N,j}^2 \xrightarrow{\mathbb{P}} 0$, which implies that $\mathbb{E}\left[\bar{U}_{N,M_N+1}^2 \,\middle|\, \mathcal{F}_{N,0}\right] \xrightarrow{\mathbb{P}} 0$, showing (a) and (b). It remains to prove (c). We have

$$\sum_{j=\tau_N+1}^{M_N} \sigma_{N,j}^2 = \sum_{j=1}^{M_N} \sigma_{N,j}^2 - 1 + \left(1 - \sum_{j=1}^{\tau_N} \sigma_{N,j}^2\right) \xrightarrow{\mathbb{P}} 0. \tag{B.14}$$

For any $\lambda > 0$,

$$\left(\mathbb{E}\left[\sum_{j=\tau_N+1}^{M_N} U_{N,j} \mathbb{1}\left\{\sum_{i=\tau_N+1}^{j} \sigma_{N,i}^2 \le \lambda\right\}\right]\right)^2 = \mathbb{E}\left[\sum_{j=\tau_N+1}^{M_N} \sigma_{N,j}^2 \mathbb{1}\left\{\sum_{i=\tau_N+1}^{j} \sigma_{N,i}^2 \le \lambda\right\}\right].$$

The term between braces converges to 0 in probability by (B.14) and its value is bounded by λ, which shows that $\sum_{j=\tau_N+1}^{M_N} U_{N,j} \mathbb{1}\left\{\sum_{i=\tau_N+1}^{j} \sigma_{N,i}^2 \le \lambda\right\} \xrightarrow{\mathbb{P}} 0$ Moreover,

$$\mathbb{P}\left(\sum_{j=\tau_N+1}^{M_N} U_{N,j} \mathbb{1}\left\{\sum_{i=\tau_N+1}^{j} \sigma_{N,i}^2 > \lambda\right\} \neq 0\right) \le \mathbb{P}\left(\sum_{i=\tau_N+1}^{M_N} \sigma_{N,i}^2 > \lambda\right),$$

which converges to 0 by (B.14). The proof is completed when $\mathbb{E}\left[U_{N,i} \,|\, \mathcal{F}_{N,i-1}\right] = 0$. To deal with the general case, it suffices to set $\bar{U}_{N,i} = U_{N,i} - \mathbb{E}\left[U_{N,i} \,|\, \mathcal{F}_{N,i-1}\right]$ and use Lemma B.17. ■

We may now specialize these results to stationary martingale increment sequences. This result was established by Billingsley (1961).

Theorem B.21. *Assume that $\{X_k,\ k \in \mathbb{N}\}$ is a strict-sense stationary, ergodic process such that $\mathbb{E}\left[X_1^2\right]$ is finite and $\mathbb{E}\left[X_k \,\middle|\, \mathcal{F}_{k-1}^X\right] = 0$, where $\{\mathcal{F}_k^X,\ k \in \mathbb{N}\}$ is the natural filtration. Then, $n^{-1/2} \sum_{k=1}^{n} X_k \xrightarrow{\mathbb{P}} \mathrm{N}(0, \mathbb{E}\left[X_1^2\right])$.*

Appendix C

Stochastic Approximation

Stochastic approximation is a set of methods that recursively search for an optimum or zero of a function $h : \mathbb{R}^d \to \mathbb{R}^d$ observed in presence of noise. If the function $\theta \mapsto h(\theta)$ is known, a simple procedure to find a root consists of using the elementary algorithm

$$\theta_n = \theta_{n-1} + \gamma_n h(\theta_{n-1}) , \tag{C.1}$$

where $\{\gamma_n,\ n \in \mathbb{N}\}$ is a sequence of positive step sizes. Under appropriate technical conditions, it may be shown that the sequence $\{\theta_k,\ k \in \mathbb{N}\}$ is bounded and eventually converges to a zero θ_* of h; see Lange (2010, Chapter 15).

In many applications, only noisy observations of the function h are available. Consider the following assumption:

Assumption AC.1. There exists a function $h : \mathbb{R}^d \to \mathbb{R}^d$ and $H : \mathbb{R}^d \times \mathsf{X} \to \mathbb{R}^d$ such that, for all $\theta \in \Theta$, $\mathbb{E}\left[|H(\theta,X)|\right] < +\infty$ and $\mathbb{E}\left[H(\theta,X)\right] = h(\theta)$.

Robbins and Monro (1951) proposed the algorithm

$$\theta_{n+1} = \theta_n + \gamma_n H(\theta_n, X_{n+1}) , \tag{C.2}$$

where $\{\gamma_n,\ n \in \mathbb{N}\}$ is a deterministic sequence of positive scalar stepsizes and $\{X_n,\ n \in \mathbb{N}\}$ is a sequence of random variables having the same law as X. The quantity $\{H(\theta_n,X_{n+1}) - h(\theta_n)\}$ is interpreted as a noisy measurement of the function h at the current estimate θ_n. Ideally, on suitable choice of $\{\gamma_n,\ n \in \mathbb{N}\}$, this recursion should converge to a zero of h with probability one. Typically, it is assumed that the sequence $\{\gamma_n,\ n \in \mathbb{N}\}$ is decreasing, $\lim_{n\to\infty} \gamma_n = 0$ and $\sum_{n=0}^{\infty} \gamma_n = \infty$. The basic idea is that decreasing step sizes $\gamma_n \to 0$ provides an averaging of the random errors committed when evaluating the function h.

Since the introduction of the now classic Robbins–Monro algorithm, stochastic approximation has been successfully used in many applications. The convergence of the stochastic approximation scheme has been established under a variety of conditions, covering most of the applications (see for instance Benveniste et al., 1990, Duflo, 1997, Kushner and Yin, 2003).

C.1 Convergence of the Robbins–Monro algorithm: Elementary results

We will briefly sketch, under strong assumptions, a proof of convergence. The Robbins–Monro recursion (C.2) may be written as

$$\theta_{n+1} = \theta_n + \gamma_{n+1} h(\theta_n) + \gamma_{n+1} \left\{ H(\theta_n, X_{n+1}) - h(\theta_n) \right\} ,$$

showing that $\{\theta_n, n \geq 0\}$ may be seen as a perturbation of the deterministic algorithm $\bar{\theta}_{n+1} = \bar{\theta}_n + \gamma_{n+1} h(\bar{\theta}_n)$ as soon as the *noise* term $\gamma_{n+1} \left\{ H(\theta_n, X_{n+1}) - h(\theta_n) \right\}$ is negligible: The deterministic sequence $\bar{\theta}_{n+1} = \bar{\theta}_n + \gamma_{n+1} h(\bar{\theta}_n)$ might be seen as a Euler discretization of the Ordinary Differential Equation (ODE) $\dot{\theta} = h(\theta)$ with a stepsize γ_n.

The stability of the ODE can be studied using Lyapunov functions. A Lyapunov function for the ordinary differential equation (ODE) $\dot{\theta} = h(\theta)$ is a function $V : \mathbb{R}^d \to \mathbb{R}^+$ such that any solution $t \mapsto \theta(t)$ of the ODE is such that $t \mapsto V(\theta(t))$ is non-increasing as t increases. If V is differentiable this is mainly equivalent to the condition $\{\nabla V(\theta)\}' h(\theta) \leq 0$ since

$$\frac{\mathrm{d}}{\mathrm{d}t} V(\theta(t)) = \{\nabla V(\theta(t))\}' \dot{\theta}(t) = \{\nabla V(\theta(t))\}' h(\theta(t)) .$$

If such a Lyapunov function exists (which is not always the case!), the ODE is said to be *dissipative*. As an example, assume that we want to minimize a function $V : \mathbb{R}^d \to \mathbb{R}$: the condition $\{\nabla V(\theta)\}' h(\theta) \leq 0$ is satisfied by setting $h = -\nabla V$.

Lemma C.2 (Robbins-Siegmund lemma). *Let $\{V_n,\ n \in \mathbb{N}\}$ and $\{W_n,\ n \in \mathbb{N}\}$ be two nonnegative adapted processes and $\{a_n,\ n \in \mathbb{N}\}$ and $\{b_n,\ n \in \mathbb{N}\}$ be two deterministic nonnegative sequences such that $\sum_{n=0}^{\infty} a_n + \sum_{n=0}^{\infty} b_n < \infty$. Assume in addition that $\mathbb{E}[V_0] < \infty$ and*

$$\mathbb{E}\left[V_{n+1} \,|\, \mathcal{F}_n \right] \leq (1 + a_n)\, V_n - W_n + b_n \,, \qquad \mathbb{P}\text{-a.s.} \tag{C.3}$$

Then

$$\sum_{n=0}^{\infty} W_n < +\infty \quad \mathbb{P}\text{-a.s.} \,, \tag{C.4}$$

$$V_n \xrightarrow{\mathbb{P}\text{-a.s.}} V_\infty \quad \textit{and} \quad \mathbb{E}[V_\infty] < +\infty \,, \tag{C.5}$$

$$\sup_{n \geq 1} \mathbb{E}[V_n] < \infty \,. \tag{C.6}$$

Proof. Set $\alpha_n = \prod_{k=1}^{n} (1 + a_k)^{-1}$ with $\alpha_0 = 1$. We will show that $\{\alpha_n,\ n \in \mathbb{N}\}$ converges to $\alpha_\infty \in (0, 1]$. By construction $\{\alpha_n,\ n \in \mathbb{N}\}$ is nonincreasing and since $1 + x \leq \exp(x)$, we get $\ln(\alpha_n) \geq -\sum_{k=1}^{n} a_k \geq -\sum_{k \geq 1} a_k$; showing that $\{\alpha_n,\ n \in \mathbb{N}\}$ is lower-bounded and therefore converges. Since $\exp\left(-\sum_{k \geq 1} a_k\right) \leq \alpha_n \leq 1$, $\alpha_\infty \in (0, 1]$. We set

$$V_n' = \alpha_{n-1} V_n \,, \quad b_n' = \alpha_n b_n \,, \quad W_n' = \alpha_n W_n \quad S_n = V_n' + \sum_{k=1}^{n-1} W_k' + \sum_{k=n}^{\infty} b_k' \,.$$

We show that $\{S_n,\ n \in \mathbb{N}\}$ converges $\mathbb{P}$-a.s. to a nonnegative random variable S_∞ satisfying $\mathbb{E}[S_\infty] < +\infty$. Using the definition of S_n and (C.3)

$$\begin{aligned}\mathbb{E}[S_{n+1} \mid \mathcal{F}_n] &\le \alpha_n \mathbb{E}[V_{n+1} \mid \mathcal{F}_n] + \sum_{k=1}^{n} W'_k + \sum_{k=n+1}^{\infty} b'_k \\ &\le \alpha_{n-1} V_n + \sum_{k=1}^{n-1} W'_k + \sum_{k=n}^{\infty} b'_k \le S_n .\end{aligned}$$

Therefore, $\{S_n,\ n \in \mathbb{N}\}$ is a non-negative supermartingale; therefore, by Theorem B.8, $S_n \xrightarrow{\mathbb{P}\text{-a.s.}} S_\infty$ where $\mathbb{E}[S_\infty] \le \mathbb{E}[S_0] = \mathbb{E}[V_0] + \sum_{k=1}^n b'_k$. Since $\sum_{k=1}^n b'_k = \sum_{k=1}^\infty \alpha_k b_k \le \sum_{k=1}^\infty b_k < \infty$, this last inequality shows that $\mathbb{E}[S_\infty] < \infty$ which implies that $S_\infty < \infty$, $\mathbb{P}$-a.s..

We first establish (C.4). The sequence of partial sums $\{\sum_{k=1}^n W'_k, n \in \mathbb{N}\}$ is increasing, and for all $n \in \mathbb{N}$, $\sum_{k=1}^n W'_k \le S_n$, which implies, $\mathbb{P}$-a.s.,

$$\lim_{n\to\infty} \sum_{k=1}^{n} W'_k \le \limsup_n S_n = S_\infty < \infty .$$

Therefore the series $\sum_{k=1}^\infty W'_k$ converges $\mathbb{P}$-a.s.. Since $\lim_{n\to\infty} \alpha_n = \alpha_\infty > 0$, this implies that $\sum_{k=1}^\infty W_k$ converges $\mathbb{P}$-a.s., since for all $n \le m$,

$$\sum_{k=n}^{m} W_k \le \alpha_m^{-1} \sum_{k=n}^{m} \alpha_k W_k = \alpha_m^{-1} \sum_{k=n}^{m} W'_k .$$

Consider now (C.5). The convergence of the series $\sum_{k=1}^\infty b_k$ implies the convergence of $\sum_{k=1}^\infty b'_k$, showing that $\lim_{n\to\infty} \sum_{k=n}^\infty b'_k = 0$. Therefore, the sequence $V'_n = S_n - \sum_{k=1}^{n-1} W'_k - \sum_{k=n}^\infty b'_k$ converges $\mathbb{P}$-a.s.. Since $V_n = \alpha_{n-1}^{-1} V'_n$ and $\lim_{n\to\infty} \alpha_n = \alpha_\infty > 0$, the sequence $\{V_n,\ n \in \mathbb{N}\}$ converges $\mathbb{P}$-a.s. to an $\mathbb{P}$-a.s. finite random variable.

Consider finally C.6. Using $\alpha_{n-1} V_n = V'_n \le S_n$, we get

$$\mathbb{E}[V_n] \le \alpha_{n-1}^{-1} \mathbb{E}[S_n] \le \alpha_\infty^{-1} \mathbb{E}[S_0] . \qquad \blacksquare$$

Consider the following assumptions:

Assumption AC.3. There exists a continuously differentiable function $V : \mathbb{R}^d \to \mathbb{R}^+$ such that

(a) ∇V is Lipshitz and $|\nabla V|^2 \le C(1+V)$,

(b) $\{\nabla V\}' h \le 0$,

(c) $\mathbb{E}\left[|H(\theta, X)|^2\right] \le C(1+V(\theta))$.

Assumption AC.4. $\{\gamma_n,\ n \in \mathbb{N}\}$ is a positive sequence satisfying $\sum_{n=0}^\infty \gamma_{n+1}^2 < +\infty$.

Assumption AC.5. The initial value θ_0 is such that $\mathbb{E}[V(\theta_0)] < +\infty$ and $\{X_n,\ n \in \mathbb{N}\}$ are i.i.d. random variables that are independent of θ_0 and have the same distribution as X.

Proposition C.6. *Assume AC.1, AC.3, AC.4 and AC.5. Then,*

$$\sup_{n\in\mathbb{N}} \mathbb{E}\left[V(\theta_n)\right] < \infty,\ V(\theta_n) \xrightarrow{\mathbb{P}\text{-a.s.}} V_\infty \text{ and } \mathbb{E}\left[V_\infty\right] < +\infty, \tag{C.7}$$

$$0 \leq -\sum_{n\geq 1} \gamma_{n+1}\{\nabla V(\theta_n)\}' h(\theta_n) < +\infty,\ \mathbb{P}\text{-a.s.}, \tag{C.8}$$

$$\theta_{n+1} - \theta_n \xrightarrow{\mathbb{P}\text{-a.s.}} 0. \tag{C.9}$$

Proof. In the proof, C is a constant that may take different values upon each appearance. Set $\mathcal{F}_n = \sigma(\theta_0, \ldots, \theta_n)$. A Taylor-Lagrange expansion at θ_n shows that

$$\begin{aligned} V(\theta_{n+1}) &= V(\theta_n + \gamma_{n+1} H(\theta_n, X_{n+1})) \\ &= V(\theta_n) + \gamma_{n+1}\{\nabla V(\theta_n)\}' H(\theta_n, X_{n+1}) \\ &\leq V(\theta_n) + \gamma_{n+1}\{\nabla V(\theta_n)\}' H(\theta_n, X_{n+1}) + \gamma_{n+1}^2 [\nabla V]_{\mathrm{Lip}} |H(\theta_n, X_{n+1})|^2 , \end{aligned} \tag{C.10}$$

where $[\nabla V]_{\mathrm{Lip}}$ denotes the Lipshitz constant. We first show that, for all $n \in \mathbb{N}$, $\mathbb{E}\left[|V(\theta_n)|\right] < \infty$. The Cauchy-Schwarz inequality implies that

$$\mathbb{E}\left[V(\theta_{n+1})\right] \leq \mathbb{E}\left[V(\theta_n)\right] + C\gamma_{n+1}(1 + \mathbb{E}\left[V(\theta_n)\right]) + C\gamma_{n+1}^2(1 + \mathbb{E}\left[V(\theta_n)\right]) .$$

Since $\mathbb{E}\left[V(\theta_0)\right] < \infty$, an easy induction shows that $\mathbb{E}\left[V(\theta_n)\right] < \infty$ for all $n \in \mathbb{N}$. Taking the conditional expectation in (C.10), we have

$$\begin{aligned} &\mathbb{E}\left[V(\theta_{n+1}) \,|\, \mathcal{F}_n\right] \\ &\leq V(\theta_n) + \gamma_{n+1}\{\nabla V(\theta_n)\}' h(\theta_n) + C\gamma_{n+1}^2 [\nabla V]_{\mathrm{Lip}} \mathbb{E}\left[|H(\theta_n, X_{n+1})|^2 \,|\, \mathcal{F}_n\right] \\ &\leq (1 + C\gamma_{n+1}^2) V(\theta_n) + \gamma_{n+1}\{\nabla V(\theta_n)\}' h(\theta_n) + C\gamma_{n+1}^2 . \end{aligned}$$

Using the fact that $\sum_{n=0}^{\infty} \gamma_n^2 < \infty$, and applying Lemma C.2 with $V_n = V(\theta_n)$, $a_n = b_n = C\gamma_{n+1}^2$ and $W_n = -\gamma_{n+1}\{\nabla V(\theta_n)\}' h(\theta_n)$, (C.7)–(C.9) follow.

Writing $\theta_{n+1} - \theta_n = \gamma_{n+1} H(\theta_n, X_{n+1})$ and using that $\sup_{n\geq 0} \mathbb{E}\left[V(\theta_n)\right] < \infty$, we get

$$\begin{aligned} \mathbb{E}\left[\sum_{n=0}^{\infty} |\theta_{n+1} - \theta_n|^2\right] &= \sum_{n=0}^{\infty} \mathbb{E}\left[|\theta_{n+1} - \theta_n|^2\right] \\ &\leq \sum_{n=0}^{\infty} \gamma_{n+1}^2 \mathbb{E}\left[|H(\theta_n, X_{n+1})|^2\right] \leq C \sum_n \gamma_{n+1}^2 (1 + \mathbb{E}\left[V(\theta_n)\right]) < \infty . \end{aligned}$$

Therefore, for any $\delta > 0$

$$\lim_{n\to\infty} \mathbb{P}\left(\sup_{m\geq n} |\theta_{m+1} - \theta_m| \geq \delta\right) \leq \delta^{-2} \lim_{n\to\infty} \sum_{m\geq n} \mathbb{E}\left[|\theta_{m+1} - \theta_m|^2\right] = 0 .$$

■

Theorem C.7. *Assume AC.1, AC.3, AC.4, AC.5. Assume in addition that*

(a) h is continuous,

(b) $\lim_{|\theta|\to\infty} V(\theta) = +\infty$,

(c) $\{\theta \in \Theta, \{\nabla V(\theta)\}' h(\theta)\} = \{\theta_*\}$ *and* $\{\theta, V(\theta) = V(\theta_*)\} = \{\theta_*\}$,

(d) $\sum_n \gamma_n = +\infty$,

then, $\theta_n \xrightarrow{\mathbb{P}\text{-a.s.}} \theta_*$.

Proof. Using Proposition C.6, we know that there exists an event $\Omega_0 \subseteq \Omega$ satisfying $\mathbb{P}(\Omega_0) = 1$ and for all $\omega \in \Omega_0$: (a) $\limsup_n V(\theta_n(\omega)) < \infty$, and (b) the series $\sum_{n=1}^{\infty} \gamma_{n+1} \{\nabla V(\theta_n(\omega))\}' h(\theta_n(\omega)) < +\infty$ converges. Since $\lim_{|\theta|\to\infty} V(\theta) = \infty$, this implies that for all $\omega \in \Omega_0$, the sequence $\{\theta_n(\omega), n \in \mathbb{N}\}$ is bounded. Therefore, the set $\Theta_\infty(\omega)$ of the limiting points of the sequence $\{\theta_n(\omega), n \in \mathbb{N}\}$ is not empty. The convergence of the series $\sum_{n\geq 1} \gamma_{n+1} \{\nabla V(\theta_n(\omega))\}' h(\theta_n(\omega)) < \infty$ and the condition $\sum_{n\geq 0} \gamma_n = \infty$ imply that

$$\liminf_{n\to\infty} \{\nabla V(\theta_n(\omega))\}' h(\theta_n(\omega)) = 0 , \quad \mathbb{P}\text{-a.s.}$$

Therefore, there is a point $\theta_\infty(\omega) \in \Theta_\infty(\omega)$ such that $\{\nabla V(\theta_\infty(\omega))\}' h(\theta_\infty(\omega)) = 0$ and since $\{\theta \in \Theta, \{\nabla V(\theta)\}' h(\theta) = 0\} = \{\theta_*\}$, we have $\theta_\infty(\omega) = \theta_*$. Since $\{V(\theta_n(\omega)), n \in \mathbb{N}\}$ converges, this implies $\lim_{n\to\infty} V(\theta_n(\omega)) = V(\theta_*)$ and for all the limiting points $\theta \in \Theta_\infty(\omega)$, $V(\theta) = V(\theta_*)$. Since $\{\theta, V(\theta) = V(\theta_*)\} = \{\theta_*\}$, the only possible stationary point is θ_*. ■

C.2 Stochastic gradient

Let $V : \mathbb{R}^d \to \mathbb{R}^+$ be such that it is a continuously differentiable function with $\{\theta \in \Theta,\ \nabla V(\theta) = 0\} = \{\theta_*\}$. Assume in addition that there exists H such that $h(\theta) = -\nabla V(\theta) = \mathbb{E}[H(\theta, X)]$.

$$h(\theta) = -\nabla V(\theta) = \mathbb{E}[H(\theta, Z)] .$$

Note that V is a Lyapunov function for the mean field h since

$$\{\nabla V(\theta)\}' h(\theta) = -\|\nabla V(\theta)\|^2 \leq 0 .$$

If the assumptions of Theorem C.7 are satisfied, then the sequence $\{\theta_n,\ n \in \mathbb{N}\}$ given by $\theta_{n+1} = \theta_n + \gamma_{n+1} H(\theta_n, X_{n+1})$ converges $\mathbb{P}$-a.s. to θ_*.

When V is strictly convex, one may always find a Lyapunov function for $h = -\nabla V$ satisfying AC.3 with compact level sets.

Lemma C.8. *Let* $G : \mathbb{R}^d \to \mathbb{R}$ *be a convex continuously differentiable function. Then,*

$$\{\{\nabla G(\theta) - \nabla G(\theta')\}\}' \{\theta - \theta'\} \geq 0 ;$$

with strict inequality if and only if $\theta \neq \theta'$ *if* G *is strictly convex.*

Proof. Define $g : [0, 1] \to \mathbb{R}$ by $g(s) = G(\theta + s(\theta' - \theta)) - G(\theta)$. Then g is a convex function (resp. strictly convex) since G is convex (resp. strictly convex) and continuously differentiable. Therefore $s \mapsto g'(s)$ is increasing (resp. strictly increasing) and $g'(1) \geq g'(0)$ (resp. $g'(1) > g'(0)$). The result follows from

$$g'(s) = \{\nabla G(\theta + s(\theta' - \theta))\}' (\theta' - \theta) .$$

■

Applying Lemma C.8 with $G = V$ and $\theta' = \theta_*$, we get

$$\{h(\theta)\}'\{\theta - \theta_*\} \le 0$$

showing that $\tilde{V} := 0.5\|\theta - \theta_*\|^2$ is a Lyapunov function for h. In addition, $\tilde{V}$ satisfies AC.3, $\lim_{\theta \to +\infty} \tilde{V}(\theta) = +\infty$, $\{\theta, \{\nabla \tilde{V}(\theta)\}' h(\theta) = 0\} = \{\theta_*\}$ and $\{\theta, \tilde{V}(\theta) = \tilde{V}(\theta_*)\} = \{\theta_*\}$. Therefore, if

- $\mathbb{E}\left[|H(\theta,X)|^2\right] \le C\{1 + \|\theta\|^2\}$
- $\{X_n,\, n \in \mathbb{N}\}$ are i.i.d. copies of X, and are independent of θ_0,
- $\{\gamma_n,\, n \in \mathbb{N}\}$ is a nonnegative sequence such that $\sum_{n=0}^{\infty} \gamma_n = +\infty$ and $\sum_{n=0}^{\infty} \gamma_n^2 < +\infty$,

the sequence $\{\theta_n,\, n \in \mathbb{N}\}$ defined by $\theta_{n+1} = \theta_n + \gamma_{n+1} H(\theta_n, X_{n+1})$ converges $\mathbb{P}$-a.s. to θ_*.

C.3 Stepsize selection and averaging

One of the main appeals of stochastic approximation is that, at least in principle, the only decision that has to be made is the choice of the step size schedule. Although in theory the method converges for a wide variety of step sizes (see Section C.1), in practice the choice of step sizes influences the actual number of simulations needed to take the parameter estimate into the neighborhood of the solution (transient regime) and its fluctuations around the solution (fluctuation near convergence). Large step sizes generally speed up convergence to a neighborhood of the solution but fail to average the noise. Small step sizes reduce noise but cause slow convergence. Heuristically, it is appropriate to use large step sizes until the algorithm reaches a neighborhood of the solution and then to switch to smaller step sizes. Whereas such a strategy appears sensible, there are serious obstacles to be dealt with when one comes to its practical implementation.

A way to alleviate the step size selection problem is to use averaging. Polyak (1990) (see also Polyak and Juditsky, 1992) showed that if the sequence of step sizes $\{\gamma_i\}$ tends to zero slower than $1/i$, yet fast enough to ensure convergence at a given rate, then the running average

$$\tilde{\theta}^i := (i - i_0 + 1)^{-1} \sum_{j=i_0}^{i} \hat{\theta}_j\,, \qquad i \ge i_0\,, \tag{C.11}$$

converges at an *optimal* rate. Here i_0 is an index at which averaging starts, so as to discard the very first steps. This result implies that one should adopt step sizes larger than usual but in conjunction with averaging (to control the increased noise due to use of the larger step sizes). The practical value of averaging has been reported in many different contexts; e.g., see Kushner and Yin (2003, Chapter 11) for a thorough investigation of averaging, as well as Delyon et al. (1999).

C.4 The Kiefer–Wolfowitz procedure

We wish to minimize the function $\mathbb{E}[H(\theta,X)] = h(\theta)$ over the $\mathbb{R}^d$-valued parameter θ, where $h(\cdot)$ is continuously differentiable and X is a random vector. Consider the following finite difference form of stochastic approximation. Let $c_n \to 0$ be a finite difference interval and let e_i be the standard unit vector in the i-th coordinate direction. Let θ_n denote the n-th estimate of the minimum. Suppose that for each i, n, and random vectors $X^+_{n,i}$, $X^-_{n,i}$, we can observe the finite difference estimate

$$Y_{n+1,i} = -\frac{[H(\theta_n + c_{n+1}e_i, X^+_{n+1,i}) - H(\theta_n - c_{n+1}e_i, X^-_{n+1,i})]}{2c_{n+1}}. \tag{C.12}$$

Define $Y_n = (Y_{n,1}, \ldots, Y_{n,r})$, and update θ_n as follows:

$$\theta_{n+1} = \theta_n + \gamma_{n+1} Y_{n+1}. \tag{C.13}$$

The algorithm (C.13) with Y_n defined by (C.12) is known as the *Kiefer-Wolfowitz* algorithm. Define

$$\psi_{n+1,i} = [h(\theta_n + c_{n+1}e_i) - H(\theta_n + c_{n+1}e_i,\ X^+_{n+1,i})] \\ - [h(\theta_n - c_n e_i) - H(\theta_n - c_n e_i,\ X^-_{n+1,i})],$$

and write

$$\frac{[h(\theta_n + c_{n+1}e_i) - h(\theta_n - c_{n+1}e_i)]}{2c_{n+1}} = \frac{\partial h}{\partial \theta^i}(\theta_n) - \beta_{n+1,i}, \tag{C.14}$$

where $-\beta_{n+1,i}$ is the bias in the finite difference estimate of $(\partial h/\partial \theta^i)(\theta_n)$, via central differences. Define $\psi_n = (\psi_{n,1}, \ldots, \psi_{n,r})$ and $\beta_n = (\beta_{n,1}, \ldots, \beta_{n,r})$. Eq. (C.13) may be rewritten as

$$\theta_{n+1} = \theta_n - \gamma_{n+1}\nabla h(\theta_n) + \gamma_{n+1}\frac{\psi_{n+1}}{2c_{n+1}} + \gamma_{n+1}\beta_{n+1}. \tag{C.15}$$

Clearly, for convergence to a local minimum to occur, it is required that $\beta_n \to 0$; under appropriate smoothness conditions, the bias is proportional to the finite difference interval $c_n \to 0$. In addition, the noise term $\gamma_n \psi_n/(2c_n)$ should average to zero. The fact that the effective noise $\psi_n/(2c_n)$ is of the order $1/c_n$ makes the Kiefer-Wolfowitz procedure less desirable than the Robbins–Monro procedure. The bias can be mitigated by computing better approximation of the derivatives via some variance reduction method or even by using an approximation to the original problem.

Suppose that $H(\cdot,\ X)$ is continuously differentiable for each value of X. Write

$$\frac{H(\theta_n - e_i c_{n+1}, X^-_{n+1,i}) - H(\theta_n + e_i c_{n+1}, X^+_{n+1,i})}{2c_{n+1}} + \frac{\partial h}{\partial \theta^i}(\theta_n) = \tilde{\psi}_{n+1,i} + \tilde{\beta}_{n+1,i}$$

where $\tilde{\beta}_{n,i}$ is the bias in the finite difference estimate.

$$\tilde{\psi}_{n+1,i} = \frac{1}{2c_{n+1}}[H(\theta_n,\ X^-_{n+1,i}) - H(\theta_n,\ X^+_{n+1,i})] \\ - \left[\frac{1}{2}\left(\frac{\partial H}{\partial \theta^i}(\theta_n,\ X^+_{n+1,i}) + \frac{\partial H}{\partial \theta^i}(\theta_n,\ X^-_{n+1,i})\right) - \frac{\partial h}{\partial \theta^i}(\theta_n)\right]$$

With these notations, the algorithm can be written as

$$\theta_{n+1} = \theta_n - \gamma_{n+1} \nabla h(\theta_n) + \gamma_{n+1} \tilde{\psi}_{n+1} + \gamma_{n+1} \tilde{\beta}_{n+1} \,. \tag{C.16}$$

This shows that it is preferable to use $X^+_{n,i} = X^-_{n,i}$ if possible, since the dominant $1/c_n$ factor cancels in the effective noise.

Appendix D

Data Augmentation

Data augmentation is a simple idea that can be traced back to the mid-twentieth century, with influences from the 1920s. Numerous researchers used the method for a variety of problems, most notably by Baum and Petrie (1966). The various approaches were synthesized in Dempster et al. (1977), which was a paper on the EM (*expectation-maximization*) algorithm that was presented and discussed at a meeting of the Royal Statistical Society. For interested readers, that paper contains an extensive presentation of the history of data augmentation prior to 1977. After Dempster et al. (1977), the development of the EM algorithm and related techniques expanded quickly. For a modern treatment, see the text McLachlan and Krishnan (2008).

The general idea is as follows. We suppose that we are interested in a parameter vector, θ, but that likelihood inference about θ based solely on the observed, but somehow "incomplete" data, Y, is intractable. The quintessential example of this problem is when some of the data are missing, as was the case in Example 2.15. We further suppose that if we had a "complete" data set, parameter estimation would be straight-forward. Hence, we postulate that there exists some *latent* variable X that is not observed, but if known, would make the estimation problem relatively simple. The pair $\{X,Y\}$ is known as the *complete data*, whereas the observed variable, Y, is called the *incomplete data*. If we assume that the joint distribution of $\{X,\ Y\}$, for a given parameter θ, admits a density $\mathrm{p}^{\theta}(x,y)$ with respect to some product measure $\lambda \times \mu$, the likelihood function of the data is given by

$$\mathrm{p}^{\theta}(y) = \int \mathrm{p}^{\theta}(x,\ y)\lambda(\mathrm{d}x)\,. \tag{D.1}$$

This relationship suggests that under certain conditions, the value of θ that maximizes the likelihood $\mathrm{p}^{\theta}(y)$ given the incomplete (but observed) data may be obtained via the complete data likelihood $\mathrm{p}^{\theta}(x,\ y)$. Recall that this idea was used in Section 2.3.2 for estimation in linear Gaussian state space models with irregularly spaced data.

In this appendix, we describe the EM algorithm, various extensions of the algorithm, some related procedures, and then present some theory regarding convergence and obtaining standard errors. For a p-dimensional parameter θ, we write $\nabla_{\theta} g(\theta) := \partial g(\theta)/\partial\theta$, a $p \times 1$ vector, and $\nabla^2_{\theta} g(\theta) := \partial^2 g(\theta)/\partial\theta\partial\theta'$, a $p \times p$ matrix, assuming existence.

D.1 The EM algorithm in the incomplete data model

Although we have already used the EM algorithm in Chapter 2, we provide a brief presentation of the technique. The log-likelihood function may be decomposed as

$$\ln p^{\theta}(Y) = \ln p^{\theta}(X,Y) - \ln p^{\theta}(X|Y) , \tag{D.2}$$

where we have introduced the conditional density of X given Y,

$$p^{\theta}(X|Y) = \frac{p^{\theta}(X,Y)}{p^{\theta}(Y)} . \tag{D.3}$$

Multiplying both sides of (D.2) by $p^{\theta'}(X|Y)$ and integrating with respect to the latent variable X, we get, $\ln p^{\theta}(Y) = \mathcal{Q}(\theta\,;\theta') - \mathcal{H}(\theta\,;\theta')$ where

$$\mathcal{Q}(\theta\,;\theta') := \mathbb{E}^{\theta'}\left[\ln p^{\theta}(X,Y) \mid Y\right] \quad \text{and} \quad \mathcal{H}(\theta\,;\theta') := \mathbb{E}^{\theta'}\left[\ln p^{\theta}(X|Y) \mid Y\right] . \tag{D.4}$$

The difference $\mathcal{H}(\theta\,;\theta') - \mathcal{H}(\theta'\,;\theta')$ is the *Kullback-Leibler divergence* (or *relative entropy*) between the conditional distribution $p(X|Y)$ indexed by θ and θ', respectively

$$\mathcal{H}(\theta\,;\theta') - \mathcal{H}(\theta'\,;\theta') = -\mathbb{E}^{\theta'}\left[\ln \frac{p^{\theta}(X|Y)}{p^{\theta'}(X|Y)} \mid Y\right] . \tag{D.5}$$

We are now ready to state the monotonicity of the EM algorithm, which is key to proving convergence. Throughout this appendix, we make the following assumptions.

Assumption AD.1. For any $\theta' \in \Theta$ and $x \in \mathsf{X}$, $\theta \mapsto \ln p^{\theta}(x|y)$ is twice continuously differentiable. For any $\theta' \in \Theta$, $\int \|\nabla_{\theta}^{k} \ln p^{\theta}(x|y)\| p^{\theta'}(x|y)\,\lambda(dx) < \infty$, for $k = 1,2$, and $\nabla_{\theta}^{k} \int \ln p^{\theta}(x|y) p^{\theta'}(x|y)\,\lambda(dx) = \int \nabla_{\theta}^{k} \ln p^{\theta}(x|y) p^{\theta'}(x|y)\,\lambda(dx)$, for $k = 1,2$.

Proposition D.2. *Under AD.1, for any $(\theta,\theta') \in \Theta \times \Theta$,*

$$\ln L(\theta) - \ln L(\theta') \geq \mathcal{Q}(\theta\,;\theta') - \mathcal{Q}(\theta'\,;\theta') , \tag{D.6}$$

where $\ln L(\theta) = \ln p^{\theta}(y)$. The inequality is strict unless $p^{\theta}(\cdot|y)$ and $p^{\theta'}(\cdot|y)$ are equal λ-a.e. For any $\theta' \in \Theta$, $\theta \mapsto \mathcal{Q}(\theta\,;\theta')$ is continuously differentiable on Θ and

$$\nabla_{\theta} \ln L(\theta) = \nabla_{\theta} \mathcal{Q}(\theta\,;\theta')\big|_{\theta'=\theta} . \tag{D.7}$$

Proof. The difference between the left-hand side and the right-hand side of (D.6) is the quantity defined in (D.5), which we already recognized as a Kullback-Leibler distance. Under AD.1 this latter term is well-defined and known to be strictly positive (by direct application of Jensen's inequality[1]) unless $p^{\theta}(\cdot|y)$ and $p^{\theta'}(\cdot|y)$ are equal $p^{\theta'}(\cdot|y)$, λ-a.e.

For (D.7), first note that $\theta \mapsto \mathcal{Q}(\theta\,;\theta')$ is a differentiable function, as the difference of two differentiable functions. Next, the previous discussion implies that $\theta \mapsto \mathcal{H}(\theta\,;\theta')$ is minimal for $\theta = \theta'$, although this may not be the only point where the minimum is achieved. Thus its gradient vanishes at θ', which proves (D.7). ■

[1] If g is a convex function and $\mathbb{E}|X| < \infty$, then $g(\mathbb{E}[X]) \leq \mathbb{E}[g(X)]$.

Proposition D.2 suggests that $\theta \mapsto \mathcal{Q}(\theta\,;\theta')$ can be used as a surrogate for the log-likelihood function to compute the maximum likelihood. This is exploited in the EM algorithm, an iterative recipe that maximizes the incomplete data likelihood $\mathrm{p}^\theta(y)$ by iteratively maximizing the auxiliary function $\theta \mapsto \mathcal{Q}(\theta\,;\theta')$. The procedure is initialized at some $\theta_0 \in \Theta$ and then iterates between two steps, expectation (E) and maximization (M)

(E) Compute the auxiliary function $\theta \mapsto \mathcal{Q}(\theta\,;\theta_{k-1})$ for the current fit of the parameter θ_{k-1}.

(M) Compute $\theta_k = \arg\max_{\theta\in\Theta} \mathcal{Q}(\theta\,;\theta_{k-1})$

Proposition D.2 provides the two decisive arguments behind the EM algorithm. First, an immediate consequence of (D.6) is that, by the very definition of the sequence $\{\theta_k,\ k \in \mathbb{N}\}$, the sequence $\{\ln \mathrm{L}(\theta_k)\}$ of log-likelihood values is non-decreasing. Hence EM is a monotone optimization algorithm. Second, if the iterations ever stop at a point $\theta_\star$, then $\mathcal{Q}(\theta\,;\theta_\star)$ has to be maximal at $\theta_\star$ (otherwise it would still be possible to improve over $\theta_\star$), and hence $\theta_\star$ is such that $\nabla_\theta \ln \mathrm{L}(\theta_\star) = 0$, that is, this is a *stationary point of the likelihood.*

Although this picture is largely correct, there is a slight flaw in the second half of the above intuitive reasoning in that the if part *(if the iterations ever stop at a point)* may indeed never happen. Stronger conditions are required to ensure that the sequence of parameter estimates produced by EM from any starting point indeed converges to a limit $\theta_\star \in \Theta$. However, it is actually true that when convergence to a point takes place, the limit has to be a stationary point of the likelihood.

The EM algorithm defined in the previous section is helpful in situations where the following general conditions hold.

E-Step: It is possible to compute, at reasonable computational cost, the intermediate quantity $\mathcal{Q}(\theta\,;\theta')$ given a value of θ'.

M-Step: $\mathcal{Q}(\theta\,;\theta')$, considered as a function of its first argument θ, is sufficiently simple to allow closed-form maximization.

A rather general context in which both of these requirements are satisfied, or at least are equivalent to easily interpretable necessary conditions, is when the complete data likelihood is a (curved) *exponential family.*

Definition D.3 (Exponential family). *The family* $\{\mathrm{p}^\theta(\cdot)\}_{\theta\in\Theta}$ *defines an* exponential family *of positive functions on* $\mathsf{X}\times\mathsf{Y}$ *if*

$$\mathrm{p}^\theta(X,Y) = \exp\{\psi(\theta,Y)'S(X,Y) - c(\theta,Y)\}h(X,Y)\,, \tag{D.8}$$

where S and ψ are vector-valued functions, c is a real-valued function and h is a non-negative real-valued function on $\mathsf{X}\times\mathsf{Y}$.

For an exponential family, the intermediate quantity of EM reduces to

$$\mathcal{Q}(\theta\,;\theta') = \psi(\theta,Y)'\mathbb{E}^{\theta'}\left[S(X,Y)\,\middle|\,Y\right] - c(\theta,Y) + \mathbb{E}^{\theta'}\left[\ln h(X,Y)\,\middle|\,Y\right]. \tag{D.9}$$

Note that the right-most term does not depend on θ and thus plays no role in the maximization. It may as well be ignored, and in practice it is not required to compute it. Except for this term, the right-hand side of (D.9) has an explicit form as soon as it is possible to evaluate the expectation of the vector of sufficient statistics S. The other important feature of (D.9), ignoring the rightmost term, is that $\mathcal{Q}(\theta\,;\theta')$, viewed as a function of θ, is similar to the logarithm of (D.8) for the particular value $S_{\theta'}(Y) = \mathbb{E}^{\theta'}\left[S(X,Y) \mid Y\right]$ of the sufficient statistic. The M-step then consists in optimizing the function $\theta \mapsto \psi(\theta)'\hat{S}_{\theta'}(Y) - c(\theta)$. In many models, the maximization of this function can be achieved in closed form.

There are variants of the EM algorithm that are handy in cases where the maximization required in the M-step is not directly feasible (McLachlan and Krishnan, 2008, Chapter 5).

D.2 The Fisher and Louis identities

It is important to consider how the gradient and Hessian of the incomplete data likelihood depends on those of the complete data likelihood. These relationships are given in the following proposition.

Proposition D.4. *Under assumption AD.1, the following identities hold:*

$$\nabla \ln \mathrm{p}^\theta(y) = \int \nabla \ln \mathrm{p}^\theta(x,y)\; \mathrm{p}^\theta(x|y)\,\lambda(\mathrm{d}x)\,, \tag{D.10}$$

$$-\nabla^2 \ln \mathrm{p}^\theta(y) = -\int \nabla^2 \ln \mathrm{p}^\theta(x,y)\,\mathrm{p}^\theta(x|y)\,\lambda(\mathrm{d}x) + \int \nabla^2 \ln \mathrm{p}^\theta(x|y)\,\mathrm{p}^\theta(x|y)\,\lambda(\mathrm{d}x)\,. \tag{D.11}$$

The second equality may be rewritten in the equivalent form

$$\nabla^2 \ln \mathrm{p}^\theta(y) + \left\{\nabla \ln \mathrm{p}^\theta(y)\right\}\left\{\nabla \ln \mathrm{p}^\theta(y)\right\}' = \int \Big[\nabla^2 \ln \mathrm{p}^\theta(x,y) + \left\{\nabla \ln \mathrm{p}^\theta(x,y)\right\}\left\{\nabla \ln \mathrm{p}^\theta(x,y)\right\}'\Big]\,\mathrm{p}^\theta(x|y)\,\lambda(\mathrm{d}x)\,. \tag{D.12}$$

Proof. Note indeed that the incomplete data likelihood is obtained by marginalizing the latent variable

$$\mathrm{p}^\theta(y) = \int \cdots \int \mathrm{p}^\theta(x,y)\lambda(\mathrm{d}x)\,.$$

By exchanging derivation and integration, the above expression implies that

$$\begin{aligned}\nabla \ln \mathrm{p}^\theta(y) &= \int \cdots \int \frac{\nabla \mathrm{p}^\theta(x,y)}{\mathrm{p}^\theta(y)}\lambda(\mathrm{d}x)\\ &= \int \cdots \int \nabla \ln \mathrm{p}^\theta(x,y)\frac{\mathrm{p}^\theta(x,y)}{\mathrm{p}^\theta(y)}\lambda(\mathrm{d}x)\\ &= \int \cdots \int \nabla \ln \mathrm{p}^\theta(x,y)\,\mathrm{p}^\theta(x|y)\lambda(\mathrm{d}x)\,.\end{aligned}$$

To prove (D.12), we start from (D.11) and note that the second term on its right-hand side is the negative of an information matrix for the parameter θ associated with the probability density function. We rewrite this second term using the well-known information matrix identity

$$-\int \nabla^2 \ln \mathrm{p}^\theta(x|y)\ \mathrm{p}^\theta(x|y)\,\lambda(\mathrm{d}x) \\ = \int \left\{\nabla \ln \mathrm{p}^\theta(x|y)\right\}\left\{\nabla \ln \mathrm{p}^\theta(x|y)\right\}' \mathrm{p}^\theta(x|y)\,\lambda(\mathrm{d}x)\,.$$

Since $\mathrm{p}^\theta(\cdot|y)$ is a probability density function for all values of θ, we have by interchanging integration and derivation

$$\int \nabla \ln \mathrm{p}^\theta(x|y)\ \mathrm{p}^\theta(x|y)\,\lambda(\mathrm{d}x) = 0\,.$$

Now use the identity $\ln \mathrm{p}^\theta(x|y) = \ln \mathrm{p}^\theta(x,y) - \ln \mathrm{p}^\theta(y)$ and (D.10) to conclude that

$$\int \left\{\nabla \ln \mathrm{p}^\theta(x|y)\right\}\left\{\nabla \ln \mathrm{p}^\theta(x|y)\right\}' \mathrm{p}^\theta(x|y)\,\lambda(\mathrm{d}x) \\ = \int \left\{\nabla \ln \mathrm{p}^\theta(x,y)\right\}\left\{\nabla \ln \mathrm{p}^\theta(x,y)\right\}' \mathrm{p}^\theta(x|y)\,\lambda(\mathrm{d}x) \\ - \left\{\nabla \ln \mathrm{p}^\theta(y)\right\}\left\{\nabla \ln \mathrm{p}^\theta(y)\right\}'\,,$$

which completes the proof. ■

Equation (D.10) is sometimes referred to as *Fisher's identity*. The left-hand side of (D.10) is the *score function* (gradient of the log-likelihood). Equation (D.10) shows that the score function may be evaluated by computing the conditional expectation [under $\mathrm{p}^{\theta'}(\cdot|y)$], of the *complete score function* $(x,y) \mapsto \nabla_\theta \ln \mathrm{p}^\theta(x,y)\big|_{\theta=\theta'}$.

Equation (D.11) is usually called the *missing information principle*, a term coined by Louis (1982). The left-hand side of (D.11) is the associated *observed information matrix*, and the second term on the right-hand side is easily recognized as the Fisher information matrix associated with the probability density function $\mathrm{p}^{\theta'}(\cdot|y)$.

Finally (D.12), which is here written in a form that highlights its symmetry, was also proved by Louis (1982) and is thus known as *Louis' identity*. Together with (D.10), it follows that the first- and second-order derivatives may be evaluated by computing the conditional expectation [under $\mathrm{p}^{\theta'}(\cdot|y)$] of the complete data likelihood.

D.3 Monte Carlo EM algorithm

In the context of HMMs, the main limitation of the EM algorithm rather appears in cases where the E-step is not feasible. This latter situation is the rule rather than the exception in models for which the state space X is not finite.

Wei and Tanner (1991) and Tanner (1993) suggest a Monte Carlo approach to approximate the intractable E-step

$$\hat{\mathcal{Q}}_m(\theta\,;\theta') := \frac{1}{m}\sum_{j=1}^{m} \ln \mathrm{p}^{\theta}(X^j, Y)\,, \tag{D.13}$$

where the missing data $\{X^i\}_{i=1}^m$ are sampled from the conditional distribution $\mathrm{p}^{\theta'}(X|Y)$. The subscript m reflects the dependence on the Monte Carlo sample size.

The EM algorithm can thus be modified into the Monte Carlo EM (MCEM) algorithm by replacing $\mathcal{Q}(\theta\,;\theta')$ by $\hat{\mathcal{Q}}_m(\theta\,;\theta')$ in the E-step.

Similar to the EM algorithm, the MCEM algorithm is well suited to problems in which the complete data likelihood belongs to the exponential family (see Definition D.3). In this case, the E-step consists in computing a Monte Carlo approximation

$$\hat{S}_{\theta'}(Y) = \frac{1}{m}\sum_{j=1}^{m} S(X^j, Y)\,. \tag{D.14}$$

In many situations, sampling from the conditional distribution $\mathrm{p}^{\theta'}(X|Y)$ may turn out difficult. One may then use Markov chain Monte Carlo techniques, in which case $\{X^j\}_{j=1}^m$ is a sequence generated by an ergodic Markov chain whose stationary distribution is $\mathrm{p}^{\theta'}(\cdot|Y)$. Another option is to use sequential Monte Carlo techniques as described in Chapter 12 for NLSS models.

Heuristically there is no need to use a large number of simulations during the initial stage of the optimization. Even rather crude estimation of $\mathcal{Q}(\theta\,;\theta')$ might suffice to drive the parameters toward the region of interest. As the EM iterations go on, the number of simulations should be increased, however, to avoid "zig-zagging" when the algorithm approaches convergence. Thus, in making the trade-off between improving accuracy and reducing the computational cost associated with a large sample size, one should favor increasing the sample size m as the parameter approaches convergence. Determining how this increase should be accomplished to produce the "best" possible result is a topic that still attracts much research interest (Booth and Hobert, 1999, Levine and Casella, 2001, Fort and Moulines, 2003, Levine and Fan, 2004).

D.3.1 Stochastic approximation EM

We now consider a variant of the MCEM algorithm that may also be interpreted as a stochastic approximation procedure. Compared to the MCEM approach, the E-step involves a weighted average of the approximations of the intermediate quantity of EM obtained in the current as well as in the previous iterations. Hence there is no need to increase the number of replications of the missing data as in MCEM.

This algorithm, called the stochastic approximation EM (SAEM) algorithm, was proposed by Delyon et al. (1999). To understand why this algorithm can be cast into the Robbins-Monro framework, consider the simple case where the complete data likelihood is from an exponential family. In this setting, the SAEM algorithm

Algorithm D.1 (Stochastic Approximation EM)

Given an initial parameter estimate $\hat{\theta}_0$ and a decreasing sequence of positive step sizes $\{\gamma_i\}_{i\geq 1}$ such that $\gamma_1 = 1$, do, for $i = 1, 2\ldots$,

Simulation: Draw $X^{i,1}, \ldots, X^{i,m}$ from the conditional distribution $\mathrm{p}^{\hat{\theta}_{i-1}}(\cdot|Y)$.

Maximization: Compute $\hat{\theta}_i$ as the maximum of the function $\hat{Q}_i(\theta)$ over the Θ, where

$$\hat{Q}_i(\theta) = \hat{Q}_{i-1}(\theta) + \gamma_i \left\{ \frac{1}{m} \sum_{j=1}^{m} \ln \mathrm{p}^{\theta}(X^{i,j}, Y) - \hat{Q}_{i-1}(\theta) \right\} . \tag{D.15}$$

updates, at each iteration, the current estimates $(\hat{S}_i, \hat{\theta}_i)$ of the complete data sufficient statistic and of the parameter. Each iteration of the algorithm is divided into two steps. In a first step, we draw $X^{i,1}, \ldots, X^{i,m}$ from the conditional density $\mathrm{p}^{\hat{\theta}^{i-1}}(\cdot|Y)$ and update $\hat{S}_i(Y)$ according to

$$\hat{S}_i(Y) = \hat{S}_{i-1}(Y) + \gamma_i \left[\frac{1}{m} \sum_{j=1}^{m} S(X^{i,j}, Y) - \hat{S}_{i-1}(Y) \right] . \tag{D.16}$$

In a second step, we compute $\hat{\theta}_i$ as the maximum of the function $\psi(\theta)'\hat{S}_i(Y) - c(\theta)$.

Assume that the function $\psi(\theta)'s - c(\theta)$ has a single global maximum, denoted $\bar{\theta}(s)$. The difference $m^{-1}\sum_{j=1}^{m} S(X^{i,j}, Y) - \hat{S}_{i-1}(Y)$ can then be considered as a noisy observation of a function $h(\hat{S}^{i-1})$, where

$$h(s) = \mathbb{E}^{\bar{\theta}(s)} \left[S(X,Y) \mid Y \right] - s . \tag{D.17}$$

Thus (D.16) fits into the Robbins-Monro framework (see Section C.1). This Robbins-Monro procedure searches for the roots of $h(s) = 0$, that is, the values of s satisfying

$$\mathbb{E}^{\bar{\theta}(s)} \left[S(X,Y) \mid Y \right] = s .$$

Assume that this equation has a solution $s_\star$ and put $\theta_\star = \bar{\theta}(s_\star)$. Now note that

$$\mathcal{Q}(\theta\,; \theta_\star) = \psi(\theta)'\mathbb{E}^{\theta_*} \left[S(X,Y) \mid Y \right] - c(\theta) = \psi(\theta)'s_* - c(\theta) ,$$

and by definition the maximum of the right-hand side of this display is obtained at $\theta_\star$. Therefore, an iteration of the EM algorithm started at θ_* will stay at θ_*, and we find that each root s_* is associated to a fixed point θ_* of the EM algorithm.

Compared to the stochastic gradient procedure, SAEM inherits most of the appealing features of the EM algorithm (among others, the invariance with respect to the parameterization). With the SAEM algorithm, the scale of the step sizes $\{\gamma_i\}$ is fixed irrespectively of the parameterization as γ_1 equals 1. As in the case of the stochastic gradient, however, the rate of decrease of the step sizes influences the practical performance of the algorithm. Here again, the use of averaging is helpful in reducing the impact of the choice of the rate of decrease of the step sizes.

D.4 Convergence of the EM algorithm

The EM algorithm is an *ascent algorithm*: the log-likelihood function, $\ln \mathrm{L}$, is monotonically increasing. Ascent algorithms (Luenberger, 1984, Chapter 6; Lange, 2010, Chapter 15) can be analyzed in a unified manner following a theory developed by Zangwill (1969). Wu (1983) showed that this general theory applies to the EM algorithm as defined above, as well as to some of its variants that he calls generalized EM (or GEM). The main result is a strong stability guarantee known as *global convergence*, which we discuss below.

We first need a mathematical formalism that describes the EM algorithm. This is done by identifying any iterative algorithm with a specific choice of a mapping M that associates θ_{k+1} to θ_k. In the theory of Zangwill (1969), one indeed considers families of algorithms by allowing for *point-to-set* maps M that associate a set $M(\theta') \subseteq \Theta$ to each parameter value $\theta' \in \Theta$. A specific algorithm in the family is such that θ_{k+1} is selected in $M(\theta_k)$. In the example of EM, we may define M as

$$M(\theta') = \left\{\theta \in \Theta : \mathcal{Q}(\theta;\theta') \geq \mathcal{Q}(\tilde{\theta};\theta') \text{ for all } \tilde{\theta} \in \Theta\right\} , \tag{D.18}$$

that is, $M(\theta')$ is the set of values θ that maximize $\mathcal{Q}(\theta;\theta')$ over Θ. In most cases $M(\theta')$ reduces to a singleton, and the mapping M is then simply a point-to-point map. But the use of point-to-set maps makes it possible to deal also with cases where the intermediate quantity of EM may have several global maxima. We next need the following definition before stating the main convergence theorem.

Definition D.5 (Closed mapping). *A map T from points of Θ to subsets of Θ is said to be* closed *on a set $\mathcal{S} \subseteq \Theta$ if for any converging sequences $\{\theta_k,\ k \in \mathbb{N}\}$ and $\{\tilde{\theta}_k,\ k \in \mathbb{N}\}$, the conditions*

(a) $\theta_k \to \theta \in \mathcal{S}$,

(b) $\tilde{\theta}_k \to \tilde{\theta}$ with $\tilde{\theta}_k \in T(\theta_k)$ for all $k \geq 0$,

imply that $\tilde{\theta} \in T(\theta)$.

Note that for point-to-point maps, that is, if $T(\theta)$ is a singleton for all θ, the definition above is equivalent to the requirement that T be continuous on $\mathcal{S}$. Definition D.5 is thus a generalization of continuity for general (point-to-set) maps. We are now ready to state the main result, which is proved in Zangwill (1969, p. 91) or Luenberger (1984, p. 187).

Theorem D.6 (Global convergence theorem). *Let Θ be a subset of $\mathbb{R}^d$ and let $\{\theta_k,\ k \in \mathbb{N}\}$ be a sequence generated by $\theta_{k+1} \in T(\theta_k)$ where T is a point-to-set map on Θ. Let $\mathcal{S} \subseteq \Theta$ be a given "solution" set and suppose that*

1. *the sequence $\{\theta_k,\ k \in \mathbb{N}\}$ is contained in a compact subset of Θ;*
2. *T is closed over $\Theta \setminus \mathcal{S}$ (the complement of $\mathcal{S}$);*
3. *there is a continuous "ascent" function s on Θ such that $s(\theta) \geq s(\theta')$ for all $\theta \in T(\theta')$, with strict inequality for points θ' that are* not *in $\mathcal{S}$.*

Then the limit of any convergent subsequence of $\{\theta_k,\ k \in \mathbb{N}\}$ is in the solution set $\mathcal{S}$. In addition, the sequence of values of the ascent function, $\{s(\theta_k), k \in \mathbb{N}\}$, converges monotonically to $s(\theta_\star)$ for some $\theta_\star \in \mathcal{S}$.

The following general convergence theorem following the proof by Wu (1983) is a direct application of the previous theory to the case of EM.

Theorem D.7. *Assume AD.1 and*

(i) $\mathcal{H}(\theta\,;\theta')$ *is continuous in its second argument,* θ', *on* Θ.

(ii) For any θ^0, *the level set* $\Theta^0 = \left\{\theta \in \Theta : \ln \mathrm{L}(\theta) \geq \ln \mathrm{L}(\theta^0)\right\}$ *is compact and contained in the interior of* Θ.

Then all limit points of any instance $\{\theta_k,\ k \in \mathbb{N}\}$ *of an EM algorithm initialized at* θ^0 *are in* $\mathcal{L}^0 = \{\theta \in \Theta^0 : \nabla_\theta \ln \mathrm{L}(\theta) = 0\}$, *the set of stationary points of* $\ln \mathrm{L}$ *with log-likelihood larger than that of* θ^0. *The sequence* $\{\ln \mathrm{L}(\theta_k)\}$ *of log-likelihoods converges monotonically to* $\ln \mathrm{L}_* = \ln \mathrm{L}(\theta_\star)$ *for some* $\theta_\star \in \mathcal{L}^0$.

Proof. This is a direct application of Theorem D.6 using $\mathcal{L}^0$ as the solution set and $\ln \mathrm{L}$ as the ascent function. The first hypothesis of Theorem D.6 follows from Theorem D.7-(ii) and the third one from Proposition D.2. The closedness assumption Theorem D.6-condition (2) follows from Proposition D.2 and the condition Theorem D.7-(i): for the EM mapping M defined in (D.18), $\tilde{\theta}_k \in M(\theta_k)$ amounts to the condition

$$\mathcal{Q}(\tilde{\theta}_k\,;\theta_k) \geq \mathcal{Q}(\theta\,;\theta_k) \quad \text{for all } \theta \in \Theta\ ,$$

which is also satisfied by the limits of the sequences $\{\tilde{\theta}_k,\ k \in \mathbb{N}\}$ and $\{\theta_k,\ k \in \mathbb{N}\}$ (if these converge) by continuity of the intermediate quantity $\mathcal{Q}$, which follows from that of $\ln \mathrm{L}$ and $\mathcal{H}$ (note that it is here important that $\mathcal{H}$ be continuous with respect to both arguments). Hence the EM mapping is indeed closed on Θ as a whole and Theorem D.7 follows. ∎

Condition Theorem D.7-(i) is very mild in typical situations. Condition Theorem D.7-(ii), however, may be restrictive, even for models in which the EM algorithm is routinely used. The practical implication of Theorem D.7-(ii) being violated is that the EM algorithm may fail to converge to the stationary points of the likelihood for some particularly badly chosen initial points θ_0.

Most importantly, the fact that θ_{k+1} maximizes the intermediate quantity $\mathcal{Q}(\cdot\,;\theta_k)$ of EM does in no way imply that, ultimately, $\ln \mathrm{L}_*$ is the global maximum of $\ln \mathrm{L}$ over Θ. There is even no guarantee that $\ln \mathrm{L}_*$ is a local maximum of the log-likelihood: it may well only be a saddle point (Wu, 1983, Section 2.1). Also, the convergence of the sequence $\ln \mathrm{L}(\theta_k)$ to $\ln \mathrm{L}_*$ does not automatically imply the convergence of $\{\theta_k\}$ to a point $\theta_\star$.

Pointwise convergence of the EM algorithm requires more stringent assumptions that are difficult to check in practice; see Wu (1983) and Boyles (1983), for an equivalent result.

Theorem D.8. *Assume A D.1 and* $\lim_{k\to\infty} \|\theta_{k+1} - \theta_k\| = 0$. *Then, all limit points of* $\{\theta_k\}$ *are in a connected and compact subset of* $\mathcal{L}_* = \{\theta \in \Theta : \ln \mathrm{L}(\theta) = \ln \mathrm{L}_*\}$, *where* $\ln \mathrm{L}_*$ *is the limit of the log-likelihood sequence* $\{\ln \mathrm{L}(\theta_k)\}$.

In particular, if the connected components of $\mathcal{L}_*$ *are singletons, then* $\{\theta_k\}$ *converges to some* $\theta_\star$ *in* $\mathcal{L}_*$.

Proof. The set of limit points of a bounded sequence $\{\theta_k\}$ with $\|\theta_{k+1} - \theta_k\| \to 0$ is connected and compact (Ostrowski, 1966, Theorem 28.1). The proof follows because under Theorem D.6, the limit points of $\{\theta_k\}$ must belong to $\mathcal{L}_*$. ■

D.5 Convergence of the MCEM algorithm

In Section D.4, the EM algorithm was analyzed by viewing each of its iterations as a mapping M on the parameter space Θ such that the EM sequence of estimates is given by the iterates $\theta_{k+1} = M(\theta_k)$. Under mild conditions, the EM sequence eventually converges to the set of fixed points, $\mathcal{L} = \{\theta \in \Theta : \theta = M(\theta)\}$, of this mapping. EM is an ascent algorithm as each iteration of M increases the observed log-likelihood $\ln \mathrm{L}$, that is, $\ln \mathrm{L} \circ M(\theta) \geq \ln \mathrm{L}(\theta)$ for any $\theta \in \Theta$ with equality if and only if $\theta \in \mathcal{L}$. This ascent property is essential in showing that the algorithm converges: it guarantees that the sequence $\{\ln \mathrm{L}(\theta^i)\}$ is non-decreasing and, hence, convergent if it is bounded.

The MCEM algorithm is an approximation of the EM algorithm. Each iteration of the MCEM algorithm is a perturbed version of an EM iteration, where the "typical size" of the perturbation is controlled by the Monte Carlo error and thus by the number of simulations. The MCEM sequence may thus be written under the form $\hat{\theta}^{i+1} = M(\hat{\theta}^i) + \zeta^{i+1}$, where ζ^{i+1} is the perturbation due to the Monte Carlo approximation. Provided that the number of simulations is increased as the algorithm approaches convergence, the perturbation ζ^i vanishes as $i \to \infty$. Whereas the MCEM algorithm is not an ascent algorithm, it is sensible, however, to expect that the behavior of the MCEM algorithm closely follows that of the EM algorithm, at least for large i, as the random perturbations vanish in the limit.

To prove that this intuition is correct, we first establish in Section D.5.1 a stability result for deterministically perturbed dynamical systems and then use this result in Section D.5.2 to deduce a set of conditions implying almost sure convergence of the MCEM algorithm.

D.5.1 Convergence of perturbed dynamical systems

Let $T : \Theta \to \Theta$ be a (point-to-point) map on Θ. We study in this section the convergence of the Θ-valued discrete time dynamical system $\theta^{i+1} = T(\theta^i)$ and the perturbed dynamical system $\theta^{i+1} = T(\theta^i) + \zeta^{i+1}$, where $\{\zeta^i\}$ is a deterministic sequence converging to zero.

To study the convergence, it is useful to introduce Lyapunov functions associated with the mapping T. A Lyapunov function, as defined below, is equivalent to the concept of ascent function that we met in Section D.4 when discussing the convergence of EM. Denote by

$$\mathcal{L} := \{\theta \in \Theta : \theta = T(\theta)\} \tag{D.19}$$

the set of fixed points of this map. A function $W : \Theta \to \mathbb{R}$ is said to be a *Lyapunov function* relative to (T, Θ) if W is continuous and $W \circ T(\theta) \geq W(\theta)$ for all $\theta \in \Theta$, with equality if and only if $\theta \in \mathcal{L}$. In other words, the map T is an ascent algorithm for the function W.

Theorem D.9. *Let Θ be an open subset of $\mathbb{R}^d$ and let $T : \Theta \to \Theta$ be a continuous map with set $\mathcal{L}$ of fixed points. Assume that there exists a Lyapunov function W relative to (T,Θ) such that $W(\mathcal{L})$ is a finite set of points. Let $\mathcal{K}$ be a compact set and $\{\theta^i\}$ a $\mathcal{K}$-valued sequence satisfying*

$$\lim_{i\to\infty} |W(\theta^{i+1}) - W \circ T(\theta^i)| = 0 . \tag{D.20}$$

Then the set $\mathcal{L} \cap \mathcal{K}$ is non-empty, the sequence $\{W(\theta^i)\}$ converges to a point $w_\star \in W(\mathcal{L} \cap \mathcal{K})$, and the sequence $\{\theta^i\}$ converges to the set $\mathcal{L}_{w_\star} = \{\theta \in \mathcal{L} \cap \mathcal{K} : W(\theta) = w_\star\}$.

The proof of the theorem is based on the following result.

Lemma D.10. *Let $\varepsilon > 0$ be a real constant, let $n \geq 1$ be an integer, and let $-\infty < a_1 < b_1 < \ldots < a_n < b_n < \infty$ be real numbers. Let $\{w_j\}$ and $\{e_j\}$ be two sequences such that* $\limsup_{j\to\infty} w_j < \infty$, $\lim_{j\to\infty} e_j = 0$ *and*

$$w_{j+1} \geq w_j + \varepsilon \mathbb{1}_{A^c}(w_j) + e_j , \quad \text{where} \quad A := \bigcup_{i=1}^{n} [a_i, b_i] . \tag{D.21}$$

Then there exists an index $k_\star \in \{1,\ldots,n\}$ such that $a_{k_\star} \leq \liminf w_j \leq \limsup w_j \leq b_{k_\star}$.

Proof. First note that (D.21) implies that the sequence $\{w_j\}$ is infinitely often in the set A (otherwise it would tend to infinity, contradicting the assumptions). Thus it visits infinitely often at least one of the intervals $[a_k, b_k]$ for some k. Choose $\eta < \varepsilon \wedge \inf_{1\leq i\leq n-1}(a_{i+1} - b_i)/2$ and set j_0 such that $|e_j| \leq \eta$ for $j \geq j_0$. Let $p \geq j_0$ such that $w_p \in [a_k, b_k]$. We will show that

$$\text{for any } j \geq p , \quad w_j \geq a_k - \eta . \tag{D.22}$$

The property is obviously true for $j = p$. Assume now that the property holds true for some $j \geq p$. If $w_j \geq a_k$, then (D.21) shows that $w_{j+1} \geq a_k - \eta$. If $a_k - \eta \leq w_j < a_k$, then $w_{j+1} \geq w_j + \varepsilon - \eta \geq a_k - \eta$. Therefore $w_{j+1} \geq a_k - \eta$, and (D.22) follows by induction. Because η was arbitrary, we find that $\liminf w_j \geq a_k$. Using a similar induction argument, one may show that $\limsup w_j \leq b_k$, which concludes the proof. ∎

Proof (of Theorem D.9). If $\mathcal{L} \cap \mathcal{K}$ was empty, then $\min_{\theta\in \mathcal{K}} W \circ T(\theta) - W(\theta) > 0$, which would contradict (D.20). Hence $\mathcal{L} \cap \mathcal{K}$ is non-empty. For simplicity, we assume in the following that $\mathcal{L} \subseteq \mathcal{K}$, if not, simply replace $\mathcal{L}$ by $\mathcal{L} \cap \mathcal{K}$.

For any $\alpha > 0$, let $[W(\mathcal{L})]_\alpha := \{x \in \mathbb{R} : \inf_{y\in W(\mathcal{L})} |x - y| < \alpha\}$. Because $W(\mathcal{L})$ is bounded, the set $[W(\mathcal{L})]_\alpha$ is a *finite* union of disjoint bounded open intervals of length at least equal to 2α. Thus there exists an integer $n_\alpha \geq 0$ and real numbers $a_\alpha(1) < b_\alpha(1) < \ldots < a_\alpha(n_\alpha) < b_\alpha(n_\alpha)$ such that

$$[W(\mathcal{L})]_\alpha = \bigcup_{k=1}^{n_\alpha} (a_\alpha(k), b_\alpha(k)) . \tag{D.23}$$

Note that $W^{-1}([W(\mathcal{L})]_\alpha)$ is an open neighborhood of $\mathcal{L}$, and define

$$\varepsilon := \inf_{\{\theta \in \mathcal{K} \setminus W^{-1}([W(\mathcal{L})]_\alpha)\}} \{W \circ T(\theta) - W(\theta)\} > 0\,. \tag{D.24}$$

Write

$$W(\theta^{i+1}) - W(\theta^i) = \left\{W \circ T(\theta^i) - W(\theta^i)\right\} + \left\{W(\theta^{i+1}) - W \circ T(\theta^i)\right\}. \tag{D.25}$$

Because $W(\theta^i) \notin [W(\mathcal{L})]_\alpha$ implies $\theta^i \notin W^{-1}([W(\mathcal{L})]_\alpha)$, we obtain

$$W(\theta^{i+1}) \geq W(\theta^i) + \varepsilon \mathbb{1}_{[W(\mathcal{L})]_\alpha^c}\left(W(\theta^i)\right) + \left\{W(\theta^{i+1}) - W \circ T(\theta^i)\right\}\,. \tag{D.26}$$

By (D.20), $W(\theta^{i+1}) - W \circ T(\theta^i) \to 0$ as $i \to \infty$. Thus by Lemma D.10, the set of limit points of the sequence $\{W(\theta^i)\}$ belongs to one of the intervals $[a_\alpha(k), b_\alpha(k)]$. Because $W(\mathcal{L}) = \bigcap_{\alpha>0}[W(\mathcal{L})]_\alpha$ and $W(\mathcal{L})$ is a finite set, the sequence $\{W(\theta^i)\}$ must be convergent with a limit that belongs to $W(\mathcal{L})$. Using (D.25) and (D.20) again, this implies that $W \circ T(\theta^i) - W(\theta^i) \to 0$ as $i \to \infty$, showing that all limit points of the sequence $\{\theta^i\}$ belongs to $\mathcal{L}$. ■

D.5.2 Convergence of the MCEM algorithm

Denote by

$$\bar{S}(\theta) := \mathbb{E}^\theta\left[S(X,Y) \mid Y\right]\,, \tag{D.27}$$

where $S(x)$ is the (vector of) sufficient statistic(s) defined below.

Assumption AD.11.

(a) Θ is an open subset of $\mathbb{R}^d$ and $\{f(\cdot;\theta)\}_{\theta\in\Theta}$ defines an exponential family of positive functions on X (see Definition D.3.

(b) The function L is positive and continuous on Θ.

(c) For any $\theta \in \Theta$, $\mathbb{E}^\theta\left[|S(X,Y)| \mid Y\right] < \infty$, and the function $\bar{S}$ is continuous on Θ.

(d) There exists a closed subset $\mathcal{S}$ that contains the convex hull of $S(\mathsf{X})$ and is such that for any $s \in \mathcal{S}$, the function $\theta \mapsto \psi^t(\theta)s - c(\theta)$ has a unique global maximum $\bar{\theta}(s) \in \Theta$. In addition, the function $\bar{\theta}(s)$ is continuous on $\mathcal{S}$.

The EM and the MCEM recursions may be expressed as

$$\text{EM:}\quad \theta^{i+1} := T(\theta^i) = \bar{\theta} \circ \bar{S}(\theta^i)\,, \qquad \text{MCEM:}\quad \hat{\theta}^{i+1} = \bar{\theta}(\hat{S}^{i+1})\,, \tag{D.28}$$

where $\{\hat{S}^i\}$ are the estimates of the complete data sufficient statistics given, for instance, by (D.14) or by an importance sampling estimate of the same quantity.

Assumption AD.12. With

$$\mathcal{L} := \{\theta \in \Theta : \bar{\theta} \circ \bar{S}(\theta) = \theta\} \tag{D.29}$$

being the set of fixed points of the EM algorithm, the image by the function L of this set $\mathcal{L}$ is a finite set of points.

Recall that if the function L is continuously differentiable, then $\mathcal{L}$ coincides with the set of stationary points of the log-likelihood. That is, $\mathcal{L} = \{\theta \in \Theta : \nabla_\theta \mathrm{L}(\theta) = 0\}$ (see in particular Theorem D.7).

To study the MCEM algorithm, we now state conditions that specify how $\hat{S}^{i+1}$ approximates $\bar{S}(\hat{\theta}^i)$.

Assumption AD.13. $\mathrm{L}[\bar{\theta}(\hat{S}^{i+1})] - \mathrm{L}[\bar{\theta} \circ \bar{S}(\hat{\theta}^i)] \to 0$ $\mathbb{P}$-a.s. as $i \to \infty$.

In other words, when $i \to \infty$, the condition implies that the Monte Carlo error $\mathrm{L}\left(\bar{\theta}(\hat{S}^i)\right) - \mathrm{L}\left(\bar{\theta} \circ \bar{S}(\hat{\theta}^i)\right)$ does not swamp the likelihood increase $\mathrm{L}\left(\bar{\theta} \circ \bar{S}(\hat{\theta}^i)\right) - \mathrm{L}(\hat{\theta}^i)$. This seems to be a minimum requirement.

Theorem D.14. *Assume AD.11, AD.12, and AD.13. Assume in addition that, almost surely, the closure of the set $\{\hat{\theta}^i\}$ is a compact subset of Θ. Then, almost surely, the sequence $\{\hat{\theta}^i\}$ converges to the set $\mathcal{L}$ and the sequence $\{\mathrm{L}(\hat{\theta}^i)\}$ has a limit.*

Proof. From Proposition D.2, each iteration of the EM algorithm increases the log-likelihood, $\mathrm{L}(\bar{\theta} \circ \bar{S}(\theta)) \geq \mathrm{L}(\theta)$, with equality if and only if $\theta \in \mathcal{L}$ (see (D.29)). Thus L is a Lyapunov function for $T = \bar{\theta} \circ \bar{S}$ on Θ. Because T is continuous by assumption, the proof follows from Theorem D.9. ∎

Lemma D.15. *Assume D.11 and that the following conditions hold.*

(i) *The sequence $\{\hat{\theta}^i\}$ is bounded* $\mathbb{P}$-a.s.

(ii) *For any $\varepsilon > 0$ and any compact set $\mathcal{K} \subseteq \Theta$,*

$$\sum_{i=1}^{\infty} \mathbb{P}\{|\hat{S}^i - \bar{S}(\hat{\theta}^{i-1})| \geq \varepsilon \mid \mathcal{F}^{i-1}\} \mathbb{1}_{\mathcal{K}}(\hat{\theta}^{i-1}) < \infty \quad \mathbb{P}\text{-a.s.}\,, \tag{D.30}$$

where $\mathcal{F}^j := \sigma(\hat{\theta}^0, \hat{S}^1, \ldots, \hat{S}^j)$.

Then AD.13 is satisfied.

Proof. We first prove that for any $\varepsilon > 0$ and any compact set $\mathcal{K} \subseteq \Theta$,

$$\sum_{i=1}^{\infty} \mathbb{P}\{|\mathrm{L}[\bar{\theta}(\hat{S}^i)] - \mathrm{L}[\bar{\theta} \circ \bar{S}(\hat{\theta}^{i-1})]| \geq \varepsilon \mid \mathcal{F}^{i-1}\} \mathbb{1}_{\mathcal{K}}(\hat{\theta}^{i-1}) < \infty \quad \mathbb{P}\text{-a.s.} \tag{D.31}$$

Note that for any $\delta > 0$ and $\varepsilon > 0$,

$$\begin{aligned}\mathbb{P}\{|\mathrm{L}[\bar{\theta}(\hat{S}^i)] - \mathrm{L}[\bar{\theta} \circ \bar{S}(\hat{\theta}^{i-1})]| \geq \varepsilon \mid \mathcal{F}^{i-1}\} &\leq \mathbb{P}\{|\hat{S}^i - \bar{S}(\hat{\theta}^{i-1})| \geq \delta \mid \mathcal{F}^{i-1}\} \\ &+ \mathbb{P}\{|\mathrm{L}[\bar{\theta}(\hat{S}^i)] - \mathrm{L}[\bar{\theta} \circ \bar{S}(\hat{\theta}^{i-1})]| \geq \varepsilon,\ |\hat{S}^i - \bar{S}(\hat{\theta}^{i-1})| \leq \delta \mid \mathcal{F}^{i-1}\}\,.\end{aligned}$$

In particular, this inequality holds in the event $\{\hat{\theta}^{i-1} \in \mathcal{K}\}$. Define the set $\mathcal{T} = \mathcal{S} \cap \{|s| \leq \sup_{\theta \in \mathcal{K}} \|\bar{S}(\theta)\| + \delta\}$. Since $\bar{S}$ is continuous, this set is compact, and therefore the function $\mathrm{L} \circ \bar{\theta}$ is uniformly continuous on $\mathcal{T}$. Hence we can find $\eta > 0$ such that $|\mathrm{L} \circ \bar{\theta}(s) - \mathrm{L} \circ \bar{\theta}(s')| \leq \varepsilon$ for any $(s, s') \in \mathcal{T} \times \mathcal{T}$ such that $|s - s'| \leq \eta$. In the event $\{\hat{\theta}^{i-1} \in \mathcal{K}\}$, we get

$$\begin{aligned}&\mathbb{P}\{|\mathrm{L}[\bar{\theta}(\hat{S}^i)] - \mathrm{L}[\bar{\theta} \circ \bar{S}(\hat{\theta}^{i-1})]| \geq \varepsilon,\ |\hat{S}^i - \bar{S}(\hat{\theta}^{i-1})| \leq \delta \mid \mathcal{F}^{i-1}\} \\ &\qquad\qquad \leq \mathbb{P}\{|\hat{S}^i - \bar{S}(\hat{\theta}^{i-1})| \geq \eta \mid \mathcal{F}^{i-1}\}\,.\end{aligned}$$

Using (D.30), (D.31) follows.

Combining (D.31) with Corollary B.14 shows that for any compact set $\mathcal{K} \subseteq \Theta$,

$$\lim_{i\to\infty} |\mathrm{L}[\bar{\theta}(\hat{S}^i)] - \mathrm{L}[\bar{\theta} \circ \bar{S}(\hat{\theta}^{i-1})]| \mathbb{1}_{\mathcal{K}}(\hat{\theta}^{i-1}) = 0 \quad \mathbb{P}\text{-a.s.}$$

The proof is concluded by noting that there exists an increasing sequence $\mathcal{K}_1 \subset \mathcal{K}_2 \subset \cdots$ of compact subsets of Θ such that $\Theta = \bigcup_{n=0}^{\infty} \mathcal{K}_n$. ■

As discussed previously, there are many different ways to approximate $\bar{S}(\theta)$. To simplify the discussion, we concentrate below on the simple situation of plain Monte Carlo approximation, assuming that

$$\hat{S}^i = m_i^{-1} \sum_{j=1}^{m_i} S(X^{i,j}), \qquad i \geq 1, \tag{D.32}$$

where m_i is the number of replications in the ith iteration and $X^{i,1}, \ldots, X^{i,m_i}$ are conditionally i.i.d. given the σ-field $\mathcal{F}^{i-1}$ with common density $p(x; \hat{\theta}^{i-1})$.

Lemma D.16. *Assume D.11 and that the closure of the set $\{\hat{\theta}^i\}$ is, almost surely, a compact subset of Θ. Assume in addition that $\sum_{i=1}^{\infty} m_i^{-r/2} < \infty$ for some $r \geq 2$ and that $\sup_{\theta \in \mathcal{K}} \int |S(x)|^r p(x;\theta)\,\lambda(dx) < \infty$ for any compact set $\mathcal{K} \subseteq \Theta$. Then the MCEM sequence $\{\hat{\theta}^i\}$ based on the estimators $\{\hat{S}^i\}$ of the sufficient statistics given by* (D.32) *satisfies Assumption D.13.*

Proof. The Markov and the Marcinkiewicz-Zygmund (see Petrov (1995, Chapter 2)) inequalities state that for any $r \geq 2$ and any $\varepsilon > 0$,

$$
\begin{aligned}
&\sum_{i=1}^{\infty} \mathbb{P}\{|\hat{S}^i - \bar{S}(\hat{\theta}^{i-1})| \geq \varepsilon \mid \mathcal{F}^{i-1}\} \mathbb{1}_{\mathcal{K}}(\hat{\theta}^{i-1}) \\
&\quad \leq \varepsilon^{-r} \sum_{i=1}^{\infty} \mathbb{E}\left[|\hat{S}^i - \bar{S}(\hat{\theta}^{i-1})|^r \mid \mathcal{F}^{i-1}\right] \mathbb{1}_{\mathcal{K}}(\hat{\theta}^{i-1}) \\
&\quad \leq C(r)\varepsilon^{-r} \sum_{i=1}^{\infty} m_i^{-r/2} \int |S(x)|^r p(x;\hat{\theta}^{i-1}) \lambda(dx)\, \mathbb{1}_{\mathcal{K}}(\hat{\theta}^{i-1}) \\
&\quad \leq C(r)\varepsilon^{-r} \sup_{\theta \in \mathcal{K}} \int |S(x)|^r p(x;\theta) \lambda(dx) \sum_{i=1}^{\infty} m_i^{-r/2},
\end{aligned}
$$

where $C(r)$ is a universal constant. The right-hand side is finite by assumption, so that the conditions of Lemma D.15 are satisfied. ■

References

Akaike, H. (1973). Information theory and an extension of the maximum likelihood principle. In *Second International Symposium on Information Theory*, volume 1, pages 267–281. Springer Verlag, New York.

Akaike, H. (1974). A new look at the statistical model identification. *IEEE Transactions on Automatic Control*, 19(6):716–723.

Akashi, H. and Kumamoto, H. (1977). Random sampling approach to state estimation in switching environment. *Automatica*, 13:429–434.

Al-Osh, M. and Alzaid, A. (1987). First order integer-valued autoregressive (INAR (1)) process. *J. Time Series Anal.*, 8:261–275.

An, H. Z. and Huang, F. C. (1996). The geometrical ergodicity of nonlinear autoregressive models. *Statist. Sinica*, 6(4):943–956.

Andel, J. (1976). Autoregressive series with random parameters. *Math. Operationsforschu. Statist.*, 7:735–741.

Anderson, B. D. O. and Moore, J. B. (1979). *Optimal Filtering*. Prentice-Hall, New Jersey.

Anderson, T. (1994). *The Statistical Analysis of Time Series.* John Wiley & Sons, New York.

Anderson, T. W. (2003). *An Introduction to Multivariate Statistical Analysis*. John Wiley & Sons, New York, 3rd edition.

Andrews, D. F. and Mallows, C. L. (1974). Scale mixtures of normal distributions. *J. Roy. Statist. Soc. B*, 36:99–102.

Andrieu, C., Doucet, A., and Holenstein, R. (2010). Particle Markov chain monte carlo methods. *J. Roy. Statist. Soc. B*, 72(Part 3):269–342.

Andrieu, C., Doucet, A., and Tadic, V. B. (2005a). Online simulation-based methods for parameter estimation in nonlinear non-Gaussian state-space models. In *Proc. IEEE Conf. Decis. Control.*

Andrieu, C., Moulines, E., and Priouret, P. (2005b). Stability of stochastic approximation under verifiable conditions. *SIAM J. Control Optim.*, 44(1):283–312 (electronic).

Andrieu, C. and Roberts, G. O. (2009). The pseudo-marginal approach for efficient Monte Carlo computations. *Ann. Statist.*, 37(2):697–725.

Andrieu, C. and Vihola, M. (2011). Markovian stochastic approximation with expanding projections. arXiv.org, arXiv:1111.5421.

Andrieu, C. and Vihola, M. (2012). Convergence properties of pseudo-marginal Markov chain Monte Carlo algorithms. arXiv.org, arXiv:1210.1484.

Andrzejak, R., Lehnertz, K., Rieke, C., Mormann, F., David, P., and Elger, C. (2001). Indications of nonlinear deterministic and finite dimensional structures in time series of brain electrical activity: Dependence on recording region and brain state. *Phys. Rev. E*, 64:061907.

Ansley, C. and Kohn, R. (1982). A geometrical derivation of the fixed interval smoothing algorithm. *Biometrika*, 69(2):486–487.

Ardia, D. and Hoogerheide, L. F. (2010). Bayesian estimation of the GARCH(1,1) model with student-t innovations. *The R Journal*, 2:41–48.

Askar, M. and Derin, H. (1981). A recursive algorithm for the Bayes solution of the smoothing problem. *IEEE Trans. Automat. Control*, 26(2):558–561.

Bahl, L., Cocke, J., Jelinek, F., and Raviv, J. (1974). Optimal decoding of linear codes for minimizing symbol error rate. *IEEE Trans. Inform. Theory*, 20(2):284–287.

Barron, A. (1985). The strong ergodic theorem for densities; generalized Shannon-McMillan-Breiman theorem. *Ann. Probab.*, 13:1292–1303.

Basawa, I. (2001). Inference in stochastic processes. In Shanbhag, D. and Rao, C., editors, *Stochastic Processes: Theory and Methods*, volume 19 of *Handbook of Statistics*, pages 55–77. Elsevier, Amsterdam.

Baum, L. E. and Petrie, T. P. (1966). Statistical inference for probabilistic functions of finite state Markov chains. *Ann. Math. Statist.*, 37:1554–1563.

Baum, L. E., Petrie, T. P., Soules, G., and Weiss, N. (1970). A maximization technique occurring in the statistical analysis of probabilistic functions of Markov chains. *Ann. Math. Statist.*, 41(1):164–171.

Baxendale, P. H. (2005). Renewal theory and computable convergence rates for geometrically ergodic Markov chains. *Ann. Appl. Probab.*, 15(1B):700–738.

Beaumont, M. A. (2003). Estimation of population growth or decline in genetically monitored populations. *Genetics*, 164(3):1139–1160.

Benveniste, A., Métivier, M., and Priouret, P. (1990). *Adaptive Algorithms and Stochastic Approximations*, volume 22. Springer, Berlin. Translated from the French by Stephen S. S. Wilson.

Bhattacharya, R. and Lee, C. (1995). On geometric ergodicity of nonlinear autoregressive models. *Statist. Probab. Lett.*, 22(4):311–315.

Billingsley, P. (1961). The Lindeberg-Lévy theorem for martingales. *Proc. Amer. Math. Soc.*, 12:788–792.

Billingsley, P. (1995). *Probability and Measure*. Wiley, New York, 3rd edition.

Billingsley, P. (1999). *Convergence of Probability Measures*. John Wiley & Sons, New York, 2nd edition.

Black, F. (1976). Studies of stock market volatility changes. In *Proceedings of the American Statistical Association*, pages 177–181.

Blackman, R. and Tukey, J. (1959). *The Measurement of Power Spectra: From the Point of View of Communications Engineering*. Dover, New York.

Block, H., Langberg, N., and Stoffer, D. (1988). Bivariate exponential and geometric autoregressive and autoregressive moving average models. *Advances in Applied Probability*, 20:798–821.

Block, H., Langberg, N., and Stoffer, D. (1990). Time series models for non-gaussian processes. *Lecture Notes-Monograph Series*, 16:69–83.

Bloomfield, P. (2004). *Fourier Analysis of Time Series: An Introduction*. Wiley-Interscience, New York.

Bollerslev, T. (1986). Generalized autoregressive conditional heteroskedasticity. *J. Econometrics*, 31:307–327.

Bollerslev, T., Engle, R. F., and Nelson, D. (1994). ARCH models. In Engle, R. F. and McFadden, D., editors, *Handbook of Econometrics*. North-Holland, Amsterdam.

Booth, J. and Hobert, J. (1999). Maximizing generalized linear mixed model likelihoods with an automated Monte Carlo EM algorithm. *J. Roy. Statist. Soc. B*, 61:265–285.

Borovkov, A. A. and Foss, S. G. (1992). Stochastically recursive sequences and their generalizations. *Siberian Adv. Math.*, 2(1):16–81.

Bougerol, P. and Picard, N. (1992). Strict stationarity of generalized autoregressive processes. *Ann. Probab.*, 20(4):1714–1730.

Box, G. and Jenkins, G. (1970). *Time Series Analysis: Forecasting and Control*. San Francisco: Holden-Day.

Box, G. E. P. and Tiao, G. C. (1973). *Bayesian Inference in Statistical Analysis*. Addison-Wesley, Reading, MA.

Boyles, R. (1983). On the convergence of the EM algorithm. *J. Roy. Statist. Soc. B*, 45(1):47–50.

Briers, M., Doucet, A., and Maskell, S. (2010). Smoothing algorithms for state-space models. *Annals Institute Statistical Mathematics*, 62(1):61–89.

Brillinger, D. (1965). An introduction to polyspectra. *The Annals of Mathematical Statistics*, 36:1351–1374.

Brillinger, D. (2001). *Time Series: Data Analysis and Theory*, volume 36. Society for Industrial Mathematics, Philadelphia.

Brockwell, P. J. and Davis, R. A. (1991). *Time Series: Theory and Methods*. New York: Springer, 2nd edition.

Brockwell, P. J., Liu, J., and Tweedie, R. L. (1992). On the existence of stationary threshold autoregressive moving-average processes. *J. Time Ser. Anal.*, 13(2):95–107.

Bunke, H. and Caelli, T., editors (2001). *Hidden Markov Models: Applications in Computer Vision*. World Scientific.

Burkholder, D. L. (1973). Distribution function inequalities for martingales. *Ann. Probability*, 1:19–42.

Caines, P. E. (1988). *Linear Stochastic Systems*. Wiley, New York.

Campbell, M. J. and Waker, A. M. (1977). A survey of statistical work on the Mackenzie River series of annual Canadian lynx trappings for the years 1821-1934 and a new analysis (with discussion). *Journal of the Royal Statistical Society, Series A: General*, 140:411–431.

Campello, R. J., Favier, G., and do Amaral, W. C. (2004). Optimal expansions of discrete-time volterra models using laguerre functions. *Automatica*, 40(5):815–822.

Cappé, O. (2001). Recursive computation of smoothed functionals of hidden Markovian processes using a particle approximation. *Monte Carlo Methods Appl.*, 7(1–2):81–92.

Cappé, O., Godsill, S. J., and Moulines, E. (2007). An overview of existing methods and recent advances in sequential Monte Carlo. *IEEE Proceedings*, 95(5):899–924.

Cappé, O., Moulines, E., and Rydén, T. (2005). *Inference in Hidden Markov Models*. Springer, New York.

Carlin, B., Polson, N., and Stoffer, D. (1992). A Monte Carlo approach to nonnormal and nonlinear state-space modeling. *J. Am. Statist. Assoc.*, 87(418):493–500.

Carter, C. K. and Kohn, R. (1994). On Gibbs sampling for state space models. *Biometrika*, 81(3):541–553.

Chan, K. S. (1993). Consistency and limiting distribution of the least squares estimator of a threshold autoregressive model. *Ann. Statist.*, 21(1):520–533.

Chan, K. S., Petruccelli, J. D., Tong, H., and Woolford, S. W. (1985). A multiple-threshold AR(1) model. *J. Appl. Probab.*, 22(2):267–279.

Chan, N. H. (2002). *Time Series: Applications to Finance*. John Wiley & Sons, New York.

Chang, R. and Hancock, J. (1966). On receiver structures for channels having memory. *IEEE Trans. Inform. Theory*, 12(4):463–468.

Chen, C. W. S. (1999). Subset selection of autoregressive time series models. *Journal of Forecasting*, 18(7):505–516.

Chen, C. W. S., Liu, F. C., and Gerlach, R. (2011). Bayesian subset selection for threshold autoregressive moving-average models. *Comput. Statist.*, 26(1):1–30.

Chen, R. and Liu, J. S. (2000). Mixture Kalman filter. *J. Roy. Statist. Soc. B*, 62(3):493–508.

Chen, R. and Tsay, R. (1993). Functional-coefficient autoregressive models. *J. Am. Statist. Assoc.*, 88:298–308.

Chopin, N. (2004). Central limit theorem for sequential Monte Carlo methods and its application to Bayesian inference. *Ann. Statist.*, 32(6):2385–2411.

Chopin, N., Jacob, P., and Papaspiliopoulos, O. (2011). SMC^2: A sequential Monte Carlo algorithm with particle Markov chain Monte Carlo updates. *arXiv:1011.1528v3*.

Chopin, N. and Singh, S. (2013). On the particle Gibbs sampler. *arXiv preprint arXiv:1304.1887*.

Chung, K. L. (2001). *A Course in Probability Theory*. Academic Press, San Diego, 3rd edition.

Cipra, B. (1993). Engineers look to Kalman filtering for guidance. *SIAM News*, 26(5):757–764.

Cline, D. B. and Pu, H.-H. (2002). A note on a simple Markov bilinear stochastic process. *Statist. Probab. Lett.*, 56(3):283–288.

Conn, A. R., Scheinberg, K., and Vicente, L. N. (2009). *Introduction to Derivative-Free Optimization*, volume 8. Siam.

Cornebise, J., Moulines, E., and Olsson, J. (2008). Adaptive methods for sequential importance sampling with application to state space models. *Stat. Comput.*, 18(4):461–480.

Cox, D. R. (1981). Statistical analysis of time series: Some recent developments. *Scand. J. Statist.*, 8(2):93–115. With discussion.

Creal, D. (2009). *A Survey of Sequential Monte Carlo Methods for Economics and Finance*. Serie research memoranda. Vrije Universiteit, Faculty of Economics and Business Administration.

Damerdji, H. (1994). Strong consistency of the variance estimator in steady-state simulation output analysis. *Math. Oper. Res.*, 19(2):494–512.

Damerdji, H. (1995). Mean-square consistency of the variance estimator in steady-state simulation output analysis. *Oper. Res.*, 43(2):282–291.

Davis, C. (1993). The computer generation of multinomial random variates. *Comput. Statist. Data Anal.*, 16(2):205–217.

Davis, R., Dunsmuir, W., and Streett, S. (2003). Observation-driven models for poisson counts. *Biometrika*, 90(4):777–790.

Dehling, H., Mikosch, T., and Sørensen, M., editors (2002). *Empirical Process Techniques for Dependent Data*. Birkhäuser, Boston.

Del Moral, P. (2004). *Feynman-Kac Formulae. Genealogical and Interacting Particle Systems with Applications*. Springer, New York.

Del Moral, P., Doucet, A., and Singh, S. (2010). A Backward Particle Interpretation of Feynman-Kac Formulae. *ESAIM M2AN*, 44(5):947–975.

Del Moral, P. and Guionnet, A. (1998). Large deviations for interacting particle systems: applications to non-linear filtering. *Stoch. Proc. App.*, 78:69–95.

Del Moral, P. and Guionnet, A. (1999). Central limit theorem for nonlinear filtering and interacting particle systems. *Ann. Appl. Probab.*, 9(2):275–297.

Del Moral, P., Hu, P., and Wu, L. (2010). On the concentration properties of interacting particle processes. *Foundations and Trends in Machine Learning*, pages 225–389.

Del Moral, P. and Miclo, L. (2000). Branching and interacting particle systems approximations of Feynman-Kac formulae with applications to non-linear filtering. In *Séminaire de Probabilités, XXXIV*, volume 1729 of *Lecture Notes in Math.*, pages 1–145. Springer, Berlin.

Del Moral, P., Patras, F., and Rubenthaler, S. (2011). A mean field theory of nonlinear filtering. In *The Oxford handbook of nonlinear filtering*, pages 705–740. Oxford Univ. Press, Oxford.

Delyon, B., Lavielle, M., and Moulines, E. (1999). On a stochastic approximation version of the EM algorithm. *Ann. Statist.*, 27(1).

Dempster, A., Laird, N., and Rubin, D. (1977). Maximum likelihood from incomplete data via the EM algorithm. *J. Roy. Statist. Soc. B*, 39(1):1–38 (with discussion).

Devroye, L., Györfi, L., and Lugosi, G. (1996). *A probabilistic theory of pattern recognition*, volume 31 of *Applications of Mathematics (New York)*. Springer-Verlag, New York.

Devroye, L. and Krzyzak, A. (2005). Random search under additive noise. In Dror, M., L'Ecuyer, P., and Szidarovszky, F., editors, *Modeling Uncertainty*, volume 46 of *International Series in Operations Research & Management Science*, pages 383–417. Springer, New York.

Diaconis, P. and Freedman, D. (1999). Iterated random functions. *SIAM Rev.*, 47(1):45–76.

Ding, Z., Granger, C., and Engle, R. (1993). A long memory property of stock market returns and a new model. *Journal of Empirical Finance*, 1(1):83–106.

Dobrushin, R. (1956). Central limit theorem for non-stationary Markov chains. I. *Teor. Veroyatnost. i Primenen.*, 1:72–89.

Doeblin, W. (1938). Sur deux problèmes de M. Kolmogoroff concernant les chaînes dénombrables. *Bull. Soc. Math. France*, 66:210–220.

Donnet, S. and Samson, A. (2011). EM algorithm coupled with particle filter for maximum likelihood parameter estimation of stochastic differential mixed-effects models. Technical Report hal-00519576, v2, Université Paris Descartes, MAP5.

Doob, J. (1953). *Stochastic Processes*. Wiley, London.

Douc, R., Doukhan, P., and Moulines, E. (2013). Ergodicity of observation-driven time series models and consistency of the maximum likelihood estimator. *Stochastic Process. Appl.*, 123(7):2620–2647.

Douc, R., Fort, G., Moulines, E., and Priouret, P. (2009a). Forgetting the initial distribution for hidden Markov models. *Stochastic Processes and their Applications*, 119(4):1235–1256.

Douc, R., Garivier, A., Moulines, E., and Olsson, J. (2010). Sequential Monte Carlo smoothing for general state space hidden Markov models. *To appear in Ann. Appl. Probab.*

Douc, R. and Matias, C. (2001). Asymptotics of the maximum likelihood estimator for general hidden Markov models. *Bernoulli*, 7(3):381–420.

Douc, R. and Moulines, E. (2008). Limit theorems for weighted samples with applications to sequential Monte Carlo methods. *Ann. Statist.*, 36(5):2344–2376.

Douc, R. and Moulines, E. (2012). Asymptotic properties of the maximum likelihood estimation in misspecified hidden Markov models. *The Annals of Statistics*, 40(5):2697–2732.

Douc, R., Moulines, E., and Olsson, J. (2009b). Optimality of the auxiliary particle filter. *Probab. Math. Statist.*, 29(1):1–28.

Douc, R., Moulines, E., Olsson, J., and van Handel, R. (2011). Consistency of the maximum likelihood estimator for general hidden Markov models. *Ann. Statist.*, 39(1):474–513.

Douc, R., Moulines, E., and Rosenthal, J. (2004a). Quantitative bounds for geometric convergence rates of Markov chains. *Ann. Appl. Probab.*, 14(4):1643–1665.

Douc, R., Moulines, E., and Rydén, T. (2004b). Asymptotic properties of the maximum likelihood estimator in autoregressive models with Markov regime. *Ann. Statist.*, 32(5):2254–2304.

Doucet, A., De Freitas, N., and Gordon, N., editors (2001). *Sequential Monte Carlo Methods in Practice*. Springer, New York.

Doucet, A., Godsill, S., and Andrieu, C. (2000). On sequential Monte-Carlo sampling methods for Bayesian filtering. *Stat. Comput.*, 10:197–208.

Doucet, A. and Johansen, A. (2009). A tutorial on particle filtering and smoothing: fifteen years later. *Oxford handbook of nonlinear filtering*.

Doucet, A., Kantas, N., Maciejowski, J., and Singh, S. (2009). An overview of sequential monte carlo methods for parameter estimation in general state-space models. *Proceedings IFAC System Identification (SySid) Meeting*.

Doucet, A., Pitt, M. K., and Kohn, R. (2012). Efficient implementation of Markov chain Monte Carlo when using an unbiased likelihood estimator. arXiv.org, arXiv:1210.1871.

Doukhan, P. and Ghindès, M. (1983). Estimation de la transition de probabilité d'une chaîne de Markov Doëblin-récurrente. Étude du cas du processus autorégressif général d'ordre 1. *Stochastic Process. Appl.*, 15(3):271–293.

Dubarry, C. and Le Corff, S. (2013). Nonasymptotic deviation inequalities for smoothed additive functionals in non-linear state-space models. *Bernoulli*. In press.

Duflo, M. (1997). *Random Iterative Models*, volume 34. Springer, Berlin. Translated from the 1990 French original by S. S. Wilson and revised by the author.

Durbin, J. (1960). Estimation of parameters in time-series regression models. *Journal of the Royal Statistical Society. Series B (Methodological)*, pages 139–153.

Durbin, J. and Koopman, S. J. (2012). *Time Series Analysis by State Space Methods*. Oxford University Press, Oxford, second edition.

Durbin, R., Eddy, S., Krogh, A., and Mitchison, G. (1998). *Biological Sequence Analysis: Probabilistic Models of Proteins and Nucleic Acids*. Cambridge University Press, Cambridge.

Durrett, R. (1999). *Essentials of Stochastic Processes*. Springer, New York.

Durrett, R. (2010). *Probability: theory and examples*. Cambridge Series in Statistical and Probabilistic Mathematics. Cambridge University Press, Cambridge, fourth edition.

Dvoretzky, A. (1972). Asymptotic normality for sums of dependent random variables. In *Proceedings of the Sixth Berkeley Symposium on Mathematical Statistics and Probability, Vol. II: Probability Theory*, pages 513–535, Berkeley, CA. Univ. California Press.

Dyk, D. A. V. and Park, T. (2008). Partially collapsed Gibbs samplers: Theory and methods. *J. Am. Statist. Assoc.*, 103(482):790–796.

Edelstein-Keshet, L. (2005). *Mathematical Models in Biology*. Society for Industrial and Applied Mathematics, Philadelphia.

Elliott, R. J. (1993). New finite dimensional filters and smoothers for Markov chains observed in Gaussian noise. *IEEE Trans. Signal Process.*, 39(1):265–271.

Elliott, R. J., Aggoun, L., and Moore, J. B. (1995). *Hidden Markov Models: Estimation and Control*. Springer, New York.

Elliott, R. J., Lau, J. W., Miao, H., and Siu, T. K. (2012). Viterbi-based estimation for Markov switching GARCH model. *Appl. Math. Finance*, 19(3):219–231.

Engle, R. and Russell, J. (1998). Autoregressive conditional duration: A new model for irregularly spaced transaction data. *Econometrica*, pages 1127–1162.

Engle, R. F. (1982). Autoregressive conditional heteroscedasticity with estimates of the variance of United Kingdom inflation. *Econometrica*, 50:987–1007.

Eubank, R. L. (1999). *Nonparametric Regression and Spline Smoothing*, volume 157. Chapman & Hall, New York.

Evans, M. and Swartz, T. (1995). Methods for approximating integrals in Statistics with special emphasis on Bayesian integration problems. *Statist. Sci.*, 10:254–272.

Evans, M. and Swartz, T. (2000). *Approximating Integrals via Monte Carlo and Deterministic Methods*. Oxford University Press.

Fama, E. (1965). The behavior of stock-market prices. *The Journal of Business*, 38(1):34–105.

Fan, J. and Yao, Q. (2003). *Nonlinear Time Series: Nonparametric and Parametric Methods*. Springer Verlag, New York.

Fearnhead, P., Wyncoll, D., and Tawn, J. (2010). A sequential smoothing algorithm with linear computational cost. *Biometrika*, 97(2):447–464.

Feigin, P. D. and Resnick, S. I. (1994). Limit distributions for linear programming time series estimators. *Stochastic Processes and their Applications*, 51(1):135–165.

Feller, W. (1943). On a general class of "contagious" distributions. *Ann. Math. Statist.*, 14:389–399.

Flegal, J. M. and Jones, G. (2010). Batch means and spectral variance estimators in Markov chain monte carlo. *Ann. Statist.*, 38(2):1034–1070.

Fokianos, K. (2009). Integer-valued time series. *Wiley Interdisciplinary Reviews: Computational Statistics*, 1(3):361–364.

Fokianos, K. and Tjøstheim, D. (2011). Log-linear Poisson autoregression. *J. Multivariate Anal.*, 102(3):563–578.

Fort, G. and Moulines, E. (2003). Convergence of the Monte Carlo expectation maximization for curved exponential families. *Ann. Statist.*, 31(4):1220–1259.

Francq, C. and Roussignol, M. (1997). On white noises driven by hidden Markov chains. *J. Time Ser. Anal.*, 18(6):553–578.

Francq, C. and Roussignol, M. (1998). Ergodicity of autoregressive processes with Markov-switching and consistency of the maximum-likelihood estimator. *Statistics*, 32:151–173.

Francq, C., Roussignol, M., and Zakoian, J.-M. (2001). Conditional heteroskedasticity driven by hidden Markov chains. *J. Time Ser. Anal.*, 2:197–220.

Franses, P. and Van Dijk, D. (2000). *Nonlinear Time Series Models in Empirical Finance*. Cambridge University Press.

Fraser, A. M. (2008). *Hidden Markov models and dynamical systems*. Society for Industrial and Applied Mathematics (SIAM), Philadelphia, PA.

Frühwirth-Schnatter, S. (1994). Data augmentation and dynamic linear models. *J. Time Ser. Anal.*, 15(2).

Frühwirth-Schnatter, S. (2006). *Finite mixture and Markov switching models*. Springer Science+ Business Media.

Fuller, W. A. (1996). *Introduction to Statistical Time Series*. John Wiley & Sons.

Fulop, A. and Li, J. (2012). Efficient learning via simulation: A marginalized resample-move approach. *Available at SSRN 1724203*.

Gao, J. (2007). *Nonlinear Time Series: Semiparametric and Nonparametric Methods*, volume 108. Chapman & Hall.

Gelman, A., Roberts, G. O., and Gilks, W. R. (1996). Efficient Metropolis jumping rules. In *Bayesian statistics, 5 (Alicante, 1994)*, Oxford Sci. Publ., pages 599–607, New York. Oxford Univ. Press.

Geman, S. and Geman, D. (1984). Stochastic relaxation, Gibbs distributions and the Bayesian restoration of images. *IEEE Trans. Pattern Anal. Mach. Intell.*, 6:721–741.

Genon-Catalot, V. and Laredo, C. (2006). Leroux's method for general hidden

Markov models. *Stochastic Process. Appl.*, 116(2):222–243.

Gerlach, R., Carter, C., and Kohn, R. (2000). Efficient Bayesian inference for dynamic mixture models. *J. Amer. Statist. Assoc.*, 95(451):819–828.

Geweke, J. (1989). Bayesian inference in econometric models using Monte-Carlo integration. *Econometrica*, 57(6):1317–1339.

Geweke, J. (1992). Evaluating the accuracy of sampling-based approaches to the calculation of posterior moments (with discussion). In Bernardo, J., Berger, J., Dawid, A., and Smith, A., editors, *Bayesian Statistics 4*, pages 169–193. Oxford University Press.

Geweke, J. (1993). Bayesian treatment of the independent student-t linear model. *Journal of Applied Econometrics*, 8(S1):S19–S40.

Geweke, J. and Tanizaki, H. (2001). Bayesian estimation of state-space models using the metropolis–hastings algorithm within gibbs sampling. *Computational Statistics & Data Analysis*, 37(2):151–170.

Geyer, C. (1994). On the convergence of Monte Carlo maximum likelihood calculations. *J. Roy. Statist. Soc. B*, 56:261–274.

Geyer, C. J. and Thompson, E. A. (1992). Constrained Monte Carlo maximum likelihood for dependent data. *J. Roy. Statist. Soc. B*, 54(3):657–699.

Gilks, W. R., Best, N. G., and Tan, K. K. C. (1995). Adaptive rejection metropolis sampling within gibbs sampling. *Journal of the Royal Statistical Society. Series C (Applied Statistics)*, 44(4):pp. 455–472.

Glynn, P. W. and Iglehart, D. (1989). Importance sampling for stochastic simulations. *Management Science*, 35(11):1367–1392.

Glynn, P. W. and Ormoneit, D. (2002). Hoeffding's inequality for uniformly ergodic Markov chains. *Statist. Probab. Lett.*, 56(2):143–146.

Godsill, S. J., Doucet, A., and West, M. (2001). Maximum *a posteriori* sequence estimation using Monte Carlo particle filters. *Ann. Inst. Stat. Math.*, 53(1):82–96.

Godsill, S. J., Doucet, A., and West, M. (2004). Monte Carlo smoothing for nonlinear time series. *Journal of the American Statistical Association*, 99(465):156–168.

Gordon, N., Salmond, D., and Smith, A. F. (1993). Novel approach to nonlinear/non-Gaussian Bayesian state estimation. *IEE Proc. F, Radar Signal Process.*, 140:107–113.

Gouriéroux, C. (1997). *ARCH Models and Financial Applications*. Springer Verlag, New York.

Granger, C. and Andersen, A. (1978). *An Introduction to Bilinear Time Series Models*. Vandenhoeck und Ruprecht, Göttingen.

Green, P. J. and Silverman, B. W. (1993). *Nonparametric Regression and Generalized Linear Models: A Roughness Penalty Approach*, volume 58. Chapman & Hall, New York.

Guégan, D. (1981). Étude d'un modéle non linéaire, le modéle superdiagonal d'ordre 1. *C. R. Acad. Sci. Paris Sér. I Math.*, 293(1):95–98.

Gupta, N. and Mehra, R. (1974a). Computational aspects of maximum likelihood estimation and reduction in sensitivity function calculations. *Automatic Control, IEEE Transactions on*, 19(6):774–783.

Gupta, N. and Mehra, R. (1974b). Computational aspects of maximum likelihood estimation and reduction in sensitivity function calculations. *IEEE Trans. Automat. Control*, 19(6):774–783.

Haggan, V. and Ozaki, T. (1981). Modelling nonlinear random vibrations using an amplitude-dependent autoregressive time series model. *Biometrika*, 68(1):189–196.

Hairer, M. and Mattingly, J. C. (2008). Spectral gaps in Wasserstein distances and the 2D stochastic Navier-Stokes equations. *Ann. Probab.*, 36(6):2050–2091.

Hairer, M. and Mattingly, J. C. (2011). Yet another look at Harris' ergodic theorem for Markov chains. In *Seminar on Stochastic Analysis, Random Fields and Applications VI*, volume 63 of *Progr. Probab.*, pages 109–117. Birkhäuser/Springer Basel AG, Basel.

Hall, P. and Heyde, C. C. (1980). *Martingale Limit Theory and its Application.* Academic Press, New York, London.

Hamilton, J. D. (1989). A new approach to the economic analysis of nonstationary time series and the business cycle. *Econometrica*, 57:357–384.

Hamilton, J. D. (1994). *Time Series Analysis.* Princeton University Press.

Hammersley, J. M. and Handscomb, D. C. (1965). *Monte Carlo Methods.* Methuen & Co., London.

Handschin, J. (1970). Monte Carlo techniques for prediction and filtering of non-linear stochastic processes. *Automatica*, 6:555–563.

Handschin, J. and Mayne, D. (1969). Monte Carlo techniques to estimate the conditionnal expectation in multi-stage non-linear filtering. In *Int. J. Control*, volume 9, pages 547–559.

Hannan, E. (1970). *Multiple Time Series.* New York: John Wiley & Sons.

Hannan, E. and Deistler, M. (1988). *The Statistical Theory of Linear Systems.* John Wiley & Sons.

Hannan, E. and Deistler, M. (2012). *The Statistical Theory of Linear Systems*, volume 70. Cambridge University Press.

Hannan, E. J. and Quinn, B. G. (1979). The determination of the order of an autoregression. *Journal of the Royal Statistical Society, Series B: Methodological*, 41:190–195.

Härdle, W., Lütkepohl, H., and Chen, R. (1997). A review of nonparametric time series analysis. *International Statistical Review*, 65(1):49–72.

Harvey, A. (1991). *Forecasting, Structural Time Series Models and the Kalman Filter.* Cambridge University Press, Cambridge.

Harvey, A. (1993). *Time Series Models.* The MIT Press, Cambridge, MA.

Hastie, T. and Tibshirani, R. (1990). *Generalized Additive Models.* Chapman & Hall, New York.

Hinich, M. and Wolinsky, M. (2005). Normalizing bispectra. *Journal of Statistical Planning and Inference*, 130(1-2):405–411.

Hitczenko, P. (1990). Best constants in martingale version of Rosenthal's inequality. *Ann. Probab.*, 18(4):1656–1668.

Ho, Y. C. and Lee, R. C. K. (1964). A Bayesian approach to problems in stochastic estimation and control. *IEEE Trans. Automat. Control*, 9(4):333–339.

Horn, R. A. and Johnson, C. R. (1985). *Matrix Analysis.* Cambridge University Press.

Hull, J. and White, A. (1987). The pricing of options on assets with stochastic volatilities. *J. Finance*, 42:281–300.

Hurvich, C. M. and Tsai, C.-L. (1993). A corrected Akaike information criterion for vector autoregressive model selection. *Journal of Time Series Analysis*, 14:271–279.

Hürzeler, M. and Künsch, H. R. (1998). Monte Carlo approximations for general state space models. *J. Comput. Graph. Statist.*, 7:175–193.

Hürzeler, M. and Künsch, H. R. (2001). Approximating and maximising the likelihood for a general state-space model. In *Sequential Monte Carlo Methods in Practice*, pages 159–175. Springer, New York.

Hwang, S. and Basawa, I. (1993). Asymptotic optimal inference for a class of nonlinear time series models. *Stochastic Processes and their Applications*, 46(1):91–113.

Hwang, S. Y. and Basawa, I. V. (1997). The local asymptotic normality of a class of generalized random coefficient autoregressive processes. *Statist. Probab. Lett.*, 34(2):165–170.

Jacobs, P. and Lewis, P. (1978a). Discrete time series generated by mixtures. i: Correlational and runs properties. *Journal of the Royal Statistical Society. Series B (Methodological)*, pages 94–105.

Jacobs, P. and Lewis, P. (1978b). Discrete time series generated by mixtures ii: Asymptotic properties. *Journal of the Royal Statistical Society. Series B (Methodological)*, pages 222–228.

Jacobs, P. and Lewis, P. (1983). Stationary discrete autoregressive-moving average time series generated by mixtures. *Journal of Time Series Analysis*, 4(1):19–36.

Jacquier, E., Polson, N. G., and Rossi, P. E. (1994). Bayesian analysis of stochastic volatility models (with discussion). *J. Bus. Econom. Statist.*, 12:371–417.

Jazwinski, A. (1970). *Stochastic processes and filtering theory.* Academic Press, New York.

Jelinek, F. (1997). *Statistical Methods for Speech Recognition.* MIT Press, Cambridge, MA.

Joe, H. and Zhu, R. (2005). Generalized poisson distribution: the property of mixture of poisson and comparison with negative binomial distribution. *Biometrical Journal*, 47(2):219–229.

Johansen, A. M. and Doucet, A. (2008). A note on auxiliary particle filters. *Statist. Probab. Lett.*, 78(12):1498–1504.

Jones, R. (1980). Maximum likelihood fitting of arma models to time series with missing observations. *Technometrics*, pages 389–395.

Jones, R. (1993). *Longitudinal Data with Serial Correlation: A State-Space Approach*, volume 47. Chapman & Hall, New York.

Jung, R. and Tremayne, A. (2011). Useful models for time series of counts or simply wrong ones? *AStA Advances in Statistical Analysis*, pages 1–33.

Kailath, T., Sayed, A., and Hassibi, B. (2000). *Linear Estimation*. Prentice Hall, Englewood Cliffs, NJ.

Kalman, R. (1960). A new approach to linear filtering and prediction problems. *Journal of Basic Engineering*, 82(Series D):35–45.

Kalman, R. E. and Bucy, R. (1961). New results in linear filtering and prediction theory. *J. Basic Eng., Trans. ASME, Series D*, 83(3):95–108.

Kantz, H. and Schreiber, T. (2004). *Nonlinear time series analysis*. Cambridge University Press, Cambridge, second edition.

Kay, S. (1988). *Modern Spectral Estimation: Theory and Applications*. Prentice Hall, Englewood Cliffs, NJ.

Kedem, B. and Fokianos, K. (2002). *Regression Models for Time Series Analysis*. John Wiley and Sons, New York.

Keenan, D. (1985). A tukey nonadditivity-type test for time series nonlinearity. *Biometrika*, 72(1):39–44.

Kim, C. and Nelson, C. (1999). *State-Space Models with Regime Switching: Classical and Gibbs-Sampling Approaches with Applications*. MIT Press, Cambridge.

Kim, J. and Stoffer, D. S. (2008). Fitting stochastic volatility models in the presence of irregular sampling via particle methods and the em algorithm. *Journal of Time Series Analysis*, 29(5):811–833.

Kitagawa, G. (1987). Non-Gaussian state space modeling of nonstationary time series. *J. Am. Statist. Assoc.*, 82(400):1023–1063.

Kitagawa, G. (1996). Monte-Carlo filter and smoother for non-Gaussian nonlinear state space models. *J. Comput. Graph. Statist.*, 1:1–25.

Kitagawa, G. (2010). *Introduction to Time Series Modeling*, volume 114. Chapman & Hall, New York.

Kitagawa, G. and Gersch, W. (1996). *Smoothness priors analysis of time series*, volume 116 of *Lecture Notes in Statistics*. Springer-Verlag, New York.

Kitagawa, G. and Sato, S. (2001). Monte Carlo smoothing and self-organising state-space model. In *Sequential Monte Carlo Methods in Practice*, pages 177–195. Springer, New York.

Klaas, M., Briers, M., De Freitas, N., Doucet, A., Maskell, S., and Lang, D. (2006). Fast particle smoothing: If I had a million particles. In *23rd Int. Conf. Machine Learning (ICML)*, Pittsburgh, Pennsylvania.

Kleptsyna, M. L. and Veretennikov, A. Y. (2008). On discrete time ergodic filters with wrong initial data. *Probab. Theory Related Fields*, 141(3-4):411–444.

Kong, A., Liu, J. S., and Wong, W. (1994). Sequential imputation and Bayesian missing data problems. *J. Am. Statist. Assoc.*, 89:590–599.

Koski, T. (2001). *Hidden Markov Models for Bioinformatics*. Kluwer Academic Publishers, Netherlands.

Krishnamurthy, V. and Rydén, T. (1998). Consistent estimation of linear and non-linear autoregressive models with Markov regime. *J. Time Ser. Anal.*, 19:291–307.

Kuhn, E. and Lavielle, M. (2004). Coupling a stochastic approximation version of EM with an MCMC procedure. *ESAIM Probab. Statist.*, 8:115–131.

Künsch, H. R. (2005). Recursive Monte-Carlo filters: algorithms and theoretical analysis. *Ann. Statist.*, 33(5):1983–2021.

Kushner, H. J. and Yin, G. G. (2003). *Stochastic Approximation and Recursive Algorithms and Applications*, volume 35. Springer, New York, 2nd edition.

Lacour, C. (2008). Nonparametric estimation of the stationary density and the transition density of a Markov chain. *Stochastic Processes and their Applications*, 118(2):232–260.

Langberg, N. and Stoffer, D. (1987). Moving-average models with bivariate exponential and geometric distributions. *Journal of Applied Probability*, pages 48–61.

Lange, K. (2010). *Numerical analysis for statisticians*. Statistics and Computing. Springer, New York, second edition.

Lauritzen, S. L. (1981). Time series analysis in 1880: A discussion of contributions made by T.N. Thiele. *International Statistical Review / Revue Internationale de Statistique*, 49(3):319–331.

Le Gland, F. and Mevel, L. (2000a). Basic properties of the projective product with application to products of column-allowable nonnegative matrices. *Math. Control Signals Systems*, 13(1):41–62.

Le Gland, F. and Mevel, L. (2000b). Exponential forgetting and geometric ergodicity in hidden Markov models. *Math. Control Signals Systems*, 13:63–93.

Lee, T., White, H., and Granger, C. (1993). Testing for neglected nonlinearity in time series models: A comparison of neural network methods and alternative tests. *Journal of Econometrics*, 56(3):269–290.

Leonov, V. P. and Shiryaev, A. N. (1959). On a method of calculation of semi-invariants *(Translated by James R. Brown)*. *Theory of Probability and its Ap-*

plications, 4:319–329.

Leroux, B. G. (1992). Maximum-likelihood estimation for hidden Markov models. *Stoch. Proc. Appl.*, 40:127–143.

Levine, R. A. and Casella, G. (2001). Implementations of the Monte Carlo EM algorithm. *J. Comput. Graph. Statist.*, 10(3):422–439.

Levine, R. A. and Fan, J. (2004). An automated (Markov chain) Monte Carlo EM algorithm. *J. Stat. Comput. Simul.*, 74(5):349–359.

Levinson, N. (1947). The Wiener RMS (root mean square) error criterion in filter design and prediction. *J. Math. Phys*, 25:261.

Lindsten, F. (2013). An efficient stochastic approximation EM algorithm using conditional particle filters. In *Proc. IEEE Int. Conf. Acoust., Speech, Signal Process. (submitted)*, Vancouver, Canada.

Lindsten, F., Jordan, M. I., and Schön, T. B. (2012). Ancestor sampling for particle Gibbs. In *Adv. Neural Inf. Process. Syst.*, Lake Tahoe, NV.

Lindsten, F. and Schön, T. B. (2012). On the use of backward simulation in the particle Gibbs sampler. In *Proc. IEEE Int. Conf. Acoust., Speech, Signal Process.*, Kyoto, Japan.

Lindsten, F. and Schön, T. B. (2013). Backward simulation methods for Monte Carlo statistical inference. *Foundations and Trends in Machine Learning (submitted)*.

Lindsten, F., Schön, T. B., and Jordan, M. I. (2013). Bayesian semiparametric wiener system identification. *Automatica (submitted)*.

Ling, S. (1999). On the probabilistic properties of a double threshold ARMA conditional heteroskedastic model. *J. Appl. Probab.*, 36(3):688–705.

Ling, S., Tong, H., and Li, D. (2007). Ergodicity and invertibility of threshold moving-average models. *Bernoulli*, 13(1):161–168.

Lipster, R. S. and Shiryaev, A. N. (2001). *Statistics of Random Processes: I. General theory*. Springer, New York, 2nd edition.

Liu, J. (2001). *Monte Carlo Strategies in Scientific Computing*. Springer, New York.

Liu, J. and Chen, R. (1995). Blind deconvolution via sequential imputations. *J. Am. Statist. Assoc.*, 90(420):567–576.

Liu, J. and Chen, R. (1998). Sequential Monte-Carlo methods for dynamic systems. *J. Am. Statist. Assoc.*, 93(443):1032–1044.

Liu, J.-C. (2012). A family of Markov-switching GARCH processes. *J. Time Series Anal.*, 33(6):892–902.

Liu, J. S. (1996). Metropolized independent sampling with comparisons to rejection sampling and importance sampling. *Stat. Comput.*, 6:113–119.

Louis, T. A. (1982). Finding the observed information matrix when using the EM algorithm. *J. Roy. Statist. Soc. B*, 44:226–233.

Luenberger, D. G. (1984). *Linear and Nonlinear Programming*. Addison-Wesley, Boston, 2nd edition.

Lütkepohl, H. (2005). *New Introduction to Multiple Time Series Analysis*. Cambridge University Press, Cambridge.

MacDonald, I. and Zucchini, W. (1997). *Hidden Markov and Other Models for Discrete-Valued Time Series*. Chapman, London.

MacDonald, I. L. and Zucchini, W. (2009). *Hidden Markov models for time series: an introduction using R*, volume 110. Chapman & Hall/CRC, London.

Mandelbrot, B. (1963). The variation of certain speculative prices. *Journal of Business*, 36(4):394–419.

Marcinkiewicz, J. (1939). Sur une propriete de la loi de Gauss. *Mathematische Zeitschrift*, 44(1):612–618.

McDiarmid, C. (1989). On the method of bounded differences. In *Surveys in combinatorics, 1989 (Norwich, 1989)*, volume 141 of *London Math. Soc. Lecture Note Ser.*, pages 148–188. Cambridge University Press, Cambridge.

McKenzie, E. (1986). Autoregressive moving-average processes with negative-binomial and geometric marginal distributions. *Advances in Applied Probability*, pages 679–705.

McLachlan, G. and Krishnan, T. (2008). *The EM Algorithm and Extensions*. Wiley, 2nd edition.

McNeil, D. (1977). *Interactive Data Analysis: A Practical Primer*. Wiley New York.

McQuarrie, A. and Tsai, C. (1998). *Regression and Time Series Model Selection*. World Scientific, London.

Meinhold, R. and Singpurwalla, N. (1983). Understanding the kalman filter. *American Statistician*, pages 123–127.

Metropolis, N., Rosenbluth, A. W., Rosenbluth, M. N., Teller, A. H., and Teller, E. (1953). Equations of state calculations by fast computing machines. *J. Chem. Phys.*, 21:1087–1092.

Mevel, L. and Finesso, L. (2004). Asymptotical statistics of misspecified hidden Markov models. *IEEE Trans. Automat. Control*, 49(7):1123–1132.

Meyn, S. P. and Tweedie, R. L. (2009). *Markov Chains and Stochastic Stability*. Cambridge University Press, London.

Mickens, R. E. (1990). *Difference Equations: Theory and Applications*. Chapman & Hall, New York.

Nakatsuma, T. (2000). Bayesian analysis of ARMA-GARCH models: a Markov chain sampling approach. *J. Econometrics*, 95(1):57–69.

Neal, R. M., Beal, M. J., and Roweis, S. T. (2003). Inferring state sequences for non-linear systems with embedded hidden Markov models. In *Adv. Neural Inf. Process. Syst.*, Vancouver, Canada.

Netto, M., Gimeno, L., and Mendes, M. (1978). On the optimal and suboptimal nonlinear filtering problem for discrete-time systems. *Automatic Control, IEEE Transactions on*, 23(6):1062–1067.

Neumann, M. H. (2011). Absolute regularity and ergodicity of Poisson count processes. *Bernoulli*, 17(4):1268–1284.

Nicholls, D. and Quinn, B. (1982). *Random Coefficient Autoregressive Models: An Introduction*. Springer-Verlag, New-York.

Nummelin, E. (1984). *General Irreducible Markov Chains and Non-Negative Operators*. Cambridge University Press, Cambridge.

Olsson, J., Cappé, O., Douc, R., and Moulines, E. (2008). Sequential Monte Carlo smoothing with application to parameter estimation in non-linear state space models. *Bernoulli*, 14(1):155–179. arXiv:math.ST/0609514.

Olsson, J. and Rydén, T. (2008). Asymptotic properties of particle filter-based maximum likelihood estimators for state space models. *Stochastic Processes and their Applications*, 118(4):649–680.

Olsson, J. and Rydén, T. (2011). Rao-Blackwellization of particle Markov chain Monte Carlo methods using forward filtering backward sampling. *IEEE Trans. Signal Process.*, 59(10):4606–4619.

Ostrowski, A. M. (1966). *Solution of Equations and Systems of Equations*. Academic Press, New York, 2nd edition.

Parzen, E. (1983). Autoregressive spectral estimation. *Handbook of statistics*, 3:221–247.

Petrie, T. (1969). Probabilistic functions of finite state Markov chains. *Ann. Math. Statist.*, 40:97–115.

Petris, G., Petrone, S., and Campagnoli, P. (2009). *Dynamic linear models with R*. Use R! Springer, New York.

Petrov, V. V. (1995). *Limit Theorems of Probability Theory*. Oxford University Press, Oxford.

Petruccelli, J. D. and Woolford, S. W. (1984). A threshold AR(1) model. *J. Appl. Probab.*, 21(2):270–286.

Pfanzagl, J. (1969). On the measurability and consistency of minimum contrast estimates. *Metrika*, 14:249–272.

Pham, T. and Tran, L. (1981). On the first-order bilinear time series model. *Journal of Applied Probability*, pages 617–627.

Philippe, A. (2006). Bayesian analysis of autoregressive moving average processes with unknown orders. *Comput. Statist. Data Anal.*, 51(3):1904–1923.

Pitt, M. K. and Shephard, N. (1999). Filtering via simulation: Auxiliary particle filters. *J. Am. Statist. Assoc.*, 94(446):590–599.

Pitt, M. K., Silva, R. S., Giordani, P., and Kohn, R. (2012). On some properties of Markov chain Monte Carlo simulation methods based on the particle filter. *Journal of Econometrics*, 171:134–151.

Pole, A. and West, M. (1989). Reference analysis of the dynamic linear model. *J. Time Ser. Anal.*, 10(2):131–147.

Polyak, B. T. (1990). A new method of stochastic approximation type. *Autom. Remote Control*, 51:98–107.

Polyak, B. T. and Juditsky, A. B. (1992). Acceleration of stochastic approximation by averaging. *SIAM J. Control Optim.*, 30(4):838–855.

Poyiadjis, G., Doucet, A., and Singh, S. (2011). Particle approximations of the score and observed information matrix in state space models with application to parameter estimation. *Biometrika*, 98(1):65–80.

Priestley, M. (1988). *Nonlinear and Nonstationary Time Series Analysis*. Academic Press, London.

Priestley, M. B. (1981). *Spectral analysis and time series. Vol. 1*. Academic Press, London.

Quinn, B. G. (1982). A note on the existence of strictly stationary solutions to bilinear equations. *J. Time Ser. Anal.*, 3(4):249–252.

Rabiner, L. R. (1989). A tutorial on hidden Markov models and selected applications in speech recognition. *Proc. IEEE*, 77(2):257–285.

Rabiner, L. R. and Juang, B.-H. (1993). *Fundamentals of Speech Recognition*. Prentice-Hall, Englewood Cliffs, NJ.

Rauch, H., Tung, F., and Striebel, C. (1965). Maximum likelihood estimates of linear dynamic systems. *AIAA Journal*, 3(8):1445–1450.

Reinsel, G. (2003). *Elements of Multivariate Time Series Analysis*. Springer Verlag, New York.

Remillard, B. (2011). Validity of the parametric bootstrap for goodness-of-fit testing in dynamic models. *Available at SSRN 1966476*.

Ricker, W. E. (1954). Stock and recruitment. *Journal of the Fisheries Board of Canada*, 11(5):559–623.

Ristic, B., Arulampalam, M., and Gordon, A. (2004). *Beyond Kalman Filters: Particle Filters for Target Tracking*. Artech House, Boston, MA.

Robbins, H. and Monro, S. (1951). A stochastic approximation method. *Ann. Math. Statist.*, 22:400–407.

Robert, C. P. and Casella, G. (2004). *Monte Carlo Statistical Methods*. Springer, New York, 2nd edition.

Roberts, G. O. and Rosenthal, J. S. (2004). General state space Markov chains and MCMC algorithms. *Probab. Surv.*, 1:20–71.

Roberts, G. O. and Tweedie, R. L. (1999). Bounds on regeneration times and convergence rates for Markov chains. *Stochastic Processes and Their Applications*, 80:211–229.

Rosenblatt, M. (1983). Cumulants and cumulant spectra. In Brillinger, D. R. and Krishnaiah, P. R., editors, *Handbook of Statistics Volume 3: Time Series in the Frequency Domain*, pages 369–382. Elsevier Science Publishing Co., New York; North-Holland Publishing Co., Amsterdam.

Rosenthal, J. S. (1995). Minorization conditions and convergence rates for Markov chain Monte Carlo. *J. Amer. Statist. Assoc.*, 90(430):558–566.

Royden, H. L. (1988). *Real Analysis*. Macmillan Publishing Company, New York, 3rd edition.

Rubin, D. B. (1987). A noniterative sampling/importance resampling alternative to the data augmentation algorithm for creating a few imputations when the fraction of missing information is modest: the SIR algorithm (discussion of Tanner and Wong). *J. Am. Statist. Assoc.*, 82:543–546.

Rubin, D. B. (1988). Using the SIR algorithm to simulate posterior distribution. In Bernardo, J. M., DeGroot, M., Lindley, D., and Smith, A., editors, *Bayesian Statistics 3*, pages 395–402. Clarendon Press, Oxford.

Rydén, T., Teräsvirta, T., and Asbrink, S. (1998). Stylized facts of daily return series and the hidden Markov model. *Journal of applied econometrics*, 13(3):217–244.

Schwarz, G. (1978). Estimating the dimension of a model. *The Annals of Statistics*, 6:461–464.

Schweppe, F. (1965). Evaluation of likelihood functions for gaussian signals. *Information Theory, IEEE Transactions on*, 11(1):61–70.

Serfling, R. J. (1980). *Approximation Theorems of Mathematical Statistics*. Wiley, New York.

Shephard, N. (1995). *Generalized Linear Autoregressions*. Working Paper, Nuffield College (University of Oxford).

Shephard, N. (1996). Statistical aspects of arch and stochastic volatility. *Monographs on Statistics and Applied Probability*, 65:1–68.

Shephard, N. and Andersen, T. (2009). Stochastic volatility: Origins and overview. In Andersen, T., Davis, R., Kreiss, J.-P., and Mikosch, T., editors, *Handbook of Financial Time Series*. Springer, Berlin.

Shiryaev, A. N. (1966). On stochastic equations in the theory of conditional Markov process. *Theory Probab. Appl.*, 11:179–184.

Shiryaev, A. N. (1996). *Probability*. Springer, New York, 2nd edition.

Shumway, R. and Stoffer, D. (1982). An approach to time series smoothing and forecasting using the em algorithm. *Journal of Time Series Analysis*, 3(4):253–264.

Shumway, R. and Stoffer, D. (2011). *Time Series Analysis and Its Applications*. Springer, New York, 3rd edition.

Spall, J. C. (2005). *Introduction to stochastic search and optimization: estimation, simulation, and control*, volume 65. Wiley-Interscience, New York, 2nd edition.

Steutel, F. and Van Harn, K. (1979). Discrete analogues of self-decomposability and stability. *Annals of Probability*, pages 893–899.

Stratonovich, R. L. (1960). Conditional Markov processes. *Theory Probab. Appl.*, 5(2):156–178.

Subba Rao, T. (1981). On the theory of bilinear time series models. *Journal of the Royal Statistical Society. Series B*, pages 244–255.

Subba Rao, T. and Gabr, M. (1984). *An Introduction to Bispectral Analysis and Bilinear Time Series Models*, volume 24. Springer-Verlag, Berlin.

Taniguchi, M. and Kakizawa, Y. (2000). *Asymptotic theory of statistical inference for time series*. Springer Series in Statistics. Springer-Verlag, New York.

Tanizaki, H. (2003). Nonlinear and non-Gaussian state-space modeling with Monte-Carlo techniques: a survey and comparative study. In Shanbhag, D. N. and Rao, C. R., editors, *Handbook of Statistics 21. Stochastic processes: Modelling and Simulation*, pages 871–929. Elsevier, Amsterdam.

Tanner, M. A. (1993). *Tools for Statistical Inference*. Springer, New York, 2nd edition.

Taylor, S. J. (1982). Financial returns modelled by the product of two stochastic processes – A study of daily sugar prices, 1961-79. In Anderson, O. D., editor, *Time Series Analysis: Theory and Practice*, volume 1, pages 203–226. Elsevier/North-Holland, Amsterdam.

Teicher, H. (1960). On the mixture of distributions. *Ann. Math. Statist.*, 31:55–73.

Teicher, H. (1963). Identifiability of finite mixtures. *Ann. Math. Statist.*, 34:1265–1269.

Teicher, H. (1967). Identifiability of mixtures of product measures. *Ann. Math. Statist.*, 38:1300–1302.

Teräsvirta, T., Tjøstheim, D., and Granger, C. (2011). *Modelling Nonlinear Economic Time Series*. Oxford University Press, Oxford.

Titterington, D. M., Smith, A. F. M., and Makov, U. E. (1985). *Statistical Analysis of Finite Mixture Distributions*. Wiley, Chichester.

Tjostheim, D. (1990). Nonlinear time series and Markov chains. *Adv. in Appl. Probab.*, 22(3):587–611.

Tjøstheim, D. (1994). Non-linear time series: a selective review. *Scand. J. Statist.*, 21(2):97–130.

Tong, H. (1978). On a threshold model. In *Pattern Recognition and Signal Processing, NATO series E, applied sciences*, pages 575–586, The Netherlands. Sijthoff & Noordhoff.

Tong, H. (1981). A note on a Markov bilinear stochastic process in discrete time. *J. Time Ser. Anal.*, 2(4):279–284.

Tong, H. (1983). *Threshold Models in Non-linear Time Series Analysis*. Springer-Verlag, New York.

Tong, H. (1990). *Non-linear Time Series: A Dynamical System Approach*. Oxford

University Press, Oxford.

Tong, H. and Lim, K. (1980). Threshold autoregression, limit cycles and cyclical data. *Journal of the Royal Statistical Society, Series B*, pages 245–292.

Tsay, R. (1986). Nonlinearity tests for time series. *Biometrika*, 73(2):461–466.

Tsay, R. (2005). *Analysis of Financial Time Series*, volume 543. Wiley-Interscience, New York.

van Dyk, D. A. and Park, T. (2008). Partially collapsed Gibbs samplers: theory and methods. *J. Amer. Statist. Assoc.*, 103(482):790–796.

Wahba, G. (1990). *Spline Models for Observational Data*, volume 59. Society for Industrial Mathematics, Philadelphia.

Wei, G. C. G. and Tanner, M. A. (1991). A Monte-Carlo implementation of the EM algorithm and the poor man's Data Augmentation algorithms. *J. Am. Statist. Assoc.*, 85:699–704.

West, M. and Harrison, J. (1997). *Bayesian Forecasting and Dynamic Models.* Springer Verlag, New York.

Whiteley, N. (2010). Discussion on Particle Markov chain Monte Carlo methods. Journal of the Royal Statistical Society: Series B, 72(3), p 306–307.

Whiteley, N., Andrieu, C., and Doucet, A. (2010). Efficient Bayesian inference for switching state-space models using discrete particle Markov chain Monte Carlo methods. Technical report, Bristol Statistics Research Report 10:04.

Whiteley, N., Andrieu, C., and Doucet, A. (2011). Bayesian computational methods for inference in multiple change-points models. *Submitted.*

Wiener, N. (1958). *Nonlinear Problems in Random Theory.* Technology Press Research Monographs. MIT Press, Boston.

Wonham, W. M. (1965). Some applications of stochastic differential equations to optimal nonlinear filtering. *SIAM J. Control*, 2(3):347–369.

Wu, C. F. J. (1983). On the convergence properties of the EM algorithm. *Ann. Statist.*, 11:95–103.

Zangwill, W. I. (1969). *Nonlinear Programming: A Unified Approach.* Prentice-Hall, Englewood Cliffs, NJ.

Zaritskii, V., Svetnik, V., and Shimelevich, L. (1975). Monte-Carlo techniques in problems of optimal data processing. *Autom. Remote Control*, 12:2015–2022.

Zeger, S. L. (1988). A regression model for time series of counts. *Biometrika*, 75(4):621–629.

Zeitouni, O. and Dembo, A. (1988). Exact filters for the estimation of the number of transitions of finite-state continuous-time Markov processes. *IEEE Trans. Inform. Theory*, 34(4):890–893.

Index

For Product Safety Concerns and Information please contact our EU
representative GPSR@taylorandfrancis.com
Taylor & Francis Verlag GmbH, Kaufingerstraße 24, 80331 München, Germany

www.ingramcontent.com/pod-product-compliance
Ingram Content Group UK Ltd.
Pitfield, Milton Keynes, MK11 3LW, UK
UKHW021114180425
457613UK00005B/88
9781466502253